Modern Systems

THIRD EDITION Analysis and Design

Technology Support

 • **Package Options** Visible Analyst or Oracle8i can be packaged with this text to provide hands-on exposure to commercial software.

 • **Companion Web site http://www.prenhall.com/hoffer** Features an Interactive Study Guide, Net Search Exercises, Destinations, PowerPoint slides, Full Glossary, Chat Facilities, and a secure, password-protected Instructor's Area.

Internet Coverage and Features

• **Coverage of Internet-based Systems** *Chapter 16* has been redesigned to address Internet-based application design topics not covered in the other chapters. Coverage includes Internet application design standards, how to maintain site consistency, security issues, and data warehousing, among other topics.

 • **Pine Valley Furniture Web Store:** PVF, a furniture company founded in 1980, now, in the *Third Edition*, explores electronic commerce as an avenue to increase its market share.

 • **Broadway Entertainment Company, Inc.:** BEC, a fictional video and record retailer, is a student project case that allows your students to study and develop a Web-based customer relationship management system.

 • **Net Search Exercises:** Margin icons for Net Search exercises on the Web site can be found in every chapter. The icon signals when a topic in the text has a corresponding Net Search exercise on the Web site.

Three Illustrative Fictional Cases

 • **Pine Valley Furniture (PVF):** Pine Valley Furniture is introduced in Chapter 5 and revisited throughout the book. As key system development life cycle concepts are presented, they are applied and illustrated with this illustrative case. A margin icon identifies the location of the case.

 • **Hoosier Burger (HB):** Starting in Chapter 2, this case illustrates how analysts would develop and implement an automated food ordering system. Hoosier Burger is a fictional fast food restaurant in Bloomington, Indiana. A margin icon identifies the location of the case segments.

 • **Broadway Entertainment Company, Inc. (BEC):** This fictional video rental and music company is used as an extended project case at the end of 15 out of 20 chapters, beginning with Chapter 4.

End-of-Chapter Material

Chapter Summary, Key Terms, Review Questions, Problems and Exercises, and Field Exercises

Supplements

• The Instructor's Resource CD-ROM features the Instructor's Manual, Test Item File, Windows PH Test Manager, PowerPoint Slides, and Image Library.

• Video Series

MODERN SYSTEMS

ANALYSIS

AND DESIGN

THIRD EDITION

JEFFREY A. HOFFER
University of Dayton

JOEY F. GEORGE
Florida State University

JOSEPH S. VALACICH
Washington State University

Prentice-Hall International, Inc.

EXECUTIVE ACQUISITIONS EDITOR: Bob Horan
PUBLISHER: Natalie Anderson
ASSOCIATE EDITOR: Lori Cerreto
EDITORIAL ASSISTANT: Erika Rusnak
MEDIA PROJECT MANAGER: Cathi Profitko
SENIOR MARKETING MANAGER: Sharon Turkovich
MARKETING ASSISTANT: Jason Smith
MANAGING EDITOR (PRODUCTION): Cynthia Regan
PRODUCTION EDITOR: Michael Reynolds
PRODUCTION ASSISTANT: Dianne Falcone
PERMISSIONS SUPERVISOR: Suzanne Grappi
ASSOCIATE DIRECTOR, MANUFACTURING: Vincent Scelta
PRODUCTION MANAGER: Arnold Vila
DESIGN DIRECTOR: Patricia Smythe
INTERIOR DESIGN: Lee Goldstein
COVER DESIGN: Steven Frim
COVER PAINTING: Steven Frim
ILLUSTRATOR (INTERIOR): Electragraphics
ASSOCIATE DIRECTOR, MULTIMEDIA PRODUCTION: Karen Goldsmith
MANAGER, MULTIMEDIA PRODUCTION: Christy Mahon
PRINT PRODUCTION LIAISON: Ashley Scattergood
COMPOSITION: UG/GGS Information Services, Inc.
FULL-SERVICE PROJECT MANAGEMENT: UG/GGS Information Services, Inc.
PRINTER/BINDER: R. R. Donnelly, Roanoke

Copyright © 2002 Prentice-Hall International, Inc.

Credits and acknowledgments borrowed form other sources and reproduced, with permission, in this textbook appear on pages 731–733.

This book may be sold only in those countries to which it is consigned by Prentice-Hall International. It is not to be re-exported and it is not for sale in the U.S.A., Mexico, or Canada.

10 9 8 7 6 5 4 3
ISBN 0-13-042363-7

Brief Contents

Contents

vii

4   Automated Tools for Systems Development   93

BROADWAY ENTERTAINMENT COMPANY, INC.: COMPANY BACKGROUND

Part II   Making the Business Case

Part III Analysis

AN OVERVIEW OF PART III

7 Determining System Requirements   202

9   Structuring System Requirements: Logic Modeling   282

BROADWAY ENTERTAINMENT COMPANY, INC.: STRUCTURING SYSTEM REQUIREMENTS: CONCEPTUAL DATA MODELING FOR THE WEB-BASED CUSTOMER RELATIONSHIP MANAGEMENT SYSTEM

11   Selecting the Best Alternative Design Strategy   348

BROADWAY ENTERTAINMENT COMPANY, INC.: FORMULATING A DESIGN STRATEGY FOR THE WEB-BASED CUSTOMER RELATIONSHIP MANAGEMENT SYSTEM

16 Designing Distributed and Internet Systems   527

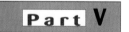

Part V   Implementation and Maintenance

17   System Implementation   570

Part VI   Advanced Analysis and Design Methods

Preface

Description

Modern Systems Analysis and Design covers the concepts, skills, methodologies, techniques, tools, and perspectives essential for systems analysts to successfully develop information systems. The primary target audience is upper division undergraduates in a management information systems or computer information systems curriculum; a secondary target audience is MIS majors in MBA and M.S. programs. Although not explicitly written for the junior college and professional development markets, this book can also be used for these programs.

We have over 50 years of combined teaching experience in systems analysis and design and have used that experience to create this newest edition of *Modern Systems Analysis and Design*. We provide a clear presentation of the concepts, skills, and techniques students need to become effective systems analysts who work with others to create information systems for businesses. We use the Systems Development Life Cycle Model as an organizing tool throughout the book to provide students with a strong conceptual and systematic framework.

The book is written assuming that students have taken an introductory course on computer systems and have experience designing programs in several programming languages. We review basic system principles for those students who have not been exposed to the material on which systems development methods are based. We also assume that students have a solid background in computing literacy and a general understanding of the core elements of a business, including basic terms associated with the production, marketing, finance, and accounting functions.

Modern Systems Analysis and Design is characterized by the following themes:

1. *Systems development is firmly rooted in an organizational context.* The successful systems analyst requires a broad understanding of organizations, organizational culture, and operation.

2. *Systems development is a practical field.* A coverage of current practices as well as accepted concepts and principles are essential in a textbook.

3. *Systems development is a profession.* Standards of practice, a sense of continuing personal development, ethics, and a respect for and collaboration with the work of others are general themes in the textbook.

4. *Systems development has significantly changed with the explosive growth in databases, data-driven architecture for systems, rapid application development, and the Internet.* Systems development and database management can be and possibly should be taught in a highly coordinated fashion. We show when rapid application development methods should and should not be applied. The Internet has rapidly become a common development platform for database driven electronic commerce systems. This text is compatible with the Hoffer, Prescott, and McFadden database text, *Modern Database Management*, Sixth Edition, also published by Prentice Hall. The proper linking of these two textbooks is a strategic opportunity to meet the needs of the IS academic field.

5. *Success in systems analysis and design requires not only skills in methodologies and techniques but also in the management of projects: time, resources, and risks.* Thus, learning systems analysis and design requires a thorough understanding of the process as well as the techniques and deliverables of the profession.

Given these themes, this textbook emphasizes the following:

- A business rather than a technology perspective.
- The role, responsibilities, and mindset of the systems analyst as well as the systems project manager rather than those of the programmer or business manager.
- The methods and principles of systems development rather than the specific tools or tool related skills of the field.

Distinctive Features

The following are some of the distinctive features of *Modern Systems Analysis and Design*:

1. This book is organized in parallel to the Hoffer, Prescott, and McFadden database text, *Modern Database Management*, Sixth Edition, which will facilitate consistency of frameworks, definitions, methods, examples, and notations to better support SA&D and database course adopting both texts. Even with the strategic compatibilities between this text and *Modern Database Management*, each of these books is designed to stand alone as a market leader.

2. The grounding of systems development in the typical architecture for systems in modern organizations, including database management and Web-based systems.

3. A clear linkage of all dimensions of systems description and modeling—process, decision, and data modeling—into a comprehensive and compatible set of systems analysis and design approaches. Such a broad coverage is necessary for students in order to understand the advanced capabilities of many systems development methodologies and tools that are automatically generating a large percentage of code from design specifications.

4. Extensive coverage of oral and written communication skills including systems documentation, project management, team management, and a variety of systems development and acquisition strategies (e.g., life cycle, prototyping, rapid application development, object orientation, joint application development, and systems reengineering).

5. Coverage of rules and principles of systems design, including decoupling, cohesion, modularity, and audits and controls.

6. Consideration of standards for the methodologies of systems analysis and the platforms on which systems are designed.

7. Discussion of systems development and implementation within the context of management of change, conversion strategies, and organizational factors in systems acceptance.

8. Careful attention to human factors in systems design that emphasize usability in both character-based and graphical user interface situations.

9. CASE technology is used throughout the text to illustrate typical systems analysis and design documents and CASE-based systems development is discussed; however, no specific CASE tool is assumed. A variety of CASE and visual development products are illustrated and the current limitations of CASE technologies are highlighted.

10. The text includes a separate chapter on systems maintenance. Given the type of job many graduates first accept and the large installed base of systems, this chapter covers an important and often neglected topic in SA&D texts.

New to the Third Edition

- *The text has been reorganized* Advanced chapters, "Rapid Application Development" and "Object-Oriented Analysis and Design," have been moved to the end of the text to form Part VI: Advanced Analysis and Design Methods. The physical database design chapter has been combined with the logical data modeling chapter to form Chapter 12. Both changes allow for a smoother flow of chapters.
- *Increased focus on make versus buy and systems integration* More and more systems development involves the use of packages in combination with legacy applications and new modules. Chapter 11 shows how companies deal with these issues.
- *Coverage of Internet-based systems* We have redesigned the distributed systems design chapter (now Chapter 16) to also address Internet-based application design topics not covered in the other chapters. We cover Internet application design standards, how to maintain site consistency, security issues, and data warehousing, among other topics. We believe that *Modern Systems Analysis and Design* now has one of the most extensive treatments of Internet application design among its competitors.
- *Integration of electronic commerce into the running cases* One of the three fictional running cases in the text, Pine Valley Furniture, is a furniture company founded in 1980, who now, in the Third Edition, has decided to explore electronic commerce as an avenue to increase its market share. Broadway Entertainment Company, Inc., BEC, a fictional video and record retailer, is a student project case that allows your students to study and develop a Web-based customer relationship management system.
- *Expanded and updated coverage of systems analysis as a profession* We have updated the coverage of codes of conduct and added new material on how systems professionals can approach business problems with ethical considerations. We have also updated information on career paths with the latest information gathered from professional societies.
- *Updated illustrations of technology* Screen captures have been updated throughout the text to show examples using the latest versions of CASE tools, programming and Internet development environments, and user interface designs. Many references to Websites are provided for students to stay current with technology trends that affect the analysis and design of information systems.
- *Expanded coverage of process modeling techniques* Chapter 8 now includes an introduction to business process modeling and functional hierarchy modeling as alternatives to data flow diagramming. These three process modeling techniques are compared so a student knows when to use each in practice.
- *Net Search Exercises* Search exercises on the Website can be found in every chapter. The icon signals when a topic in the text has a corresponding Net Search exercise on the Website. Students can access the exercise from http://www.prenhall.com/hoffer and email their findings to their instructors.

Pedagogical Features

The pedagogical features of *Modern Systems Analysis and Design* reinforce and apply the key content of the book.

Three Illustrative Fictional Cases

PINE VALLEY FURNITURE

Pine Valley Furniture (*PVF*): In addition to an electronic business-to-consumer shopping Website, several other systems development activities from Pine Valley Furniture are used to illustrate key points. Pine Valley Furniture is introduced in Chapter 3 and revisited throughout the book. As key system development life cycle concepts are presented, they are applied and illustrated with this illustrative case. For example, in Chapter 6, we explore how PVF plans a development project for a customer tracking system. A margin icon identifies the location of the case.

Hoosier Burger (HB): This second illustrative case is introduced in Chapter 2 and revisited throughout the book. Hoosier Burger is a fictional fast food restaurant in Bloomington, Indiana. We use this case to illustrate how analysts would develop and implement an automated food ordering system. A margin icon identifies the location of the case segments.

Broadway Entertainment Company, Inc. (BEC): This fictional video rental and music company is used as an extended project case at the end of fifteen out of twenty chapters, beginning with Chapter 4. Designed to bring the chapter concepts to life, this case illustrates how a company initiates, plans, models, designs, and implements a web-based customer relationship management system. Discussion questions are included to promote critical thinking and class participation. Suggested solutions to the discussion questions are provided in the Instructor's Manual.

End-of-Chapter Material We developed an extensive selection of end-of-chapter material designed to accommodate various learning and teaching styles.

- *Chapter Summary* Reviews the major topics of the chapter and previews the connection of the current chapter to future chapters.
- *Key Terms* Designed as a self-test feature, students match each key term in the chapter with a definition.
- *Review Questions* Test students' understanding of key concepts.
- *Problems and Exercises* Test students' analytical skills and require them to apply key concepts.
- *Field Exercises* Give students the opportunity to explore the practice of SA&D in organizations.

Margin Term Definitions Each key term and its definition appear in the margin. Glossaries of terms and acronyms appear in the back of the book.

References Located at the end of each chapter, references together amount to over 100 books, journals, and Websites that can provide students and faculty with additional coverage of topics.

Using This Text

As stated earlier, the book is intended for mainstream SA&D courses. It may be used in a one semester course on SA&D or over two quarters (first in a systems analysis and then in a systems design course). Because of the consistency with *Modern Database Management*, chapters from this book and from *Modern Database Management* can be used in various sequences suitable for your curriculum. The book will be adopted typically in business schools or departments, not in computer science programs. Applied computer science or computer technology programs may adopt the book.

The typical faculty member who will find this book most interesting is someone

- with a practical, rather than technical or theoretical, orientation
- with an understanding of databases and systems that use databases
- who uses practical projects and exercises in the course.

More specifically, academic programs that are trying to better relate their SA&D and database courses as part of a comprehensive understanding of systems development will be especially attracted to this book.

The outline of the book generally follows the systems development life cycle, which allows for a logical progression of topics. However, the book emphasizes that various approaches (e.g., prototyping and iterative development) are also used, so what appears to be a logical progression often is a more cyclic process. Part I of the book provides an overview of systems development and previews the remainder of the book. Part I also covers those skills and concepts that are applied throughout systems development, including systems concepts, project management, and CASE and other automated development technologies. The remaining five sections provide thorough coverage of the six phases of a generic systems development life cycle, interspersing coverage of alternatives to the SDLC as appropriate. Some chapters may be skipped depending on the orientation of the instructor or the students' background. For example, Chapters 1 (environment of SA&D) and 2 (critical success factors for SA&D) cover topics that are emphasized in some introductory MIS courses. Chapter 5 (project identification and selection) can be skipped if the instructor wants to emphasize systems development once projects are identified or if there are fewer than 15 weeks available for the course. Chapters 10 (conceptual data modeling) and 12 (database design) can be skipped or quickly scanned (as a refresher) if students have already had a thorough coverage of these topics in a previous database or data structures course. Finally, Chapter 18 (maintenance) can be skipped if these topics are beyond the scope of your course.

Because the material is presented within the flow of a systems development project, it is not recommended that you attempt to use the chapters out of sequence, with a few exceptions: Chapters 8 (process modeling), 9 (logic modeling), and 10 (conceptual data modeling) can be taught in any sequence; and Chapter 12 (database design) can be taught after Chapters 13 (output design) and 14 (interface design), but Chapters 13 and 14 should be taught in sequence.

Software Packaging Options

- Visible Analyst
- Oracle8i

To enhance the hands-on learning process, Prentice Hall can package this text with Visible Analyst or Oracle8i software. Your Prentice Hall sales representative can provide you with additional information on pricing and ordering.

The Supplement Package

A comprehensive and flexible technology support package is available to enhance the teaching experience:

Instructor's Resource CD-ROM　The Instructor's Resource CD features the following:

- *Instructor's Resource Manual,* by Jeffrey A. Hoffer, Joey F. George, Joseph S. Valacich, and Lisa Miller, with teaching suggestions and answers to all text review questions, problems, and exercises. Lecture notes on how to use the video series (described below) are also included. The Instructor's Resource Manual is also available in print and from the faculty area of the text's Website.

- *Test Item File and Windows PH Test Manager,* by Lisa Miller, University of Central Oklahoma, includes over 3,000 test questions including multiple choice, true/false, completion, and essay questions. The Test Item File is available in Microsoft Word and as the computerized Prentice Hall Test Manager, which is a comprehensive suite of tools for testing and assessment. Test Manager allows instructors to easily create and distribute tests for their courses, either by printing and distributing through traditional methods or by on-line delivery via a Local Area Network (LAN) server. Test Manager features Screen Wizards to assist you as you move through the program, and the software is backed with full technical support.

- *PowerPoint Presentation Slides* feature lecture notes that highlight key text terms and concepts. Professors can customize the presentation by adding their own slides, or editing the existing ones.

- *Image Library* is a collection of the text art organized by chapter. This includes all figures, tables, and screenshots, as permission allows.

Companion Website (http://prenhall.com/hoffer)   The Companion Website accompanying *Modern Systems Analysis and Design* includes

1. An interactive study guide with multiple choice, true/false, and essay questions. Students receive automatic feedback to their answers. Responses to the essay questions, and results from the multiple choice and true/false questions can be emailed to the instructor after a student finishes a quiz.

2. Web-based exploratory exercises, referenced in the text margin as "Net Search" features are developed on the site.

3. Destinations module (links) includes many useful Web links to help students explore systems analysis and design, CASE tools, and information systems on the Web.

4. PowerPoint presentations for each chapter are available in the student area of the site.

5. A full glossary is available both alphabetically and by chapter, along with a glossary of acronyms.

6. Chat facilities include Message Board and Live Chat. Message Board allows users to post messages and check back periodically for responses. Live Chat allows users to discuss course topics in real-time, and enables professors to host on-line classes.

7. A secure, password-protected Instructor's area features downloads of the Instructor's Resource Manual and data sets to accompany the text case studies.

Video Series

Four of the five clips on this video were prepared by Electronic Data Systems Corporation (EDS) and cover topics such as joint application design and application engineering; the fifth clip covers the application of object-oriented analysis and design in a municipal government agency. Each clip is approximately 15 minutes in length and includes an introduction and prologue from the text authors. Lecture notes and suggestions on how to use the videos are included in the Instructor's Resource Manual.

Acknowledgments

The authors have been blessed by considerable assistance from many people on all aspects of preparation of this text and its supplements. We are, of course, responsible for what eventually appears between the covers, but the insights, corrections, contributions, and proddings of others have greatly improved our manuscript. The people we recognize here all have a strong commitment to students, to the IS field, and to excellence. Their contributions have stimulated us, and frequently stimulated the inclusion of new topics and innovative pedagogy.

We would like to recognize the efforts of the many faculty and practicing systems analysts who have been reviewers of this and its associated text, *Essentials of Systems Analysis and Design*. We have tried to deal with each reviewer comment, and although we did not always agree with specific points (within the approach we wanted to take with this book), all reviewers made us stop and think carefully about what and how we were writing. The reviewers were:

Bonnie C. Glassberg, *University of Buffalo*

Gene Klawikowski, *Nicolet Area Technical College*

Roger McHaney, *Kansas State University*

Steven Ross, *Western Washington University*

Harry Reif, *James Madison University*

Stephen Priest, *Daniel Webster College*

Barbara Allen, *Douglas College*

Jay E. Aronson, *University of Georgia*

Susan Athey, *Colorado State University*

Bill Boroski, *Trident Technical College*

Penny Brunner, *University of North Carolina, Asheville*

Pedro Cabrejos, *Champlain College*

Donald Chand, *Bentley College*

Amir Dabirian, *California State University, Fullerton*

Mark Dishaw, *University of Wisconsin at Oshkosh*

Jerry Dubyk, *Northern Alabama Institute of Technology*

Bob Foley, *DeVry Institute of Technology*

Barry Frew, *Naval Post-Graduate School*

Jim Gifford, *University of Wisconsin*

Mike Godfrey, *California State University, Long Beach*

Dale Gust, *Central Michigan University*

John Haney, *Walla Walla College*

Alexander Hars, *University of Southern California*

Ellen Hoadley, *Loyola College-Baltimore*

Monica Holmes, *Central Michigan University*

Robert Jackson, *Brigham Young University*

Murray Jennex, *University of Phoenix*

Len Jessup, *Washington State University*

Robert Keim, *Arizona State University*

Mat Klempa, *California State University at Los Angeles*

Ned Kock, *Temple University*

Rebecca Koop, *Wright State University*

Sophie Lee, *University of Massachusetts at Boston*

Chang-Yang Lin, *Eastern Kentucky University*

Nancy Martin, *USA Group, Indianapolis, Indiana*

Nancy Melone, *University of Oregon*

David Paper, *Utah State University*

G. Prernkumar, *Iowa State University*

Mary Prescott, *University of South Florida*

Terence Ryan, *Southern Illinois University*

Robert Saldarini, *Bergen Community College*

Elaine Seeman, *Pitt Community College*

Eugene Stafford, *Iona College*

Sultan Bhimjee, *San Francisco State*

Bob Tucker, *Antares Alliance, Plano, Texas*

Merrill Warkentin, *Northeastern University*

Cheryl Welch, *Ashland University*

Connie Wells, *Nicholls State University*

Chris Westland, *University of Southern California*

Charles Winton, *University of North Florida*

Terry Zuechow, *EDS Corporation, Piano, Texas*

We extend a special note of thanks to Jeremy Alexander of Web-X.com. Jeremy was instrumental in conceptualizing and writing the Pine Valley Furniture WebStore feature that appears in Chapters 6–18. The addition of this feature has helped make those chapters more modern and innovative. Jeremy also built the installation procedures on the Website for Oracle and Saonee Sarker of Washington State University developed the Oracle tutorial modules.

Lisa Miller from the University of Central Oklahoma has worked with us on several projects and has once again provided us with thoughtful and timely content that has improved the pedagogy of our book. Lisa prepared an extensive test bank, and revised the Instructor's Manual for this text. Meikin Clark (a graduate of the University of Dayton), Melissa Koenig (a graduate of Indiana University), and Sara DiMaio (University of Dayton) contributed the solutions for the Broadway Entertainment Company student project case studies.

We also wish to thank Atish Sinha of the University of Wisconsin-Milwaukee for writing the original version of Chapter 20 on object-oriented analysis and design. Dr. Sinha, who has been teaching this topic for several years to both undergraduates and MBA students, executed a challenging assignment with creativity and cooperation.

We are also indebted to our undergraduate and MBA students at the University of Dayton, Florida State University, and Washington State University, who have given us many helpful comments as they worked with drafts of this text.

Our unique supplement to this text is a series of five videotapes that illustrate common activities and situations encountered by systems analysts. We are very excited about the pedagogical value of these tapes, and compliment EDS Corporation for the sizable commitment of human and financial resources to develop and produce four of these tapes for exclusive use with our book. Specifically, we thank Stu Bailey, Michael Cummings, Vern Olsen, Chris Ryan, and Terry Zuechow of EDS, Bob Tucker of Antares Alliance, and Bill Satterwhite of Whitecap Productions for all of their work on this project. The fifth tape was scripted and produced by the Center for Business and Economics Research at the University of Dayton, and addresses the analysis of needs for a new information system using object-oriented principles. We thank Mike Kurtz and the rest of the CBER staff for their outstanding work.

Thanks also go to Fred McFadden (University of Colorado, Colorado Springs) and Mary Prescott (University of South Florida) for their assistance in coordinating this text with its companion book, *Modern Database Management*, also by Prentice Hall.

Finally, we have been fortunate to work with a large number of creative and insightful people at Prentice Hall, who have added much to the development, format, and production of this text. We have been thoroughly impressed with their commitment to this text and to the IS education market. These people include: Bob Horan (Executive Editor), Lori Cerreto (Associate Editor), Mike Reynolds (Production Editor), Cheryl Asherman (Senior Designer), Erika Rusnak (Editorial Assistant), Sharon Turkovich (Marketing Manager), and Jason Smith (Marketing Assistant).

The writing of this text has involved thousands of hours of time from the authors and from all of the people listed above. Although our names will be visibly associated with this book, we know that much of the credit goes to the individuals and organizations listed here for any success this book might achieve. It is important for the reader to recognize all the individuals and organizations that have been committed to the preparation and production of this book.

Jeffrey A. Hoffer, Dayton, Ohio
Joey F. George, Tallahassee, Florida
Joseph S. Valacich, Pullman, Washington

Part ONE

Foundations for Systems Development

An Overview of Part ONE

Foundations for Systems Development

You are beginning a journey that will allow you to build on every aspect of your education and experience. Becoming a systems analyst is not a goal, it is a path to a rich and diverse career that will allow you to exercise and continue to develop a wide range of talents. We hope that this introductory part of the text helps open your mind to the opportunities of the systems analysis and design field and to the engaging nature of systems work.

Chapter 1 shows that what you do as a systems analyst occurs within a multifaceted organizational process involving other organizational members and external parties. Understanding systems development requires an understanding not only of each technique, tool, and method, but also of how these elements cooperate, complement, and support each other within an organizational setting.

You will discover how the general approach to systems development has changed and continues to change as technology evolves and users' expectations grow. Today, the modern approach to systems analysis and design integrates several views of systems with special emphasis on data as the core material. You'll also discover that the systems analysis and design field is constantly adapting to new situations due to a strong commitment to constantly improve. Our goal in this book is to provide you with a mosaic of the skills needed to effectively work in whichever environment you find yourself, armed with the knowledge to determine the best practices for that situation and argue for them effectively.

In systems development, certain fundamental principles and concepts play a role in every phase of the systems development life cycle and, in a sense, hold the systems development process together. The purpose of Chapters 2, 3, and 4 is to outline those aspects of systems development that are independent of but applicable to each life cycle step. We cover these principles now since they will be used repeatedly in subsequent chapters, and it can be confusing to introduce them in a piecemeal fashion.

Individual factors for a systems analyst that contribute to successful systems development are the topic of Chapter 2. Various studies of successful systems analysis and design indicate that the systems analyst must possess analytical, technical, management, and interpersonal skills. Because many systems analysis and design courses require significant project work, either in groups or in a real organizational setting, you will soon have an opportunity to practice the critical success factors. As Chapter 2 ends, we develop an appreciation for systems analysis as a profession and show how these critical skills lead to standards, ethics, and career paths for the field.

Chapter 3 addresses a fundamental characteristic of life as a systems analyst: working within the framework of projects with constrained resources. All systems work demands attention to deadlines, working within budgets, and coordinating the work of various people. The very nature of the systems development life cycle (SDLC) implies a systematic approach involving a project—a group of related activities leading to a final deliverable. Projects must be planned, started, executed, and completed. The planned work of the project must be represented in such a way that all interested parties can review and understand it. In your first job as a systems analyst, you will have to work within the schedule and other project plans and thus it is important to understand the management process controlling your work.

Chapter 4 overviews the effort to provide automated support for the systems analysis and design process. The focus of this effort today is on computer-aided software engineering, or CASE, and emerging development tools such as object-oriented development and visual programming. CASE provides an engineering-style discipline with associated tools to enforce standards of practice and to greatly increase the speed by which systems are developed and maintained. Likewise, object-oriented development tools are helping to alleviate maintenance problems plaguing organizations that have multiple systems with redundant functions through the use of software objects that can be infinitely reused, becoming the building blocks for all systems. Visual development tools allow systems developers to quickly build new user interfaces, reports, and other features into new and existing systems in a fraction of the time previously required by allowing systems designers the ability to "draw" the design using predefined objects.

Finally, Part I introduces Broadway Entertainment Company, Inc. (BEC). The BEC case helps demonstrate how what you learn in each chapter might fit into a practical, organizational situation. Two BEC case sections are included after Chapter 4; the remaining book chapters through Chapter 17 each have an associated BEC case. The first BEC section introduces the company. The second BEC section provides more detail on existing systems to aid in understanding BEC operations when we look at the requirements and design for new systems in later BEC case sections.

It is time to begin your journey.

1

The Systems Development Environment

LEARNING OBJECTIVES

After studying this chapter, you should be able to:

● Define information systems analysis and design.

● Discuss the modern approach to systems analysis and design that combines both process and data views of systems.

● Describe the organizational roles, including systems analyst, involved in information systems development.

● Describe the different types of information systems.

● Describe the information systems development life cycle (SDLC).

● List alternative approaches to the systems development life cycle and compare the advantages and deficiencies of the SDLC and these approaches.

● Explain briefly the role of computer-aided software engineering (CASE) tools in systems development.

INTRODUCTION

Information systems analysis and design is a complex, challenging, and stimulating organizational process that a team of business and systems professionals uses to develop and maintain computer-based information systems. Although advances in information technology continually give us new capabilities, the analysis and design of information systems is driven from an organizational perspective. An organization might consist of a whole enterprise, specific departments, or individual work groups. Organizations can respond to and anticipate problems and opportunities through innovative uses of information technology. Information systems analysis and design is, therefore, an organizational improvement process. Systems are built and rebuilt for organizational benefits. Benefits result from adding value during the process of creating, producing, and supporting the organization's products and services. Thus, the analysis and design of information systems is based on your understanding of the organization's objectives, structure, and processes as well as your knowledge of how to exploit information technology for advantage.

In the current business environment, the trend is to incorporate the Internet, especially the World Wide Web, more and more into an organization's way of doing business. Although you are probably most familiar with marketing done on the Web and Web-based retailing sites, like CDNow or Amazon.com, the overwhelming majority of business use for the Web is business-to-business applications. These applications run the gamut of everything businesses do, from transmitting orders and payments to suppliers, to fulfilling orders and collecting payment from customers, to maintaining business relationships, to establishing electronic marketplaces where businesses can shop online for the best deals on resources they need for assembling their products and services. Although the Internet seems to pervade business these days, it is important to remember that many of the key aspects of business—offering a product or service for sale, collecting payment, paying employees, maintaining supplier and client relationships—have not changed in the Internet Age. Understanding the business and the way it functions is still the key to successful systems development, even in the fast-paced, technology-driven environment organizations find themselves in today.

Few business careers present a greater opportunity for significant and visible impact on business as do careers in systems development. One reason for this is that there is a tremendous demand for information technology workers in business. According to the Information Technology Association of America (ITAA), over 1.6 million new information technology jobs were to have been created in 2000 (ITAA, 2000). ITAA projected that more than half of those jobs would not be filled by year's end. Not all of these jobs were in systems development. Many were in technical support and network administration. Yet demand for systems developers remains strong, accounting for 20 percent of all the new jobs. That's over 300,000 new jobs in systems development created in just one year! Information technology workers with Web-related skills accounted for another 13 percent of the new jobs. With such promising career prospects, combined with the challenges and opportunities of dealing with the rapid advances in information technologies, it is difficult to imagine a more exciting career choice than systems analysis and design. Furthermore, analyzing and designing information systems will give you the chance to understand organizations at a depth and breadth that might take many more years to accomplish in other careers.

An important (but not the only) result of systems analysis and design is **application software**; that is, software designed to support a specific organizational function or process, such as inventory management, payroll, or market analysis. In addition to application software, the total information system includes the hardware and systems software on which the application software runs, documentation and training materials, the specific job roles associated with the overall system, controls, and the people who use the software along with their work methods. Although we will address all these various dimensions of the overall system, we will emphasize application software development—your primary responsibility as a systems analyst.

In the early years of computing, analysis and design was considered to be an art. Now that the need for systems and software has become so great, people in industry and academia have developed work methods that make analysis and design a disciplined process (similar to processes followed in engineering fields). Our goal is to help you develop the knowledge and skills needed to understand and follow such software engineering processes. Central to software engineering processes (and to this book) are various *methodologies*, *techniques*, and *tools* that have been developed, tested, and widely used over the years to assist people like you during systems analysis and design.

Methodologies are comprehensive, multiple-step approaches to systems development that will guide your work and influence the quality of your final product: the information system. A methodology adopted by an organization will be consistent with its general management style (for example, an organization's orientation

Information systems analysis and design: The complex organizational process whereby computer-based information systems are developed and maintained.

Application software: Computer software designed to support organizational functions or processes.

toward consensus management will influence its choice of systems development methodology). Most methodologies incorporate several development techniques.

Techniques are particular processes that you, as an analyst, will follow to help ensure that your work is well thought-out, complete, and comprehensible to others on your project team. Techniques provide support for a wide range of tasks including conducting thorough interviews to determine what your system should do, planning and managing the activities in a systems development project, diagramming the system's logic, and designing the reports your system will generate.

Tools are typically computer programs that make it easy to use and benefit from the techniques and to faithfully follow the guidelines of the overall development methodology. To be effective, both techniques and tools must be consistent with an organization's systems development methodology. Techniques and tools must make it easy for systems developers to conduct the steps called for in the methodology. These three elements—methodologies, techniques, and tools—work together to form an organizational approach to systems analysis and design.

Although many people in organizations are responsible for systems analysis and design, in most organizations the **systems analyst** has the primary responsibility. When you begin your career in systems development, you will most likely begin as a systems analyst or as a programmer with some systems analyst responsibilities. The primary role of a systems analyst is to study the problems and needs of an organization in order to determine how people, methods, and information technology can best be combined to bring about improvements in the organization. A systems analyst helps system users and other business managers define their requirements for new or enhanced information services. As such, a systems analyst is an agent of change and innovation.

In the rest of this chapter, we will examine the systems approach to studying organizations and the systems approach to studying analysis and design. You will learn about the dominant complementary approaches to systems development—the data- and process-oriented approaches. You will also identify the various people who develop systems and the different types of systems they develop. The chapter ends with a discussion of some of the methodologies, techniques, and tools created to support the systems development process.

Systems analyst: The organizational role most responsible for the analysis and design of information systems.

N ET   S E A R C H
The number of new terms and words that appear each year related to information systems and new technologies is incredible. Visit http://www.prenhall.com/hoffer to complete an exercise on this topic.

A MODERN APPROACH TO SYSTEMS ANALYSIS AND DESIGN

The analysis and design of computer-based information systems began in the 1950s. Since then, the development environment has changed dramatically, driven by organizational needs as well as by rapid changes in the technological capabilities of computers. In the 1950s, the focus of the development effort was on the processes the software performed. Since computer power was a critical resource, efficiency of processing became the main goal. Computers were large, expensive, and not very reliable. Emphasis was placed on automating existing processes, such as purchasing or paying, often within single departments. All applications had to be developed in machine language or assembly language, and they had to be developed from scratch, as there was no software industry. Because computers were so expensive, computer memory was also at a premium, so system developers conserved as much memory for data storage as they could.

The first procedural or third-generation computer programming languages did not become available until the beginning of the 1960s. Computers were still large and expensive. But the 1960s saw important breakthroughs in technology that enabled the development of smaller, faster, less-expensive computers—mini-computers—and the beginnings of the software industry. Most organizations still developed their applications from scratch, using their in-house development staffs. System development was more of an art than a science. This view of systems development began to change in the 1970s, however, as organizations began to realize how expensive it was to develop customized information systems for every application. Systems development came to be more disciplined as many people worked to make it more like engineering. Early database management systems, using hierarchical and network models, helped bring discipline to the storage and retrieval of data. The development of database management systems helped shift the focus of systems development from processes first to data first.

The 1980s were also marked by major breakthroughs in computing in organizations, as microcomputers became key organizational tools. The software industry expanded greatly as more and more people began to write off-the-shelf software for microcomputers. Developers began to write more and more applications in fourth-generation languages, which unlike procedural languages, instructed a computer on what to do instead of how to do it. Computer-aided software engineering (CASE) tools were developed to make systems developers' work easier and more consistent. As computers continued to get smaller, faster, and cheaper, and as the operating systems for computers moved away from line prompt interfaces to windows- and icon-based interfaces, organizations moved to applications with more graphics. Organizations developed less software in-house and bought relatively more from software vendors. The systems developer's job went through a transition from builder to integrator.

The systems development environment in the late 1990s focused on systems integration. Developers used visual programming environments, like PowerBuilder or VisualBasic, to design the user interfaces for systems that run on client/server platforms. The database, which may be relational or object-oriented, and which may have been developed using software from firms such as Oracle, Microsoft, or Ingres, resided on the server. In many cases, the application logic resided on the same server. Alternatively, an organization may have decided to purchase their entire enterprise-wide system from companies such as SAP AG or PowerSoft Inc. Enterprise-wide systems are large, complex systems that consist of a series of independent system modules. Developers assemble systems by choosing and implementing specific modules. Starting in the middle years of the 1990s, more and more systems development efforts focused on the Internet, especially the Web.

Today, in the first years of the new century, there continues to be a lot of focus on developing systems for the Internet and for firms' Intranets and Extranets. As happened with traditional systems, Internet developers now rely on computer-based tools, such as Cold Fusion, to speed and simplify the development of Web-based systems. Many CASE tools, such as those developed by Oracle, now directly support Web application development. More and more, systems implementation involves a three-tier design, with the database on one server, the application on a second server, and client logic located on user machines. The other determining factor in the early years of the new century is the move to wireless components in systems. Wireless devices, such as cell phones and personal digital assistant devices, like the Palm Pilot, are increasingly able to access Web-based applications from almost anywhere. Finally, the trend continues toward assembling systems from programs and components purchased off the shelf. In many cases, organizations not only don't develop the application in-house; they don't even run the application in-house, choosing instead to use the application on a per-use basis by accessing through an application service provider (ASP).

Although the systems development environment has changed dramatically since its beginnings in the 1950s, some basic principles that govern the approach to systems development have been applicable through all the different eras of computing in organizations. One principle is the distinction between data, data flows, and processing logic, a distinction that will be considered in the next sections.

Separating Data and Processes That Handle Data

Every information system consists of three key components that must be clearly understood by anyone who analyzes and designs systems: data, data flows, and processing logic (see Figure 1-1). **Data** are raw facts that describe people, objects, and events in an organization, such as a customer's account number, the number of boxes of cereal bought, and whether someone is a Democrat or a Republican. Every information system depends on data in order to produce **information**, which is processed data presented in a form suitable for human interpretation. Systems developers must understand what kind of data a system uses and where the data originate. Data and the relationships among data may be described using various techniques, as we will see later. Figure 1-1 shows the structure of employee data as a simple table of rows (records about different employees) and columns (attributes describing each employee).

Data flows are groups of data that move and flow through a system and include a description of the sources and destinations for each data flow. For example, a customer's account number may be captured when he or she uses a credit card to pay for a purchased item. The account number may then be stored in a file within the system until needed to compile a billing statement or prepare a mailing address for a sales circular. When needed, the account number can be extracted from storage and used to complete a system function. Figure 1-1 illustrates data flows with directional lines that connect rounded rectangles, which represent the processing steps that accept input data flows and produce output data flows.

Processing logic, the third component, describes the steps in the transformation of the data and the events that trigger these steps. For example, processing logic in a credit card application will explain how to compute available credit given the current credit balance and the amount of the current transaction. Processing logic will also indicate that the computation of the new credit balance will occur when a clerk presses a key on a credit card scanner to confirm the sales transaction. Figure 1-1, using an English-like language, illustrates the rules for calculating an employee's pay and the event (receipt of a new hours-worked value) that causes this calculation to be made.

Traditionally, an information system's design was based upon what the system was supposed to do, such as billing and inventory control: the focus was on output and pro-

Data: Raw facts about people, objects, and events in an organization.

Information: Data that have been processed and presented in a form suitable for human interpretation, often with the purpose of revealing trends or patterns.

Data flow: Data in motion, moving from one place in a system to another.

Processing logic: The steps by which data are transformed or moved and a description of the events that trigger these steps.

Figure 1-1
Differences among data, data flow, and processing logic

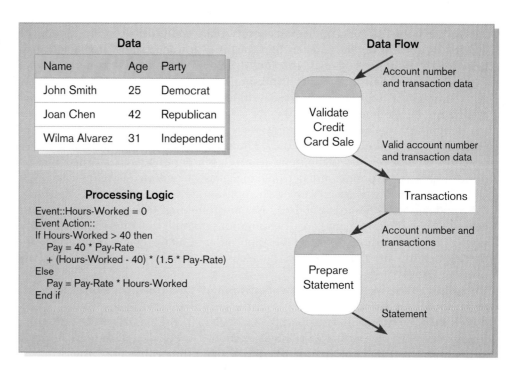

Data

Name	Age	Party
John Smith	25	Democrat
Joan Chen	42	Republican
Wilma Alvarez	31	Independent

Processing Logic

```
Event::Hours-Worked = 0
Event Action::
If Hours-Worked > 40 then
    Pay = 40 * Pay-Rate
    + (Hours-Worked - 40) * (1.5 * Pay-Rate)
Else
    Pay = Pay-Rate * Hours-Worked
End if
```

Data Flow

Account number and transaction data

Validate Credit Card Sale

Valid account number and transaction data

Transactions

Account number and transactions

Prepare Statement

Statement

cessing logic. Although the data the system used as input were important, data were subordinate to the application. The assumption was that we could anticipate all outputs and the proper processing steps with their need for data. Therefore, we could easily derive all data requirements from all known system deliverables. Furthermore, each application contained its own files and data storage capacity. The data had to match the specifications established in each application, and each application was considered separately.

This concentration on the flow, use, and transformation of data in an information system typified the **process-oriented approach** to systems development. The techniques and notations developed from this approach track the movement of data from their sources, through intermediate processing steps, and on to final destinations. Since various parts of an information system work on different schedules and at different speeds, the process-oriented approach also shows where data are temporarily stored until needed for processing. The natural structure of the data is, however, not specified within the traditional process-oriented approach. Until recently, techniques for the process-oriented approach did not address the timing or triggering of processing steps, only their sequence.

Data processing managers soon realized that there were problems with analyzing and designing systems using only a process-oriented approach. One result was the existence of several specialized files of data, each locked within different applications and programs. Many of the files in these different applications contained the same data elements (see Figure 1-2a). When a single data element changed, it had to be

Process-oriented approach: An overall strategy to information systems development that focuses on how and when data are moved through and changed by an information system.

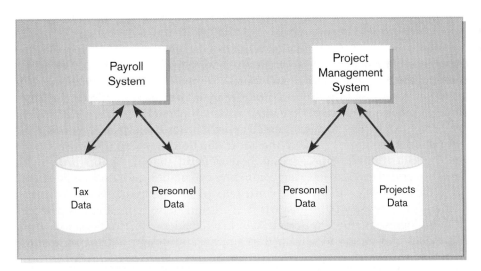

Figure 1-2
Traditional relationship between data and applications, with redundant data, versus the database approach
(a) Traditional approach

(b) Database approach

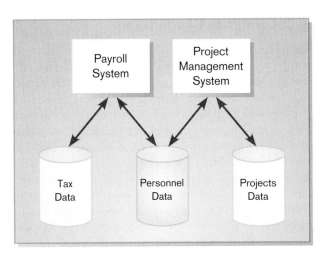

TABLE 1-1   Key Differences Between the Process-Oriented and Data-Oriented Approaches to Systems Development

Characteristic	Process-Orientation	Data-Orientation
System focus	What the system is supposed to do and when	Data the system needs to operate
Design stability	Limited, as business processes and the applications that support them change constantly	More enduring, as the data needs of an organization do not change rapidly
Data organization	Data files designed for each individual application	Data files designed for the enterprise
State of the data	Much uncontrolled duplication	Limited, controlled duplication

changed in each of these files. If, for example, such a system were in effect at your university and your address changed, it would have to be changed in the files of the library, the registrar's office, the financial aid office, and every other place your address was stored. It also became difficult to combine specialized data files. Even if the files contained the same data elements, each file might use a different name and format for the data. Since it was important to standardize how data elements were represented, data processing managers gradually came to separate the application programs and the data these programs used.

Data-oriented approach: An overall strategy of information systems development that focuses on the ideal organization of data rather than where and how data are used.

This focus on data typified the **data-oriented approach** to information systems development. The data-oriented approach depicts the ideal organization of data, independent of where and how data are used within a system (see Figure 1-2b). The techniques used for data orientation result in a data model that describes the kinds of data needed in systems and the business relationships among the data. A data model describes the rules and policies of a business. Some people believe a data model to be more permanent than a process model since a data model reflects the inherent nature of a business instead of the way a business operates, which is constantly changing. Some people refer to data-oriented approaches as *information engineering*.

Table 1-1 highlights some of the key distinctions between the process-oriented and the data-oriented approaches to systems development. Although we highlight these two approaches as separate competing orientations, we do so only to emphasize their differences and unique contributions to systems analysis and design and to give you a sense of the historical evolution of systems analysis and design methodologies. As either approach is, by itself, inadequate, this book will cover the techniques and tools you will need to analyze and design both process and data aspects of systems.

Separating Databases and Applications

As data storage management technology advanced, it became possible to represent data not in separate files for each application but in coherent and shared databases. A **database** is a shared collection of logically related data organized in ways that facilitate capture, storage, and retrieval for multiple users in an organization. Databases involve methods of data organization that allow data to be centrally managed, standardized, and consistent. Instead of a proliferation of separate and distinct data files, the database approach allows central databases to be the sole source of data for many varied applications.

Database: A shared collection of logically related data designed to meet the information needs of multiple users in an organization.

Application independence: The separation of data and the definition of data from the applications that use these data.

Under the data-oriented approach to systems development, databases are designed around *subjects*, such as customers, suppliers, and parts. Designing databases around subjects enables you to use and revise databases for many different independent applications. This focus results in **application independence**, the separation of data and the definition of data from applications.

The central point of application independence is that data and applications are separate. For the data-oriented approach to be effective, however, another change in the system design is needed: Organizations that have centrally managed repositories of organizational data must design new applications to work with existing databases. Organizations that do not have centrally managed repositories of organizational data must design databases that will support both current and future applications.

YOUR ROLE AND OTHER ORGANIZATIONAL RESPONSIBILITIES IN SYSTEMS DEVELOPMENT

In an organization that develops its own information systems internally, there are several types of jobs involved. In medium to large organizations, there is usually a separate Information Systems (IS) department. Depending on how the organization is set up, the IS department may be a relatively independent unit, reporting to the organization's top manager. Alternatively, the IS department may be part of another functional department, such as Finance, or there may even be an IS department in several major business units. In any of these cases, the manager of an IS department will be involved in systems development. If the department is large enough, there will be a separate division for systems development, which would be homebase for systems analysts, and another division for programming, where programmers would be based (see Figure 1-3). The people for whom the systems are designed are located in the functional departments and are referred to as users or *end users*.

Some organizations use a different structure for their IS departments. Following this model, analysts are assigned and may report to functional departments. In this way, analysts learn more about the business they support. This approach is supposed to result in better systems, since the analyst becomes an expert in both systems development and the business area.

Regardless of how an organization structures its information systems department, systems development is a *team effort*. Systems analysts work together in a team, usually organized on a project basis. Team membership can be expanded to include IS managers, programmers, users, and other specialists who may be involved (throughout or at specific points) in the systems development project. It is rare to

Figure 1-3
Organization chart for typical IS department

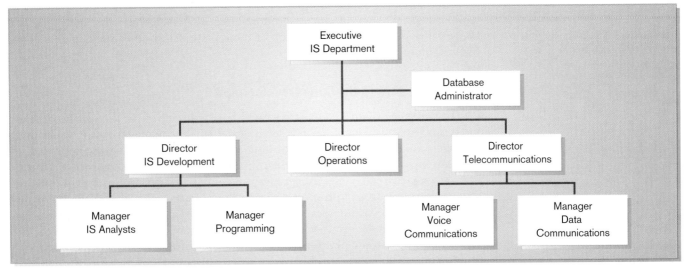

TABLE 1-2 Characteristics of Successful Teams

- Diversity in backgrounds, skills, and goals
- Tolerance of diversity, uncertainty, ambiguity
- Clear and complete communication
- Trust
- Mutual respect and putting one's own views second to the team
- Reward structure that promotes shared responsibility and accountability

find an organizational information system project that involves only one person. Thus, learning how to work with others in teams is an important skill for any IS professional, and we will stress team skills throughout this book.

A good team has certain characteristics, some that are a result of how the group is assembled and others that must be acquired through effort on the part of team members (see Table 1-2). A good team is diverse and tolerant of diversity:

- A diverse team has representation from all the different groups interested in a system, and the representation of these groups on the team increases the likelihood of acceptance of the changes a new system will cause.

- Diversity exposes team members to new and different ideas, ideas they might never think of were all team members from the same background, with the same skills and goals.

- New and different ideas can help a team generate better solutions to its problems and defend the course of action it chooses.

- Team members must be able to entertain new ideas without being overly critical, without dismissing new ideas out of hand simply because they are new.

- Team members must be able to deal with ambiguous information as well as with complexity and must learn to play a role on a team (and different roles on different teams) so that the talents of all team members can best be utilized.

In order to work well together, a good team must strive to communicate clearly and completely with its members. Team members will communicate more effectively if they trust each other. Trust, in turn, is built on mutual respect and an ability to place one's own goals and views secondary to the goals and views of the group. To help ensure that a team will work well together, management needs to develop a reward structure that promotes shared responsibility and accountability within the team. In addition to rewards for individual efforts, team members must be rewarded by IS managers for their work as members of an effective work unit.

Team success depends not only on how a team is assembled or the efforts of the group but also on the management of the team. Reward systems are one part of good team management. Effective project management is another key element of successful teams. Project management includes devising a feasible and realistic work plan and schedule, monitoring progress against this schedule, coordinating the project with its sponsors, allocating resources to the project, and sometimes even deciding whether and when a project should be terminated before completing the system.

The characteristics of each systems analysis and design project will dictate which types of individuals should be on the project team. In general, those involved in systems development include IS managers, systems analysts, programmers, end users, and business managers as well as additional IS managers, technicians, and specialists. We will now preview the role of each of these players and other **stakeholders** in systems development.

Stakeholder: A person who has an interest in an existing or new information system.

IS Managers in Systems Development

The manager of an IS department may have a direct role in the systems development process if the organization is small or if that is the manager's style. Typically, IS managers are more involved in allocating resources to and overseeing approved system development projects rather than in the actual development process. Thus, IS managers may attend some project review meetings and certainly will expect written status reports on project progress covering their areas of concern. IS managers may prescribe what methodologies, techniques, and tools are to be used and the procedure for reporting the status of projects. As department leaders, IS managers are also responsible for career planning and development for systems analysts and other employees and for solving problems that arise in the course of development projects.

There are, of course, several IS managers in any medium to large IS department (see Figure 1-3). The manager of an entire IS department may have the title Chief Information Officer and may report to the president or chairman of the firm. Each division of the IS department will also have a manager. Typical titles for these managers are Director of IS Development, IS Operations Manager, and IS Programming Director. The Director of IS Development may be responsible for several development projects at any given time, each of which has a project manager. The responsibilities and focus of any particular IS manager depend on his or her level in the department and on how the organization manages and supports the systems development process.

Systems Analysts in Systems Development

Systems analysts are the key individuals in the systems development process. To succeed as a systems analyst, you will need to develop four skills: analytical, technical, managerial, and interpersonal. *Analytical skills* enable you to understand the organization and its functions, to identify opportunities and problems, and to analyze and solve problems. One of the most important analytical skills you can develop is systems thinking, or the ability to see organizations and information systems as systems. Systems thinking provides a framework from which to see the important relationships among information systems, the organizations they exist in, and the environment in which the organizations themselves exist. *Technical skills* help you understand the potential and the limitations of information technology. As an analyst, you must be able to envision an information system that will help users solve problems and that will guide the system's design and development. You must also be able to work with programming languages, various operating systems, and computer hardware platforms. *Management skills* help you manage projects, resources, risk, and change. *Interpersonal skills* help you work with end users as well as with other analysts and programmers. As a systems analyst, you will play a major role as a liaison among users, programmers, and other systems professionals. Effective written and oral communication, including competence in leading meetings, interviewing, and listening, is a key skill analysts must master. Effective analysts successfully combine these four skills, as Figure 1-4, a typical advertisement for a systems analyst position, illustrates.

As with any profession, becoming a good systems analyst takes years of study and experience. Once hired by an organization, you will generally be trained in the development methodology used by the organization. There is usually a career path for systems analysts that allows them to gain experience and advance into project management and further IS or business management. Many academic IS departments train their undergraduate students to be systems analysts. As your career progresses, you may get the chance to become a manager inside or outside the IS area. In some organizations, you can opt to follow a technical career advancement ladder. As an analyst, you will become aware of a consistent set of professional practices, many of which are governed by a professional code of ethics, similar to other professions.

Programmers in Systems Development

Programmers convert the system specifications given to them by the analysts into instructions the computer can understand. Writing a computer program is sometimes called writing code, or *coding*. Programmers also write program documentation and programs for testing systems. For many years, programming was considered an art. However, computer scientists found that code could be improved if it were structured, so they introduced what is now called *structured programming* (Bohm and Jacopini, 1966). In structured programming, all computing instructions can be represented through the use of three simple structures: sequence, repetition, and selection. Becoming a skilled programmer takes years of training and experience. Many computer information systems undergraduates begin work as programmers or programmer/analysts.

Figure 1-4
Typical job ad for a systems analyst

SIMON & TAYLOR, INC.
SYSTEMS ANALYST

Simon & Taylor, Inc., a candy manufacturer, has an immediate opening for a systems analyst in its Montana-based office.

The ideal candidate will have:

• A Bachelor's degree in management information systems and/or computer science.

• Two to three years' UNIX/RDBMS programming experience.

• Experience with the HP/UX Operating System or Linux, and HTML. Experience with Cold Fusion and knowledge of XML are desired but not essential.

• Familiarity with distribution and manufacturing concepts (allocation, replenishment, shop floor control, and production schedule).

• Working knowledge of project management and all phases of software development life cycle.

• Strong analytical and organizational skills.

We offer a competitive salary, a signing bonus, relocation assistance, and the challenges of working in a state-of-the-art IT environment.

E-mail your résumé to www.human_resources@simontaylor.com with salary requirement.

Simon & Taylor, Inc., is an Equal Opportunity Employer

Programming is very labor-intensive, therefore, special-purpose computing tools called *code generators* have been developed to generate reasonably good code from specifications, saving an organization time and money. Code generators do not put programmers out of work; rather, these tools change the nature of programming. Where code generators are in use, programmers take the generated code and fix problems with it, optimize it, and integrate it with other parts of the system. The goal of some computer-aided software engineering (CASE) tools is to provide a variety of code generators that can automatically produce 90 percent or more of code directly from the system specifications normally given a programmer. When this goal is achieved, the role of programmers on systems development teams will be changed further.

Business Managers in Systems Development

Another group important to systems development efforts is business managers, such as functional department heads and corporate executives. These managers are important to systems development because they have the power to fund development projects and to allocate the resources necessary for the projects' success. Because of their decision-making authority and knowledge of the firm's lines of business, department heads and executives are also able to set general requirements and constraints for development projects. In larger companies where the relative importance of systems projects is determined by a steering committee, these executives have additional power as they are usually members of the steering committees or systems planning groups. Business managers, therefore, have the power to set the direction for systems development, to propose and approve projects, and to determine the relative importance of projects that have already been approved and assigned to other people in the organization.

Other IS Managers/Technicians in Systems Development

In larger organizations where IS roles are more differentiated, there may be several additional IS professionals involved in the systems development effort. A firm with an existing set of databases will most likely have a *database administrator* who is usually involved in any systems project affecting the firm's databases. Network and *telecommunications experts* help develop systems involving data and/or voice communication, either internal or external to the organization. Some organizations have *human factors* departments that are concerned with system interfaces and ease-of-use issues, training users, and writing user documentation and manuals. Overseeing much of the development effort, especially for large or sensitive systems, are an organization's *internal auditors* who ensure that required controls are built into the system. In many organizations, auditors also have responsibility for keeping track of changes in the system's design. The necessary interaction of all these individuals makes systems development very much a team effort.

TYPES OF INFORMATION SYSTEMS AND SYSTEMS DEVELOPMENT

As you can see, several different people in an organization can be involved in developing information systems. Given the broad range of people and interests represented in systems development, you might assume that it could take several different types of information systems to satisfy all an organization's information system needs. Your assumption would be correct.

Up until now we have been talking about information systems in generic terms, but there are actually several different types or classes of information systems. In general, these types are distinguished from each other on the basis of what the system does or by the technology used to construct the system. As a systems analyst, part of your job will be to determine which kind of system will best address the organizational problem or opportunity on which you are focusing. In addition, different classes of systems may require different methodologies, techniques, and tools for development.

From your prior studies and experiences with information systems, you are probably aware of at least four classes of information systems:

- Transaction processing systems
- Management information systems
- Decision support systems (for individuals, groups, and executives)
- Expert systems

In addition, many organizations recognize scientific (or technical) computing and office automation systems. To preview the diversity of systems development approaches, the following sections briefly highlight how systems analysis and design methods differ across the four major types of systems.

Transaction Processing Systems

Transaction processing systems (TPS) automate the handling of data about business activities or transactions, which can be thought of as simple, discrete events in the life of an organization. Data about each transaction are captured, transactions are verified and accepted or rejected, and validated transactions are stored for later aggregation. Reports may be produced immediately to provide standard summarizations of transactions, and transactions may be moved from process to process in order to handle all aspects of the business activity.

The analysis and design of a TPS means focusing on the firm's current procedures for processing transactions, whether those procedures are manual or automated. The focus on current procedures implies a careful tracking of data capture, flow, processing, and output. The goal of TPS development is to improve transaction processing by speeding it up, using fewer people, improving efficiency and accuracy, integrating it with other organizational information systems, or providing information not previously available.

Management Information Systems

A management information system (MIS) takes the relatively raw data available through a TPS and converts them into a meaningful aggregated form that managers need to conduct their responsibilities. Developing an MIS calls for a good understanding of what kind of information managers require and how managers use information in their jobs. Sometimes managers themselves may not know precisely what they need or how they will use information. Thus, the analyst must also develop a good understanding of the business and the transaction processing systems that provide data for an MIS.

Management information systems often require data from several transaction processing systems (for example, customer order processing, raw material purchasing, and employee timekeeping). Development of an MIS can, therefore, benefit from a data-orientation, in which data are considered an organization resource separate from the TPS in which they are captured. Because it is important to be able to draw on data from various subject areas, developing a comprehensive and accurate model of data is essential in building an MIS.

Decision Support Systems

Decision support systems (DSS) are designed to help organizational decision makers make decisions. Instead of providing summaries of data, as with an MIS, a DSS provides an interactive environment in which decision makers can quickly manipulate data and models of business operations. A DSS is composed of a database (which may be extracted from a TPS or MIS), mathematical or graphical models of business processes, and a user interface (or dialogue module) that provides a way for the decision maker, usually a nontechnical manager, to communicate with the DSS. A DSS may use both hard historical data as well as judgments (or "what if" scenarios) about alternative histories or possible futures. In many cases, the historical data come from a firm's data warehouse. A data warehouse is a collection of integrated, subject-oriented databases, designed to support the decision support function, where each unit of data is relevant to some moment in time (Bischoff, 1997). One form of a DSS, an executive information system (EIS), emphasizes the unstructured capability for senior management to explore data starting at a high level of aggregation and selectively drilling down into specific areas where more detailed understandings of the business are required. In either case, a DSS is characterized by less structured and predictable use; rather, a DSS is a software resource intended to support a certain scope of decision-making activities (from problem finding to choosing a course of action).

The systems analysis and design for a DSS often concentrates on the three main DSS components: database, model base, and user dialogue. As with an MIS, a data-orientation is most often used for understanding user requirements. In addition, the systems analysis and design project will carefully document the mathematical rules that define interrelationships among different data. These relationships are used to predict future data or to find the best solutions to decision problems. Thus, decision logic must be carefully understood and documented. Also, since a decision maker typically interacts with a DSS, the design of easy-to-use yet thorough user dialogues and screens is important. Because a DSS often deals with situations not encountered every day or situations that can be handled in many different ways, there can be considerable uncertainty on what a DSS should actually do. Thus, systems developers often use methods that prototype the system and iteratively and rapidly redevelop the system based on trial use. The development of a DSS, hence, often does not follow as formal a project plan as is done for a TPS or MIS, since the software deliverable is more uncertain at the beginning of the project.

Expert Systems

Different from any of the other classes of systems we have discussed so far, an expert system (ES) attempts to codify and manipulate knowledge rather than information. If-then-else rules or other knowledge representation forms describe the way an expert would approach situations in a specific domain of problems. Typically, users communicate with an ES through an interactive dialogue. The ES asks questions (that an expert would ask) and the end user supplies the answers. The answers are then used to determine which rules apply and the ES provides a recommendation based on the rules.

The focus on developing an ES is acquiring the knowledge of the expert in the particular problem domain. Knowledge engineers perform knowledge acquisition; they are similar to systems analysts but are trained to use different techniques, as determining knowledge is considered more difficult than determining data.

Summary of Information Systems Types

Many information systems you build or maintain will contain aspects of each of the four major types of information systems. Thus, as a systems analyst, you will likely employ specific methodologies, techniques, and tools associated with each of the four information system types. Table 1-3 summarizes the general characteristics and development methods for each type.

So far, we have concentrated on the context of information systems development, looking at the different organizations where software is developed, the people involved in development efforts, and the different types of information systems that exist in organizations. Now that we have a good idea of context, we can turn to the actual process by which many information systems are developed in organizations, the *systems development life cycle.*

TABLE 1-3   **Systems Development for Different IS Types**

IS Type	IS Characteristics	Systems Development Methods
Transaction processing system	High-volume, data capture focus; goal is efficiency of data movement and processing and interfacing different TPSs	Process-orientation; concern with capturing, validating, and storing data and with moving data between each required step
Management information system	Draws on diverse yet predictable data resources to aggregate and summarize data; may involve forecasting future data from historical trends and business knowledge	Data-orientation; concern with understanding relationships between data so data can be accessed and summarized in a variety of ways; builds a model of data that supports a variety of uses
Decision support system	Provides guidance in identifying problems, finding and evaluating alternative solutions, and selecting or comparing alternatives; potentially involves groups of decision makers; often involves semi-structured problems and the need to access data at different levels of detail	Data- and decision logic-orientations; design of user dialogue; group communication may also be key and access to unpredictable data may be necessary; nature of systems require iterative development and almost constant updating
Expert system	Provides expert advice by asking users a sequence of questions dependent on prior answers that lead to a conclusion or recommendation	A specialized decision logic-orientation in which knowledge is elicited from experts and described by rules or other forms

DEVELOPING INFORMATION SYSTEMS AND THE SYSTEMS DEVELOPMENT LIFE CYCLE

Systems development methodology:
A standard process followed in an organization to conduct all the steps necessary to analyze, design, implement, and maintain information systems.

Systems development life cycle (SDLC): The traditional methodology used to develop, maintain, and replace information systems.

N E T   S E A R C H
Different organizations and authors have not agreed on a single SDLC. Visit http://www.prenhall.com/hoffer to complete an exercise on this topic.

Most organizations find it beneficial to use a standard set of steps, called a **systems development methodology**, to develop and support their information systems. Like many processes, the development of information systems often follows a life cycle. For example, a commercial product follows a life cycle in that it is created, tested, and introduced to the market. Its sales increase, peak, and decline. Finally, the product is removed from the market and replaced by something else. The **systems development life cycle (SDLC)** is a common methodology for systems development in many organizations, featuring several phases that mark the progress of the systems analysis and design effort. Every textbook author and information system development organization uses a slightly different life cycle model, with anywhere from three to almost twenty identifiable phases.

Although any life cycle appears at first glance to be a sequentially ordered set of phases, it actually is not (see Figure 1-5). The specific steps and their sequence are meant to be adapted as required for a project, consistent with management approaches. For example, in any given SDLC phase, the project can return to an earlier phase if necessary. Similarly, if a commercial product does not perform well just after its introduction, it may be temporarily removed from the market and improved before being reintroduced. In the systems development life cycle, it is also possible to complete some activities in one phase in parallel with some activities of another phase. Sometimes the life cycle is iterative; that is, phases are repeated as required until an acceptable system is found. Such an iterative approach is especially characteristic of rapid application development methods, such as prototyping, which we introduce later in the chapter (see Figure 1-6). Some people consider the life cycle to be a spiral, in which we constantly cycle through the phases at different levels of detail (see Figure 1-7). The life cycle can also be thought of as a circular process in

Figure 1-5
The systems development life cycle

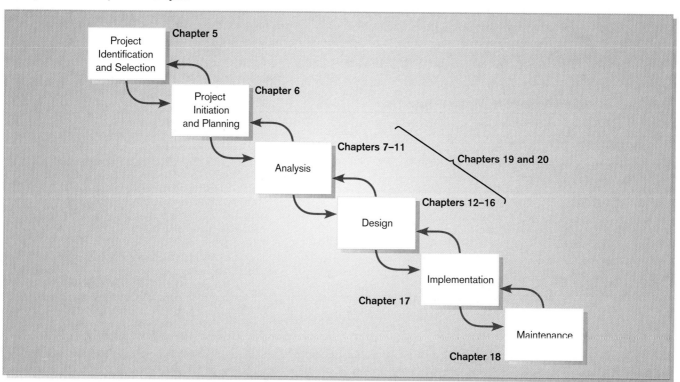

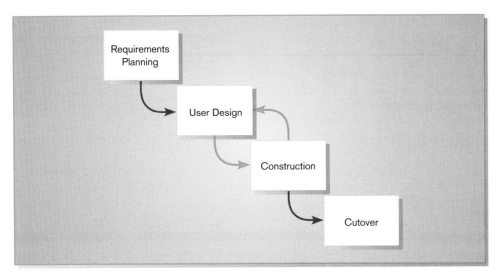

Figure 1-6
Rapid application development
SDLC

which the end of the useful life of one system leads to the beginning of another project that will develop a new version or replace an existing system altogether. However conceived, the systems development life cycle used in an organization is an orderly set of activities conducted and planned for each development project. The skills required of a systems analyst apply to all life cycle models. Software is the most obvious end product of the life cycle; other essential outputs include documentation about the system and how it was developed as well as training for users.

Every medium to large corporation and every custom software producer will have its own specific, detailed life cycle or systems development methodology in place (see Figure 1-8). Even if a particular methodology does not look like a cycle, you will probably discover that many of the SDLC steps are performed and SDLC techniques and tools are used. Learning about systems analysis and design from the life cycle approach will serve you well no matter which systems development methodology you use.

When you begin your first job, you are likely to spend several weeks or months learning your organization's SDLC and its associated methodologies, techniques, and tools. In order to make this book as general as possible, we follow a rather generic life

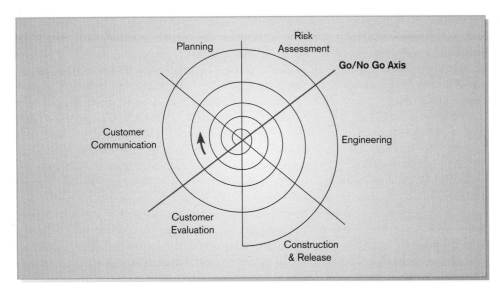

Figure 1-7
Evolutionary model SDLC

Figure 1-8
Seer Technologies, Inc., SDLC
*(Source: The Systems Development Life Cycle,
© 1990–2000 Level 8 Technologies, Inc.
Reprinted by permission.)*

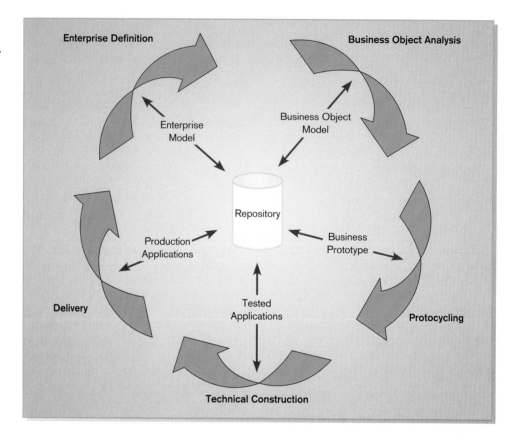

cycle model, as described in more detail in Figure 1-5. Notice how our model resembles a staircase with arrows connecting each step to the step before it and to the step after it. This representation of the SDLC is sometimes referred to as the waterfall model. We use this SDLC as one example of a methodology but, more importantly, as a way to arrange the topics of systems analysis and design. Thus, what you learn in this book you can apply to almost any life cycle you might follow. As we describe this SDLC throughout the book, you will see that each phase has specific outcomes and deliverables that feed important information to other phases. At the end of each phase (and sometimes within phases for intermediate steps), a systems development project reaches a *milestone* and, as deliverables are produced, they are often reviewed by parties outside the project team. In the rest of this section we provide a brief overview of each SDLC phase. At the end of the section we summarize this discussion in a table listing the main deliverables or outputs from each SDLC phase.

Project identification and selection:
The first phase of the SDLC in which an organization's total information system needs are identified, analyzed, prioritized, and arranged.

The first phase in the SDLC is called **project identification and selection**. In this phase, someone identifies the need for a new or enhanced system. In larger organizations, this recognition may be part of a corporate and systems planning process. Information needs of the organization as a whole are examined, and projects to meet these needs are proactively identified. The organization's information system needs may result from requests to deal with problems in current procedures, from the desire to perform additional tasks, or from the realization that information technology could be used to capitalize on an existing opportunity. These needs can then be prioritized and translated into a plan for the IS department, including a schedule for developing new major systems. In smaller organizations (as well as in large ones), determination of which systems to develop may be affected by ad hoc user requests submitted as the need for new or enhanced systems arises as well as from a formalized information planning process. In either case, during project identification and selection, an organization determines whether or not resources should be devoted to the development or enhancement of each information system under consideration. The outcome of the project identification and selection process is a determination of which systems development projects should be undertaken by the organization, at least in terms of an initial study.

The second phase is **project initiation and planning**. The two major activities in this phase are the formal, yet still preliminary, investigation of the system problem or opportunity at hand and the presentation of reasons why the system should or should not be developed by the organization. A critical step at this point is determining the scope of the proposed system. The project leader and initial team of systems analysts also produce a specific plan for the proposed project the team will follow using the remaining SDLC steps. This baseline project plan customizes the standardized SDLC and specifies the time and resources needed for its execution. The formal definition of a project is based on the likelihood that the organization's IS department is able to develop a system that will solve the problem or exploit the opportunity and determine whether the costs of developing the system outweigh the benefits it could provide. The final presentation of the business case for proceeding with the subsequent project phases is usually made by the project leader and other team members to someone in management or to a special management committee with the job of deciding which projects the organization will undertake.

Project initiation and planning: The second phase of the SDLC in which a potential information systems project is explained and an argument for continuing or not continuing with the project is presented; a detailed plan is also developed for conducting the remaining phases of the SDLC for the proposed system.

The next phase is **analysis**. During this phase, the analyst thoroughly studies the organization's current procedures and the information systems used to perform organizational tasks. Analysis has several subphases. The first is requirements determination. In this subphase, you and other analysts work with users to determine what the users want from a proposed system. This subphase usually involves a careful study of any current systems, manual and computerized, that might be replaced or enhanced as part of this project. Next, you study the requirements and structure them according to their interrelationships and eliminate any redundancies. Third, you generate alternative initial designs to match the requirements. Then you compare these alternatives to determine which best meets the requirements within the cost, labor, and technical levels the organization is willing to commit to the development process. The output of the analysis phase is a description of (but not a detailed design for) the alternative solution recommended by the analysis team. Once the recommendation is accepted by those with funding authority, you can begin to make plans to acquire any hardware and system software necessary to build or operate the system as proposed.

Analysis: The third phase of the SDLC in which the current system is studied and alternative replacement systems are proposed.

The fourth phase is devoted to designing the new or enhanced system. During **design**, you and the other analysts convert the description of the recommended alternative solution into logical and then physical system specifications. You must design all aspects of the system from input and output screens to reports, databases, and computer processes. You must then provide the physical specifics of the system you have designed, either as a model or as detailed documentation, to guide those who will build the new system.

Design: The fourth phase of the SDLC in which the description of the recommended solution is converted into logical and then physical system specifications.

That part of the design process that is independent of any specific hardware or software platform is referred to as **logical design**. Theoretically, the system could be implemented on any hardware and systems software. The idea is to make sure that the system functions as intended. Logical design concentrates on the business aspects of the system and tends to be oriented to a high level of specificity.

Logical design: The part of the design phase of the SDLC in which all functional features of the system chosen for development in analysis are described independently of any computer platform.

Once the overall high-level design of the system is worked out, you begin turning logical specifications into physical ones. This process is referred to as **physical design**. As part of physical design, you design the various parts of the system to perform the physical operations necessary to facilitate data capture, processing, and information output. This can be done in many ways, from creating a working model of the system to be implemented, to writing detailed specifications describing all the different parts of the system and how they should be built. In many cases, the working model becomes the basis for the actual system to be used. During physical design, the analyst team must determine many of the physical details necessary to build the final sys-

Physical design: The part of the design phase of the SDLC in which the logical specifications of the system from logical design are transformed into technology-specific details from which all programming and system construction can be accomplished.

tem, from the programming language the system will be written in to the database system that will store the data to the hardware platform on which the system will run. Often the choices of language and database and platform are already decided by the organization or by the client, and at this point, these information technologies must be taken into account in the physical design of the system. The final product of the design phase is the physical system specifications in a form ready to be turned over to programmers and other system builders for construction. Figure 1-9 illustrates the difference between logical and physical design.

The physical system specifications, whether in the form of a detailed model or as detailed written specifications, are turned over to programmers as the first part of the **implementation** phase. During implementation, you turn system specifications into a working system that is tested and then put into use. Implementation includes coding, testing, and installation. During *coding*, programmers write the programs that make up the system. Sometimes the code is generated by the same system used to build the detailed model of the system. During *testing*, programmers and analysts test individual programs and the entire system in order to find and correct errors. During *installation*, the new system becomes a part of the daily activities of the organization. Application software is installed, or loaded, on existing or new hardware and users are

Implementation: The fifth phase of the SDLC in which the information system is coded, tested, installed, and supported in the organization.

Figure 1-9
The difference between logical design and physical design

(a) A skateboard ramp blueprint (logical design)

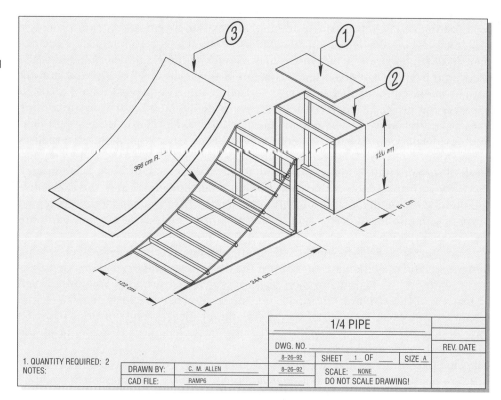

(b) A skateboard ramp (physical design)
(Sources: www.tumyeto.com/tydu/skatebrd/ organizations/plans/14pipe.jpg; www.tumyeto.com/tydu/skatebrd/ organizations/iuscblue.html; accessed September 16, 1999. Reprinted by permission of the International Association of Skateboard Companies.)

introduced to the new system and trained. Begin planning for both testing and installation early as the project initiation and planning phase, since both testing and installation require extensive analysis in order to develop exactly the right approach.

Implementation activities also include initial user support such as the finalization of documentation, training programs, and ongoing user assistance. Note that documentation and training programs are finalized during implementation; documentation is produced throughout the life cycle, and training (and education) occur from the inception of a project. Implementation can continue for as long as the system exists since ongoing user support is also part of implementation. Despite the best efforts of analysts, managers, and programmers, however, installation is not always a simple process. Many well-designed systems have failed because the installation process was faulty. Our point is that even a well-designed system can fail if implementation is not well managed. Since the management of implementation is usually done by the project team, we stress implementation issues throughout this book.

The final phase is **maintenance**. When a system (including its training, documentation, and support) is operating in an organization, users sometimes find problems with how it works and often think of better ways to perform its functions. Also, the organization's needs with respect to the system change over time. In maintenance, programmers make the changes that users ask for and modify the system to reflect changing business conditions. These changes are necessary to keep the system running and useful. In a sense, maintenance is not a separate phase but a repetition of the other life cycle phases required to study and implement the needed changes. Thus, you might think of maintenance as an overlay to the life cycle rather than a separate phase. The amount of time and effort devoted to maintenance depends a great deal on the performance of the previous phases of the life cycle. There inevitably comes a time, however, when an information system is no longer performing as desired, when maintenance costs become prohibitive, or when an organization's needs have changed substantially. Such problems indicate that it is time to begin designing the system's replacement, thereby completing the loop and starting the life cycle over again. Often the distinction between major maintenance and new development is not clear, which is another reason maintenance often resembles the life cycle itself.

The SDLC is a highly linked set of phases whose products feed the activities in subsequent phases. Table 1-4 summarizes the outputs or products of each phase

Maintenance: The final phase of the SDLC in which an information system is systematically repaired and improved.

TABLE 1-4   Products of SDLC Phases

Phase	Products, Outputs, or Deliverables
Project identification and selection	Priorities for systems and projects; an architecture for data, networks, hardware, and IS management is the result of associated systems planning activities
Project initiation and planning	Detailed steps, or work plan, for project; specification of system scope and high-level system requirements or features; assignment of team members and other resources; system justification or business case
Analysis	Description of current system and where problems or opportunities are with a general recommendation on how to fix, enhance, or replace current system; explanation of alternative systems and justification for chosen alternative
Logical design	Functional, detailed specifications of all system elements (data, processes, inputs, and outputs)
Physical design	Technical, detailed specifications of all system elements (programs, files, network, system software, etc.); acquisition plan for new technology
Implementation	Code, documentation, training procedures, and support capabilities
Maintenance	New versions or releases of software with associated updates to documentation, training, and support

based on the in-text descriptions. The chapters on the SDLC phases will elaborate on the products of each phase as well as on how the products are developed.

Throughout the systems development life cycle, the systems development project itself needs to be carefully planned and managed. The larger the systems project, the greater the need for project management. Several project management techniques have been developed in this century and many have been made more useful through automation. Chapter 3 contains a more detailed treatment of project planning and management techniques. Next, we will discuss some of the criticisms of the systems development life cycle and alternatives developed to address those criticisms.

The Traditional SDLC

There are several criticisms of the traditional life cycle approach to systems development as followed exactly as outlined in Figure 1-5. One criticism relates to the way the life cycle is organized. Although we know that phases of the life cycle can sometimes overlap, traditionally one phase ended and another began once a *milestone* had been reached. The milestone usually took the form of some deliverable or pre-specified output from the phase. For example, the design deliverable is the set of detailed physical design specifications. Once the milestone had been reached and the new phase initiated, it became difficult to go back. Even though business conditions continued to change during the development process and analysts were pressured by users and others to alter the design to match changing conditions, it was necessary for the analysts to freeze the design at a particular point and go forward. The enormous amount of effort and time necessary to implement a specific design meant that it would be very expensive to make changes in a system once it was developed. There were no CASE tools, no code generators, and no fourth-generation languages when the SDLC was popularized in the 1960s. If the design was not frozen, the system would never be completed, as programmers would no sooner be done with coding one design than they would receive requests for major changes. The traditional life cycle, then, had the property of locking in users to requirements that had been previously determined, even though those requirements might have changed.

Another criticism of the way the traditional life cycle is often used is that it tends to focus too little time on good analysis and design. The result is a system that does not match users' needs and one that requires extensive maintenance, unnecessarily increasing development costs. According to some estimates, maintenance costs account for 40 to 70 percent of the system development costs (Dorfman and Thayer, 1997). Given these problems, people working in systems development began to look for better ways to conduct systems analysis and design.

Structured Analysis and Structured Design

Ed Yourdon and his colleagues developed *structured analysis* and *structured design* in the early 1970s as a way to address some of the problems with the traditional SDLC (Yourdon and Constantine, 1979). By making analysis and design more disciplined, similar to engineering fields, through the use of tools such as data flow diagrams and transform analysis, Yourdon and colleagues sought to emphasize and improve the analysis and design phases of the life cycle. The goal was to reduce maintenance time and effort. Structured analysis and design makes it easier to go back to earlier phases in the life cycle when necessary—for example, when requirements change. Finally, there is also an emphasis on partitioning or dividing a problem into smaller, more manageable units and on making a clear distinction between physical and logical design (DeMarco, 1979; Yourdon and Constantine, 1979). The life cycle used in this book is faithful to these structured principles.

Object-Oriented Analysis and Design

A more recent approach to systems development that is becoming more and more popular is **object-oriented analysis and design (OOAD)** (we elaborate on this approach in Chapter 20). OOAD is often called the third approach to systems development, after the process-oriented and data-oriented approaches. The object-oriented approach combines data and processes (called *methods*) into single entities called **objects**. Objects usually correspond to the real things an information system deals with, such as customers, suppliers, contracts, and rental agreements. Putting data and processes together in one place recognizes the fact that there are a limited number of operations for any given data structure. Putting data and processes together makes sense even though typical systems development keeps data and processes independent of each other. The goal of OOAD is to make system elements more reusable, thus improving system quality and the productivity of systems analysis and design.

Another key idea behind object-orientation is that of **inheritance**. Objects are organized into **object classes**, which are groups of objects sharing structural and behavioral characteristics. Inheritance allows the creation of new classes that share some of the characteristics of existing classes. For example, from a class of objects called "person," you can use inheritance to define another class of objects called "customer." Objects of the class "customer" would share certain characteristics with objects of the class "person": they would both have names, addresses, phone numbers, and so on. Since "person" is the more general class and "customer" is more specific, every customer is a person but not every person is a customer.

As you might expect, you need a computer programming language that can create and manipulate objects and classes of objects in order to create object-oriented information systems. Several object-oriented programming languages have been created (e.g., C++, Eiffel, and ObjectPAL—for Paradox for Windows). In fact, object-oriented languages were developed first and object-oriented analysis and design techniques followed. Because OOAD is still relatively new, there is little consensus or standardization among the many OOAD techniques available. In general, the primary task of object-oriented analysis is identifying objects, defining their structure and behavior, and defining their relationships. The primary tasks of object-oriented design are modeling the details of the objects' behavior and communication with other objects so that system requirements are met and reexamining and redefining objects to better take advantage of inheritance and other benefits of object-orientation.

Object-oriented analysis and design (OOAD): Systems development methodologies and techniques based on objects rather than data or processes.

Object: A structure that encapsulates (or packages) attributes and methods that operate on those attributes. An object is an abstraction of a real-world thing in which data and processes are placed together to model the structure and behavior of the real-world object.

Inheritance: The property that occurs when entity types or object classes are arranged in a hierarchy and each entity type or object class assumes the attributes and methods of its ancestors; that is, those higher up in the hierarchy. Inheritance allows new but related classes to be derived from existing classes.

Object class: A logical grouping of objects that have the same (or similar) attributes and behaviors (methods).

DIFFERENT APPROACHES TO IMPROVING DEVELOPMENT

In the continuing effort to improve the systems analysis and design process, several different approaches have been developed. We will describe the more important approaches in more detail in later chapters. Attempts to make system development less of an art and more of a science are usually referred to as *systems engineering* or *software engineering*. As the names indicate, rigorous engineering techniques are applied to systems development. A very influential practice borrowed from engineering is called prototyping. We will discuss prototyping next, followed by an introduction to *Joint Application Design (JAD)*. Both prototyping and JAD are fast becoming standard parts of the typical systems analysis and design process.

Prototyping

Designing and building a scaled-down but functional version of a desired system is the process known as **prototyping**. You can build a prototype with any computer language or development tool, but special prototyping tools have been developed to

Prototyping: An iterative process of systems development in which requirements are converted to a working system that is continually revised through close work between an analyst and users.

Figure 1-10
The prototyping methodology
(Adapted from "Prototyping: The New Paradigm for Systems Development," by J. D. Naumann and A. M. Jenkins, MIS Quarterly 6 (3), pp. 29–44.)

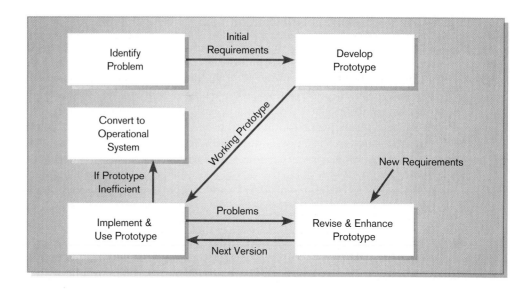

N E T   S E A R C H
CASE tools are used to automate the SDLC. Visit http://www.prenhall.com/hoffer to complete an exercise on this topic.

simplify the process. A prototype can be developed with some fourth-generation languages (4GLs), with the query and screen and report design tools of a database management system, and with tools called *computer-aided software engineering (CASE)* tools.

Using prototyping as a development technique (see Figure 1-10), the analyst works with users to determine the initial or basic requirements for the system. The analyst then quickly builds a prototype. When the prototype is completed, the users work with it and tell the analyst what they like and do not like about it. The analyst uses this feedback to improve the prototype and takes the new version back to the users. This iterative process continues until the users are relatively satisfied with what they have seen. Two key advantages of the prototyping technique are the large extent to which prototyping involves the user in analysis and design and its ability to capture requirements in concrete, rather than verbal or abstract, form. In addition to being used stand-alone, prototyping may also be used to augment the SDLC. For example, a prototype of the final system may be developed early in analysis to help the analysts identify what users want. Then the final system is developed based on the specifications of the prototype. We discuss prototyping in greater detail in Chapter 7 and use various prototyping tools in Chapters 13 and 14 to illustrate the design of system outputs.

Prototyping is a form of rapid application development, or RAD. The fundamental principle of any RAD methodology is to delay producing detailed system design documents until after user requirements are clear. The prototype serves as the working description of needs. RAD methodologies emphasize gaining user acceptance of the human-system interface and developing core capabilities as quickly as possible, sacrificing computer efficiency for gains in human efficiency in rapidly building and rebuilding working systems. On the other hand, RAD methodologies can overlook important software engineering principles, the result of which are inconsistencies between system modules, noncompliance with standards, and lack of reusability of system components (Bourne, 1994). Chapter 19 addresses RAD in detail.

Joint Application Design

Joint Application Design (JAD): A structured process in which users, managers, and analysts work together for several days in a series of intensive meetings to specify or review system requirements.

In the late 1970s, systems development personnel at IBM developed a new process for collecting information system requirements and reviewing system designs. The process is called **Joint Application Design (JAD)**. The basic idea behind JAD is to bring structure to the requirements determination phase of analysis and to the reviews that occur as part of design. Users, managers, and systems developers are brought together for a series of intensive structured meetings run by a JAD session leader who maintains the structure and adheres to the agenda. By gathering the people directly affected by an IS in one room at the same time to work together to agree on system requirements and design details, time and organizational resources are better managed. As an added plus, group members are more likely to develop a shared understanding of what the IS is supposed to do. We will discuss JAD in more detail in Chapter 7.

IMPROVING IS DEVELOPMENT PRODUCTIVITY

Other efforts to improve the system development process have taken advantage of the benefits offered by computing technology itself. The result has been the creation and fairly widespread use of *computer-aided software engineering* or *CASE* tools. CASE tools have been developed for internal use and for sale by several leading firms, including Oracle (Designer), and Computer Associates (COOL: Gen), to name a couple.

CASE tools are built around a central repository for system descriptions and specifications, including information about data names, format, uses, and locations. The idea of a central repository of information about the project is not new—the manual form of such a repository is called a project dictionary or workbook. The difference is that CASE tools automate the repository for easier updating and for consistency. CASE tools also include diagramming tools for data flow diagrams and other graphical aids, screen and report design tools, and other special-purpose tools. CASE helps programmers and analysts do their jobs more efficiently and more effectively by automating routine tasks. There is more information on CASE in Chapter 4, and we relate many examples of the use of CASE throughout this book.

Summary

This chapter introduced you to information systems analysis and design, the complex organizational process whereby computer-based information systems are developed and maintained. You read about the differences between the process-oriented and data-oriented approaches to systems analysis and design: Process-orientation focuses on what the system is supposed to do while data-orientation focuses on the data the system needs to operate. Process-orientation provides a less stable design than does data-orientation, as business processes change faster than do the data an organization uses. With process-orientation, data files are designed for specific applications whereas data files are designed for the whole enterprise with data-orientation; process-orientation leads to much uncontrolled data redundancy whereas data redundancy is controlled under data-orientation. You also learned about application independence, the separation of data from the computer applications that use the data. Data-orientation and application independence frame the way you learn about systems analysis and design in this book.

A major part of this chapter was devoted to examining the context of systems analysis and design. You read about the various people in organizations who develop systems, including systems analysts, programmers, IS managers, business managers, end users, database administrators, human factors experts, telecommunications experts, and auditors. You also learned that there are many different kinds of information systems used in organizations, from transaction processing systems to expert systems to office systems. Development techniques vary with system type.

Finally, you learned about the basic framework that guides systems analysis and design, the systems development life cycle, with its six major phases: project identification and selection, project initiation and planning, analysis, design, implementation, and maintenance. The life cycle has had its share of criticism, which you read about, and other frameworks have been developed to address the life cycle's problems. These frameworks include prototyping (a Rapid Application Development approach) and Joint Application Design.

Key Terms

1. Analysis
2. Application independence
3. Application software
4. Data
5. Database
6. Data flow
7. Data-oriented approach
8. Design
9. Implementation
10. Information
11. Information systems analysis and design

12. Inheritance
13. Joint Application Design (JAD)
14. Logical design
15. Maintenance
16. Object
17. Object class
18. Object-oriented analysis and design (OOAD)
19. Physical design

20. Process-oriented approach
21. Processing logic
22. Project identification and selection
23. Project initiation and planning
24. Prototyping
25. Stakeholder
26. Systems analyst
27. Systems development life cycle (SDLC)
28. Systems development methodology

Match each of the key terms above with the definition that best fits it.

_____ Systems development methodologies and techniques based on objects rather than data or processes.

_____ The first phase of the SDLC, in which an organization's total information system needs are identified, analyzed, prioritized, and arranged.

_____ The second phase of the SDLC in which a potential information systems project is explained and an argument for continuing or not continuing with the project is presented; a detailed plan is also developed for conducting the remaining phases of the SDLC for the proposed system.

_____ The fourth phase of the SDLC in which the description of the recommended solution is converted into logical and then physical system specifications.

_____ The complex organizational process whereby computer-based information systems are developed and maintained.

_____ Computer software designed to support organizational functions or processes.

_____ The organizational role most responsible for the analysis and design of information systems.

_____ An entity that has a well-defined role in the application domain and has state, behavior, and identity.

_____ A structured process in which users, managers, and analysts work together for several days in a series of intensive meetings to specify or review system requirements.

_____ An iterative process of systems development in which requirements are converted to a working system that is continually revised through close work between an analyst and users.

_____ The part of the design phase of the SDLC in which all functional features of the system chosen for development in analysis are described independent of any computer platform.

_____ The part of the design phase of the SDLC in which the logical specifications of the system from logical design are transformed into technology-specific details from which all programming and system construction can be accomplished.

_____ A set of objects that share a common structure and a common behavior.

_____ The third phase of the SDLC in which the current system is studied and alternative replacement systems are proposed.

_____ The fifth phase of the SDLC in which the information system is coded, tested, installed, and supported in the organization.

_____ The final phase of the SDLC in which an information system is systematically repaired and improved; or changes made to a system to fix or enhance its functionality.

_____ A standard process followed in an organization to conduct all the steps necessary to analyze, design, implement, and maintain information systems.

_____ The property that occurs when entity types or object classes are arranged in a hierarchy and each entity type or object class assumes the attributes and methods of its ancestors, that is, those higher up in the hierarchy.

_____ The traditional methodology used to develop, maintain, and replace information systems.

_____ The separation of data and the definition of data from the applications that use these data.

_____ A shared collection of logically related data designed to meet the information needs of multiple users in an organization.

_____ Data in motion, moving from one place in a system to another.

_____ The steps by which data are transformed or moved and a description of the events that trigger these steps.

_____ An overall strategy for information systems development that focuses on how and when data are moved through and changed by an information system.

_____ An overall strategy of information systems development that focuses on the ideal organization of data rather than on where and how they are used.

_____ Data that have been processed and presented in a form suitable for human interpretation, often with the purpose of revealing trends or patterns.

_____ Raw facts about people, objects, and events in an organization.

_____ A person who has an interest in an existing or new information system. Someone who is involved in the development of a system, in the use of a system, or someone who has authority over the parts of the organization affected by the system.

Review Questions

1. What is information systems analysis and design?

2. Explain the traditional application-based approach to systems development. How is this different from the data-based approach?

3. What are the organizational roles associated with systems development? Describe the responsibilities of each role.

4. List the different classes of information systems described in this chapter. How do they differ from each other?

5. List and explain the different phases in the systems development life cycle.

6. What are structured analysis and structured design?

7. What is prototyping?

8. What is JAD?

9. What is object-oriented analysis and design?

10. Explain how systems analysis and design has changed from 1950–2000.

11. What are the characteristics of successful systems development teams?

Problems and Exercises

1. Why is it important to use systems analysis and design methodologies when building a system? Why not just build the system in whatever way seems to be "quick and easy"? What value is provided by using an "engineering" approach?

2. Choose a business transaction you undertake regularly, such as using an ATM machine, buying groceries at the supermarket, or buying a ticket for a university's basketball game. For this transaction, define the data, draw the data flow diagram, and describe processing logic.

3. How would you organize a project team of students to work with a small business client? How would you organize a project team if you were working for a professional consulting organization? How might these two methods of organization differ? Why?

4. How might prototyping be used as part of the SDLC?

5. Describe the difference in the role of a systems analyst in the SDLC versus prototyping.

6. Contrast process-oriented and data-oriented approaches to systems analysis and design. Why does this book make the point that these are complementary, not competing, approaches to systems development?

7. Compare Figures 1-7 and 1-10. What similarities and differences do you see?

8. Compare Figures 1-5 and 1-8. Can you match steps in Figure 1-8 with phases in Figure 1-5? How might you explain the differences?

9. Construct a table with the six phases of the SDLC as columns and the various people/roles involved in a systems development project as the rows. Place a P in a cell if that person participates in that phase, and place an L in a cell if that person plays a leadership role in that phase. Explain your responses.

Field Exercises

1. Choose an organization with a fairly extensive information systems department. Get a copy of its organization chart (or draw one). At what level in the organization is the highest ranking information systems employee? What is his or her title? To whom does he or she report? What are his or her primary duties? Does the nature of this person's job description, responsibilities, and authority help or hinder this person in doing a good job?

2. Choose an organization and identify personnel who fulfill each of the following roles: IS manager, systems analyst, programmer, end user, business manager, database administrator, network and/or telecommunications manager, information system security manager. Are these roles filled formally or informally? Draw an organization chart linking these people. When these people build, use, and maintain the information systems, do they work together as a team or is

their work fairly independent of each other? Why? Are they effective? Why or why not?

3. Choose an organization that you interact with regularly and list as many different "systems" (whether computer-based or not) as you can that are used to process transactions, provide information to managers and executives, help managers and executives make decisions, aid groups to make decisions together, capture knowledge and provide expertise, help design products and/or facilities, and assist people in communicating with each other. Draw a diagram that shows how each of these systems interacts (or should interact) with each other. Are these systems well integrated?

4. Imagine an information system built without using a systems analysis and design methodology and without any thinking about the SDLC. Use your imagination and describe any and all problems that might occur, even if they seem a bit extreme and absurd. Surprisingly, the problems you will describe have probably already happened in one setting or another.

5. Choose a relatively small organization that is just beginning to use information systems. What types of systems are being used? for what purposes? To what extent are these systems integrated with each other? with systems outside the organization? How are these systems developed and controlled? Who is involved in systems development, use, and control?

6. You may want to keep a personal journal of ideas and observations about systems analysis and design while you are studying this book. Use this journal to record comments you hear, summaries of news stories or professional articles you read, original ideas or hypotheses you create, and questions that require further analysis. Keep your eyes and ears open for anything related to systems analysis and design. Your instructor may ask you to turn in a copy of your journal from time to time in order to provide feedback and reactions. The journal is an unstructured set of personal notes that will supplement your class notes and can stimulate you to think beyond the topics covered within the time limitations of most courses.

References

Bischoff, J. 1997. *Data Warehouse*. Upper Saddle River, NJ: Prentice-Hall, Inc.

Bohm, C., and I. Jacopini. 1966. "Flow Diagrams, Turing Machines, and Languages with Only Two Formation Rules." *Communications of the ACM* 9 (May): 366–71.

Bourne, K. C. 1994. "Putting Rigor Back in RAD." *Database Programming & Design* 7(8) (Aug.): 25–30.

DeMarco, T. 1979. *Structured Analysis and System Specification*. Englewood Cliffs, NJ: Prentice-Hall.

Dorfman, M. and R. M. Thayer (eds). 1997. *Software Engineering*. Los Alamitos, CA: IEEE Computer Society Press.

Information Technology Association of America. 2000. *Bridging the Gap: Information Technology Skills for a New Millennium*. Arlington, VA: ITAA.

Naumann, J. D., and A. M. Jenkins. 1982. "Prototyping: The New Paradigm for Systems Development." *MIS Quarterly* 6 (3): 29–44.

Yourdon, E., and L. L. Constantine. 1979. *Structured Design*. Englewood Cliffs, NJ: Prentice-Hall.

Chapter 2

Succeeding as a Systems Analyst

LEARNING OBJECTIVES

After studying this chapter, you should be able to:

- Discuss the analytical skills, including systems thinking, needed for a systems analyst to be successful.
- Describe the technical skills required of a systems analyst.
- Discuss the management skills required of a systems analyst.
- Identify the interpersonal skills required of a systems analyst.
- Describe the systems analysis profession.

INTRODUCTION

In the first chapter, you learned about the different types of information systems developed in organizations, the people who develop them, and the project environment in which systems are developed. Before we explore the systems development life cycle in more detail, however, we need to examine the skills needed to succeed as a systems analyst. You will first examine the analytical skills a systems analyst needs, then discuss the technical, management, and interpersonal skills required of a good analyst. One of the key analytical skills you will study is systems thinking, or the ability to see things as *systems*. You probably learned about systems and systems thinking in your introductory information systems class, so we will review here the highlights of systems and systems thinking that directly affect the design of information systems and how a systems analyst develops systems.

As illustrated in Figure 2-1, an analyst works throughout all phases of the systems development life cycle. The life cycle model represents the process of developing information systems, the same process you read about in Chapter 1. The skills the analyst needs to be successful are represented by the objects placed in the diagram. The laptop computer represents technical skills; the briefcase represents management skills; the magnifying glass represents analytical skills; and the telephone represents interpersonal skills. As you can see, to follow the guidelines established by any development methodology, an analyst needs to rely on many skills. Although we cannot possibly provide thorough coverage of these skills in this chapter, some will be covered in considerable depth in later chapters while others are discussed more generally. Our goal for these general

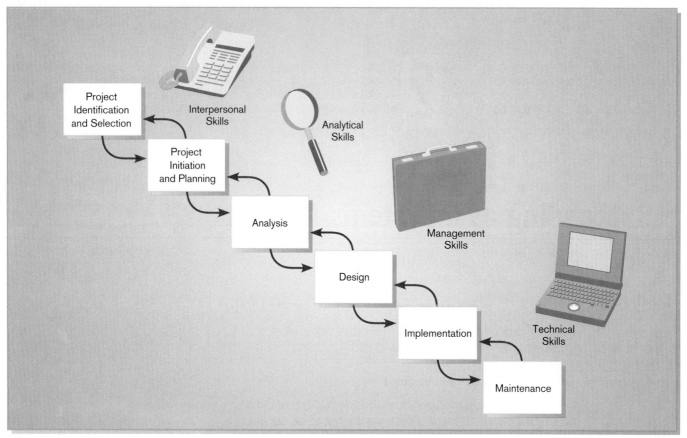

Figure 2-1
The relationship between a systems analyst's skills and the systems development life cycle

skills is to sensitize you to abilities that you need to develop from other courses and materials in order to become a successful systems analyst. The chapter ends by stepping back from these specific skills to examine systems analysis as a profession, with its own standards of practice, ethics, and career paths.

ANALYTICAL SKILLS FOR SYSTEMS ANALYSTS

Given the title systems analyst, you might think that analytical skills are the most important. While there is no question that analytical skills are essential, other skills are equally required. First, however, we will focus on the four sets of analytical skills: systems thinking, organizational knowledge, problem identification, and problem analyzing and solving.

Systems Thinking: A Review

If you counted the number of times each key term is used in this book, the key term used most frequently would undoubtedly be *system*. Let's take the time now to examine systems in general and information systems in particular. (For a more thorough treatment of system concepts, see Martin, et al., 1999). Let's start by examining what we mean by a system and identify the characteristics that define a system.

System: An interrelated set of components, with an identifiable boundary, working together for some purpose.

Definitions of a System and Its Parts   A **system** is an interrelated set of components with an identifiable boundary, working together for some purpose. A system has nine characteristics (see Figure 2 2):

1. Components
2. Interrelated components

3. A boundary
4. A purpose
5. An environment
6. Interfaces
7. Input
8. Output
9. Constraints

A system is made up of components. A **component** is either an irreducible part or an aggregate of parts, also called a *subsystem*. The simple concept of a component is very powerful. For example, just as with an automobile or a stereo system with proper design, we can repair or upgrade the system by changing individual components without having to make changes throughout the entire system. The components are **interrelated**; that is, the function of one is somehow tied to the functions of the others. For example, the work of one component, such as producing a daily report of customer orders received, may not progress successfully until the work of another component is finished, such as sorting customer orders by date of receipt. A system has a **boundary**, within which all of its components are contained and which establishes the limits of a system, separating the system from other systems. Components within the boundary can be changed whereas things outside the boundary cannot be changed. All of the components work together to achieve some overall **purpose** for the larger system: the system's reason for existing.

A system exists within an **environment**—everything outside the system's boundary. For example, we might consider the environment of a state university to include the

Component: An irreducible part or aggregation of parts that make up a system, also called a subsystem.

Interrelated components: Dependence of one subsystem on one or more subsystems.

Boundary: The line that marks the inside and outside of a system and that sets off the system from its environment.

Purpose: The overall goal or function of a system.

Environment: Everything external to a system that interacts with the system.

Figure 2-2
A general depiction of a system

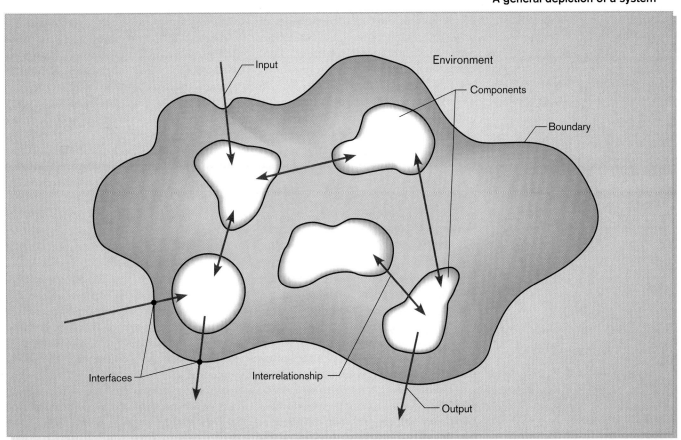

Figure 2-3
Special characteristics of interfaces

> **INTERFACE FUNCTIONS**
>
> Because an interface exists at the point where a system meets its environment, the interface has several special, important functions. An interface provides
>
> - **Security,** protecting the system from undesirable elements that may want to infiltrate it
> - **Filtering** unwanted data, both for elements leaving the system and entering it
> - **Coding and decoding** incoming and outgoing messages
> - **Detecting and correcting errors** in its interaction with the environment
> - **Buffering,** providing a layer of slack between the system and its environment, so that the system and its environment can work on different cycles and at different speeds
> - **Summarizing** raw data and transforming them into the level of detail and format needed throughout the system (for an input interface) or in the environment (for an output interface)
>
> Because interface functions are critical in communication between system components or a system and its environment, interfaces receive much attention in the design of information systems (see Chapters 13 and 14).

Interface: Point of contact where a system meets its environment or where subsystems meet each other.

Constraint: A limit to what a system can accomplish.

Input: Whatever a system takes from its environment in order to fulfill its purpose.

Output: Whatever a system returns to its environment in order to fulfill its purpose.

legislature, prospective students, foundations and funding agencies, and the news media. Usually the system interacts with its environment, exchanging, in the case of an information system, data and information. The points at which the system meets its environment are called **interfaces**, and there are also interfaces between subsystems (Figure 2-3 provides a list of functions performed by interfaces). An example of a subsystem interface is the clutch subsystem, which acts as the point of interaction between the engine and transmission subsystems of a car. As can be seen from Figure 2-3, interfaces may include much functionality. You will spend a considerable portion of time in systems development dealing with interfaces, especially interfaces between an automated system and its users (manual systems) and interfaces between different information systems. It is the design of good interfaces that permits different systems to work together without being too dependent on each other.

A system must face **constraints** in its functioning because there are limits (in terms of capacity, speed, or capabilities) to what it can do and how it can achieve its purpose within its environment. Some of these constraints are imposed inside the system (for example, a limited number of staff available) and others are imposed by the environment (for example, due dates or regulations). A system takes **input** from its environment in order to function. Mammals, for example, take in food, oxygen, and water from the environment as input. Finally, a system returns **output** to its environment as a result of its functioning and thus achieves its purpose.

Now that you know the definition of a system and its nine important characteristics, let's take an example of a system and use it to illustrate the definition and each system characteristic. Consider a system that is familiar to you: a fast-food restaurant (see Figure 2-4).

How is a fast-food restaurant a system? Let's take a look at the fictional Hoosier Burger Restaurant in Bloomington, Indiana. First, Hoosier Burger has components or subsystems. We can figure out what the subsystems are in many ways but, for the sake of illustration, let's focus on Hoosier Burger's physical subsystems as follows: kitchen, dining room, counter, storage, and office. As you might expect, the subsystems are interrelated and work together to prepare food and deliver it to customers, one purpose for the restaurant's existence. Food is delivered to Hoosier Burger early in the morning, kept in storage, prepared in the kitchen, sold at the counter, and often eaten in the dining room. The boundary of Hoosier Burger is represented by its physical walls and the primary purpose for the restaurant's existence is to make a profit for its owners, Bob and Thelma Mellankamp.

Hoosier Burger's environment consists of those external elements that interact with the restaurant, such as customers (many of whom come from nearby Indiana University), the local labor supply, food distributors (much of the produce is grown locally), banks, and neighborhood fast-food competitors. Hoosier Burger has one

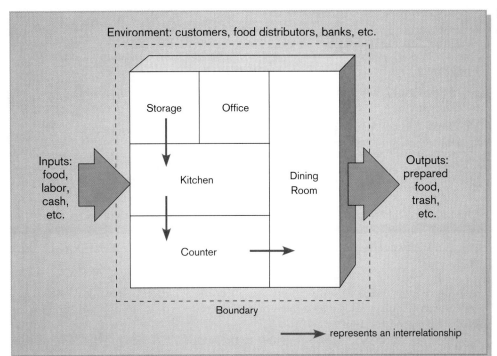

Figure 2-4
A fast-food restaurant as a system

interface at the counter where customers place orders and another at the back door where food and supplies are delivered. Still another interface is the telephone managers use regularly to talk with bankers and food distributors. The restaurant faces several constraints. It is designed for the easy and cost-effective preparation of certain popular foods, such as hamburgers and milk shakes, which constrains the restaurant in the foods it may offer for sale. Hoosier Burger's size and its location in the university neighborhood constrain how much money it can make on any given day. The Monroe County Health Department also imposes constraints, such as rules governing food storage. Inputs include, but are not limited to, ingredients for the burgers and other food as well as cash and labor. Outputs include, but are not limited to, prepared food, bank deposits, and trash.

Important System Concepts   Once we have recognized something as a system and identified the system's characteristics, how do we understand the system? Further, what principles or concepts about systems guide the design of information systems? A key aspect of a system for building systems is the system's relationship with its environment. Some systems, called **open systems**, interact freely with their environments, taking in input and returning output. As the environment changes, an open system must adapt to the changes or suffer the consequences. A **closed system** does not interact with the environment; changes in the environment and adaptability are not issues for a closed system. However, all business information systems are open, and in order to understand a system and its relationships to other information systems, to the organization, and to the larger environment, you must always think of information systems as open and constantly interacting with the environment.

There are several other important systems concepts with which systems analysts need to become familiar:

- Decomposition
- Modularity
- Coupling
- Cohesion

Open system: A system that interacts freely with its environment, taking input and returning output.

Closed system: A system that is cut off from its environment and does not interact with it.

Figure 2-5
Purposes of decomposition

DECOMPOSITION FUNCTIONS

Decomposition aids a systems analyst and other systems development project team members by

- Breaking a system into smaller, more manageable, and understandable subsystems
- Facilitating the focusing of attention on one area (subsystem) at a time without interference from other parts
- Allowing attention to concentrate on the part of the system pertinent to a particular audience, without confusing people with details irrelevant to their interests
- Permitting different parts of the system to be built at independent times and/or by different people

---

In addition, you need to understand the differences between viewing a system at a logical and at a physical level, each with associated descriptions concentrating on different aspects of a system.

Decomposition deals with being able to break down a system into its components. These components may themselves be systems (subsystems) and can be broken down into their components as well. How does decomposition aid understanding of a system? Decomposition results in smaller and less complex pieces that are easier to understand than larger, complex pieces. Decomposing a system also allows us to focus on one particular part of a system, making it easier to think of how to modify that one part independently of the entire system (Figure 2-5). Figure 2-6 shows the decomposition of a portable compact disc (CD) player. At the highest level of abstraction, this system simply accepts CDs and settings of the volume and tone controls as input and produces music as output. Decomposing the system into subsystems reveals the system's inner workings: There are separate systems for reading the digital signals from the CDs, for amplifying the signals, for turning the signals into sound waves, and for controlling the volume and tone of the sound. Breaking the subsystems down into their components would reveal even more about the inner workings of the system and greatly enhance our understanding of how the overall system works.

Modularity, a direct result of decomposition, refers to dividing a system up into chunks or modules of a relatively *uniform* size. Modules can represent a system simply, making it not only easier to understand but also easier to redesign and rebuild.

Coupling is the extent to which subsystems are dependent on each other. Subsystems should be as independent as possible. If one subsystem fails and other subsystems are highly dependent on it, the others will either fail themselves or have

Modularity: Dividing a system up into chunks or modules of a relatively uniform size.

Coupling: The extent to which subsystems depend on each other.

Figure 2-6
An example of system decomposition

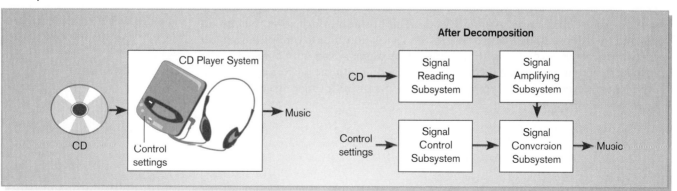

problems functioning. Looking at Figure 2-6, we would say the components of a portable CD player are tightly coupled. The amplifier and the unit that reads the CD signals are wired together in the same container, and the boundaries between these two subsystems may be difficult to draw clearly. If one subsystem fails, the entire CD player must be sent off for repair. In a home stereo system, the components are loosely coupled since the subsystems, such as the speakers, the amplifier, the receiver, and the CD player, are all physically separate and function independently. For example, if the amplifier in a home stereo system fails, only the amplifier needs to be repaired.

Finally, **cohesion** is the extent to which a subsystem performs a single function. In biological systems, subsystems tend to be well differentiated and thus very cohesive. In man-made systems, subsystems are not always as cohesive as they should be.

One final key systems concept with which you should be familiar is the difference between logical and physical systems. Any description of a system is abstract since the definition is not the system itself. When we talk about logical and physical systems, we are actually talking about logical and physical system descriptions.

A **logical system description** portrays the purpose and function of the system without tying the description to any specific physical implementation. For example, in developing a logical description of the portable CD player, we describe the basic components of the player (signal reader, amplifier, speakers, controls) and their relations to each other, focusing on the function of playing CDs using a self-contained, portable unit. We do not specify whether the earphone jack contains aluminum or gold, where we could buy the laser that reads the CDs, or how much the jack or the laser cost to produce.

The **physical system description**, on the other hand, is a material depiction of the system, a central concern of which is building the system. A physical description of the portable CD player would provide details on the construction of each subunit, such as the design of the laser, the composition of the earphones, and whether the controls feature digital readouts. A systems analyst should deal with function (logical system description) before form (physical system description), just as an architect does for the analysis and design of buildings.

Benefiting from Systems Thinking

The first step in systems thinking is to be able to identify something as a system. This identification also involves recognizing each of the system's characteristics, for example, identifying where the boundary lies and all of the relevant inputs. But once you have identified a system, what is the value of thinking of something as a system? Visualizing a set of things and their interrelationships as a system allows you to translate a specific physical situation into more general, abstract terms. From this abstraction, you can think about the essential characteristics of a specific situation. This in turn allows you to gain insights you might never get from focusing too much on the details of the specific situation. Also, you can question assumptions, provide documentation, and manipulate the abstract system without disrupting the real situation.

Let's look again at Hoosier Burger. How can visualizing a fast-food restaurant as a system help us gain insights about the restaurant that we might not get otherwise? Let's imagine that Hoosier Burger is facing more demand for its food than it can handle. Some people are convinced that its hamburgers are the best in Bloomington, maybe even in southern Indiana. Many people, especially IU students and faculty, frequently eat at Hoosier Burger, and the staff is having a difficult time keeping up with the demand. For the owner-managers, Bob and Thelma Mellankamp, the high level of demand is both a problem and an opportunity. The problem is that if the restaurant can't keep up with demand, people will stop coming to eat here, and the owners will lose money. The opportunity is to capitalize on Hoosier Burger's popularity and serve even more customers every day, making larger profits for the owners (which is the purpose of their system).

Cohesion: The extent to which a system or a subsystem performs a single function.

Logical system description: Description of a system that focuses on the system's function and purpose without regard to how the system will be physically implemented.

Physical system description: Description of a system that focuses on how the system will be materially constructed.

How does looking at Hoosier Burger as a system help? By decomposing the restaurant into subsystems, we can analyze each subsystem separately and discover if one or more subsystems is at capacity. Capacity is a general problem common to many systems. Let's say, after careful study, we discover that the kitchen, storage, and dining room subsystems have plenty of available capacity. However, the counter is unable to handle the rush of people. Customers have to wait in line for several minutes to place and receive their orders. The counter is the restaurant's bottleneck; thus the capacity of the counter needs to be increased. If we redesign the counter area or the procedures for taking customer orders, then we can increase the counter's capacity and better match it to the kitchen's capacity. Customers will have to wait in line less time to place their orders and they will get their food faster. Fewer customers will turn away because of long lines, which should translate into more food sold and higher profits.

There are other aspects of the system we could have examined, such as outputs, inputs, or environmental conditions, but to make the example more clear and concise, we looked only at subsystems. For this particular problem, decomposing Hoosier Burger into its subsystems enabled us to determine its problem with demand. Other problems may have required an examination of all aspects of the restaurant system.

N E T   S E A R C H

The concept of "system" is central to systems analysis and design. Visit http://www.prenhall.com/hoffer to complete an exercise on this topic.

Applying Systems Thinking to Information Systems   None of the examples of systems we have examined so far in this chapter have been information systems, even though information systems are the focus of this book. There are two reasons why we have looked at other types of systems first. One is so that you will become accustomed to thinking of some of the many different things you encounter daily as systems and realize how useful systems thinking can be. The second is that thinking of organizations as systems is a useful perspective from which to begin developing information systems. *Information systems can be seen as subsystems in larger organizational systems, taking input from, and returning output to, their organizational environments.*

Let's examine a simplified version of an information system as a special kind of system. In our fast-food restaurant example, Hoosier Burger uses an information system to take customer orders, send the orders to the kitchen, monitor goods sold and inventory, and generate reports for management. The information system is depicted as a data flow diagram in Figure 2-7 (you will learn how to draw data flow diagrams in Chapter 8).

As the diagram illustrates, Hoosier Burger's customer order system contains four components or subsystems: Process Customer Food Order, Update Goods Sold File, Update Inventory File, and Produce Management Reports. The arrows in the diagram show how these subsystems are interrelated. For example, the first process produces four outputs: a Kitchen Order, a Receipt, Goods Sold data, and Inventory Data. The latter two outputs serve as input for other subsystems. The dotted line illustrates the boundary of the system. Notice that the Customer, the Kitchen, and the Restaurant Manager (Bob Mellankamp) are all considered to be outside the customer order system. The specific purpose of the system is to facilitate customer orders, monitor inventory, and generate reports; the system's general purpose is to improve the efficiency of the restaurant's operations.

Since this information system is smaller in scope and purpose than the Hoosier Burger system itself, its environment is also smaller. For our purposes, we can limit the environment to those entities that interact with the system: Customers, the Kitchen, and the Restaurant Manager. Constraints on the system may or may not be apparent from the diagram. For example, the diagram implicitly shows (by omission) that there is no direct data exchange between the customer order system and information systems used by the restaurant's suppliers; this prevents the system from automatically issuing an order for supplies directly to the suppliers when inventory falls below a certain level. We do not know, however, if any other Hoosier Burger system

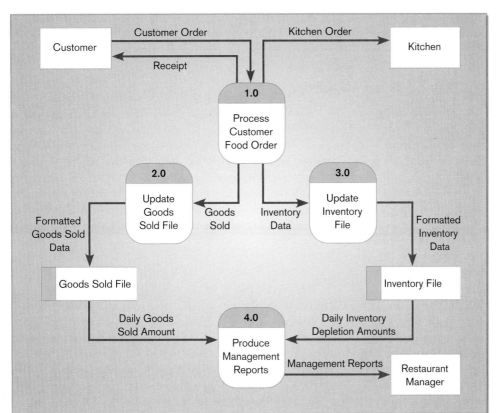

Figure 2-7
A fast-food restaurant's customer order information system depicted in a data flow diagram

supports such direct data exchange. Another constraint may be the system's inability to provide on-line, real-time information on inventory levels, limiting Bob Mellankamp to receiving nightly batched reports. This is not at all clear from Figure 2-7. In contrast, system input and output are very clear. The only system input is the Customer Order and there are three overall system outputs: a Receipt for the customer, a Kitchen Order, and Management Reports.

On one level of analysis and description, Hoosier Burger's customer order system is a physical system that takes input, processes data, and returns output. The physical system consists of a computerized cash register that a clerk uses to enter a customer order and return a paper receipt to the customer. Another piece of paper, the kitchen order, is generated from a printer in the restaurant's kitchen. The cash register sends data on the order about goods sold and inventory to a computer in Hoosier Burger's office, where computer files on goods sold and inventory are updated by applications software. Other application software uses data in the Goods Sold and Inventory files to generate and print reports on a laser printer in the office.

On another level of analysis and description, Hoosier Burger's customer order system can be explained using a logical description of an information system that focuses on the flow and transformation of data. The physical system is one possible implementation of the more abstract, logical information system description. For the logical information system description, it is irrelevant whether the customer's order shows up in the kitchen as a piece of paper or as lines of text on a monitor screen. What's important is the information that is sent to Hoosier Burger's kitchen. For every logical information system description, there can be several different physical implementations of it.

The way we draw information systems shows how we think of them as systems. Data flow diagrams clearly illustrate inputs, outputs, system boundaries, the environment, subsystems, and interrelationships. Purpose and constraints are much more difficult to illustrate and must therefore be documented using other notations. In total, all elements of the logical system description must address all nine characteristics of a system.

Organizational Knowledge

As a systems analyst, you will work in organizations. Whether you are an in-house or contract custom software developer, you must understand how organizations work. In addition, you must understand the functions and procedures of the particular organization (or enterprise) you are working for. Furthermore, many of the systems you will build or maintain serve one organizational department and you must understand how that department operates, its purpose, its relationships with other departments and, if applicable, its relationships with customers and suppliers. Table 2-1 lists various kinds of organizational knowledge that a systems analyst must acquire in order to be successful.

Problem Identification

What is a problem? Pounds (1969) defines a problem as the difference between an existing situation and a desired situation. For him, the process of identifying problems is the process of defining differences, so problem solving is the process of finding a way to reduce differences. According to Pounds, a manager defines differences by comparing the current situation to the output of a model that predicts what the output should be. For example, at Hoosier Burger, a certain portion of the food ordered from local produce distributors is expected to go bad before it can be used. Comparing a current food spoilage rate of 10 percent to a desired spoilage rate of 5 percent defines a difference and therefore identifies a problem. In this case, Bob Mellankamp has used a model to determine the desired spoilage rate of 5 percent. The particular model used, showing how fast produce ripens after harvesting, typical delivery times, and how long produce will stay fresh in a refrigerator, has come from

---

TABLE 2-1   Selected Areas of Organizational Knowledge for a Systems Analyst

How Work Officially Gets Done in a Particular Organization
Terminology, abbreviations, and acronyms
Policies
Standards and procedures
Standards of practice
Formal organization structure
Job descriptions
Understanding the Organization's Internal Politics
Influence and inclinations of key personnel
Who the experts are in different subject areas
Critical incidents in the organization's history
Informal organization structure
Coalition membership and power structures
Understanding the Organization's Competitive and Regulatory Environment
Government regulations
Competitors, domestic and international
Products, services, and markets
Role of technology
Understanding the Organization's Strategies and Tactics
Short- and long-term strategy and plans
Values and mission

research carried out at Purdue University's College of Agriculture. Based on the research, the Mellankamps have set a standard of a 5 percent spoilage rate, with an acceptable variance of 2 percent in either direction. According to this standard, a 5 percent variance between desired and actual is clearly out of line and merits attention. Another model might have indicated that a 10 percent spoilage rate was acceptable. You can see that understanding how managers identify problems is understanding the models they use to define differences.

In order to identify problems that need solving, you must be able to compare the current situation in an organization to the desired situation. You must develop a repertoire of models to define the differences between what is and what ought to be. It is also important that you appreciate the models that information systems users rely on to identify problems. Every functional area of the organization will use different models to find problems; what is helpful in accounting will not necessarily work well in manufacturing. Often you must be able to see problems from a broader perspective. By relying on models from their own particular functional areas, users may not see the real problem from an organizational view.

Problem Analyzing and Solving

Once a problem has been identified, you must analyze the problem and determine how to solve it. Analysis entails finding out more about the problem. Systems analysts learn through experience, with guidance from proven methods, how to get the needed information from people as well as from organizational files and documents. As you seek out additional information, you also begin to formulate alternative solutions to the problem. Devising solutions leads to a search for more information, which in turn leads to improvements in the alternatives. Obviously, such a process could continue indefinitely, but at some point, the alternatives are compared and typically one is chosen as the best solution. Once the analyst, users, and management agree on the general suitability of the solution, they devise a plan for implementing it.

The approach for analyzing and solving problems we describe was formally described by Herbert Simon and colleagues (Simon, 1960). The approach has four phases: intelligence, design, choice, and implementation. During the intelligence phase, all information relevant to the problem is collected. During the design phase, alternatives are formulated and, during choice the best alternative solution is chosen. The solution is put into practice during the implementation phase.

This problem-analysis and -solving approach should be familiar to you: It is essentially the same general process as that described earlier in the systems development life cycle (see Figure 2-8). Simon's intelligence phase corresponds roughly to the first three phases in the life cycle: project identification and selection, project initiation and planning, and analysis. Simon's design phase corresponds to that part of analysis where alternative solutions are formulated. The detailed solution formulation (once the solution is chosen), however, would be performed in the life cycle's latter parts of the design phase. Choice of the best solution is made in stages, first at the end of the analysis phase and then during design. In our life cycle model, activities that occur in design, implementation, and maintenance correspond to Simon's implementation phase.

Simon's problem-solving model is a useful one that lends insight into how people solve certain kinds of problems, but there are other factors in organizations that influence how problems are solved. Among these are personal interests, political considerations, and limits in time and cognitive ability that affect how much information people can gather and process. We will say more about these factors in later chapters; now we will turn to an examination of the technical skills required of systems analysts.

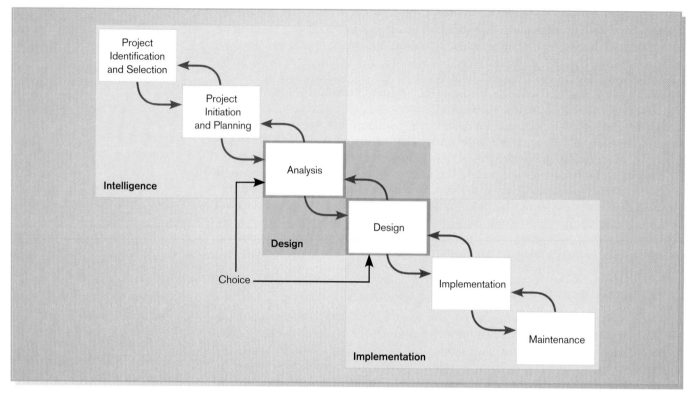

Figure 2-8
The systems development life cycle and Simon's problem-solving model

TECHNICAL SKILLS

Many aspects of your job as a systems analyst are technically oriented. In order to develop computer-based information systems, you must understand how computers, data networks, database management and operating systems, and a host of other technologies work as well as their potential and limitations. Further, you must be technically adept with different notations for representing, or modeling, various aspects of information systems. You need these technical skills not only to perform tasks assigned to you but also to communicate with the other people with whom you work in systems development (see Chapter 1 for a discussion of the roles of various people in systems analysis and design). Rather than develop a single set of technical skills to use throughout your career, you must constantly re-educate yourself about information technology, techniques, and methodologies. These information technology, techniques, and methodologies change quickly, and you must keep up with the changes. You need to understand alternative technologies (like Microsoft Windows, Linux, and UNIX operating environments) as organizational preferences, since choices vary across companies and over time. Versatility, based on a sound understanding of technical concepts rather than specific tools, gives you the flexibility needed for such a changing skill set.

The following activities will help you stay versatile and up-to-date:

- Read trade publications (for example, *Computerworld* or *PCWeek*) and books.
- Join professional societies (for example, the *Association of Information Technology Professionals* or the *Association for Computing Machinery*) or other clubs and their meetings.
- Attend classes or teach at a local college. Teaching is a wonderful way to force yourself to stay current and to learn from others.
- Attend any courses or training sessions offered by your organization.

- Attend professional conferences, seminars, or trade shows.
- Participate in electronic bulletin boards, news groups, or conferences on local, national, or international networks.
- Regularly browse Websites that focus on industry news, such as CNET. Many trade publications, like *Computerworld*, also have Websites.

Maybe you have seen the cartoon of the person wearing tattered clothes, looking thin, sitting on a park bench feeding the birds. The caption reads, "He was an outstanding systems analyst, but he took a six-month vacation and fell too far behind in his field." Being a systems analyst working in the systems field requires continuous learning.

Because of the rapid changes that occur in technology, we do not dwell on specifics in this section. For example, when this book was being written, object-oriented database technology was considered new and experimental. It is quite possible, however, that this technology may be popular and widespread when you read this book. In general, you should be as familiar as possible with such families of technologies as

- Microcomputers, workstations, minicomputers, and mainframe computers
- Programming languages
- Operating systems, both for single machines and networks
- Database and file management systems
- Data communication standards and software for local and wide area networks
- Systems development tools and environments (such as form and report generators and graphical interface design tools)
- Web development languages and tools, such as HTML, Cold Fusion, and Microsoft's Front Page
- Decision support system generators and data analysis tools

as well as modern methods and techniques for describing, modeling, and building systems. How technical you must be will vary by job assignment and where you are in your career. Often, you will be asked to be more technical in the early stages of your career, and then you will assume more managerial responsibilities as you gain experience. We discuss career progression later in this chapter.

MANAGEMENT SKILLS

Systems analysts are almost always members of project teams and are frequently asked to lead teams. Management skills are very useful for anyone in a leadership role. As an analyst, you also need to know how to manage your own work and how to use organizational resources in the most productive ways possible. Self-management, then, is an important skill for an analyst. In this section, we describe four categories of management skills: resource, project, risk, and change management.

Resource Management

Any organizational worker must know how to obtain and work effectively with organizational resources. A systems analyst must know how to get the most out of a wide range of resources: system documentation, information technology, and money. For an analyst leading a team, the most important resource is people. A team leader must learn how to best utilize the particular talents of other team members. He or she must also be able to delegate responsibility, empowering people to do the tasks they have been assigned.

Resource management includes the following capabilities:

- Predicting resource usage (budgeting)
- Tracking and accounting for resource consumption
- Learning how to use resources effectively
- Evaluating the quality of resources used
- Securing resources from abusive use
- Relinquishing resources when no longer needed and obsoleting resources when they can no longer be useful

Project Management

Effectively managing projects is crucial to a systems analyst's job. Information systems development projects range from one-person projects that take very little time and effort to multi-person, multi-year efforts costing millions of dollars. The goal of project management is to prevent projects from coming in late and going over budget. In addition, project management is designed to help managers keep track of the project's progress.

Even if you are not a project leader, you will be given responsibilities for parts of a project, or subprojects. In the role of project or subproject manager, you first need to decompose a (sub)project into several independent tasks. The next step is to determine how the tasks are related to each other and who will be responsible for each task. As we will see in Chapter 3, analysts use established tools and techniques to help manage projects. The most important element, however, is managing the people working on the project. Successful analysts motivate people to work together and instill a sense of trust and interdependence among them. Project management extends beyond the organization to any vendors or contractors working on the project.

Often, in today's development environment, many aspects of a project may be farmed out to various contractors outside the organization. Using independent contractors has many advantages. A particular contractor may be more skilled than internal personnel in a technology or may be less expensive. If a project is short on time, it may also make sense to contract out some parts of a development project to help speed up the overall process. Many times, however, contractors deliver work that is late or of low quality, or that does not meet requirements. If the system requirements are unstable or not well defined, the potential problems with contractors can be exaggerated. For these reasons, it is just as important to manage outside contractors as it is to manage everyone else involved in a project. Two mechanisms that help manage contractors are contracts and relationship managers. Very well-specified contracts that spell out just exactly what is expected and when, and that lay out explicit sanctions for nonperformance, may motivate contractors to perform up to expectations. On the other hand, very explicit contracts may scare off contractors who know that they cannot live up to such a contract's terms. Relationship managers act as liaisons between your firm and the contractors. By establishing personal relationships with the parties involved, relationship managers may be in a position to sense trouble before it happens and work with both parties toward reasonable settlements.

Risk Management

Risk management is the ability to anticipate what might go wrong in a project. Once risks to the project have been identified, you must be able to minimize the likelihood that those risks will actually occur. If minimizing risk is not possible, then you try to minimize the damage that might result. Risk management also includes knowing where to place resources (such as people) where they can do the most good and prioritizing activities to achieve the greatest gain. We discuss a key part of risk management that is carried out during project justification—risk assessment—in Chapter 6.

Change Management

Introducing a new or improved information system into an organization is a change process. In general, people do not like change and tend to resist it; therefore, any change in how people perform their work in an organization must be carefully managed. Change management, then, is a very important skill for systems analysts, who are organizational *change agents*. You must know how to get people to make a smooth transition from one information system to another, giving up their old ways of doing things and accepting new ways. Change management also includes the ability to deal with technical issues related to change, such as obsolescence and reusability. You will learn more about managing the change that accompanies a new information system in Chapter 17.

INTERPERSONAL SKILLS

Although, as a systems analyst, you will be working in the technical area of designing and building computer-based information systems, you will also work extensively with all types of people. Perhaps the most important skills you will need to master are interpersonal. In this part of the chapter, we will discuss the various interpersonal skills necessary for successful systems analysis work: communication skills; working alone and with a team; facilitating groups; and managing expectations of users and managers.

Communication Skills

The single most important interpersonal skill for an analyst, as well as for any professional, is the ability to communicate clearly and effectively with others. Analysts should be able to successfully communicate with users, other information systems professionals, and management. Analysts must establish a good, open working relationship with clients early in the project and maintain it throughout by communicating effectively.

Communication takes many forms, from written (memos, reports) to verbal (phone calls, face-to-face conversations) to visual (presentation slides, diagrams). The analyst must be able to master as many forms of communication as possible. Oral communication and listening skills are considered by many information system professionals as the most important communication skills analysts need to succeed. Interviewing skills are not far behind. All types of communication, however, have one thing in common: They improve with experience. The more you practice, the better you get. Some of the specific types of communication we will mention are interviewing and listening, the use of questionnaires, and written and oral presentations.

Interviewing, Listening, and Questionnaires
Interviewing is one of the primary ways analysts gather information about an information systems project. Early in a project, you may spend a large amount of time interviewing users about their work and the information they use. There are many ways to effectively interview someone, and becoming a good interviewer takes practice. We will discuss interviewing in more detail in Chapter 7, but it is important to point out now that asking questions is only one part of interviewing. Listening to the answers is just as important, if not more so. Careful listening helps you understand the problem you're investigating and, many times, the answers to your questions lead to additional questions that may be even more revealing and probing than the questions you prepared before your interview.

Although interviews are very effective ways of communicating with people and obtaining important information from them, interviews can also be very expensive and time-consuming. Because questionnaires provide no direct means by which to ask follow-up questions, they are generally less effective than interviews. It is possible, however, though time-consuming, to call respondents and ask them follow-up questions. Questionnaires are less expensive to conduct because the questioner does not have to invest the same amount of time and effort to collect the same information using a questionnaire as he or she does in conducting an interview. For example, using a written questionnaire that respondents complete themselves, you could gather the same information from 100 people in one hour that you could collect from only one person in a one-hour interview. In addition, questionnaires have the advantage of being less biased in how the results are interpreted because the questions and answers are standardized. Creating good questionnaires is a skill that comes only with practice and experience. You will learn more about questionnaire design in Chapter 7.

Written and Oral Presentations   At many points during the systems development process, you must document the progress of the project and communicate that progress to others. This communication takes the following forms:

- Meeting agenda
- Meeting minutes
- Interview summaries
- Project schedules and descriptions
- Memoranda requesting information, an interview, participation in a project activity, or the status of a project
- Requests for proposal from contractors and vendors

and a host of other documents. This documentation is essential to provide a written, not just oral, history for the project, to convey information clearly, to provide details needed by those who will maintain the system after you are off the project team, and to obtain commitments and approvals at key project milestones.

The larger the organization and the more complicated the systems development project, the more writing you will have to do. You and your team members will have to complete and file a report at the end of each stage of the systems development life cycle. The first report will be the business case for getting approval to start the project. The last report may be an audit of the entire development process. And at each phase, the analysis team will have to document the system as it evolves. To be effective, you need to write both clearly and persuasively.

As there are often many different parties involved in the development of a system, there are many opportunities to inform people of the project's status. Periodic written status reports are one way to keep people informed, but there will also be unscheduled calls for ad hoc reports. Many projects will also involve scheduled and unscheduled oral presentations. Part of oral presentations involves preparing slides, overhead transparencies, or multimedia presentations, including system demonstrations. Another part involves being able to field and answer questions from the audience.

How can you improve your communication skills? We have four simple yet powerful suggestions:

1. Take every opportunity to practice. Speak to a civic organization about trends in computing. Such groups often look for local speakers to present talks on topics of general interest. Conduct a training class on some topic on which you have special expertise. Some people have found participation in Toastmasters, an international organization with local chapters, a very helpful way to improve oral communication skills.

2. Videotape your presentations and do a critical self-appraisal of your skills. You can view videotapes of other speakers and share your assessments with each other.

3. Make use of writing centers located at many colleges as a way to critique your writing.

4. Take classes on business and technical writing from colleges and professional organizations.

Working Alone and with a Team

As a systems analyst, you must often work alone on certain aspects of any systems development project. To this end, you must be able to organize and manage your own schedule, commitments, and deadlines. Many people in the organization will depend on your individual performance, yet you are almost always a member of a team and must work with the team toward achieving project goals. As we saw in Chapter 1, working with a team entails a certain amount of give and take. You need to know when to trust the judgment of other team members as well as when to question it. For example, when team members are speaking or acting from their base of experience and expertise, you are more likely to trust their judgment than when they are talking about something beyond their knowledge. For this reason, the analyst leading the team must understand the strengths and weaknesses of the other team members. To work together effectively and to ensure the quality of the group product, the team must establish standards of cooperation and coordination that guide their work (review Table 1-2 for the characteristics of a successful team).

There are several dimensions to the cooperation and coordination that influence team work. Table 2-2 lists the twelve characteristics of a high-performance team (McConnell, 1996). The first characteristic is a shared vision, which allows each team member to have a clear understanding of the project's objectives. A shared vision helps team members keep their priorities straight and not allow small items of little significance to become overwhelming and distracting. To provide motivation for team members, the vision also needs to present a challenge to team members. The second characteristic, team identity, emerges as team members work together closely and begin to share a common language and sense of humor. Team identity can lead to the synergy of effort only possible when groups work together well.

Shared vision and team identity are important but they alone may not be enough for a team to actually accomplish something. The third characteristic of high-performance teams is how the teams are organized. A result-driven structure is one that depends on clear roles, effective communication systems, means of monitoring individual performance, and decision making based on facts rather than emotions. Choosing the right people for the team is the fourth characteristic. McConnell (1996) reports that team performance may differ by as much as a factor of 5, depend-

---

TABLE 2-2   Characteristics of a High-Performance Team (McConnell, 1996)

1. Shared, elevated vision or goal
2. Sense of team identity
3. Result-driven structure
4. Competent team members
5. Commitment to the team
6. Mutual trust
7. Interdependence among team members
8. Effective communication
9. Sense of autonomy
10. Sense of empowerment
11. Small team size
12. High level of enjoyment

---

ing only on the skills and attitudes of a team's members. Although the skills of each team member are important determinants of how well the team will perform, all members must be committed to the team, the fifth characteristic of high-performance teams. A group of the best and brightest individuals, committed only to their own self-interests, cannot outperform a true team of lesser talents who are genuinely committed to each other and to their joint effort.

The next five characteristics of high-performance teams all have to do with how the team members interact with each other. It is very important that team members develop genuine trust for each other. The need for trust is why you see so many team-building exercises; for example, an individual falls backwards into the arms of a fellow team member, not knowing if the other person is really there but trusting that he or she will be. Similarly, members of high-performance teams work interdependently, relying on each others' strengths; develop effective means of communication; give each team member the autonomy to do whatever he or she believes is best for the team and for the project; and empower each team member.

All of these high-performance characteristics seem to work best, according to McConnell (1996), in small teams no larger than eight to ten people. Finally, it is important that teams have fun. Enjoying working together leads to increased team cohesiveness, which has been shown to be a key ingredient of team productivity (Lakhanpal, 1993).

Facilitating Groups

Sometimes you need to interact with a group in order to communicate and receive information. In Chapter 1, we introduced you to the Joint Application Design (JAD) process in which analysts actively work with groups during systems development. Analysts use JAD sessions to gather systems requirements and to conduct design reviews. The assembled group is the most important resource the analyst has access to during a JAD and you must get the most out of that resource; successful group facilitation is one way to do that. In a typical JAD, there is a trained session leader running the show. He or she has been specially trained to facilitate groups, to help them work together, and to help them achieve their common goals. Facilitation necessarily involves a certain amount of neutrality on the part of the facilitator. The facilitator must guide the group without being part of the group and must work to keep the effort on track by ferreting out disagreements and helping the group resolve differences. Obviously, group facilitation requires training. Many organizations that rely on group facilitation train their own facilitators. Figure 2-9 lists some guidelines for running an effective meeting, a task that is fundamental to facilitating groups.

Managing Expectations

Systems development is a change process, and any organizational change is greeted with anticipation and uncertainty by organization members. Organization members will have certain ideas, perhaps based on their hopes and wishes, about what a new information system will be able to do for them; these expectations about the new system can easily run out of control. Ginzberg (1981) found that successfully managing user expectations is related to successful systems implementation. For you to successfully manage expectations, you need to understand the technology and what it can do. You must understand the work flows that the technology will support and how the new system will affect them. More important than understanding, however, is your ability to communicate a realistic picture of the new system and what it will do for users and managers. Managing expectations begins with the development of the business case for the system and extends all the way through training people to use the finished system. You need to educate those who have few expectations as well as temper the optimism of those who expect the new system to perform miracles.

Figure 2-9
**Some guidelines for running
effective meetings**
*(Adapted from Option Technologies, Inc.
[1992])*

- Become comfortable with your role as facilitator by gaining confidence in your ability, being clear about your purpose, and finding a style that is right for you.
- At the beginning of the meeting, make sure the group understands what is expected of them and of you.
- Use physical movement to focus on yourself or on the group, depending on which is called for at the time.
- Reward group member participation with thanks and respect.
- Ask questions instead of making statements.
- Be willing to wait patiently for group members to answer the questions you ask them.
- Be a good listener.
- Keep the group focused.
- Encourage group members to feel ownership of the group's goals and of their attempts to reach those goals.

SYSTEMS ANALYSIS AS A PROFESSION

Even though systems analysis is a relatively new field, those in the field have established standards for education, training, certification, and practice. Such standards are required for any profession.

Whether or not systems analysis is a profession is open to debate. Some feel systems analysis is not a profession because it simply has not been around long enough to have established the rigorous standards that define a profession. Others feel that at least some standards are already in place. There are guidelines for college curricula and there are standard ways of analyzing, designing, and implementing systems. Professional societies that systems analysts may join include the Society for Information Management, the Association of Information Technology Professionals, and the Association for Computing Machinery (ACM). There is a Certified Computing Professional (CCP) exam, much like the Certified Public Accountant (CPA) exam, that you can take to prove your competency in the field, although, unlike the CPA certificate, very few jobs and employers in the IS field require you to have the CCP certificate. Codes of ethics to govern behavior also exist. In this section, we will discuss several aspects of a systems analyst's job: standards of practice, the ACM code of ethics, and career paths for those choosing to become systems analysts.

Standards of Practice

Standard methods or practices of performing systems development are emerging that make systems development less of an art and more of a science. Standards are developed through education and practice and spread as systems analysts move from one organization to another. We will focus here on four standards of practice: an endorsed development methodology, approved development platforms, well-defined roles for people in the development process, and a common language.

There are several different development methodologies now being used in organizations. Although there is no standardization of a single methodology across all organizations, a few prominent methodologies are in common use. An *endorsed development methodology* lays out specific procedures and techniques to be used during the development process. These standards are central to promoting consistency and reliability in methods across all of an organization's development projects. Some methodologies are spread through the work of well-known consultants; others are spread through major consulting firms.

Closely associated with endorsed methodologies are approved development platforms. Some methodologies are closely tied to platforms, but other methodologies are more adaptable and can work in close accordance with development platforms that exist in the organization, such as database management systems and 4GLs. The point is that organizations, and hence the analysts who work for them, are standardizing around specific platforms, and standards for development emerge from this standardization.

Roles for the various people involved in the development process are also becoming standardized. End users, managers, and analysts are each assigned certain responsibilities for development projects. The training that analysts receive in college, on their first jobs, and during their interactions with other analysts, combine to create a gestalt of the analyst's job. For example, as you study this book and talk about systems development in your class, you are forming certain ideas about what systems analysts do and how systems are developed in organizations. Your ideas are also shaped and reinforced by the other IS courses you take in college. Once you get your first job, you will receive additional training and you will adjust your understanding of systems analysis accordingly. As you gain experience working on projects and interacting with other analysts, who may have been trained at other universities and in other organizations, your ideas will continue to change and grow, but the basic core of what systems analysis means to you will have been established. Many of the experiences you have on the job will reinforce much of what you have already learned about systems analysis. When you leave an organization and go to work elsewhere, you will carry your understanding of systems analysis with you. Over time, as you and other analysts change jobs and move from one organization to another, what it means to be an analyst becomes standardized across organizations, and the standards of practice in the field help define what it means to be an analyst.

Another factor moving the job of the systems analyst toward professionalism is the development of a common language analysts use to talk to each other. Analysts communicate on the job, at meetings of professional societies, and through publications. As analysts develop a special language for communication among themselves, their language becomes standardized. One example is the Unified Modeling Language (UML), which has emerged as a common way to specify and design information systems based on the object-oriented approach (see Chapter 20). Other examples of communication becoming standardized include the widespread use of common programming languages such as COBOL and C and the spread of SQL as the language of choice for data definition and manipulation for relational databases. As their common language develops, analysts become more cohesive as a group—a characteristic of professions.

Ethics

The ACM is a large professional society made up of information system professionals and academics. It has over 85,000 members. Founded in 1947, the ACM is dedicated to promoting information processing as an academic discipline and to encouraging the responsible use of computers in a wide range of applications. Because of its size and membership, it has much influence in the information systems community. The ACM has developed a code of ethics for its members called the "ACM Code of Ethics and Professional Conduct." The full statement is reproduced in Figure 2-10. The code applies to all ACM members and directly applies to systems analysts.

Note the emphasis in the Code on personal responsibility, on honesty, and on respect for relevant laws. Notice also that compliance with a code of ethics such as this one is voluntary, although article 4.2 calls for, at a minimum, peer pressure for compliance. No one can force an information systems professional to follow these guidelines. However, it is voluntary compliance with the guidelines that makes someone a professional in the first place. Notice that for leaders there is the burden of educating non-IS professionals about computing—about what computing can and

Figure 2-10
ACM Code of Ethics and Professional Conduct, Revision Draft No. 19 (9/19/91).
(Copyright © Association for Computing Machinery, reprinted with permission.)

Association for Computing Machinery Professional Code of Ethics

Preamble

Commitment to ethical professional conduct is expected of every member (voting members, associate members, and student members) of the Association for Computing Machinery (ACM).

This Code, consisting of 24 imperatives formulated as statements of personal responsibility, identifies the elements of such a commitment. It contains many, but not all, issues professionals are likely to face. Section1 outlines fundamental ethical considerations, while Section 2 addresses additional, more specific considerations of professional conduct. Statements in Section 3 pertain more specifically to individuals who have a leadership role, whether in the workplace or in a volunteer capacity such as with organizations like ACM. Principles involving compliance with this Code are given in Section 4.

(1.0) General Moral Imperatives

(As an ACM member I will . . .)
(1.1) Contribute to society and human well-being.
(1.2) Avoid harm to others.
(1.3) Be honest and trustworthy.
(1.4) Be fair and take action not to discriminate.
(1.5) Honor property rights including copyrights and patent.
(1.6) Give proper credit for intellectual property.
(1.7) Respect the privacy of others.
(1.8) Honor confidentiality.

(2.0) More Specific Professional Responsibilities

(As an ACM computing professional I will . . .)
(2.1) Strive to achieve the highest quality, effectiveness and dignity in both the process and products of professional work.
(2.2) Acquire and maintain professional competence.
(2.3) Know and respect existing laws pertaining to professional work.
(2.4) Accept and provide appropriate professional review.
(2.5) Give comprehensive and thorough evaluations of computer systems and their impacts, including analysis of possible risks.
(2.6) Honor contracts, agreements, and assigned responsibilities.
(2.7) Improve public understanding of computing and its consequences.
(2.8) Access computing and communication resources only when authorized to do so.

(3.0) Organizational Leadership Imperatives

(As an ACM member and an organizational leader I will . . .)
(3.1) Articulate social responsibilities of members of an organizational unit and encourage full acceptance of those responsibilities.
(3.2) Manage personnel and resources to design and build information systems that enhance the quality of working life.
(3.3) Acknowledge and support proper and authorized uses of an organization's computing and communication resources.
(3.4) Ensure that users and those who will be affected by a system have their needs clearly articulated during the assessment and design of requirements; later the system must be validated to meet requirements.
(3.5) Articulate and support policies that protect the dignity of users and others affected by a computing system.
(3.6) Create opportunities for members of the organization to learn the principles and limitations of computer systems.

(4.0) Compliance with the Code

(As an ACM member I will . . .)
(4.1) Uphold and promote the principles of this Code.
(4.2) Treat violations of this code as inconsistent with membership in the ACM.

cannot do. The Code also expresses concern for the quality of work life and for protecting the dignity and privacy of others when performing professional work, such as developing information systems.

Though not written specifically for systems analysts, the ACM Code of Ethics can easily be adapted to the systems analysis job. Many systems development projects deal directly with many of the issues addressed in the Code: privacy, quality of work life, user participation, and managing expectations. When an analyst must confront one or more of these issues, the Code can be used as a guide for professional conduct.

It is also important to remember that systems analysts work within organizations. Codes of ethics, such as that approved by the ACM, may not provide all of the guidance analysts need for dealing with ethically questionable situations in business organizations. The study of ethics, however, is a complex and sometimes bewildering exercise, so simplified and targeted approaches can be very helpful. One such approach has been developed by Smith and Hasnas (1999) for business managers, but it can be usefully applied by information systems professionals as well.

Smith and Hasnas describe three different ways to view business problems with ethical considerations. The first is the stockholder approach, which holds that any action taken by a business is ethically acceptable as long as it is legal, not deceptive, and maximizes profits for stockholders.

The second view is the stakeholder approach. A stakeholder is not the same as a stockholder. A stakeholder, like a stockholder, may own part of the firm, but a stakeholder typically has a greater involvement with the firm than does a stockholder. A stakeholder is either vital to the ongoing operation of the firm or is vitally affected by the actions of the firm. According to the stakeholder approach, you first have to determine who your stakeholders are. Then every action you are considering that violates the rights of any one of these stakeholders must be rejected. Only actions that best balance the rights of the different stakeholder groups can be taken by the firm.

The third approach is called the social contract approach. The focus of this approach is much broader than the other two, as it extends beyond stockholders and stakeholders to members of society at large. Any actions, potentially taken by the firm, that are deceptive, that could dehumanize employees, or that could discriminate, must be rejected outright. Further, any potential actions that could reduce the welfare of the members of society must also be eliminated. Only then can actions that would enhance the financial liability of the firm be considered. Table 2-3 lists the key ethical obligations of anyone employing one of these three approaches.

The best way to compare these three approaches is to look at how they could be applied to a business situation. Smith and Hasnas (1999) supply just such an exam-

TABLE 2-3   Comparison of Ethical Obligations for Three Different Approaches to Business Ethics (from Smith and Hasnas, 1999). Adapted with permission of *MIS Quarterly*.

Stockholder	Stakeholder	Social Contract
• Conform to laws and regulations • Avoid fraud and deception • Maximize profits	• Determine who are relevant stakeholders • Determine rights of each; reject options that violate these • Accept remaining option that best balances interests of stakeholders	• Reject actions that are fraudulent/deceptive, dehumanize employees, or involve discrimination • Eliminate options that reduce welfare of society's members • Choose remaining option that maximizes probability of financial success

ple. In December 1990, Blockbuster Entertainment Corporation announced a plan to sell the rental history of its customers to direct marketers. (Federal law prohibits the disclosure of the titles of videos that people rent, but it is legal to disclose the categories of videos that one rents.) Blockbuster scuttled the plan in response to public outcry. However, the Blockbuster case provides a good example for comparing the three approaches to business ethics.

Under the stockholder approach, Smith and Hasnas say the Blockbuster plan would have been ethically acceptable. It was legal and not deceptive, and the income from selling rental histories would have added to Blockbuster's profits. Under the stakeholder approach, the different stakeholder groups would have to be identified first, and then the effects of the plan on each group would have to be determined and compared to the effects on other groups. If, for example, the plan resulted in limited income and severely inconvenienced customers, say, for example, in terms of all the junk mail they would receive from the direct marketers who purchased their rental histories, then the plan would probably not be ethically acceptable. However, if the plan generated lots of new revenue and customers were only slightly inconvenienced, then the plan might well be ethically acceptable. Finally, using the social contract approach, the plan to sell rental histories would probably not be ethically acceptable. The primary reason for this determination, according to Smith and Hasnas (1999), is that neither Blockbuster employees nor customers would benefit in any material way from the implementation of the plan.

Although this example is greatly simplified, it does illustrate well the types of ethical dilemmas confronting information technology professionals. It also does a very good job in showing how the ethics of a situation depends greatly on the ethical approach taken to examine the issue in the first place.

NET SEARCH
Most professional societies and some organizations have codes of ethics for their members. Visit http://www.prenhall.com/hoffer to complete an exercise on this topic.

Career Paths

Currently, there are many different opportunities for a recent college graduate with a degree in management information systems (MIS). Traditionally, most recent graduates took jobs as systems analysts or programmer/analysts with large consulting firms. Many still do. But the traditional path has changed in the past few years. We explore below some of the many alternatives available to information technology specialists, but first we discuss the consulting option, which remains viable and attractive.

The information systems business at consulting firms has been growing at double-digit levels over the past decade, creating a need for hundreds if not thousands of MIS graduates. Typically, if you are hired by a consulting firm, the first thing you do is to report for extensive and intensive training in the tools and technologies the firm uses. Most of these firms have their own campuses where they train new recruits. After training, you would be assigned to a project. The project may or may not be close to the city you have chosen as your base.

As a junior consultant, your job would involve lots of travel, and you would be involved in many projects over the years. You would be exposed to many different organizations, technologies, industries, and systems. One type of system consulting firms have been heavily involved with lately has been enterprise resource planning or ERP systems. These systems are so large and complex, organizations seeking to implement them need the expert help only experienced consulting firms can provide. Once you had been working for the firm for a while, if you were successful, you would have to decide if you wanted to compete for a partner position. Unlike corporations, where stockholders own the company, most consulting firms are organized as partnerships, where the partners own the company. As you might imagine, there are not many partnerships available, and competition for them is fierce. Many consulting firm employees decide not to compete for a partner position. Instead, many go to work for client firms, or start their own small consulting firms.

Another opportunity available to you is to work in the information systems shop of a corporation. The work is very similar to what you would do as an analyst for a consulting firm, except that your clients all work for the same corporation as you do. This does not mean you won't have to travel. Large corporations have plants and offices all over the world, so you might have to travel to these distant locations as part of your job. You would not be exposed to as many different types of systems and technologies as would be the case if you worked for a consulting firm, but instead you would have the chance to gain deep expertise in the technologies the corporation has chosen to use. You would also have the opportunity, if you chose, to become a division or department manager in the information systems shop. You might also decide you wanted to become the corporation's Chief Information Officer or CIO. Until very recently, most corporations chose not to hire new graduates right out of school for their information systems shops, but that is no longer the case for many organizations.

Finally, considering that so many systems organizations use application packages developed by software vendors, there are many opportunities available for the recent MIS graduate in the software industry. The software industry is a massive, multi-billion dollar a year industry, so the range of opportunities is vast, ranging from a job with a large established firm like Microsoft, to working for an Internet start-up. Working for a software vendor, you would work on developing and testing information systems, just as would be the case in a consulting firm or corporation, but unlike those jobs, you would rarely ever see or talk with the end user of the system you develop. Opportunities exist for you to move up in the company or to use what you have learned to start your own software development firm.

Not all recent MIS graduates become systems analysts, however. There are many other types of information technology jobs that make use of the skills you will acquire as part of completing an MIS degree. Among the other opportunities now available for recent MIS graduates are the following:

- Network administration, which involves installing, managing, monitoring, and upgrading the firm's internal data and communication networks

- Technical support specialist, which involves troubleshooting, customer service, hardware and/or software installation, and systems maintenance (ITAA, 2000)

- Help desk support, in which you attempt to solve user problems and answer user questions about systems they rely on

- E-business and multimedia product and service development, where you help migrate existing systems to the Internet as well as develop new applications that take advantage of trends in electronic business

- Decision support analyst, in which you design database queries and data analysis routines to support business analysis and decision making, often for one department, such as market research or investments

- Data warehouse specialist, which involves converting massive amounts of historical data to aggregated data useful for decision support

- Quality assurance specialist, in which you review and test software to make sure it is as error-free as possible

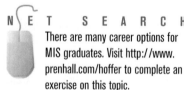

N E T   S E A R C H

There are many career options for MIS graduates. Visit http://www.prenhall.com/hoffer to complete an exercise on this topic.

Obviously, not all of these opportunities will exist in every company that uses information technology. Not every firm will have need for data warehouse specialists or systems analysts. Yet almost every firm needs network administrators and technical support specialists. These positions account for half of the new information technology jobs created in 2000 (ITAA, 2000). Even though every firm does not employ analysts or database developers, 20 percent of the new jobs created in 2000 fall into these categories. An additional 13 percent of new information technology jobs are related to E-business and Internet development. In short, the opportunities are widespread, regardless of which particular area of the information technology profession you decide to pursue.

Summary

In this chapter, we have surveyed the skills necessary for success as a systems analyst. The requisite skills are analytical, technical, management, and interpersonal. Analytical skills include the concept of systems thinking, which is one of the most important skills an analyst can learn. Systems thinking provides a disciplined foundation on which all other analyst skills can build. In addition, an analyst needs to understand the nature of business and of the particular enterprise he or she serves and to be able to identify, analyze, and solve problems.

Technical skills change over time as technology changes and analysts need to keep current with changing information technology. This can be accomplished through reading trade journals, joining professional societies, attending or teaching classes, attending conferences, and participating in electronic bulletin boards and news groups. Some technology areas that play a continuing important role are programming languages, operating systems, database management systems, data communications, and systems development techniques and tools.

A useful skill is the ability to manage resources, projects, risk, and change. Interpersonal skills, especially clear communication, are also important. Analysts communicate with team members in interviews, with questionnaires, through written and oral presentations, and through facilitating groups. A key component of communicating about information systems is managing the expectations of both users and managers.

The chapter concluded with an examination of the system analyst's position, the standards of practice, the ACM Code of Ethics, and possible career paths. Systems analysis is becoming more of a science and less of an art as the systems analysis field becomes a profession.

Key Terms

1. Boundary
2. Closed system
3. Cohesion
4. Components
5. Constraints
6. Coupling
7. Environment
8. Input
9. Interface
10. Interrelated components
11. Logical system description
12. Modularity
13. Open system
14. Output
15. Physical system description
16. Purpose
17. System

Match each of the key terms above with the definition that best fits it.

_____ A system that is cut off from its environment and does not interact with it.

_____ An interrelated set of components, with an identifiable boundary, working together for some purpose.

_____ An irreducible part or aggregation of parts that make up a system, also called a subsystem.

_____ Dependence of one part of the system on one or more other system parts.

_____ The line that marks the inside and outside of a system, and that sets off the system from its environment.

_____ The overall goal or function of a system.

_____ Whatever a system returns to its environment in order to fulfill its purpose.

_____ Everything external to a system that interacts with the system.

_____ Point of contact where a system meets its environment or where subsystems meet each other.

_____ A limit to what a system can accomplish.

_____ Dividing a system up into chunks or modules of a relatively uniform size.

_____ The extent to which subsystems depend on each other.

_____ The extent to which a system or subsystem performs a single function.

_____ Whatever a system takes from its environment in order to fulfill its purpose.

_____ Description of a system that focuses on the system's function and purpose without regard to how the system will be physically implemented.

_____ A system that interacts freely with its environment, taking input and returning output.

_____ Description of a system that focuses on how the system will be materially constructed.

Review Questions

1. What is systems thinking? How is it useful for thinking about computer-based information systems?

2. What is decomposition? coupling? cohesion?

3. In what way are organizations systems?

4. What are the differences between problem identification and problem solving?

5. How can a systems analyst determine if his or her technical skills are up-to-date?

6. Explain the management skills needed by systems analysts.

7. Which communication skills are important for analysts? Why?

8. Is systems analysis a profession? Why or why not?

9. What is a code of ethics?

10. What's the difference between a logical system description and a physical system description?

11. Which areas of organizational knowledge are important for a systems analyst to know?

12. What's the difference between an open and a closed system?

13. What kinds of tasks are included in resource management?

14. Why is the development of an information sometimes done by independent contractors?

15. What are the twelve characteristics of high performance teams? Compare Table 2-2 with Table 1-2. What differences do you see between these two tables?

Problems and Exercises

1. Describe your university or college as a system. What is the input? the output? the boundary? the components? their interrelationships? the constraints? the purpose? the interfaces? the environment? Draw a diagram of this system.

2. a. A car is a system with several subsystems, including the braking subsystem, the electrical subsystem, the engine, the fuel subsystem, the climate control subsystem, and the passenger subsystem. Draw a diagram of a car as a system and label all of its system characteristics.

   b. Your personal computer is a system. Draw and label a personal computer as a system as you did for a car in part (a).

3. Describe yourself in terms of your abilities at resource, project, risk, and change management. Among these categories, what are your strengths and weaknesses? Why? How can you best capitalize on your strengths and strengthen areas where you are weak? If you do not have managerial or supervisory experience, answer these questions as if you were generalizing from your experiences thus far to your performance later as a manager.

4. Describe yourself in terms of your abilities at each of the following interpersonal skills: working alone versus working with a team, interviewing, listening, writing, presenting, facilitating a group, and managing expectations. Where are your strengths and weaknesses? Why? What can you do to capitalize on your strengths and strengthen areas where you are weak?

5. Use your imagination and hypothesize what a systems analyst would be like if he or she were a person with no personal or professional ethics. What types of systems would that person help to create, and how might they go about building such systems? What would the consequences of these actions be, with what implications for the analyst, for his or her information systems department, for his or her users, for the organization? Specifically, how would a code of ethics and

professional conduct help curb the behavior of this person? This may seem like a silly exercise, but even your wildest guesses about the things an unethical analyst might do have probably happened in some setting.

6. You likely receive (and pay) one or more bills each month or semester (for example, tuition, rent, utilities, or telephone). Describe one of these billing systems as an information system. Be sure to list at least one example of each of the nine characteristics of a system for your example billing system.

7. The chapter mentioned that choosing the boundary for a system is a crucial step in analyzing and studying a system. What criteria would you use to determine where to draw a system boundary? What are the ramifications of setting too broad a boundary? Too narrow a boundary?

8. Make a list of the technical skills you have developed at school, as part of any job you've held, and on your own. Using newspaper want ads, trade journals, and other sources, determine if your technical skills are up-to-date. If not, devise a plan to update your technical skills.

9. Recall a team on which you have worked in a job or course project. How well did this team follow the twelve characteristics of high-performance teams? How could you have improved the performance of this team?

10. Chapter 1 outlined the role of systems analyst and others in systems development. How might these roles change if contractors are used on a project? Which of these roles might a contractor provide?

11. Figure 2-11 contains the code of ethics for the Association of Information Technology Professionals (AITP). Compare it to the ACM code of ethics (Figure 2-10) on a point-by-point basis. You may want to go to the Websites for each group to get more detailed information on their respective codes of ethics (www.acm.org & www.aitp.org).

Figure 2-11
The Code of Ethics of the AITP.
(Used with permission.)

Code of Ethics

I acknowledge:

That I have an obligation to management, therefore, I shall promote the understanding of information processing methods and procedures to management using every resource at my command.

That I have an obligation to my fellow members, therefore, I shall uphold the high ideals of AITP as outlined in the Association Bylaws. Further, I shall cooperate with my fellow members and shall treat them with honesty and respect at all times.

That I have an obligation to society and will participate to the best of my ability in the dissemination of knowledge pertaining to the general development and understanding of information processing. Further, I shall not use knowledge of a confidential nature to further my personal interest, nor shall I violate the privacy and confidentiality of information entrusted to me or to which I may gain access.

That I have an obligation to my College or University, therefore, I shall uphold its ethical and moral principles.

That I have an obligation to my employer whose trust I hold, therefore, I shall endeavor to discharge this obligation to the best of my ability, to guard my employer's interests, and to advise him or her wisely and honestly.

That I have an obligation to my country, therefore, in my personal, business, and social contacts, I shall uphold my nation and shall honor the chosen way of life of my fellow citizens.

I accept these obligations as a personal responsibility and as a member of this Association. I shall actively discharge these obligations and I dedicate myself to that end.

Field Exercises

1. Describe an organization of your choice as an open system. What factors lead you to believe that this system is open? Describe the organization in terms of decomposition, coupling, cohesion, and modularity. What is beneficial about thinking of the organization in this way?

2. Think about a problem you have, perhaps with a grade in a class, with a job you're not satisfied with, or with a co-worker on the job. Describe the problem as a difference between "what is" and "what should be." What must happen to shift your situation from "what is" to "what should be" to bring about a situation that you are satisfied with? What specific actionable steps must you take to make this change happen? What information, if any, will you need to gather about this situation? From where, and/or from whom, must the information come? How can you get this information?

3. Choose a manager you know in any area and describe this person in terms of his or her abilities at resource, project, risk, and change management. Overall, is he or she successful or not? Why or why not?

4. Investigate where on your campus and community you could go to get help and practice with public speaking. Talk with other students, contact your instructors, look in the telephone book and directory of services at your college, and explore avenues to uncover as many sources of public speaking help you can find.

5. Many organizations have an approved technology list from which units are free to purchase hardware and software for application development. Contact the computing services unit at your college (or other organization) and find out what hardware and software are supported on your campus for administrative computing. Given this list of supported technologies, what would you infer are the technical skills required for a systems analyst at your college (or other organization)?

References

Ginzberg, M. J. 1981. "Early Diagnosis of MIS Implementation Failure: Promising Results and Unanswered Questions." *Management Science* 27 (April): 459–78.

Information Technology Association of America. 2000. *Bridging the Gap: Information Technology Skills for a New Millennium.* Arlington, VA: ITAA.

Lakhanpal, B. 1993. "Understanding the Factors Influencing the Performance of Software Development Groups: An Exploratory Group-level Analysis." *Information & Software Technology* 35(8): 468–73.

Martin. E. W., D. W. DeHayes, J. A. Hoffer, W. C. Perkins, and C. V. Brown. 1999. *Managing Information Technology: What Managers Need to Know.* 3rd ed. Upper Saddle River, NJ: Prentice Hall Inc.

McConnell, S. 1996. *Rapid Development.* Redmond, WA: Microsoft Press.

Option Technologies, Inc. 1992. *Just-In-Time Knowledge for Teams.* Mendotta Heights, MN.

Pounds, W. F. 1969. "The Process of Problem Finding." *Industrial Management Review* (Fall): 1–19.

Simon, H. A. 1960. *The New Science of Management Decision.* New York: Harper & Row.

Smith, H. J., and J. Hasnas. 1999. "Ethics and Information Systems: The Corporate Domain." *MIS Quarterly* 23(1), 109–27.

Managing the Information Systems Project

After studying this chapter, you should be able to:

- Explain the process of managing an information systems project.

- Describe the skills required to be an effective project manager.

- List and describe the skills and activities of a project manager during project initiation, project planning, project execution, and project close-down.

- Explain what is meant by critical path scheduling and describe the process of creating Gantt and PERT charts.

- Explain how commercial project management software packages can be used to assist in representing and managing project schedules.

INTRODUCTION

Many aspects of information technology in general and the development of information systems in particular are more glamorous than the management of development projects. This view is underscored by a quote from a classic book that focuses on the management of information systems projects:

> Project management has rarely received the attention it deserves, and particularly within the computing profession, it has been overshadowed by the battles within the technological arena: Manufacturer versus manufacturer, development language versus development language, mainframe versus micro, operating system versus operating system are the stuff from

which great legends are born and great leaders emerge as role models . . . Pity the humble project manager who manages to bring the general ledger system in on time, within budget, and working to the users' satisfaction (Thomsett, 1985).

As the above quotation typifies, some may not view project management to be a glamorous occupation. Yet, project management is an important aspect of the development of information systems and a critical skill for a systems analyst. The focus of project management is to assure that system development projects meet customer expectations and are delivered within budget and time constraints. This chapter describes how you can wear many different hats while managing all or part of a project.

The project manager is responsible for virtually all aspects of a systems development project: What you experience as a project manager is an environment of continual change and problem solving. In some organizations the project manager is a senior systems analyst who "has been around the block" a time or two. In others, both junior and senior analysts are expected to take on this role, managing parts of a project or actively supporting a more senior colleague who is assuming this role. In addition, there is a shift in the types of projects most firms are undertaking, which makes project management much more difficult and even more critical to project success (Kirsch, 2000). For example, in the past, organizations focused much of their development on very large custom-designed stand-alone applications. Today, much of the systems development effort in organizations focuses on implementing packaged software such as enterprise resource planning (ERP) and data warehousing systems. Existing legacy applications are also being modified so that business-to-business transactions can seamlessly occur over the Internet. New Web-based interfaces are being added to existing legacy systems so that a broader range of users, often distributed globally, can access corporate information and systems. Working with vendors to supply applications, with customers or suppliers to integrate systems, or with a broader and diverse user community requires that project managers be highly skilled. Consequently, it is important that you gain an understanding of the project management process; this will become a critical skill for your future success.

In this chapter we focus on the systems analyst's role in managing information systems projects and will refer to this role as the project manager. The next section will provide the background for Pine Valley Furniture, a manufacturing company that we will visit throughout the remainder of the book. The following section will provide you with an understanding of the project manager's role and the project management process. The subsequent section examines techniques for reporting project plans using Gantt and PERT charts. The chapter will conclude by discussing the use of commercially available project management software that can be used to assist with a wide variety of project management activities.

PINE VALLEY FURNITURE COMPANY BACKGROUND

Pine Valley Furniture Company (PVF) manufactures high-quality wood furniture and distributes it to retail stores within the United States. Its product lines include dinette sets, stereo cabinets, wall units, living room furniture, and bedroom furniture. In the early 1980s, PVF's founder, Alex Schuster, started to make and sell custom furniture in his garage. Alex managed invoices and kept track of customers by using file folders and a filing cabinet. By 1984, business expanded and Alex had to rent a warehouse and hire a part-time bookkeeper. PVF's product line had multiplied, sales volume had doubled, and staff had increased to 50 employees. By 1990, PVF moved into its third and present location. Due to the added complexity of the company's operations, Alex reorganized the company into the following functional areas:

- Manufacturing, which was further subdivided into three separate functions—Fabrication, Assembling, and Finishing
- Sales
- Orders
- Accounting
- Purchasing

Alex and the heads of the functional areas established manual information systems, such as accounting ledgers and file folders, which worked well for a time. Eventually, however, PVF selected and installed a minicomputer to automate invoicing, accounts receivable, and inventory control applications.

When the applications were first computerized, each separate application had its own individual data files tailored to the needs of each functional area. As is typical in such situations, the applications closely resembled the manual systems on which they were based. Three computer applications at Pine Valley Furniture are depicted in Figure 3-1: order filling, invoicing, and payroll. In the late 1990s, PVF formed a task force to study the possibility of moving to a database approach. After a preliminary study, management decided to convert its information systems to such an approach. The company upgraded its minicomputer and implemented a database management system. By the time we catch up with Pine Valley Furniture, it has successfully designed and populated a company-wide database and has converted its applications to work with the database. However, PVF is continuing to grow at a rapid rate, putting pressure on its current application systems.

The computer-based applications at PVF support its business processes. When customers order furniture, their orders must be processed appropriately: Furniture must be built and shipped to the right customer and the right invoice mailed to the right address. Employees have to be paid for their work. Given these tasks, most of PVF's computer-based applications are located in the accounting and financial areas. The applications include order filling, invoicing, accounts receivable, inventory control, accounts payable, payroll, and general ledger. At one time, each application had its own data files. For example, there was a customer master file, an inventory master file, a back order file, an inventory pricing file, and an employee master file. The order filling system used data from three files: customer master, inventory master, and back order. With PVF's new centralized database, data are organized around entities, or subjects, such as customers, invoices, and orders.

Pine Valley Furniture, like many firms, decided to develop its application software in-house; that is, it hired staff and bought computer hardware and software nec-

Figure 3-1
Three computer applications at Pine Valley Furniture: Order Filling, Invoicing, and Payroll
(Source: Hoffer, Prescott, and McFadden, 2002)

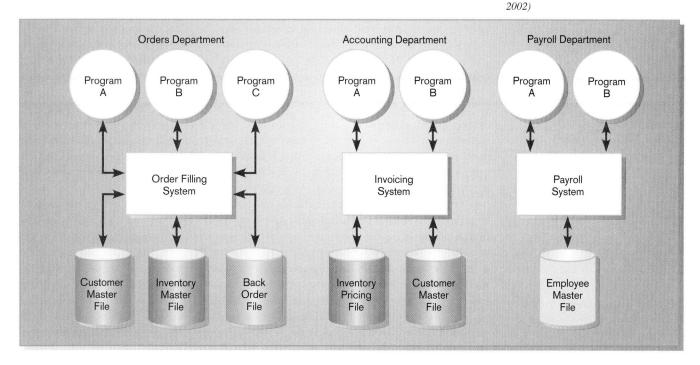

essary to build application software suited to its own needs. (Other methods used to obtain application software, are explained in Chapter 11.) Let's see how a project manager plays a key role in developing a new information system for Pine Valley Furniture.

N E T   S E A R C H
You may not be aware that you can become a certified project manager. Visit http://www.prenhall.com/hoffer to complete an exercise related to this topic.

Project manager: A systems analyst with a diverse set of skills—management, leadership, technical, conflict management, and customer relationship—who is responsible for initiating, planning, executing, and closing down a project.

Project: A planned undertaking of related activities to reach an objective that has a beginning and an end.

PINE VALLEY FURNITURE

Deliverable: An end product in a phase of the SDLC.

Feasibility study: Determines if the information system makes sense for the organization from an economic and operational standpoint.

MANAGING THE INFORMATION SYSTEMS PROJECT

Project management is an important aspect of the development of information systems and a critical skill for a systems analyst. The focus of project management is to assure that system development projects meet customer expectations and are delivered within budget and time constraints.

The **project manager** is a systems analyst with a diverse set of skills—management, leadership, technical, conflict management, and customer relationship—who is responsible for initiating, planning, executing, and closing down a project. As a project manager, your environment is one of continual change and problem solving. In some organizations the project manager is a senior systems analyst who "has been around the block" a time or two. In others, both junior and senior analysts are expected to take on this role, managing parts of a project or actively supporting a more senior colleague who is assuming this role. Understanding the project management process is a critical skill for your future success.

Creating and implementing successful projects require managing resources, activities, and tasks needed to complete the information systems project. A **project** is a planned undertaking of a series of related activities to reach an objective that has a beginning and end. The first question you might ask yourself is "Where do projects come from?" and, after considering all the different things that you could be asked to work on within an organization, "How do I know which projects to work on?" The ways in which each organization answers these questions vary.

In the rest of this section, we describe the process followed by Juanita Lopez and Chris Martin during the development of Pine Valley Furniture's Purchasing Fulfillment System. Juanita works in the order department, and Chris is a systems analyst.

Juanita observed problems with the way orders were processed and reported: Sales growth had increased the workload for the Manufacturing department, and the current systems no longer adequately supported the tracking of orders. It was becoming more difficult to track orders and get the right furniture and invoice to the right customers. Juanita contacted Chris, and together they developed a system that corrected these Ordering department problems.

The first **deliverable**, or end product, produced by Chris and Juanita was a System Service Request (SSR), a standard form PVF uses for requesting systems development work. Figure 3-2 shows an SSR for a purchasing fulfillment system. The form includes the name and contact information of the person requesting the system, a statement of the problem, and the name and contact information of the liaison and sponsor.

This request was then evaluated by the Systems Priority Board of PVF. Because all organizations have limited time and resources, not all requests can be approved. The board evaluates development requests in relation to the business problems or opportunities the system will solve or create, It also considers how the proposed project fits within the organization's information systems architecture and long-range development plans. The review board selects those projects that best meet overall organizational objectives (we learn more about organizational objectives in Chapter 5). In the case of the Purchasing Fulfillment System request, the board found merit in the request and approved a more detailed **feasibility study**. A feasibility study, conducted by the project manager, involves determining if the information system makes sense

Pine Valley Furniture
System Service Request

REQUESTED BY   Juanita Lopez   DATE   November 1, 2001

DEPARTMENT   Purchasing, Manufacturing Support

LOCATION   Headquarters, 1-322

CONTACT   Tel: 4-3267   FAX: 4-3270   e-mail: jlopez

TYPE OF REQUEST   URGENCY

[X]   New System   [   ]   Immediate – Operations are impaired or
                                                           opportunity lost

[   ]   System Enhancement   [   ]   Problems exist, but can be worked around
[   ]   System Error Correction   [X]   Business losses can be tolerated until new
                                                           system installed

PROBLEM STATEMENT

Sales growth at PVF has caused greater volume of work for the manufacturing support unit within Purchasing. Further,
more concentration on customer service has reduced manufacturing lead times, which puts more pressure on purchasing
activities. In addition, cost-cutting measures force Purchasing to be more agressive in negotiating terms with vendors,
improving delivery times, and lowering our investments in inventory. The current modest systems support for
manufacturing purchasing is not responsive to these new business conditions. Data are not available, information cannot
be summarized, supplier orders cannot be adequately tracked, and commodity buying is not well supported. PVF is
spending too much on raw materials and not being responsive to manufacturing needs.

SERVICE REQUEST

I request a thorough analysis of our current operations with the intent to design and build a completely new information
system. This system should handle all purchasing transactions, support display and reporting of critical purchasing data,
and assist purchasing agents in commodity buying.

IS LIAISON   Chris Martin (Tel: 4-6204   FAX: 4-6200   e-mail: cmartin)

SPONSOR   Sal Divario, Director, Purchasing

------------------------ TO BE COMPLETED BY SYSTEMS PRIORITY BOARD ------------------------

[   ]   Request approved   Assigned to   _____
                                  Start date   _____
[   ]   Recommend revision
[   ]   Suggest user development
[   ]   Reject for reason   _____

Figure 3-2
**System Service Request for
Purchasing Fulfillment System with
name and contact information of
the person requesting the system,
a statement of the problem, and
the name and contact information
of the liaison and sponsor**

for the organization from an economic and operational standpoint. The study takes place before the system is constructed. Figure 3-3 is a graphical view of the steps followed during the project initiation of the Purchasing Fulfillment System.

In summary, systems development projects are undertaken for two primary reasons: to take advantage of business opportunities and to solve business problems. Taking advantage of an opportunity might mean providing an innovative service to customers through the creation of a new system. For example, PVF may want to create a Web page so that customers can easily access its catalog and place orders at any time. Solving a business problem could involve modifying how an existing system processes data so that more accurate or timely information is provided to users. For example, a company such as PVF may create a password-protected Intranet site that contains important announcements and budget information.

Projects are not always initiated for the rational reasons (taking advantage of business opportunities or solving business problems) stated above. For example, in some instances organizations and government undertake projects to spend resources, attain or pad budgets, keep people busy, or help train people and develop their skills. Our focus in this chapter is not on how and why organizations identify projects but on the management of projects once they have been identified.

Figure 3-3
A graphical view of the five steps followed during the project initiation of the purchasing fulfillment system

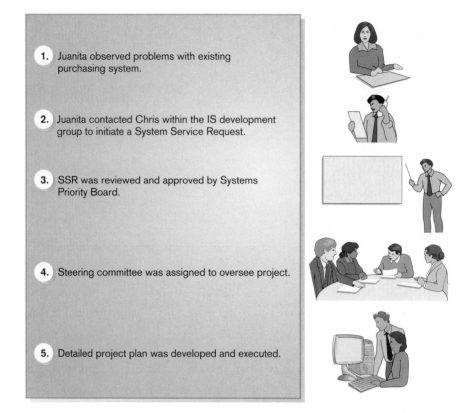

1. Juanita observed problems with existing purchasing system.

2. Juanita contacted Chris within the IS development group to initiate a System Service Request.

3. SSR was reviewed and approved by Systems Priority Board.

4. Steering committee was assigned to oversee project.

5. Detailed project plan was developed and executed.

Once a potential project has been identified, an organization must determine the resources required for its completion. This is done by analyzing the scope of the project and determining the probability of successful completion. After getting this information, the organization can then determine whether taking advantage of an opportunity or solving a particular problem is feasible within time and resource constraints. If deemed feasible, a more detailed project analysis is then conducted.

As you will see, determining the size, scope, and resource requirements for a project are just a few of the many skills that a project manager must possess. A project manager is often referred to as a juggler keeping aloft many balls, which reflect the various aspects of a project's development, as depicted in Figure 3-4.

To successfully orchestrate the construction of a complex information system, a project manager must have interpersonal, leadership, and technical skills. Table 3-1 lists the project manager's common skills and activities. Note that many of the skills are related to personnel or general management, not simply technical skills. Table 3-1 shows that not only does an effective project manager have varied skills, but he or she is also the most instrumental person to the successful completion of any project.

Project management: A controlled process of initiating, planning, executing, and closing down a project.

The remainder of this chapter will focus on the **project management** process, which involves four phases:

1. Initiating the project
2. Planning the project
3. Executing the project
4. Closing down the project

Several activities must be performed during each of these four phases. Following this formal project management process greatly increases the likelihood of project success.

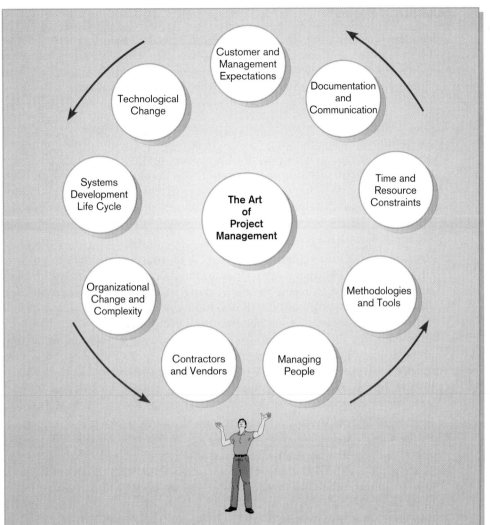

Figure 3-4
A project manager juggles
numerous activities

TABLE 3-1   Common Activities and Skills of a Project Manager

Activity	Description	Skill
Leadership	Influencing the activities of others toward the attainment of a common goal through the use of intelligence, personality, and abilities	Communication; liaison between management, users, and developers; assigning activities; monitoring progress
Management	Getting projects completed through the effective utilization of resources	Defining and sequencing activities; communicating expectations; assigning resources to activities; monitoring outcomes
Customer relations	Working closely with customers to assure project deliverables meet expectations	Interpreting system requests and specifications; site preparation and user training; contact point for customers
Technical problem solving	Designing and sequencing activities to attain project goals	Interpreting system requests and specifications; defining activities and their sequence; making trade-offs between alternative solutions; designing solutions to problems
Conflict management	Managing conflict within a project team to assure that conflict is not too high or too low	Problem solving; smoothing out personality differences; compromising; goal setting
Team management	Managing the project team for effective team performance	Communication within and between teams; peer evaluations; conflict resolution; team building; self-management
Risk and change management	Identifying, assessing, and managing the risks and day-to-day changes that occur during a project	Environmental scanning; risk and opportunity identification and assessment; forecasting; resource redeployment

Initiating a Project

Project initiation: The first phase of the project management process in which activities are performed to assess the size, scope, and complexity of the project and to establish procedures to support later project activities.

During **project initiation** the project manager performs several activities that assess the size, scope, and complexity of the project, and establishes procedures to support subsequent activities. Depending on the project, some initiation activities may be unnecessary and some may be very involved. The types of activities you will perform when initiating a project are summarized in Figure 3-5 and described next.

1. *Establishing the project initiation team.* This activity involves organizing an initial core of project team members to assist in accomplishing the project initiation activities. For example, during the Purchasing Fulfillment System project at PVF, Chris Martin was assigned to support the Purchasing department. It is a PVF policy that all initiation teams consist of at least one user representative, in this case Juanita Lopez, and one member of the IS development group. Therefore, the project initiation team consisted of Chris and Juanita; Chris was the project manager.

2. *Establishing a relationship with the customer.* A thorough understanding of your customer builds stronger partnerships and higher levels of trust. At PVF, management has tried to foster strong working relationships between business units (like Purchasing) and the IS development group by assigning a specific individual to work as a liaison between both groups. Because Chris had been assigned to the Purchasing unit for some time, he was already aware of some of the problems with the existing purchasing systems. PVF's policy of assigning specific individuals to each business unit helped to assure that both Chris and Juanita were comfortable working together prior to the initiation of the project. Many organizations use a similar mechanism for establishing relationships with customers.

3. *Establishing the project initiation plan.* This step defines the activities required to organize the initiation team while it is working to define the goals and scope of the project (Abdel-Hamid, Sengupta, and Swett, 1999). Chris's role was to help Juanita translate her business requirements into a written request for an improved information system. This required the collection, analysis, organization, and transformation of a lot of information. Because Chris and Juanita were already familiar with each other and their roles within a development project, they next needed to define when and how they would communicate, define deliverables and project steps, and set deadlines. Their initiation plan included agendas for several meetings. These steps eventually led to the creation of their System Service Request (SSR) form.

4. *Establishing management procedures.* Successful projects require the development of effective management procedures. Within PVF, many of these management procedures had been established as standard operating procedures

Figure 3-5
Five project initiation activities

Project Initiation

1. Establishing the Project Initiation Team
2. Establishing a Relationship with the Customer
3. Establishing the Project Initiation Plan
4. Establishing Management Procedures
5. Establishing the Project Management Environment and Project Workbook

by the Systems Priority Board and the IS development group. For example, all project development work is charged back to the functional unit requesting the work. In other organizations, each project may have unique procedures tailored to its needs. Yet, in general when establishing procedures, you are concerned with developing team communication and reporting procedures, job assignments and roles, project change procedures, and determining how project funding and billing will be handled. It was fortunate for Chris and Juanita that most of these procedures were already established at PVF, allowing them to move on to other project activities.

5. *Establishing the project management environment and project workbook.* The focus of this activity is to collect and organize the tools that you will use while managing the project and to construct the **project workbook**. For example, most diagrams, charts, and system descriptions provide much of the project workbook contents. Thus, the project workbook serves as a repository for all project correspondence, inputs, outputs, deliverables, procedures, and standards established by the project team (Rettig, 1990). The project workbook can be stored as an on-line electronic document or in a large three-ring binder. The project workbook is used by all team members and is useful for project audits, orientation of new team members, communication with management and customers, identifying future projects, and performing postproject reviews. The establishment and diligent recording of all project information in the workbook are two of the most important activities you will perform as project manager.

Project workbook: An on-line or hard-copy repository for all project correspondence, inputs, outputs, deliverables, procedures, and standards that is used for performing project audits, orientating new team members, communicating with management and customers, identifying future projects, and performing post-project reviews.

Figure 3-6 shows the project workbook for the Purchasing Fulfillment System. It consists of both a large hard-copy binder and electronic diskettes where the system data dictionary, a catalog of data stored in the database, and diagrams are stored. For this system, all project documents can fit into a single binder. It is not unusual, however, for project documentation to be spread over several binders. As more information is captured and recorded electronically, however, fewer hard-copy binders may be needed. Many project teams keep their project workbooks on the Web. A Web site

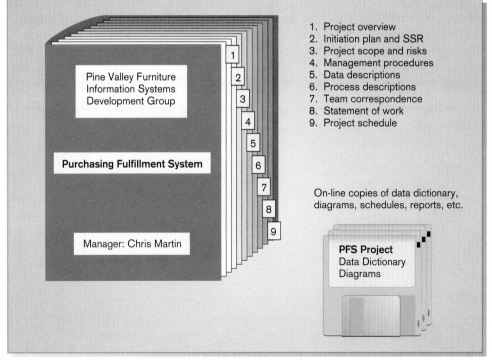

Figure 3-6
The project workbook for the Purchase Fulfillment System project contains nine key documents in both hard-copy and electronic form

Pine Valley Furniture
Information Systems
Development Group

Purchasing Fulfillment System

Manager: Chris Martin

1. Project overview
2. Initiation plan and SSR
3. Project scope and risks
4. Management procedures
5. Data descriptions
6. Process descriptions
7. Team correspondence
8. Statement of work
9. Project schedule

On-line copies of data dictionary, diagrams, schedules, reports, etc.

PFS Project
Data Dictionary
Diagrams

can be created so that all project members can easily access all project documents. This Web site can be a simple repository of documents or an elaborate site with password protection and security levels. The best feature of using the Web as your repository is that it allows all project members and customers to review a project's status and all related information continually.

Project initiation is complete once these five activities have been performed. Before moving on to the next phase of the project, the work performed during project initiation is reviewed at a meeting attended by management, customers, and project team members. An outcome of this meeting is a decision to continue the project, modify it, or abandon it. In the case of the Purchasing Fulfillment System project at Pine Valley Furniture, the board accepted the SSR and selected a project steering committee to monitor project progress and to provide guidance to the team members during subsequent activities. If the scope of the project is modified, it may be necessary to return to project initiation activities and collect additional information. Once a decision is made to continue the project, a much more detailed project plan is developed during the project planning phase.

Planning the Project

Project planning: The second phase of the project management process, which focuses on defining clear, discrete activities and the work needed to complete each activity within a single project.

The next step in the project management process is **project planning** where prior research has found a positive relationship between effective project planning and positive project outcomes (Guinan, Cooprider, and Faraj, 1998; Kirsch, 2000). Project planning involves defining clear, discrete activities and the work needed to complete each activity within a single project. It often requires you to make numerous assumptions about the availability of resources such as hardware, software, and personnel. It is much easier to plan nearer-term activities than those occurring in the future. In actual fact, you often have to construct longer-term plans that are more general in scope and nearer-term plans that are more detailed. The repetitive nature of the project management process requires that plans be constantly monitored throughout the project and periodically updated (usually after each phase) based upon the most recent information.

Figure 3-7 illustrates the principle that nearer-term plans are typically more specific and firmer than longer-term plans. For example, it is virtually impossible to rigorously plan activities late in the project without first completing the earlier activities. Also, the outcome of activities performed earlier in the project are likely to have impact on later activities. This means that it is very difficult, and very likely inefficient, to try to plan detailed solutions for activities that will occur far into the future.

Figure 3-7
Level of project planning detail should be high in the short term, with less detail as time goes on

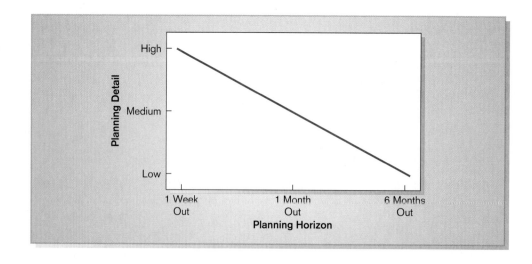

Project Planning

1. Describing Project Scope, Alternatives, and Feasibility

2. Dividing the Project into Manageable Tasks

3. Estimating Resources and Creating a Resource Plan

4. Developing a Preliminary Schedule

5. Developing a Communication Plan

6. Determining Project Standards and Procedures

7. Identifying and Assessing Risk

8. Creating a Preliminary Budget

9. Developing a Statement of Work

10. Setting a Baseline Project Plan

Figure 3-8
Ten project management activities

As with the project initiation process, varied and numerous activities must be performed during project planning. For example, during the Purchasing Fulfillment System project, Chris and Juanita developed a 10-page plan. However, project plans for very large systems may be several hundred pages in length. The types of activities that you can perform during project planning are summarized in Figure 3-8 and are described below:

1. *Describing project scope, alternatives, and feasibility.* The purpose of this activity is to understand the content and complexity of the project. Within PVF's system development methodology, one of the first meetings must focus on defining a project's scope. Although project scope information was not included in the SSR developed by Chris and Juanita, it was important that both shared the same vision for the project before moving too far along. During this activity, you should reach agreement on the following questions:

   - What problem or opportunity does the project address?
   - What are the quantifiable results to be achieved?
   - What needs to be done?
   - How will success be measured?
   - How will we know when we are finished?

   After defining the scope of the project, your next objective is to identify and document general alternative solutions for the current business problem or opportunity. You must then assess the feasibility of each alternative solution and choose which to consider during subsequent SDLC phases. In some instances, off-the-shelf software can be found. It is also important that any unique problems, constraints, and assumptions about the project be clearly stated.

2. *Dividing the project into manageable tasks.* This is a critical activity during the project planning process. Here, you must divide the entire project into manageable tasks and then logically order them to ensure a smooth evolution between tasks. The definition of tasks and their sequence is referred to as the **work breakdown structure**. Some tasks may be performed in parallel whereas others must follow one another sequentially. Task sequence depends on which tasks produce deliverables needed in other tasks, when critical

Work breakdown structure: The process of dividing the project into manageable tasks and logically ordering them to ensure a smooth evolution between tasks.

Figure 3-9
Gantt chart showing project tasks, duration times for those tasks (d= days), and predecessors.

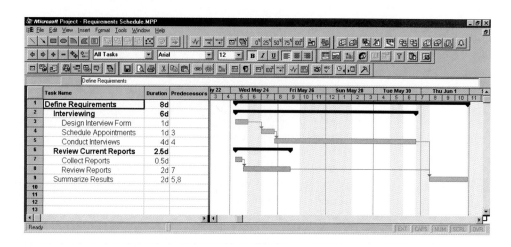

Gantt chart: A graphical representation of a project that shows each task as a horizontal bar whose length is proportional to its time for completion.

resources are available, the constraints placed on the project by the client, and the process outlined in the SDLC.

For example, suppose that you are working on a new development project and need to collect system requirements by interviewing users of the new system and reviewing reports they currently use to do their job. A work breakdown for these activities is represented in a Gantt chart in Figure 3-9. A **Gantt chart** is a graphical representation of a project that shows each task as a horizontal bar whose length is proportional to its time for completion. Different colors, shades, or shapes can be used to highlight each kind of task. For example, those activities on the critical path (defined below) may be in red and a summary task could have a special bar. Note that the black horizontal bars—rows 1, 2, and 6 in Figure 3-9—represent summary tasks. Planned versus actual times or progress for an activity can be compared by parallel bars of different colors, shades, or shapes. Gantt charts do not (typically) show how tasks must be ordered (precedence) but simply show when an activity should begin and end. In Figure 3-9, the task duration is shown in the second column by days, "d," and necessary prior tasks are noted in the third column as predecessors. Most project management software tools support a broad range of task durations including minutes, hours, days, weeks, and months. As you will learn in later chapters, the SDLC consists of several phases, which you need to break down into activities. Creating a work breakdown structure requires that you decompose phases into activities—summary tasks—and activities into specific tasks. For example, Figure 3-9 shows that the activity Interviewing consists of three tasks: design interview form, schedule appointments, and conduct interviews.

Defining tasks in too much detail will make the management of the project unnecessarily complex. You will develop the skill of discovering the optimal level of detail for representing tasks through experience. For example, it may be very difficult to list tasks that require less than one hour of time to complete in a final work breakdown structure. Alternatively, choosing tasks that are too large in scope (e.g., several weeks long) will not provide you with a clear sense of the status of the project or of the interdependencies between tasks. What are the characteristics of a "task"? A task

- Can be done by one person or a well-defined group
- Has a single and identifiable deliverable (The task is, however, the process of creating the deliverable.)
- Has a known method or technique
- Has well-accepted predecessor and successor steps
- Is measurable so that percent completed can be determined

3. *Estimating resources and creating a resource plan.* The goal of this activity is to estimate resource requirements for each project activity and use this information to create a project resource plan. The resource plan helps assemble and deploy resources in the most effective manner. For example, you would not want to bring additional programmers onto the project at a rate faster than you could prepare work for them.

   People are the most important, and expensive, part of project resource planning. Project time estimates for task completion and overall system quality are significantly influenced by the assignment of people to tasks. It is important to give people tasks that allow them to learn new skills. It is equally important to make sure that project members are not in "over their heads" or working on a task that is not well suited to their skills. Resource estimates may need to be revised based upon the skills of the actual person (or people) assigned to a particular activity. Figure 3-10 indicates the relative programming speed versus the relative programming quality of three programmers. The figure suggests that Carl should not be assigned tasks in which completion time is critical and that Brenda should be assigned to tasks in which high quality is most vital.

   One approach to assigning tasks is to assign a single task type (or only a few task types) to each worker for the duration of the project. For example, you could assign one worker to create all computer displays and another to create all system reports. Such specialization ensures that both workers become efficient at their own particular tasks. A worker may become bored if the task is too specialized or is long in duration, so you could assign workers to a wider variety of tasks. However, this approach may lead to lowered task efficiency. A middle ground would be to make assignments with a balance of both specialization and task variety. Assignments depend upon the size of the development project and the skills of the project team. Regardless of the manner in which you assign tasks, make sure that each team member works only on one task at a time. Exceptions to this rule can occur when a task occupies only a small portion of a team member's time (e.g., testing the programs developed by another team member) or during an emergency.

4. *Developing a preliminary schedule.* During this activity, you use the information on tasks and resource availability to assign time estimates to each activity in the work breakdown structure. These time estimates will allow you to create target starting and ending dates for the project. Target dates can be revisited and modified until a schedule produced is acceptable to the customer. Determining an acceptable schedule may require that you find additional or different resources or that the scope of the project be changed. The schedule may be represented as a Gantt chart, as illustrated in Figure 3-9, or as a PERT

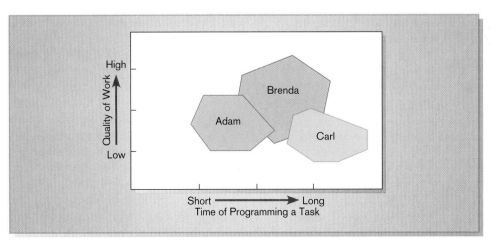

Figure 3-10
Trade-offs between the quality of the program code versus the speed of programming

Figure 3-11
A PERT chart illustrates tasks with rectangles (or ovals) and the relationships and sequences of those activities with arrows

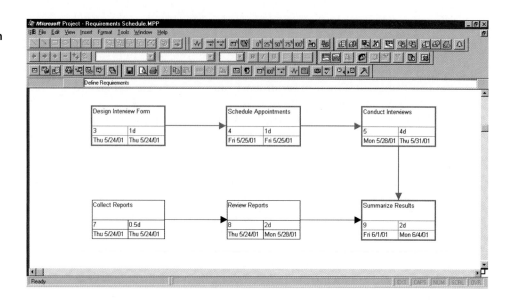

PERT chart: A diagram that depicts project tasks and their interrelationships. PERT stands for Program Evaluation Review Technique.

chart, as illustrated in Figure 3-11. A **PERT** (Program Evaluation Review Technique) **chart** is a graphical depiction of project tasks and their interrelationships. As with a Gantt chart, each type of task can be highlighted by different features on the PERT chart. The distinguishing feature of a PERT chart is that the ordering of tasks is shown by connecting tasks—depicted as rectangles or ovals—with its predecessor and successor tasks. However, the relative size of a node (representing a task) or a gap between nodes does not imply the task's duration. Only the individual task items are drawn on a PERT chart, which is why the summary tasks 1, 2, and 6—the black bars—from Figure 3-9 are not shown in Figure 3-11. We describe both of these charts later in this chapter.

5. *Developing a communication plan.* The goal of this activity is to outline the communication procedures among management, project team members, and the customer. The communication plan includes when and how written and oral reports will be provided by the team, how team members will coordinate work, what messages will be sent to announce the project to interested parties, and what kinds of information will be shared with vendors and external contractors involved with the project. It is important that free and open communication occur among all parties, with respect for proprietary information and confidentiality with the customer (Kettelhut, 1991; Kirsch, 2000).

6. *Determining project standards and procedures.* During this activity, you will specify how various deliverables are produced and tested by you and your project team. For example, the team must decide on which tools to use, how the standard SDLC might be modified, which SDLC methods will be used, documentation styles (e.g., type fonts and margins for user manuals), how team members will report the status of their assigned activities, and terminology. Setting project standards and procedures for work acceptance is a way to assure the development of a high-quality system. Also, it is much easier to train new team members when clear standards are in place. Organizational standards for project management and conduct make the determination of individual project standards easier and the interchange or sharing of personnel among different projects feasible.

7. *Identifying and assessing risk.* The goal of this activity is to identify sources of project risk and to estimate the consequences of those risks. Risks might arise

from the use of new technology, prospective users' resistance to change, availability of critical resources, competitive reactions or changes in regulatory actions due to the construction of a system, or team member inexperience with technology or the business area. You should continually try to identify and assess project risk.

The identification of project risks is required to develop PVF's new Purchasing Fulfillment System. Chris and Juanita met to identify and describe possible negative outcomes of the project and their probabilities of occurrence. Although we list the identification of risks and the outline of project scope as two discrete activities, they are highly related and often concurrently discussed.

8. *Creating a preliminary budget.* During this phase, you need to create a preliminary budget that outlines the planned expenses and revenues associated with your project. The project justification will demonstrate that the benefits are worth these costs. Figure 3-12 shows a cost-benefit analysis for a new development project. This analysis shows net present value calculations of the project's benefits and costs as well as a return on investment and cash flow analysis. We discuss project budgets fully in Chapter 6.

9. *Developing a Statement of Work.* An important activity that occurs near the end of the project planning phase is the development of the Statement of Work. Developed primarily for the customer, this document outlines work that will be done and clearly describes what the project will deliver. The Statement of Work is useful to make sure that you, the customer, and other project team members have a clear understanding of the intended project size, duration, and outcomes.

10. *Setting a Baseline Project Plan.* Once all of the prior project planning activities have been completed, you will be able to develop a Baseline Project Plan. This baseline plan provides an estimate of the project's tasks and resource requirements and is used to guide the next project phase—execution. As new information is acquired during project execution, the baseline plan will continue to be updated.

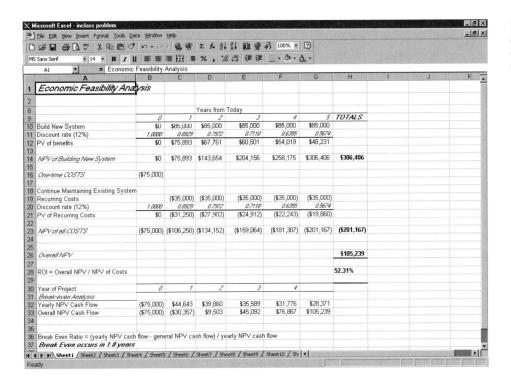

Figure 3-12
A financial cost and benefit analysis for a systems development project

At the end of the project planning phase, a review of the Baseline Project Plan is conducted to double-check all information in the plan. As with the project initiation phase, it may be necessary to modify the plan, which means returning to prior project planning activities before proceeding. As with the Purchasing Fulfillment System project, you may submit the plan and make a brief presentation to the project steering committee at this time. The committee can endorse the plan, ask for modifications, or determine that it is not wise to continue the project as currently outlined.

Executing the Project

Project execution: The third phase of the project management process in which the plans created in the prior phases (project initiation and planning) are put into action.

Project execution puts the Baseline Project Plan into action. Within the context of the SDLC, project execution occurs primarily during the analysis, design, and implementation phases. During the development of the Purchasing Fulfillment System, Chris Martin was responsible for five key activities during project execution. These activities are summarized in Figure 3-13 and described in the remainder of this section:

1. *Executing the Baseline Project Plan.* As project manager, you oversee the execution of the baseline plan. This means that you initiate the execution of project activities, acquire and assign resources, orient and train new team members, keep the project on schedule, and assure the quality of project deliverables. This is a formidable task, but a task made much easier through the use of sound project management techniques. For example, as tasks are completed during a project, they can be "marked" as completed on the project schedule. In Figure 3-14, tasks 3 and 7 are marked as completed by showing 100 percent in the "% Complete" column. Members of the project team will come and go. You are responsible for initiating new team members by providing them with the resources they need and helping them assimilate into the team. You may want to plan social events, regular team project status meetings, team-level reviews of project deliverables, and other group events to mold the group into an effective team.

2. *Monitoring project progress against the Baseline Project Plan.* While you execute the Baseline Project Plan, you should monitor your progress. If the project gets ahead of (or behind) schedule, you may have to adjust resources, activities, and budgets. Monitoring project activities can result in modifications to the current plan. Measuring the time and effort expended on each activity will help you improve the accuracy of estimations for future projects. It is possible with project schedule charts, like Gantt, to show progress against a plan; and it is easy with PERT charts to understand the ramifications of delays in an activity. Monitoring progress also means that the team leader

Figure 3-13
Five project execution activities

Project Execution

1. Executing the Baseline Project Plan
2. Monitoring Project Progress against the Baseline Project Plan
3. Managing Changes to the Baseline Project Plan
4. Maintaining the Project Workbook
5. Communicating the Project Status

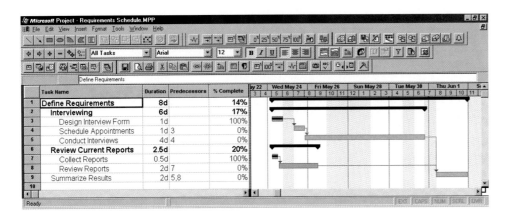

Figure 3-14
Gantt chart with tasks 3 and 7 completed

must evaluate and appraise each team member, occasionally change work assignments or request changes in personnel, and provide feedback to the employee's supervisor.

3. *Managing changes to the Baseline Project Plan.* You will encounter pressure to make changes to the baseline plan. At PVF, policies dictate that only approved changes to the project specification can be made and all changes must be reflected in the baseline plan and project workbook, including all charts. For example, if Juanita suggests a significant change to the existing design of the Purchasing Fulfillment System, a formal change request must be approved by the steering committee. The request should explain why changes are desired and describe all possible impacts on prior and subsequent activities, project resources, and the overall project schedule. Chris would have to help Juanita develop such a request. This information allows the project steering committee to more easily evaluate the costs and benefits of a significant midcourse change.

In addition to changes occurring through formal request, changes may also occur from events outside your control. In fact, numerous events may initiate a change to the Baseline Project Plan, including the following possibilities:

- A slipped completion date for an activity

- A bungled activity that must be redone

- The identification of a new activity that becomes evident later in the project

- An unforeseen change in personnel due to sickness, resignation, or termination

When an event occurs that delays the completion of an activity, you typically have two choices: Devise a way to get back on schedule or revise the plan. Devising a way to get back on schedule is the preferred approach because no changes to the plan will have to be made. The ability to head off and smoothly work around problems is a critical skill that you need to master.

As you see later in the chapter, project schedule charts are very helpful in assessing the impact of change. Using such charts, you can quickly see if the completion time of other activities will be affected by changes in the duration of a given activity or if the whole project completion date will change. Often you will have to find a way to rearrange the activities because the ultimate project completion data may be rather fixed. There may be a penalty to the organization (even legal action) if the expected completion date is not met.

TABLE 3-2   **Project Team Communication Methods**

Procedure	Formality	Use
Project workbook	High	Inform Permanent record
Meetings	Medium to high	Resolve issues
Seminars and workshops	Low to medium	Inform
Project newsletters	Medium to high	Inform
Status reports	High	Inform
Specification documents	High	Inform Permanent record
Minutes of meetings	High	Inform Permanent record
Bulletin boards	Low	Inform
Memos	Medium to high	Inform
Brown bag lunches	Low	Inform
Hallway discussions	Low	Inform Resolve issues

4. *Maintaining the project workbook.* As in all project phases, maintaining complete records of all project events is necessary. The workbook provides the documentation new team members require to assimilate project tasks quickly. It explains why design decisions were made and is a primary source of information for producing all project reports.

5. *Communicating the project status.* The project manager is responsible for keeping all team members—system developers, managers, and customers—abreast of the project status. Clear communication is required to create a shared understanding of the activities and goals of the project; such an understanding assures better coordination of activities. This means that the entire project plan should be shared with the entire project team and any revisions to the plan should be communicated to all interested parties so that everyone understands how the plan is evolving. Procedures for communicating project activities vary from formal meetings to informal hallway discussions. Some procedures are useful for informing others of project status, others for resolving issues, and others for keeping permanent records of information and events. Table 3-2 lists numerous communication procedures, their level of formality, and most likely use. Whichever procedure you use, frequent communication helps to assure project success (Kettelhut, 1991; Kirsch 2000).

This section outlined your role as the project manager during the execution of the Baseline Project Plan. The ease with which the project can be managed is significantly influenced by the quality of prior project phases. If you develop a high-quality project plan, it is much more likely that the project will be successfully executed. The next section describes your role during project closedown, the final phase of the project management process.

Closing Down the Project

Project closedown: The final phase of the project management process that focuses on bringing a project to an end.

The focus of **project closedown** is to bring the project to an end. Projects can conclude with a natural or unnatural termination. A natural termination occurs when the requirements of the project have been met—the project has been completed and is a success. An unnatural termination occurs when the project is stopped before completion (Keil et al., 2000). Several events can cause an unnatural termination to a proj-

Project Closedown

1. Closing Down the Project

2. Conducting Postproject Reviews

3. Closing the Customer Contract

Figure 3-15
Three project closedown activities

ect. For example, it may be learned that the assumption used to guide the project proved to be false or that the performance of the system or development group was somehow inadequate or that the requirements are no longer relevant or valid in the customer's business environment. The most likely reasons for the unnatural termination of a project relate to running out of time or money, or both. Regardless of the project termination outcome, several activities must be performed: closing down the project, conducting postproject reviews, and closing the customer contract. Within the context of the SDLC, project closedown occurs after the implementation phase. The system maintenance phase typically represents an ongoing series of projects, each needing to be individually managed. Figure 3-15 summarizes the project closedown activities that are described more fully in the remainder of this section:

1. *Closing down the project.* During closedown, you perform several diverse activities. For example, you have several team members working with you, project completion may signify job and assignment changes for some members. You will likely be required to assess each team member and provide an appraisal for personnel files and salary determination. You may also want to provide career advice to team members, write letters to superiors praising special accomplishments of team members, and send thank-you letters to those who helped but were not team members. As project manager, you must be prepared to handle possible negative personnel issues such as job termination, especially if the project was not successful. When closing down the project, it is also important to notify all interested parties that the project has been completed and to finalize all project documentation and financial records so that a final review of the project can be conducted. You should also celebrate the accomplishments of the team. Some teams will hold a party, and each team member may receive memorabilia (e.g., a T-shirt with "I survived the X project"). The goal is to celebrate the team's effort to bring a difficult task to a successful conclusion.

2. *Conducting postproject reviews.* Once you have closed down the project, final reviews of the project should be conducted with management and customers. The objective of these reviews is to determine the strengths and weaknesses of project deliverables, the processes used to create them, and the project management process. It is important that everyone understands what went right and what went wrong in order to improve the process for the next project. Remember, the systems development methodology adopted by an organization is a living guideline that must undergo continual improvement.

3. *Closing the customer contract.* The focus of this final activity is to ensure that all contractual terms of the project have been met. A project governed by a contractual agreement is typically not completed until agreed to by both parties, often in writing. Thus, it is paramount that you gain agreement from your customer that all contractual obligations have been met and that further work is either their responsibility or covered under another System Service Request or contract.

N E T   S E A R C H

There is ample information available to help you become a better project manager. Visit http://www.prenhall. com/hoffer to complete an exercise related to this topic.

Closedown is a very important activity. A project is not complete until it is closed, and it is at closedown that projects are deemed a success or failure. Completion also signifies the chance to begin a new project and apply what you have learned. Now that you have an understanding of the project management process, the next section describes specific techniques used in systems development for representing and scheduling activities and resources.

Representing and Scheduling Project Plans

A project manager has a wide variety of techniques available for depicting and documenting project plans. These planning documents can take the form of graphical or textual reports, although graphical reports have become most popular for depicting project plans. The most commonly used methods are Gantt and PERT charts. Because Gantt charts do not (typically) show how tasks must be ordered (precedence) but simply show when a task should begin and when it should end, they are often more useful for depicting relatively simple projects or subparts of a larger project, the activities of a single worker, or for monitoring the progress of activities compared to scheduled completion dates (Figure 3-16). Recall that a PERT chart shows the ordering of activities by connect-

Figure 3-16
Graphical diagrams that depict project plans
(a) A Gantt Chart

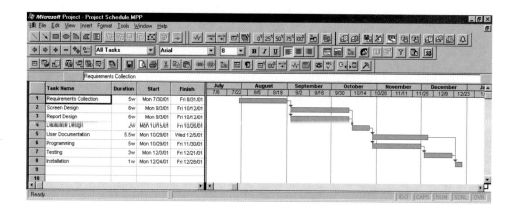

(b) A PERT chart

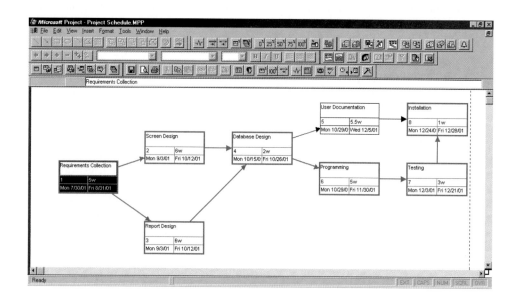

ing a task to its predecessor and successor tasks (see Figure 3-16). Sometimes a PERT chart is preferable; other times a Gantt chart more easily shows certain aspects of a project. Here are the key differences between these two charts.

- Gantt visually shows the duration of tasks whereas PERT visually shows the sequence dependencies between tasks.

- Gantt visually shows the time overlap of tasks whereas PERT does not show time overlap but does show which tasks could be done in parallel.

- Some forms of Gantt charts can visually show slack time available within an earliest start and latest finish duration. PERT shows this by data within activity rectangles.

Project managers also use textual reports that depict resource utilization by tasks, complexity of the project, and cost distributions to control activities. For example, Figure 3-17 shows a screen from Microsoft Project for Windows that summarizes all project activities, their durations in weeks, and their scheduled starting and ending dates. Most project managers use computer-based systems to help develop their graphical and textual reports. Later in this chapter, we discuss these automated systems in more detail.

A project manager will periodically review the status of all ongoing project task activities to assess whether the activities will be completed early, on time, or late. If early or late, the duration of the activity, depicted in column 2 of Figure 3-17, can be updated. Once changed, the scheduled start and finish times of all subsequent tasks will also change. Making such a change will also alter a Gantt or PERT chart used to represent the project tasks. The ability to easily make changes to a project is a very powerful feature of most project management environments. It allows the project manager to determine easily how changes in task duration impact the project completion date. It is also useful for examining the impact of "what if" scenarios of adding or reducing resources, such as personnel, for an activity.

Resources: Any person, group of people, piece of equipment, or material used in accomplishing an activity.

Representing Project Plans

Project scheduling and management require that time, costs, and resources be controlled. **Resources** are any person, group of people, piece of equipment, or material used in accomplishing an activity. PERT is a **critical path scheduling** technique used for controlling resources. A critical path refers to a sequence of task activities whose order and durations directly affect the completion date of a project. PERT is

Critical path scheduling: A scheduling technique whose order and duration of a sequence of task activities directly affect the completion date of a project.

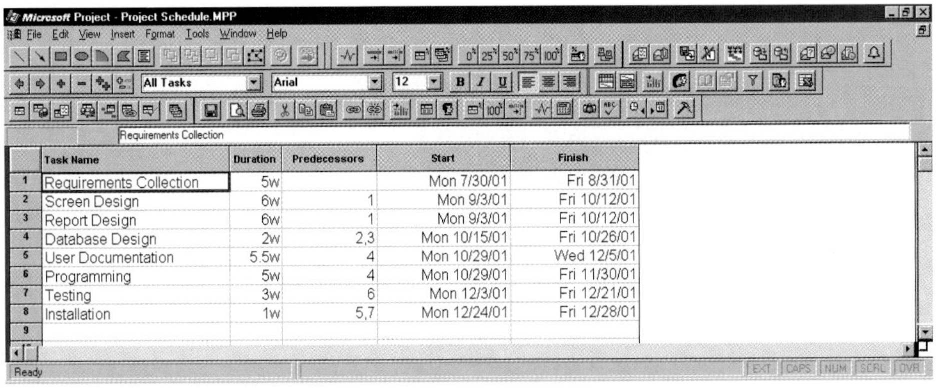

	Task Name	Duration	Predecessors	Start	Finish
1	Requirements Collection	5w		Mon 7/30/01	Fri 8/31/01
2	Screen Design	6w	1	Mon 9/3/01	Fri 10/12/01
3	Report Design	6w	1	Mon 9/3/01	Fri 10/12/01
4	Database Design	2w	2,3	Mon 10/15/01	Fri 10/26/01
5	User Documentation	5.5w	4	Mon 10/29/01	Wed 12/5/01
6	Programming	5w	4	Mon 10/29/01	Fri 11/30/01
7	Testing	3w	6	Mon 12/3/01	Fri 12/21/01
8	Installation	1w	5,7	Mon 12/24/01	Fri 12/28/01
9					

Figure 3-17
A screen from Microsoft Project for Windows summarizes all project activities, their durations in weeks, and their scheduled starting and ending dates.

one of the most widely used and best-known scheduling methods. You would use a PERT chart when tasks

- Are well-defined and have a clear beginning and endpoint
- Can be worked on independently of other tasks
- Are ordered
- Serve the purpose of the project

A major strength of the PERT technique is its ability to represent how completion times vary for activities. Because of this, it is more often used than Gantt charts to manage projects such as information systems development where variability in the duration of activities is the norm. PERT charts use a graphical network diagram composed of circles or rectangles representing activities and connecting arrows showing required work flows, as illustrated in Figure 3-18.

Constructing a Gantt Chart and PERT Chart at Pine Valley Furniture

Although Pine Valley Furniture has historically been a manufacturing company, it has recently entered the direct sales market for selected target markets. One of the fastest growing of these markets is economically priced furniture suitable for college students. Management has requested that a new Sales Promotion Tracking System (SPTS) be developed. This project has already successfully moved through project initiation and is currently in the detailed project planning stage, which corresponds to the SDLC phase of project initiation and planning. The SPTS will be used to track the sales purchases by college students for the next fall semester. Students typically purchase low-priced beds, bookcases, desks, tables, chairs, and dressers. Because PVF does not normally stock a large quantity of lower-priced items, management feels that a tracking system will help provide information about the college student market that can be used for follow-up sales promotions (e.g., a midterm futon sale).

The project is to design, develop, and implement this information system before the start of the fall term in order to collect sales data at the next major buying period. This deadline gives the project team 24 weeks to develop and implement the system. The Systems Priority Board at PVF wants to make a decision this week based on the feasibility of completing the project within the 24-week deadline. Using PVF's project planning methodology, the project manager, Jim Woo, knows that the next step is to construct Gantt and PERT charts of the

Figure 3-18
PERT chart showing activities (represented by circles) and sequence of those activities (represented by arrows)

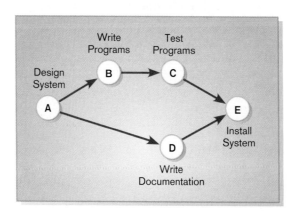

project to represent the Baseline Project Plan so that he can use these charts to estimate the likelihood of completing the project within 24 weeks. A major activity of project planning focuses on dividing the project into manageable activities, estimating times for each, and sequencing their order. Here are the steps Jim followed to do this.

1. *Identify each activity to be completed in the project.* After discussing the new Sales Promotion Tracking System with PVF's management, sales, and development staff, Jim identified the following major activities for the project:

   - Requirements collection
   - Screen design
   - Report design
   - Database construction
   - User documentation creation
   - Software programming
   - System testing
   - System installation

2. *Determine time estimates and calculate the expected completion time for each activity.* Three estimates are used to determine the expected completion time for an activity: optimistic time, realistic time, and pessimistic time. The optimistic (o) and pessimistic (p) times reflect the minimum and maximum possible periods of time for an activity to be completed. The realistic time (r), or most likely time, reflects the project manager's "best guess" of the amount of time the activity actually will require for completion. Once each of these estimates is made for an activity, an expected time (ET) can be calculated. Because the completion time should be closest to the realistic time (r), it is weighted four times more than the optimistic (o) and expected (p) times. Once you add these values together, it must be divided by 6 to determine the ET. This equation is shown in the following formula:

$$ET = \frac{o + 4r + p}{6}$$

Figure 3-19
Estimated time calculations for the SPTS project

ACTIVITY	TIME ESTIMATE (in weeks)			EXPECTED TIME (ET) $\dfrac{o + 4r + p}{6}$
	o	r	p	
1. Requirements Collection	1	5	9	5
2. Screen Design	5	6	7	6
3. Report Design	3	6	9	6
4. Database Design	1	2	3	2
5. User Documentation	3	6	7	5.5
6. Programming	4	5	6	5
7. Testing	1	3	5	3
8. Installation	1	1	1	1

Figure 3-20
Sequence of Activities within the SPTS Project

ACTIVITY	PRECEDING ACTIVITY
1. Requirements Collection	—
2. Screen Design	1
3. Report Design	1
4. Database Design	2,3
5. User Documentation	4
6. Programming	4
7. Testing	6
8. Installation	5,7

where

ET = expected time for the completion for an activity

o = optimistic completion time for an activity

r = most likely completion time for an activity

p = pessimistic completion time for an activity

After identifying the major project activities, Jim established optimistic, realistic, and pessimistic time estimates for each activity. These numbers were then used to calculate the expected completion times for all project activities. Figure 3-19 shows the estimated time calculations for each activity of the Sales Promotion Tracking System project.

3. *Determine the sequence of the activities and precedence relationships among all activities by constructing Gantt and PERT charts.* This step helps you understand how various activities are related. Jim starts by determining the order in which activities should take place. The results of this analysis for the SPTS project are shown in Figure 3-20. The first row of this figure shows that no activities precede requirements collection. Row 2 shows that screen design must be preceded by requirements collection. Row 4 shows that both screen and report design must precede database construction. Thus, activities may be preceded by zero, one, or more activities.

Figure 3-21
Gantt chart that illustrates the sequence and duration of each activity of the SPTS project

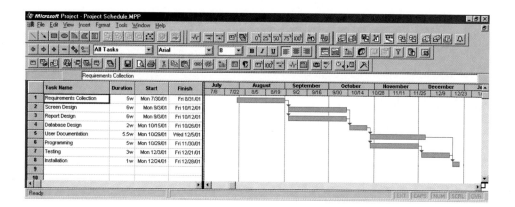

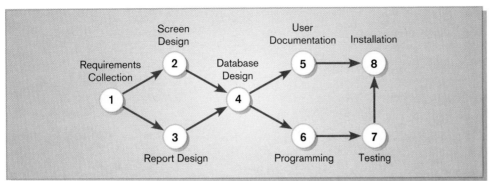

Figure 3-22
PERT chart that illustrates the
activities (circles) and the
sequence (arrows) of those
activities

Using the estimated time and activity sequencing information from Figures 3-19 and 3-20, Jim can now construct Gantt and PERT charts of the project's activities. To construct the Gantt chart, a horizontal bar is drawn for each activity that reflects its sequence and duration, as shown in Figure 3-21. The Gantt chart may not, however, show direct interrelationships between activities. For example, just because the database design activity begins right after the screen design and report design bars finish does not imply that these two activities must finish before database design can begin. To show such precedence relationships, a PERT chart must be used. The Gantt chart in Figure 3-21 does, however, show precedence relationships.

PERT charts have two major components: arrows and nodes. Arrows reflect the sequence of activities whereas nodes reflect activities that consume time and resources. A PERT chart for the SPTS project is shown in Figure 3-22. This diagram has eight nodes labeled 1 through 8.

4. *Determine the critical path.* The critical path of a PERT network is represented by the sequence of connected activities that produce the longest overall time period. All nodes and activities within this sequence are referred to as being "on" the **critical path**. The critical path represents the shortest time in which a project can be completed. In other words, any activity on the critical path that is delayed in completion delays the entire project. Nodes not on the critical path, however, can be delayed (for some amount of time) without delaying the final completion of the project.

Critical path: The shortest time in which a project can be completed.

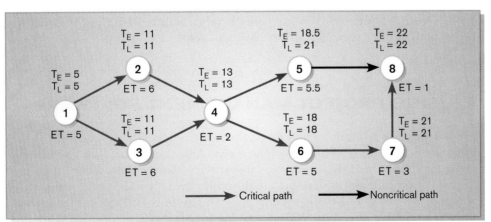

Figure 3-23
PERT chart for the SPTS project
showing estimated times for each
activity and the earliest and latest
expected completion time for each
activity

Figure 3-24
Activity slack time calculations for the SPTS project; all activities except number 5 are on the critical path

ACTIVITY	T_E	T_L	SLACK $T_L - T_E$	ON CRITICAL PATH
1	5	5	0	✓
2	11	11	0	✓
3	11	11	0	✓
4	13	13	0	✓
5	18.5	21	2.5	
6	18	18	0	✓
7	21	21	0	✓
8	22	22	0	✓

Slack time: The amount of time that an activity can be delayed without delaying the project.

Nodes not on the critical path contain **slack time** and allow the project manager some flexibility in scheduling.

Figure 3-23 shows the PERT chart that Jim constructed to determine the critical path and expected completion time for the SPTS project. To determine the critical path, Jim calculated the earliest and latest expected completion time for each activity. He found each activity's earliest expected completion time (T_E) by summing the estimated time (ET) for each activity from left to right (i.e., in precedence order), starting at activity 1 and working toward activity 8. In this case, T_E for activity 8 is equal to 22 weeks. If two or more activities precede an activity, the largest expected completion time of these activities is used in calculating the new activity's expected completion time. For example, because activity 8 is preceded by both activities 5 and 7, the largest expected completion time between 5 and 7 is 21, so T_E for activity 8 is 21 + 1, or 22. The earliest expected completion time for the last activity of the project represents the amount of time the project should take to complete. Because the time of each activity can vary, however, the projected completion time represents only an estimate. The project may in fact require more or less time for completion.

The latest expected completion time (T_L) refers to the time in which an activity can be completed without delaying the project. To find the values for each activity's T_L, Jim started at activity 8 and set T_L equal to the final T_E (22 weeks). Next, he worked right to left toward activity 1 and subtracted the expected time for each activity. The slack time for each activity is equal to the difference between its latest and earliest expected completion times ($T_L - T_E$). Figure 3-24 shows the slack time calculations for all activities of the SPTS project. All activities with a slack time equal to zero are on the critical path. Thus, all activities except 5 are on the critical path. Part of the diagram in Figure 3-23 shows two critical paths, between activities 1-2-4 and 1-3-4, because both of these parallel activities have zero slack.

USING PROJECT MANAGEMENT SOFTWARE

N E T   S E A R C H
There is powerful software available to support the management of a project's activities. Visit http://www.prenhall.com/hoffer to complete an exercise on this topic.

A wide variety of automated project management tools are available to help you manage a development project. New versions of these tools are continuously being developed and released by software vendors. Most of the available tools have a common set of features that include the ability to define and order tasks, assign resources to tasks, and easily modify tasks and resources. Project management tools are available to run on IBM-compatible personal computers, the Macintosh, and larger mainframe and workstation-based systems. These systems vary in the number of task activities supported, the complexity of relationships, system processing and storage requirements, and, of course, cost. Prices for these systems can range from a few hundred dollars for personal computer-based systems to more than $100,000 for large-scale multiproject systems. Yet a lot can be done with systems like Microsoft Project as well as public domain and shareware systems. For example, numerous shareware project management programs (e.g., Calendar Quick, Delegator, and SureTrak) can be downloaded from the World Wide Web (e.g., at www.download.com) and on-line bulletin boards. Because these systems are continuously changing, you should comparison shop before choosing a particular package.

We now illustrate the types of activities you would perform when using project management software. Microsoft Project for Windows is a project management system that has had consistent high marks in computer publication reviews. When using this system to manage a project, you need to perform at least the following activities:

- Establish a project starting or ending date
- Enter tasks and assign task relationships
- Select a scheduling method to review project reports

Establishing a Project Starting Date

Defining the general project information includes obtaining the name of the project and project manager and the starting or ending date of the project. Starting and ending dates are used to schedule future activities or backdate others (see below) based upon their duration and relationships to other activities. An example from Microsoft Project for Windows of the data entry screen for establishing a project starting or ending date is shown in Figure 3-25. This screen shows PVF's Purchasing Fulfillment System project. Here, the starting date for the project is Monday, November 19, 2001.

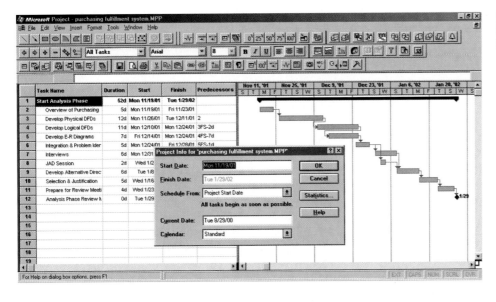

Figure 3-25
Establishing a project starting date in Microsoft Project for Windows

Figure 3-26
Entering tasks and assigning task relationships in Microsoft Project for Windows

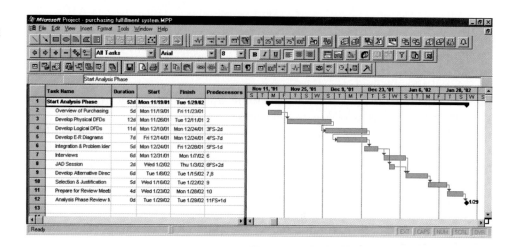

Entering Tasks and Assigning Task Relationships

The next step in defining a project is to define project tasks and their relationships. For the Purchasing Fulfillment System project, Chris defined 11 tasks to be completed when he performed the initial system analysis activities of the project (Task 1—Start Analysis Phase—is a summary task that is used to group related tasks). The task entry screen, shown in Figure 3-26, is similar to a financial spreadsheet program. The user moves the cursor to a cell with arrow keys or the mouse and then simply enters a textual Name and a numeric Duration for each activity. Scheduled Start and Scheduled Finish are automatically entered based upon the project start date and duration. To set an activity relationship, the ID number (or numbers) of the activity that must be completed before the start of the current activity is entered in the Predecessors column. Additional codes under this column make the precedence relationships more precise. For example, consider the Predecessor column for ID 6. The entry in this cell says that activity 6 cannot start until one day before the finish of activity 5. (Microsoft Project provides many different options for precedence and delays such as in this example, but discussion of these is beyond the scope of our coverage.) The project management software uses this information to construct Gantt, PERT, and other project-related reports.

Selecting a Scheduling Method to Review Project Reports

Once information about all the activities for a project has been entered, it is very easy to review the information in a variety of graphical and textual formats using displays or printed reports. For example, Figure 3-26 shows the project information in a Gantt chart screen whereas Figure 3-27 shows the project information in a condensed PERT chart display. You can easily change how you view the information by making a selection from the View menu shown in Figure 3-27.

As mentioned in the chapter, interim project reports to management will often compare actual progress to plans. Figure 3-28 illustrates how Microsoft Project shows progress with a solid line within the activity bar. In this figure, task 2 is completed and task 3 is almost completed, but there remains a small percentage of work, as shown by the incomplete solid lines within the bar for this task. Assuming that this screen represents the status of the project on Friday, December 7, 2001, the third activity is approximately on schedule, but the second

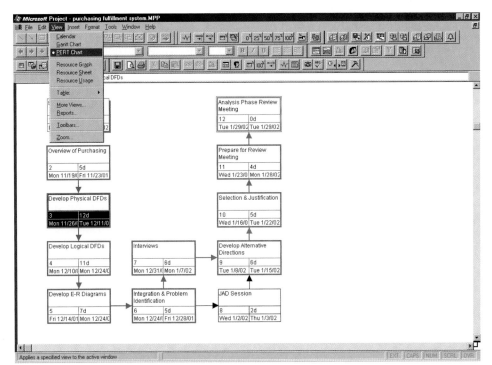

Figure 3-27
Viewing project information as a
PERT chart in Microsoft Project for
Windows

activity is behind its expected completion date. Tabular reports can summarize the same information.

This brief introduction to project management software has only scratched the surface to show you the power and the features of these systems. Other features that are widely available and especially useful for multiperson projects relate to resource usage and utilization. Resource-related features allow you to define characteristics such as standard costing rates and daily availability via a calendar that records holidays, working hours, and vacations. These features are particularly useful for billing and estimating project costs. Often, resources are shared across multiple projects, which could significantly affect a project's schedule.

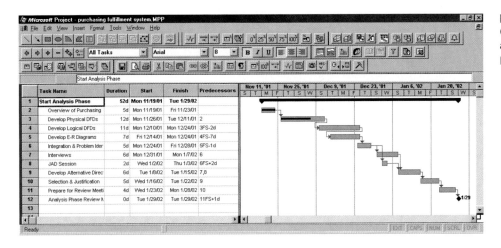

Figure 3-28
Gantt chart showing progress of
activities (right frame) versus
planned activities (left frame)

Depending upon how projects are billed within an organization, assigning and billing resources to tasks is a very time-consuming activity for most project managers. The features provided in these powerful tools can greatly ease both the planning and managing of projects so that both project and management resources are effectively utilized.

Summary

The focus of this chapter was on managing information system projects and the role of the project manager in this process. A project manager has both technical and managerial skills and is ultimately responsible for determining the size, scope, and resource requirements for a project. Once a project is deemed feasible by an organization, the project manager ensures that the project meets the customer's needs and is delivered within budget and time constraints. To manage the project, the project manager must execute four primary activities: project initiation, project planning, project execution, and project closedown. The focus of project initiation is on assessing the size, scope, and complexity of a project and establishing procedures to support later project activities. The focus of project planning is on defining clear, discrete activities and the work needed to complete each activity. The focus of project execution is on putting the plans developed in project initiation and planning into action. Project closedown focuses on bringing the project to an end.

Gantt and PERT charts are powerful graphical techniques used in planning and controlling projects. Both Gantt and PERT scheduling techniques require that a project have activities that can be defined as having a clear beginning and end, can be worked on independently of other activities, are ordered, and are such that their completion signifies the end of the project. Gantt charts use horizontal bars to represent the beginning, duration, and ending of an activity. PERT is a critical path scheduling method that shows the interrelationships between activities. Critical path scheduling refers to planning methods whereby the order and duration of the project's activities directly affect the completion date of the project. These charts show when activities can begin and end, which activities cannot be delayed without delaying the whole project, how much slack time each activity has, and progress against planned activities. PERT's ability to use probability estimates in determining critical paths and deadlines makes it a widely used technique for very complex projects.

A wide variety of automated tools for assisting the project manager are available. Most tools have common features including the ability to define and order tasks, assign resources to tasks, and modify tasks and resources. Systems vary regarding the number of activities supported, the complexity of relationships, processing and storage requirements, and cost.

Key Terms

1. Critical path
2. Critical path scheduling
3. Deliverable
4. Feasibility study
5. Gantt chart
6. PERT chart
7. Project
8. Project closedown
9. Project execution
10. Project initiation
11. Project management
12. Project manager
13. Project planning
14. Project workbook
15. Resources
16. Slack time
17. Work breakdown structure

Match each of the key terms above with the definition that best fits it.

_____ A systems analyst with a diverse set of skills—management, leadership, technical, conflict management, and customer relationship—who is responsible for initiating, planning, executing, and closing down a project.

_____ A planned undertaking of related activities to reach an objective that has a beginning and an end.

_____ An end product in a phase of the SCLC.

_____ Determines if the information system makes sense for the organization from an economic and operational standpoint.

_____ A controlled process of initiating, planning, executing, and closing down a project.

_____ The first phase of the project management process in which activities are performed to assess the size, scope, and complexity of the project and to establish procedures to support later project activities.

_____ An on-line or hard-copy repository for all project correspondence, inputs, outputs, deliverables, procedures, and standards.

_____ The second phase of the project management process, which focuses on defining clear, discrete activities and the work needed to complete each activity within a single project.

_____ The process of dividing the project into manageable tasks and logically ordering them to ensure a smooth evolution between tasks.

_____ A graphical representation of a project that shows each task as a horizontal bar whose length is proportional to its time for completion.

_____ A diagram that depicts project tasks and their interrelationships.

_____ The third phase of the project management process in which the plans created in the prior phases are put into action.

_____ The final phase of the project management process that focuses on bringing a project to an end.

_____ Any person, group of people, piece of equipment, or material used in accomplishing an activity.

_____ A scheduling technique whose order and duration of a sequence of task activities directly affect the completion date of a project.

_____ The shortest time in which a project can be completed.

_____ The amount of time that an activity can be delayed without delaying the project.

Review Questions

1. Contrast the following terms:
   a. critical path scheduling, Gantt, PERT, slack time
   b. project, project management, project manager
   c. project initiation, project planning, project execution, project closedown
   d. project workbook, resources, work breakdown structure

2. Discuss the reasons why organizations undertake information system projects.

3. List and describe the common skills and activities of a project manager. Which skill do you think is most important? Why?

4. Describe the activities performed by the project manager during project initiation.

5. Describe the activities performed by the project manager during project planning.

6. Describe the activities performed by the project manager during project execution.

7. List various project team communication methods and describe an example of the type of information that might be shared among team members using each method.

8. Describe the activities performed by the project manager during project closedown.

9. What characteristics must a project have in order for critical path scheduling to be applicable?

10. Describe the steps involved in making a Gantt chart.

11. Describe the steps involved in making a PERT chart.

12. In which phase of the systems development life cycle does project planning typically occur? In which phase does project management occur?

13. What are some reasons why one activity may have to precede another activity before the second activity can begin? In other words, what causes precedence relationships between project activities?

Problems and Exercises

1. Which of the four phases of the project management process do you feel is most challenging? Why?

2. What are some sources of risk in a systems analysis and design project and how does a project manager cope with risk during the stages of project management?

3. Search computer magazines or the Web for recent reviews of project management software. Which packages seem to be most popular? What are the relative strengths and weaknesses of each package software? What advice would you give to someone intending to buy project management software for their PC? Why?

4. How are information system projects similar to other types of projects? How are they different? Are the project management packages you evaluated in Problem and Exercise 3

suited for all types of projects or for particular types of projects? Which package is best suited for information systems projects? Why?

5. If given the chance, would you become the manager of an information systems project? If so, why? Prepare a list of the strengths that you would bring to the project as its manager. If not, why not? What would it take for you to feel more comfortable managing an information systems project? Prepare a list and timetable for the necessary training you would need to feel more comfortable about managing an information systems project.

6. Calculate the expected time for the following tasks.

Task	Optimistic Time	Most Likely Time	Pessimistic Time	Expected Time
A	3	7	11	
B	5	9	13	
C	1	2	9	
D	2	3	16	
E	2	4	18	
F	3	4	11	
G	1	4	7	
H	3	4	5	
I	2	4	12	
J	4	7	9	

7. A project has been defined to contain the following list of activities along with their required times for completion.

Activity	Time (weeks)	Immediate Predecessors
1 – collect requirements	2	–
2 – analyze processes	3	1
3 – analyze data	3	2
4 – design processes	7	2
5 – design data	6	2
6 – design screens	1	3, 4
7 – design reports	5	4, 5
8 – program	4	6, 7
9 – test and document	8	7
10 – install	2	8, 9

a. Draw a PERT chart for the activities.
b. Calculate the earliest expected completion time.
c. Show the critical path.
d. What would happen if Activity 6 was revised to take six weeks instead of one week?

8. Construct a Gantt chart for the project defined in Problem and Exercise 7 above.

9. Look again at the activities outlined in Problem and Exercise 7. Assume that your team is in its first week of the project and has discovered that each of the activity duration estimates is wrong. Activity 2 will take only two weeks to complete. Activities 4 and 7 will each take three times longer than anticipated. All other activities will take twice as long to complete as previously estimated. In addition, a new activity, number 11, has been added. It will take one week to com-

plete and its immediate predecessors are Activities 10 and 9. Adjust the PERT chart and recalculate the earliest expected completion times.

10. Construct Gantt and PERT charts for a project you are or will be involved in. Choose a project of sufficient depth at either work, home, or school. Identify the activities to be completed, determine the sequence of the activities, and construct a diagram reflecting the starting, ending, duration, and precedence (PERT only) relationships among all activities. For your PERT chart, use the procedure in this chapter to determine time estimates for each activity and calculate the expected time for each activity. Now determine the critical path and the early and late starting and finishing times for each activity. Which activities have slack time?

11. For the project you described in Problem and Exercise 10, assume that the worst has happened. A key team member has dropped out of the project and has been assigned to another project in another part of the country. The remaining team members are having personality clashes. Key deliverables for the project are now due much earlier than expected. In addition, you have just determined that a key phase in the early life of the project will now take much longer than you had originally expected. To make matters worse, your boss absolutely will not accept that this project cannot be completed by this new deadline. What will you do to account for these project changes and problems? Begin by reconstructing your Gantt and PERT charts and determine a strategy for dealing with the specific changes and problems described above. If new resources are needed to meet the new deadline, outline the rationale that you will use to convince your boss that these additional resources are critical to the success of the project.

12. Assume you have a project with seven activities labeled A–G (below). Derive the earliest completion time (or early finish—EF), latest completion time (or late finish—LF), and slack for each of the following tasks (begin at time = 0). Which tasks are on the critical path? Draw a Gantt chart for these tasks.

Task	Preceding Event	Expected Duration	EF	LF	Slack	Critical Path?
A	–	5				
B	A	3				
C	A	4				
D	C	6				
E	B, C	4				
F	D	1				
G	D, E, F	5				

13. Draw a PERT chart for the tasks shown in Problem and Exercise 12. Highlight the critical path.

14. Assume you have a project with 10 activities labeled A–J (immediately following). Derive the earliest completion time (or early finish—EF), latest completion time (or late finish—LF), and slack for each of the following tasks (begin at time = 0). Which tasks are on the critical path? Highlight the critical path on your PERT chart.

Activity	Preceding Event	Expected Duration	EF	LF	Slack	Critical Path?
A	–	4				
B	A	5				
C	A	6				
D	A	7				
E	A, D	6				
F	C, E	5				
G	D, E	4				
H	E	3				
I	F, G	4				
J	H, I	5				

Activity	Preceding Event	Expected Duration	EF	LF	Slack	Critical Path?
A	–	2				
B	A	3				
C	B	4				
D	C	5				
E	C	4				
F	D, E	3				
G	F	4				
H	F	6				
I	G, H	5				
J	G	2				
K	I, J	4				

15. Draw a Gantt chart for the tasks shown in Problem and Exercise 14.

16. Assume you have a project with 11 activities labeled A–K (immediately following). Derive the earliest completion time (or early finish—EF), latest completion time (or late finish—LF), and slack for each of the following tasks (begin at time = 0). Which tasks are on the critical path? Draw both Gantt and PERT charts for these tasks and make sure you highlight the critical path on your PERT chart.

17. Make a list of the tasks that you performed when designing your schedule of classes for this term. Develop a table showing each task, its duration, preceding event(s), and expected duration. Develop a PERT chart for these tasks. Highlight the critical path on your PERT chart.

Field Exercises

1. Identify someone who manages an information systems project in an organization. Describe to him or her each of the skills and activities listed in Table 3-1. Determine which items they are responsible for on the project. Of those they are responsible for, determine which are the more challenging and why. Of those they are not responsible for, determine why not and who is responsible for these activities. What other skills and activities, not listed in Table 3-1, is this person responsible for in managing this project?

2. Identify someone who manages an information systems project in an organization. Describe to him or her each of the project planning elements in Figure 3-8. Determine the extent to which each of these elements is part of that person's project planning process. If that person is not able to perform some of these planning activities, or if he or she cannot spend as much time on any of these activities as he or she would like, determine what barriers are prohibitive for proper project planning.

3. Identify someone who manages an information systems project (or other team-based project) in an organization. Describe to him or her each of the project team communication methods listed in Table 3-2. Determine which types of communication methods are used for team communication and describe which he or she feels are best for communicating various types of information.

4. Identify someone who manages an information systems project in an organization. Describe to them each of the project execution elements in Figure 3-13. Determine the extent to which each of these elements is part of that person's project execution process. If that person does not perform some of these activities, or if he or she cannot spend much time on any of these activities, determine what barriers or reasons prevent performing all project execution activities.

5. Interview a sample of project managers. Divide your sample into two small subsamples, one for managers of information systems projects and one for managers of other types of projects. Ask each respondent to identify personal leadership attributes that contribute to successful project management and explain why these are important. Summarize your results. What seem to be the attributes most often cited as leading to successful project management, regardless of the type of project? Are there any consistent differences between the responses in the two subsamples? If so, what are these differences? Do they make sense to you? If there are no apparent differences between the responses of the two subsamples, why not? Are there no differences in the skill sets necessary for managing information system projects versus managing other types of projects?

6. Observe a real information systems project team in action for an extended period of time. Keep a notebook as you watch individual members performing their individual tasks, as you review the project management techniques used by the team's leader, and as you sit in on some of their meetings. What seem to be the team's strengths and weaknesses? What are some areas in which the team can improve?

References

Abdel-Hamid, T. K., Sengupta, K., and Swett, C. 1999. "The Impact of Goals on Software Project Management: An Experimental Investigation." *MIS Quarterly* (23)4.

Guinan, P. J., Cooprider, J. G., and Faraj, S. (1998). "Enabling Software Development Team Performance During Requirements Definition: A Behavioral Versus Technical Approach." *Information Systems Research* 9(2): 101–25.

Hoffer, J. A., Prescott, M. B., and McFadden, F.R. 2002. *Modern Database Management.* Upper Saddle River, NJ: Prentice Hall.

Keil, M. Tan, B. C. Y, Wei, K. K., Saarinen, T., Tuunainen, V., and Wassenaar, A. 2000. "A Cross-Cultural Study on Escalation of Commitment Behavior in Software Projects." *MIS Quarterly* (24)2.

Kettelhut, M. C. 1991. "Avoiding group-induced errors in systems development." *Journal of Systems Management* 42 (December): 13–17.

Kirsch, L. J. 2000. "Software Project Management: An Integrated Perspective for and Emerging Paradigm." In *Framing the Domains of IT Management: Projecting the Future from the Past,* Cincinatti: Pinnaflex Educational Resources. ed. R. W. Zmud, Chapter 15, 285–304.

Page-Jones, M. 1985. *Practical Project Management.* New York: Dorset House.

Rettig, M. 1990. "Software Teams." *Communications of the ACM* 33 (10): 23–27.

Thomsett, R. 1985. Foreword to *Practical Project Management,* by M. Page-Jones. New York: Dorset House.

Chapter 4

Automated Tools for Systems Development

After studying this chapter, you should be able to:

- Identify the trade-offs when using CASE to support systems development activities.

- Describe organizational forces for and against the adoption of CASE tools.

- Describe the role of CASE tools and how they are used to support activities within the SDLC.

- List and describe the typical components of a comprehensive CASE environment.

- Describe the general functions of upper CASE tools, lower CASE tools, cross life-cycle CASE tools, and the CASE repository.

- Describe visual and emerging development tools and how they are being used.

INTRODUCTION

In the past, system development was viewed by many as an art that only a few skilled individuals could master. Within many organizations, the techniques employed by each developer could also vary substantially. This lack of consistency in technique and methodology often made it difficult to integrate systems and data or to quickly construct new systems. As a result, many organizations faced a growing backlog of applications to be developed; once developed, many of these systems were error-ridden, over budget, and late. Lack of standards also made maintenance difficult.

To address these problems, information systems professionals concluded that software development needed an engineering-type discipline (Nunamaker, 1992). The goal was to concentrate on developing common techniques, standard methodologies, and automated tools in a manner similar to the traditional engineering field. This chapter covers the evolution and use of automated tools to support the information systems development process. **Computer-aided software engineering (CASE)** refers to automated software tools used by systems analysts to develop information systems.

Computer-aided software engineering (CASE): Software tools that provide automated support for some portion of the systems development process.

These tools can be used to automate or support activities throughout the systems development process with the objective of increasing productivity and improving the overall quality of systems. Additionally, systems development and programming environments are continuously evolving to include *visual* and CASE-like features, especially for tools to build Internet and Electronic Commerce applications. This evolution is occurring rapidly and people with skill in using these tools are in hot demand.

In the next section, we will describe how CASE is being used within organizations and examine the typical components of a comprehensive CASE environment. Next, visual and emerging development tools will be described with a focus on object-oriented tools, visual programming, Internet development, and systems development tools of the future.

THE USE OF CASE IN ORGANIZATIONS

The purpose of CASE is to make it much easier to enact a single design philosophy within an organization with many projects, systems, and people. CASE can support most of the system development activities; Figure 4-1 highlights selected CASE facilities for each life-cycle phase. Although CASE tools run on a variety of mini and mainframe systems, recent advances in microcomputers have made the PC the predominant CASE workstation. CASE helps provide an engineering-type discipline to software development and to the automation of the entire software life-cycle process,

Figure 4-1
CASE can provide effective support for most system development activities

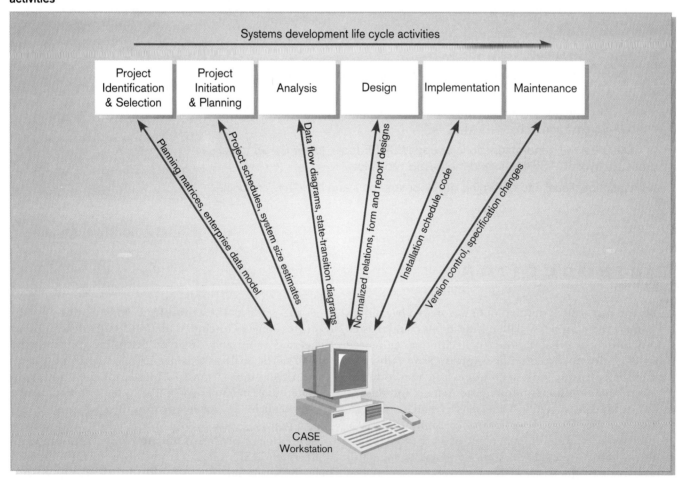

sometimes with a single family of integrated software tools. In general, CASE assists systems builders in managing the complexities of information system projects and helps assure that high-quality systems are constructed on time and within budget (Hapgood, 2000).

CASE and System Quality

From a systems analyst's perspective, the reason for using CASE may stem from a very straightforward and practical decision such as, "It makes my life easier." However, from an organizational perspective, the reasons for using CASE are much broader (see Table 4-1). Organizations primarily adopt CASE to improve the quality and speed of the systems development process. It is easy to argue that all organizations should openly embrace the adoption and deployment of CASE, since the objectives of CASE are consistent with the objectives of most organizations.

For example, Collect America, Ltd. (CA), a company that collects long overdue customer and commercial debts, uses Oracle Designer and Developer CASE tools for developing their information systems (Oracle, 1998). Based in Denver, Colorado, CA tracks and monitors debtors using an Oracle database environment. To make collections, CA has franchise agreements with regional law firms in the United States, Canada, and Mexico. As their business has grown, so too have their information management needs. CA's use of Oracle's CASE tools to design, construct, and maintain their applications has improved system quality and speed over their previously non-CASE developed systems. The main reason is that systems developed using CASE allow applications to be easily moved from one hardware platform to another. Consequently, as the hardware infrastructure in CA has evolved, their applications have also evolved without major changes. For example, with the rapid growth and easy accessibility of the World Wide Web, CA could save a lot of money if franchise offices around North America used the Web to access CA's database. Using Oracle's CASE tools, system developers reconstructed applications so that they would operate via Java Applets inside a standard Web browser. Today, because access to this information is through the Web, huge cost savings are being realized by CA; the company no longer has to invest in routers, dedicated lines, and all the other technology needed to manage a dedicated network. For CA, CASE has really paid off.

Yet, adopting CASE can have numerous and widespread effects on an organization and its information systems development process beyond quality and speed improvements. As a result, the deployment of CASE within organizations has been slower than expected, with several factors inhibiting widespread deployment (Richman, 1996).

TABLE 4-1   Objectives of CASE

Most organizations use CASE to

- Improve the quality of the systems developed
- Increase the speed with which systems are designed and developed
- Ease and improve the testing process through the use of automated checking
- Improve the integration of development activities via common methodologies
- Improve the quality and completeness of documentation
- Help standardize the development process
- Improve the management of the project
- Simplify program maintenance
- Promote reusability of modules and documentation
- Improve software portability across environments

The Cost of Case

Most agree that the start-up cost of using CASE is the greatest single factor. Integrated CASE environments range in price from less than $5,000 per analyst to more than $50,000! Lower-end systems are often low in functionality and fail to provide substantial productivity benefits. Furthermore, without adequate CASE training, most people are not able to gain the expertise needed to fully use CASE. CASE, an expensive technology, often leaves only large-scale system builders capable of making the investment required for organization-wide adoption. For example, it has been estimated that equipping a 150-person IS organization with CASE technology, including software, hardware, training, and maintenance, could exceed $3,000,000 over a five-year period (*I/S Analyzer,* 1993). A practical guideline is that it costs between $5,000 and $15,000 per year to provide CASE to one systems analyst. In spite of this, smaller organizations have effectively deployed CASE by using fewer and less-sophisticated tools to automate only a subset of all development activities, thus substantially reducing deployment costs. Nonetheless, adopting CASE is an important decision that must be made at the highest levels of the organization.

Another factor influencing CASE adoption relates to how organizations evaluate their return on investments. The big benefits to using CASE come in the late stages of the SDLC: system construction, testing, implementation, and especially maintenance. Additionally, CASE often lengthens the duration of early stages of the project, sometimes by more than 40 percent (Stone, 1993). This increase in front-end effort is necessitated by the need to completely finish the system design before using automated code generators during system construction. In essence, although CASE provides the potential to significantly shorten the *overall* process, many users and managers are often frustrated by the seemingly long duration of planning, analysis, and design when using CASE. As a result, enthusiasm for CASE can dwindle. Organizations must be patient when making CASE investments—it is likely that the long-term payback from CASE takes more time than some organizations are willing to accept.

Other factors can also influence CASE adoption. One factor, cited as a significant productivity bottleneck, is that some CASE tools cannot easily share information between tools. A related issue is the extent to which CASE can support all SDLC activities. Providing a complete set of tools for all aspects of the SDLC turns out to be more difficult than first imagined. This has left many organizations with distinct tools that "refuse to talk to each other."

The adoption of CASE is highly related to use of a formal systems development process. Many CASE products force or encourage analysts to follow a specific methodology or philosophy for systems development. Thus, an organization without a widely used methodology or one that is not compatible with CASE tools will find it difficult to use CASE. Without a match between a methodology and CASE, CASE is simply another graphical drawing, word-processing, and reporting package.

The Outlook for CASE

Despite these issues, the long-term prognosis for CASE is very good. The functionality of CASE tools is increasing and the costs are coming down (Richman, 1996). During the next several years, CASE technologies and the market for CASE will begin to mature. This should help improve product offerings and reduce system costs. Additionally, by exposing more systems analysts to CASE technology earlier in their education and career, adoption with less training and better results should result.

Another factor that should stimulate the market for CASE products is the desire of organizations to extend the life of existing systems. Categories of CASE products referred to as *reverse engineering* and *reengineering* tools are "breathing new life" into existing systems by allowing old programs to be more easily modified to run on new hardware configurations (Pfrenzinger, 1992).

Reverse engineering refers to the process of creating design specifications for a system or program module from program code and data definitions. For example, CASE tools that support reverse engineering read program source code as input, perform an analysis, and extract information such as program control structures, data structures, logic, and data flow. Once a program is represented at a design level using both graphical and textual representations, the systems analyst can more effectively restructure the code to current business needs or programming practices. Figure 4-2 shows a screen from Imagix 4D that graphically displays how symbols within an information system are related to one another. This tool provides analysts with a powerful method to quickly explore and understand a system. For example, Figure 4-2 shows a mapping between variables and program procedures. This high-level view of a program allows programmers to see more quickly the interrelationships and structure of a program, making it easier to understand and maintain. As with many legacy systems, only the source code may exist, yet additional documentation is necessary to make program maintenance productive.

Reengineering tools are similar to reverse engineering tools but include analysis features that can automatically, or interactively with a systems analyst, alter an existing system in an effort to improve its quality or performance. Although most organizations may have numerous systems that are candidates for reverse engineering or reengineering, the complexity and effort in using these tools have limited their widespread use. Additionally, most CASE environments do not yet have reverse or reengineering capabilities. However, as automated development environments evolve to support these features, CASE should evolve to have greater impact beyond what the technology has so far experienced.

Besides financial and productivity issues, the culture of an organization can significantly influence the success of CASE adoption. IS personnel with different career orientations have different attitudes toward CASE (Orlikowski, 1989). IS personnel with a managerial orientation welcome CASE because they believe it helps reduce the risk and uncertainty in managing the SDLC. IS personnel with a more technical orientation tend to resist CASE because they feel threatened by the technology's ability to replace some skills they have taken years to master. Table 4-2 lists several possi-

Reverse engineering: Automated tools that read program source code as input and create graphical and textual representations of program design-level information such as program control structures, data structures, logical flow, and data flow.

Reengineering: Automated tools that read program source code as input, perform an analysis of the program's data and logic, and then automatically, or interactively with a systems analyst, alter an existing system in an effort to improve its quality or performance.

NET SEARCH
Reengineering an existing database so that you could move it from one platform to another could be very time-consuming unless you use a reengineering tool that can automate this process. Visit http://www.prenhall.com/hoffer to complete an exercise related to this topic.

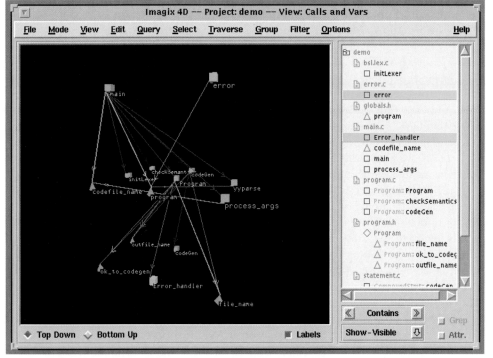

Figure 4-2
Sample screen of reverse engineering tool from Imagix
(Source: Imagix, 2000)

TABLE 4-2 Common Impacts of CASE on Individuals within Organizations

Individuals	Common Impact
Systems Analysts	CASE automates many routine tasks of the analyst, making the communication skills (rather than analytical skills) of the analyst most critical.
Programmers	Programmers will piece together objects created by code generators and fourth-generation languages. Their role will include more of maintaining designs using diagramming tools rather than source code.
Users	Users will be much more active in the systems development process through the use of upper CASE tools.
Top Managers	Top managers will play a more active role in setting priorities and strategic directions for IS by using CASE-based planning and through user-oriented system development methods.
Functional Managers	Functional managers will play a greater role in leading development projects by using CASE to reengineer their business processes.
IS Project Managers	IS project managers will have greater control over development projects and resources.

Adapted from Chen and Norman, 1992

TABLE 4-3 Driving Organizational Forces for the Adoption of CASE

Organizations adopt CASE to
- Provide new systems with shorter development time
- Improve the productivity of the systems development process
- Improve the quality of the systems development process
- Improve worker skills
- Improve the portability of new systems
- Improve the management of the systems development process

TABLE 4-4 Resisting Organizational Forces for the Adoption of CASE

Organizations reject CASE because of
- The high cost of purchasing CASE
- The high cost of training personnel
- Low organizational confidence in the IS department to deliver high-quality systems on time and within budget
- Lack of methodology standards within the organization
- Viewing CASE as a threat to job security
- Lack of confidence in CASE products

ble impacts of CASE on the roles of individuals within organizations. It should be clear after reviewing this table that the adoption of CASE should be a well thought-out and highly orchestrated activity.

Driving and Resisting Forces for CASE

A report prepared by a large accounting firm identified several driving and resisting forces that influence CASE adoption (Chen and Norman, 1992). Forces were identified from both an organization-wide and an individual perspective (see Tables 4-3 and 4-4). Organizational forces came from the pressures to develop and market information-intensive products and services, the increasingly shorter time-to-market, the transition to information-based organizations, and the potential for using information systems to support the organization's competitive strategy. Other factors included the increasing pressure to improve developer productivity and system quality. CASE was also viewed as an attractive vehicle for training, for building systems independent of specific implementation platforms, and for more easily managing projects. From the individual perspective, IS professionals who viewed CASE favorably felt that understanding how to use CASE was a valuable and marketable skill, that CASE automated many routine and boring tasks, and that it was an effective method for enforcing a common development methodology.

There were several resisting forces that had acted to preclude many organizations from making the investment in CASE. From an organizational perspective, it was reported that many business managers were losing confidence in the IS department's ability to deliver quality systems on time. Thus, organizations were investing in other options such as end-user computing and outsourcing.

Effective CASE adoption requires using a common design methodology. Thus, if an organization does not *first* have a standard methodology for developing systems, it is unlikely that CASE will be successfully adopted. Additionally, individuals within the IS organization may view CASE as a career threat if no coherent strategy for adoption and deployment is instituted.

This section has described the use of CASE within organizations and discussed several issues surrounding the adoption of CASE. It should now be clear to you that adopting and using CASE is a significant event in any organization. In view of this,

TABLE 4-5   **CASE Implementation Issues and Strategies**

Top Management Support	CASE requires a substantial long-term investment. Top management support is essential for success.
Contribution	CASE adoption is a business decision that must be justified as bringing business value to the organization by addressing or overcoming some identifiable problem in the development of information systems.
Manage Expectation	History has shown that the benefits of CASE are often long-term (for example, easier maintenance). IS managers must be careful not to oversell CASE as a cure-all for the development of information systems.
Prevent Resistance	The resistance to CASE can come from both inside and outside the IS department. Effectively managing expectations, providing extensive training, and selecting the right pilot projects and personnel will ease adoption and diffusion.
Deploy Carefully	CASE deployment should be done with care. Initial projects and personnel should be selected carefully. Personnel should be respected individuals with high credibility, training, and expertise.
Evaluate Continuously	The effects of CASE (and all other substantial investments) on the organization should be continuously monitored so that timely adjustments in strategy can be made.

Adapted from Chen and Norman, 1992

the adoption of CASE, or any other technology with so many potential impacts, should be guided with a clear strategy. Table 4-5 outlines many critical issues and implementation strategies. Although all items listed in this table are important, our personal experience is that without top management support, any significant systems project is destined for failure.

COMPONENTS OF CASE

CASE tools are used to support a wide variety of SDLC activities. CASE tools can be used to help in the project identification and selection, project initiation and planning, analysis, and design phases (**upper CASE**) and/or in the implementation and maintenance phases (**lower CASE**) of the SDLC (see Figure 4-3). A third category of CASE, **cross life-cycle CASE**, is tools used to support activities that occur *across* multiple phases of the SDLC. For example, tools used to assist in ongoing activities such as managing the project, developing time estimates for activities, and creating documentation are often considered cross life-cycle tools. Over the past several years, vendors of upper, lower, and cross life-cycle CASE products have "opened up" their systems through the use of standard databases and data conversion utilities to more easily share information across products and tools. An integrated and standard database called a *repository* is the common method for providing product and tool integration and has been a key factor in enabling CASE to more easily manage larger, more complex projects and to seamlessly *integrate* data across various tools and products. Integrated CASE or *I-CASE* will be described in more detail later in the chapter. User interface standards such as Microsoft's Windows^{TM} have also greatly eased the integration and deployment of these systems. The general types of CASE tools are listed below:

- *Diagramming tools* that enable system process, data, and control structures to be represented graphically.

Upper CASE: CASE tools designed to support the information planning and the project identification and selection, project initiation and planning, analysis, and design phases of the systems development life cycle.

Lower CASE: CASE tools designed to support the implementation and maintenance phases of the systems development life cycle.

Cross life-cycle CASE: CASE tools designed to support activities that occur *across* multiple phases of the systems development life cycle.

Figure 4-3
The relationship between CASE tools and the systems development life cycle

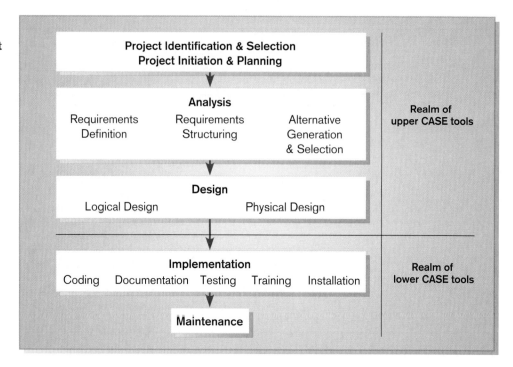

- *Computer display and report generators* that help prototype how systems "look and feel" to users. Display (or form) and report generators also make it easier for the systems analyst to identify data requirements and relationships.

- *Analysis tools* that automatically check for incomplete, inconsistent, or incorrect specifications in diagrams, forms, and reports.

- A central *repository* that enables the integrated storage of specification, diagrams, reports, and project management information.

- *Documentation generators* that help produce both technical and user documentation in standard formats.

- *Code generators* that enable the automatic generation of program and database definition code directly from the design documents, diagrams, forms, and reports.

Besides providing an array of tools, most CASE products also support ad hoc inquiry into and extraction from the repository. Security features, which may be important in some development environments, are also widely available. For example, if you contract with a custom software developer, you may expect them to secure your system specifications so that other project teams may not access your system requirements, design, and code. Some more advanced CASE products also support version control, which allows one repository to contain the description of several versions or releases of the same application system. Also, some CASE products provide import and export facilities to automatically move data between the CASE repository and other software development tools such as word processors, software libraries, and testing environments. Finally, as a shared development database, CASE environments should provide facilities for backup and recovery, user account management, and usage accounting. In other words, to provide the greatest benefits, CASE should be used to support all activities within the SDLC.

Nonetheless, many organizations that use CASE tools do not use them to support all phases of the SDLC. Some organizations may extensively use the diagramming features but not use code generators. Table 4-6 summarizes how CASE is commonly used within each SDLC phase. There are a variety of reasons why organizations

TABLE 4-6   Examples of CASE Usage Within the SDLC

SDLC Phase	Key Activities	CASE Tool Usage
Project identification and selection	Display and structure high-level organizational information	Diagramming and matrix tools to create and structure information
Project initiation and planning	Develop project scope and feasibility	Repository and documentation generators to develop project plans
Analysis	Determine and structure system requirements	Diagramming to create process, logic, and data models
Logical and physical design	Create new system designs	Form and report generators to prototype designs; analysis and documentation generators to define specifications
Implementation	Translate designs into an information system	Code generators and analysis, form, and report generators to develop system; documentation generators to develop system and user documentation
Maintenance	Evolve information system	All tools are used (repeat life-cycle)

choose to adopt CASE partially or to not use it at all. These reasons range from a lack of vision for applying CASE to all aspects of the SDLC to the belief that CASE technology will fail to meet an organization's unique system development needs. Several key differences between using traditional nonintegrated system development processes versus CASE-based development are summarized in Table 4-7.

In traditional systems development, much of the time is spent on coding and testing. When software changes are approved, often the code is first changed and then tested. Once the functionality of the code is assured, the documentation and specification documents are updated to reflect system changes. In effect, this means the systems analyst's job is that of *maintaining code and documentation*. Changes to the system "ripple" through all aspects of the system and supporting documentation and maintaining systems requires that all discrete system documents be updated. For example, if you use a word-processing system for logging documentation changes, these files need to be updated. If you use a separate program to graphically reflect process flow and data, changes to these diagrams must occur. Additionally, all other documents must be updated to reflect even the most minuscule changes to the system. The process of keeping all system documentation current can be a very boring and time-consuming activity that is often neglected. This neglect makes future maintenance by the same or (more likely) different programmers difficult at best. Experienced analysts know that complete and consistent documentation is the most important determinant to maintaining large computer systems. Thus, a primary drawback to the traditional development approach is the lack of integration among specification documents, program code, and supporting documentation.

TABLE 4-7   Traditional Systems Development versus CASE-Based Development

Traditional Systems Development	CASE-Based Systems Development
Emphasis on coding and testing	Emphasis on analysis and design
Paper-based specification	Rapid interactive prototyping
Manual coding of programs	Automated code generation
Manual documenting	Automated documentation generation
Intensive software testing	Automated design checking
Maintain code and documentation	Maintain design specifications

Adapted from McClure, 1989

A primary objective of the CASE-based approach to systems development is to overcome the drawbacks of the traditional approach. When using an integrated CASE environment, your primary role is to maintain the design documents, as most other aspects of the system can "flow" directly from these diagrams, forms, and report templates. In the remainder of this section, we discuss the components of CASE.

CASE Diagramming Tools

Diagramming tools: CASE tools that support the creation of graphical representations of various system elements such as process flow, data relationships, and program structures.

CASE **diagramming tools** allow you to represent a system and its various components visually. Diagrams are very effective for representing process flows, data structures, and program structures. For example, a diagramming technique called data flow diagramming is often used to represent the movement of data between business processes.

An example of a business process is customer billing within a company when invoices are received from a customer, a bill is produced, and this bill is sent to the customer; each of these activities is a business process. A data flow diagram (DFD) is used to model business processes as information flows through an organization. For example, Figure 4-4 shows a DFD from Oracle's Designer CASE system. This view shows some of the business processes within a department of motor vehicles, specifically the processes related to that of issuing a driver's license to an applicant. The diagram shows process 1.2, Issue License, as consisting of three subprocesses, (1.2.3) Update DMV database, (1.2.1) Get photograph of applicant, and (1.2.2) Create drivers license. Arrows, called data flows, move information between processes, data stores (e.g., DMV database), and between sources and sinks of information. In this figure, the DMV Evaluator is a source that provides the data flow, Successful test results. Likewise, the Applicant is a sink that receives information from the data flow, Drivers License. Note that processes could be automated or nonautomated. Additionally, the CASE tool allows the system designer to easily decompose higher-level processes into more specific subprocesses. CASE allows the designer to examine the system processes at a high aggregate level or at a highly detailed level. For example, Figure 4-5 provides a high-level view of the motor vehicle system in a functional hierarchy diagram.

Figure 4-4
A data flow diagram (DFD) from Oracle's Designer CASE environment

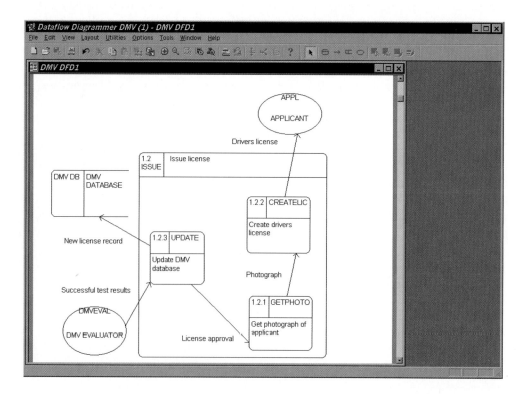

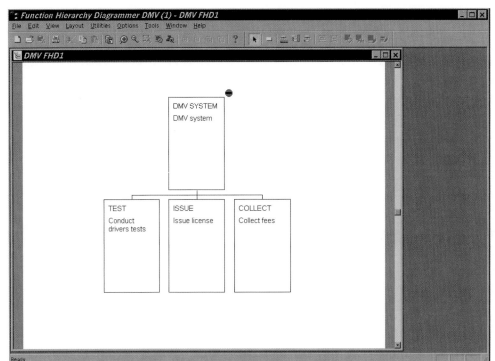

Figure 4-5
A functional hierarchy diagram from Oracle's Designer CASE environment

In addition to providing methods for modeling business processes, most CASE systems provide numerous options for representing other types of system-related information. A common method for representing data is the entity-relationship diagram (ERD). An example of a CASE-drawn ERD is shown in Figure 4-6 and is again from Oracle's Designer. In this diagram, there are four separate *entities*—supplier, shipment, product, and task. The lines connecting these entities represent relation-

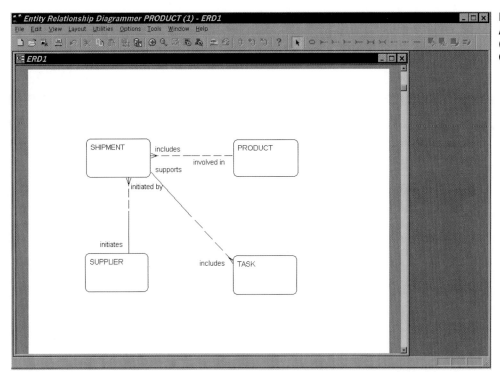

Figure 4-6
An entity-relationship diagram (ERD) from Oracle's Designer CASE environment

ships between entities, which means these entities can be joined together to provide information. For example, the product entity lists all of a company's products. When joined with the shipment entity, and when the supplier entity is joined with the shipment entity, it is easy to identify the products contained in a given shipment from a given supplier. Systems designers use these high-level models to gain a clear understanding of the interrelationships among different entities. Most CASE tools can generate the database schema—the separate tables needed to physically store the data in the database—directly from the ERD.

As with DFDs, most CASE environments allow you to display data models at higher or lower levels of detail. In addition, not only can you "link" higher and lower level views, you can also associate items in one diagram with items in another. For example, an entity of an ERD can be linked to a data store on a DFD, and the data elements in a data flow on a DFD can be linked to data elements of an entity and defined in the repository. The types of diagrams you will use depend upon the methodology standards within your organization and the type of information you are trying to represent. Most IS professionals believe the old proverb that "a picture is worth a thousand words" and have found this to be especially true when representing business processes and complex data relationships. Diagrams are an effective method for developing a common language that users and analysts can use to discuss system requirements. This has resulted in making the diagramming capabilities of most CASE environments a fundamental and indispensable component.

CASE Form and Report Generator Tools

Automated tools for developing computer displays (forms) and reports help the systems analyst design how the user will interact with the new system. These tools, referred to as **form and report generators**, are most commonly used for two purposes: (1) to create, modify, and test prototypes of computer display forms and reports and (2) to identify which data items to display or collect for each form or report. Figure 4-7 shows an example of a form layout design, along with the resulting form, using Oracle's Developer form design facility. This facility has numerous features to help you quickly design forms and windows that look and feel consistent to your users. For example, Developer has a template feature that allows you to define common headings, footers, and function key assignments. Once defined, all forms in the system inherit these template definitions. Also, if a common change is desired for all forms, you can simply change the template definition and all system forms will automatically inherit this change. Once you are happy with the design, you can quickly test its usability by converting the design template into working prototype forms. Figure 4-7 shows both the form layout design and the form it generates.

You will find that using automated tools for developing forms and reports is useful for both you and the eventual users of your system. For users, interacting with you during the early stages of the SDLC as forms and reports are outlined may help to ease system implementation. Involved users will be more familiar with the system when it is completed. As a result, these users may require less training than uninvolved users. Additionally, they may feel more positive that the system will meet their needs. From your perspective as the analyst, close interaction with users will help you develop a common frame of reference and enable you to better understand their data and processing requirements.

CASE Analysis Tools

One important objective of CASE is to help you handle the complexities of building large systems. We have described how CASE environments automate the creation of diagrams to represent system process flows, data, and structures in addi-

Form and report generators: CASE tools that support the creation of system forms and reports in order to prototype how systems will "look and feel" to users.

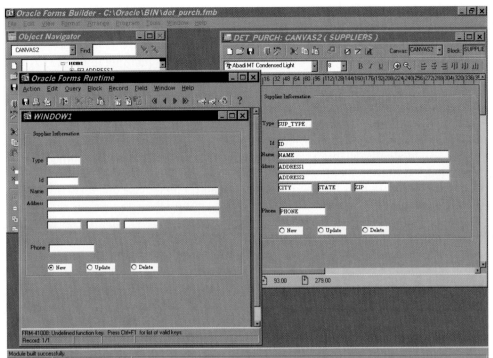

Figure 4-7
Form design tool from Oracle's
Developer CASE environment

tion to constructing forms and reports. **Analysis tools** generate reports that help you identify possible inconsistencies, redundancies, and omissions in these diagrams, forms, and reports. For example, many analysis activities can be performed on the graphical diagrams created by the analyst. Each general diagramming technique has numerous rules that govern how a diagram can be drawn; for example, the "balancing" rule must be followed when creating a lower-level DFD from a higher-level DFD. This rule requires that the number of data flows or arrows flowing into and out of a high-level process must equal those flowing into and out of all this process' lower-level subprocesses. Figure 4-8 shows a violation of the balancing rule: Data flow "C" flows out of process 1.2 on the Level-1 diagram, but it does not flow out of process 1 on the Level-0 diagram—dataflow X is a product produced

Analysis tools: CASE tools that enable automatic checking for incomplete, inconsistent, or incorrect specifications in diagrams, forms, and reports.

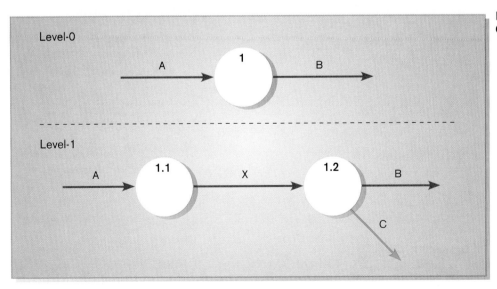

Figure 4-8
Out of balance data flow diagrams

within process 1 and would not be identified as an error. Thus, these diagrams are out of balance. An analysis performed on these Level-0 and Level-1 diagrams would alert you to this rule violation.

It is important to note that the types of analyses vary depending upon the organization's development methodology and the features of the CASE environment in use. In addition, many CASE systems support the creation of customized analysis reports. Typically, however, CASE analysis functions focus primarily on data structures and usage and on diagram completeness and consistency. For example, an analysis of all data input forms for a system could be performed. One report generated from this analysis could identify all input forms where the data elements contained on one form were identical to those contained on another (completely redundant forms). A second report could identify those forms where the data elements of one form were completely contained as a subset of the elements on another form (a partially redundant form). Such an analysis may identify that data are being entered into the system from more than one source—a possible data control violation. Such analysis capabilities are especially useful after the work of several systems analysts is combined.

CASE Repository

Substantial benefits when using CASE can only be achieved through the *integration* of various CASE tools and their data. Integrated CASE, or **I-CASE**, tools rely on common terminology, notations, and methods for systems development across all tools. Furthermore, all integrated CASE tools have a common user interface and can share system representations without systems analysts having to convert between different formats used by different tools. Hence, central to I-CASE is the idea of using a common repository for all tools so that this information can be easily shared between tools and SDLC activities. The **repository**, a centralized database, is the nucleus of a comprehensive I-CASE environment and is paramount to the smooth integration of the tools used during the various SDLC phases (Hanna, 1996). This means that the repository holds the complete information needed to create, modify, and evolve a software system from project initiation and planning to code generation and maintenance (see Figure 4-9). With a true I-CASE product, all tools throughout the entire life cycle will use a common

I-CASE: An automated systems development environment that provides numerous tools to create diagrams, forms, and reports; provides analysis, reporting, and code generation facilities; and seamlessly shares and integrates data across and between tools.

Repository: A centralized database that contains all diagrams, forms and report definitions, data structure, data definitions, process flows and logic, and definitions of other organizational and system components; it provides a set of mechanisms and structures to achieve seamless data-to-tool and data-to-data integration.

Figure 4-9
System development items stored in the CASE repository

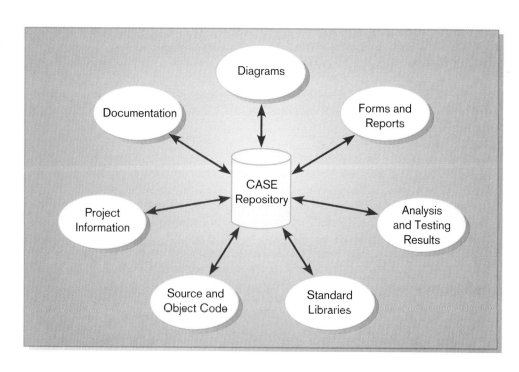

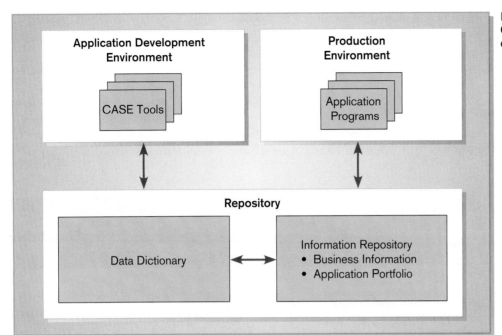

Figure 4-10
Common components of a
comprehensive CASE repository

repository. In this book, we will interchangeably use both CASE and I-CASE to refer to automated environments used to support the creation of information systems.

For years common development repositories have been used to create information systems independent of CASE. Figure 4-10 reflects the common components of a comprehensive CASE repository. The application development environment is one in which either information specialists or end users use CASE tools, high-level languages, and other tools to develop new applications. The production environment is one in which these same people use applications to build databases, keep the data current, and extract data from databases.

Within a repository there are two primary segments: the information repository and the data dictionary. The **information repository** combines information about an organization's business information and its application portfolio and provides automated tools to manage and control access to the repository (Bruce, Fuller, and Moriarty, 1989). Business information is the data stored in the corporate databases while the application portfolio consists of the application programs used to manage business information.

The **data dictionary** is a computer software tool used to manage and control access to the information repository. It provides facilities for recording, storing, and processing descriptions of an organization's significant data and data processing resources (Lefkovitz, 1985). Data dictionary features within a CASE repository are especially valuable for the systems analyst when cross referencing data items. **Cross referencing** enables one description of a data item to be stored and accessed by all individuals (systems analysts and end users) so that a single definition for a data item is established and used. Such a description helps to avoid data duplication and makes systems development and maintenance more efficient. For example, if the field length of a data element is changed, the data dictionary can produce a report identifying all programs affected by this change. Within an I-CASE environment, all diagrams, forms, reports, and programs can be automatically updated by the single change to the data dictionary definition. Each entry in a data dictionary has a standard "definition" that can include information such as the following attributes:

1. Element name and any aliases (can include both data items and programs)
2. Textual description of the element

Information repository: Automated tools used to manage and control access to organizational business information and application portfolio as components within a comprehensive repository.

Data dictionary: The repository of all data definitions for all organizational applications.

Cross referencing: A feature performed by a data dictionary that enables one description of a data item to be stored and accessed by all individuals so that a single definition for a data item is established and used.

3. List of related elements

4. Element type and format (for example, a calendar date might be of data type "date" and be of the format 12-Jan-01)

5. Range of acceptable values

6. Other information unique to the proper processing of this element

Figure 4-11 shows two computer screens that display information from Oracle's repository tool within the Designer and Developer CASE environment. Figure 4-11a shows a screen from the standard reporting system. Using this tool, one can view

Figure 4-11
Sample CASE repository contents
(a) Screen from Oracle's information repository showing standard reports

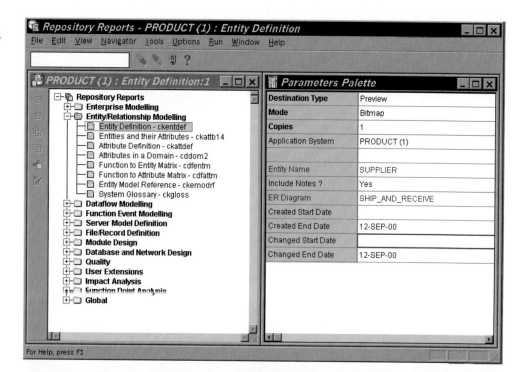

(b) Screen from Oracle's repository showing data dictionary details

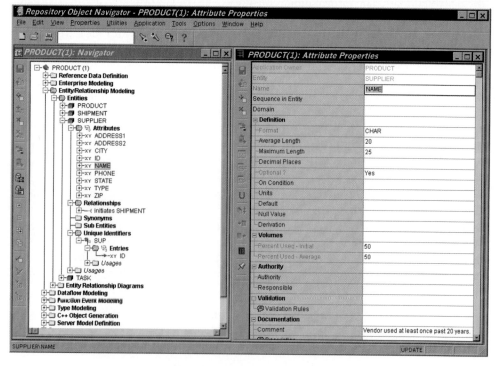

exhaustive information on all repository contents. In this figure, a report on the SHIP__AND__RECEIVE ERD is being viewed. The details on any individual item in the repository can be viewed by moving sequentially through each item or by locating an item through an interactive query. For example, Figure 4-11b shows a screen examining the data dictionary details of the NAME attribute of the SUPPLIER entity. In these details, we see that supplier names are a required field within a maximum length of 25 characters.

CASE Repository and the SDLC

During the project initiation and planning phase, the repository is used to store all information, both textual and graphical, related to the problem being solved. Details such as the problem domain, project resources and history, and organizational context are stored in the repository. As the project evolves, the repository becomes the basis for the integration of the various SDLC activities and phases.

During the analysis and design phases of the SDLC, the CASE repository is used to store graphical diagrams and prototype forms and reports. When completeness and consistency analyses are done, the repository allows the data from all diagrams, forms, and reports to be accessed simultaneously for comprehensive analysis. The data stored in the repository are also used as the foundation for the generation of code and documentation. Thus, the CASE repository is the integrating mechanism on all cross life-cycle tools and activities.

Additional Advantages of a CASE Repository

Besides specific tool integration, there are two additional advantages of using a comprehensive CASE repository that relate to project management and reusability. The development of most software systems requires that more than one person work on the project; therefore, to coordinate most effectively the activities of multiple developers, project management techniques should be used (Chapter 3). The CASE repository provides a wealth of information to the project manager and allows the manager to exert an appropriate amount of control on the project. For example, on large software development projects, it is customary (and more efficient) to partition the development into distinct subprojects where one or a few people (that is, a team) have primary responsibility for their development. Through the CASE repository, the project manager can restrict members' access to only those aspects of the system for which they are responsible. This reduces the complexity of the system for a given team and provides security such that data are not inadvertently changed or deleted. Partitioning allows multiple teams to work in parallel on different aspects of a single system, potentially reducing total system development time.

Another important use of the CASE repository relates to software **reusability.** In a large organization with many software systems, up to 75 percent of the application programs contain a significant amount of identical functions (Jones, 1986). In addition, as much as 50 percent of systems-level programs and upwards of 70 percent of telecommunications programs contain a significant amount of identical functions. Thus, one easy way for systems developers to enhance their productivity is to stop "reinventing the wheel" (or here, reinventing the function). If all organizational systems were created using CASE technology with a common repository, it would be possible to reuse significant portions of prior systems (or the design of prior systems) in the development of new ones. There are many items that can be reused besides programming code such as design documents (diagrams, specification documents, form and report layouts) and project management modules (schedules, assignments, report formats, project plans). The benefits of reusability are reduced development time and cost and improved software quality by using time-tested modules.

Reusability: The ability to design software modules in a manner so that they can be used again and again in different systems without significant modification.

NET   SEARCH

One of the strengths of using CASE-based development relates to how the repository integrates information across design activities and between tools. Visit http://www.prenhall.com/hoffer to complete an exercise related to this topic.

CASE without a Common Repository

Organizations that do not adopt a single integrated CASE environment must share design and development information among tools. Sharing may be necessary since different, nonintegrated tools may be used on the same project or on different projects for coupled systems. Besides using a single CASE repository, you can share data between CASE tools by the following methods:

- Manually entering specifications contained in one repository into another repository (obviously, not a very desirable approach).

- Converting repository contents into some neutral format, like an ASCII file, and then importing these into another repository. This is a better solution in terms of human effort but there still may be considerable loss of information or extra manual effort since the structure of the specifications may be lost in such a conversion.

- Converting the specifications in different repositories by using vendor or third-party utilities that translate either directly or through an industry standard exchange format among the repository formats of different CASE tools (IBM External Source Format and the CASE Data Interchange Format provide this type of passive repository sharing, if supported by the CASE tool).

- Allowing one CASE tool to directly read the repository of another CASE tool. This more active type of sharing across CASE tools is possible only if a CASE vendor opens up its database format, as many DBMS vendors have done.

CASE Documentation Generator Tools

Documentation generators: CASE tools that enable the easy production of both technical and user documentation in standard formats.

Each phase of the SDLC produces documentation. The types of documentation that flow from one phase to the next vary depending upon the organization, methodologies employed, and type of system being built. **Documentation generators** are modules that can create standard reports based upon the contents of the repository. Typically, SDLC documentation includes textual descriptions of needs, solution trade-offs, diagrams of data and processes, prototype forms and reports, program specifications, and user documentation including application and reference materials. A system that does not have adequate documentation is virtually impossible to use and maintain (Brooks, 1995).

A common problem when developing systems is that ". . . programmers concentrate on getting the application software up and running, rather than producing a document at the end of each development phase" (Hanna, 1992). Thus, documentation is a task that is often left to be dealt with *after* the programs have been completed. This time lag between when development activities occur and when documentation of these activities is produced often results in lower-quality documentation. The value of good documentation in relation to system maintenance is shown in Figure 4-12. This figure shows that the system maintenance effort takes 400 percent longer with poor-quality documentation. High-quality documentation leads to an 80 percent reduction in the system maintenance effort when compared to average-quality documentation. The practical implication of this is that there are some benefits for taking steps to improve the quality of system documentation and severe disadvantages for producing less-than-average-quality documentation.

Documentation generators within a CASE environment provide a method for managing the vast amounts of documentation created during the SDLC. Documentation generators allow the creation of master templates that can be used to verify that the documentation created for each SDLC phase conforms to a standard and that all required documents have been produced. Documentation is an often overlooked aspect of systems development yet, as pointed out in Figure 4-12, is decidedly the most important aspect to building *maintainable* systems. (The interested reader is encouraged to read the classic discussion of systems development and documentation by Brooks [1995] in *The Mythical Man-Month*.)

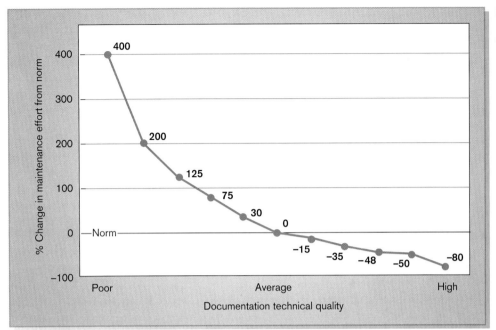

Figure 4-12
Impact of documentation quality on system maintenance
(Source: Hanna, 1992)

CASE Code Generation Tools

Code generators are automated systems that produce high-level program source code from diagrams and forms used to represent the system. As target environments vary on several dimensions, such as hardware and operating system platforms, many code generators are designed to be special-purpose systems that produce source code for a particular environment in a particular programming language. Most CASE tools that generate source code take a more flexible approach by producing standard source code and database definitions. Using standard language conventions, CASE-generated code can typically be compiled and executed on numerous hardware and operating system platforms with no, or very minor, changes. Yet standard code and definitions may not take advantage of special hardware or operating system features of specific environments.

Code generators: CASE tools that enable the automatic generation of program and database definition code directly from the design documents, diagrams, forms, and reports stored in the repository.

VISUAL AND EMERGING DEVELOPMENT TOOLS

Systems development tools are rapidly evolving. In particular, programming tools are undergoing radical changes to make programming much faster and easier. Without these powerful tools, programming would remain a slow and tedious process. The following three subsections describe some of these tools: The first describes object-oriented development, the second describes visual development, and the third examines the evolution and future of systems development tools.

Object-Oriented Development Tools

Development tools with object-oriented capabilities are among the newest type of programming languages. Object-oriented development tools are helping to alleviate maintenance problems plaguing organizations that have multiple systems with redundant functions. To understand how object-oriented tools work, let's start by describing an *object*. In simple terms, an object is a chunk of program and data that is built to perform common functions within a system. The idea is to make objects that can be *easily* reused and literally become the building blocks for all systems. This is very different from traditional programming languages where variables, procedures, and data are managed separately. This process of grouping the data and instructions together into a single object is called *encapsulation* (see Chapter 20). By encapsulating the instructions and data together, programs are easier to maintain because the objects that are grouped together are protected or isolated from other parts of the program. An example of an object might include employee identification and payroll information with a set of corresponding rules for calculating monthly payroll for a variety of job classifications and tax rules. Another example might be a printing object that can be used in multiple applications. Then, when a modification to the print object needs to be made—for example, let's suppose we want to modify the print object to support color printing—*one* module can be changed and *all* systems that use this common object can include the enhanced color printing functionality.

Consequently, one of the major advantages of object-oriented development is that each object contains easily reusable code. This means that objects used in one program can simply be inserted into a different program without having to be recreated or reprogrammed. In other words, once an object is created, it can be plugged into a number of different applications, avoiding the time (and expense!) of having to reprogram that particular set of instructions. Just as a radio made by Pioneer can be plugged into several different cars, an object can be plugged into several different applications. For example, when using an object-oriented development environment like Visual C++, a developer has numerous controls (that is, objects) that are built directly into the development tool. This means that when developing a user interface or input form, the programmer can use predefined objects—each object has a predefined look and numerous predefined properties that are associated with the object. Figure 4-13 shows a form from

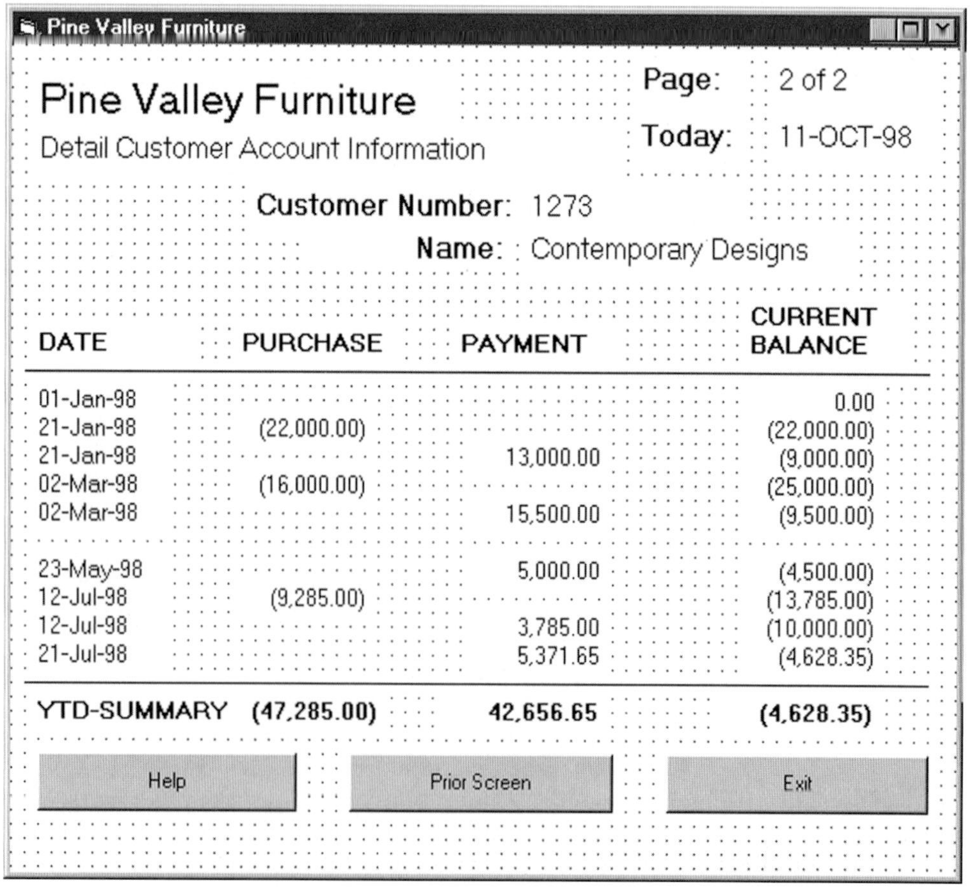

Figure 4-13
A form designed in Visual C++

DATE	PURCHASE	PAYMENT	CURRENT BALANCE
01-Jan-98			0.00
21-Jan-98	(22,000.00)		(22,000.00)
21-Jan-98		13,000.00	(9,000.00)
02-Mar-98	(16,000.00)		(25,000.00)
02-Mar-98		15,500.00	(9,500.00)
23-May-98		5,000.00	(4,500.00)
12-Jul-98	(9,285.00)		(13,785.00)
12-Jul-98		3,785.00	(10,000.00)
21-Jul-98		5,371.65	(4,628.35)
YTD-SUMMARY	(47,285.00)	42,656.65	(4,628.35)

Pine Valley Furniture
Detail Customer Account Information
Page: 2 of 2
Today: 11-OCT-98
Customer Number: 1273
Name: Contemporary Designs

Help   Prior Screen   Exit

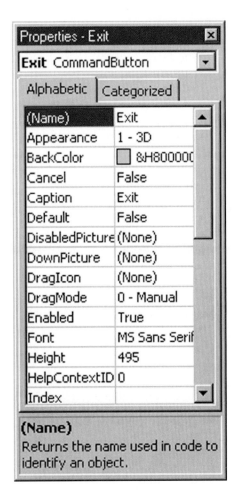

Figure 4-14
Button property customization in Visual C++

PINE VALLEY FURNITURE

Pine Valley Furniture with three command buttons at the bottom of the form—Help, Prior Screen, and Exit. Figure 4-14 shows the button property customization form for the Exit button of Figure 4-13 where the programmer can set specific properties for that object. Using this method of development, the object can be customized and be easily reused in other systems. Thus, the goal of object-oriented development is to make software easier to create, simpler and more consistent to use, and far more reliable (Verity and Schwartz, 1994). An easy way to think about object-oriented development for building information systems is to think about using Lego™ Bricks for building model bridges, planes, and buildings. The Lego Bricks are reusable in an infinite number of structures. Likewise, the well-designed software object is also infinitely reusable.

Object-oriented languages are currently among the most popular and their use continues to grow. For example, C++ is an object-oriented enhancement of the original C programming language. Java has become a very popular object-oriented development language for building Internet Web applications. As you might expect, there are object-oriented extensions to many other traditional programming languages including COBOL, FORTRAN, and BASIC. The biggest growth area in object-oriented development is the area of visual development tools. This will be described next.

Visual Development Tools

Visual development tools are a relatively new and extremely powerful way to rapidly develop systems (Hapgood, 2000). Visual Studio by Microsoft, PowerBuilder by Powersoft, Delphi by Borland International, ColdFusion by Allaire Corporation, and other visual development tools are some of the most widely used development environments today. These visual tools allow systems developers to quickly build new user interfaces, reports, and other features into new and existing systems in a fraction of the time previously required. Instead of building a screen, report, or menu by typing crude commands, designers use visual tools to "draw" the design using predefined objects. For example, to build a menu system in a visual programming environment is simple. Analysts can quickly list the order of menu commands in a development module called the Menu Editor (see Figure 4-15a) and instantly test the look of their design (see Figure 4-15b). In fact, once designed, these systems convert the design into the appropriate computer instructions, but all these details are hidden from the developer.

Although the popularity and the capabilities of most visual programming environments continue to expand, each environment has its own way of doing things. Yet, all have similar functions and capabilities. The market for these tools is very competitive, therefore, if one tool adds a new feature, most others quickly follow. Consequently, each tool has its strengths and weaknesses; it would be impossible to generalize that one system or approach to visual development is best.

Microsoft's Visual Studio is arguably the most popular visual development environment and is very powerful and easy to use. The Visual Studio environment helps to integrate Internet development with traditional systems development. Using this environment, developers can work on common database stored proce-

Figure 4-15
Building a menu with a visual development tool
(a) Menu Editor commands to build a menu in Visual Basic

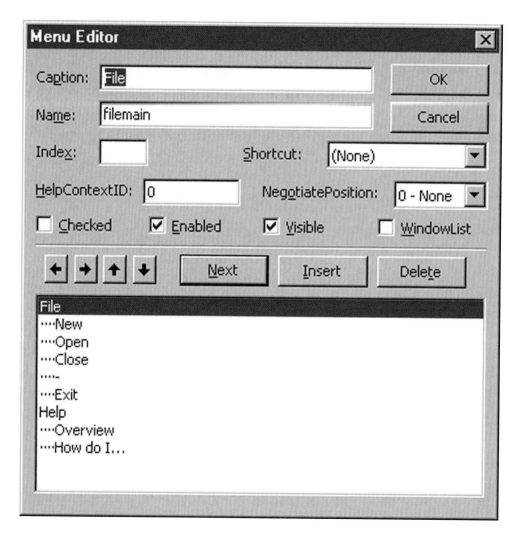

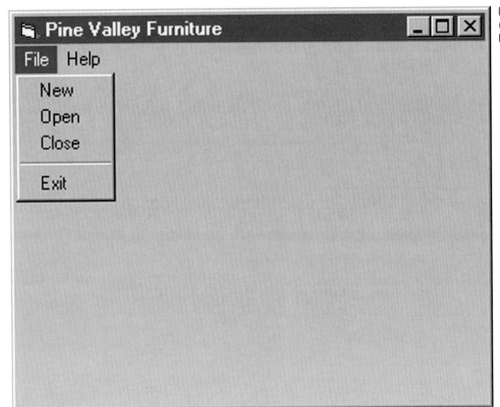

Figure 4-15 (continued)
(b) Menu system created using Visual Basic's Menu Editor

dures, Web server components, Active Server Pages, and client applets from any of the development tools in an integrated and consistent approach. Visual Studio includes several popular development tools—Visual Basic, Visual C++, Visual J++, Visual FoxPro, and Visual InterDev—where each tool works as a stand-alone application, but accesses common data and component libraries that are managed through the Visual Studio environment. This allows developers to pick and choose which tools to use to speed and ease the development process. For example, Visual Basic, with its strong reputation for ease of use, might be used to rapidly develop the user interface (see Figure 4-16), while Visual C++ could be used to create efficient and compact code for processing information. This flexibility has made Visual Studio a very popular development environment with countless organizations including Boeing, Qualcomm, Baan, and even Comedy Central (see www.microsoft.com).

PowerBuilder is another visual development environment that is extremely powerful and somewhat more expensive than Visual Basic (see Figure 4-17 for an example of this development environment). PowerBuilder's greatest strength is that it runs and develops applications for a broad range of hardware, computing platforms, and databases. PowerBuilder is primarily used by professional systems developers in larger organizations such as American Express, Fox Broadcasting, Nissan Motors, and Sega Video Games (see www.powersoft.com). Delphi is another powerful visual programming tool that has a large and loyal following among professional developers (see Figure 4-18). Delphi, like Visual Basic and PowerBuilder, has extensive object-oriented capabilities that developers can quickly and easily include in their systems. Organizations such as American Airlines, BMW, Coors, NASA, and countless others have reported great success when using Delphi (see www.inprise.com). For Internet development, ColdFusion

Figure 4-16
Visual Basic development environment within Visual Studio

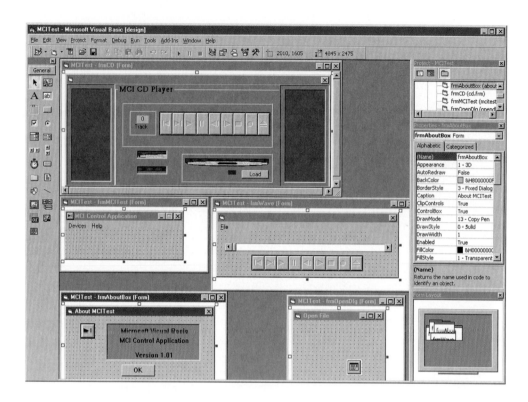

Figure 4-17
PowerBuilder development environment

(Source: Sybase, 2000; www.sybase.com)

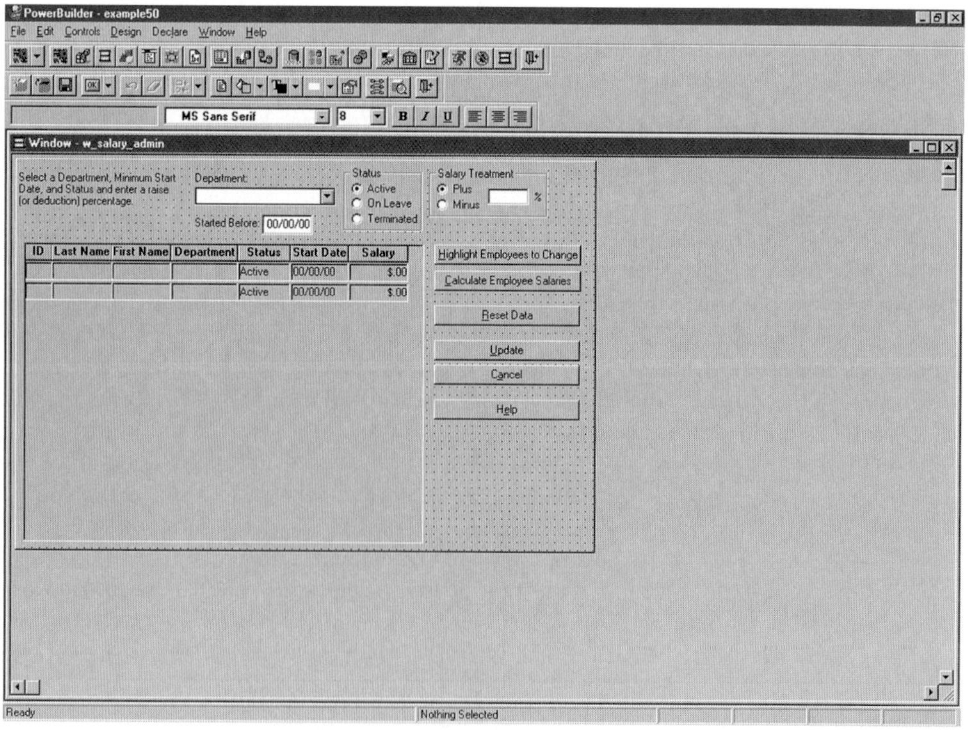

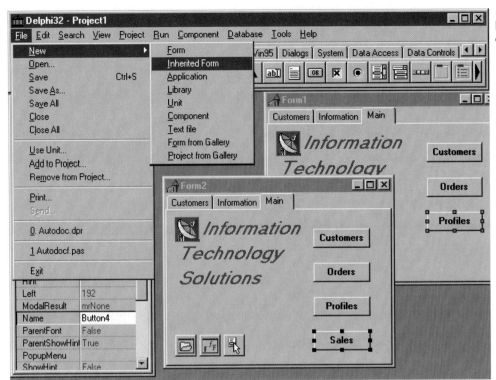

Figure 4-18
Borland's Delphi® development environment
(Source: Borland Software Corporation; 2000, 2001, www.borland.com)

from Allaire Corporation (www.allaire.com) is a highly acclaimed tool that provides an integrated development environment, allowing easy authoring of pages with fields that can be visually connected to server-based databases (see Figure 4-19a). ColdFusion supports quality assurance in development by providing useful utilities like an interactive debugging feature that can greatly speed development (see Figure 4-19b). Like the other products described here, ColdFusion has an impressive list of customers including Swiss Army, Casio, Pacific Bell, and the Internal Revenue Service. Given the popularity of visual development, if you haven't yet used these tools, you soon will!

Evolution and Future of Development Tools

We have already described two other approaches—object-oriented and visual development tools—which are rapidly being deployed to ease and empower the development process. A third way that is also emerging is to embed artificial intelligence into development environments. For example, in the near future, developers will be able to go well beyond 4GLs and develop large-scale information systems by simply telling the computer what they want it to do by using intelligent agents, created by other programmers, that will reside in the computer. A user or programmer would make a request to the agent through a conversational dialogue. The agents would use preestablished objects and visually display the results of the evolving system. In sum, the evolution and future of development tools will include extensive use of object-oriented modules and the visual display of results. As you will experience in your career, the evolution of development tools and techniques will be ongoing and continuous. This will result in more capable and complex systems for users and more capable and complex development environments. Hang on!

N E T   S E A R C H

Microsoft has a vision for the future of Internet-based application development that goes far beyond the capabilities of current leading tools like Visual Studio. Visit http://www.prenhall.com/hoffer to complete an exercise related to this topic.

Figure 4-19
**ColdFusion development
environment**
(Courtesy of Allaire.)
(a) Code to connect Web page to
server-based database

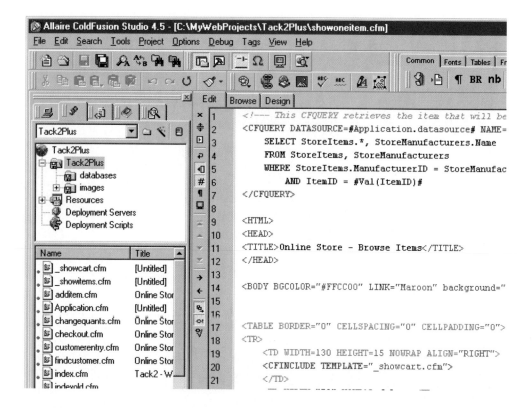

(b) Debugging feature

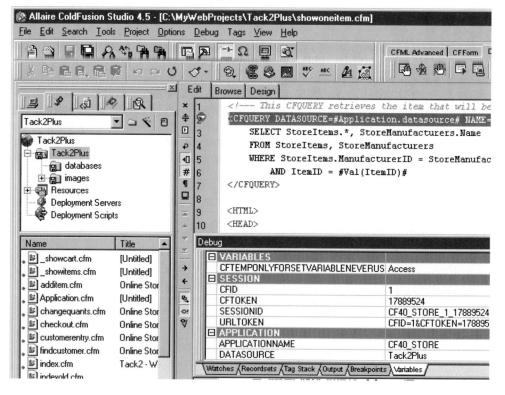

Summary

The purpose of this chapter has been to introduce you to the use of automated tools to support the systems development process. First, we examined how CASE is being used within organizations. The objectives of CASE are to aid the systems analyst with automated tools so that higher-quality systems are constructed on time and within budget, maintained economically, and changed rapidly.

Yet, with all the potential benefits of adoption, CASE can affect many individuals within the organization in a variety of ways. Systems development personnel will focus more on their interpersonal and communication skills than solely on their analytical skills. Likewise, users will be much more involved in the development process by actively working with development personnel when defining requirements, finalizing designs, and testing features and capabilities. Additionally, managers will play a more active role in setting project and design priorities. Overall, CASE acts to make the development process more collaborative. It should be clear that implementing CASE is not a small undertaking for most organizations. A long-term perspective must be taken or the potential benefits of CASE may not be realized.

Different categories of CASE tools were introduced, such as reverse and reengineering tools for rebuilding legacy systems. The components of a comprehensive CASE system are divided into upper, lower, and cross life-cycle CASE tools, covering different segments of the SDLC. Upper CASE tools—diagramming tools, form and report generators, and analysis tools—are used primarily to support project identification and selection, project initiation and planning, analysis, and design. Lower CASE tools—code generators—are used primarily to support system implementation and maintenance. Cross life-cycle tools—project management tools—coordinate project activities. The repository and documentation generators are used across multiple life-cycle phases to support project management, activity estimation, and documentation creation.

The final major section of the chapter examined visual and emerging development tools, focusing primarily on object-oriented and visual development. These modern tools are being rapidly deployed because they are enabling organizations to more quickly and easily develop and maintain systems. In short, object-oriented tools allow the use of software objects that encapsulate the programming instructions and data together, making programs easier to develop and maintain. A big benefit of object-oriented development is the ease with which software objects can be reused in other systems. Many development environments and programming languages use object-orientated capabilities, including C++, COBOL, FORTRAN, and BASIC. Visual development tools allow systems developers to quickly build new user interfaces, reports, and other features into new and existing systems in a fraction of the time previously required. Through the use of development tools like Visual Studio by Microsoft, PowerBuilder by Powersoft, Delphi by Borland International, or ColdFusion by Allaire, new systems are constructed by piecing together predefined "visual" objects rather than typing crude programming commands as was done in the past. This evolution in development is radically changing how systems are designed, developed, and maintained. As a result, more powerful systems can be constructed in a fraction of the time previously required. In the next chapter, we look in detail at the SDLC by focusing on project identification and selection.

Key Terms

1. Analysis tools
2. Code generators
3. Computer-aided software engineering (CASE)
4. Cross life-cycle CASE
5. Cross referencing
6. Data dictionary
7. Diagramming tools
8. Documentation generators
9. Form and report generators
10. I-CASE
11. Information repository
12. Lower CASE
13. Reengineering
14. Repository
15. Reusability
16. Reverse engineering
17. Upper CASE

Match each of the key terms above with the definition that best fits it.

_____ Software tools that provide automated support for some portion of the systems development process.

_____ Automated tools that read program source code as input and create graphical and textual representations of program design-level information such as program control structures, data structures, logical flow, and data flow.

_____ Automated tools that read program source code as input, perform an analysis of the program's data and logic, and then automatically, or interactively with a systems analyst, alter an existing system in an effort to improve its quality or performance.

_____ CASE tools designed to support the information planning and the project identification and selection, project initiation and planning, analysis, and design phases of the systems development life cycle.

_____ CASE tools designed to support the implementation and maintenance phases of the systems development life cycle.

_____ CASE tools designed to support the activities that occur *across* multiple phases of the systems development life cycle.

_____ CASE tools that support the creation of graphical representations of various system elements such as process flow, data relationships, and program structures.

_____ CASE tools that support the creation of system forms and reports in order to prototype how systems will "look and feel" to users.

_____ CASE tools that enable automatic checking for incomplete, inconsistent, or incorrect specifications in diagrams, forms, and reports.

_____ An automated systems development environment that provides numerous tools to create diagrams, forms, and reports; provides analysis, reporting, and code generation facilities; and seamlessly shares and integrates data across and between tools.

_____ A centralized database that contains all diagrams, forms and reports definitions, data structure, data definitions, process flows and logic, and definitions of other organizational and system components; it provides a set of mechanisms and structures to achieve seamless data-to-tool and data-to-data integration.

_____ Automated tools used to manage and control access to organizational business information and application portfolio as components within a comprehensive repository.

_____ The repository of all data definitions for all organizational applications.

_____ A feature performed by a data dictionary that enables one description of a data item to be stored and accessed by all individuals so that a single definition for a data item is established an used.

_____ The ability to design software so that they can be used again and again in different systems without significant modification.

_____ CASE tools that enable the easy production of both technical and user documentation in standard formats.

_____ CASE tools that enable the automatic generation of program and database definitions code directly from the design documents, diagrams, forms, and reports stored in the repository.

Review Questions

1. Describe the evolution of CASE and its outlook for the future.

2. List five objectives for CASE within organizations.

3. How does or can the role of CASE change in relation to the size of the organization?

4. How does or can the role of CASE change in relation to the type of information systems developed by an organization?

5. Who are the individuals impacted by the adoption of CASE within an organization, and how are they impacted?

6. What are the driving forces behind the adoption of CASE? What are the resisting forces?

7. Describe each major component of a comprehensive CASE system. Is any component more important than any other?

8. Contrast the differences between a data dictionary and repository.

9. Contrast the difference between traditional nonautomated systems development and CASE-based systems development.

10. Describe how CASE is used to support each phase of the SDLC.

11. Describe the concept of software reusability. Is reusability possible without CASE?

Problems and Exercises

1. Review the driving and resisting forces for CASE described in Tables 4-3 and 4-4 and the CASE implementation issues listed in Table 4-5. What do you forecast for CASE evolution and adoption in the future? Why? Would you recommend to colleagues that they adopt CASE tools? If so, under what circumstances? Why? If you wouldn't recommend using CASE tools now, why not, and what must change to cause you to recommend CASE tool adoption?

2. Review the sample computer forms in Figure 4-7. How is a form generator different from a standard graphics package like, say, Microsoft Windows Paintbrush® or PowerPoint? Why not simply use one of these graphics packages instead of using a CASE tool? What is gained by using a CASE tool rather than a graphics package? To answer these questions adequately, you may need to call a CASE tool vendor directly or find CASE tool product evaluations in the popular press.

3. Using a university or your workplace as the setting, list as many data elements (e.g., student identification number or customer identification number) as you can. Imagine how much information would be gathered and organized in this organization's database files and data dictionary. Estimate how much computer-based storage space would be needed to store, process, and back up this information. In an environment with computers, but without CASE tools, how

would this organization store, organize, and retrieve this information? What are the limitations, weaknesses, and potential problems in trying to do this without the data dictionary component of a CASE tool?

4. What forms of user documentation, either hard copy or on-line, came with your PC, operating system, network operating system, and applications software at work, home, or school? Is this documentation accessible and helpful? At what level would you rate the quality of the documentation—high, average or low? Why? What could be done to improve the documentation? How could CASE help to improve documentation?

5. A goal stated by many vendors of CASE products is to have CASE ultimately be able to automatically generate (and regenerate to any platform and with any changes) 100 percent of the code, error-free, for a new or modified informa-

tion system. This goal is considered important in order to achieve systems development productivity gains necessary to deal with systems backlog, to improve system quality, and to enhance our ability to maintain systems. Do you think this goal is possible? Why or why not?

6. What problems might occur during systems development if an organization used different CASE tools that did not share a common repository? What parallels can you make between the purpose of shared databases and the purpose of I-CASE?

7. Review the data dictionary entries on the SUPPLIER data entity in Figure 4-11b. What other data about an entity would you suggest be included in such a data dictionary entry? Why?

8. Contrast the differences between CASE, object-oriented, and visual development tools. How are these tools similar? How are they different?

Field Exercises

1. Interview a systems analyst, programmer, or an IS project manager to elicit their views of CASE, object-oriented, or visual development tools. Are they or their organization using any of these tools? If so, what are their evaluations thus far? What effects have these tools had on their jobs? If they use CASE, do their perceptions fit with those summarized in Table 4-2? If they are not using these development tools, why not? What must happen before they adopt one of these tools?

2. Find a detailed description of a CASE, object-oriented, or visual development tool and determine which of the functions and/or capabilities discussed in the chapter this tool supports. To do this, you may need to call a tool vendor directly, find a product evaluation in the popular press, or search a company's Website.

3. Interview information systems professionals who use CASE, object-oriented, or visual development tools and find out how they use the tools throughout the SDLC process. Ask them what advantages and disadvantages they see in using the tools that they do.

4. The systems development tool market is rapidly changing. Search through recent trade publications, like *Computerworld*,

PCWeek, *Intelligent Enterprise*, and *Software Magazine* (many of these publications can now be found on the Web). Find ads or articles that list and/or evaluate tools. Develop a list of these products. What are the benefits claimed for each in these ads and articles? What features do the vendors promote?

5. Choose an organization that is developing information systems without CASE tools. Talk to the people involved in building the systems and find out more about how they build them. Determine the parts of the systems development process that could be better done using CASE tools. Estimate the time and expense involved in those parts of the process that could be better done using CASE tools. Does it make sense for this organization to adopt CASE tools? Why or why not?

6. Go to a CASE tool vendor's Website and determine the product's price, functionality, and advantages. Try to find information related to any future plans for the product. If changes are planned, what changes and/or enhancements are planned for future versions? Why are these changes being made?

References

Allaire. 2000. Information from: *www.allaire.com*. Information verified: December 14, 2000.

Brooks, F. P., Jr. 1995. *The Mythical Man-Month: Essays on Software Engineering.* Reading, MA: Addison-Wesley.

Bruce, T., J. Fuller, and T. Moriarty. 1989. "So You Want a Repository." *Database Programming & Design* 2(May): 60–69.

Chen M., and R. J. Norman. 1992. "Integrated Computer-aided Software Engineering (CASE): Adoption, Implementation, and Impacts." *Proceedings of the Hawaii International Conference on System Sciences*, edited by J. F. Nunamaker, Jr.

Los Alamitos, CA: IEEE Computer Society Press, Vol. 3: 362–73.

Hanna, M. 1992. "Using Documentation as a Life-cycle Tool." *Software Magazine* 12(12) (Dec.): 41–51.

Hapgood, F. 2000. "CASE Closed." *CIO Magazine*, April 1, 2000 (*www.cio.com*).

Inprise. 2000. Information from: *www.inprise.com*. Information verified: December 14, 2000.

I/S Analyzer. 1993. "The Cost and Benefits of CASE." Rockville, MD: United Communications Group. 31(6) (June).

Jones, C. 1986. *Programming Productivity.* New York, NY: McGraw-Hill.

Lefkovitz, H. C. 1985. *Proposed American National Standards Information Resource Dictionary System.* Wellesley, MA: QED Information Sciences.

McClure, C. L. 1989. *CASE is Software Automation.* Englewood Cliffs, NJ: Prentice-Hall.

Microsoft, 2000. Information from: *www.microsoft.com.* Information verified: December 14, 2000.

Nunamaker, J. F. 1992. "Build and Learn, Evaluate and Learn." *Informatica* 1(1): 1–6.

Oracle, 1998. Information from: *www.oracle.com.* Information verified: February 14, 1998.

Orlikowski, W. J. 1989. "Division Among the Ranks: The Social Implications of CASE Tools for System Developers." *Proceedings of the Tenth International Conference on Information Systems.* 199–210.

Pfrenzinger, S. 1992. "Reengineering Goals Shift Toward Analysis, Transition." *Software Magazine* 12(10) (Oct.): 44–57.

Powersoft, 2000. Information from: *www.powersoft.com.* Information verified: December 14, 2000.

Richman, D. 1996. "CASE Reincarnation." *ComputerWorld,* July 29, 1996 *(www.computerworld.com).*

Stone, J. 1993. *Inside ADW and IEF: The Promise and Reality of CASE.* New York, NY: McGraw-Hill.

Sybase. 2000. Information from: *www.sybase.com.* Information verified: December 14, 2000.

Verity, J. W., and E. I. Schwartz. 1994. Software Made Simple. In P. Gray, W. R. King, E. R. McLean, and H. J. Watson (eds.), *Management of Information Systems,* 2nd ed., pp. 293–99. Fort Worth, TX: The Dryden Press.

BROADWAY ENTERTAINMENT COMPANY, INC.
Company Background

CASE INTRODUCTION

Broadway Entertainment Company, Inc. (BEC) is a fictional company in the video rental and recorded music retail industry, but its size, strategies, and business problems (and opportunities) are comparable to those of real businesses in this fast-growing industry.

In this section we'll introduce you to the company, the people who work for it, and the company's information systems. At the end of most subsequent chapters we'll revisit BEC to illustrate the phase of the life cycle discussed in that chapter. Our aim is to provide you with a realistic case example of how the systems development life cycle moves through its phases and how analysts, managers, and users work together to develop an information system. Through this example, you practice working on tasks and discussing issues related to each phase in an ongoing systems development project.

THE COMPANY

As of January 2001, Broadway Entertainment Company, or BEC, owned 2,443 outlets across the United States, Canada, Mexico, and Costa Rica. There is at least one BEC outlet in every state (except Montana) and in each Canadian province. There are 58 Canadian stores, 25 in Mexico, and six in Costa Rica. The company is currently struggling to open a retail outlet in Japan and plans to expand into the European Union (EU) within a year. United States Broadway operations are headquartered in Spartanburg, South Carolina; Canadian operations are headquartered in Vancouver, British Columbia; and Latin American operations are based in Mexico City, Mexico.

Each BEC outlet offers for sale two product lines, recorded music (on CDs and cassette tapes) and video games. Each outlet also rents two product lines, recorded videos (on VHS tape and DVDs) and video games. In calendar year 2000, music sales and video rentals together accounted for 85 percent of Broadway's U.S. revenues (see BEC Table 4–1). Foreign operations added another $24,500,000 to company revenues.

BEC TABLE 4-1 BEC Domestic Revenue, by Category, Calendar Year 2000

Category	Revenue (in $000s)	Percent
Music Sales	637,020	36
Compact Disks	426,500	24
Cassettes	210,520	12
Video Game Sales	104,760	6
Video Game Rentals	159,140	9
Video Rentals	862,080	49
Video Tapes	766,780	43
DVDs	95,300	5
Total	1,763,000	100

The home video and music retail industries are strong and growing, both domestically and internationally. For several years, home video has generated more revenue than either theatrical box office or movie pay-per-view.

To get a good idea of the industry in which Broadway competes, we look at five key elements of the home video and music retail industries:

1. Suppliers—all of the major distributors of recorded music (Sony, Matsushita, Time Warner), video games (Nintendo, Sega), and recorded videos (CBS, Fox, Viacom)

2. Buyers—individual consumers

3. Substitutes—television (broadcast, cable, satellite), first-run movies, Internet-based multimedia, theater, radio, concerts, and sporting events

4. Barriers to entry—few barriers and many threats, including alliances between telecommunications and entertainment companies to create cable television and Web TV, which lets consumers choose from a large number and variety of videos, music, and other home entertainment products from a computerized menu system in their homes

5. Rivalries among competing firms—large music chains (such as Musicland and Tower Records, all smaller than BEC) and large video chains (such as Blockbuster Entertainment, which is larger and more globally competitive than BEC)

Company History

The first BEC outlet opened in the Westgate Mall in Spartanburg, South Carolina, in 1977 as a music (record) sales store. The first store exclusively sold recorded music, primarily in vinyl format, but also stocked cassette tapes. Broadway's founder and current chairman of the board, Nigel Broad, had immigrated to South Carolina from his native Great Britain in 1968. After nine years of playing in a band in jazz clubs, Nigel used the money he had been left by his mother to form Broadway Entertainment Company, Inc., and opened the first BEC outlet.

Sales were steady and profits increased. Soon Nigel was able to open a second outlet and then a third. Predicting that his BEC stores had already met Spartanburg's demand for recorded music, Nigel decided to open his fourth store in nearby Greenville in 1981. At about the same time, he added a new product line—Atari video game cartridges. Atari's release of its Space Invaders game cartridge resulted in huge profits for Nigel. The company continued to grow and Broadway expanded beyond South Carolina into neighboring states.

In the early 1980s, Nigel saw the potential in videotapes. A few video rental outlets had opened in some of Broadway's markets, but they were all small independent operations. Nigel saw the opportunity to combine video rentals with music sales in one place. He also decided that he could rent more videos to customers if he changed some of the typical video store rules such as eliminating the heavy membership fee and allowing customers to keep videos more than one night. Nigel also wanted to offer the best selection of videos anywhere.

Nigel opened his first joint music and video store at the original BEC outlet in Spartanburg in 1985. Customer response was overwhelming. In 1986, Nigel decided to turn all 17 BEC outlets into joint music and video stores. To move into the video rental business in a big way, Nigel and his chief financial officer, Bill Patton, decided to have a public offering. They were happily surprised when all 1 million shares sold at $7 per share. The proceeds also allowed Broadway to revive the dying video game line by dropping Atari and adding the newly released Nintendo game cartridges.

Profits from BEC outlets continued to grow throughout the 1980s, and Broadway further expanded by acquiring existing music and video store chains including Music World. From 1987 through 1993, the number of BEC outlets roughly doubled each year. The decision to go international, made in 1991, resulted in 12 Canadian stores that year. The initial 3 Latin American stores were opened in mid-1994. From its beginnings in 1977, with 10 employees and $398,000 in revenues, Broadway Entertainment Company, Inc., grew to 24,225 employees and worldwide revenues of $1,787,500,000 by January 1, 2001.

Company Organization

In 1992, when the company opened its one-thousandth store, Nigel decided that he no longer wanted to be chief executive officer of the company. Nigel decided to fill only the position of chairman and he promoted his close friend Ira Abramowitz to the offices of president and CEO (see BEC Figure 4-1).

Most of Broadway's other senior officers have also been promoted from within. Bill Patton, the chief financial officer, started as the fledgling company's first bookkeeper and accountant. Karen Gardner had been part of the outside consulting team that built Broadway's first information system in 1986 and 1987. She became the vice president in charge of IS for BEC in 1990. Bob Panofsky,

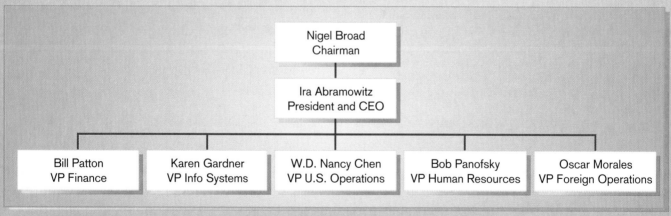

BEC Figure 4-1
Broadway Entertainment Company, Inc. organization chart

the vice president for human resources, had been with the company since 1981. An exception to the promote-from-within tendency, W. D. Nancy Chen, the vice president for domestic operations, had been recruited from Music World in 1991, shortly before the chain was purchased by Broadway. Oscar Morales had been hired in 1992 from Blockbuster Entertainment, where he had been in charge of Latin American expansion.

Development of Information Systems

Broadway Entertainment Company operated from 1977 until 1984 without any computer-based information systems support. As the company grew, accounting ledgers, files, and customer account information became unruly. Like many businesses this size, the owner did not have the expertise or the capital for developing the company's own information systems. For example, Bill Patton, managed inventory by hand until he bought an IBM AT in 1984. Computerizing the company made the expansion to 10 stores in 1984 much easier.

In 1985 BEC had nobody trained in information systems on staff, and all the BEC managers were quite busy coping with the business expansion. Nigel and Bill considered hiring a small staff of experienced IS professionals, but they did not know how to manage such a group, how to select quality staff, or what to expect from such employees. Nigel and Bill realized that computer software could be quite complicated, and building systems for a rapidly changing organization could be quite a challenge. Nigel and Bill also knew that building information systems required discipline. So Nigel, after talking with leaders of several other South Carolina businesses, contacted the information consulting firm of Fitzgerald McNally, Inc., about designing and building a custom computer-based information system for Broadway. In 1985, no prewritten programs were available to help run the still relatively new business of video and music rental and sales stores.

Nigel and Bill wanted the new system to perform accounting, payroll, and inventory control. Nigel wanted the system to be readily expandable as he was planning for Broadway's rapid growth. At the operational level, Nigel realized that the video rental business would require unique features in its information system. For one thing, rental customers would not only be taking product from the store, they would also be returning it at the end of the rental period. Further, customers would be required to register with Broadway and attach some kind of deposit to their account in order to help ensure that videos would be returned.

At a managerial level, Nigel wanted the movement of videos in and out of the stores and all customer accounts computerized. Nigel also wanted to be able to search through the data on Broadway's customers describing their rental habits. He wanted to know which videos were the most popular, and he wanted to know who Broadway's most frequent customers were, not only in South Carolina but also in every location where Broadway did business.

Fitzgerald McNally, Inc. was happy to get Broadway's account. It assigned Karen Gardner to head the development team. Karen led a team of her own staff of analysts and programmers, along with several BEC managers, in a thorough analysis and design study. The methodology applied in this study provided the discipline needed for such a major systems development effort. The methodology began with information planning and continued through all phases of the systems development life cycle.

Karen and her team delivered and installed the system at the end of the two-year project. The system was centralized, with an IBM 4381 mainframe installed at headquarters in Spartanburg and three terminals, three light pens, and three dot-matrix printers installed in each BEC outlet. The light pens recorded, for example, when the tapes were rented and when they were returned by reading the bar code on the cassette. The light pens were also used to read the customer's account number, which was recorded in a bar code on the customer's BEC account card. The printers generated receipts. In addition, the system included a small personal computer and printer to handle a few office functions such as the ordering and receiving of goods. The software monitored and updated inventory levels. Another software product generated and updated the customer database, whereas other parts of the final software package were designed for accounting and payroll.

In 1990, Karen Gardner left Fitzgerald McNally and joined Broadway as the head of its information systems group. Karen led the effort to expand and enhance Broadway's information systems as the company grew to over 2,000 company-owned stores in 1995. Broadway now uses a client/server network of computers at headquarters and in-store point-of-sale (POS) computer systems to handle the transaction volume generated by millions of customers at all BEC outlets.

INFORMATION SYSTEMS AT BEC TODAY

BEC has two systems development and support groups, one for in-store applications and the other for corporate, regional, and country-specific applications. The corporate development group has liaison staff with the in-store group, because data in many corporate systems feed or are fed by in-store applications (e.g., market analysis systems depend on transaction data collected by the in-store systems). BEC creates both one-year and three-year IS plans that encompass both store and corporate functions.

The functions of the original in-store systems at BEC have changed very little since they were installed in 1987—for example, customer and inventory tracking are still done by pen-based, bar code scanning of product labels and membership cards. Rentals and returns, sales, and other changes in inventory as well as employee time in and out are all captured at the store in electronic form via a local point-of-sale (POS) computer system. These data are transmitted in batches at night using modems and regular telephone connections to corporate headquarters where all records are stored in a network of IBM AS/400 computers (see BEC Figure 4-2).

As shown in BEC Figure 4-2, each BEC store has an NCR computer that serves as a host for a number of POS terminals at checkout counters and is used for generating reports. Some managers have also learned how to use spreadsheet, word processing, and other packages to handle functions not supported by systems provided by BEC. The front-end communications processor offloads traffic from the IBM AS/400 network so that the servers can concentrate on data processing applications. BEC's communication protocol is SNA (System Network Architecture), an IBM standard. Corporate databases are managed by IBM's relational DBMS DB2. BEC uses a variety of programming environments, including C, COBOL, SQL (as part of DB2), and code generators.

Inventory control and purchasing are done centrally and employees are paid by the corporation. Each store has electronic records of only its own activity, including inventory and personnel. Profit and loss, balance sheets, and other financial statements are produced for each store by centralized systems. In the following sections we'll review the applications that exist in the stores and at the corporate level.

In-Store Systems

BEC Table 4-2 lists the application systems installed in each store. BEC has developed a turnkey package of hardware and software (called Entertainment Tracker—ET), which is installed in each store worldwide. Besides English, the system also works in Spanish and French.

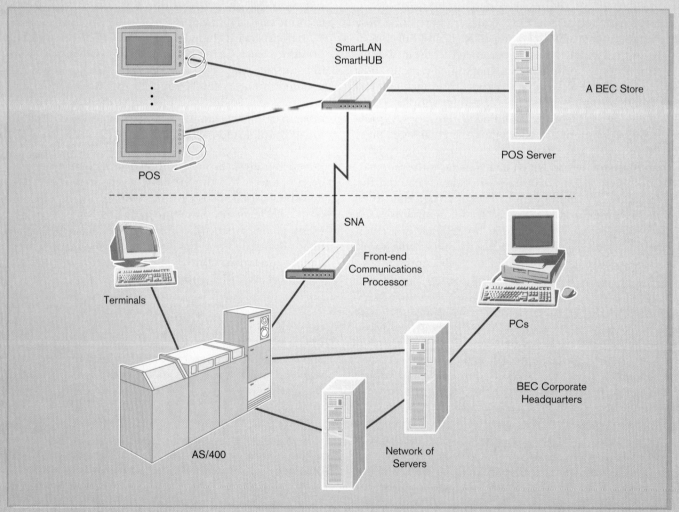

BEC Figure 4-2
BEC hardware and network architecture

BEC TABLE 4-2 List of BEC In-Store
(Entertainment Tracker) Applications

System Name	Description
Membership	Supports enrollment of new members, issuing membership cards, reinstatement of inactive members, and local data management for transient members
Rental	Supports rentals and returns of all products and outstanding rental reports
Sales	Supports sales and returns of all products (including videos, music, snack food, BEC apparel, and gift certificates)
Inventory control	Supports all changes in rental and sales inventory that are not sales based (e.g., receipt of a new tape for rental, rejection of goods damaged in shipment, and transfer of an item from rental to sales categories)
Employee	Supports hiring and terminating hourly employees, as well as all time-reporting activities

As you can see from BEC Table 4-2, all of these applications are transaction processing systems. In fact, there is a master screen on the POS terminals from which each ET application is activated. These systems work off a local decentralized database, and there is a similarly structured database for each store. Various batched data transfers occur between corporate and store systems at night (store transactions, price and membership data updates, etc.). The local database contains data on members, products, sales, rentals, returns, employees, and work assignments. The database contains only current data—the history of customer sales and rentals is retained in a corporate database. Thus, local stores do not retain any customer sales and rental activity (except for open rentals).

Data for those members who have had no activity at a local store for more than one year are purged from the local database. When members use a BEC membership card and no member record exists in the local database, members are asked to provide a local address and phone number where they can be contacted.

All store employees, except the store manager who is on salary, are paid on an hourly basis, so clock-in and -out times are entered as a transaction, using employee badges with a bar code strip, on the same POS terminal used for member transactions. Paychecks are delivered by express mail twice a month. Employee reports (e.g., attendance, payroll, and productivity) are produced by corporate systems and sent to store managers.

All other store recordkeeping is manual and corporation offices handle accounts receivables and payables. The local store manager is responsible for contacting via phone or mail members who are late in returning rented

items. Each night a file of delinquent members is transmitted to each store and, if a member tries to use a delinquent membership, the member is asked to return all outstanding rentals before renting any more items and the current transaction is invalidated. When terminated members try to use their cards, a BEC store clerk keeps the membership card and members are given a printed form that explains their rights at that point. Stolen membership cards are handled similarly, except that the store manager deals personally with people using cards that have been reported stolen.

Corporate Systems

Corporate systems run on IBM servers using IBM's DB2 relational database management system, although some run on PCs. Application software is written in COBOL, C, SQL (a database processing language), and several 4GLs, and all systems are developed by BEC. Clerks and managers use PCs for interactive access into corporate systems as well as for stand-alone, end-user applications such as word processing, spreadsheets, specialized databases, and business graphics.

There are more than 20 major corporate systems with over 350 programs and approximately 500,000 lines of code. There are many more specialized systems, often developed for individual managers, projects, or special events. BEC Table 4-3 lists some of the most active and largest of the major corporate systems.

One interesting aspect of the banking application is that because stores have no financial responsibilities, BEC uses a local bank only for daily deposits and getting change. BEC's corporate bank, NCNB, arranges correspondent banking relationships for BEC so that local deposits are electronically transferred to BEC's corporate accounts with NCNB.

BEC's applications are still expanding, and they are under constant revision. For example, in cooperation with several hotel and motel chains that provide VCRs for rental, BEC is undertaking a new marketing campaign aimed at frequent travelers. At any one time, there are approximately 10 major system changes or new systems under development for corporate applications with over 250 change requests received annually covering requirements from minor bug fixes to reformatting or creating new reports to whole new systems.

Status of Systems

A rapidly expanding business, BEC has created significant growth for the information systems group managers. Karen Gardner is considering reorganizing her staff to provide more focused attention to the international area. BEC still uses the services of Fitzgerald McNally when Karen's resources are fully committed. Karen's depart-

BEC TABLE 4-3 List of BEC Corporate Applications

System Name	Description
Human resources	Supports all employee functions, including payroll, benefits, employment and evaluation history, training, and career development (including a college scholarship for employees and dependents)
Accounts receivable	Supports notification of overdue fees and collection of payment from delinquent customers
Banking	Supports interactions with banking institutions, including account management and electronic funds transfers
Accounts payable, purchasing, and shipping	Supports ordering products and all other purchased items used internally and resold/rented, distribution of products to stores, and payment to vendors
General ledger and financial accounting	Supports all financial statement and reporting functions
Property management	Supports the purchasing, rental, and management of all properties and real estate used by BEC
Member tracking	Supports recordkeeping on all BEC members and transmits and receives member data between corporate and in-store systems
Inventory management	Supports tracking inventory of items in stores and elsewhere and reordering those items that must be replenished
Sales tracking and analysis	Supports a variety of sales analysis activities for marketing and product purchasing functions based on sales and rental transaction data transmitted nightly from stores
Store contact	Supports transmittal of data between corporate headquarters and stores nightly, and the transfer of data to and from corporate and store systems
Fraud	Supports monitoring abuse of membership privileges
Shareholder services	Supports all shareholder activities, including recording stock purchases and transfers, disbursement of dividends, and reporting
Store and site analysis	Supports the activity and profit analysis of stores and the analysis of potential sites for stores

ment includes 33 developers (programmers, analysts, and other specialists in database, networking, etc.) plus data center staff, which is now large and technically skilled enough to handle almost all requests.

Karen's current challenge in managing the IS group is keeping her staff current in the skills they need to successfully support the systems in a rapidly changing and competitive business environment. In addition, Karen's staff needs to be excellent project managers, to understand the business completely, and to exhibit excellent communication with clients and each other. Karen is also concerned about information systems literacy among BEC management and that technology is not being as thoroughly exploited as it could be.

To deal with this situation, Karen is considering several initiatives. First, she has requested a sizable increase in her training budget, including expanding the benefits of the college tuition reimbursement program. Second, Karen is considering instituting a development program that will better develop junior staff members and will involve user departments. As part of this program, BEC personnel will rotate in and out of the IS group as part of normal career progression. This program should greatly improve relationships with user departments and increase end-user understanding of technology. The development of this set of technical, managerial, business, and interpersonal skills in and outside IS is a critical success factor for Karen's group in responding to the significant demands and opportunities of the IS area.

CASE SUMMARY

Broadway Entertainment Company is a $1.79-billion international chain of music, video, and game rental and sales outlets. BEC started with one store in Spartanburg, South Carolina, in 1977 and has grown through astute management of expansion and acquisitions into over 2,000 stores in four countries.

BEC's hardware and software environment is similar to that used by many national retail chains. Each store has a computer system with point-of-sale terminals that run mainly sales and rental transaction processing applications, such as product sales and rental, membership, store-level inventory, and employee pay activities. Corporate systems are executed on a network of computers at a corporate data center. Corporate systems handle all accounting, banking, property, sales and member tracking, and other applications that involve data from all stores.

BEC is a rapidly growing business with significant demand for information services. To build and maintain systems, BEC has divided its staff into functional area groups for both domestic and international needs. BEC uses modern database management and programming language technologies. The BEC IS organization is challenged by keeping current in both business and technology areas. We will see in case studies in subsequent chapters how BEC responds to a request for a new system within this business and technology environment.

CASE QUESTIONS

1. What qualities have led to BEC's success so far?

2. Is the IS organization at BEC poised to undertake significant systems development in the near future?

3. What specific management skills do systems analysts at BEC need?

4. What specific communication skills do systems analysts at BEC need?

5. What specific areas of organizational knowledge do systems analysts at BEC need beyond the information provided in this case?

6. Why did BEC decide to originally use an outside contractor, Fitzgerald McNally, to develop its first computer applications?

7. What has BEC done to facilitate the global utilization of their application systems?

8. Do corporate and in-store systems seem to be tightly or loosely related at BEC? Why do you think this is so?

9. What challenges and limitations will affect what and how systems are developed in the future at BEC?

10. Chapter 1 of this book identified roles associated with systems development. This BEC case lists the types of jobs held by people in the IS organization at BEC. Do you see any IS development roles missing at BEC?

Part TWO

Making the Business Case

● **Chapter 5**
Identifying and Selecting Systems
Development Projects

● **Chapter 6**
Initiating and Planning Systems Development Projects

An Overview of Part TWO

Making the Business Case

The demand for new or replacement systems exceeds the ability and resources of most organizations to conduct systems development projects by either themselves or consultants. This means that organizations must set priorities and a direction for systems development that will yield development projects with the greatest net benefits. As a systems analyst, you must not only analyze user information requirements but also help make the business case, or justify why the system should be built and the development project conducted.

The reason for any new or improved information system is to add value to the organization. As systems analysts, we must choose to use systems development resources to build the mix of systems that adds the greatest value to the organization. How can we determine the business value of systems and identify those applications that provide the most critical gains? Part II addresses this topic, which we call making the business case. Business value comes from supporting the most critical business goals and helping the organization deliver on its business strategy. All systems, whether supporting operation or strategic functions, must be linked to business goals. The chapters in this part of the book show how to make this linkage.

The source of systems projects is either initiatives from information systems planning (proactive identification of systems) or request from users or IS professionals (reactions to problems or opportunities) for new or enhanced systems. In Chapter 5 we outline the linkage among corporate planning, information systems planning, and the identification and selection of projects. We do not include IS planning as part of the SDLC, but the results of IS planning greatly influence the birth and conduct of systems projects. Chapter 5 makes a strong argument that IS planning provides not only insights into choosing which systems an organization needs, but also describes the strategies necessary for evaluating the viability of any potential systems project.

A more frequent source of project identification originates from system service requests (SSRs) from business managers and IS professionals, usually for very focused systems or incremental improvements in existing systems. User managers request a new or replacement system when they believe that improved information services will help them do their jobs. IS professionals may request system updates when technological changes make current system implementations obsolete or when the performance of an existing system needs improvement. In either case, the request for service must be understood by management and a justification for the system and associated project must be developed.

We continue with the Broadway Entertainment Company (BEC) case following Chapter 5. In this case, we show how an idea for a new information system was stimulated by a synergy of corporate strategic planning and the creativity of an individual business manager. We also show how this idea is initially evaluated and how it leads to the initiation of a systems development project.

Chapter 6 focuses on what happens after a project has been identified and selected: the next step in making the business case, initiating and planning the proposed system request. This plan develops a better understanding of the scope of the potential system change and the nature of the needed system features. From this preliminary understanding of system requirements, a project plan is developed that shows both the detailed steps and resources needed in order to conduct the analysis phase of the life cycle and the more general steps for subsequent phases. The feasibility and potential risks of the requested system are also outlined and an economic cost-benefit analysis is conducted to show the potential impact of the system change. In addition to the economic feasibility or justification of the system, technical, organizational, political, legal, schedule, and other feasibilities are assessed. Potential risks—unwanted outcomes—are identified and plans for dealing with these possibilities are identified. Project initiation and planning ends when a formal proposal for the systems development project is completed and submitted for approval to whomever must commit the resources to sys-

tems development. If approved, the project moves into the analysis phase.

We illustrate a typical project initiation and planning phase in a BEC case following Chapter 6. In this case, we show how BEC identified one critically important business goal, which provided the motivation for a requested system. The case further shows how an analysis of this business goal leads to the justification for a system with a competitive advantage for BEC and then to the associated development project plan.

5

Identifying and Selecting Systems Development Projects

LEARNING OBJECTIVES

After studying this chapter, you should be able to:

● Describe the project identification and selection process.

● Describe the corporate strategic planning and information systems planning process.

● Explain the relationship between corporate strategic planning and information systems planning.

● Describe how information systems planning can be used to assist in identifying and selecting systems development projects.

● Analyze information systems planning matrices to determine affinity between information systems and IS projects and to forecast the impact of IS projects on business objectives.

● Describe the three classes of Internet electronic commerce applications: Internet, Intranets, and Extranets.

INTRODUCTION

The scope of information systems today is the whole enterprise. Managers, knowledge workers, and all other organizational members expect to easily access and retrieve information, regardless of its location. Nonintegrated systems used in the past—often referred to as "islands of information"—are being replaced with cooperative, integrated enterprise systems that can easily support information sharing. While the goal of building bridges between these "islands" will take some time to achieve, it represents a clear direction for information systems development. The use of enterprise resource planning (ERP) systems like SAP R/3 (www.sap.com), PeopleSoft (www.peoplesoft.com), Oracle (www.oracle.com), and Baan (www.baan.com) have enabled the linking of these "islands" in many organizations. Additionally, as the use of the Internet continues to evolve to support business activities, system integration has become a paramount concern of organizations (Hasselbring, 2000).

Obtaining integrated enterprise-wide computing presents significant challenges for both corporate and information systems management. For example, given the proliferation of personal and departmental computing wherein disparate systems and databases have been created, how can the organization possibly control and maintain all of these systems and data? In many cases they simply cannot because it is nearly impossible to track who has which systems and what data, where there are overlaps or inconsistencies, and how accurate the information is. The reason that personal and departmental systems and databases abound is that users are either unaware of the information that exists in corporate databases or they cannot easily get at it, so they create and maintain their own information and systems. Intelligent identification and selection of system projects, for both new and replacement systems, are critical steps in gaining control of systems and data. It is the hope of many chief information officers (CIOs) that with the advent of ERP systems, improved system integration, and the rapid deployment of corporate Internet solutions, these islands will be reduced or eliminated (Ross and Feeny, 2000).

The acquisition, development, and maintenance of information systems consume substantial resources for most organizations. This suggests that organizations can benefit from following a formal process for identifying and selecting projects. The first phase of the systems development life cycle—project identification and selection—deals with this issue. In the next section, you will learn about a general method for identifying and selecting projects and the deliverables and outcomes from this process. This is followed by brief descriptions of corporate strategic planning and information systems planning, two activities that can greatly improve the project identification and selection process.

IDENTIFYING AND SELECTING SYSTEMS DEVELOPMENT PROJECTS

The first phase of the SDLC is project identification and selection. During this activity, a senior manager, a business group, an IS manager, or a steering committee identify and assess all possible systems development projects that an organization unit could undertake. Next, those projects deemed most likely to yield significant organizational benefits, given available resources, are selected for subsequent development activities. Organizations vary in their approach to identifying and selecting projects. In some organizations, project identification and selection is a very formal process in which projects are outcomes of a larger overall planning process. For example, a large organization may follow a formal project identification process whereby a proposed project is rigorously compared to all competing projects. Alternatively, a small organization may use informal project selection processes that allow the highest-ranking IS manager to independently select projects or allow individual business units to decide on projects after agreeing to provide project funding.

There is a variety of sources for information systems development requests. One source is requests by managers and business units for replacing or extending an existing system to gain needed information or to provide a new service to customers. Another source for requests is IS managers who want to make a system more efficient, less costly to operate, or want to move it to a new operating environment. A final source of projects is a formal planning group that identifies projects for improvement to help the organization meet its corporate objectives (for example, a new system to provide better customer service). Regardless of how a given organization actually executes the project identification and selection process, there is a common sequence of activities that occurs. In the following sections, we describe a general process for identifying and selecting projects and producing the deliverables and outcomes of this process.

The Process of Identifying and Selecting IS Development Projects

As shown in Figure 5-1, project identification and selection consists of three primary activities:

1. Identifying potential development projects
2. Classifying and ranking projects
3. Selecting projects for development

Each of these steps is described below:

1. *Identifying Potential Development Projects.* Organizations vary as to how they identify projects. This process can be performed by

    - A key member of top management, either the CEO of a small- or medium-sized organization or a senior executive in a larger organization
    - A steering committee, composed of a cross section of managers with an interest in systems
    - User departments, in which either the head of the requesting unit or a committee from the requesting department decides which projects to submit (often you, as a systems analyst, will help users prepare such requests)
    - The development group or a senior IS manager

    All methods of identification have been found to have strengths and weaknesses. Research has found, for example, that projects identified by top management more often have a strategic organizational focus. Alternatively, projects identified by steering committees more often reflect the diversity of the committee and therefore have a cross-functional focus. Projects identified by individual departments or business units most often have a narrow, tactical focus. Finally, a dominant characteristic of projects

Figure 5-1
Systems development life cycle with project identification and selection highlighted

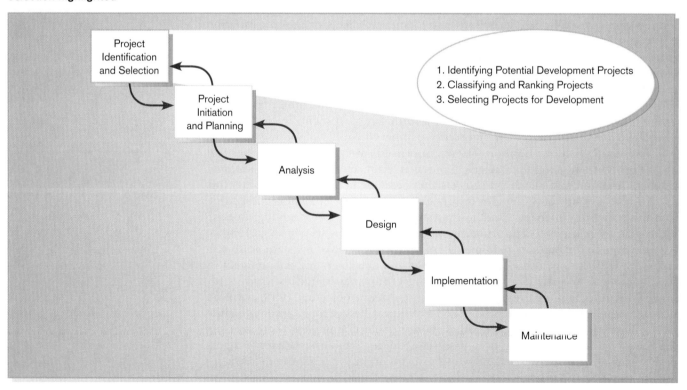

identified by the development group is the ease with which existing hardware and systems will integrate with the proposed project. Other factors, such as project cost, duration, complexity, and risk, are also influenced by the source of a given project. Characteristics of each selection method are briefly summarized in Table 5-1. In addition to who makes the decision, characteristics specific to the organization—such as the level of firm diversification, level of vertical integration, or extent of growth opportunities—can also influence any investment or project selection decision (Dewan, Michael, and Min, 1998).

Of all the possible project sources, those identified by top management and steering committees most often reflect the broader needs of the organization. This occurs because top management and steering committees are likely to have a broader understanding of overall business objectives and constraints. Projects identified by top management or by a diverse steering committee are therefore referred to as coming from a top-down source.

Projects identified by a functional manager, business unit, or by the information systems development group are often designed for a particular business need within a given business unit. In other words, these projects may not reflect the overall objectives of the organization. This does not mean that projects identified by individual managers, business units, or the IS development group are deficient, only that they may not consider broader organizational issues. Project initiatives stemming from managers, business units, or the development group are generally referred to as coming from a bottom-up source. These are the types of projects in which you, as a systems analyst, will have the earliest role in the life cycle as part of your ongoing support of users. You will help user managers provide the description of information needs and the reasons for doing the project that will be evaluated in selecting, among all submitted projects, which ones will be approved to move into the project initiation and planning phase of the SDLC.

In sum, projects are identified by both top-down and bottom-up initiatives. The formality of the process of identifying and selecting projects can vary substantially across organizations. Also, since limited resources preclude the development of all proposed systems, most organizations have some process of classifying and ranking the merit of each project. Those projects

TABLE 5-1 Characteristics of Alternative Methods for Making Information Systems Identification and Selection Decisions

Selection Method	Characteristics
Top Management	Greater strategic focus
	Largest project size
	Longest project duration
Steering Committee	Cross-functional focus
	Greater organizational change
	Formal cost-benefit analysis
	Larger and riskier projects
User Department	Narrow, nonstrategic focus
	Faster development
	Fewer users, management layers, and business functions
Development Group	Integration with existing systems focus
	Fewer development delays
	Less concern on cost-benefit analysis

Adapted from McKeen, Guimaraes, and Wetherbe, 1994

deemed to be inconsistent with overall organizational objectives, redundant in functionality to some existing system, or unnecessary will thus be removed from consideration. This topic is discussed next.

2. *Classifying and Ranking IS Development Projects.* The second major activity in the project identification and selection process focuses on assessing the relative merit of potential projects. As with the project identification process, classifying and ranking projects can be performed by top managers, a steering committee, business units, or the IS development group. Additionally, the criteria used when assigning the relative merit of a given project can vary. Commonly used criteria for assessing projects are summarized in Table 5-2. In any given organization, one or several criteria might be used during the classifying and ranking process.

As with the project identification and selection process, the actual criteria used to assess projects will vary by organization. If, for example, an organization uses a steering committee, it may choose to meet monthly or quarterly to review projects and use a wide variety of evaluation criteria. At these meetings, new project requests will be reviewed relative to projects already identified, and ongoing projects are monitored. The relative ratings of projects are used to guide the final activity of this identification process — project selection.

An important project evaluation method that is widely used for assessing information systems development projects is called **value chain analysis** (Porter, 1985; Shank and Govindarajan, 1993). Value chain analysis is the process of analyzing an organization's activities for making products and/or services to determine where value is added and costs are incurred. Once an organization gains a clear understanding of its value chain, improvements in the organization's operations and performance can be achieved. Information systems projects providing the greatest benefit to the value chain will be given priority over those with fewer benefits.

As you might have guessed, information systems have become one of the primary ways for organizations to make changes and improvements in their value chains. Many organizations, for example, are using the Internet to exchange important business information with suppliers and customers such as orders, invoices, and receipts. To conduct a value chain analysis for an organization, think about an organization as a big input/output process (see Figure 5-2). At one end are the inputs to the organization, for example, supplies that are purchased. Within the organizations, those supplies and resources are integrated in some way to produce products and services. At the other end are the outputs, which represent the products and services that are

Value chain analysis: Analyzing an organization's activities to determine where value is added to products and/or services and the costs incurred for doing so; usually also includes a comparison with the activities, added value, and costs of other organizations for the purpose of making improvements in the organization's operations and performance.

TABLE 5-2 Possible Evaluation Criteria When Classifying and Ranking Projects

Evaluation Criteria	Description
Value Chain Analysis	Extent to which activities add value and costs when developing products and/or services
Strategic Alignment	Extent to which the project is viewed as helping the organization achieve its strategic objectives and long-term goals
Potential Benefits	Extent to which the project is viewed as improving profits, customer service, and so forth and the duration of these benefits
Resource Availability	Amount and type of resources the project requires and their availability
Project Size/Duration	Number of individuals and the length of time needed to complete the project
Technical Difficulty/Risks	Level of technical difficulty to successfully complete the project within given time and resource constraints

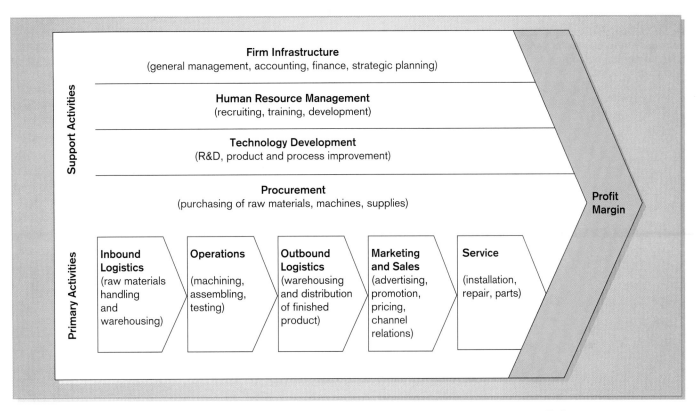

Figure 5-2
Organizational value chain
(Source: Reprinted with the permission of The Free Press, a Division of Simon & Schuster, Inc. from Competitive Advantage: Creating and Sustaining Superior Performance *by Michael E. Porter. Copyright © 1985 by Michael Porter.)*

marketed, sold, and then distributed to customers. In value chain analysis, you must first understand each activity, function, and process where value is or should be added. Next, determine the costs (and the factors that drive costs or cause them to fluctuate) within each of the areas. After understanding your value chain and costs, you can benchmark (compare) your value chain and associated costs with those of other organizations, preferably your competitors. By making these comparisons, you can identify priorities for applying information systems projects.

3. *Selecting IS Development Projects.* The final activity in the project identification and selection process is the actual selection of projects for further development. Project selection is a process of considering both short- and long-term projects and selecting those most likely to achieve business objectives. Additionally, as business conditions change over time, the relative importance of any single project may substantially change. Thus, the identification and selection of projects is a very important and ongoing activity.

   Numerous factors must be considered when making project selection decisions. Figure 5-3 shows that a selection decision requires that the perceived needs of the organization, existing systems and ongoing projects, resource availability, evaluation criteria, current business conditions, and the perspectives of the decision makers will all play a role in project selection decisions. Numerous outcomes can occur from this decision process. Of course, projects can be accepted or rejected. Acceptance of a project usually means that funding to conduct the next phase of the SDLC has been approved. Rejection means that the project will no longer be considered for development. However, projects may also be conditionally accepted; projects may be accepted pending the approval or availability of needed resources or the demonstration that a particularly difficult aspect of the system *can* be developed. Projects may also be returned to the original requesters who are told to

N E T   S E A R C H
Value chain management has become increasingly important as corporations strive to leverage the Internet for business advantage. Visit http://www.prenhall.com/hoffer to complete an exercise on this topic.

Figure 5-3
Project selection decisions must consider numerous factors and can have numerous outcomes

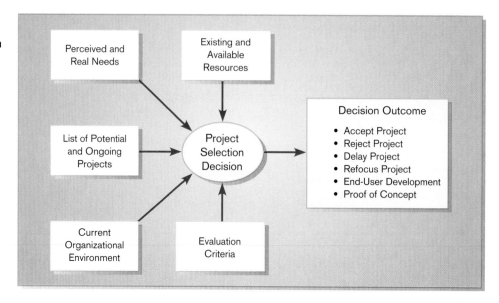

develop or purchase the requested system themselves. Finally, the requesters of a project may be asked to modify and resubmit their request after making suggested changes or clarifications.

Deliverables and Outcomes

The primary deliverable from the first SDLC phase is a schedule of specific IS development projects, coming from both top-down and bottom-up sources, to move into the next SDLC phase—project initiation and planning (see Figure 5-4). An outcome of this phase is the assurance that careful consideration was given to project selection, with a clear understanding of how each project can help the organization reach its objectives. Due to the principle of **incremental commitment**, a selected project does not necessarily result in a working system. After each subsequent SDLC

Incremental Commitment: A strategy in systems analysis and design in which the project is reviewed after each phase and continuation of the project is rejustified in each of these reviews.

Figure 5-4
Information systems development projects come from both top-down and bottom-up initiatives

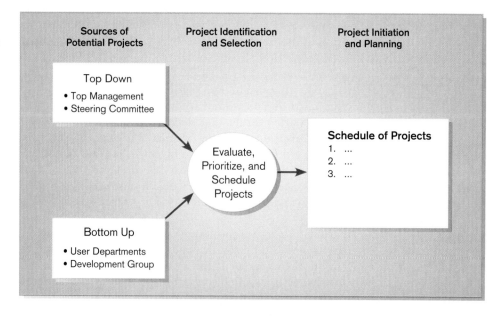

phase, you, other members of the project team, and organizational officials will reassess your project to determine whether the business conditions have changed or whether a more detailed understanding of a system's costs, benefits, and risks would suggest that the project is not as worthy as previously thought.

Many organizations have found that in order to make good project selection decisions and to provide sound guidance as issues arise in your work as a systems analyst on a project, a clear understanding of overall organizational business strategy and objectives is required. This means that a clear understanding of the business and the desired role of information systems in achieving organizational goals is a precondition to improving the identification and selection process. In the next section we provide a brief overview of the process many organizations follow, involving corporate strategic planning and information systems planning, when setting their business strategy and objectives and when defining the role of information systems in their plans.

CORPORATE AND INFORMATION SYSTEMS PLANNING

Although there are numerous motivations for carefully planning the identification and selection of projects (see Atkinson, 1990; Ross and Feeny, 2000), organizations have not traditionally used a systematic planning process when determining how to allocate IS resources. Instead, projects have often resulted from attempts to solve isolated organizational problems. In effect, organizations have asked the question: "What procedure (application program) is required to solve *this particular problem* as it exists today?" The difficulty with this approach is that the required organizational procedures are likely to change over time as the environment changes. For example, a company may decide to change its method of billing customers or a university may change its procedures for registering students. When such changes occur, it is usually necessary to again modify existing information systems.

In contrast, planning-based approaches essentially ask the question: "What information (or data) requirements will satisfy the decision-making needs or business processes of the enterprise today and well into the future?" A major advantage of this approach is that an organization's informational needs are less likely to change (or will change more slowly) than its business processes. For example, unless an organization fundamentally changes its business, its underlying data structures may remain reasonably stable for more than ten years. However, the procedures used to access and process the data may change many times during that period. Thus, the challenge of most organizations is to design comprehensive information models containing data that are relatively independent from the languages and programs used to access, create, and update them.

To benefit from a planning-based approach for identifying and selecting projects, an organization must analyze its information needs and plan its projects carefully. Without careful planning, organizations may construct databases and systems that support individual processes but do not provide a resource that can be easily shared throughout the organization. Further, as business processes change, lack of data and system integration will hamper the speed at which the organization can effectively make business strategy or process changes.

The need for improved information systems project identification and selection is readily apparent when we consider factors such as the following:

1. The cost of information systems has risen steadily and approaches 40 percent of total expenses in some organizations.

2. Many systems cannot handle applications that cross organizational boundaries.

3. Many systems often do not address the critical problems of the business as a whole nor support strategic applications.

4. Data redundancy is often out of control and users may have little confidence in the quality of data.

5. Systems maintenance costs are out of control as old, poorly planned systems must constantly be revised.

6. Application backlogs often extend three years or more and frustrated end users are forced to create (or purchase) their own systems, often creating redundant databases and incompatible systems in the process.

Careful planning and selection of projects alone will certainly not solve all of these problems. We believe, however, that a disciplined approach, driven by top management commitment, is a prerequisite to most effectively apply information systems in order to reach organizational objectives. The focus of this section is to provide you with a clear understanding of how specific development projects with a broader organizational focus can be identified and selected. Specifically, we describe corporate strategic planning and information systems planning, two processes that can significantly improve the quality of project identification and selection decisions. This section also outlines the types of information about business direction and general systems requirements that can influence the selection decisions and guide the direction of approved projects.

Corporate Strategic Planning

Corporate strategic planning: An ongoing process that defines the mission, objectives, and strategies of an organization.

A prerequisite to making effective project selection decisions is to gain a clear idea of where an organization is, its vision of where it wants to be in the future, and how to make the transition to its desired future state. Figure 5-5 represents this as a three-step process. The first step focuses on gaining an understanding of the current enterprise. In other words, if you don't know where you are, it is impossible to tell where you are going. Next, top management must determine where it wants the enterprise to be in the future. Finally, after gaining an understanding of the current and future enterprise, a strategic plan can be developed to guide this transition. The process of developing and refining models of the current and future enterprise as well as a transition strategy is often referred to as **corporate strategic planning**. During corporate strategic planning, executives typically develop a mission statement, statements of future corporate objectives, and strategies designed to help the organization reach its objectives.

Mission statement: A statement that makes it clear what business a company is in.

All successful organizations have a mission. The **mission statement** of a company typically states in very simple terms what business the company is in. For example, the mission statement for Pine Valley Furniture (PVF) is shown in Figure

Figure 5-5
Corporate strategic planning is a three-step process

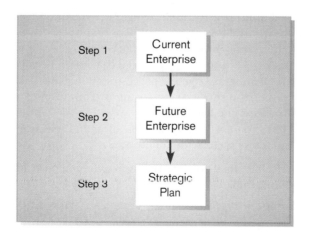

Figure 5-6
Mission statement (Pine Valley Furniture)

5-6. After reviewing the mission statement from PVF, it becomes clear that they are in the business of constructing and selling high-quality wood furniture to the general public, businesses, and institutions such as universities and hospitals. It is also clear that PVF is not in the business of fabricating steel file cabinets or selling their products through wholesale distributors. Based on this mission statement, you could conclude that PVF does not need a retail sales information system; instead, a high-quality human resource information system would be consistent with their goal.

After defining its mission, an organization can then define its objectives. The **objective statements** refer to "broad and timeless" goals for the organization. These goals can be expressed as a series of statements that are either qualitative or quantitative but that typically do not contain details likely to change substantially over time. Objectives are often referred to as "critical success factors." Here, we will simply use the term "objectives." The objectives for PVF are shown in Figure 5-7, with most relating to some aspect of the organizational mission. For example, objective number two relates to how PVF views its relationships with customers.

Objective statements: A series of statements that express an organization's qualitative and quantitative goals for reaching a desired future position.

Figure 5-7
Statement of corporate objectives (Pine Valley Furniture)

Pine Valley Furniture
Statement of Objectives

1. PVF will strive to increase market share and profitability (prime objective).
2. PVF will be considered a market leader in customer service.
3. PVF will be innovative in the use of technology to help bring new products to market faster than our competition.
4. PVF will employ the fewest number of the highest-quality people necessary to accomplish our prime objective.
5. PVF will create an environment that values diversity in gender, race, values, and culture among employees, suppliers, and customers.

TABLE 5-3   Generic Competitive Strategies

Strategy	Description
Low-Cost Producer	This strategy reflects competing in an industry on the basis of product or service cost to the consumer. For example, in the automobile industry, the South Korean-produced Hyundai is a product line that competes on the basis of low cost.
Product Differentiation	This competitive strategy reflects capitalizing on a key product criterion requested by the market (for example, high quality, style, performance, roominess). In the automobile industry, many manufacturers are trying to differentiate their products on the basis of quality (for example, "At Ford, quality is job one.").
Product Focus or Niche	This strategy is similar to both the low-cost and differentiation strategies but with a much narrower market focus. For example, a niche market in the automobile industry is the convertible sports car market. Within this market, some manufacturers may employ a low-cost strategy while others may employ a differentiation strategy on performance or style.

Adapted from Porter, 1980

Competitive strategy: The method by which an organization attempts to achieve its mission and objectives.

This goal would suggest that PVF might want to invest in electronic data interchange or on-line order status systems that would contribute to high-quality customer service. Once a company has defined its mission and objectives, a competitive strategy can be formulated.

A **competitive strategy** is the method by which an organization attempts to achieve its mission and objectives. In essence, the strategy is an organization's game plan for playing in the competitive business world. In his classic book on competitive strategy, Michael Porter (1980) defined three generic strategies—low-cost producer, product differentiation, and product focus or niche—for reaching corporate objectives (see Table 5-3). These generic strategies allow you to more easily compare two companies in the same industry that may not employ the same competitive strategy. In addition, organizations employing different competitive strategies often have different informational needs to aid decision making. For example, Rolls Royce and GEO are two car lines with different strategies: One is a high-prestige line in the ultra-luxury *niche* while the other is a relatively *low-priced* line for the general automobile market. Rolls Royce may build information systems to collect and analyze information on customer satisfaction to help manage a key company objective. Alternatively, GEO may build systems to track plant and material utilization in order to manage activities related to their low-cost strategy.

To effectively deploy resources such as the creation of a marketing and sales organization, *or to build the most effective information systems*, an organization must clearly understand its mission, objectives, and strategy. A lack of understanding will make it impossible to know which activities are essential to achieving business objectives. From an information systems development perspective, by understanding which activities are most critical for achieving business objectives, an organization has a much greater chance to identify those activities that need to be supported by information systems. In other words, **it is only through the clear understanding of the organizational mission, objectives, and strategies that IS development projects should be identified and selected**. The process of planning how information systems can be employed to assist organizations to reach their objectives is the focus of the next section.

N E T   S E A R C H

Establishing a Web strategy has become a hot topic for most organizations. Visit http://www.prenhall.com/hoffer to complete an exercise on this topic.

Information systems planning (ISP): An orderly means of assessing the information needs of an organization and defining the systems, databases, and technologies that will best satisfy those needs.

Information Systems Planning

The second planning process that can play a significant role in the quality of project identification and selection decisions is called **information systems planning (ISP)**. Information systems planning is an orderly means of assessing the information needs

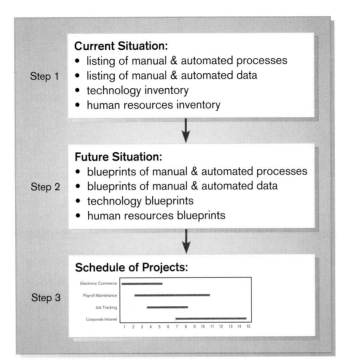

Figure 5-8
Information systems planning is a three-step process

of an organization and defining the information systems, databases, and technologies that will best satisfy those needs (Carlson, Gardner, and Ruth, 1989; Parker and Benson, 1989; Segars and Grover, 1999). This means that during ISP you (or, more likely, senior IS managers responsible for the IS plan) must model current and future organization informational needs, and develop strategies and *project plans* to migrate the current information systems and technologies to their desired future state. ISP is a top-down process that takes into account the outside forces—industry, economic, relative size, geographic region, and so on—critical to the success of the firm. This means that ISP must look at information systems and technologies in terms of how they help the business achieve its objectives defined during corporate strategic planning.

The three key activities of this modeling process are represented in Figure 5-8. Like corporate strategic planning, ISP is a three-step process in which the first step is to assess current IS-related assets—human resources, data, processes, and technologies. Next, target blueprints of these resources are developed. These blueprints reflect the desired future state of resources needed by the organization to reach its objectives as defined during strategic planning. Finally, a series of scheduled projects is defined to help move the organization from its current to its future desired state. (Of course, scheduled projects from the ISP process are just one source for projects. Others include bottom-up requests from managers and business units like the System Service Request in Figure 3-2.)

For example, a project may focus on reconfiguration of a telecommunications network to speed data communications, or may restructure work and data flows between business areas. Projects can include not only the development of new information systems or the modification of existing ones but also the acquisition and management of new systems, technologies, and platforms. These three activities parallel those of corporate strategic planning and this relationship is shown in Figure 5-9. Numerous methodologies such as Business Systems Planning (BSP) and Information Engineering (IE) have been developed to support the ISP process (see Segars and Grover, 1999); most contain the three key activities described below:

1. *Describing the Current Situation.* The most widely used approach for describing the current organizational situation is generically referred to as top-down

Figure 5-9
Parallel activities of corporate strategic planning and information systems planning

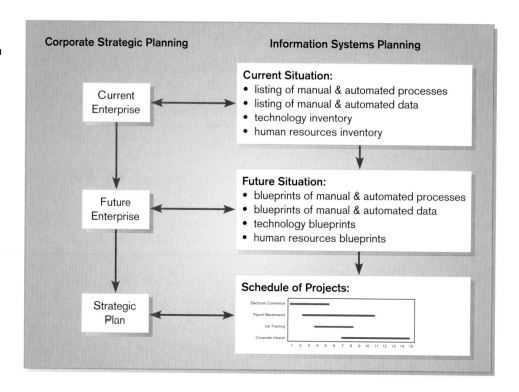

Top-down planning: A generic information systems planning methodology that attempts to gain a broad understanding of the information system needs of the entire organization.

Bottom-up planning: A generic information systems planning methodology that identifies and defines IS development projects based upon solving operational business problems or taking advantage of some business opportunities.

planning. **Top-down planning** attempts to gain a broad understanding of the informational needs of the entire organization. The approach begins by conducting an extensive analysis of the organization's mission, objectives, and strategy and determining the information requirements needed to meet each objective. This approach to ISP implies by its name a high-level organizational perspective with active involvement of top-level management. The top-down approach to ISP has several advantages over other planning approaches, and these are summarized in Table 5-4.

In contrast to the top-down planning approach, a **bottom-up planning** approach requires the identification of business problems and opportunities that are used to define projects. Using the bottom-up approach for creating IS

TABLE 5-4 Advantages to the Top-Down Planning Approach over Other Planning Approaches

Advantage	Description
Broader Perspective	If not viewed from the top, information systems may be implemented without first understanding the business from general management's viewpoint.
Improved Integration	If not viewed from the top, totally new management information systems may be implemented rather than planning how to evolve existing systems.
Improved Management Support	If not viewed from the top, planners may lack sufficient management acceptance of the role of information systems in helping them achieve business objectives.
Better Understanding	If not viewed from the top, planners may lack the understanding necessary to implement information systems across the entire business rather than simply to individual operating units.

IBM, 1982; pp. 236–37

plans can be faster and less costly to develop than using the top-down approach and also has the advantage of identifying pressing organizational problems. Yet, the bottom-up approach often fails to view the informational needs of the *entire* organization. This can result in the creation of disparate information systems and databases that are redundant or not easily integrated without substantial rework.

The process of describing the current situation begins by selecting a planning team that includes executives chartered to model the existing situation. To gain this understanding, the team will need to review corporate documents; interview managers, executives, and customers; and conduct detailed reviews of competitors, markets, products, and finances. The type of information that must be collected to represent the current situation includes the identification of all organizational locations, units, functions, processes, data (or data entities), and information systems.

Within Pine Valley Furniture, for example, organizational locations would consist of a list of all geographic areas in which the organization operates (for example, the locations of the home and branch offices). Organizational units represent a list of people or business units that operate within the organization. Thus, organizational units would include vice president manufacturing, sales manager, sales person, and clerk. Functions are cross-organizational collections of activities used to perform day-to-day business operations. Examples of business functions might include research and development, employee development, purchasing, and sales. Processes represent a list of manual or automated procedures designed to support business functions. Examples of business processes might include payroll processing, customer billing, and product shipping. Data entities represent a list of the information items generated, updated, deleted, or used within business processes. Information systems represent automated and nonautomated systems used to transform data into information to support business processes. For example, Figure 5-10 shows portions of the business functions, data entities, and information systems of PVF. Once high-level information is collected, each item can typically be decomposed into smaller units as more detailed planning is performed. Figure 5-11 shows the decomposition of several of PVF's high-level business functions into more detailed supporting functions.

PINE VALLEY FURNITURE

After creating these lists, a series of matrices can be developed to cross-reference various elements of the organization. The types of matrices typically developed include the following:

- *Location-to-Function:* This matrix identifies which business functions are being performed at various organizational locations.

- *Location-to-Unit:* This matrix identifies which organizational units are located in or interact with a specific business location.

N E T   S E A R C H
The Web is requiring businesses to think more globally about their Web strategy. Visit http://www.prenhall.com/hoffer to complete an exercise on this topic.

FUNCTIONS:	DATA ENTITIES:	INFORMATION SYSTEMS:
• business planning	• customer	• payroll processing
• product development	• product	• accounts payable
• marketing and sales	• vendor	• accounts receivable
• production operations	• raw material	• time card processing
• finance and accounting	• order	• inventory management
• human resources	• invoice	...
...	• equipment	
	...	

Figure 5-10
Information systems planning information (Pine Valley Furniture)

Figure 5-11
Functional decomposition of
information systems planning
information (Pine Valley Furniture)

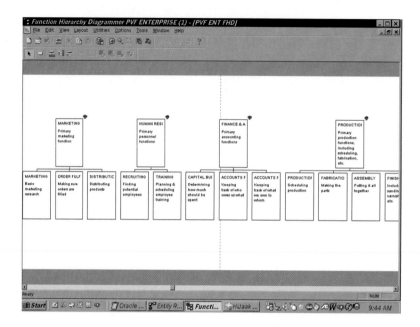

- *Unit-to-Function:* This matrix identifies the relationships between organizational entities and each business function.
- *Function-to-Objective:* This matrix identifies which functions are essential or desirable in achieving each organizational objective.
- *Function-to-Process:* This matrix identifies which processes are used to support each business function.
- *Function-to-Data Entity:* This matrix identifies which business functions utilize which data entities.
- *Process-to-Data Entity:* This matrix identifies which data are captured, used, updated, or deleted within each process.
- *Process-to-Information System:* This matrix identifies which information systems are used to support each process.
- *Data Entity-to-Information System:* This matrix identifies which data are created, updated, accessed, or deleted in each system.
- *Information System-to-Objective:* This matrix identifies which information systems support each business objective as identified during organizational planning.

   Different matrices will have different relationships depending upon what is being represented. For example, Figure 5-12 shows a portion of the Data Entity-to-Function matrix for Pine Valley Furniture. The "X" in various cells of the matrix represents which business functions utilize which data entities. A more detailed picture of data utilization would be shown in the Process-to-Data Entity matrix (not shown here), where the cells would be coded as "C" for the associated process that creates or captures data for the associated data entity, "R" for retrieve (or used), "U" for update, and "D" for delete. This means that different matrices can have different relationships depending upon what is being represented. Because of this flexibility and ease for representing information, analysts use a broad range of matrices to gain a clear understanding of an organization's current situation and to plan for its future (Kerr, 1990). A primer on using matrices for information systems planning is provided in Figure 5-13.

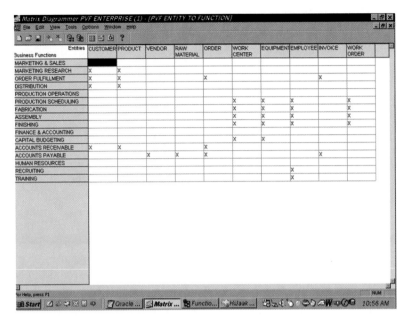

Figure 5-12
Data Entity-to-Function matrix (Pine Valley Furniture)

Business Functions	CUSTOMER	PRODUCT	VENDOR	RAW MATERIAL	ORDER	WORK CENTER	EQUIPMENT	EMPLOYEE	INVOICE	WORK ORDER
MARKETING & SALES										
MARKETING RESEARCH	X	X								
ORDER FULFILLMENT	X	X			X			X		
DISTRIBUTION	X	X								
PRODUCTION OPERATIONS										
PRODUCTION SCHEDULING						X	X	X		X
FABRICATION						X	X	X		X
ASSEMBLY						X	X	X		X
FINISHING						X	X	X		X
FINANCE & ACCOUNTING										
CAPITAL BUDGETING						X	X			
ACCOUNTS RECEIVABLE	X	X			X					
ACCOUNTS PAYABLE			X	X	X				X	
HUMAN RESOURCES										
RECRUITING								X		
TRAINING								X		

Figure 5-13
Making sense out of planning matrices

During the information systems planning process, before individual projects are identified and selected, a great deal of "behind the scenes" analysis takes place. During this planning period, which can span from six months to a year, IS planning team members develop and analyze numerous matrices like those described in the associated text. Matrices are developed to represent the current and the future views of the organization. Matrices of the "current" situation are called "as is" matrices. In other words, they describe the world "as" it currently "is." Matrices of the target or "future" situation are called "to be" matrices. Contrasting the current and future views provides insights into the relationships existing in important business information, and most importantly, forms the basis for the identification and selection of specific development projects. Many CASE tools provide features that will help you make sense out of these matrices in at least three ways:

1. **Management of Information.** A big part of working with complex matrices is managing the information. Using the dictionary features of the CASE tool repository, terms (such as business functions and process and data entities) can be defined or modified in a single location. All planners will therefore have the most recent information.
2. **Matrix Construction.** The reporting system within the CASE repository allows matrix reports to be easily produced. Since planning information can be changed at any time by many team members, an easy method to record changes and produce the most up-to-date reports is invaluable to the planning process.
3. **Matrix Analysis.** Possibly the most important feature CASE tools provide to planners is the ability to perform complex analyses within and across matrices. This analysis is often referred to as **affinity clustering.** Affinity refers to the extent to which information holds things in common. Thus, affinity clustering is the process of arranging matrix information so that clusters of information with some predetermined level or type of affinity are placed next to each other on a matrix report. For example, an affinity clustering of a Process-to-Data Entry matrix would create roughly a block diagonal matrix with processes that use similar data entities appearing in adjacent rows and data entities used in common by the same processes grouped into adjacent columns. This general form of analysis can be used by planners to identify items that often appear together (or should!). Such information can be used by planners to most effectively group and relate information (e.g., data to processes, functions to locations, and so on). For example, those data entities used by a common set of processes are candidates for a specific database. And those business processes that relate to a strategically important objective will likely receive more attention when managers from those areas request system changes.

Affinity clustering: The process of arranging planning matrix information so the clusters of information with some predetermined level or type of affinity are placed next to each other on a matrix report.

2. *Describing the Target Situation, Trends, and Constraints.* After describing the current situation, the next step in the ISP process is to define the target situation that reflects the desired future state of the organization. This means that the target situation consists of the desired state of the locations, units, functions, processes, data, and information systems (see Figure 5-8). For example, if a desired future state of the organization is to have several new branch offices or a new product line that require several new employee positions, functions, processes, and data, then most lists and matrices will need to be updated to reflect this vision. The target situation must be developed in light of technology and business trends, in addition to organizational constraints. This means that lists of business trends and constraints should also be constructed in order to help assure that the target situation reflects these issues.

In summary, to create the target situation, planners must first edit their initial lists and record the *desired* locations, units, functions, processes, data, and information systems within the constraints and trends of the organization environment (for example, time, resources, technological evolution, competition, and so on). Next, matrices are updated to relate information in a manner consistent with the desired future state. Planners then focus on the *differences* between the current and future lists and matrices to identify projects and transition strategies.

3. *Developing a Transition Strategy and Plans.* Once the creation of the current and target situations is complete, a detailed transition strategy and plan are developed by the IS planning team. This plan should be very comprehensive, reflecting both broad, long-range issues in addition to providing sufficient detail to guide all levels of management concerning what needs doing, how, when, and by whom in the organization. The components of a typical information systems plan are outlined in Figure 5-14.

The IS plan is typically a very comprehensive document that looks at both short- and long-term organizational development needs. The short- and long-term developmental needs identified in the plan are typically expressed as a series of projects (see Figure 5-15). Projects from the long-term plan tend to build a foundation for later projects (such as transforming databases from old technology into newer technology). Projects from the short-term plan consist of specific steps to fill the gap between current and desired systems or respond to dynamic business conditions. The top-down (or plan-driven) projects join a set of bottom-up or needs-driven projects submitted as system service requests from managers to form the short-term systems development plan. Collectively, the short- and long-term projects set clear directions for the project selection process. The short-term plan includes not only those projects identified from the planning process but also those selected from among bottom-up requests. The overall IS plan may also influence all development projects. For example, the IS mission and IS constraints may cause projects to choose certain technologies or emphasize certain application features as systems are designed.

In this section, we outlined a general process for developing an IS plan. ISP is a detailed process and an integral part of deciding how to best deploy information systems and technologies to help reach organizational goals. It is beyond the scope of this chapter, however, to extensively discuss ISP, yet it should be clear from our discussion that planning-based project identification and selection will yield substantial benefits to an organization. It is probably also clear to you that, as a systems analyst, you are not usually involved in IS planning, since this process requires senior IS and corporate management participation. On the other hand, the results of IS planning,

I. **Organizational Mission, Objectives, and Strategy**
   Briefly describes the mission, objectives, and strategy of the organization. The current and future views of the company are also briefly presented (i.e., where are we, where we want to be).

II. **Informational Inventory**
   This section provides a summary of the various business processes, functions, data entities, and information needs of the enterprise. This inventory will view both current and future needs.

III. **Mission and Objectives of Information Systems**
   Description of the primary role IS will play in the organization to transform the enterprise from its current to future state. While it may later be revised, it represents the current best estimate of the overall role for IS within the organization. This role may be as a necessary cost, an investment, or a strategic advantage, for example.

IV. **Constraints on IS Development**
   Briefly describes limitations imposed by technology and current level of resources within the company—financial, technological, and personnel.

V. **Overall Systems Needs and Long-Range IS Strategies**
   Presents a summary of the overall systems needed within the company and the set of long-range (2–5 years) strategies chosen by the IS department to fill the needs.

VI. **The Short-Term Plan**
   Shows a detailed inventory of present projects and systems and a detailed plan of projects to be developed or advanced during the current year. These projects may be the result of the long-range IS strategies or of requests from managers that have already been approved and are in some stage of the life cycle.

VII. **Conclusions**
   Contains likely but not-yet-certain events that may affect the plan, an inventroy of business change elements as presently known, and a description of their estimated impact on the plan.

Figure 5-14
Outline of an information systems plan

such as planning matrices like that in Figure 5-12, can be a source of very valuable information as you identify and justify projects.

The planning process that we have described requires collecting, organizing, and analyzing vast amounts of data. This process can be eased by using upper CASE tools to more effectively deal with the vast amounts of information and to assist in providing (and enforcing) a common methodology throughout the process. In short, as information is collected during information systems planning, for example, it can be stored in the CASE repository. This information can then be summarized as needed to help identify and select specific development projects. For example, a particular matrix, like Information System-to-Objective, could be generated from the repository to help identify projects that address upgrading or replacing a specific system in order to better achieve prime objectives. In this simple way CASE can be used to improve the project identification and selection process. For more information on the role of CASE during the project identification and selection process, see Chapter 4.

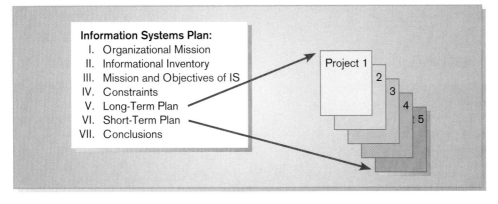

Figure 5-15
Systems development projects flow from the information systems plan

Information Systems Plan:
 I. Organizational Mission
 II. Informational Inventory
 III. Mission and Objectives of IS
 IV. Constraints
 V. Long-Term Plan
 VI. Short-Term Plan
 VII. Conclusions

Project 1
2
3
4
5

ELECTRONIC COMMERCE APPLICATION: IDENTIFYING AND SELECTING SYSTEMS DEVELOPMENT PROJECTS

Identifying and selecting systems development projects for an Internet-based electronic commerce application is no different than the process followed for more traditional applications. Nonetheless, there are some special considerations when developing an Internet-based application. In this section, we highlight some of those issues that relate directly to the process of identifying and selecting systems development projects.

Internet Basics

The name **"Internet"** is derived from the concept of "internetworking," that is, connecting host computers and their networks to form an even larger, global network. And that is essentially what the Internet is—a large worldwide network of networks that use a common protocol to communicate with each other. The interconnected networks include UNIX, IBM, Novell, Apple, and many other network and computer types. The Internet stands as the most prominent representation of global networking. Using the Internet to support day-to-day business activities is broadly referred to as **electronic commerce** (EC). However, not all Internet EC applications are the same. For example, there are three general classes of Internet EC applications; typically referred to as Internets, Intranets, and Extranets. Figure 5-16 shows three possible modes of EC using the Internet. The term used to describe transactions between individuals and businesses is *Internet*-based EC. So, the term Internet is used to refer to both the global computing network and to business-to-consumer EC applications, or commonly as just "B-to-C." **Intranet** refers to the use of the Internet within the same business and **Extranet** refers to the use of the Internet between firms. Extranet EC is commonly referred to as "B-to-B" since it is business-to-business EC.

Intranets and Extranets are examples of two ways organizations have been communicating via technology for years. For example, Intranets are a lot like having a "global" local area network. Organizations with Intranets dictate what applications will run over the Intranet—such as electronic mail or an inventory control system—as well as dictate the speed and quality of the hardware connected to the Intranet. In other words, Intranets are a new twist—a global twist—to an old way of using information systems to support business activities within a single organization. Likewise, Extranets are also similar to an established computing model, **electronic data interchange (EDI)**. EDI refers to the use of telecommunications technologies to directly transfer business documents between organizations. Using EDI, trading partners (suppliers, manufacturers, customers, etc.) establish computer-to-computer links that allow them to exchange data electronically. For example, a company using EDI may send an electronic purchase order to a supplier instead of a paper request. The paper order may take several days to arrive at

Internet: A large worldwide network of networks that use a common protocol to communicate with each other; a global computing network to support business-to-consumer electronic commerce.

Electronic commerce: Internet-based communication to support day-to-day business activities.

Intranet: Internet-based communication to support business activities within a single organization.

Extranet: Internet-based communication to support business-to-business activities.

Electronic data interchange (EDI): The use of telecommunications technologies to directly transfer business documents between organizations.

Figure 5-16
Three possible modes of electronic commerce

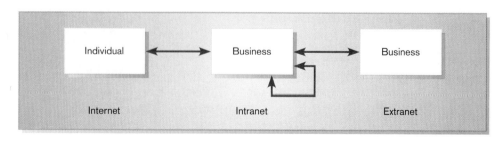

TABLE 5-5 Unknowns That Must Be Dealt with When Designing and Building Internet Applications

User	• Concern: Who is the user?
	• Example: Where is the user located? What is their expertise, education, or expectations?
Connection Speed	• Concern: What is the speed of the connection and what information can be effectively displayed?
	• Example: Modem, Cable Modem, DSL, Satellite, Broadband, Cellular.
Access Method	• Concern: What is the method of accessing the net?
	• Example: Web browser, Personal Digital Assistant (PDA), Web-enabled Cellular Phone, Web-enabled Television.

the supplier, whereas an EDI purchase order will only take a few seconds. EDI is fast becoming the standard by which organizations will communicate with each other in the world of electronic commerce.

When developing either an Intranet or Extranet, developers know who the users are, what applications will be used, the speed of the network connection, and the type of communication devices (e.g., Web browsers like Netscape or Internet Explorer, personal digital assistants like a Palm Pilot, or a Web-enabled cellular phone). On the other hand, when developing an Internet EC application (hereafter, simply EC), there are countless unknowns that developers have to discern in order to build a useful system. Table 5-5 lists a sample of the numerous unknowns to be dealt with when designing and building an EC application. These unknowns may result in making trade-offs based on a careful analysis of who the users are likely to be, where they are likely to be located, and how they are likely to be connected to the Internet. Even with all these difficulties to contend with, there is no shortage of Internet EC applications springing up all across the world. One company that has decided to get onto the Web with their own EC site is Pine Valley Furniture.

Pine Valley Furniture WebStore

PINE VALLEY FURNITURE

The board of directors of PVF has requested that a project team be created to explore the opportunity to develop an EC system. Specifically, market research has found that there is a good opportunity for online furniture purchases, especially in the areas of:

- Corporate furniture buying
- Home office furniture purchasing
- Student furniture purchasing

The board wants to incorporate all three target markets into their long-term EC plan, but wants to initially focus on the corporate furniture buying system. They feel that this segment has the greatest potential to provide an adequate return on investment and would be a good building block for moving into the customer-based markets. Because the corporate furniture buying system will be specifically targeted to the business furniture market, it will be easier to define the system's operational requirements. Additionally, this EC system should integrate nicely with two currently existing systems, Purchasing Fulfillment and Customer Tracking. Together, these attributes make it an ideal candidate for initiating PVF's Web strategy. Throughout the remainder of the book, we will follow the evolution of the WebStore project until it becomes operational for PVF.

Summary

In this chapter, we described the first phase of the SDLC—project identification and selection. Project identification and selection consists of three primary activities: identifying potential development projects, classifying and ranking projects, and selecting projects for development. A variety of organizational members or units can be assigned to perform this process including top management, a diverse steering committee, business units and functional managers, the development group, or the most senior IS executive. Potential projects can be evaluated and selected using a broad range of criteria such as value chain analysis, alignment with business strategy, potential benefits, resource availability and requirements, and risks.

The quality of the project identification and selection process can be improved if decisions are guided by corporate strategic planning and information systems planning. Corporate strategic planning is the process of identifying the mission, objectives, and strategies of an organization. Crucial in this process is selecting a competitive strategy that states how the organization plans to achieve its objectives.

Information systems planning is an orderly means for assessing the information needs of an organization and defining the systems and databases that will best satisfy those needs. ISP is a top-down process that takes into account outside forces that drive the business and the factors critical to the success of the firm. ISP evaluates the current inventory of systems, the desired future state of the organization and its system, and determines which projects are needed to transform systems to meet the future needs of the organization.

Corporate and IS planning are highly interrelated. Conceptually, these relationships can be viewed via various matrices that show how organizational objectives, locations, units, functions, processes, data entities, and systems relate to one another. Selected projects will be those viewed to be most important in supporting the organizational strategy.

The Internet is a global network consisting of thousands of interconnected individual networks that communicate with each other using TCP/IP. Electronic commerce (EC) refers to the use of the Internet to support day-to-day business activities. Internet-based EC are transactions between individuals and businesses. Intranet refers to the use of the Intranet within the same organization. Extranet refers to the use of the Internet between firms.

The focus of this chapter was to provide you with a clearer understanding of how organizations identify and select projects. The need for improved project identification and selection is apparent for reasons such as the following: the cost of information systems is rising rapidly, systems cannot handle applications that cross organizational boundaries, systems often do not address critical organizational objectives, data redundancy is often out of control, and system maintenance costs continue to rise. Thus, effective project identification and selection is essential if organizations are to realize the greatest benefits from information systems.

Key Terms

1. Affinity clustering
2. Bottom-up planning
3. Competitive strategy
4. Corporate strategic planning
5. Electronic commerce
6. Electronic Data Interchange (EDI)
7. Extranet
8. Incremental commitment
9. Information systems planning (ISP)
10. Internet
11. Intranet
12. Mission statement
13. Objective statements
14. Top-down planning
15. Value chain analysis

Match each of the key terms above with the definition that best fits it.

_____ Analyzing an organization's activities to determine where value is added to products and/or services and the costs incurred for doing so.

_____ A strategy in systems analysis and design in which the project is reviewed after each phase and continuation of the project is rejustified in each of these reviews.

_____ An ongoing process that defines the mission, objectives, and strategies of an organization.

_____ A statement that makes it clear what business a company is in.

_____ A series of statements that express an organization's qualitative and quantitative goals for reaching a desired future position.

_____ The method by which an organization attempts to achieve its mission and objectives.

_____ An orderly means of assessing the information needs of an organization and defining the systems, databases, and technologies that will best satisfy those needs.

_____ A generic information systems planning methodology that attempts to gain a broad understanding of the information system needs of the entire organization.

_____ A generic information systems planning methodology that identifies and defines IS development projects based upon solving operational business problems or taking advantage of some business opportunities.

_____ The process of arranging planning matrix information so the clusters of information with some predetermined level or type of affinity are placed next to each other on a matrix report.

_____ A large worldwide network of networks that use a common protocol to communicate with each other.

_____ Internet-based communication to support day-to-day business activities.

_____ Internet-based communication to support business activities within a single organization.

_____ Internet-based communication to support business-to-business activities.

_____ The use of telecommunications technologies to directly transfer business documents between organizations.

Review Questions

1. Contrast the following terms:
   a. mission; objective statements; competitive strategy
   b. corporate strategic planning; information systems planning
   c. top-down planning; bottom-up planning
   d. low-cost producer; product differentiation; product focus or niche
   e. data entity; information system

2. Describe the project identification and selection process.

3. Describe several project evaluation criteria.

4. Describe value chain analysis and how organizations use this technique to evaluate and compare projects.

5. Discuss several factors that provide evidence for the need for improved information systems planning today.

6. Describe the steps involved in corporate strategic planning.

7. What are three generic competitive strategies?

8. Describe what is meant by information systems planning and the steps involved in the process.

9. List and describe the advantages of top-down planning over other planning approaches.

10. Briefly describe nine planning matrices that are used in information systems planning and project identification and selection.

11. Discuss some of the factors that must be considered when designing and building Internet applications.

Problems and Exercises

1. Write a mission statement for a business that you would like to start. The mission statement should state the area of business you will be in and what aspect of the business you value highly.

2. When you are happy with the mission statement you have developed in response to the prior question, describe the objectives and competitive strategy for achieving that mission.

3. Consider an organization that you believe does not conduct adequate strategic IS planning. List at least six reasons why this type of planning is not done appropriately (or is not done at all). Are these reasons justifiable? What are the implications of this inadequate strategic IS planning? What limits, problems, weaknesses, and barriers might this present?

4. IS planning, as depicted in this chapter, is highly related to corporate strategic planning. What might those responsible for IS planning have to do if they operate in an organization without a formal corporate planning process?

5. The economic analysis carried out during the project identification and selection phase of the systems development life cycle is rather cursory. Why is this? Consequently, what factors do you think tend to be most important for a potential project to survive this first phase of the life cycle?

6. In those organizations that do an excellent job of IS planning, why might projects identified from a bottom-up process still find their way into the project initiation and planning phase of the life cycle?

7. Figure 5-14 introduces the concept of affinity clustering. Suppose that through affinity clustering it was found that three business functions provided the bulk of the use of five data entities. What implications might this have for project identification and subsequent steps in the systems development life cycle?

8. Timberline Technology manufactures membrane circuits in its northern California plant. In addition, all circuit design and R&D work occur at this site. All finance, accounting, and

human resource functions are headquartered at the parent company in the upper-Midwest. Sales take place through six sales representatives located in various cities across the country. Information systems for payroll processing, accounts payable, and accounts receivable are located at the parent office while systems for inventory management and computer-integrated manufacturing are at the California plant. As best you can, list the locations, units, functions, processes, data entities, and information systems for this company.

9. For each of the following categories, create the most plausible planning matrices for Timberline Technology, described in Problem and Exercise 8: function-to-data entity, process-to-data entity, process-to-information system, data entity-to-information system. What other information systems not listed is Timberline likely to need?

10. The owners of Timberline Technology (described in Problem and Exercise 8) are considering adding a plant in Idaho and one in Arizona and six more sales representatives at various sites across the country. Update the matrices from Problem and Exercise 9 so that the matrices account for these changes.

Field Exercises

1. Obtain a copy of an organization's mission statement. (One can typically be found in the organization's Annual Report, which are often available in university libraries or in corporate marketing brochures. If you are finding it difficult to locate this material, write or call the organization directly and ask for a copy of the mission statement.) What is this organization's area of business? What does the organization value highly (e.g., high-quality products and services, low cost to consumers, employee growth and development, etc.)? If the mission statement is well written, these concepts should be clear. Do you know anything about the information systems in this company that would demonstrate that the types of systems in place reflect the organization's mission? Explain.

2. Interview the managers of the information systems department of an organization to determine the level and nature of their strategic information systems planning. Does it appear to be adequate? Why or why not? Obtain a copy of that organization's mission statement. To what degree do the strategic IS plan and the organizational strategic plan fit together? What are the areas where the two plans fit and do not fit? If there is not a good fit, what are the implications for the success of the organization? For the usefulness of their information systems?

3. Choose an organization that you have contact with, perhaps your employer or university. Follow the "Outline of an information systems plan" shown in Figure 5-15 and complete a short information systems plan for the organization you chose. Write at least a brief paragraph for each of the seven categories in the outline. If IS personnel and managers are available, interview them to obtain information you need. Present your mock plan to the organization's IS manager and ask for feedback on whether or not your plan fits the IS reality for that organization.

4. Choose an organization that you have contact with, perhaps your employer or university. List significant examples for each of the items used to create planning matrices. Next, list possible relationships among various items and display these relationships in a series of planning matrices.

5. Write separate mission statements that you believe fit well for Microsoft, IBM, and AT&T. Compare your mission statements with the real mission statements of these companies. Their mission statements can typically be found in their Annual Reports. Were your mission statements comparable to the real mission statements? Why or why not? What differences and similarities are there among these three mission statements? What information systems are necessary to help these companies deliver on their mission statements?

6. Choose an organization that you have contact with, perhaps your employer or university. Determine how information systems projects are identified. Are projects identified adequately? Are they identified as part of the information systems planning or the corporate strategic planning process? Why or why not?

References

Atkinson, R. A. 1990. "The Motivations for Strategic Planning." *Journal of Information Systems Management* 7 (4): 53–56.

Carlson, C. K., E. P. Gardner, and S. R. Ruth. 1989. "Technology-Driven Long-Range Planning." *Journal of Information Systems Management* 6 (3): 24–29.

Dewan, S., Michael, S. C., and Min, C-K. 1998. "Firm Characteristics and Investments in Information Technology: Scale and Scope Effects," *Information Systems Research* 9(3): 219–232.

Hasselbring, W. 2000. "Information System Integration," *Communications of the ACM* 43(6): 33–38.

IBM. 1982. "Business Systems Planning." In *Advanced System Development/Feasibility Techniques*, edited by J. D. Couger, M. A. Colter, and R. W. Knapp. New York, NY: Wiley: 236–314.

Kerr, J. 1990. "The Power of Information Systems Planning." *Database Programming & Design* 3 (December): 60–66.

McKeen, J. D., T. Guimaraes, and J. C. Wetherbe. 1994. "A Comparative Analysis of MIS Project Selection Mechanisms." *Data Base* 25 (February): 43–59.

Parker, M. M., and R. J. Benson. 1989. "Enterprisewide Information Management: State-of-the-Art Strategic Planning." *Journal of Information Systems Management* 6 (Summer): 14–23.

Porter, M. 1980. *Competitive Strategy: Techniques for Analyzing Industries and Competitors.* New York, NY: Free Press.

Porter, M. 1985. *Competitive Advantage.* New York, NY: Free Press.

Ross, J., and Feeny, D. 2000. "The Evolving Role of the CIO." In *Framing the Domains of IT Management: Projecting the Future from the Past,* edited by R. W. Zmud. Cincinnati, OH: Pinnaflex Educational Resources: Chapter 19: 385–402.

Segars, A. H., and Grover, V. 1999. "Profiles of Strategic Information Systems Planning," *Information Systems Planning* 10(3): 199–232.

Shank, J. K., and V. Govindarajan. 1993. *Strategic Cost Management.* New York, NY: Free Press.

BROADWAY ENTERTAINMENT COMPANY, INC.

Identifying and Selecting the Customer Relationship Management System

CASE INTRODUCTION

Carrie Douglass graduated from Stillwater State University with a Bachelors of Arts degree in Marketing. Among the courses Carrie took at Stillwater were several on information technology in marketing, including one on electronic commerce. While at Stillwater, Carrie worked part-time as an assistant manager at the Broadway Entertainment Company (BEC) store in Centerville, Ohio, a suburb of Dayton. After graduation, Carrie was recruited by BEC for a full-time position because of her excellent job experience at BEC and her outstanding record in classes and student organizations at Stillwater. Carrie immediately entered the BEC Manager Development Program, which consisted of three months of training, observation of experienced managers at several stores, and work experience.

The first week of training was held at the BEC regional headquarters in Columbus, Ohio. Carrie learned about company procedures and policies, trends in the home entertainment industry, and personnel practices used in BEC stores. It was during this week that Carrie was introduced to the BEC Blueprint for the Decade, a vision statement for the firm, as shown in BEC Figure 5-1.

The Blueprint, as it is called, seemed rather abstract to Carrie while in training. Carrie saw a video in which Nigel Broad, BEC's chairman, explained the importance of the Blueprint. Nigel was very sincere and clearly passionate about BEC's future hinging on every employee finding innovative ways for BEC to achieve the vision outlined in the Blueprint.

After the three-month development program was over, Carrie was surprised to be appointed manager of the Centerville store. The previous manager was promoted to a marketing position in Columbus, which created this opportunity. Carrie started her job with enthusiasm, wanting to apply what she had learned at Stillwater and in the Management Development Program.

THE IDEA FOR A NEW SYSTEM

Although confident in her skills, Carrie believes that learning never stops. So, she logged onto the Amazon.com Web site one night from her home computer to look for some books on trends in retail marketing. While on the Web site, Carrie saw that Amazon.com was selling some of the same products BEC sells and rents in its stores. She had visited the BEC Website often. Although a rich source of information about the company (she had found her first job with BEC from a job posting on the company's Website), BEC was not engaged in electronic commerce with customers. All of a sudden, the words of the BEC Blueprint for the Decade started to come to life for Carrie. The Blueprint said that "BEC will be a leader in all areas of our business—human resources, *technology, operations,* and *marketing*." And, "BEC will be innovative in the use of technology . . . to provide better service to our customers." These statements caused Carie to recall a conversation she had in the store just that day with a mother of several young children.

The mother, a frequent BEC customer, had complimented Carrie on the cleanliness of the store and efficiency of checkout. The mother added, however, that she wished BEC better understood all her needs. For example, she allowed her children to pick out movies and games, but she found that the industry rating system was not always consistent with her wishes. It would be great if she and other parents could submit and view comments about videos and games. This way, parents would be more aware of the content of the products and the reactions of other children to these products. Carrie wondered why this kind of information couldn't be placed on a Website for anyone to use. Probably the comments made by parents shopping at the Centerville store would be different from those of parents shopping at other stores, so it seemed to make sense that this information service should be a part of local store operations.

One of the books Carrie found on Amazon.com discussed customer relationship marketing. This seemed like exactly what the mother wanted from BEC. The mother didn't want just products and services; rather she wanted a store that understood and supported all of her needs for home entertainment. She wanted the store to relate to her, not just sell and rent products to her and her children.

BLUEPRINT FOR THE DECADE

FOREWORD

This blueprint provides guidance to Broadway Entertainment Company (BEC) for this coming decade. It shows our vision for the firm—our mission, objectives, and strategy fit together—and provides direction for all individuals and decisions of the firm.

OUR MISSION

BEC is a publicly held, for-profit organization focusing on the home entertainment industry that has a global focus for operations. BEC exists to serve customers with a primary goal of enhancing shareholders' investment through the pursuit of excellence in everything we do. BEC will operate under the highest ethical standards; will respect the dignity, rights, and contributions of all employees; and will strive to better society.

OUR OBJECTIVES

1. BEC will strive to increase market share and profitability (prime objective).
2. BEC will be a leader in all areas of our business—human resources, technology, operations, and marketing.
3. BEC will be cost-effective in the use of all resources.
4. BEC will rank among industry leaders in both profitability and growth.
5. BEC will be innovative in the use of technology to help bring new products and services to market faster than our competition and to provide better service to our customers.
6. BEC will create an environment that values diversity in gender, race, values, and culture among employees, suppliers, and customers.

OUR STRATEGY

BEC will be a **global** provider of home entertainment products and services by providing the highest-quality **customer service**, the **broadest range of products and services**, at the **lowest possible price.**

BEC Figure 5-1
Broadway Entertainment Company's mission, objectives, and strategy

---

As a new store manager, Carrie was quite busy, but she was excited to do something about her idea. She still did not understand how all aspects of BEC worked (e.g., the Manager Development Program had not discussed how to work with BEC's IS organization). Carrie thought that maybe she should call someone in the IS organization in Spartanburg to discuss her idea. She found the corporate Subject Matter Expert list and saw the name of Karen Gardner, VP for Information Systems. Calling a vice president didn't seem like a smart move at this point, because without a more thorough plan for her idea about a customer information service, there was no way she could get BEC management to pay attention to it. Carrie also knew the help desk phone number, which she called when there were problems with Entertainment Tracker, the BEC computer system that store employees used. This, too, did not seem like the right call to make, since her idea did not relate directly to Entertainment Tracker. Carrie knew a way, however, to better develop her idea while still giving all the attention she needed to her new job. All she needed to do is to make one phone call, and she thought her idea could take shape.

FORMALIZING A PROJECT PROPOSAL

Carrie's call was to Professor Martha Tann, head of the management information systems (MIS) program at Stillwater State University. Carrie had taken Professor Tann's course on MIS required of all business students at Stillwater. Professor Tann also supervises a two-semester capstone course for MIS majors in which student teams work in local organizations to do systems analysis, design, and development for a new or replacement information system. Carrie's idea was to have an MIS student team develop a prototype of the system and use this prototype to sell the concept of the system to BEC management.

Over the next few weeks Carrie and Professor Tann discussed Carrie's idea and how projects are conducted by MIS students. Students in the course indicate which projects they want to work on among a set of projects submitted for the course by local organizations. There are always more requests submitted by local organizations than can be handled by the course, just like most organizations have more demand for information systems than can be satisfied by the available resources. Projects are presented to the students via a System Service Request form, typical of what would be used inside an organization for a user to request the IS group to undertake a systems development project. Once a group of students is assigned by Professor Tann to a project of their choice, the student team proceeds as if they were a group of systems analysts employed by the sponsoring organization or an outside consulting firm. Within any limitations imposed by the sponsoring organization, the students may conduct the project using any methodology or techniques appropriate for the situation.

The initial System Service Request that Carrie submitted for review by Professor Tann appears in BEC Figure 5-2. This request appears in a standard format used for all project submissions for the MIS project course at Stillwater State University. Professor Tann reviews initial requests for understandability by the students and gives submitters guidance on how to make the project more appealing to students.

When selecting among final System Service Requests, the students look for the projects that will give them the

BEC Figure 5-2
System Service Request from Carrie Douglass

Systems Service Request
Stillwater State University
Capstone MIS Project Course

STILLWATER STATE UNIVERSITY

REQUESTED BY ___Carrie Douglass___ DATE ___August 12, 2001___

ORGANIZATION ___Broadway Entertainment Company, Store OH-84___

ADDRESS ___4600 So. Main Street___

CONTACT ___Tel: 422-7700 FAX: 422-7760 e-mail: CarrieDoug@aol.com___

TYPE OF SYSTEM   URGENCY
[X] New System   [ ] Immediate—Operations are impaired or opportunity lost
[ ] System Enhancement   [ ] Problems exist, but can be worked around
[ ] System Error Correction   [X] Business losses can be tolerated until new system installed

PROBLEM STATEMENT

Today, Broadway Entertainment Company (BEC) sells and rents videos, music, and games to customers. BEC is profitable and growing. Increased competition from existing and emerging competitors requires BEC to constantly consider better ways to meet the needs of its customers. Increasingly, customers want information services as well as products as part of the relationship with our store. Customers want us to be aware of their likes, dislikes, and preferences, and want us to create a sense of community for the exchange of information among customers. The vision of BEC is to be a market leader in the use of technology to provide the highest-quality customer service with the broadest range of products and services. Even though providing information services as part of our relationship with our customers is consistent with this vision, no such services are provided today. The purpose of the proposed project is to prove (or disprove) that such customer information services will improve customer satisfaction and lead to increased revenue and potentially increased market share. A sustainable competitive advantage would be desirable, but is not necessary at this stage.

Specifically, the proposed system will provide information services such as: (1) ability for customers to submit unstructured and structured comments about movies, music, and games they have bought or rented; (2) submit requests for new products for sale and rent; (3) check on due dates for a customer's outstanding rentals; (4) extend a rental without penalty for a minor fee to be applied when the item is returned; (5) review the inventory of items carried in the store; (6) parents can monitor (see a list of) items rented or purchased by their children. This project should conduct a thorough analysis of such information services desired by customers, design a Web-based system to provide such services, and implement and test a prototype of this system.

SERVICE REQUEST

I request a thorough analysis of this idea be conducted. I need a working prototype of the system that could be tested with a selected group of actual customers. The prototype should include major system functions. A survey of users should be conducted to gather evidence to support (or possibly not support) my subsequent request to BEC to build such a system for all stores.

IS LIAISON ___Student team leader, assigned when a team is selected for this project___

SPONSOR ___Carrie Douglass, Manager BEC Store OH-84___

- - - - - - - - - - - - TO BE COMPLETED BY SYSTEMS PRIORITY BOARD - - - - - - - - - - - -
[ ] Request approved   Assigned to _____
                              Start date _____
[ ] Recommend revision
[ ] Suggest user development
[ ] Reject for reason _____
                      _____

best opportunity to learn and integrate the skills needed to manage and conduct a systems analysis and design project. Professor Tann also asks the students to pretend to be a steering committee (sometimes called a Systems Priority Board) to select projects that appear to be well justified and of value to the sponsoring organization. So, Carrie knew that she would have to make the case for the project succinctly and persuasively, even before a preliminary study of the situation could be conducted. Her project idea would have to compete with other submissions, just as it will when she proposes it later within BEC. At least by then, she will have the experience from the prototype to prove the value of her ideas—if the students at Stillwater accept her request.

Carrie and Professor Tann discussed how she might make a more persuasive argument for Carrie's project idea. Professor Tann suggested that students would have many reasons to be motivated to work on submitted projects, including convenience, opportunity to work with interesting technologies, doability, and a sense that their work might make a difference for the organization. The first three reasons seemed to be satisfied with this project, but Professor Tann asked Carrie to think more about the last reason. Obviously, the more likely the system the students might develop would be accepted by the BEC MIS group for final development, the more positive the students would feel that their work would make a difference. Professor Tann suggested that Carrie investigate what the BEC IS priorities were, and try to link her project idea to those priorities.

Carrie called the former manager of the Centerville store, Steve Tettau, to see if he, in his new position in the regional office, had an idea about the corporate IS priorities. Carrie got lucky. Steve had just been assigned to a team of IS staff and business managers who were charged with conducting a thorough review of Entertainment Tracker. The leader of this project had asked the team to analyze how well Entertainment Tracker aligned with the IS plan for BEC. What this meant was to relate Entertainment Tracker to the objectives for BEC found in the Blueprint for the Decade and to specific IS strategic objectives:

- Better align IS development with corporate objectives
- Deliver global system solutions
- Reduce systems development backlog
- Increase skill level of IS staff

The team leader had told Steve that the BEC Systems Priority Board, which decided which IS projects were funded, would need to see a high-level analysis of how their project, and all other projects competing for limited corporate support, related to corporate and IS strategic plans.

With this new information, Carrie prepared a chart to relate her idea for a customer relationship management system to business and IS objectives; this chart appears in BEC Figure 5-3. Although at a very high level, this information nonetheless might help the Stillwater

BEC Figure 5-3
Web-based customer relationship management system alignment with IS plan

	Rating	Brief Explanation
Blueprint Objectives		
1. Increase market share and profitability	High	Loyalty through system increases market share and retains customers, increasing profitability
2. Leader in all areas of business	Moderate	System improves marketing
3. Cost-effective use of resources	Low	Saves some time of store employees
4. Industry leader in profitability and growth	High	See 1
5. Innovative use of technology	High	We have a chance to leapfrog our competition
6. Value diversity	High	Every customer's input will be recorded
IS Plan Objectives		
1. Better align IS development with corporate objectives	High	This project scores well on all six objectives above
2. Deliver global system solutions	High	Web system can easily be deployed globally, with multiple language sites
3. Reduce systems development backlog	None	No apparent impact
4. Increase skill level of IS staff	Moderate	This is a leading-edge application, requiring leading-edge skills
Impact on Current Systems	Low	Some interaction with Entertainment Tracker to share or exchange data, but proposed system does not change architecture of existing systems

students to see the potential scope of the new system and the potential areas where the system might have impact.

CASE SUMMARY

Ideas for new or improved information systems come from a variety of sources, including: the need to fix a broken system, the need to improve the performance of an existing system, competitive pressures or new/changed government regulations, requirements generated from top-down organizational initiatives, and creative ideas by individual managers. The request for a Web-based customer information system submitted by Carrie Douglass is an example of this common, last category. Often an organization is overwhelmed by such requests. An organization must determine which ideas are the most worthy, and what action should be taken in response to each request.

Carrie's proposal creates an opportunity for students at Stillwater State University to engage in an actual systems development project. Although Carrie is not expecting a final, professional, and complete system, a working prototype that will be used by actual customers can serve as an example of the type of system that could be built by Broadway Entertainment. The project as proposed requests that all the typical steps in the analysis and design of an information system be conducted. Carrie Douglass could be rewarded for her creativity if the system proves to be worthwhile, or her idea could flop. The success of her idea depends on the quality of the work done by students at Stillwater.

CASE QUESTIONS

1. The System Service Request (SSR) submitted by Carrie Douglass (BEC Figure 5-2) has not been reviewed by Professor Tann. If you were Professor Tann, would you ask for any changes to the request as submitted? If so, what changes, and if no changes, why? Remember, an SSR is a call for a preliminary study, not a thorough problem statement.

2. If you were a student in Professor Tann's class, would you want to work on this project? Why or why not?

3. If you were a member of BEC's steering committee, the Systems Priority Board, what action would you recommend for this project request if you had received it? Justify your answer.

4. Should Carrie have contacted Karen Gardner? Should Carrie have accepted only Steve Tettau's suggestions, or should she have talked to others at this point? Justify your answer.

5. One of the ideas presented in the chapter is to relate system requests to satisfying the competitive strategy of the organization. What is BEC's competitive strategy (address at least the items in Table 5-3), and how would you position Carrie's project request with respect to this competitive strategy?

6. If you were a systems analyst in the BEC corporate IS department and you had received a call from Carrie Douglass about her project idea, what would you recommend to Carrie? What do you think Carrie would need to prepare or do, in addition to what she has already prepared, to submit a request to the System Priority Board?

7. Do you question any of the ratings and explanations in BEC Figure 5-3 (think like a member of the System Priority Board who might see this table)? Explain. Based on your reading of the introduction to the BEC case in Chapter 4, do you suggest adding into this table any other high-level summary information about the proposed system? Consider what other information could further relate the proposed system to corporate and IS strategy or help to motivate the potential worth of the proposed system (review ideas presented in this chapter about how projects are selected to generate possible responses to this question)?

8. If you were an account representative with a small consulting firm that had received a request for proposal from Carrie Douglass to conduct the project she outlines, what would be your response? Is the System Service Request and project alignment document sufficient as a request for proposal? If so, why? If not, what is missing?

Chapter 6

Initiating and Planning Systems Development Projects

LEARNING OBJECTIVES

After studying this chapter, you should be able to:

- Describe the steps involved in the project initiation and planning process.

- Explain the need for and the contents of a Statement of Work and Baseline Project Plan.

- List and describe various methods for assessing project feasibility.

- Describe the differences between tangible and intangible benefits and costs and between one-time and recurring benefits and costs.

- Perform cost-benefit analysis and describe what is meant by the time value of money, present value, discount rate, net present value, return on investment, and break-even analysis.

- Describe the general rules for evaluating the technical risks associated with a systems development project.

- Describe the activities and participant roles within a structured walkthrough.

INTRODUCTION

The first phase of the systems development life cycle is project identification and selection, during which the need for a new or enhanced system is recognized. This first life cycle phase does not deal with a specific project but rather identifies the portfolio of projects to be undertaken by the organization. Thus, project identification and selection is a pre-project step in the life cycle. This recognition of potential projects may come as part of a larger planning process, informa-tion systems planning, or from requests from managers and business units. Regardless of how a project is iden-tified and selected, the next step is to conduct a more detailed assessment of one particular project selected during the first phase. This assessment does not focus on how the proposed system will operate but rather on understanding the scope of a proposed project and its feasibility of completion given the available resources. It is crucial that organizations understand whether resources should be devoted to a project, otherwise very expensive mistakes can be made. The focus of this chapter is on this process. In other words, project initiation and planning is where projects are either accepted for development, rejected, or redirected. This is also where you, as a systems analyst, begin to play a major role in the systems development process.

In the next section, the project initiation and planning process is briefly reviewed. Next, numerous techniques for assessing project feasibility are described. The information uncovered during feasibility analysis is organized into a document called a Baseline Project Plan. Once this plan is developed, a formal review of the project can be conducted. The process of building this plan is discussed next. Yet, before the project can evolve to the next phase of the systems development life cycle—analysis—the project plan must be reviewed and accepted. In the final major section of the chapter, we provide an overview of the project review process.

INITIATING AND PLANNING SYSTEMS DEVELOPMENT PROJECTS

A key consideration when conducting project initiation and planning (PIP) is deciding when PIP ends and when analysis, the next phase of the SDLC, begins. This is a concern since many activities performed during PIP could also be completed during analysis. Pressman (2001) speaks of three important questions that must be considered when making this decision on the division between PIP and analysis:

1. *How much effort should be expended on the project initiation and planning process?*
2. *Who is responsible for performing the project initiation and planning process?*
3. *Why is project initiation and planning such a challenging activity?*

Finding an answer to the first question, how much effort should be expended on the PIP process, is often difficult. Practical experience has found, however, that the time and effort spent on initiation and planning activities easily pay for themselves later in the project. Proper and insightful project planning, including determining project scope as well as identifying project activities, can easily reduce time in later project phases. For example, a careful feasibility analysis that leads to deciding that a project is not worth pursuing can save a considerable expenditure of resources. The actual amount of time expended will be affected by the size and complexity of the project as well as by the experience of your organization in building similar systems. A rule of thumb is that between 10 and 20 percent of the entire development effort should be expended on the PIP study. Thus, you should not be reluctant to spend considerable time in PIP in order to fully understand the motivation for the requested system.

For the second question, who is responsible for performing the PIP, most organizations assign an experienced systems analyst, or team of analysts for large projects, to perform PIP. The analyst will work with the proposed customers (managers and users) of the system and other technical development staff in preparing the final plan. Experienced analysts working with customers who well understand their information services needs should be able to perform PIP without the detailed analysis typical of the analysis phase of the life cycle. Less-experienced analysts with customers who only vaguely understand their needs will likely expend more effort during PIP in order to be certain that the project scope and work plan are feasible.

Third, the project initiation and planning process is viewed as a challenging activity because the objective of the PIP study is to transform a vague system request document into a tangible project description. This is an open-ended process. The analyst must clearly understand the motivation for and objectives of the proposed system. Therefore, effective communication among the systems analyst, users, and management is crucial to the creation of a meaningful project plan. Getting all parties to agree on the direction of a project may be difficult for cross-department projects when different parties have different business objectives. Thus, more complex organizational settings for projects will result in more time required for analysis of the current and proposed systems during PIP.

In the remainder of this chapter, we will describe the necessary activities used to answer these questions. In the next section, we will revisit the project initiation and planning activities originally outlined in Chapter 3 in the section on Managing the Information Systems Project. This is followed by a brief description of the deliverables and outcomes from this process.

The Process of Initiating and Planning IS Development Projects

As its name implies, two major activities occur during the second phase of the SDLC, project initiation and planning (Figure 6-1). As the steps in the project initiation and planning process were explained in Chapter 3, our primary focus in this chapter is to describe several techniques that are used when performing this process. Therefore, we will only briefly review the PIP process.

Project initiation focuses on activities designed to assist in *organizing* a team to conduct project planning. During initiation, one or more analysts are assigned to work with a customer—that is, a member of the business group that requested or will be impacted by the project—to establish work standards and communication procedures. Examples of the types of activities performed are shown in Table 6-1. Depending upon the size, scope, and complexity of the project, some project initiation activities may be unnecessary or may be very involved. Also, many organizations have established procedures for assisting with common initiation activities.

Figure 6-1
Systems development life cycle with project initiation and planning highlighted

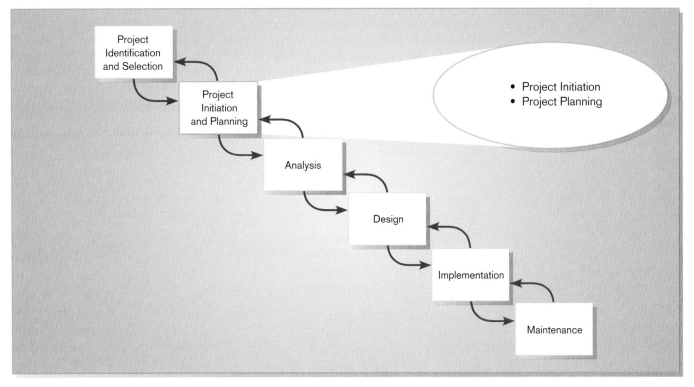

TABLE 6-1   Elements of Project Initiation

- Establishing the Project Initiation Team
- Establishing a Relationship with the Customer
- Establishing the Project Initiation Plan
- Establishing Management Procedures
- Establishing the Project Management Environment and Project Workbook

Project planning, the second activity within PIP, is distinct from general information systems planning, which focuses on assessing the information systems needs of the *entire* organization (discussed in Chapter 5). Project planning is the process of defining clear, discrete activities and the work needed to complete each activity within a *single* project. The objective of the project planning process is the development of a *Baseline Project Plan* (*BPP*) and the *Statement of Work* (*SOW*). The BPP becomes the foundation for the remainder of the development project. The SOW produced by the team clearly outlines the objectives and constraints of the project for the customer. As with the project initiation process, the size, scope, and complexity of a project will dictate the comprehensiveness of the project planning process and resulting documents. Further, numerous assumptions about resource availability and potential problems will have to be made. Analysis of these assumptions and system costs and benefits forms a **business case**. The range of activities performed during project planning are listed in Table 6-2.

Business case: The justification for an information system, presented in terms of the tangible and intangible economic benefits and costs, and the technical and organizational feasibility of the proposed system.

Baseline Project Plan (BPP). A major outcome and deliverable from the project initiation and planning phase, which contains the best estimate of a project's scope, benefits, costs, risks, and resource requirements.

Deliverables and Outcomes

The major outcomes and deliverables from the project initiation and planning phase are the Baseline Project Plan and the Statement of Work. The **Baseline Project Plan (BPP)** contains all information collected and analyzed during project initiation and planning. The plan reflects the best estimate of the project's scope, benefits, costs, risks, and resource requirements given the current understanding of the project. The BPP specifies detailed project activities for the next life cycle phase—analysis—and less detail for subsequent project phases (since these depend on the results of the analysis phase). Similarly, benefits, costs, risks, and resource requirements will become more specific and quantifiable as the project progresses. The BPP is used by the project selection committee to help decide whether the project should be accepted, redirected, or canceled. If selected, the BPP becomes the foundation document for all subsequent SDLC activities; however, it is also expected to evolve as the project evolves. That is, as new information is learned during subsequent SDLC phases, the baseline plan will be updated. Later in the chapter we describe how to construct the BPP.

TABLE 6-2   Elements of Project Planning

- Describing the Project Scope, Alternatives, and Feasibility
- Dividing the Project into Manageable Tasks
- Estimating Resources and Creating a Resource Plan
- Developing a Preliminary Schedule
- Developing a Communication Plan
- Determining Project Standards and Procedures
- Identifying and Assessing Risk
- Creating a Preliminary Budget
- Developing a Statement of Work
- Setting a Baseline Project Plan

The **Statement of Work (SOW)** is a short document prepared for the customer that describes what the project will deliver and outlines all work required to complete the project. The SOW assures that both you and your customer gain a common understanding of the project and is a very useful communication tool. The SOW is a very easy document to create because it typically consists of a high-level summary of the BPP information (described later). A sample SOW is shown in Figure 6-2. Depending upon your relationship with your customer, the role of the SOW may vary. At one extreme, the SOW can be used as the basis of a formal contractual agreement outlining firm deadlines, costs, and specifications. At the other extreme, the SOW can simply be used as a communication vehicle to outline the current best estimates of what the project will deliver, when it will be completed, and the resources it may consume. A contract programming or consulting firm, for example, may establish a very formal relationship with a customer and use a SOW that is extensive and

Statement of Work (SOW):
Document prepared for the customer during project initiation and planning that describes what the project will deliver and outlines generally at a high level all work required to complete the project.

Pine Valley Furniture	Prepared: 9/20/01
Statement of Work	

Project Name: Customer Tracking Systems
PVF Project Manager: Jim Woo

Customer: Marketing
Project Sponsor: Jackie Judson

Project Start/End (projected):   10/1/01–2/1/02

PVF Development Staff Estimates (man-months):

Programmers:	2.0
Jr. Analysts:	1.5
Sr. Analysts:	0.3
Supervisors:	0.1
Consultants:	0.0
Librarian:	0.1

TOTAL:	**4.0**

Project Description

Goal

This project will implement a customer tracking system for the marketing department. The purpose of this system is to automate the ... to save employee time, reduce errors, have more timely information, ...

Objective

- minimize data entry errors
- provide more timely information
- ...

Phases of Work

The following tasks and deliverables reflect the current understanding of the project:
In Analysis, ...
In Design, ...
In Implementation, ...

Figure 6-2
Statement of Work for the Customer Tracking System (Pine Valley Furniture)

formal. Alternatively, an internal development group may develop a SOW that is only one-to-two pages in length and is intended to inform customers rather than to set contractual obligations and deadlines.

ASSESSING PROJECT FEASIBILITY

All projects are feasible given unlimited resources and infinite time (Pressman, 2001). Unfortunately, most projects must be developed within tight budgetary and time constraints. This means that assessing project feasibility is a required activity for all information systems projects and is potentially a large undertaking. It requires that you, as a systems analyst, evaluate a wide range of factors. Typically, some of these factors will be more important than others for some projects and relatively unimportant for other projects. Although the specifics of a given project will dictate which factors are most important, most feasibility factors are represented by the following categories:

- *Economic*
- *Technical*
- *Operational*
- *Schedule*
- *Legal and Contractual*
- *Political*

Together, the culmination of these feasibility analyses form the business case that justifies the expenditure of resources on the project. In the remainder of this section, we will examine various feasibility issues. We begin by examining issues related to economic feasibility and demonstrate techniques for conducting this analysis. This is followed by a discussion of techniques for assessing technical project risk. Finally, issues not directly associated with economic and technical feasibility, but no less important to assuring project success, are discussed.

PINE VALLEY FURNITURE

To help you better understand the feasibility assessment process, we will examine a project at Pine Valley Furniture. For this project, a Systems Service Request (SSR) was submitted by Pine Valley Furniture's (PVF) vice-president of Marketing, Jackie Judson, to develop a Customer Tracking System (Figure 6-3). Jackie feels that this system would allow PVF's marketing group to better track customer purchase activity and sales trends. She also feels that, if constructed, the Customer Tracking System (CTS) would provide many tangible and intangible benefits to PVF. This project was selected by PVF's Systems Priority Board for a project initiation and planning study. During project initiation, senior systems analyst Jim Woo was assigned to work with Jackie to initiate and plan the project. At this point in the project, all project initiation activities have been completed. Jackie and Jim are now focusing on project planning activities in order to complete the BPP.

Assessing Economic Feasibility

Economic feasibility: A process of identifying the financial benefits and costs associated with a development project.

The purpose for assessing **economic feasibility** is to identify the financial benefits and costs associated with the development project; economic feasibility is often referred to as cost-benefit analysis. During project initiation and planning, it will be impossible for you to precisely define all benefits and costs related to a particular project. Yet, it is important that you spend adequate time identifying and quantifying these items or it will be impossible for you to conduct an adequate economic analysis and make meaningful comparisons between rival projects. Here we will describe typical benefits and costs resulting from the development of an information system and provide several useful worksheets for recording costs and benefits. Additionally, sev-

Pine Valley Furniture
System Service Request

REQUESTED BY ___Jackie Judson___ DATE: ___August 23, 2001___

DEPARTMENT ___Marketing___

LOCATION ___Headquarters, 570c___

CONTACT ___Tel: 4-3290   FAX: 4-3270   e-mail: jjudson___

TYPE OF REQUEST   URGENCY
[X] New System   [   ] Immediate – Operations are impaired or opportunity lost
[   ] System Enhancement   [   ] Problems exist, but can be worked around
[   ] System Error Correction   [X] Business losses can be tolerated until new system installed

PROBLEM STATEMENT

Sales growth at PVF has caused a greater volume of work for the marketing department. This volume
of work has greatly increased the volume and complexity of the data we need to deal with and
understand. We are currently using manual methods and a complex PC-based electronic spreadsheet
to track and forecast customer buying patterns. This method of analysis has many problems: (1) we
are slow to catch buying trends as there is often a week or more delay before data can be taken from
point of sales system and manually enter it into our spreadsheet; (2) the process of manual data entry is
prone to errors (which makes the results of our subsequent analysis suspect); and (3) the volume of
data and the complexity of analyses conducted in the system seem to be overwhelming our current
system—sometimes the program starts recalculating and never returns while for others it returns
information that we know cannot be correct.

SERVICE REQUEST

I request a thorough analysis of our current method of tracking and analysis of customer purchasing
activity with the intent to design and build a completely new information system. This system should
handle all customer purchasing activity, support display and reporting of critical sales information, and
assist marketing personnel in understanding the increasingly complex and competitive business
environment. I feel that such a system will improve the competitiveness of PVF, particularly in our
ability to better serve our customers.

IS LIAISON ___Jim Woo, 4-6207   FAX: 4-6200 e-mail: jwoo___

SPONSOR ___Jackie Judson, Vice-President, Marketing___

-------------------- TO BE COMPLETED BY SYSTEMS PRIORITY BOARD ----------------------
[   ] Request approved   Assigned to _____
                             Start date _____
[   ] Recommend revision
[   ] Suggest user development
[   ] Reject for reason _____

Figure 6-3
System service request for Customer Tracking System (Pine Valley Furniture)

eral common techniques for making cost-benefit calculations are presented. These worksheets and techniques are used after each SDLC phase as the project is reviewed in order to decide whether to continue, redirect, or kill a project.

Determining Project Benefits   An information system can provide many benefits to an organization. For example, a new or renovated IS can automate monotonous jobs, reduce errors, provide innovative services to customers and suppliers, and improve organizational efficiency, speed, flexibility, and morale. In general, the benefits can be viewed as being both tangible and intangible. **Tangible benefits** refer to items that can be measured in dollars and with certainty. Examples of tangible benefits might include reduced personnel expenses, lower transaction costs, or higher profit margins. It is important to note that not all tangible benefits can be easily quantified. For example, a tangible benefit that allows a company to perform a task in 50 percent of the time may be difficult to quantify in terms of hard dollar savings. Most tangible benefits will fit within the following categories:

Tangible benefit: A benefit derived from the creation of an information system that can be measured in dollars and with certainty.

- Cost reduction and avoidance
- Error reduction
- Increased flexibility
- Increased speed of activity
- Improvement of management planning and control
- Opening new markets and increasing sales opportunities

Within the Customer Tracking System at PVF, Jim and Jackie identified several tangible benefits, summarized on a tangible benefits worksheet shown in Figure 6-4. Jackie and Jim had to establish the values in Figure 6-4 after collecting information from users of the current customer tracking system. They first interviewed the person responsible for collecting, entering, and analyzing the correctness of the current customer tracking data. This person estimated that they spent 10 percent of their time correcting data entry error. Given that this person's salary is $25,000, Jackie and Jim estimated an *error reduction* benefit of $2,500. Jackie and Jim also interviewed managers who used the current customer tracking reports. Using this information they were able to estimate other tangible benefits. They learned that *cost reduction or avoidance* benefits could be gained due to better inventory management. Also, *increased flexibility* would likely occur from a reduction in the time normally taken to manually reorganize data for different purposes. Further, *improvements in management planning*

Figure 6-4
Tangible benefits for Customer Tracking System (Pine Valley Furniture)

TANGIBLE BENEFITS WORKSHEET Customer Tracking System Project	
	Year 1 through 5
A. Cost reduction or avoidance	$ 4,500
B. Error reduction	2,500
C. Increased flexibility	7,500
D. Increased speed of activity	10,500
E. Improvement in management planning or control	25,000
F. Other _____	0
TOTAL tangible benefits	**$50,000**

TABLE 6-3   Intangible Benefits from the Development of an Information System

- Competitive necessity
- More timely information
- Improved organizational planning
- Increased organizational flexibility
- Promotion of organizational learning and understanding
- Availability of new, better, or more information
- Ability to investigate more alternatives

- Faster decision making
- Information processing efficiency
- Improved asset utilization
- Improved resource control
- Increased accuracy in clerical operations
- Improved work process that can improve employee morale
- Positive impacts on society

Adapted from Parker and Benson, 1988

or control should result from a broader range of analyses in the new system. Overall, this analysis forecasts that benefits from the system would be approximately $50,000 per year.

Jim and Jackie also identified several intangible benefits of the system. Although they could not quantify these benefits, they will still be described in the final BPP. **Intangible benefits** refer to items that *cannot* be easily measured in dollars or with certainty. Intangible benefits may have direct organizational benefits such as the improvement of employee morale or they may have broader societal implications such as the reduction of waste creation or resource consumption. Potential tangible benefits may have to be considered intangible during project initiation and planning since you may not be able to quantify them in dollars or with certainty at this stage in the life cycle. During later stages, such intangibles can become tangible benefits as you better understand the ramifications of the system you are designing. In this case, the BPP is updated and the business case revised to justify continuation of the project to the next phase. Table 6-3 lists numerous intangible benefits often associated with the development of an information system. Actual benefits will vary from system to system. After determining project benefits, project costs must be identified.

Determining Project Costs   Similar to benefits, an information system can have both tangible and intangible costs. **Tangible costs** refer to items that you can easily measure in dollars and with certainty. From an IS development perspective, tangible costs include items such as hardware costs, labor costs, and operational costs such as employee training and building renovations. Alternatively, intangible costs are those items that you cannot easily measure in terms of dollars or with certainty. **Intangible costs** can include loss of customer goodwill, employee morale, or operational inefficiency. Table 6-4 provides a summary of common costs associated with the development and operation of an information system. Predicting the costs associated with the development of an information system is an inexact science. IS researchers, however, have identified several guidelines for improving the cost-estimating process (see Table 6-5). Both underestimating and overestimating costs are problems you must avoid (Lederer and Prasad, 1992). Underestimation results in cost overruns while overestimation results in unnecessary allocation of resources that might be better utilized.

Besides tangible and intangible costs, you can distinguish IS-related development costs as either one-time or recurring (the same is true for benefits although we do not discuss this difference for benefits). **One-time costs** refer to those associated with project initiation and development and the start-up of the system. These costs typically encompass activities such as system development, new hardware and software purchases, user training, site preparation, and data or system conversion. When conducting an economic cost-benefit analysis, a worksheet should be created for capturing these expenses. For very large projects, one-time costs may be staged over one or more years. In these cases, a separate one-time cost worksheet should be created

Intangible benefit: A benefit derived from the creation of an information system that cannot be easily measured in dollars or with certainty.

Tangible cost: A cost associated with an information system that can be measured in dollars and with certainty.

Intangible cost: A cost associated with an information system that cannot be easily measured in terms of dollars or with certainty.

One-time cost: A cost associated with project start-up and development, or system start-up.

TABLE 6-4   Possible Information Systems Costs

Types of Costs	Examples	Types of Costs	Examples
Procurement	Consulting costs	Project-related	Application software
	Equipment purchase or lease		Software modifications to fit local systems
	Equipment installation costs		
	Site preparation and modifications		Personnel, overhead, et al., from in-house development
	Capital costs		
	Management and staff time		Training users in application use
Start-up	Operating system software		Collecting and analyzing data
	Communications equipment installation		Preparing documentation
	Start-up personnel		Managing development
	Personnel searches and hiring activities	Operating	System maintenance costs (hardware, software, and facilities)
	Disruption to the rest of the organization		Rental of space and equipment
	Management to direct start-up activity		Asset depreciation
			Management, operation, and planning personnel

Adapted with the permission of the Association for Computing Machinery. Copyright © 1978 J. L. King and E. Schrems

Recurring cost: A cost resulting from the ongoing evolution and use of a system.

for each year. This separation will make it easier to perform present value calculations (see below). **Recurring costs** refer to those costs resulting from the ongoing evolution and use of the system. Examples of these costs typically include

- Application software maintenance
- Incremental data storage expense
- Incremental communications
- New software and hardware leases
- Supplies and other expenses (for example, paper, forms, data center personnel)

Both one-time and recurring costs can consist of items that are fixed or variable in nature. Fixed costs refer to costs that are billed or incurred at a regular interval and usually at a fixed rate (a facility lease payment). Variable costs refer to items that vary in relation to usage (long distance phone charges).

TABLE 6-5   Guidelines for Better Cost Estimating

1. Assign the initial estimating task to the final developers.
2. Delay finalizing the initial estimate until the end of a thorough study.
3. Anticipate and control user changes.
4. Monitor the progress of the proposed project.
5. Evaluate proposed project progress by using independent auditors.
6. Use the estimate to evaluate project personnel.
7. Study the cost estimate carefully before approving it.
8. Rely on documented facts, standards, and simple arithmetic formulas rather than guessing, intuition, personal memory, and complex formulas.
9. Don't rely on cost-estimating software for an accurate estimate.

Adapted with the permission of the Association for Computing Machinery. Copyright © 1992 A. L. Lederer and J. Prasad.

```
                ONE-TIME COSTS WORKSHEET
                Customer Tracking System Project

                                                    Year 0

A. Development costs                                 $20,000

B. New hardware                                      15,000

C. New (purchased) software, if any
     1. Packaged applications software                5,000
     2. Other _____                 0

D. User training                                      2,500

E. Site preparation                                       0

F. Other _____                   0

TOTAL one-time cost                                 $42,500
```

Figure 6-5
One-time costs for Customer Tracking System (Pine Valley Furniture)

During the process of determining project costs, Jim and Jackie identified both one-time and recurring costs for the project. These costs are summarized in Figures 6-5 and 6-6. These figures show that this project will incur a one-time cost of $42,500 and a recurring cost of $28,500 per year. One-time costs were established by discussing the system with Jim's boss who felt that the system would require approximately four months to develop (at $5,000 per month). To effectively run the new system, the Marketing department would need to upgrade at least five of their current workstations (at $3,000 each). Additionally, software licenses for each workstation (at $1,000 each) and modest user training fees (ten users at $250 each) would be necessary.

As you can see from Figure 6-6, Jim and Jackie believe the proposed system will be highly dynamic and will require, on average, five months of annual maintenance, primarily for enhancements as users expect more from the system. Other ongoing expenses such as increased data storage, communications equipment, and supplies should also be expected. You should now have an understanding of the types of benefit and cost categories associated with an information systems project. It should be clear that there are many potential benefits and costs associated with a given project. Additionally, since the development and useful life of a system may span several

```
                RECURRING COSTS WORKSHEET
                Customer Tracking System Project

                                              Year 1 through 5

A. Application software maintenance                 $25,000

B. Incremental data storage required: 20 MB × $50.    1,000
   (estimated cost/MB = $50)

C. Incremental communications (lines, messages, . . .)  2,000

D. New software or hardware leases                        0

E. Supplies                                             500

F. Other _____                    0

TOTAL recurring costs                               $28,500
```

Figure 6-6
Recurring costs for Customer Tracking System (Pine Valley Furniture)

years, these benefits and costs must be normalized into present-day values in order to perform meaningful cost-benefit comparisons. In the next section, we address the relationship between time and money.

The Time Value of Money   Most techniques used to determine economic feasibility encompass the concept of the *time value of money* (TVM). TVM refers to the concept of comparing present cash outlays to future expected returns. As previously discussed, the development of an information system has both one-time and recurring costs. Furthermore, benefits from systems development will likely occur sometime in the future. Since many projects may be competing for the same investment dollars and may have different useful life expectancies, all costs and benefits must be viewed in relation to their *present value* when comparing investment options.

A simple example will help in understanding the TVM. Suppose you want to buy a used car from an acquaintance and she asks that you make three payments of $1,500 for three years, beginning next year, for a total of $4,500. If she would agree to a single lump sum payment at the time of sale (and if you had the money!), what amount do you think she would agree to? Should the single payment be $4,500? Should it be more or less? To answer this question, we must consider the time value of money. Most of us would gladly accept $4,500 today rather than three payments of $1,500, because a dollar today (or $4,500 for that matter) is worth more than a dollar tomorrow or next year, because money can be invested. The rate at which money can be borrowed or invested is called the *cost of capital*, and is called the **discount rate** for TVM calculations. Let's suppose that the seller could put the money received for the sale of the car in the bank and receive a 10 percent return on her investment. A simple formula can be used when figuring out the **present value** of the three $1,500 payments:

Discount rate: The rate of return used to compute the present value of future cash flows.

Present value: The current value of a future cash flow.

$$PV_n = Y \times \left[\frac{1}{(1 + i)^n} \right]$$

where PV_n is the present value of Y dollars n years from now when i is the discount rate.

From our example, the present value of the three payments of $1,500 can be calculated as

$$PV_1 = 1500 \times \left[\frac{1}{(1 + .10)^1} \right] = 1500 \times .9091 = 1363.65$$

$$PV_2 = 1500 \times \left[\frac{1}{(1 + .10)^2} \right] = 1500 \times .8264 = 1239.60$$

$$PV_3 = 1500 \times \left[\frac{1}{(1 + .10)^3} \right] = 1500 \times .7513 = 1126.95$$

where PV_1, PV_2, and PV_3 reflect the present value of each $1,500 payment in year 1, 2, and 3, respectively.

To calculate the *net present value* (NPV) of the three $1,500 payments, simply add the present values calculated above (NPV = PV_1 + PV_2 + PV_3 = 1363.65 + 1239.60 + 1126.95 = $3730.20). In other words, the seller could accept a lump sum payment of $3,730.20 as equivalent to the three payments of $1,500, given a discount rate of 10 percent.

Given that we now know the relationship between time and money, the next step in performing the economic analysis is to create a summary worksheet reflecting the present values of all benefits and costs as well as all pertinent analyses. Due to the fast pace of the business world, PVF's System Priority Board feels that the useful life of many information systems may not exceed five years. Therefore, all cost-benefit analysis calculations will be made using a five-year time horizon as the upper boundary on all time-related analyses. In addition, the management of PVF has set their cost of capital to be 12 percent (that is, PVF's discount rate). The worksheet constructed by Jim is shown in Figure 6-7.

Cell H11 of the worksheet displayed in Figure 6-7 summarizes the NPV of the total tangible benefits from the project. Cell H19 summarizes the NPV of the total costs from the project. The NPV for the project ($35,003) shows that, overall, benefits from the project exceed costs (see cell H22).

The overall return on investment (ROI) for the project is also shown on the worksheet in cell H25. Since alternative projects will likely have different benefit and cost values and, possibly, different life expectancies, the overall ROI value is very useful for making project comparisons on an economic basis. Of course, this example shows ROI for the overall project. An ROI analysis could be calculated for each year of the project.

The last analysis shown in Figure 6-7 is a break-even analysis. The objective of the break-even analysis is to discover at what point (if ever) benefits equal costs (that is,

	A	B	C	D	E	F	G	H
1	**Pine Valley Furniture**							
2	**Economic Feasibility Analysis**							
3	**Customer Tracking System Project**							
4								
5					Year of Project			
6		Year 0	Year 1	Year 2	Year 3	Year 4	Year 5	TOTALS
7	Net economic benefit	$0	$50,000	$50,000	$50,000	$50,000	$50,000	
8	Discount rate (12%)	1.0000	0.8929	0.7972	0.7118	0.6355	0.5674	
9	PV of benefits	$0	$44,643	$39,860	$35,589	$31,776	$28,371	
10								
11	NPV of all BENEFITS	$0	$44,643	$84,503	$120,092	$151,867	$180,239	$180,239
12								
13	One-time COSTS	($42,500)						
14								
15	Recurring Costs	$0	($28,500)	($28,500)	($28,500)	($28,500)	($28,500)	
16	Discount rate (12%)	1.0000	0.8929	0.7972	0.7118	0.6355	0.5674	
17	PV of Recurring Costs	$0	($25,446)	($22,720)	($20,286)	($18,112)	($16,172)	
18								
19	NPV of all COSTS	($42,500)	($67,946)	($90,666)	($110,952)	($129,064)	($145,236)	($145,236)
20								
21								
22	Overall NPV							$35,003
23								
24								
25	Overall ROI - (Overall NPV / NPV of all COSTS)							0.24
26								
27								
28	Break-even Analysis							
29	Yearly NPV Cash Flow	($42,500)	$19,196	$17,140	$15,303	$13,664	$12,200	
30	Overall NPV Cash Flow	($42,500)	($23,304)	($6,164)	$9,139	$22,803	$35,003	
31								
32	Project break-even occurs between years 2 and 3							
33	Use first year of positive cash flow to calculate break-even fraction - ((15303 - 9139) / 15303) = .403							
34	Actual break-even occurred at 2.4 years							
35								
36	Note: All dollar values have been rounded to the nearest dollar							

Figure 6-7
Summary spreadsheet reflecting the present value calculations of all benefits and costs for the Customer Tracking System (Pine Valley Furniture)

when break-even occurs). To conduct this analysis, the NPV of the yearly cash flows are determined. Here, the yearly cash flows are calculated by subtracting both the one-time cost and the present values of the recurring costs from the present value of the yearly benefits. The overall NPV of the cash flow reflects the total cash flows for all preceding years. Examination of line 30 of the worksheet shows that break-even occurs between years 2 and 3. Since year 3 is the first in which the overall NPV cash flows figure is nonnegative, the identification of what point during the year break-even occurs can be derived as follows:

$$\text{Break-Even Ratio} = \frac{\text{Yearly NPV Cash Flow} - \text{Overall NPV Cash Flow}}{\text{Yearly NPV Cash Flow}}$$

Using data from Figure 6-7,

$$\text{Break-Even Ratio} = \frac{15,303 - 9,139}{15,303} = .404$$

Therefore, project break-even occurs at approximately 2.4 years. A graphical representation of this analysis is shown in Figure 6-8. Using the information from the economic analysis, PVF's Systems Priority Board will be in a much better position to understand the potential economic impact of the Customer Tracking System. It should be clear from this analysis that, without such information, it would be virtually impossible to know the cost-benefits of a proposed system and impossible to make an informed decision regarding approval or rejection of the service request.

There are many techniques that you can use to compute a project's economic feasibility. Because most information systems have a useful life of more than one year and will provide benefits and incur expenses for more than one year, most techniques for analyzing economic feasibility employ the concept of the TVM. Some of these cost-benefit analysis techniques are quite simple while others are more sophisticated. Table 6-6 describes three commonly used techniques for conducting economic feasibility analysis. For a more detailed discussion of TVM or cost-benefit analysis techniques in general, the interested reader is encouraged to review an introductory finance or managerial accounting textbook.

A systems project, to be approved for continuation, may not have to achieve break-even or have an ROI above some organizational threshold as estimated during

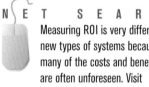

N E T   S E A R C H
Measuring ROI is very different for new types of systems because many of the costs and benefits are often unforeseen. Visit http://www.prenhall.com/hoffer to complete an exercise related to this topic.

Figure 6-8
Break-even analysis for Customer Tracking System (Pine Valley Furniture)

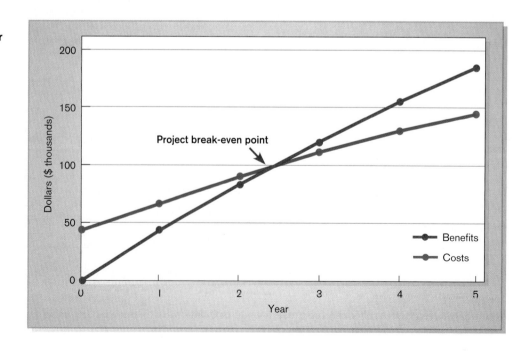

TABLE 6-6   Commonly Used Economic Cost-Benefit Analysis Techniques

Analysis Technique	Description
Net Present Value (NPV)	NPV uses a discount rate determined from the company's cost of capital to establish the present value of a project. The discount rate is used to determine the present value of both cash receipts and outlays.
Return on Investment (ROI)	ROI is the ratio of the net cash receipts of the project divided by the cash outlays of the project. Trade-off analysis can be made among projects competing for investment by comparing their representative ROI ratios.
Break-Even Analysis (BEA)	BEA finds the amount of time required for the cumulative cash flow from a project to equal its initial and ongoing investment.

project initiation and planning. Since you may not be able to quantify many benefits or costs at this point in a project, such financial hurdles for a project may be unattainable. In this case, simply doing as thorough an economic analysis as possible, including producing a long list of intangibles, may be sufficient for the project to progress. One other option is to run the type of economic analysis shown in Figure 6-7 using pessimistic, optimistic, and expected benefit and cost estimates during project initiation and planning. This range of possible outcomes, along with the list of intangible benefits and the support of the requesting business unit, will often be enough to allow the project to continue to the analysis phase. You must, however, be as precise as you can with the economic analysis, especially when investment capital is scarce. In this case, it may be necessary to conduct some typical analysis phase activities during project initiation and planning in order to clearly identify inefficiencies and shortcomings with the existing system and to explain how a new system will overcome these problems. Thus, building the economic case for a systems project is an open-ended activity; how much analysis is needed depends on the particular project, stakeholders, and business conditions. Also, conducting economic feasibility analyses for new types of information systems is often very difficult.

Assessing Technical Feasibility

The purpose of assessing **technical feasibility** is to gain an understanding of the organization's ability to construct the proposed system. This analysis should include an assessment of the development group's understanding of the possible target hardware, software, and operating environments to be used as well as system size, complexity, and the group's experience with similar systems. In this section, we will discuss a framework you can use for assessing the technical feasibility of a project in which a level of project risk can be determined after answering a few fundamental questions.

Technical feasibility: A process of assessing the development organization's ability to construct a proposed system.

It is important to note that all projects have risk and that risk is not necessarily something to avoid. Yet it is also true that, because organizations typically expect a greater return on their investment for riskier projects, understanding the sources and types of technical risks proves to be a valuable tool when you assess a project. Also, risks need to be managed in order to be minimized; you should, therefore, identify potential risks as early as possible in a project. The potential consequences of not assessing and managing risks can include the following outcomes:

1. Failure to attain expected benefits from the project
2. Inaccurate project cost estimates
3. Inaccurate project duration estimates
4. Failure to achieve adequate system performance levels
5. Failure to adequately integrate the new system with existing hardware, software, or organizational procedures

You can manage risk on a project by changing the project plan to avoid risky factors, assigning project team members to carefully manage the risky aspects, and setting up monitoring methods to determine whether or not potential risk is, in fact, materializing.

The amount of technical risk associated with a given project is contingent on four primary factors: project size, project structure, the development group's experience with the application and technology area, and the user group's experience with development projects and application area (see also Kirsch, 2000). Aspects of each of these risk areas are summarized in Table 6-7. Using these factors for conducting a technical risk assessment, four general rules emerge:

1. *Large projects are riskier than small projects.* Project size, of course, relates to the relative project size that the development group is familiar working with. A "small" project for one development group may be relatively "large" for another. The types of factors that influence project size are listed in Table 6-7.

2. *A system in which the requirements are easily obtained and highly structured will be less risky than one in which requirements are messy, ill-structured, ill-defined, or subject to the judgment of an individual.* For example, the development of a payroll system has requirements that may be easy to obtain due to legal reporting requirements and standard accounting procedures. On the other hand, the development of an executive support system would need to be customized to the particular executive decision style and critical success factors of the organization, thus making its development more risky (see Table 6-7).

3. *The development of a system employing commonly used or standard technology will be less risky than one employing novel or nonstandard technology.* A project has a greater likelihood of experiencing unforeseen technical problems when the development group lacks knowledge related to some aspect of the technology environment. A less risky approach is to use standard development tools and hardware environments. It is not uncommon for experienced system developers to talk of the difficulty of using leading-edge (or in their words, *bleeding* edge) technology (see Table 6-7).

TABLE 6-7   Project Risk Assessment Factors

Risk Factor	Examples
Project Size	Number of members on the project team
	Project duration time
	Number of organizational departments involved in project
	Size of programming effort (e.g., hours, function points)
Project Structure	New system or renovation of existing system(s)
	Organizational, procedural, structural, or personnel changes resulting from system
	User perceptions and willingness to participate in effort
	Management commitment to system
	Amount of user information in system development effort
Development Group	Familiarity with target-hardware, software development environment, tools, and operating system
	Familiarity with proposed application area
	Familiarity with building similar systems of similar size
User Group	Familiarity with information systems development process
	Familiarity with proposed application area
	Familiarity with using similar systems

Adapted from Applegate, McFarlan, and McKenny, 1999

4. *A project is less risky when the user group is familiar with the systems development process and application area than if unfamiliar.* Successful IS projects require active involvement and cooperation between the user and development groups. Users familiar with the application area and the systems development process are more likely to understand the need for their involvement and how this involvement can influence the success of the project (see Table 6-7).

A project with high risk may still be conducted. Many organizations look at risk as a portfolio issue: Considering all projects, it is okay to have a reasonable percentage of high-, medium-, and low-risk projects. Given that some high-risk projects will get into trouble, an organization cannot afford to have too many of these. Having too many low-risk projects may not be aggressive enough to make major breakthroughs in innovative uses of systems. Each organization must decide on its acceptable mix of projects of varying risk.

A matrix for assessing the relative risks related to the general rules described above is shown in Figure 6-9. Using the risk factor rules to assess the technical risk level of the Customer Tracking System, Jim and Jackie concluded the following about their project:

1. The project is a relatively *small project* for PVF's development organization. The basic data for the system is readily available so the creation of the system will not be a large undertaking.

2. The requirements for the project are *highly structured* and easily obtainable. In fact, an existing spreadsheet-based system is available for analysts to examine and study.

3. The *development group is familiar* with the technology that will likely be used to construct the system, as the system will simply extend current system capabilities.

4. The *user group is familiar* with the application area since they are already using the PC-based spreadsheet system described in Figure 6-3.

Given this risk assessment, Jim and Jackie mapped their information into the risk framework of Figure 6-9. They concluded that this project should be viewed as having "very low" technical risk (cell 4 of the figure). Although this method is useful for gaining an understanding of technical feasibility, numerous other issues can influence the success of the project. These nonfinancial and nontechnical issues are described in the following section.

		Low Structure	High Structure
High Familiarity with Technology or Application Area	Large Project	(1) Low risk (very susceptible to mismanagement)	(2) Low risk
	Small Project	(3) Very low risk (very susceptible to mismanagement)	(4) Very low risk
Low Familiarity with Technology or Application Area	Large Project	(5) Very high risk	(6) Medium risk
	Small Project	(7) High risk	(8) Medium-low risk

Figure 6-9
Effects of degree of project structure, project size, and familiarity with application area on project implementation risk
(Adapted from: Corporate Information Systems Management: The Challenges of Managing in an Information Age, 5th ed., F.11.1, p. 284, by L. M. Applegate and F. W. McFarlan. Copyright © 1999. Reprinted by permission of The McGraw-Hill Companies.

Assessing Other Feasibility Concerns

Operational feasibility: The process of assessing the degree to which a proposed system solves business problems or takes advantage of business opportunities.

In this section, we will briefly conclude our discussion of project feasibility issues by reviewing other forms of feasibility that you may need to consider when formulating the business case for a system during project planning. The first relates to examining the likelihood that the project will attain its desired objectives, called **operational feasibility**. Its purpose is to gain an understanding of the degree to which the proposed system will likely solve the business problems or take advantage of the opportunities outlined in the systems service request or project identification study. For a project motivated from information system planning, operational feasibility includes justifying the project on the basis of being consistent with or necessary for accomplishing the IS plan. In fact, the business case for any project can be enhanced by showing a link to the business or information systems plan. Your assessment of operational feasibility should also include an analysis of how the proposed system will affect organizational structures and procedures. Systems that have substantial and widespread impact on an organization's structure or procedures are typically riskier projects to undertake. Thus, it is important for you to have a clear understanding of how an IS will fit into the current day-to-day operations of the organization.

Schedule feasibility: The process of assessing the degree to which the potential time frame and completion dates for all major activities within a project meet organizational deadlines and constraints for affecting change.

Another feasibility concern relates to project duration and is referred to as assessing schedule feasibility. The purpose of assessing **schedule feasibility** is for you, as a systems analyst, to gain an understanding of the likelihood that all potential time frames and completion date schedules can be met and that meeting these dates will be sufficient for dealing with the needs of the organization. For example, a system may have to be operational by a government-imposed deadline, by a particular point in the business cycle (such as the beginning of the season when new products are introduced), or at least by the time a competitor is expected to introduce a similar system. Further, detailed activities may only be feasible if resources are available when called for in the schedule. For example, the schedule should not call for system testing during rushed business periods or for key project meetings during annual vacation or holiday periods. The schedule of activities produced during project initiation and planning will be very precise and detailed for the analysis phase. The estimated activities and associated times for activities after the analysis phase are typically not as detailed (e.g., it will take two weeks to program the payroll report module) but are rather at the life-cycle phase level (e.g., it will take six weeks for physical design, four months for programming, and so on). This means that assessing schedule feasibility during project initiation and planning is more of a "rough-cut" analysis of whether the system can be completed within the constraints of the business opportunity or the desires of the users. While assessing schedule feasibility you should also evaluate scheduling trade-offs. For example, factors such as project team size, availability of key personnel, subcontracting or outsourcing activities, and changes in development environments may all be considered as having possible impact on the eventual schedule. As with all forms of feasibility, schedule feasibility will be reassessed after each phase, when you can specify with greater certainty the detailed steps and their duration for the next phase.

Legal and contractual feasibility: The process of assessing potential legal and contractual ramifications due to the construction of a system.

A third concern relates to assessing **legal and contractual feasibility** issues. In this area, you need to gain an understanding of any potential legal ramifications due to the construction of the system. Possible considerations might include copyright or nondisclosure infringements, labor laws, antitrust legislation (which might limit the creation of systems to share data with other organizations), foreign trade regulations (for example, some countries limit access to employee data by foreign corporations), and financial reporting standards as well as current or pending contractual obligations. Contractual obligations may involve ownership of software used in joint ventures, license agreements for use of hardware or software, nondisclosure agreements with partners, or elements of a labor agreement (for example, a union agreement may preclude certain compensation or work-monitoring capabilities a user may want in a system). A common situation is that development of a new application system for use on new computers may require new or expanded, and more costly, system software licenses. Typically, legal and contractual feasibility is a greater consideration if your organization has historically used an outside organization for specific systems or services that you now are considering handling yourself. In this case, ownership of program source code by another party may make it difficult to extend an existing system or link a new system with an existing, purchased system.

A final feasibility concern focuses on assessing **political feasibility** in which you attempt to gain an understanding of how key stakeholders within the organization view the proposed system. Since an information system may affect the distribution of information within the organization, and thus the distribution of power, the construction of an IS can have political ramifications. Those stakeholders not supporting the project may take steps to block, disrupt, or change the intended focus of the project.

In summary, depending upon the given situation, numerous feasibility issues must be considered when planning a project. This analysis should consider economic, technical, operational, schedule, legal, contractual, and political issues related to the project. In addition to these considerations, project selection by an organization may be influenced by issues beyond those discussed here. For example, projects may be selected for construction given high project costs and high technical risk if the system is viewed as a strategic necessity; that is, a project viewed by the organization as being critical to its survival. Alternatively, projects may be selected because they are deemed to require few resources and have little risk. Projects may also be selected due to the power or persuasiveness of the manager proposing the system. This means that project selection may be influenced by factors beyond those discussed here and beyond items that can be analyzed. Understanding the reality that projects may be selected based on factors beyond analysis, your role as a systems analyst is to provide a thorough examination of the items that can be assessed. Your analysis will ensure that a project review committee has as much information as possible when making project approval decisions. In the next section, we discuss how project plans are typically reviewed.

> **Political feasibility:** The process of evaluating how key stakeholders within the organization view the proposed system.

BUILDING THE BASELINE PROJECT PLAN

All the information collected during project initiation and planning is collected and organized into a document called the Baseline Project Plan. Once the BPP is completed, a formal review of the project can be conducted with project clients and other interested parties. This presentation is called a *walkthrough* and is discussed later in the chapter. The focus of this review is to verify all information and assumptions in the baseline plan before moving ahead with the project. As mentioned above, the project size and organizational standards will dictate the comprehensiveness of the project initiation and planning process as well as the BPP. Yet, most experienced systems builders have found project planning and a clear project plan to be invaluable to project success. An outline of a Baseline Project Plan is provided in Figure 6-10, which shows that it contains four major sections:

1. Introduction
2. System Description
3. Feasibility Assessment
4. Management Issues

The purpose of the *Introduction* is to provide a brief overview of the entire document and outline a recommended course of action for the project. The entire Introduction section is often limited to only a few pages. Although the Introduction section is sequenced as the first section of the BPP, it is often the final section to be written. It is only after performing most of the project planning activities that a clear overview and recommendation can be created. One activity that should be performed initially is the definition of project scope.

BASELINE PROJECT PLAN REPORT

1.0 Introduction

A. Project Overview—Provides an executive summary that specifies the project's scope, feasibility, justification, resource requirements, and schedules. Additionally, a brief statement of the problem, the environment in which the system is to be implemented, and constraints that affect the project are provided.

B. Recommendation—Provides a summary of important findings from the planning process and recommendations for subsequent activities.

2.0 System Description

A. Alternatives—Provides a brief presentation of alternative system configurations.

B. System Description—Provides a description of the selected configuration and a narrative of input information, tasks performed, and resultant information.

3.0 Feasibility Assessment

A. Economic Analysis—Provides an economic justification for the system using cost-benefit analysis.

B. Technical Analysis—Provides a discussion of relevant technical risk factors and an overall risk rating of the project.

C. Operational Analysis—Provides an analysis of how the proposed system solves business problems or takes advantage of business opportunities in addition to an assessment of how current day-to-day activities will be changed by the system.

D. Legal and Contractual Analysis—Provides a description of any legal or contractual risks related to the project (e.g., copyright or nondisclosure issues, data capture or transferring, and so on).

E. Political Analysis—Provides a description of how key stakeholders within the organization view the proposed system.

F. Schedules, Timeline, and Resource Analysis—Provides a description of potential time frame and completion date scenarios using various resource allocation schemes.

4.0 Management Issues

A. Team Configuration and Management—Provides a description of the team member roles and reporting relationships.

B. Communication Plan—Provides a description of the communication procedures to be followed by management, team members, and the customer.

C. Project Standards and Procedures—Provides a description of how deliverables will be evaluated and accepted by the customer.

D. Other Project-Specific Topics—Provides a description of any other relevant issues related to the project uncovered during planning.

Figure 6-10
Outline of a Baseline Project Plan

When defining scope for the Customer Tracking System within PVF, Jim Woo first needed to gain a clear understanding of the project's objectives. To do this, Jim briefly interviewed Jackie Judson and several of her colleagues to gain a clear idea of their needs. He also spent a few hours reviewing the existing system's functionality, processes, and data use requirements for performing customer tracking activities. These activities provided him with the information needed to define the project scope and to identify possible alternative solutions. Alternative system solutions can relate to different system scopes, platforms for deployment, or approaches to acquiring the system. We elaborate on the idea of alternative solutions, called design strategies, when we discuss the analysis phase of the life cycle. During project initiation and planning, the most crucial element of the design strategy is the system's scope. In sum, a determination of scope will depend on these factors:

- Which organizational units (business functions and divisions) might be affected by or use the proposed system or system change?

- With which current systems might the proposed system need to interact or be consistent, or which current systems might be changed due to a replacement system?

- Who inside and outside the requesting organization (or the organization as a whole) might care about the proposed system?

- What range of potential system capabilities are to be considered?

The statement of project scope for the Customer Tracking System project is shown in Figure 6-11.

For the Customer Tracking System (CTS), project scope was defined using only textual information. It is not uncommon, however, to define project scope using diagrams such as data flow diagrams and entity-relationship models. For example, Figure 6-12 shows a context-level data flow diagram used to define system scope for PVF's Purchasing Fulfillment System described on the *Modern Systems Analysis and Design*'s World Wide Web site (http://www.prenhall.com/hoffer/student/supp.html). The other items in the Introduction section of the BPP are simply executive summaries of the other sections of the document.

The second section of the BPP is the *System Description* where you outline possible alternative solutions in addition to the one deemed most appropriate for the given situation. Note that this description is at a very high level, mostly narrative in form. For example, alternatives may be stated as simply as this:

1. Mainframe with central database
2. Distributed with decentralized databases
3. Batch data input with on-line retrieval
4. Purchasing of a prewritten package

If the project is approved for construction or purchase, you will need to collect and structure information in a more detailed and rigorous manner during the analysis phase and evaluate in greater depth these and other alternative directions for the system. At this point in the project, your objective is only to identify the most obvious alternative solutions.

When Jim and Jackie were considering system alternatives for the CTS, they focused on two primary issues. First, they discussed how the system would be acquired and considered three options: *purchase* the system if one could be found that met PVF's needs, *outsource* the development of the system to an outside organization, or *build* the system within PVF. The second issue focused on defining the comprehensiveness of the system's functionality. To complete this task, Jim asked Jackie to write a series of statements listing the types of tasks that she envisioned marketing personnel would be able to accomplish when using the CTS. This list of statements became the basis of the system description and was instrumental in helping them make their acquisition decision. After considering the unique needs of the marketing group, both decided that the best decision was to build the system within PVF.

In the third section, *Feasibility Assessment*, issues related to project costs and benefits, technical difficulties, and other such concerns are outlined. This is also the section where high-level project schedules are specified using PERT and Gantt charts. Recall from Chapter 3 that this process is referred to as a work breakdown structure. During project initiation and planning, task and activity estimates are not generally detailed. An accurate work breakdown can only be done for the next one or two life-cycle activities. After defining the primary tasks for the project, an estimate of the

PINE VALLEY FURNITURE

Pine Valley Furniture   Prepared by: Jim Woo
Statement of Project Scope   Date: September 18, 2001

General Project Information
    Project Name: Customer Tracking System
    Sponsor: Jackie Judson, VP Marketing
    Project Manager: Jim Woo

Problem/Opportunity Statement:
    Sales growth has outpaced the marketing department's ability to accurately track and forecast
    customer buying trends. An improved method for performing this process must be found in
    order to reach company objectives.

Project Objectives:
    To enable the marketing department to accurately track and forecast customer buying patterns
    in order to better serve customers with the best mix of products. This will also enable PVF to
    identify the proper application of production and material resources.

Project Description:
    A new information system will be constructed that will collect all customer purchasing activity,
    support display and reporting of sales information, aggregate data, and show trends in order to
    assist marketing personnel in understanding dynamic market conditions. The project will follow
    PVF's systems development life cycle.

Business Benefits:
    Improved understanding of customer buying patterns
    Improved utilization of marketing and sales personnel
    Improved utilization of production and materials

Project Deliverables:
    Customer tracking system analysis and design
    Customer tracking system programs
    Customer tracking documentation
    Training procedures

Estimated Project Duration:
    5 months

Figure 6-11
Statement of project scope (Pine Valley Furniture)

resource requirements can be made. As with defining tasks and activities, this activity
is primarily concerned with gaining rough estimates of the human resource requirements, since people are the most expensive resource element. Once you define the
major tasks and resource requirements, a preliminary schedule can be developed.
Defining an acceptable schedule may require that you find additional or different
resources or that the scope of the project be changed. The greatest amount of project planning effort is typically expended on these Feasibility Assessment activities.

    The final section, *Management Issues*, outlines a number of managerial con-
cerns related to the project. This will be a very short section if the proposed proj-

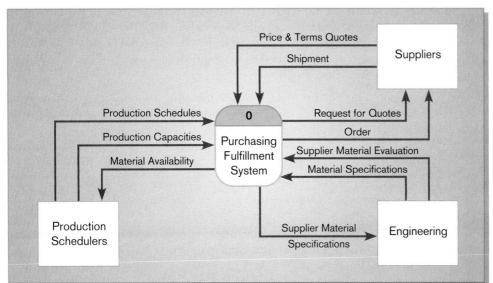

Figure 6-12
Context-level data flow diagram showing project scope for **Purchasing Fulfillment System (Pine Valley Furniture)**

ect is going to be conducted exactly as prescribed by the organization's standard systems development methodology. Most projects, however, have some unique characteristics that require minor to major deviation from the standard methodology. In the Team Configuration and Management portion, you identify the types of people to work on the project, who will be responsible for which tasks, and how work will be supervised and reviewed. In the Communications Plan portion, you explain how the user will be kept informed about project progress (such as periodic review meetings or even a newsletter) and which mechanisms will be used to foster sharing of ideas among team members, such as some form of computer-based conference facility. An example of the type of information contained in the Project Standards and Procedures portion would be procedures for submitting and approving project change requests and any other issues deemed important for the project's success.

You should now have a feel for how a BPP is constructed and the types of information it contains. Its creation is not meant to be a project in and of itself but rather a step in the overall systems development process. Developing the BPP has two primary objectives. First, it helps to assure that the customer and development group share a common understanding of the project. Second, it helps to provide the sponsoring organization with a clear idea of the scope, benefits, and duration of the project.

REVIEWING THE BASELINE PROJECT PLAN

Before the next phase of the SDLC can begin, the users, management, and development group must review the Baseline Project Plan in order to verify that it makes sense. This review takes place before the BPP is submitted or presented to some project approval body, such as an IS steering committee or the person who must fund the project. The objective of this review is to assure that the proposed system conforms to organizational standards and to make sure that all relevant parties understand and agree with the information contained in the Baseline Project Plan. A common method for performing this review (as well as reviews during subsequent life-cycle phases) is called a structured **walkthrough**. Walkthroughs are peer group reviews of any product created during the systems development process and are widely used by professional development organizations. Experience has shown that walkthroughs are a very effective way to ensure the quality of an information system and have become a common day-to-day activity for many system analysts.

Walkthrough: A peer group review of any product created during the systems development process. Also called structured walkthrough.

Most walkthroughs are not rigidly formal or exceeding long in duration. It is important, however, that a specific agenda be established for the walkthrough so that all attendees understand what is to be covered and the expected completion time. At walkthrough meetings, there is a need to have individuals play specific roles. These roles are as follows (Yourdon, 1989):

- *Coordinator.* This person plans the meeting and facilitates a smooth meeting process. This person may be the project leader or a lead analyst responsible for the current life-cycle step.
- *Presenter.* This person describes the work product to the group. The presenter is usually an analyst who has done all or some of the work being presented.
- *User.* This person (or group) makes sure that the work product meets the needs of the project's customers. This user would usually be someone not on the project team.
- *Secretary.* This person takes notes and records decisions or recommendations made by the group. This may be a clerk assigned to the project team or it may be one of the analysts on the team.
- *Standards bearer.* The role of this person is to ensure that the work product adheres to organizational technical standards. Many larger organizations have staff groups within the unit responsible for establishing standard procedures, methods, and documentation formats. These standards bearers validate the work so that it can be used by others in the development organization.
- *Maintenance oracle.* This person reviews the work product in terms of future maintenance activities. The goal is to make the system and its documentation easy to maintain.

N E T S E A R C H
Managing a structured walkthrough with a very large group can be difficult. Visit http://www.prenhall.com/hoffer to complete an exercise related to this topic.

After Jim and Jackie completed their BPP for the Customer Tracking System, Jim approached his boss and requested that a walkthrough meeting be scheduled and that a walkthrough coordinator be assigned to the project. PVF assists the coordinator by providing a Walkthrough Review Form, shown in Figure 6-13. Using this form, the coordinator can more easily make sure that a qualified individual is assigned to each walkthrough role, that each member has been given a copy of the review materials, and that each member knows the agenda, date, time, and location of the meeting. At the meeting, Jim presented the BPP and Jackie added comments from a user perspective. Once the walkthrough presentation was completed, the coordinator polled each representative for his or her recommendation concerning the work product. The results of this voting may result in validation of the work product, validation pending changes suggested during the meeting, or a suggestion that the work product requires major revision before being presented for approval. In this latter case, substantial changes to the work product are usually requested after which another walkthrough must be scheduled before the project can be proposed to the Systems Priority Board (steering committee). In the case of the Customer Tracking System, the BPP was supported by the walkthrough panel pending some minor changes to the duration estimates of the schedule. These suggested changes were recorded by the secretary on a Walkthrough Action List (see Figure 6-14) and given to Jim to incorporate into a final version of the baseline plan presented to the steering committee.

As suggested above, walkthrough meetings are a common occurrence in most systems development groups and can be used for more activities than reviewing the BPP, including the following:

- System specifications
- Logical and physical designs
- Code or program segments
- Test procedures and results
- Manuals and documentation

Pine Valley Furniture
Walkthrough Review Form

Session Coordinator:

Project/Segment:

Coordinator's Checklist:

1. Confirmation with producer(s) that material is ready and stable: _____
2. Issue invitations, assign responsibilities, distribute materials: [ ] Y   [ ] N
3. Set date, time, and location for meeting:

Date: ___ / ___ / ___   Time: _____ A.M. / P.M. (circle one)

Location: _____

Responsibilities	Participants	Can Attend	Received Materials
Coordinator	_____	[ ] Y [ ] N	[ ] Y [ ] N
Presenter	_____	[ ] Y [ ] N	[ ] Y [ ] N
User	_____	[ ] Y [ ] N	[ ] Y [ ] N
Secretary	_____	[ ] Y [ ] N	[ ] Y [ ] N
Standards	_____	[ ] Y [ ] N	[ ] Y [ ] N
Maintenance	_____	[ ] Y [ ] N	[ ] Y [ ] N

Agenda:
____ 1. All participants agree to follow PVF's Rules of a Walkthrough
____ 2. New material: walkthrough of all material
____ 3. Old material: item-by-item checkoff of previous action list
____ 4. Creation of new action list (contribution by each participant)
____ 5. Group decision (see below)
____ 6. Deliver copy of this form to the project control manager

Group Decision:
_____ Accept product as-is
_____ Revise (no further walkthrough)
_____ Review and schedule another walkthrough

Signatures		

Figure 6-13
Walkthrough review form (Pine Valley Furniture)

Pine Valley Furniture
Walkthrough Action List

Session Coordinator:

Project/Segment:

Date and Time of Walkthrough:

Date: ____ / ____ / ____   Time: _____ A.M. / P.M. (circle one)

Fixed (✓)	Issues raised in review:

Figure 6-14
Walkthrough action list (Pine Valley Furniture)

One of the key advantages to using a structured review process is to ensure that formal review points occur during the project. At each subsequent phase of the project, a formal review should be conducted (and shown on the project schedule) to make sure that all aspects of the project are satisfactorily accomplished before assigning additional resources to the project. This conservative approach of reviewing each major project activity with continuation contingent on successful completion of the prior phase is called *incremental commitment.* It is much easier to stop or redirect a project at any point when using this approach.

ELECTRONIC COMMERCE APPLICATION: INITIATING AND PLANNING SYSTEMS DEVELOPMENT PROJECTS

PINE VALLEY FURNITURE

Initiating and planning systems development projects for an Internet-based electronic commerce application is very similar to the process followed for more traditional applications. In the last chapter, you read how Pine Valley Furniture's management began the WebStore project—to sell furniture products over the Internet. In this section, we highlight some of the issues that relate directly to the process of identifying and selecting systems development projects.

TABLE 6-8   Web-Based System Costs

Cost Category	Examples
Platform costs	• Web-hosting service • Web server • Server software • Software plug-ins • Firewall server • Router • Internet connection
Content and service	• Creative design and development • Ongoing design fees • Web project manager • Technical site manager • Content staff • Graphics staff • Support staff • Site enhancement funds • Fees to license outside content • Programming, consulting, and research • Training and travel
Marketing	• Direct mail • Launch and ongoing public relations • Print advertisement • Paid links to other Websites • Promotions • Marketing staff • Advertising sales staff

TABLE 6-9   PVF WebStore: Project Benefits and Costs

Tangible Benefits	Intangible Benefits
• Lower per-transaction overhead cost • Repeat business	• First to market • Foundation for complete Web-based IS • Simplicity for customers
Tangible Costs (one-time)	*Intangible Costs*
• Internet service setup fee • Hardware • Development cost • Data entry	• No face-to-face interaction • Not all customers use Internet
Tangible Costs (recurring)	
• Internet service hosting fee • Software • Support • Maintenance • Decreased sales via traditional channels	

Initiating and Planning Systems Development Projects for Pine Valley Furniture's WebStore

Given the high priority of the WebStore project, Jackie Judson, vice-president of Marketing, and senior systems analyst Jim Woo were assigned to work on this project. Like the Customer Tracking System described earlier in the chapter, their initial activity was to begin the project's initiation and planning activities.

Initiating and Planning PVF's E-Commerce System   To start the initiation and planning process, Jim and Jackie held several meetings over several days. At the first meeting they agreed that "WebStore" would be the proposed system project name. Next, they worked on identifying potential benefits, costs, and feasibility concerns. To assist in this process, Jim developed a list of potential costs from developing Web-based systems that he shared with Jackie and the other project team members (see Table 6-8).

WebStore Project Walkthrough   After meeting with the project team, Jim and Jackie established an initial list of benefits and costs (see Table 6-9) as well as several feasibility concerns (see Table 6-10). Next, Jim worked with several of PVF's technical specialists to develop an initial project schedule. Figure 6-15 shows the Gantt chart for this 84 day schedule. Finally, Jim and Jackie presented their initial project plans in a walkthrough to PVF's Board of Directors and senior management. All were excited about the project plan and approval was given to move the WebStore project onto the analysis phase.

TABLE 6-10   PVF WebStore: Feasibility Concerns

Feasibility Concern	Description
Operational	On-line store is open 24/7/365 Returns/customer support
Technical	New skill set for development, maintenance and operation
Schedule	Must be open for business by Q3
Legal	Credit card fraud
Political	Traditional distribution channel loses business

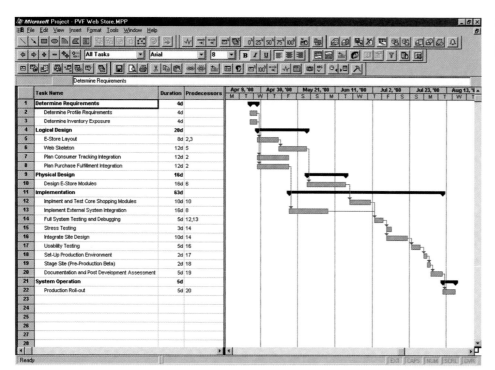

Figure 6-15
Schedule for WebStore project at Pine Valley Furniture

Summary

The project initiation and planning phase is a critical activity in the life of a project. It is at this point that projects are accepted for development, rejected as infeasible, or redirected. The objective of this process is to transform a vague system request into a tangible system description clearly outlining the objectives, feasibility issues, benefits, costs, and time schedules for the project.

Project initiation includes forming the project initiation team, establishing customer relationships, developing a plan to get the project started, setting project management procedures, and creating an overall project management environment. A key activity in project planning is the assessment of numerous feasibility issues associated with the project. Feasibilities include economic, technical, operational, schedule, legal and contractual, and political. These issues are influenced by the project size, the type of system proposed, and the collective experience of the development group and potential customers of the system. High project costs and risks are not necessarily bad; rather it is more important that the organization understands the costs and risks associated with a project and with the portfolio of active projects before proceeding.

After completing all analyses, a Baseline Project Plan can be created. A BPP includes a high-level description of the proposed system or system change, an outline of the various feasibilities, and an overview of management issues specific to the project. Before the development of an information system can begin, the users, management, and development group must review and agree on this specification. The focus of this walkthrough review is to assess the merits of the project and to assure that the project, if accepted for development, conforms to organizational standards and goals. An objective of this process is to also make sure that all relevant parties understand and agree with the information contained in the plan before subsequent development activities begin.

Project initiation and planning is a challenging and time-consuming activity that requires active involvement from many organizational participants. The eventual success of development projects, and the MIS function in general, hinges on the effective use of disciplined, rational approaches such as the techniques outlined in this chapter. In subsequent chapters you will be exposed to numerous other tools that will equip you to become an effective designer and developer of information systems.

Key Terms

1. Baseline Project Plan (BPP)
2. Business case
3. Discount rate
4. Economic feasibility
5. Intangible benefit
6. Intangible cost
7. Legal and contractual feasibility
8. One-time cost
9. Operational feasibility
10. Present value
11. Political feasibility
12. Recurring cost
13. Schedule feasibility
14. Statement of Work (SOW)
15. Tangible benefit
16. Tangible cost
17. Technical feasibility
18. Walkthrough

Match each of the key terms above with the definition that best fits it.

_____ The process of evaluating how key stakeholders within the organization view the proposed system.

_____ Document prepared for the customer during project initiation and planning that describes what the project will deliver and outlines generally at a high level all work required to complete the project.

_____ The justification for an information system, presented in terms of the tangible and intangible economic benefits and costs, and the technical and organizational feasibility of the proposed system.

_____ A process of identifying the financial benefits and costs associated with a development project.

_____ The process of assessing the degree to which a proposed system solves business problems or takes advantage of business opportunities.

_____ A cost resulting from the ongoing evolution and use of a system.

_____ The rate of return used to compute the present value of future cash flows.

_____ A benefit derived from the creation of an information system that cannot be easily measured in dollars or with certainty.

_____ The process of assessing the degree to which the potential time frame and completion dates for all major activities within a project meet organizational deadlines and constraints for affecting change.

_____ A cost associated with an information system that can be easily measured in dollars and certainty.

_____ A peer group review of any product created during the system development process.

_____ A process of assessing the development organization's ability to construct a proposed system.

_____ A cost associated with project start-up and development, or system start-up.

_____ The current value of a future cash flow.

_____ A benefit derived from the creation on an information system that can be measured in dollars and with certainty.

_____ The process of assessing potential legal and contractual ramifications due to the construction of a system.

_____ A cost associated with an information system that cannot be easily measured in terms of dollars or with certainty.

_____ The plan is the major outcome and deliverable from the project initiation and planning phase and contains the best estimate of the project's scope, benefits, costs, risks, and resource requirements.

Review Questions

1. Contrast the following terms:
   a. break-even analysis, present value, net present value, return on investment
   b. economic feasibility, legal and contractual feasibility, operational feasibility, political feasibility, schedule feasibility
   c. intangible benefit, tangible benefit
   d. intangible cost, tangible cost

2. List and describe the steps in the project initiation and planning process.

3. What is contained in a Baseline Project Plan? Is the content and format of all baseline plans the same? Why or why not?

4. Describe three commonly used methods for performing economic cost-benefit analysis.

5. List and discuss the different types of project feasibility factors. Is any factor most important? Why or why not?

6. What are the potential consequences of not assessing the technical risks associated with an information systems development project?

7. In what ways could you identify an IS project that was riskier than another?

8. What are the types or categories of benefits from an IS project?

9. What intangible benefits might an organization obtain from the development of an IS?

10. Describe the concept of the time value of money. How does the discount rate affect the value of $1 today versus one year from today?

11. Describe the structured walkthrough process. What roles need to be performed during a walkthrough?

Problems and Exercises

1. Consider the purchase of a PC and laser printer for use at your home and assess risk for this project using the project risk assessment factors in Table 6-7.

2. Consider your use of a PC at either home or work and list tangible benefits from an information system. Based on this list, does your use of a PC seem to be beneficial? Why or why not? Now do the same using Table 6-3, the intangible benefits from an information system. Does this analysis support or contradict your previous analysis? Based on both analyses, does your use of a PC seem to be beneficial?

3. Consider, as an example, buying a network of PCs for a department at your workplace or, alternatively, consider outfitting a laboratory of PCs for students at a university. For your example, estimate the costs outlined in Table 6-4, one-time and recurring costs.

4. Assuming monetary benefits of an information system at $85,000 per year, one-time costs of $75,000, recurring costs of $35,000 per year, a discount rate of 12 percent , and a five-year time horizon, calculate the Net Present Value of these costs and benefits of an information system. Also calculate the overall Return on Investment of the project and then present a Break-Even Analysis. At what point does break-even occur?

5. Choose as an example one of the information systems you described in Problem and Exercise 3 above, either buying a network of PCs for a department at your workplace or outfitting a laboratory of PCs for students at a university. Estimate the costs and benefits for your system and calculate the Net Present Value, Return on Investment and present a Break-Even Analysis. Assume a discount rate of 12 percent and a five-year time horizon.

6. Use the outline for the Baseline Project Plan provided in Figure 6-10 to present the system specifications for the information system you chose for Problems and Exercises 3 and 5 above.

7. Change the discount rate for Problem and Exercise 4 to 10 percent and redo the analysis.

8. Change the recurring costs in Problem and Exercise 4 to $40,000 and redo the analysis.

9. Change the time horizon in Problem and Exercise 4 to three years and redo the analysis.

10. Assume monetary benefits of an information system of $50,000 the first year and increasing benefits of $5,000 a year for the next four years (Year 1 = 50,000; Year 2 = 55,000; Year 3 = 60,000; Year 4 = 65,000; Year 5 = 70,000). One-time development costs were $90,000 and recurring costs beginning in Year 1 were $40,000 over the duration of the system's life. The discount rate for the company was 10 percent. Using a five-year horizon, calculate the Net Present Value of these costs and benefits. Also calculate the overall Return on Investment of the project and then present a Break-Even analysis. At what point does break-even occur?

11. Change the discount rate for Problem and Exercise 10 to 12 percent and redo the analysis.

12. Change the recurring costs in Problem and Exercise 10 to $60,000 and redo the analysis.

13. Change the time horizon in Problem and Exercise 4 to three years and redo the analysis.

14. For the system you chose for Problems and Exercises 3 and 5, complete section 1.0, A, Project Overview, of the Baseline Project Plan Report. How important is it that this initial section of the Baseline Project Plan Report be done well? What could go wrong if this section is incomplete or incorrect?

15. For the system you chose for Problems and Exercises 3 and 5, complete section 2.0, A, Alternatives, of the Baseline Project Plan Report. Without conducting a full-blown feasibility analysis, what is your gut feeling as to the feasibility of this system?

16. For the system you chose for Problems and Exercises 3 and 5, complete section 3.0, A–F, Feasibility Analysis, of the Baseline Project Plan Report. How does this feasibility analysis compare with your gut feeling from the previous question? What might go wrong if you rely on your gut feeling in determining system feasibility?

17. For the system you chose for Problems and Exercises 3 and 5, complete section 4.0, A–C, Management Issues, of the Baseline Project Plan Report. Why might people sometimes feel that these additional steps in the project plan are a waste of time? What would you say to them to convince them that these steps are important?

Field Exercises

1. Describe several projects you are involved in or plan to undertake, whether they be related to your education or to your professional or personal life. Some examples are purchasing a new vehicle, learning a new language, renovating a home, and so on. For each, sketch out a Baseline Project Plan like that outlined in Figure 6-10. Focus your efforts on item number 1.0 (Introduction) and 2.0 (System Description).

2. For each project from the previous question, assess the feasibility in terms of economic, operational, technical, schedule, legal and contractual, as well as political aspects.

3. Network with a contact you have in some organization that conducts projects (these might be IS projects, but they could be construction, product development, research and development, or any type of project). Interview a project

manager and find out what type of Baseline Project Plan is constructed. For a typical project, in what ways are baseline plans modified during the life of a project? Why are plans modified after the project begins? What does this tell you about project planning?

4. Through a contact you have in some organization that uses packaged software, interview an IS manager responsible for systems in an area that uses packaged application software. What contractual limitations, if any, has the organization encountered with using the package? If possible, review the license agreement for the software and make a list of all the restrictions placed on a user of this software.

5. Choose an organization that you are familiar with and determine what is done to initiate information systems projects. Who is responsible for performing this? Is this process formal or informal? Does this appear to be a top-down or bottom-up process? How could this process be improved?

6. Find an organization that does not use Baseline Project Plans for their IS projects. Why doesn't this organization use this method? What are the advantages and disadvantages of not using this method? What benefits could be gained from implementing the use of Baseline Project Plans? What barriers are there to implementing this method?

References

Applegate, L. M., McFarlan, F. W., and McKenny, J. L. 1999. *Corporate Information Systems Management: The Challenges of Managing in an Information Age*, 5th ed. Boston, MA: Irwin/McGraw-Hill.

King, J. L. and E. Schrems. 1978. "Cost Benefit Analysis in Information Systems Development and Operation. *ACM Computing Surveys* 10(1): 19–34.

Kirsch, L. J. 2000. "Software Project Management: An Integrated Perspective for an Emerging Paradigm." In *Framing the Domains of IT Management: Projecting the Future from the Past*, edited by R. W. Zmud. Cincinnati: Pinnaflex Educational Resources, Chapter 15, 285–304.

Lederer, A. L., and J. Prasad. 1992. "Nine Management Guidelines for Better Cost Estimating." *Communications of the ACM* 35(2): 51–59.

Parker, M. M. and R. J. Benson. 1988. *Information Economics*, Englewood Cliffs, NJ: Prentice-Hall.

Pressman, R. S. 2001. *Software Engineering*. 5th ed. New York, NY: McGraw-Hill.

Yourdon, E. 1989. *Structured Walkthroughs*, 4th ed. Englewood Cliffs, NJ: Prentice-Hall.

BROADWAY ENTERTAINMENT COMPANY, INC.

Initiating and Planning the Customer Relationship Management System

CASE INTRODUCTION

Carrie Douglass, manager of the Broadway Entertainment Company store in Centerville, Ohio, was pleased when the Management Information System (MIS) students at Stillwater State University accepted her request to design a customer information system. The students saw the development of this system as a unique opportunity. This system deals with one of the hottest topics in business today—customer relationship management—and is a simple form of one of the most active areas of information systems development—electronic commerce. Many of the MIS students wanted to work on this project, but Professor Tann limits each team to four members. Professor Tann selected a team of Tracey Wesley, John Whitman, Missi Davies, and Aaron Sharp to work on the BEC project.

The BEC student team had never taken on such a large, open-ended project. Each team member had been on many teams in other classes, and each had some practical work experience from part-time jobs, summer internships, or cooperative education terms. None of the team members had ever worked in a store like those operated by BEC, although all of them were regular BEC customers. And, this group of students had never worked on the same team. Both Tracey and John are parents of children and, hence, have some personal interest in an information system such as Carrie has proposed. Professor Tann selected Tracey, John, Missi, and Aaron, in part, for their diversity. Tracey is very interested in programming, and spent a busy fall term in 1999 in an internship working on Y2K conversions for the local electric utility company. John works full-time for the Dayton Public Schools as a computer applications trainer. Missi was a marketing major before she transferred to MIS after her sophomore year. Missi worked part-time in customer service at a local department store before her switch into MIS. Aaron, the most technical member of the team, actually began taking MIS classes at Stillwater while in high school. Aaron was in charge of the Website for his high school. Aaron and another MIS student formed their own firm during their junior year to do Website consulting with several small businesses in the Dayton area.

INITIATING AND PLANNING THE PROJECT

The students were eager to meet their clients. The team met after class one day to plan the first meeting with Carrie. They decided that this first meeting should be short (no more than an hour), informal, and at the Centerville store. There would be no presentation and no formal interview. The agenda included getting to know Carrie, asking if she had any new ideas since submitting the SSR, setting a schedule for the next couple of weeks, and seeking from Carrie some resources they would need.

The student team agreed to meet Carrie one morning before the store opened for business. This was a good time for Carrie because she would not have any distractions. Also, since there was no meeting room in the store, their discussion would not interfere with store operations. More formal meetings might have to be on campus or at an office center near the store.

The team shared with Carrie information about their project course requirements, including a tentative schedule of when the team was expected to submit deliverables for each system development phase to Professor Tann (see BEC Figure 6-1). Although Carrie was eager to get the project done, she was aware of this schedule from talking with Professor Tann. Each team member explained his or her background and skills, and stated personal goals for the project. Carrie explained her background. The team members were surprised to discover that Carrie had only recently graduated from Stillwater, however, they felt that they had found a kindred spirit for a project sponsor.

Before developing a plan for the detailed steps of the analysis phase and general steps for subsequent phases, the team wanted to determine if anything had changed in Carrie's mind since she submitted the request. Carrie, not too surprisingly, had been very busy since she submitted the request and had only a few new ideas. First, Carrie had become even more excited about the system she proposed. She explained that once the project started she would probably generate new ideas every few days. The team explained that although this would be helpful, at some point the system requirements would have to be

Date (Week)	Deliverable	Format
Every 2 weeks	Status Report	One-page memo
4	Baseline Project Plan	Written report
8	Requirements Statement	Written report, CASE repository, BPP update
9	Requirements Walkthrough	Oral presentation to class and client
11	Functional Design Specifications	Written report, CASE repository, BPP update
14	Physical Design Specifications	Written report, CASE repository, BPP update
15	Testing and Installation Plan	Written report
19	Code Walkthrough	Oral presentation to one other class team
21	Preliminary System Demonstration	Oral presentation to class and client
25	User Documentation	Written
27	Preliminary Final Report	Written report
29	Final Report	Written report, all system documentation
29	Installation and Pass-off to Client	Status report on result of installation
29	Practice Final Presentation	Oral presentation in class
Week 30	Final Presentation	Oral presentation to MIS program Advisory Board

BEC Figure 6-1
Tentative schedule for project deliverables in MIS senior project course

fixed so that a detailed design could be completed and the system prototype built. Further ideas could be incorporated in later enhancements. Second, Carrie raised a concern about what would happen once the course was over and an initial system was built. How would she get any further help? The team suggested that they would consider how follow-up might be handled. Missi offered the option that if Carrie thought that the final product of the project was good enough, it might be time to involve BEC corporate IS people near the end of the project in a hand-off meeting. Carrie suggested that this would be an alternative to readdress later in the project.

Third, Carrie also asked how the team would interact with her during the project. The team responded by saying that they would provide to Carrie in the next two weeks a detailed schedule for the next phase of the project (the whole analysis phase) as part of a comprehensive plan for the project. Along with this schedule would be a statement of when there would be face-to-face review meetings, the nature of written status reports, and other elements of a communication plan for the project. Finally, Carrie had one new idea about the requirements for the system. She suggested that a useful feature would be a page that would change weekly with comments from a store employee concerning his or her favorite picks for the week in several categories: adventure, mystery, documentary, children's, and so on. Carrie emphasized that she wanted to provide these comments from her own employees, not from outside sources or to provide links to outside movie and music review Websites.

As this first meeting was coming to a close, the team asked Carrie to do one thing for them. They asked Carrie to send a note to all store employees to introduce the project. The team wanted employees to know that a project was underway, what the reasons and goals were for the project, that there would be students from Stillwater asking questions and observing, and that the students needed the cooperation of all employees. Carrie said she would be glad to distribute such a note, but asked that the team draft the note for her. Carrie reserved the right to make changes to it before giving it to employees.

Carrie then took the students on a quick tour of the Centerville BEC store. As she walked by different areas of the store she related activities at those areas to the proposed Web customer information system. At the checkout counter, she mentioned how the system could save time of employees from looking up various information about customers that the customers would be able to view on-line themselves. She introduced the team to one of the employees as she entered the store, and explained how the employee could devote time to more important tasks rather than give advice on what products to buy or rent. And near the entrance to the store, Carrie pointed out a location where she thought an in-store kiosk could be placed where customers could do everything with the system that they could do at home via the Internet.

As the quick tour came to an end, the team members each thanked Carrie for her time and for the project. The students vowed to take the project seriously and to use the system to further the career of a Stillwater graduate.

Carrie offered to provide any other information they might need over the next few weeks as they formulate the initial project plan.

DEVELOPING THE BASELINE PROJECT PLAN

The next day the Stillwater student team met to review the meeting with Carrie. They were pleased. Carrie had been easy to talk with, very interested in the project (maybe too eager, Missi thought), and supportive. They had not learned a great deal more about the proposed system from this first visit, but they had not wanted this first meeting to drill down into details. Possibly, they would have to talk with Carrie again before the project plan would be submitted.

There would be a lot of work to do over the next few weeks to develop the Baseline Project Plan (BPP). Using a whiteboard, the team began to identify the major parts of the BPP so that they could divide up the work to get everything completed in time. What they identified included:

- *A detailed requirements statement.* Of course, the requirements will be verified and elaborated during the analysis phase, but the team needed something more complete than the SSR Carrie had submitted.

- *A model of the relationship of the Web system to other BEC information systems.* This was an area of real uncertainty to the team. Carrie had mentioned the point-of-sale (POS) system used in the store, Entertainment Tracker (ET), during their brief tour of the store. The team members would have to investigate further how closely the Web system they were to design and build would have to work with ET.

- *A macro-level model of the proposed system.* The team decided that one of the best ways to summarize the system they thought they were being asked to develop was to present in the BPP a general picture of what the system could do. The students had learned various system modeling notations in classes, and thought that there were several that could be used at such an early stage of the project. The students wanted to make sure that they were on the same page as the client with system expectations.

- *A list of tangible and intangible benefits and costs.* The team probably cannot quantify many of the tangible benefits and costs now, but the analysis phase will need to capture the information necessary for this essential part of the product of the analysis phase. The team noted that the Blueprint for the Decade and the document on aligning the BEC IS plan with the proposed system would be good starting points, but a much more detailed analysis would be needed to convince Carrie to spend money required to deploy the new system.

- *A list of project and system risks.* The BPP should identify technical, operational, schedule, legal, or political risks and how the team intended to cope with these.

- *A project schedule.* The schedule should be detailed for the analysis phase, and more macro for subsequent phases. The project milestones needed to occur roughly when major deliverables were to be turned in to Professor Tann, but the team could propose a different schedule for course deliverables than outlined in the course syllabus.

- *A statement of resource requirements to do the project.* Human resources were limited, and nobody from the client organization could be devoted to the project. Professor Tann expected the team to do all the work, and besides Carrie, it was unclear who else at the BEC store could help them. This was especially true for knowledge about Entertainment Tracker (ET). Hopefully documentation on ET would provide the information they needed, or maybe there would be someone at the BEC corporate IS department to whom they could direct some questions. The team thought that they would likely interview some employees and contact customers to discover desired system features and to better calibrate benefits and costs.

- *A list of technology resources.* The team members discussed how they could use Stillwater computer resources for most of the development work, but at some point Carrie would have to invest in some computer equipment, an Internet connection, and possibly a contract with an Internet Service Provider (ISP). How much of these resources were needed for the project and how much could be delayed until when Carrie accepted their system needed to be determined.

- *Management and communication plans.* The team members debated how to organize the team. They had discussed each other's strengths and weaknesses when they first met. Some of the team members thought they should divide up by strengths to make the workload easier. Others thought that it was important to take advantage of the project and gain experience in areas of weakness. Another proposal was to rotate jobs among the team members each time there was a major project phase change. A good argument could be made for each team member to become the project leader. A solution on team organization was not clear. In terms of a communication plan, Professor Tann wanted biweekly status reports. The team members thought that a slightly different version of these status reports would also work as a regular report to Carrie. They would use e-mail with each other and with Carrie. Someone on the team would have to decide on a format for all status reports and major written deliverables. They wanted all reports and presentations to have a consistent look, which would make a more professional impression on Professor Tann and anyone from BEC.

After identifying what they thought needed to go into the BPP, the team members still had difficulty deciding who should take on which tasks. So, the members decided to have each person develop a bullet list of reasons why he or she should be in charge of each item the team had written on the whiteboard. They would meet again the next day and try to resolve this first dilemma for the team.

CASE SUMMARY

The project initiation and planning phase of the project demonstrates the need to carefully develop a Baseline Project Plan. The Web-based customer information system for the Centerville, Ohio, Broadway Entertainment Company store has potential benefit in increased sales and rentals and many potential intangible benefits. However both one-time and ongoing costs as well as the risks of implementing a new system make the success of such projects dubious. The motivation for the project is linked to several important BEC objectives, and an enthusiastic client (Carrie) exists. However, since this is the first large project for which the Stillwater MIS students have had total responsibility, they are being cautious to get the project started with a solid plan.

CASE QUESTIONS

1. Carrie agreed to send a note to employees announcing the project if the team would draft the note. Prepare this note for Carrie's review. Give some thought to what employees need to know so that they are not surprised about the project, they are cooperative, and they are ready to give the team members information needed for the project.

2. When the student team met with Carrie, she said that she would "generate new ideas every few days" for the project. Suggest some project management procedures that the Stillwater team could use to deal with this prediction from Carrie.

3. The Stillwater students stated that a detailed requirements statement should be included in the Baseline Project Plan. With the information you have so far from the BEC cases in previous chapters, you really cannot prepare much more of a statement that was given in the SSR. What activities should the students do to prepare this statement for the BPP? Remember, this statement is done before the analysis phase, but will be revised based upon the results of analysis.

4. The list of items the student team identified should go into a Baseline Project Plan included a macro level model of the system. From your experience with systems analysis and design techniques so far in the course you are taking and from what you have seen in prior courses and work experience, what type of models might be suitable? If possible, draw a high-level model of the proposed Web-based customer relationship information system, including how this system interacts with Entertainment Tracker and other existing BEC corporate information systems.

5. In this case, the Stillwater team members were struggling with assigning tasks to individuals. Another approach would be to define team roles and then assign tasks to roles and people to roles. List and define the roles you believe should exist on the Stillwater student team for the proposed project.

6. There are nine bulleted items that the Stillwater students have identified as elements of the Baseline Project Plan for their project. Did they miss anything? If so, give a brief explanation of each missing element.

7. One of the BPP items the Stillwater team members listed is the set of tangible and intangible benefits and costs of the system. Prepare this list as you think it should be presented as part of the BPP. How do you propose collecting the information to quantify each potential tangible benefit?

8. What do you consider to be the risks of the project as you currently understand it? Is this a low-, medium-, or high-risk project? Justify your answer. How would you propose dealing with the risks?

9. Develop a tentative schedule for this project. As stated in this case, make the schedule detailed for the next phase, analysis, and more macro for all subsequent phases. Assume that you have one academic year in which to complete the project and that the project must result in a working system.

10. What additional activities should the Stillwater team members conduct in order to prepare the details of the BPP? Describe each activity.

Part THREE

Analysis

An Overview of Part THREE

Analysis

Analysis is the first SDLC phase where you begin to understand, in depth, the needs for system changes. Systems analysis involves a substantial amount of effort and cost and is therefore undertaken only after management has decided that the systems development project under consideration has merit and should be pursued through this phase. The analysis team should not take the analysis process for granted or attempt to speed through it. Most observers would agree that many of the errors in developed systems are directly traceable to inadequate efforts in the analysis and design phases of the life cycle. As analysis is a large and involved process, we divide it into three main activities to make the overall process easier to understand:

- *Requirements determination.* This is primarily a fact-finding activity.

- *Requirements structuring.* This activity creates a thorough and clear description of current business operations and new information processing services.

- *Alternative generation and selection.* This process results in a choice among alternative strategies for subsequent systems design.

The purpose of analysis is to determine what information and information processing services are needed to support selected objectives and functions of the organization. Gathering this information is called requirements determination, the subject of Chapter 7. The fact-finding techniques in Chapter 7 are used to learn about the current system, the organization that the replacement system will support, and user requirements or expectations for the replacement system.

In Chapter 7, we also discuss a major source of new systems, Business Process Reengineering (BPR). In contrast to the incremental improvements that drive many systems development projects, BPR results in radical redesign of the processes that information systems are designed to support. We show how BPR relates to information systems analysis in Chapter 8, where we use data flow diagrams to support the reengineering process. At the end of Chapter 7, the Broadway Entertainment Company (BEC) running case shows how the results of an interview can be used to better understand the requirements for a new system and how different requirements determination techniques can be used in combination to gain a thorough understanding of requirements.

Information about current operations and requirements for a replacement system must be organized for analysis and design. Organizing, or structuring, system requirements results in diagrams and descriptions (models) that can be analyzed to show deficiencies, inefficiencies, missing elements, and illogical components of the current business operation and information systems. Along with user requirements, they are used to determine the strategy for the replacement system.

The results of the requirements determination can be structured according to three essential views of the current and replacement information systems:

- *Process*—the sequence of data movement and handling operations within the system

- *Logic and timing*—the rules by which data are transformed and manipulated and an indication of what triggers data transformation

- *Data*— the inherent structure of data independent of how or when it is processed

The *process* view of a system can be represented by data flow diagrams, the subject of Chapter 8. *System logic and timing*, the subject of Chapter 9, describes what goes on inside the process boxes in data flow diagrams and when these processes occur. System logic and timing can be represented in many ways, including Structured English, decision tables, and decision trees. Finally, the *data* view of a system, discussed in Chapter 10, shows the rules that govern the structure and integrity of data and concentrates on what data about business entities and relationships among these entities must be accessed within the system. Broadway Entertainment Company (BEC) cases appear following Chapters 8, 9, and 10 to illustrate the process, logic, and data models that describe a new system. The cases also illustrate how diagrams and models for each of these three views of a system relate to one another to form a consistent and thorough structured description of a proposed system.

Chapter 11 discusses the final step within analysis—how to choose among alternative design strategies. The systems development team realizes that there are many possible strategies for providing the features, technology, and acquisition of the desired system. Analysts will first generate several competing alternative strategies. When alternatives have been identified and documented, the systems development team presents its best choice to management, and the analysis phase ends. At this point, management has to decide whether to continue the project and proceed to design and, if so, how the design phase of the project should be structured. Chapter 11 also discusses updating the Baseline Project Plan and the transition from analysis to design. Chapter 11 concludes with a BEC case that shows how alternative design strategies are generated and evaluated and the best strategy selected. The case also explains how the Baseline Project Plan is updated and presented to management to determine if and how the project should continue.

Chapter 7

Determining System Requirements

LEARNING OBJECTIVES

After studying this chapter, you should be able to:

- Describe options for designing and conducting interviews and develop a plan for conducting an interview to determine system requirements.

- Design, distribute, and analyze questionnaires to determine system requirements.

- Explain the advantages and pitfalls of observing workers and analyzing business documents to determine system requirements.

- Explain how computing can provide support for requirements determination.

- Participate in and help plan a Joint Application Design session.

- Use prototyping during requirements determination.

- Select the appropriate methods to elicit system requirements.

- Understand how requirements determination techniques apply to development for Internet applications.

INTRODUCTION

Systems analysis is the part of the systems development life cycle in which you determine how the current information system functions and assess what users would like to see in a new system. As you learned in Chapter 1, there are three subphases in analysis: requirements determination, requirements structuring, and alternative generation and choice.

In this chapter, you will learn about the beginning subphase of analysis—determining system require-

ments. Techniques used in requirements determination have evolved over time to become more structured and, as we will see in this chapter, current methods increasingly rely on the computer for support. We will first study the more traditional requirements determination methods including interviewing, using questionnaires, observing users in their work environment, and collecting procedures and other written documents. We will then discuss modern methods for collecting system requirements. The first of these methods is Joint Application Design (JAD), which you first read about in Chapter 1. Next, you will read about how analysts rely more and more on information systems to help them perform analysis. As you will see, group support systems have been used to support systems analysis, especially as part of the JAD process. CASE tools, discussed in Chapter 4, are also very useful in requirements determination. You will learn how prototyping has become a key tool for some requirements determination efforts. Finally, you will learn how requirements analysis continues to be an important part of systems analysis and design whether the approach is radical, as with business process redesign, or new, as with developing Internet applications.

PERFORMING REQUIREMENTS DETERMINATION

As stated earlier and shown in Figure 7-1, there are three subphases to systems analysis: requirements determination, requirements structuring, and generating alternative system design strategies and selecting the best one. We will address these as three separate steps, but you should consider these steps as somewhat parallel and iterative. For example, as you determine some aspects of the current and desired system(s), you begin to structure these requirements or to build prototypes to show users how a system might behave. Inconsistencies and deficiencies discovered through structuring and prototyping lead you to explore further the operation of current system(s) and the future needs of the organization. Eventually your ideas and discoveries converge on a thorough and accurate depiction of current operations and what the requirements are for the new system. As you think about beginning the analysis phase, you probably wonder what exactly is involved in requirements determination. We discuss this process in the next section.

Figure 7-1
Systems development life cycle with analysis phase highlighted

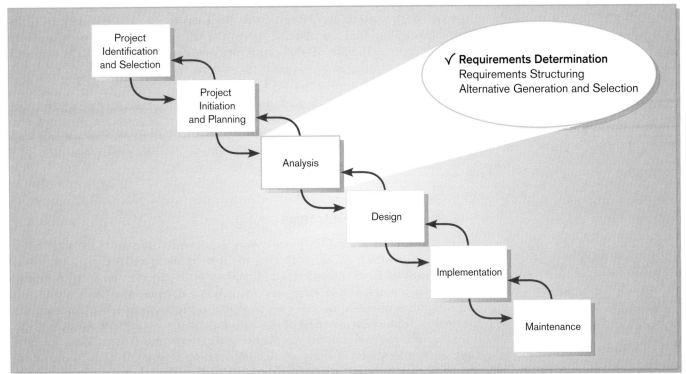

The Process of Determining Requirements

Once management has granted permission to pursue development of a new system (this was done at the end of the project identification and selection phase of the SDLC) and a project is initiated and planned (see Chapter 6), you begin determining what the new system should do. During requirements determination, you and other analysts gather information on what the system should do from as many sources as possible: from users of the current system, from observing users, and from reports, forms, and procedures. All of the system requirements are carefully documented and made ready for structuring, the subject of Chapters 8 through 10.

In many ways, gathering system requirements is like conducting any investigation. Have you read any of the Sherlock Holmes or similar mystery stories? Do you enjoy solving puzzles? From these experiences, we can detect some similar characteristics for a good systems analyst during the requirements determination subphase. These characteristics include

- *Impertinence.* You should question everything. You need to ask such questions as: Are all transactions processed the same way? Could anyone be charged something other than the standard price? Might we someday want to allow and encourage employees to work for more than one department?

- *Impartiality*: Your role is to find the best solution to a business problem or opportunity. It is not, for example, to find a way to justify the purchase of new hardware or to insist on incorporating what users think they want into the new system requirements. You must consider issues raised by all parties and try to find the best organizational solution.

- *Relax constraints*: Assume anything is possible and eliminate the infeasible. For example, do not accept this statement: "We've always done it that way, so we have to continue the practice." Traditions are different from rules and policies. Traditions probably started for a good reason but, as the organization and its environment change, traditions may turn into habits rather than sensible procedures.

- *Attention to details*: Every fact must fit with every other fact. One element out of place means that the ultimate system will fail at some time. For example, an imprecise definition of who a customer is may mean that you purge customer data when a customer has no active orders; yet these past customers may be vital contacts for future sales.

- *Reframing*: Analysis is, in part, a creative process. You must challenge yourself to look at the organization in new ways. You must consider how each user views his or her requirements. You must be careful not to jump to this conclusion: "I worked on a system like that once—this new system must work the same way as the one I built before."

Deliverables and Outcomes

The primary deliverables from requirements determination are the various forms of information gathered during the determination process: transcripts of interviews; notes from observation and analysis of documents; analyzed responses from questionnaires; sets of forms, reports, job descriptions, and other documents; and computer-generated output such as system prototypes. In short, anything that the analysis team collects as part of determining system requirements is included in the deliverables resulting from this subphase of the systems development life cycle. Table 7-1 lists examples of some specific information that might be gathered during requirements determination.

TABLE 7-1 *Deliverables for Requirements Determination*

1. **Information collected from conversations with or observations of users:** interview transcripts, questionnaire responses, notes from observation, meeting minutes
2. **Existing written information:** business mission and strategy statements, sample business forms and reports and computer displays, procedure manuals, job descriptions, training manuals, flowcharts and documentation of existing systems, consultant reports
3. **Computer-based information:** results from Joint Application Design sessions, transcripts or files from group support system sessions, CASE repository contents and reports of existing systems, and displays and reports from system prototypes.

These deliverables contain the information you need for systems analysis within the scope of the system you are developing. In addition, you need to understand the following components of an organization:

- The business objectives that drive what and how work is done
- The information people need to do their jobs
- The data (definition, volume, size, etc.) handled within the organization to support the jobs
- When, how, and by whom or what the data are moved, transformed, and stored
- The sequence and other dependencies among different data-handling activities
- The rules governing how data are handled and processed
- Policies and guidelines that describe the nature of the business and the market and environment in which it operates
- Key events affecting data values and when these events occur

As should be obvious, such a large amount of information must be organized in order to be useful. This is the purpose of the next subphase—requirements structuring.

From just this subphase of analysis, you probably already realize that the amount of information to be gathered could be huge, especially if the scope of the system under development is broad. The time required to collect and structure a great deal of information can be extensive and, because it involves so much human effort, quite expensive. Too much analysis is not productive and the term "analysis paralysis" has been coined to describe a systems development project that has bogged down in an abundance of analysis work. Because of the dangers of excessive analysis, today's systems analysts focus more on the system to be developed than on the current system. The techniques you will learn about later in the chapter, JAD and prototyping, were developed to keep the analysis effort at a minimum yet still effective. Other processes have been developed to limit the analysis commitment even more, providing an alternative to the SDLC. One of these processes is called *Rapid Application Development (RAD)* and is the subject of Chapter 19. As you will learn, RAD relies on JAD, prototyping, and integrated CASE tools to be effective. Even RAD, as well as the structured analysis methods discussed in subsequent chapters in the analysis section of this book, rely on a basic understanding of the business area served by an information system. Thus, before you can fully appreciate RAD, you need to learn about traditional fact-gathering techniques. These techniques are the subject of the next section.

TRADITIONAL METHODS FOR DETERMINING REQUIREMENTS

At the core of systems analysis is the collection of information. At the outset, you must collect information about the information systems that are currently being used and how users would like to improve the current systems and organizational operations

TABLE 7-2   Traditional Methods of Collecting System Requirements

- Individually *interview* people informed about the operation and issues of the current system and needs for systems in future organizational activities
- Survey people via *questionnaires* to discover issues and requirements
- *Interview groups* of people with diverse needs to find synergies and contrasts among system requirements
- *Observe workers* at selected times to see how data are handled and what information people need to do their jobs
- *Study business documents* to discover reported issues, policies, rules, and directions as well as concrete examples of the use of data and information in the organization

with new or replacement information systems. One of the best ways to get this information is to talk to the people who are directly or indirectly involved in the different parts of the organizations affected by the possible system changes: users, managers, funders, etc. Another way to find out about the current system is to gather copies of documentation relevant to current systems and business processes. In this chapter, you will learn about various ways to get information directly from stakeholders: interviews, questionnaires, group interviews, the Nominal Group Technique, and direct observation. You will learn about collecting documentation on the current system and organizational operation in the form of written procedures, forms, reports, and other hard copy. These traditional methods of collecting system requirements are listed in Table 7-2.

Interviewing and Listening

Interviewing is one of the primary ways analysts gather information about an information systems project. Early in a project, an analyst may spend a large amount of time interviewing people about their work, the information they use to do it, and the types of information processing that might supplement their work. Other stakeholders are interviewed to understand organizational direction, policies, expectations managers have on the units they supervise, and other nonroutine aspects of organizational operations. During interviewing you will gather facts, opinions, and speculation and observe body language, emotions, and other signs of what people want and how they assess current systems.

There are many ways to effectively interview someone and no one method is necessarily better than another. Some guidelines you should keep in mind when you interview, summarized in Table 7-3, are discussed next.

First, you should prepare thoroughly before the interview. Set up an appointment at a time and for a duration convenient for the interviewee. The general nature of the interview should be explained to the interviewee in advance. You may ask the interviewee to think about specific questions or issues or to review certain documentation to prepare for the interview. You should spend some time thinking about what you need to find out and write down your questions. Do not assume that you can anticipate all possible questions. You want the interview to be natural and, to some degree, you want to spontaneously direct the interview as you discover what expertise the interviewee brings to the session.

You should prepare an interview guide or checklist so that you know in which sequence you intend to ask your questions and how much time you want to spend in each area of the interview. The checklist might include some probing questions to ask as follow-up if you receive certain anticipated responses. You can, to some degree, integrate your interview guide with the notes you take during the interview, as depicted in a sample guide in Figure 7-2. This same guide can serve as an outline for a summary of what you discover during an interview.

The first page of the sample interview guide contains a general outline of the interview. Besides basic information on who is being interviewed and when, you list major

TABLE 7-3   Guidelines for Effective Interviewing

Plan the Interview
- Prepare interviewee: appointment, priming questions
- Prepare checklist, agenda, and questions

Listen carefully and take notes (tape record if permitted)

Review notes within 48 hours of interview

Be neutral

Seek diverse views

objectives for the interview. These objectives typically cover the most important data you need to collect, a list of issues on which you need to seek agreement (for example, content for certain system reports), and which areas you need to explore, not necessarily with specific questions. You also include reminder notes to yourself on key information about the interviewee (for example, job history, known positions taken on issues, and role with current system). This information helps you to be personal, shows that you consider the interviewee important, and may assist you in interpreting some answers. Also included is an agenda for the interview with approximate time limits for different sections of the interview. You may not follow the time limits precisely but the schedule helps you cover all areas during the time the interviewee is available. Space is also allotted for general observations that do not fit under specific questions and for notes taken during the interview about topics skipped or issues raised that could not be resolved.

On subsequent pages you list specific questions; the sample form in Figure 7-2 includes space for taking notes on these questions. Because unanticipated information arises, you will not strictly follow the guide in sequence. You can, however, check off the questions you have asked and write reminders to yourself to return to or skip certain questions as the dynamics of the interview unfold.

Interview Outline	

Interviewee:
  Name of person being interviewed
	Interviewer:
	  Name of person leading interview

Location/Medium:
  Office, conference room,
  or phone number
	Appointment Date:
	Start Time:
	End Time:

Objectives:
  What data to collect
  On what to gain agreement
  What areas to explore
	Reminders:
	  Background/experience of interviewee
	  Known opinions of interviewee

Agenda:
Introduction
Background on Project
Overview of Interview
  Topics To Be Covered
  Permission to Tape Record
Topic 1 Questions
Topic 2 Questions
...
Summary of Major Points
Questions from Interviewee
Closing
	Approximate Time:
	1 minute
	2 minutes
	
	1 minute
	
	5 minutes
	7 minutes
	...
	2 minutes
	5 minutes
	1 minute

General Observations: *Interviewee seemed busy–probably need to call in a few days for follow-up questions since he gave only short answers. PC was turned off–probably not a regular PC user.*	
Unresolved Issues, Topics not Covered: *He needs to look up sales figures from 1999. He raised the issue of how to handle returned goods, but we did not have time to discuss.*	
	(continues)

Figure 7-2
Typical interview guide

Figure 7-2 (continued)
Typical interview guide

Interviewee:	Date:
Questions:	Notes:

When to ask question, if conditional
Question number: 1
    Have you used the current sales
tracking system? If so, how often?
	Answer
	    Yes, I ask for a report on my
	    product line weekly
	
	Observations
	    Seemed anxious—may be
	    overestimating usage frequency

If yes, go to Question 2	

Question: 2
    What do you like least about the
system?
	Answer
	    Sales are shown in units, not
	    dollars
	
	Observations
	    System can show sales in dollars,
	    but user does not know this.

Open-ended questions: Questions in interviews and on questionnaires that have no prespecified answers.

Closed-ended questions: Questions in interviews and on questionnaires that ask those responding to choose from among a set of specified responses.

Choosing Interview Questions You need to decide what mix and sequence of open-ended and closed-ended questions you will use. **Open-ended questions** are usually used to probe for information for which you cannot anticipate all possible responses or for which you do not know the precise question to ask. The person being interviewed is encouraged to talk about whatever interests him or her within the general bounds of the question. An example is, "What would you say is the best thing about the information system you currently use to do your job?" or "List the three most frequently used menu options." You must react quickly to answers and determine whether or not any follow-up questions are needed for clarification or elaboration. Sometimes body language will suggest that a user has given an incomplete answer or is reluctant to divulge some information; a follow-up question might yield additional insight. One advantage of open-ended questions in an interview is that previously unknown information can surface. You can then continue exploring along unexpected lines of inquiry to reveal even more new information. Open-ended questions also often put the interviewees at ease since they are able to respond in their own words using their own structure; open-ended questions give interviewees more of a sense of involvement and control in the interview. A major disadvantage of open-ended questions is the length of time it can take for the questions to be answered. In addition, open-ended questions can be difficult to summarize.

Closed-ended questions provide a range of answers from which the interviewee may choose. Here is an example:

Which of the following would you say is the one best thing about the information system you currently use to do your job (pick only one):
a. Having easy access to all of the data you need
b. The system's response time
c. The ability to access the system from remote locations

Closed-ended questions work well when the major answers to questions are well known. Another plus is that interviews based on closed-ended questions do not necessarily require a large time commitment—more topics can be covered. As opposed to collecting such information via questionnaires, you can see body language and hear voice tone, which can aid in interpreting the interviewee's responses. Closed-ended questions can also be an easy way to begin an interview and to determine which line of open-ended questions to pursue. You can include an "other" option to encourage the interviewee to add unanticipated responses. A major disadvantage of closed-ended questions is that useful information that does not quite fit into the defined answers may be overlooked as the respondent tries to make a choice instead of providing his or her best answer.

Closed-ended questions, like objective questions on an examination, can follow several forms, including these choices:

- True or false.

- Multiple choice (with only one response or selecting all relevant choices).

- Rating a response or idea on some scale, say from bad to good or strongly agree to strongly disagree. Each point on the scale should have a clear and consistent meaning to each person and there is usually a neutral point in the middle of the scale.

- Ranking items in order of importance.

Interview Guidelines First, with either open- or closed-ended questions, do not phrase a question in a way that implies a right or wrong answer. The respondent must feel that he or she can state his or her true opinion and perspective and that his or her idea will be considered equally with those of others. Questions such as, "Should the system continue to provide the ability to override the default value, even though most users now do not like the feature?" should be avoided as such wording predefines a socially acceptable answer.

The second guideline to remember about interviews is to listen very carefully to what is being said. Take careful notes or, if possible, record the interview on a tape recorder (be sure to ask permission first!). The answers may contain extremely important information for the project. Also, this may be the only chance you have to get information from this particular person. If you run out of time and still need to get information from the person you are talking to, ask to schedule a follow-up interview.

Third, once the interview is over, go back to your office and type up your notes within 48 hours. If you recorded the interview, use the recording to verify the material in your notes. After 48 hours, your memory of the interview will fade quickly. As you type and organize your notes, write down any additional questions that might arise from lapses in your notes or from ambiguous information. Separate facts from your opinions and interpretations. Make a list of unclear points that need clarification. Call the person you interviewed and get answers to these new questions. Use the phone call as an opportunity to verify the accuracy of your notes. You may also want to send a written copy of your notes to the person you interviewed so the person can check your notes for accuracy. Finally, make sure you thank the person for his or her time. You may need to talk to your respondent again. If the interviewee will be a user of your system or is involved in some other way in the system's success, you want to leave a good impression.

Fourth, be careful during the interview not to set expectations about the new or replacement system unless you are sure these features will be part of the delivered system. Let the interviewee know that there are many steps to the project and the perspectives of many people need to be considered, along with what is technically possible. Let respondents know that their ideas will be carefully considered, but that due to the iterative nature of the systems development process, it is premature to say now exactly what the ultimate system will or will not do.

Fifth, seek a variety of perspectives from the interviews. Find out what potential users of the system, users of other systems that might be affected by changes, man-

N E T   S E A R C H
There are commercially available services that can help you conduct interviews over the phone. Visit http://www.prenhall.com/hoffer to complete an exercise related to this topic.

agers and superiors, information systems staff who have experience with the current system, and others think the current problems and opportunities are and what new information services might better serve the organization. You want to understand all possible perspectives so that in a later approval step you will have information on which to base a recommendation or design decision that all stakeholders can accept.

Administering Questionnaires

Interviews are very effective ways of communicating with people and obtaining important information from them. However, interviews are also very expensive and time-consuming to conduct. Thus, a limited number of questions can be covered and people contacted. In contrast, questionnaires are passive and often yield less depth of understanding than interviews; however, questionnaires are not as expensive to administer per respondent. In addition, questionnaires have the advantage of gathering information from many people in a relatively short time and of being less biased in the interpretation of their results.

Choosing Questionnaire Respondents Sometimes there are more people to survey than you can handle and you must decide which set of people to send the questionnaire to or which questionnaire to send to which group of people. Whichever group of respondents you choose, it should be representative of all users. In general, you can achieve a representative sample by any one or any combination of these four methods:

1. Those *convenient to* sample: these may be people at a local site, those willing to be surveyed, or those most motivated to respond

2. A *random* group: if you get a list of all users of the current system, simply choose every nth person on the list; or, you could select people by skipping names on the list based on numbers from a random number table

3. A *purposeful* sample: here you may specify only people who satisfy certain criteria, such as users of the system for more than two years or users who use the system most often

4. A *stratified* sample: in this case, you have several categories of people whom you definitely want to include—choose a random set from each category (e.g., users, managers, foreign business unit users)

Samples that combine characteristics of several approaches are also common. In any case, once the questionnaires are returned, you should check for nonresponse bias; that is, a systematic bias in the results since those who responded are different from those who did not respond. You can refer to books on survey research to find out how to determine if your results are confounded by nonresponse bias.

Designing Questionnaires Questionnaires are usually administered on paper although they can be administered in person (resembling a structured interview), over the phone (computer-assisted telephone interviewing), or even on diskette or Website. Questionnaires are less expensive, however, if they do not require a person to administer them directly; that is, if the people answering the questions can complete the questionnaire without help. Also, answers can be provided at the convenience of the respondent, as long as the answers are returned by a specific date.

Questionnaires typically include closed-ended questions, more than can be effectively asked in an interview, and sometimes contain open-ended questions as well. Closed-ended questions are preferable because they are easier to complete and they define the exact coverage required. A few open-ended questions give the person being surveyed an opportunity to add insights not anticipated by the designer of the questionnaire. In general, questionnaires take less time to complete than interviews structured to obtain the same information. In addition, questionnaires are given to many people simultaneously whereas interviews are usually limited to one person at a time.

Questionnaires are generally less rich in information content than interviews, however, because they provide no direct means by which to ask follow-up questions (although it is possible, though time-consuming, to contact respondents after they have returned their completed questionnaires to ask for further information). Also, since questionnaires are written, they do not provide the opportunity to judge the accuracy of the responses. In an interview, you can sometimes determine if people are answering truthfully or fully by the words they use, whether they make direct eye contact, the tone of voice they use, or their body language.

The ability to create good questionnaires is a skill that improves with practice and experience. Because the questions are written, they must be extremely clear in meaning and logical in sequence. When a person is completing a questionnaire, he or she only has the written questions to interpret and answer. You are not there to clarify each question's meaning. For example, what if a closed-ended question were phrased in this way:

How often do you back up your computer files?
a. Frequently
b. Sometimes
c. Hardly at all
d. Never

There are at least two sources of ambiguity in the wording of the question. The first source of ambiguity is the categories offered for the answer: the only nonambiguous answer is "never." "Hardly at all" could mean anything from once per year to once per month, depending on who is answering the question. "Sometimes" could cover the same range of possibilities as "Hardly at all." "Frequently" could be anything from once per hour to once per week. The second source of ambiguity is in the question itself. Does the term "computer files" pertain only to those on my hard disk? Or does it also mean the files I have stored on floppy disk? What if I have more than one PC in my office? And what about the files I have stored on the minicomputer I use for certain applications? I don't back up those files; the system operator does it on a regular basis for all minicomputer files, not just mine. With no questioner present to explain the ambiguities, the respondent is at a loss and must try to answer the question in the best way he or she knows how. Whether the respondent's interpretation is the same as other respondents' is anyone's guess. The respondent cannot be there when the data are analyzed to tell exactly what was meant.

A less ambiguous way to phrase the question and its response categories would be something like this:

How often do you back up the computer files stored on the hard disk on the PC you use for over 50%
of your work time?
a. Frequently (at least once per week)
b. Sometimes (from one to three times per month)
c. Hardly at all (once per month or less)
d. Never

As you can see from the wording of the question, the phrasing is a bit awkward, but it avoids ambiguity. You may want to break up a single question into multiple questions, or a set of questions and statements, to avoid awkward phrasing. Notice also that the possible responses are much clearer now that they have been specifically defined, and they cover the full range of possibilities, from never to at least once per week with no overlapping time periods.

Obviously, care must be taken in the task of composing closed-ended and open-ended questions. Further, you should be as careful in composing questions for interviews as for questionnaires, since sloppily worded questions cannot be identified every time in an interview unless the interviewee asks for clarification. For both interviews and questionnaires, it is wise to pretest your questions. Pose the questions in a simulated interview and ask the interviewee to rephrase each question as he or she interprets the question. Check responses for reasonableness. You can even ask the same question in what you think are several different ways to see if you receive a materially different response. Use this feedback to adjust the questions to make them less ambiguous.

Questionnaires are most useful in the requirements determination process when used for very specific purposes rather than for more general information gathering. For example, one useful application of questionnaires is to measure levels of user satisfaction with a system or with particular aspects of it. Another useful application is to have several users choose from among a list of system features available in many off-the-shelf software packages. You could ask users to choose the features they most want and quickly tabulate the results to find out which features are most in demand. You could then recommend a system solution based on a particular software package to meet the demands of most of the users.

Choosing Between Interviews and Questionnaires

To summarize the previous sections, you can see that interviews are good tools for collecting rich, detailed information and that interviews allow exploration and follow-up (see Table 7-4). On the other hand, interviews are quite time-intensive and expensive. In comparison, questionnaires are inexpensive and take less time, as specific information can be gathered from many people at once without the personal intervention of an interviewer. The information collected from a questionnaire is less rich, however, and is potentially ambiguous if questions are not phrased precisely. In addition, follow-up to a questionnaire is more difficult as it often involves interviews or phone calls, adding to the expense of the process.

These differences and others are important for you to remember during the analysis phase. Deciding which method to use and what strategy to employ to gather information will vary with the system being studied and its organizational context. For example, if the organization is large and the system being studied is vast and complex, then there will probably be dozens of affected users and stakeholders. If you know little about the system or the organization, a good strategy is to identify key users and stakeholders and interview them. You would then use the information gathered in the interviews to create a questionnaire that would be distributed to a large number of users. You could then schedule follow-up interviews with a few users. At the other extreme, if the system and organization are small and you understand them well, the best strategy may be to interview only one or two key users or stakeholders.

TABLE 7-4   Comparison of Interviews and Questionnaires

Characteristic	Interviews	Questionnaires
Information Richness	High (many channels)	Medium to low (only responses)
Time Required	Can be extensive	Low to moderate
Expense	Can be high	Moderate
Chance for Follow-up and Probing	Good: probing and clarification questions can be asked by either interviewer or interviewee	Limited: probing and follow-up done after original data collection
Confidentiality	Interviewee is known to interviewer	Respondent can be unknown
Involvement of Subject	Interviewee is involved and committed	Respondent is passive, no clear commitment
Potential Audience	Limited numbers, but complete responses from those interviewed	Can be quite large, but lack of response from some can bias results

Interviewing Groups

One drawback to using interviews and questionnaires to collect systems requirements is the need for the analyst to reconcile apparent contradictions in the information collected. A series of interviews may turn up inconsistent information about the current system or its replacement. You must work through all of these inconsistencies to figure out what the most accurate representation of current and future systems might be. Such a process requires several follow-up phone calls and additional interviews. Catching important people in their offices is often difficult and frustrating, and scheduling new interviews may become very time-consuming. In addition, new interviews may reveal new questions that in turn require additional interviews with those interviewed earlier. Clearly, gathering information about an information system through a series of individual interviews and follow-up calls is not an efficient process.

Another option available to you is the group interview. In a group interview, you interview several key people at once. To make sure all of the important information is collected, you may conduct the interview with one or more analysts. In the case of multiple interviewers, one analyst may ask questions while another takes notes, or different analysts might concentrate on different kinds of information. For example, one analyst may listen for data requirements while another notes the timing and triggering of key events. The number of interviewees involved in the process may range from two to however many you believe can be comfortably accommodated.

A group interview has a few advantages. One, it is a much more effective use of your time than is a series of interviews with individuals (although the time commitment of the interviewees may be more of a concern). Two, interviewing several people together allows them to hear the opinions of other key people and gives them the opportunity to agree or disagree with their peers. Synergies also often occur. For example, the comments of one person might cause another person to say, "That reminds me of . . ." or "I didn't know that was a problem." You can benefit from such a discussion as it helps you identify issues on which there is general agreement and areas where views diverge widely.

The primary disadvantage of a group interview is the difficulty in scheduling it. The more people involved, the more difficult it will be finding a convenient time and place for everyone. Modern technology such as video conferences and video phones can minimize the geographical dispersion factors that make scheduling meetings so difficult. Group interviews are at the core of the Joint Application Design process, which we discuss in a later section in this chapter. A specific technique for working with groups, Nominal Group Technique, is discussed next.

Nominal Group Technique

Many different techniques have been developed over the years to improve the process of working with groups. One of the more popular techniques for generating ideas among group members is called **Nominal Group Technique** (NGT). NGT is exactly what the name indicates—the individuals working together to solve a problem are a group in name only, or nominally. Group members may be all gathered in the same room for NGT, but they all work alone for a period of time. Typically, group members make a written list of their ideas. At the end of the idea generation time, group members pool their individual ideas under the guidance of a trained facilitator. Pooling usually involves having the facilitator ask each person in turn for an idea that has not been presented before. As the person reads the idea aloud, someone else writes down the idea on a blackboard or flip chart. After all of the ideas have been introduced, the facilitator will then ask for the group to openly discuss each idea, primarily for clarification.

Once all of the ideas are understood by all of the participants, the facilitator will try to reduce the number of ideas the group will carry forward for additional consid-

Nominal Group Technique: A facilitated process that supports idea generation by groups. At the beginning of the process, group members work alone to generate ideas, which are then pooled under the guidance of a trained facilitator.

eration. There are many ways to reduce the number of ideas. The facilitator may ask participants to choose only a subset of ideas that they believe are important. Then the facilitator will go around the room, asking each person to read aloud an idea important to him or her, which has not yet been identified by someone else. Or the facilitator may work with the group to identify and either eliminate or combine ideas that are very similar to others. At some point, the facilitator and the group end up with a tractable set of ideas, which can be further prioritized.

In a requirements determination context, the ideas being sought in an NGT exercise would typically apply to problems with the existing system or ideas for new features in the system being developed. The end result would be a list of either problems or features that group members themselves had generated and prioritized. There should be a high level of ownership of such a list, at least for the group that took part in the NGT exercise.

There is some evidence to support the use of NGT to help focus and refine the work of a group, in that the number and quality of ideas that result from an NGT may be higher than what would normally be obtained from an unfacilitated group meeting. An NGT exercise could be used to complement the work done in a typical group interview, or a part of a Joint Application Design effort, described in more detail later in the chapter.

Directly Observing Users

All the methods of collecting information that we have been discussing up until now involve getting people to recall and convey information they have about an organizational area and the information systems that support these processes. People, however, are not always very reliable informants, even when they try to be reliable and tell what they think is the truth. As odd as it may sound, people often do not have a completely accurate appreciation of what they do or how they do it. This is especially true concerning infrequent events, issues from the past, or issues for which people have considerable passion. Since people cannot always be trusted to reliably interpret and report their own actions, you can supplement and corroborate what people tell you by watching what they do or by obtaining relatively objective measures of how people behave in work situations. (See the box "Lost Soft Drink Sales" for an example of the importance of systems analysts learning firsthand about the business for which they are designing systems.)

For example, one possible view of how a hypothetical manager does her job is that a manager carefully plans her activities, works long and consistently on solving problems, and controls the pace of her work. A manager might tell you that is how she spends her day. When Mintzberg (1973) observed how managers work, however, he found that a manager's day is actually punctuated by many, many interruptions. Managers work in a fragmented manner, focusing on a problem or on a communication for only a short time before they are interrupted by phone calls or visits from their subordinates and other managers. An information system designed to fit the work environment described by our hypothetical manager would not effectively support the actual work environment in which that manager finds herself.

As another example, consider the difference between what another employee might tell you about how much he uses electronic mail and how much electronic mail use you might discover through more objective means. An employee might tell you he is swamped with e-mail messages and that he spends a significant proportion of his time responding to e-mail messages. However, if you were able to check electronic mail records, you might find that this employee receives only three e-mail messages per day on average, and that the most messages he has ever received during one eight-hour period is ten. In this case, you were able to obtain an accurate behavioral measure of how much e-mail this employee copes with without having to watch him read his e-mail.

Lost Soft Drink Sales

A systems analyst was quite surprised to read that sales of all soft drink products were lower, instead of higher, after a new delivery truck routing system was installed. The software was designed to reduce stock-outs at customer sites by allowing drivers to visit each customer more often using more efficient delivery routes.

Confused by the results, management asked the analyst to delay a scheduled vacation, but he insisted that he could look afresh at the system only after a few overdue days of rest and relaxation.

Instead of taking a vacation, however, the analyst called a delivery dispatcher he had interviewed during the design of the system and asked to be given a route for a few days. The analyst drove a route (for a regular driver actually on vacation), following the schedule developed from the new system. What the analyst discovered was that the route was very efficient, as expected; so at first the analyst could not see any reason for lost sales.

During the third and last day of his "vacation" the analyst stayed overtime at one store to ask the manager if she had any ideas why sales might have dropped off in recent weeks. The manager had no explanation, but did make a seemingly unrelated observation that the regular route driver appeared to have less time to spend in the store. He did not seem to take as much interest in where the products were displayed and did not ask for promotional signs to be displayed, as he had often done in the past.

From this conversation, the analyst concluded that the new delivery truck routing system was, in one sense, too good. It placed the driver on such a tight schedule that a driver had no time left for the "schmoozing" required to get special treatment, that gave the company's products an edge over the competition.

Without firsthand observation of the system in action participating as a system user, the analyst might never have discovered the true problem with the system design. Once time was allotted for not only stocking new products but also for necessary marketing work, product sales returned to and exceeded levels achieved before the new system had been introduced.

The intent behind obtaining system records and direct observation is the same, however, and that is to obtain more firsthand and objective measures of employee interaction with information systems. In some cases, behavioral measures will be a more accurate reflection of reality than what employees themselves believe. In other cases, the behavioral information will substantiate what employees have told you directly. Although observation and obtaining objective measures are desirable ways to collect pertinent information, such methods are not always possible in real organizational settings. Thus, these methods are not totally unbiased, just as no other one data-gathering method is unbiased.

For example, observation can cause people to change their normal operating behavior. Employees who know they are being observed may be nervous and make more mistakes than normal, may be careful to follow exact procedures they do not typically follow, and may work faster or slower than normal. Moreover, since observation typically cannot be continuous, you receive only a snapshot image of the person or task you observe, which may not include important events or activities. Since observation is very time-consuming, you will not only observe for a limited time but also a limited number of people and at a limited number of sites. Again, observation yields only a small segment of data from a possibly vast variety of data sources. Exactly which people or sites to observe is a difficult selection problem. You want to pick both typical and atypical people and sites and observe during normal and abnormal conditions and times to receive the richest possible data from observation.

Analyzing Procedures and Other Documents

As noted above, asking questions of the people who use a system every day or who have an interest in a system is an effective way to gather information about current and future systems. Observing current system users is a more direct way of seeing how an existing system operates, but even this method provides limited exposure to all aspects of current operations. These methods of determining system requirements can be enhanced by examining system and organizational documentation to discover more details about current systems and the organization these systems support.

Although we discuss here several important types of documents that are useful in understanding possible future system requirements, our discussion does not exhaust all possibilities. You should attempt to find all written documents about the organizational areas relevant to the systems under redesign. Besides the few specific documents we discuss, organizational mission statements, business plans, organization charts, business policy manuals, job descriptions, internal and external correspondence, and reports from prior organizational studies can all provide valuable insight.

What can the analysis of documents tell you about the requirements for a new system? In documents you can find information about

- Problems with existing systems (e.g., missing information or redundant steps)
- Opportunities to meet new needs if only certain information or information processing were available (e.g., analysis of sales based on customer type)
- Organizational direction that can influence information system requirements (e.g., trying to link customers and suppliers more closely to the organization)
- Titles and names of key individuals who have an interest in relevant existing systems (e.g., the name of a sales manager who led a study of buying behavior of key customers)
- Values of the organization or individuals who can help determine priorities for different capabilities desired by different users (e.g., maintaining market share even if it means lower short-term profits)
- Special information processing circumstances that occur irregularly that may not be identified by any other requirements determination technique (e.g., special handling needed for a few very large-volume customers and which requires use of customized customer ordering procedures)
- The reason why current systems are designed as they are, which can suggest features left out of current software, which may now be feasible and more desirable (e.g., data about a customer's purchase of competitors' products were not available when the current system was designed; these data are now available from several sources)
- Data, rules for processing data, and principles by which the organization operates that must be enforced by the information system (e.g., each customer is assigned exactly one sales department staff member as a primary contact if the customer has any questions)

One type of useful document is a written work procedure for an individual or a work group. The procedure describes how a particular job or task is performed, including data and information that are used and created in the process of performing the job. For example, the procedure shown in Figure 7-3 includes data (list of features and advantages, drawings, inventor name, and witness names) required to prepare an invention disclosure. It also indicates that besides the inventor, the vice president for research and department head and dean must review the material, and that a witness is required for any filing of an invention disclosure. These insights clearly affect what data must be kept, to whom information must be sent, and the rules that govern valid forms.

Procedures are not trouble-free sources of information, however. Sometimes your analysis of several written procedures will reveal a duplication of effort in two or more jobs. You should call such duplication to the attention of management as an issue to be resolved before system design can proceed. That is, it may be necessary to redesign the organization before the redesign of an information system can achieve its full benefits. Another problem you may encounter with a procedure occurs when the procedure is missing. Again, it is not your job to create a document for a missing procedure—that is up to management. A third and common problem with a written procedure happens when the procedure is out of date. You may realize the proce-

Figure 7-3
Example of a procedure

GUIDE FOR PREPARATION OF INVENTION DISCLOSURE
(See FACULTY and STAFF MANUALS for detailed
Patent Policy and routing procedures.)

(1) DISCLOSE ONLY ONE INVENTION PER FORM.

(2) PREPARE COMPLETE DISCLOSURE.

The disclosure of your invention is adequate for patent purposes ONLY if it enables a person skilled in the art to understand the invention.

(3) CONSIDER THE FOLLOWING IN PREPARING A COMPLETE DISCLOSURE:

(a) All essential elements of the invention, their relationship to one another, and their mode of operation.

(b) Equivalents that can be substituted for any elements.

(c) List of features believed to be new.

(d) Advantages this invention has over the prior art.

(e) Whether the invention has been built and/or tested.

(4) PROVIDE APPROPRIATE ADDITIONAL MATERIAL.

Drawings and descriptive material should be provided as needed to clarify the disclosure. Each page of this material must be signed and dated by each inventor and properly witnessed. A copy of any current and/or planned publication relating to the invention should be included.

(4) INDICATE PRIOR KNOWLEDGE AND INFORMATION.

Pertinent publications, patents or previous devices, and related research or engineering activities should be identified.

(5) HAVE DISCLOSURE WITNESSED.

Persons other than co-inventors should serve as witnesses and should sign each sheet of the disclosure only after reading and understanding the disclosure.

(7) FORWARD ORIGINAL PLUS ONE COPY (two copies if supported by grant/contract) TO VICE PRESIDENT FOR RESEARCH VIA DEPARTMENT HEAD AND DEAN.

dure is out of date when you interview the person responsible for performing the task described in the procedure. Once again, the decision to rewrite the procedure so that it matches reality is made by management, but you may make suggestions based upon your understanding of the organization. A fourth problem often encountered with written procedures is that the formal procedures may contradict information you collected from interviews, questionnaires, and observation about

Formal system: The official way a system works as described in organizational documentation.

Informal system: The way a system actually works.

how the organization operates and what information is required. As in the other cases, resolution rests with management.

All of these problems illustrate the difference between **formal systems** and **informal systems**. Formal systems are systems recognized by the official documentation of the organization; informal systems are the way in which the organization actually works. Informal systems develop because of inadequacies of formal procedures, individual work habits and preferences, resistance to control, and other factors. It is important to understand both formal and informal systems since each provides insight into information requirements and what will be required to convert from present to future information services.

A second type of document useful to systems analysts is a business form (see Figure 7-4). Forms are used for all types of business functions, from recording an

Figure 7-4
A blank invoice form
(Source: http://www.giraffeonline.com) (Used by permission.)

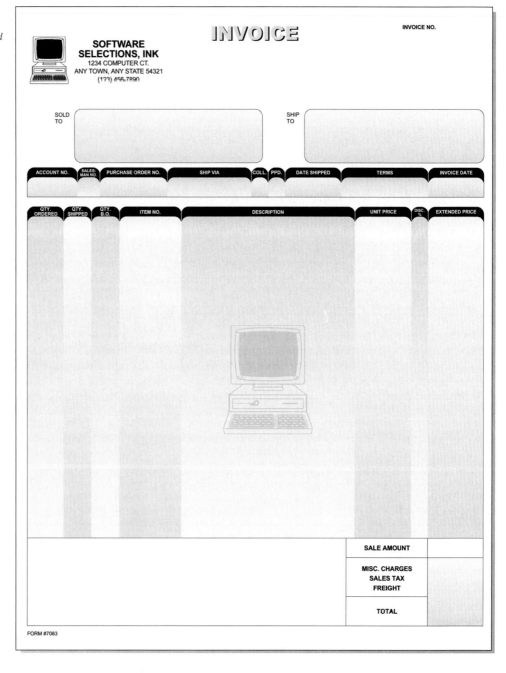

order acknowledging the payment of a bill to indicating what goods have been shipped. Forms are important for understanding a system because they explicitly indicate what data flow in or out of a system and which are necessary for the system to function. In the sample invoice form in Figure 7-4, we see locations for data such as the name of the customer, the customer's sold to and ship to addresses, data (item number, quantity, etc.) about each line item on the invoice, and calculated data such as tax, freight, and totals.

The form gives us crucial information about the nature of the organization. For example, the company can ship and bill to different addresses; customers can have discounts applied; and the freight expense is charged to the customer. A printed form may correspond to a computer display that the system will generate for someone to enter and maintain data or to display data to on-line users. Forms are most useful to you when they contain actual organizational data, as this allows you to determine the characteristics of data that are actually used by the application. The ways in which people use forms change over time, and data that were needed when a form was designed may no longer be required. You can use the systems analysis techniques presented in Chapters 8 through 10 to help you determine which data are no longer required.

A third type of useful document is a report generated by current systems. As the primary output for some types of systems, a report enables you to work backwards from the information on the report to the data that must have been necessary to generate them. Figure 7-5 presents an example of a typical financial report. This report shows

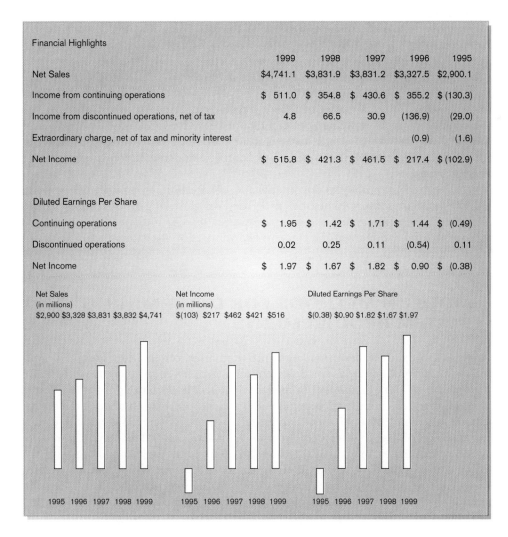

Figure 7-5
Financial highlights report from Corning
(*Source: http://www.corning.com/investor/annual_report/99_annual/financial_highlights.htm*) (*Reprinted by permission of Corning, Inc.*)

Financial Highlights

	1999	1998	1997	1996	1995
Net Sales	$4,741.1	$3,831.9	$3,831.2	$3,327.5	$2,900.1
Income from continuing operations	$ 511.0	$ 354.8	$ 430.6	$ 355.2	$ (130.3)
Income from discontinued operations, net of tax	4.8	66.5	30.9	(136.9)	(29.0)
Extraordinary charge, net of tax and minority interest				(0.9)	(1.6)
Net Income	$ 515.8	$ 421.3	$ 461.5	$ 217.4	$ (102.9)

Diluted Earnings Per Share

	1999	1998	1997	1996	1995
Continuing operations	$ 1.95	$ 1.42	$ 1.71	$ 1.44	$ (0.49)
Discontinued operations	0.02	0.25	0.11	(0.54)	0.11
Net Income	$ 1.97	$ 1.67	$ 1.82	$ 0.90	$ (0.38)

Net Sales (in millions)
$2,900 $3,328 $3,831 $3,832 $4,741

Net Income (in millions)
$(103) $217 $462 $421 $516

Diluted Earnings Per Share
$(0.38) $0.90 $1.82 $1.67 $1.97

1995 1996 1997 1998 1999   1995 1996 1997 1998 1999   1995 1996 1997 1998 1999

TABLE 7-5   Comparison of Observation and Document Analysis

Characteristic	Observation	Document Analysis
Information Richness	High (many channels)	Low (passive) and old
Time Required	Can be extensive	Low to moderate
Expense	Can be high	Low to moderate
Chance for Follow-up and Probing	Good: probing and clarification questions can be asked during or after observation	Limited: probing possible only if original author is available
Confidentiality	Observee is known to interviewer; observee may change behavior when observed	Depends on nature of document; does not change simply by being read
Involvement of Subject	Interviewees may or may not be involved and committed depending on whether they know if they are being observed	None, no clear commitment
Potential Audience	Limited numbers and limited time (snapshot) of each	Potentially biased by which documents were kept or because document not created for this purpose

the financial highlights for a corporation for five consecutive years. The report shows the information in both tabular and graphical formats. You would analyze such reports to determine which data need to be captured over what time period and what manipulation of these raw data would be necessary to produce each field on the report.

If the current system is computer-based, a fourth set of useful documents are those that describe the current information systems—how they were designed and how they work. There are a lot of different types of documents that fit this description, everything from flowcharts to data dictionaries and CASE tool reports to user manuals. An analyst who has access to such documents is lucky, as many in-house–developed information systems lack complete documentation (unless a CASE tool has been used).

Analysis of organizational documents and observation, along with interviewing and questionnaires, are the methods most used for gathering system requirements. In Table 7-4 we summarized the comparative features of interviews and questionnaires. Table 7-5 summarizes the comparative features of observation and analysis of organizational documents.

MODERN METHODS FOR DETERMINING SYSTEM REQUIREMENTS

Even though we called interviews, questionnaires, observation, and document analysis traditional methods for determining a system's requirements, all of these methods are still very much used by analysts to collect important information. Today, however, there are additional techniques to collect information about the current system, the organizational area requesting the new system, and what the new system should be like. In this section, you will learn about several modern information-gathering techniques for analysis (listed in Table 7-6): Joint Application Design (JAD), group support systems, CASE tools, and prototyping. As we said earlier, these techniques can support effective information collection and structuring while reducing the amount of time required for analysis. An alternative to the SDLC, RAD, which combines JAD, CASE tools, and prototyping, is described in more detail in Chapter 19.

TABLE 7-6   Modern Methods for Collecting System Requirements

- Bringing together in a *Joint Application Design* (*JAD*) session users, sponsors, analysts, and others to discuss and review system requirements
- Using *group support systems* to facilitate the sharing of ideas and voicing opinions about system requirements
- Using *CASE tools* to analyze current systems to discover requirements to meet changing business conditions
- Iteratively developing system *prototypes* that refine the understanding of system requirements in concrete by showing working versions of system features

Joint Application Design

You were introduced to Joint Application Design, or JAD, in Chapter 1. There you learned JAD started in the late 1970s at IBM and that since then the practice of JAD has spread throughout many companies and industries. For example, it is quite popular in the insurance industry in Connecticut where a JAD users group has been formed. In fact, several generic approaches to JAD have been documented and popularized (see Wood and Silver, 1995, for an example). You also learned in Chapter 1 that the main idea behind JAD is to bring together the key users, managers, and systems analysts involved in the analysis of a current system. In that respect, JAD is similar to a group interview; a JAD, however, follows a particular structure of roles and agenda that is quite different from a group interview during which analysts control the sequence of questions answered by users. The primary purpose of using JAD in the analysis phase is to collect systems requirements simultaneously from the key people involved with the system. The result is an intense and structured, but highly effective, process. As with a group interview, having all the key people together in one place at one time allows analysts to see where there are areas of agreement and where there are conflicts. Meeting with all these important people for over a week of intense sessions allows you the opportunity to resolve conflicts, or at least to understand why a conflict may not be simple to resolve.

JAD sessions are usually conducted in a location other than the place where the people involved normally work. The idea behind such a practice is to keep participants away from as many distractions as possible so that they can concentrate on systems analysis. A JAD may last anywhere from four hours to an entire week and may consist of several sessions. A JAD employs thousands of dollars of corporate resources, the most expensive of which is the time of the people involved. Other expenses include the costs associated with flying people to a remote site and putting them up in hotels and feeding them for several days.

The typical participants in a JAD are listed below:

- **JAD session leader:** The JAD leader organizes and runs the JAD. This person has been trained in group management and facilitation as well as in systems analysis. The JAD leader sets the agenda and sees that it is met. The JAD leader remains neutral on issues and does not contribute ideas or opinions but rather concentrates on keeping the group on the agenda, resolving conflicts and disagreements, and soliciting all ideas.

- Users: The key users of the system under consideration are vital participants in a JAD. They are the only ones who have a clear understanding of what it means to use the system on a daily basis.

- Managers: Managers of the work groups who use the system in question provide insight into new organizational directions, motivations for and organizational impacts of systems, and support for requirements determined during the JAD.

JAD session leader: The trained individual who plans and leads Joint Application Design sessions.

- Sponsor: As a major undertaking due to its expense, a JAD must be sponsored by someone at a relatively high level in the company. If the sponsor attends any sessions, it is usually only at the very beginning or the end.

- Systems Analysts: Members of the systems analysis team attend the JAD although their actual participation may be limited. Analysts are there to learn from users and managers, not to run or dominate the process.

- **Scribe:** The scribe takes notes during the JAD sessions. This is usually done on a personal computer or laptop. Notes may be taken using a word processor, or notes and diagrams may be entered directly into a CASE tool.

- IS staff: Besides systems analysts, other IS staff, such as programmers, database analysts, IS planners, and data center personnel, may attend to learn from the discussion and possibly to contribute their ideas on the technical feasibility of proposed ideas or on technical limitations of current systems.

Scribe: The person who makes detailed notes of the happenings at a Joint Application Design session.

JAD sessions are usually held in special-purpose rooms where participants sit around horseshoe-shaped tables, as in Figure 7-6. These rooms are typically equipped with whiteboards (possibly electronic, with a printer to make copies of what is written on the board). Other audiovisual tools may be used, such as transparencies and overhead projectors, magnetic symbols that can be easily rearranged on a whiteboard, flip charts, and computer-generated displays. Flip chart paper is typically used for keeping track of issues that cannot be resolved during the JAD or for those issues requiring additional information that can be gathered during breaks in the proceedings. Computers may be used to create and display form or report designs, for diagramming existing or replacement systems, and for prototyping.

Figure 7-6
Illustration of the typical room layout for a JAD (Adapted from Wood and Silver, 1995)

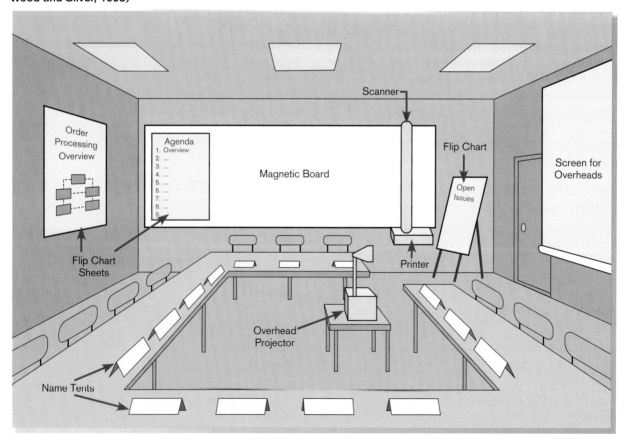

When a JAD is completed, the end result is a set of documents that detail the workings of the current system related to the study of a replacement system. Depending on the exact purpose of the JAD, analysts may also walk away from the JAD with some detailed information on what is desired of the replacement system.

Taking Part in a JAD Imagine that you are a systems analyst taking part in your first JAD. What might participating in a JAD be like? Typically, JADs are held off site, in comfortable conference facilities. On the first morning of the JAD, you and your fellow analysts walk into a room that looks much like the one depicted in Figure 7-6. The JAD facilitator is already there; she is finishing writing the day's agenda on a flip chart. The scribe is seated in a corner at a microcomputer, preparing to take notes on the day's activities. Users and managers begin to enter in groups and seat themselves around the U-shaped table. You and the other analysts review your notes describing what you have learned so far about the information system you are all here to discuss. The session leader opens the meeting with a welcome and a brief rundown of the agenda. The first day will be devoted to a general overview of the current system and major problems associated with it. The next two days will be devoted to an analysis of current system screens. The last two days will be devoted to analysis of reports.

The session leader introduces the corporate sponsor who talks about the organizational unit and current system related to the systems analysis study and the importance of upgrading the current system to meet changing business conditions. He leaves, and the JAD session leader takes over. She yields the floor to the senior analyst who begins a presentation on key problems with the system that have already been identified. After the presentation, the session leader opens the discussion to the users and managers in the room.

After a few minutes of talk, a heated discussion begins between two users from different corporate locations. One user, who represents the office that served as the model for the original systems design, argues that the system's perceived lack of flexibility is really an asset, not a problem. The other user, who represents an office that was part of another company before a merger, argues that the current system is so inflexible as to be virtually unusable. The session leader intervenes and tries to help the users isolate particular aspects of the system that may contribute to the system's perceived lack of flexibility.

Questions arise about the intent of the original developers. The session leader asks the analysis team about their impressions of the original system design. Because these questions cannot be answered during this meeting, as none of the original designers are present and none of the original design documents are readily available, the session leader assigns the question about intent to the "to do" list. This question becomes the first one on a flip chart sheet of "to do" items and the session leader gives you the assignment of finding out about the intent of the original designers. She writes your name next to the "to do" item on the list and continues with the session. Before the end of the JAD, you must get an answer to this question.

The JAD will continue like this for its duration. Analysts will make presentations, help lead discussions of form and report design, answer questions from users and managers, and take notes on what is being said. After each meeting, the analysis team will meet, usually informally, to discuss what has occurred that day and to consolidate what they have learned. Users will continue to contribute during the meetings and the session leader will facilitate, intervening in conflicts, seeing that the group follows the agenda. When the JAD is over, the session leader and her assistants must prepare a report that documents the findings in the JAD and is circulated among users and analysts.

CASE Tools During JAD The CASE tools most useful to analysis during a JAD are those referred to as upper CASE (see Chapter 4), as they apply most directly to activities occurring early in the systems development life cycle. Upper CASE tools usually include planning tools (see Chapter 5), diagramming tools (see Chapter 8 for examples of diagrams used during systems analysis), and prototyping tools, such as computer form and report generators. For requirements determination and structuring, the most useful CASE tools are for diagramming and for form and report generation. The more interaction analysts have with users during this phase, the more useful this set of tools. The analyst can use diagramming and prototyping tools to give graphic form to system requirements, show the tools to users, and make changes based on the users' reactions. The same tools are very valuable for requirements structuring as well. Using common CASE tools during requirements determination and structuring makes the transition between these two subphases easier and reduces the time spent. In structuring, CASE tools that analyze requirements information for correctness, completeness, and consistency are also useful. Finally, for alternative generation and selection, diagramming and prototyping tools are key to presenting users with graphic illustrations of what the alternative systems will look like. Such a practice provides users and analysts with better information to select the most desirable alternative system.

Some observers advocate using CASE tools during JADs (Lucas, 1993). Running a CASE tool during a JAD allows analysts to enter system models directly into a CASE tool, providing consistency and reliability in the joint model-building process. The CASE tool captures system requirements in a more flexible and useful way than can a scribe or an analysis team making notes. Further, the CASE tool can be used to project menu, display, and report designs, so users can directly observe old and new designs and evaluate their usefulness for the analysis team.

Supporting JAD with GSS Even though CASE tools can greatly augment a JAD, the group interaction process is typically not well supported by computing. Other than CASE tools, most of the computer use at a JAD is by one person, the scribe taking notes. Since JAD is a structured group process, JAD can benefit from the same computer-based support that can be applied to any group process. Group support systems (GSS) can be used to support group meetings. Here we will discuss how JAD can benefit from GSS use.

One disadvantage to a JAD session is that it suffers from many of the same problems as any group meeting. For example, the more people in a group, the less time there is for all of them to speak and state their views. Even if you assumed that they all spoke for an equal amount of time, no one would have much time to talk in a one-hour meeting for 12 people (only five minutes each!). The assumption about speaking equally points out a second problem with meetings—one or a few people always dominate the discussion. On the other hand, some people will say absolutely nothing. Whatever outcome the meeting produces tends to be tilted toward those who spoke the most during the meeting and others may not be fully committed to the conclusions reached. A third problem with group meetings is that some people are afraid to speak out for fear they will be criticized. A fourth problem is that most people are not willing to criticize or challenge their bosses in a meeting, even if what the boss is saying is wrong.

JADs suffer from all of these problems with meetings. The result is that important views often are not aired. Such an outcome is unfortunate as the design of the new system could be adversely impacted and the system may have to be reworked at great expense when those important views finally become known.

GSSs have been designed specifically to help alleviate some of the problems with group meetings. In order to provide everyone in the meeting with the same chance to contribute, group members type their comments into computers rather than speak them. The GSS is set up so that all members of the group can see what every other member has been typing. In the one-hour meeting for 12 people mentioned earlier, all 12 can contribute for the full hour, instead of just for five minutes, using a GSS. If everyone in the meeting is typing, not talking, and everyone has the same chance to contribute, then the chances of domination of the meeting by any one individual are greatly reduced. Also, comments typed into a GSS are anonymous. Anonymity helps those who fear criticism because only the comment, and not the person, can be criticized, since no one knows who typed what. Anonymity also provides the ability to criticize your boss.

Supporting a JAD with a GSS has many potential benefits. Using a GSS, a JAD session leader is more likely to obtain contributions from everyone, rather than from just a few. Important ideas are less likely to be missed. Similarly, poor ideas are more likely to be criticized. A study comparing traditional JAD to JAD supported with GSS found that using a GSS did lead to certain enhancements in the JAD process (Carmel, George, and Nunamaker, 1992). Among the findings were that GSS-supported JADs tended to be more time-efficient than traditional JAD and participation was more equal because there was less domination by certain individuals than in traditional JAD. The study also found that introducing a GSS into a JAD session had other, less desirable, effects. GSS-supported JADs tended to be less structured and it was more difficult to identify and resolve conflicts when a GSS was used, due in part to the anonymity of interaction. Supporting a JAD with GSS, then, does seem to provide some benefits through altering how the group works together. Yet a reduction in the JAD leader's ability to resolve conflicts could be a problem, especially since JAD was designed to help uncover and resolve conflicts.

Using Prototyping During Requirements Determination

You were introduced to prototyping in Chapter 1 (see Figure 1–9 for an overview of prototyping). There you learned that prototyping is an iterative process involving analysts and users whereby a rudimentary version of an information system is built and rebuilt according to user feedback. You also learned that prototyping could replace the systems development life cycle or augment it. What we are interested in here is how prototyping can augment the requirements determination process.

In order to gather an initial basic set of requirements, you will still have to interview users and collect documentation. Prototyping, however, will allow you to quickly convert basic requirements into a working, though limited, version of the desired information system. The prototype will then be viewed and tested by the user. Typically, seeing verbal descriptions of requirements converted into a physical system will prompt the user to modify existing requirements and generate new ones. For example, in the initial interviews, a user might have said that he wanted all relevant utility billing information on a single computer display form, such as the client's name and address, the service record, and payment history. Once the same user sees how crowded and confusing such a design would be in the prototype, he might change his mind and instead ask for the information to be organized on several screens, but with easy transitions from one screen to another. He might also be reminded of some important requirements (data, calculations, etc.) that had not surfaced during the initial interviews.

You would then redesign the prototype to incorporate the suggested changes. Once modified, users would again view and test the prototype. And, once again, you would incorporate their suggestions for change. Through such an iterative process, the chances are good that you will be able to better capture a system's requirements. The goal with using prototyping to support requirements determination is to develop concrete specifications for the ultimate system, not to build the ultimate system from prototyping.

Prototyping is possible with several 4GLs and with CASE tools, as pointed out in Chapter 4 and the earlier section on CASE tools and analysis in this chapter. As we saw there, you can use CASE tools as part of a JAD to provide a type of limited prototyping with a group of users.

N E T   S E A R C H
There is additional information on the Web on Joint Application Design. Visit http://www.prenhall.com/hoffer to complete an exercise related to this topic.

Prototyping is most useful for requirements determination when

- User requirements are not clear or well understood, which is often the case for totally new systems or systems that support decision making

- One or a few users and other stakeholders are involved with the system

- Possible designs are complex and require concrete form to fully evaluate

- Communication problems have existed in the past between users and analysts and both parties want to be sure that system requirements are as specific as possible

- Tools (such as form and report generators) and data are readily available to rapidly build working systems

Prototyping also has some drawbacks as a tool for requirements determination. These include

- A tendency to avoid creating formal documentation of system requirements, which can then make the system more difficult to develop into a fully working system

- Prototypes can become very idiosyncratic to the initial user and difficult to diffuse or adapt to other potential users

- Prototypes are often built as stand-alone systems, thus ignoring issues of sharing data and interactions with other existing systems, as well as issues with scaling up applications

- Checks in the SDLC are bypassed so that some more subtle, but still important, system requirements might be forgotten (e.g., security, some data entry controls, or standardization of data across systems)

N ET   S E A R C H
Prototyping is a word that is used in a broad variety of contexts. Visit http://www.prenhall.com/hoffer to complete an exercise related to this topic.

RADICAL METHODS FOR DETERMINING SYSTEM REQUIREMENTS

Whether traditional or modern, the methods for determining system requirements that you have read about in this chapter apply to any requirements determination effort, regardless of its motivation. But most of what you have learned has traditionally been applied to systems development projects that involve automating existing processes. Analysts use system requirements determination to understand current problems and opportunities, as well as what is needed and desired in future systems. Typically, the current way of doing things has a large impact on the new system. In some organizations, though, management is looking for new ways to perform current tasks. These new ways may be radically different from how things are done now, but the payoffs may be enormous: fewer people may be needed to do the same work, relationships with customers may improve dramatically, and processes may become much more efficient and effective, all of which can result in increased profits. The overall process by which current methods are replaced with radically new methods is generally referred to as **business process reengineering** or **BPR**.

To better understand BPR, consider the following analogy. Suppose you are a successful European golfer who has tuned your game to fit the style of golf courses and weather in Europe. You have learned how to control the flight of the ball in heavy winds, roll the ball on wide open greens, putt on large and undulating greens, and aim at a target without the aid of the landscaping common on North American courses. When you come to the United States to make your fortune on the U.S. tour, you discover that incrementally improving your putting, driving accuracy, and sand shots will help, but the new competitive environment is simply not suited to your style of the game. You

Business Process Reengineering (BPR): The search for, and implementation of, radical change in business processes to achieve breakthrough improvements in products and services.

need to reengineer your whole approach, learning how to aim at targets, spin and stop a ball on the green, and manage the distractions of crowds and press. If you are good enough, you may survive, but without reengineering, you will never be a winner.

Just as the competitiveness of golf forces good players to adapt their games to changing conditions, the competitiveness of our global economy has driven most companies into a mode of continuously improving the quality of their products and services (Dobyns and Crawford-Mason, 1991). Organizations realize that creatively using information technologies can yield significant improvements in most business processes. The idea behind BPR is not just to improve each business process but, in a systems modeling sense, to reorganize the complete flow of data in major sections of an organization to eliminate unnecessary steps, achieve synergies among previously separate steps, and become more responsive to future changes. Companies such as IBM, Procter & Gamble, Wal-Mart, and Ford are actively pursuing BPR efforts and have had great success. Yet, many other companies have found difficulty in applying BPR principles (Moad, 1994). Nonetheless, BPR concepts are actively applied in both corporate strategic planning and information systems planning as a way to radically improve business processes (as described in Chapter 5).

BPR advocates suggest that radical increases in the quality of business processes can be achieved through creative application of information technologies. BPR advocates also suggest that radical improvement cannot be achieved by tweaking existing processes but rather by using a clean sheet of paper and asking "If we were a new organization, how would we accomplish this activity?" Changing the way work is performed also changes the way information is shared and stored, which means that the results of many BPR efforts are the development of information system maintenance requests or requests for system replacement. It is likely that you will encounter or have encountered BPR initiatives in your own organization.

Identifying Processes to Reengineer

A first step in any BPR effort relates to understanding what processes to change. To do this, you must first understand which processes represent the **key business processes** for the organization. Key business processes are the structured set of measurable activities designed to produce a specific output for a particular customer or market. The important aspect of this definition is that key processes are focused on some type of organizational outcome such as the creation of a product or the delivery of a service. Key business processes are also customer-focused. In other words, key business processes would include all activities used to design, build, deliver, support, and service a particular product for a particular customer. BPR efforts, therefore, first try to understand those activities that are part of the organization's key business processes and then alter the sequence and structure of activities to achieve radical improvements in speed, quality, and customer satisfaction. The same techniques you learned to use for systems requirement determination can be used to discover and understand key business processes. Interviewing key individuals, observing activities, reading and studying organizational documents, and conducting JADs can all be used to find and fathom key business processes.

After identifying key business processes, the next step is to identify specific activities that can be radically improved through reengineering. Hammer and Champy (1993), the two people most identified with the term BPR, suggest that three questions be asked to identify activities for radical change:

1. How important is the activity to delivering an outcome?
2. How feasible is changing the activity?
3. How dysfunctional is the activity?

The answers to these questions provide guidance for selecting which activities to change. Those activities deemed important, changeable, yet dysfunctional, are pri-

Key business processes: The structured, measured set of activities designed to produce a specific output for a particular customer or market.

mary candidates. To identify dysfunctional activities, they suggest you look for activities where there are excessive information exchanges between individuals, where information is redundantly recorded or needs to be rekeyed, where there are excessive inventory buffers or inspections, and where there is a lot of rework or complexity. Many of the tools and techniques for modeling data, processes, events, and logic within the IS development process are also being applied to model business processes within BPR efforts (see Davenport, 1993). Thus, the skills of a systems analyst are often central to many BPR efforts.

Disruptive Technologies

Once key business processes and activities have been identified, information technologies must be applied to radically improve business processes. To do this, Hammer and Champy (1993) suggest that organizations think "inductively" about information technology. Induction is the process of reasoning from the specific to the general, which means that managers must *learn* the power of new technologies and *think* of innovative ways to alter the way work is done. This is contrary to deductive thinking where problems are first identified and solutions are then formulated.

Disruptive technologies:
Technologies that enable the breaking of long-held business rules that inhibit organizations from making radical business changes.

Hammer and Champy suggest that managers especially consider **disruptive technologies** when applying deductive thinking. Disruptive technologies are those that enable the breaking of long-held business rules that inhibit organizations from making radical business changes. For example, Saturn is using production schedule databases and electronic data interchange (EDI) to work with its suppliers as if they and Saturn were one company. Suppliers do not wait until Saturn sends them a purchase order for more parts but simply monitor inventory levels and automatically send shipments as needed (Hammer and Champy, 1993: 90). Table 7-7 shows several long-held business rules and beliefs that constrain organizations from making radical process improvements. For example, the first rule suggests that information can only appear in one place at a time. However, the advent of distributed databases (see Chapter 18) has "disrupted" this long-held business belief.

In this section, we discussed how BPR is increasingly being used to identify ways to adapt existing information systems to changing organizational information needs and processes. It was our intent to provide a brief introduction to this topic, as the specific tools and techniques for performing BPR are still evolving. For more information on this exciting topic, the interested reader is encouraged to see the books by Davenport (1993), Hammer and Champy(1993), and Hammer (1996).

TABLE 7-7   Long-Held Organizational Rules That Are Being Eliminated Through Disruptive Technologies

Rule	Disruptive Technology
Information can appear in only one place at a time.	Distributed databases allow the sharing of information.
Only experts can perform complex work.	Expert systems can aid nonexperts.
Businesses must choose between centralization and decentralization.	Advanced telecommunications networks can support dynamic organizational structures.
Managers must make all decisions.	Decision-support tools can aid nonmanagers.
Field personnel need offices where they can receive, store, retrieve, and transmit information.	Wireless data communication and portable computers provide a "virtual" office for workers.
The best contact with a potential buyer is personal contact.	Interactive communication technologies allow complex messaging capabilities.
You have to find out where things are.	Automatic identification and tracking technology know where things are.
Plans get revised periodically.	High-performance computing can provide real-time updating.

INTERNET DEVELOPMENT: DETERMINING SYSTEM REQUIREMENTS

PINE VALLEY FURNITURE

Determining systems requirements for an Internet-based electronic commerce application is no different than the process followed for other applications. In the last chapter, you read how Pine Valley Furniture's management began the WebStore project, a project to sell furniture products over the Internet. In this section, we examine the process followed by PVF to determine system requirements and highlight some of the issues and capabilities that you may want to consider when developing your own Internet-based application.

Determining System Requirements for Pine Valley Furniture's WebStore

To collect system requirements as quickly as possible, Jim and Jackie decided to hold a three-day JAD session. In order to get the most out of these sessions, they invited a broad range of people, including representatives from Sales and Marketing, Operations, and Information Systems. Additionally, they asked an experienced JAD facilitator, Cheri Morris, to conduct the session. Together with Cheri, Jim and Jackie developed a very ambitious and detailed agenda for the session. Their goal was to collect requirements on the following items:

- System Layout and Navigation Characteristics
- WebStore and Site Management System Capabilities
- Customer and Inventory Information
- System Prototype Evolution

In the remainder of this section, we briefly highlight the outcomes of the JAD session.

System Layout and Navigation Characteristics As part of the process of preparing for the JAD session, all participants were asked to visit several established retail Websites, including *www.amazon.com*, *www.landsend.com*, *www.sony.com*, and *www.pier1.com*. At the JAD session, participants were asked to identify characteristics of these sites that they found appealing and those characteristics they found cumbersome. This allowed participants to identify and discuss those features that they wanted the WebStore to possess. The outcomes of this activity are summarized in Table 7-8.

WebStore and Site Management System Capabilities After agreeing to the general layout and navigational characteristics of the WebStore, the session then turned its focus onto the basic system capabilities. To assist in this process, systems analysts from the Information Systems Department developed a draft skeleton of the WebStore. This skeleton was based on the types of screens common to and capabilities of popular retail Websites. For example, many retail Websites have a "shopping cart" feature that allows customers to accumulate multiple items before checking out rather than buying a single item at a time. After some discussion, the participants

TABLE 7-8   Desired Layout and Navigation Feature of WebStore

Layout and Design	• Navigation menu and logo placement should remain consistent throughout the entire site (this allows users to maintain familiarity while using the site and minimizes users who get "lost" in the site) • Graphics should be lightweight to allow for quick page display • Text should be used over graphics whenever possible
Navigation	• Any section of the store should be accessible from any other section via the navigation menu • Users should always be aware of what section they are currently in

TABLE 7-9   System Structure of the WebStore and Site Management Systems

WebStore System	Site Management System
❑ Main Page	❑ User Profile Manager
• Product Line (Catalog)	❑ Order Maintenance Manager
✓ Desks	❑ Content (Catalog) Manager
✓ Chairs	❑ Reports
✓ Tables	• Total Hits
✓ File Cabinets	• Most Frequent Page Views
• Shopping Cart	• Users/Time of Day
• Checkout	• Users/Day of Week
• Account Profile	• Shoppers not purchasing (used
• Order Status/History	shopping cart—did not checkout)
• Customer Comments	• Feedback Analysis
❑ Company Info	
❑ Feedback	
❑ Contact Information	

agreed that the system structure shown in Table 7-9 would form the foundation for the WebStore system.

In addition to the WebStore capabilities, members of the Marketing and Sales Department described several reports that would be necessary to effectively manage customer accounts and sales transactions. In addition, the department wants to be able to conduct detailed analyses of site visitors, sales tracking, and so on. Members of the Operations Department expressed a need to easily update the product catalog. These collective requests and activities were organized into a system design structure, called the Site Management System, which is summarized in Table 7-9. The structures of both the WebStore and Site Management Systems will be given to the Information Systems Department as the baseline for further analysis and design activities.

Customer and Inventory Information   The WebStore will be designed to support the furniture purchases of three distinct types of customers:

- Corporate customers
- Home office customers
- Student customers

TABLE 7-10   Customer and Inventory Information for WebStore

Corporate Customer	Home Office Customer	Student Customer
• Company Name	• Name	• Name
• Company Address	• Doing Business As	• School
• Company Phone	(Company Name)	• Address
• Company Fax	• Address	• Phone
• Company Preferred	• Phone	• E-Mail
Shipping Method	• Fax	
• Buyer Name	• E-Mail	
• Buyer Phone		
• Buyer E-Mail		
Inventory Information		
• SKU	• Finished Product Size	• Available Colors
• Name	• Finished Product Weight	• Price
• Description	• Available Materials	• Lead Time

TABLE 7-11   Stages of System Implementation of WebStore

Stage 1 (Basic Functionality):
- Simple catalog navigation; 2 products per section—limited attributes set
- 25 sample users
- Simulated credit card transaction
- Full shopping cart functionality

Stage 2 (Look and Feel):
- Full product attribute set and media (images, video)—commonly referred to as the "product data catalog"
- Full site layout
- Simulated integration with Purchasing Fulfillment and Customer Tracking systems

Stage 3 (Staging/Preproduction):
- Full integration with Purchasing Fulfillment and Customer Tracking systems
- Full credit card processing integration
- Full product data catalog

To effectively track the sales to these different types of customers, distinct information must be captured and stored by the system. Table 7-10 summarizes this information for each customer type that was identified during the JAD session. In addition to the customer information, information about the products ordered must also be captured and stored. Orders reflect the range of product information that must be specified to execute a sales transaction. Thus, in addition to capturing the customer information, product and sales data must also be captured and stored. Table 7-10 lists the results of this analysis.

System Prototype Evolution As a final activity, the JAD participants, benefiting from extensive input from the Information Systems staff, discussed how the system implementation should evolve. After completing analysis and design activities, it was agreed to that the system implementation should progress in three main stages so that changes to the requirements could more easily be identified and implemented. Table 7-11 summarizes these stages and the functionality that would be incorporated at each stage of the implementation.

At the conclusion of the JAD session there was a good feeling among the participants. All felt that a lot of progress had been made and that clear requirements had been identified. With these requirements in hand, Jim and the Information Systems staff could now begin to turn these lists of requirements into formal analysis and design specifications. To show how information flows through the WebStore, Data Flow Diagrams (Chapter 8) will be produced. To show a conceptual model of the data used within WebStore, an Entity-Relationship Diagram (Chapter 10) will be produced. Both of these analysis documents will become part of the foundation for detailed system design and implementation.

Summary

As we saw in Chapter 1, there are three subphases in the systems analysis phase of the systems development life cycle: requirements determination, requirements structuring, and alternative generation and choice. Chapter 7 has focused on requirements determination, the gathering of information about current systems and the need for replacement systems. Chapters 8 through 10 will address techniques for structuring the requirements elicited during requirements determination. Chapter 11 closes Part III of the book by explaining how analysts generate alternative design strategies for replacement systems and choose the best one.

For requirements determination, the traditional sources of information about a system include interviews, questionnaires, observation, group interviews, and procedures, forms, and other useful documents. Often many or even all of these sources are used to gather perspectives on the adequacy of current systems and the requirements for replacement systems. Each form of information collection has its advantages and disadvantages, which were summarized in Tables 7-4 and 7-5. Selecting the methods to use depends on the need for rich or thorough information, the time and budget available, the need to probe deeper once initial information is collected, the need for confidentiality for those providing assessments of system requirements, the desire to get people involved and committed to a project, and the potential audience from which requirements should be collected.

Both open- and closed-ended questions can be posed during interviews or in questionnaires. In either case, you must be very precise in formulating a question in order to avoid ambiguity and to insure a proper response. During observation you must try not to intrude or interfere with normal business activities so that the people being observed do not modify their activities from normal processes. The results of all requirements gathering methods should be compared, since there may be differences between the formal or official system and the way people actually work, the informal system.

You also learned about alternative methods to collect requirements information, many of which themselves make use of information systems. Joint Application Design (JAD) begins with the idea of the group interview and adds structure and a JAD session leader to it. Typical JAD participants include the session leader, a scribe, key users, managers, a sponsor, and systems analysts. JAD sessions are usually held off-site and may last as long as one week.

Although JAD sessions typically rely little on computer support, systems analysis is increasingly performed with computer assistance, such as group support systems. You also read how information systems can support requirements determination with CASE tools and for prototyping. As part of the prototyping process, users and analysts work closely together to determine requirements the analyst then builds into a model. The analyst and user then work together on revising the model until it is close to what the user desires.

Business process reengineering (BPR) is an approach to radically changing business processes. BPR efforts are a source of new information requirements. Information systems and technologies often enable BPR by allowing an organization to eliminate or relax constraints on traditional business rules. Most of the same techniques used for requirements determination for traditional systems can also be fruitfully applied to the development of Internet applications. Accurately capturing requirements in a timely manner for Internet applications is just as important as for more traditional systems.

The result of requirements determination is a thorough set of information, including some charts, that describes the current systems being studied and the need for new and different capabilities to be included in the replacement systems. This information, however, is not in a form that makes analysis of true problems and clear statements of new features possible. Thus, you and other analysts will study this information and structure it into standard formats suitable for identifying problems and unambiguously describing the specifications for new systems. With modern information systems, structuring requires documenting the flow or movement of data throughout the organization and information systems, the logic of transforming data into information (including the timing of events that cause data to be transformed or processed), and the rules that govern the relationships between different data handled by the system. We discuss a variety of popular techniques for structuring requirements in the next three chapters.

Key Terms

1. Business process reengineering
2. Closed-ended questions
3. Disruptive technologies
4. Formal system
5. Informal system
6. JAD session leader
7. Key business processes
8. Nominal Group Technique
9. Open-ended questions
10. Scribe

Match each of the key terms above with the definition that best fits it.

_____ Questions in interviews and on questionnaires that ask those responding to choose from among a set of specified responses.

_____ Technologies that enable the breaking of long-held business rules that inhibit organizations from making radical business changes.

_____ A facilitated process that supports idea generation by groups. At the beginning of the process, group members work alone to generate ideas, which are then pooled under the guidance of a trained facilitator.

_____ The structured, measured set of activities designed to produce a specific output for a particular customer or market.

_____ The official way a system works as described in organizational documentation.

_____ The search for, and implementation of, radical change in business processes to achieve breakthrough improvements in products and services.

_____ The way a system actually works.

_____ The person who makes detailed notes of the happenings at a Joint Application Design session.

_____ Questions in interviews and on questionnaires that have no prespecified answers.

_____ The trained individual who plans and leads Joint Application Design sessions.

Review Questions

1. Describe systems analysis and the major activities that occur during this phase of the systems development life cycle.

2. Describe four traditional techniques for collecting information during analysis. When might one be better than another?

3. Compare collecting information by interview and by questionnaire. Describe a hypothetical situation in which each of these methods would be an effective way to collect information system requirements.

4. What is JAD? How is it better than traditional information-gathering techniques? What are its weaknesses?

5. How has computing been used to support requirements determination?

6. How can NGT be used for requirements determination?

7. How can CASE tools be used to support requirements determination? Which type of CASE tools are appropriate for use during requirements determination?

8. Describe how prototyping can be used during requirements determination. How is it better or worse than traditional methods?

9. What unique benefits does a GSS provide for group methods of requirements determination, such as group interviews or JAD?

10. When conducting a business process reengineering study, what should you look for when trying to identify business processes to change? Why?

11. What are disruptive technologies and how do they enable organizations to radically change their business processes?

Problems and Exercises

1. Choose either CASE or GSS as a topic and review a related article from the popular press and from the academic research literature. Summarize the two articles and, based on your reading, prepare a list of arguments for why this type of system would be useful in a JAD session. Also address the limits for applying this type of system in a JAD setting.

2. One of the potential problems with gathering information requirements by observing potential system users mentioned in the chapter is that people may change their behavior when observed. What could you do to overcome this potential confounding factor in accurately determining information requirements?

3. Summarize the problems with the reliability and usefulness of analyzing business documents as a method for gathering information requirements. How could you cope with these problems to effectively use business documents as a source of insights on system requirements?

4. Suppose you were asked to lead a JAD session. List 10 guidelines you would follow to assist you in playing the proper role of a JAD session leader.

5. Prepare a plan, similar to Figure 7-2, for an interview with your academic advisor to determine which courses you should take to develop the skills you need to be hired as a programmer/analyst.

6. Write at least three closed-ended questions that you might use on a questionnaire that would be sent to users of a word-processing package in order to develop ideas for the next version of the package. Test these questions by asking a friend to answer the questions; then interview your friend to determine why she responded as she did. From this interview, determine if she misunderstood any of your questions and, if so, rewrite the questions to be less ambiguous.

7. An interview lends itself easily to asking probing questions, or asking different questions depending on the answers provided by the interviewee. Although not impossible, probing and alternative questions can be handled in a questionnaire. Discuss how you could include probing or alternative sets of questions in a questionnaire.

8. Figure 7-2 shows part of a guide for an interview. How might an interview guide differ when a group interview is to be conducted?

9. Group interviews and JADs are very powerful ways to collect system requirements but special problems arise during group requirements collection sessions. Summarize the special interviewing and group problems that arise in such group sessions, and suggest ways that you, as a group interviewer or group facilitator, might deal with these problems.

10. Review the material in Chapter 5 on corporate and information systems strategic planning. How are these processes different from BPR? What new perspectives might BPR bring that classical strategic planning methods may not have?

Field Exercises

1. Effective interviewing is not something that you can learn from just reading about it. You must first do some interviewing, preferably a lot of it, as interviewing skills only improve with experience. To get an idea of what interviewing is like, try the following: Find three friends or classmates to help you complete this exercise. Organize yourselves into pairs. Write down a series of questions you can use to find out about a job your partner now has or once held. You decide what questions to use, but at a minimum, you must find out the following: (1) the job's title, (2) the job's responsibilities, (3) whom your partner reported to, (4) who reported to your partner, if anyone did, and (5) what information your partner used to do his or her job. At the same time, your partner should be preparing questions to ask you about a job you have had. Now conduct the interview. Take careful notes. Organize what you find into a clear form that another person could understand (you might want to use a systems diagram like Figure 2-7, with boundaries, inputs, and outputs). Now repeat the process, but this time, your partner interviews you.

   While the two of you have been interviewing each other, your two other friends should have been doing the same thing. When all four of you are done, switch partners and repeat the entire process. When you are all done, each of you should have interviewed two people, and each of you should have been interviewed by two people. Now, you and the person who interviewed your original partner should compare your findings. Most likely, your findings will not be identical to what the other person found. If your findings differ, discover why. Did you use the same questions? Did the other person do a more thorough job of interviewing your first partner, since it was the second time he or she had conducted an interview? Did you both ask follow-up questions? Did you both spend about the same amount of time on the interview? Prepare a report with this person about why your findings differed. Now find both of the people who interviewed you. Does one set of findings differ from the other? Try and figure out why. Did one of them (or both of them) misrepresent or misunderstand what you told them? Each of you should now write a report on your experience, using it to explain why interviews are sometimes inconsistent and inaccurate and why having two people interview someone on a topic is better than having just one person do the interview. Explain the implications of what you have learned for the requirements determination subphase of the systems development life cycle.

2. Choose a work team at your work or university and interview them in a group setting. Ask them about their current system (whether computer-based or not) for performing their work. Ask each of them what information they use and/or need and from where/whom they get it. Was this a useful method for you to learn about their work? Why or why not? What comparative advantages does this method provide as compared to one-on-one interviews with each team member? What comparative disadvantages?

3. For the same work team you used in Field Exercise 2, examine copies of any relevant written documentation (e.g., written procedures, forms, reports, system documentation). Are any of these forms of written documentation missing? Why? With what consequences? To what extent does this written documentation fit with the information you received in the group interview?

4. Interview systems analysts, users, and managers who have been involved in JAD sessions. Determine the location, structure and outcomes of each of their JAD sessions. Elicit their evaluations of their sessions. Were they productive? Why or why not?

5. Survey the literature on JAD in the academic and popular press and determine the "state of the art." How is JAD being used to help determine system requirements? Is using JAD for this process beneficial? Why or why not? Present your analysis to the IS manager at your work or at your university. Does your analysis of JAD fit with his or her perception? Why or why not? Is he or she currently using JAD, or a JAD-like method, for determining system requirements? Why or why not?

6. With the help of other students or your instructor, contact someone in an organization that has carried out a BPR study. What effects did this study have on information systems? In what ways did IT (especially disruptive technologies) facilitate making the radical changes discovered in the BPR study?

References

Carmel, E., J. F. George, and J. F. Nunamaker, Jr. 1992. "Supporting Joint Application Development (JAD) with Electronic Meeting Systems: A Field Study." *Proceedings of the Thirteenth International Conference on Information Systems.* Dallas, TX, December: 223–32.

Davenport, T. H. 1993. *Process Innovation: Reengineering Work through Information Technology.* Boston, MA: Harvard Business School Press.

Dobyns, L., and C. Crawford-Mason. 1991. *Quality or Else.* Boston, MA: Houghton-Mifflin.

Hammer, M. 1996. *Beyond Reengineering.* New York: Harper Business.

Hammer, M., and J. Champy. 1993. *Reengineering the Corporation.* New York, NY: Harper Business.

Lucas, M.A. 1993. "The Way of JAD." *Database Programming & Design* 6 (July): 42–49.

Mintzberg, H. 1973. *The Nature of Managerial Work.* New York, NY: Harper & Row.

Moad, J. 1994. "After Reengineering: Taking Care of Business." *Datamation* 40 (20): 40–44.

Wood, J., and D. Silver. 1995. *Joint Application Development,* 2/e. New York, NY: John Wiley & Sons.

BROADWAY ENTERTAINMENT COMPANY, INC.

Determining Requirements for the Web-Based Customer Relationship Management System

CASE INTRODUCTION

Carrie Douglass, manager of the Broadway Entertainment Company store in Centerville, Ohio, was pleased with the progress of the team of Management Information Systems (MIS) students from Stillwater State University. The first meeting with Tracey Wesley, John Whitman, Missi Davies, and Aaron Sharp had been productive, and the plan they developed for the BEC Web-based customer relationship management project seemed workable. Carrie was eager for the students to begin to show her possible designs for the proposed system. But, the plan indicated that several activities for gathering system requirements needed to be conducted before a possible system would be presented to her. Carrie hoped these activities could be done quickly and without much of a drain on employee time.

GETTING STARTED ON REQUIREMENTS DETERMINATION

The Stillwater team of students was almost ready to conduct a detailed analysis of the requirements for the customer relationship management system. Carrie's comments from the first meeting had been helpful; however, the team members had a few more questions to help them orient the requirements determination portion of the analysis phase. Two of the team members, Missi and Aaron, scheduled a second, brief visit with Carrie to gather some additional background information.

First, they asked Carrie what the business goals were for her and her store. Carrie explained that each BEC store had three main goals: (1) to increase dollar income volume by at least 1.5 percent each month, (2) to increase profit by at least 1 percent each month, and (3) to maintain customer overall satisfaction above 95 percent. The store manager has bonus pay incentives to achieve these objectives. Customer satisfaction is measured monthly by random telephone calls placed by an independent market research firm. A sample of customers from each store and in each area where there is a BEC store are contacted and asked to answer a list of 10 questions about their experience at BEC or competing stores. Personally, Carrie wants to perform as an above-average store manager during her first year, which means that she expects to clearly beat the goals given her by BEC. She also wants to be viewed as an innovative store manager, someone with potential for more significant positions later in her career. She likes working for BEC and sees a long-term career within the organization.

Second, the team asked Carrie what expectations she had about how the requirements determination phase of the project would be conducted. Carrie expects the team to act independently without much direction or supervision from her; she does not have the time to work closely with them. She also expects that they will ask any questions of her or her employees and that everyone will cooperate with the project. Missi and Aaron confirmed with Carrie that she had no concerns about the team meeting with store customers, who could give insights on possible system features. Carrie agreed, and suggested that a flyer could be given to customers with each purchase or rental asking them to participate in whatever way the team chose to capture their ideas (whether that be individual or group interviews, surveys, or whatever). Next, Carrie asked about the time line for the project. The team members reminded Carrie that their project plan outlined this—their course included 27 more weeks of work over the following 30 calendar weeks.

At the end of the meeting the team members asked Carrie for a copy of a description of the computer capabilities available to them through in-store technology. Carrie agreed to send them a copy of an overview of the BEC information systems she obtained in the training program (see the BEC case at the end of Chapter 4). If they need more details, they should ask for specific information, and Carrie will try to get those details from corporate staff.

CONDUCTING REQUIREMENTS DETERMINATION

After Missi and Aaron shared with their other team members the results of their meeting with Carrie, the students from Stillwater State felt that they were ready to conduct various requirements determination activities. The information Missi and Aaron collected in their meeting with Carrie would help the team prepare good questions to employees, customers, and prospective customers. Carrie's answers to the question about goals for her store suggested to them that including prospective customers—those who typically shopped at other home entertainment stores—could provide helpful insights. They had decided and outlined in the Baseline Project Plan three steps for gathering requirements (see BEC Figure 7-1).

At the same time these sessions aimed at generating system features are being conducted, Tracey and John will study the Entertainment Tracker (ET) documentation to determine what data can be transferred from ET and their format. This analysis will be essential to determine the information flows needed to coordinate ET and the customer relationship management system. This analysis of ET as well as the results of the feature analysis sessions will help the team to determine the critical events of the business, the logic of reacting to these events and to making choices in the system, and the rules that govern the integrity of system data. Determining these system characteristics are essential for proposing and documenting a structure and design for the system, which will be the following step in their project. Their work in requirements determination must capture a great deal of the information they need not only to develop a prototype to show the system features and user interface, but also need for subsequent systems development steps involving system design, implementation, and installation.

CASE SUMMARY

The Stillwater team of students is off to an enthusiastic start to the BEC customer relationship management system project. The two initial meetings with the client seemed to have gone well. Team members liked Carrie, and she seemed to like them and spoke frankly to them. But, there were also signs of risk for the project. First, Carrie might be overly enthusiastic and naïve about the system. As an inexperienced store manager, she may not have well-seasoned or definitive ideas. Second, other store employees, many of whom are part-time help with no long-term commitment to BEC, have not been involved in the development of the project's ideas. The team will have to assess whether the employees or other people are critical stakeholders in the project. Third, Carrie needs the project to be conducted with minimal costs. Until the team can develop a clear understanding of the system requirements and the available technology in the store, additional costs are unclear. Certainly the team can develop the design and proof-of-concept prototype of a system on computers at Stillwater, but if the system is to be used in the store, there may be significant start-up costs. Finally, the team is a little concerned about Carrie's reluctance for corporate IS staff to be involved in the project. Carrie seemed hesitant about the handoff meeting as a way to provide follow-up after the team's work is done, and she wants to be a buffer with corporate about IS details. Being able to coordinate the new system with Entertainment Tracker is a critical success factor for this project.

BEC Figure 7-1
Stillwater MIS student team plan for requirements determination

1. Conduct a group interview of four BEC store employees	In this interview, the team will obtain the reactions from employees to the system requirements Carrie had outlined in the System Service Request (see BEC Figure 5-2) and seek their ideas, based on their interactions with customers, about what features would be desirable. The team also wanted to find out more from the employees about the possible relationships between Entertainment Tracker, the in-store point-of-sale system, and the new system. Employees can also help the team to identify potential hazards to the successful implementation of the system, and acceptance of a system by employees and customers.
2. Conduct individual interviews with at least four BEC and non-BEC customers	In these interviews, the team will explain the general purpose of the proposed system, ask the participants for ideas about system features, and get reactions from the interviewees to the features of other similar Web-systems for BEC competitors and noncompetitors.
3. Conduct a structured walk-through session with students from their MIS project class	In this session, the team members will demonstrate a prototype of the system and get reactions from their colleagues before showing a revised prototype to Carrie, store employees, and customers.

The team members agree that the project looks like a great learning experience. There is original analysis and design work to be done. The project schedule and techniques to be used are wide open, within the constraints of their course requirements. There are some interesting stakeholder issues to be handled. The benefits of the system are still vague, as are the costs for a complete implementation. And the team is diverse, with a variety of skills and experience, but with the unknowns of how members will react and work together when critical deadlines must be met.

CASE QUESTIONS

1. From what you know so far about the customer relationship management system project at BEC, whom do you consider to be the stakeholders in this system? Are these stakeholders involved in the early stages of requirements determination as outlined in this case? How would you suggest involving each stakeholder in the project in order to gain the greatest insights during requirements determination and to achieve success for the project?

2. Develop a detailed project schedule for the analysis phase. This schedule should follow from answers to questions in BEC cases from prior chapters and from any class project guidelines given to you by your instructor.

3. Your answer to Case Question 2 likely included review points with Carrie and other stakeholders. These review points are part of the project's overall communication plan. Explain the overall communication

activities you would suggest for this project. How should team members communicate questions, findings, and results to one another? How should the team communicate with stakeholders?

4. Evaluate the overall plan for the three requirements determination activities the Stillwater students have planned. Would you suggest different steps to gather system requirements? Why or why not?

5. Independent of your answer to Case Question 4, prepare detailed interview plans for the two interview sessions the Stillwater students have outlined. Use Figure 7-2 as a template for documenting an interview plan. How does Carrie's list of three goals for her BEC store affect your interview plans? Prepare a draft of the flyer discussed in the case for inviting customers to be interviewed. How would you propose finding non-BEC store customers to be interviewed?

6. How would you propose involving BEC store customers (actual and potential) in requirements determination, possibly in ways other than are planned by the Stillwater students? Do they need to be involved? If not, why not? If so, prepare a plan for questionnaires, interviews, focus group or JAD session, observation, or whatever means you suggest using to elicit their requirements for the system.

7. How would you propose involving BEC store employees in requirements determination, possibly in ways other than are planned by the Stillwater students? Do they need to be involved? If not, why not? If so, prepare a plan for questionnaires, interviews, focus group or JAD session, observation, or whatever means you suggest using to elicit their requirements for the system.

8

Structuring System

Requirements: Process Modeling

LEARNING OBJECTIVES

After studying this chapter, you should be able to:

- Understand the logical modeling of processes through studying examples of data flow diagrams.

- Draw data flow diagrams following specific rules and guidelines that lead to accurate and well-structured process models.

- Decompose data flow diagrams into lower-level diagrams.

- Balance higher-level and lower-level data flow diagrams.

- Explain the differences among four types of DFDs: current physical, current logical, new physical, and new logical.

- Use data flow diagrams as a tool to support the analysis of information systems.

- Compare and contrast data flow diagrams with Oracle's process modeling tool and with functional hierarchy diagrams.

- Discuss process modeling for Internet applications.

INTRODUCTION

In the last chapter, you learned about various methods used by systems analysts to collect the information necessary to determine information systems requirements. In this chapter, our focus will be on one tool used to coherently represent the information gathered as part of requirements determination—data flow diagrams. Data flow diagrams allow you to model how data flow through an information system, the relationships among the data flows, and how data come to be stored at specific locations. Data flow diagrams also show the processes that change or transform data. Because data flow diagrams concentrate on the movement of data between processes, these diagrams are called process models.

As the name indicates, a data flow diagram is a graphical tool that allows analysts (and users, for that matter) to depict the flow of data in an information system. The system can be physical or logical, manual or computer-based. In this chapter, you will learn the basic mechanics of drawing and revising data flow diagrams and you will learn the basic symbols and a set of rules for drawing them. You will learn about what to do and what *not* to do when drawing data flow diagrams. You will learn two important concepts related to data flow diagrams: *balancing* and *decomposition*. You will also learn the differences between four different types of data flow diagrams: current physical, current logical, new logical, and new physical. Toward the end of the chapter, you will learn how to use data flow diagrams as part of the analysis of an information system and as a tool for supporting business process reengineering. You will learn about two other tools available for process modeling: Oracle's process modeling tool, and functional hierarchy diagrams. Finally, you will learn how process modeling is also important for the analysis of Internet-based systems.

PROCESS MODELING

Process modeling involves graphically representing the functions, or processes, which capture, manipulate, store, and distribute data between a system and its environment and between components within a system. A common form of a process model is a **data flow diagram**. Over the years, several different tools have been developed for process modeling. In this chapter, we focus on data flow diagrams, the traditional process modeling technique of structured analysis and design and the technique most often used today for process modeling.

Data flow diagramming is one of several notations that are called structured analysis techniques. Although not all organizations use each structured analysis technique, collectively techniques like data flow diagrams have had a significant impact on the quality of the systems development process. For example, Raytheon (Gibbs, 1994) has reported a savings from 1988 through 1994 of $17.2 million in software costs by applying structured analysis techniques, due mainly to avoiding rework to fix requirements flaws. This represents a doubling of systems developers' productivity and helped them avoid costly system mistakes.

Data flow diagram: A picture of the movement of data between external entities and the processes and data stores within a system.

Modeling a System's Process

As Figure 8-1 shows, there are three subphases of the analysis phase of the systems development life cycle: requirements determination, requirements structuring, and generating alternative systems and selecting the best one. The analysis team enters requirements structuring with an abundance of information gathered during requirements determination. During requirements structuring, you and the other team members must organize the information into a meaningful representation of the information system that exists and of the requirements desired in a replacement system. In addition to modeling the processing elements of an information system and how data are transformed in the system, you must also model the processing logic and the timing of events in the system (Chapter 9) and the structure of data within the system (Chapter 10). Thus, a process model is only one of three major complementary views of an information system. Together, process, logic and timing, and data models provide a thorough specification of an information system and, with the proper supporting tools, also provide the basis for the automatic generation of many working information system components.

N E T   S E A R C H
In addition to data flow diagrams, there are numerous ways to model business processes. Visit http://www.prenhall.com/hoffer to complete an exercise related to this topic.

Deliverables and Outcomes

In structured analysis, the primary deliverables from process modeling are a set of coherent, interrelated data flow diagrams. Table 8-1 provides a more detailed list of the deliverables that result from studying and documenting a system's process,

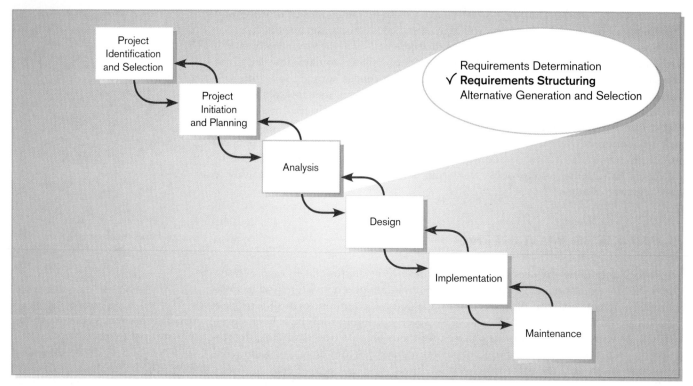

Figure 8-1
Systems development life cycle
with the analysis phase highlighted

using data flow diagrams as the documentation tool. First, a context diagram shows the scope of the system, indicating which elements are inside and which are outside the system. Second, data flow diagrams of the current physical system specify which people and technologies are used in which processes to move and transform data, accepting inputs and producing outputs. These diagrams are developed into sufficient detail to understand the current system and to eventually determine how to convert the current system into its replacement. Third, technology-independent, or logical, data flow diagrams of the current system show what data processing functions are performed by the current information system. Fourth, the data movement, or flow, structure, and functional requirements of the new system are represented in logical data flow diagrams. Finally, entries for all of the objects included in all diagrams are included in the project dictionary or CASE repository. This logical progression of deliverables allows you to understand the existing system. You can then abstract this system into its essential elements to show the way in which the new system should meet its information processing requirements identified during requirements determination. Remember, the deliverables of process modeling are simply stating *what* you learned during requirements determination; in later steps in the systems development life cycle, you and other project team members will make decisions on exactly *how* the new system will deliver these new requirements in specific manual and automated functions. Since requirements determination and structur-

TABLE 8-1   Deliverables for Process Modeling

1. Context data flow diagram (DFD)
2. DFDs of current physical system (adequate detail only)
3. DFDs of current logical system
4. DFDs of new logical system
5. Thorough descriptions of each DFD component

ing are often parallel steps, data flow diagrams evolve from the more general to the more detailed as current and replacement systems are better understood.

Even though data flow diagrams remain popular tools for process modeling and can significantly increase software development productivity, as reported earlier for Raytheon, data flow diagrams are not used in all systems development methodologies. Some organizations, like Electronic Data Systems, have developed their own type of diagrams to model processes. Other organizations rely on the process modeling tool in Oracle's Designer CASE tool, or on more widely available functional hierarchy diagrams, which are available in many CASE toolsets. Some methodologies, such as Rapid Application Development (RAD), do not model process separately at all. Instead RAD builds process into the prototypes that are created as the core of its development life cycle (see Chapter 19). However, even if you never formally use data flow diagrams in your professional career, they remain a part of system development's history. DFDs give you a notation as well as illustrate important concepts about the movement of data between manual and automated steps and a way to depict work flow in an organization. DFDs continue to be beneficial to information systems professionals as tools for both analysis and communication. For that reason, we devote almost an entire chapter to DFDs.

DATA FLOW DIAGRAMMING MECHANICS

Data flow diagrams are versatile diagramming tools. With only four symbols, you can use data flow diagrams to represent both physical and logical information systems. Data flow diagrams (DFDs) are not as good as flowcharts for depicting the details of physical systems; on the other hand, flowcharts are not very useful for depicting purely logical information flows. In fact, flowcharting has been criticized by proponents of structured analysis and structured design because it is too physically oriented. Flowcharting symbols primarily represent physical computing equipment, such as punch cards, terminals, and tape reels. One continual criticism of system flowcharts has been that reliance on them tends to result in premature physical system design. Consistent with the incremental commitment philosophy of the SDLC, you should wait to make technology choices and to decide on physical characteristics of an information system until you are sure all functional requirements are right and accepted by users and other stakeholders.

DFDs do not share this problem of premature physical design because they do not rely on any symbols to represent specific physical computing equipment. They are also easier to use than flow charts as they involve only four different symbols.

NET SEARCH
In this book, we use the DeMarco and Yourdon conventions for data flow diagrams. Visit http://www.prenhall.com/hoffer to complete an exercise related to this topic.

Definitions and Symbols

There are two different standard sets of data flow diagram symbols, but each set consists of four symbols that represent the same things: data flows, data stores, processes, and sources/sinks (or external entities). The set of symbols we will use in this book was devised by Gane and Sarson (1979). The other standard set was developed by DeMarco and Yourdon (DeMarco, 1979; Yourdon and Constantine, 1979).

A *data flow* can be best understood as data in motion, moving from one place in a system to another. A data flow could represent data on a customer order form or a payroll check. A data flow could also represent the results of a query to a database, the contents of a printed report, or data on a data entry computer display form. A data flow is data that move together. Thus, a data flow can be composed of many individual pieces of data that are generated at the same time and flow together to common destinations. A **data store** is data at rest. A data store may represent one of many different physical locations for data, for example, a file folder, one or more com-

Data store: Data at rest, which may take the form of many different physical representations.

Process: The work or actions performed on data so that they are transformed, stored, or distributed.

Source/sink: The origin and/or destination of data, sometimes referred to as external entities.

puter-based file(s), or a notebook. To understand data movement and handling in a system, the physical configuration is not really important. A data store might contain data about customers, students, customer orders, or supplier invoices. A **process** is the work or actions performed on data so that they are transformed, stored, or distributed. When modeling the data processing of a system, it doesn't matter whether a process is performed manually or by a computer. Finally, a **source/sink** is the origin and/or destination of the data. Source/sinks are sometimes referred to as external entities because they are outside the system. Once processed, data or information leave the system and go to some other place. Since sources and sinks are outside the system we are studying, there are many characteristics of sources and sinks that are of no interest to us. In particular, we do not consider the following:

- Interactions that occur between sources and sinks
- What a source or sink does with information or how it operates (that is, a source or sink is a "black box")
- How to control or redesign a source or sink since, from the perspective of the system we are studying, the data a sink receives and often what data a source provides are fixed
- How to provide sources and sinks direct access to stored data since, as external agents, they cannot directly access or manipulate data stored within the system; that is, processes within the system must receive or distribute data between the system and its environment

These principles are consistent with the concepts of system, boundaries, and environment presented in Chapter 2.

The symbols for each set of DFD conventions are presented in Figure 8-2. For both conventions, a data flow is depicted as an arrow. The arrow is labeled with a meaningful name for the data in motion; for example, customer order, sales receipt, or paycheck. The name represents the aggregation of all the individual elements of data moving as part of one packet, that is, all the data moving together at the same time. A square is used in both conventions for sources/sinks and has a name

Figure 8-2
Comparison of DeMarco & Yourdan and Gane & Sarson DFD symbol sets

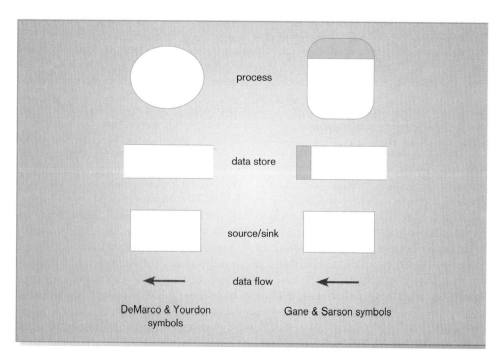

process

data store

source/sink

data flow

DeMarco & Yourdon
symbols

Gane & Sarson symbols

that states what the external agent is, such as customer, teller, EPA office, or inventory control system. The Gane & Sarson symbol for a process is a rectangle with rounded corners; it is a circle for DeMarco & Yourdon. The Gane & Sarson rounded rectangle has a line drawn through the top. The upper portion is used to indicate the number of the process. Inside the lower portion is a name for the process, such as generate paycheck, calculate overtime pay, or compute grade point average. The Gane & Sarson symbol for a data store is a rectangle that is missing its right vertical side. At the left end is a small box used to number the data store and inside the main part of the rectangle is a meaningful label for the data store, such as student file, transcripts, or roster of classes. The DeMarco data store symbol consists of two parallel lines, which may be depicted horizontally or vertically.

As stated earlier, sources/sinks are *always* outside the information system and define the boundaries of the system. Data must originate outside a system from one or more sources and the system must produce information to one or more sinks (these are principles of open systems, and almost every information system is an example of an open system). If any data processing takes place inside the source/sink, we are not interested in it, as this processing takes place outside of the system we are diagramming. A source/sink might consist of the following:

- Another organization or organization unit that sends data to or receives information from the system you are analyzing (for example, a supplier or an academic department—in either case, this organization is external to the system you are studying)

- A person inside or outside the business unit supported by the system you are analyzing and who interacts with the system (for example, a customer or loan officer)

- Another information system with which the system you are analyzing exchanges information

Many times students who are just learning how to use DFDs will be confused about whether something is a source/sink or a process within a system. This dilemma occurs most often when the data flows in a system cross office or departmental boundaries so that some processing occurs in one office and the processed data is moved to another office where additional processing occurs. Students are tempted to identify the second office as a source/sink to emphasize the fact that the data have been moved from one physical location to another (Figure 8-3a). However, we are not concerned with where the data are physically located. We are more interested in how they are moving through the system and how they are being processed. If the processing of data in the other office may be automated by your system or the handling of data there may be subject for redesign, then you should represent the second office as one or more processes rather than as a source/sink (Figure 8-3b).

Developing DFDs: An Example

To illustrate how DFDs are used to model the logic of data flows in information systems, we will present and work through an example. Consider Hoosier Burger's food ordering system, which you first saw in Chapter 2. The highest-level view of this system, a **context diagram**, is shown in Figure 8-4. You will notice that this context diagram contains only one process, no data stores, four data flows, and three sources/sinks. The single process, labeled "0," represents the entire system; all context diagrams have only one process labeled "0." The sources/sinks represent its environmental boundaries. Since the data stores of the system are conceptually inside the one process, no data stores appear on a context diagram.

Context diagram: An overview of an organizational system that shows the system boundaries, external entities that interact with the system, and the major information flows between the entities and the system.

Figure 8-3
Differences between sources/sinks and processes
(a) An improperly drawn DFD showing a process as a source/sink

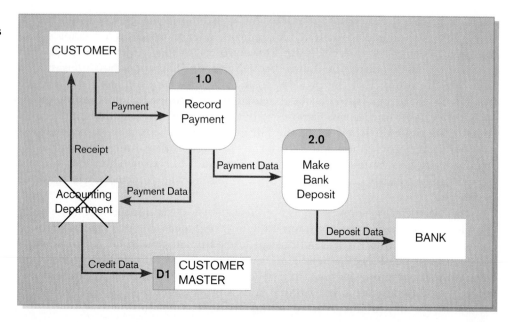

(b) A DFD showing proper use of a process

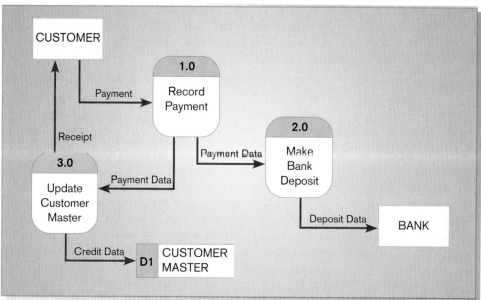

The next step for the analyst is to think about which processes are represented by the single process in the context diagram. As you can see in Figure 8-5, we have identified four separate processes, providing more detail of the system we are studying. The main processes represent the major functions of the system and these major functions correspond to such actions as the following:

1. Capturing data from different sources (e.g., Process 1.0)
2. Maintaining data stores (e.g. Processes 2.0 and 3.0)
3. Producing and distributing data to different sinks (e.g., Process 4.0)
4. High-level descriptions of data transformation operations (e.g., Process 1.0).

Often, these major functions correspond to the selections of activities on the main system menu.

We see the system begins with an order from a customer, as was the case with the context diagram. In the first process, labeled "1.0," we see that the customer order is

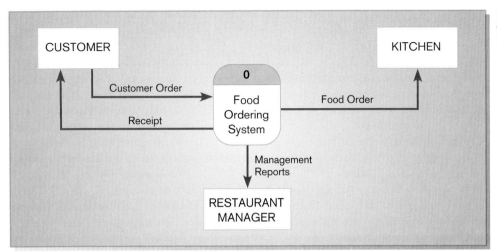

Figure 8-4
Context diagram of Hoosier
Burger's food ordering system

processed. The results are four streams or flows of data: (1) the food order is transmitted to the kitchen, (2) the customer order is transformed into a list of goods sold, (3) the customer order is transformed into inventory data, and (4) the process generates a receipt for the customer.

Notice that the sources/sinks are the same in the context diagram and in this diagram: the customer, the kitchen, and the restaurant's manager. This diagram is called a **level-0 diagram** as it represents the primary individual processes in the system at the

Level-0 diagram: A data flow diagram that represents a system's major processes, data flows, and data stores at a high level of detail.

Figure 8-5
Level-0 DFD of Hoosier Burger's
food ordering system

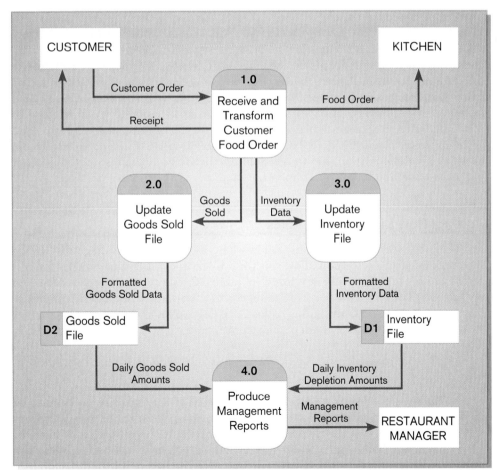

highest possible level. Each process has a number that ends in .0 (corresponding to the level number of the DFD).

Two of the data flows generated by the first process, "Receive and Transform Customer Food Order," go to external entities, so we no longer have to worry about them. We are not concerned about what happens outside of our system. Let's trace the flow of the data represented in the other two data flows. First, the data labeled Goods Sold goes to Process 2.0, Update Goods Sold File. The output for this process is labeled Formatted Goods Sold Data. This output updates a data store labeled Goods Sold File. If the customer order was for two cheeseburgers, one order of fries, and a large soft drink, each of these categories of goods sold in the data store would be incremented appropriately. The daily goods sold amounts are then used as input to Process 4.0, Produce Management Reports. Similarly, the remaining data flow generated by Process 1.0, called Inventory Data, serves as input for Process 3.0, Update Inventory File. This process updates the Inventory File data store, based on the inventory that would have been used to create the customer order. For example, an order of two cheeseburgers would mean that Hoosier Burger now has two fewer hamburger patties, two fewer burger buns, and four fewer slices of American cheese. The Daily Inventory Depletion Amounts are then used as input to Process 4. The data flow leaving Process 4.0, Management Reports, goes to the sink Restaurant Manager.

Figure 8-5 illustrates several important concepts about information movement. Consider the data flow Inventory Data moving from Process 1.0 to Process 3.0. We know from this diagram that Process 1.0 produces this data flow and that Process 3.0 receives it. However, we do not know the timing of when this data flow is produced, how frequently it is produced, or what volume of data is sent. Thus, this DFD hides many physical characteristics of the system it describes. We do know, however, that this data flow is needed by Process 3.0 and that Process 1.0 provides this needed data.

Also implied by the Inventory Data data flow is that whenever Process 1.0 produces this flow, Process 3.0 must be ready to accept it. Thus, Processes 1.0 and 3.0 are coupled to each other. In contrast, consider the link between Process 2.0 and Process 4.0. The output from Process 2.0, Formatted Goods Sold Data, is placed in a data store and, later, when Process 4.0 needs such data, it reads Daily Goods Sold Amounts from this data store. In this case, Processes 2.0 and 4.0 are decoupled by placing a buffer, a data store, between them. Now, each of these processes can work at their own pace and Process 4.0 does not have to be vigilant by being able to accept input at any time. Further, the Goods Sold File becomes a data resource that other processes could potentially draw upon for data.

N E T   S E A R C H
Fast-food restaurants like Hoosier Burger have a broad range of special-purpose software available to them to support the operation and management of the restaurant. Visit http://www.prenhall.com/hoffer to complete an exercise related to this topic.

Data Flow Diagramming Rules

There is a set of rules you must follow when drawing data flow diagrams. Unlike system flowcharts, these rules allow you (or a CASE tool) to evaluate DFDs for correctness. The rules for DFDs are listed in Table 8-2. Figure 8-6 illustrates incorrect ways to draw DFDs and the corresponding correct application of the rules. The rules that prescribe naming conventions (rules C, G, I, and P) and those that explain how to interpret data flows in and out of data stores (rules N and O) are not illustrated in Figure 8-6.

Besides the rules of Table 8-2, there are two DFD guidelines that apply most of the time:

- *The inputs to a process are different from the outputs of that process*: The reason is that processes, to have a purpose, typically transform inputs into outputs, rather than simply pass the data through without some manipulation. What may happen is that the same input goes in and out of a process but the process also produces other new data flows that are the result of manipulating the inputs.

TABLE 8-2   Rules Governing Data Flow Diagramming

Process:

A. No process can have only outputs. It is making data from nothing (a miracle). If an object has only outputs, then it must be a source.

B. No process can have only inputs (a black hole). If an object has only inputs, then it must be a sink.

C. A process has a verb phrase label.

Data Store:

D. Data cannot move directly from one data store to another data store. Data must be moved by a process.

E. Data cannot move directly from an outside source to a data store. Data must be moved by a process that receives data from the source and places the data into the data store.

F. Data cannot move directly to an outside sink from a data store. Data must be moved by a process.

G. A data store has a noun phrase label.

Source/Sink:

H. Data cannot move directly from a source to a sink. It must be moved by a process if the data are of any concern to our system. Otherwise, the data flow is not shown on the DFD.

I. A source/sink has a noun phrase label.

Data Flow:

J. A data flow has only one direction of flow between symbols. It may flow in both directions between a process and a data store to show a read before an update. The latter is usually indicated, however, by two separate arrows since these happen at different times.

K. A fork in a data flow means that exactly the same data goes from a common location to two or more different processes, data stores, or sources/sinks (this usually indicates different copies of the same data going to different locations).

L. A join in a data flow means that exactly the same data comes from any of two or more different processes, data stores, or sources/sinks to a common location.

M. A data flow cannot go directly back to the same process it leaves. There must be at least one other process that handles the data flow, produces some other data flow, and returns the original data flow to the beginning process.

N. A data flow to a data store means update (delete or change).

O. A data flow from a data store means retrieve or use.

P. A data flow has a noun phrase label. More than one data flow noun phrase can appear on a single arrow as long as all of the flows on the same arrow move together as one package.

Adapted from Celko, 1987

- *Objects on a DFD have unique names*: Every process has a unique name. There is no reason to have two processes with the same name. To keep a DFD uncluttered, however, you may repeat data stores and sources/sinks. When two arrows have the same data flow name, you must be careful that these flows are exactly the same. It is easy to reuse the same data flow name when two packets of data are almost the same, but not identical. A data flow name represents a specific set of data, and another data flow that has even one more or one less piece of data must be given a different, unique name.

Decomposition of DFDs

In the earlier example of Hoosier Burger's food ordering system, we started with a high-level context diagram. Upon thinking more about the system, we saw that the larger system consisted of four processes. The act of going from a single system to four component processes is called (*functional*) *decomposition*. **Functional decomposition** is an iterative process of breaking the description or perspective of a system down into finer and finer detail. This process creates a set of hierarchically related

Functional decomposition: An iterative process of breaking the description of a system down into finer and finer detail, which creates a set of charts in which one process on a given chart is explained in greater detail on another chart.

Figure 8-6
Incorrect and correct ways to draw data flow diagrams

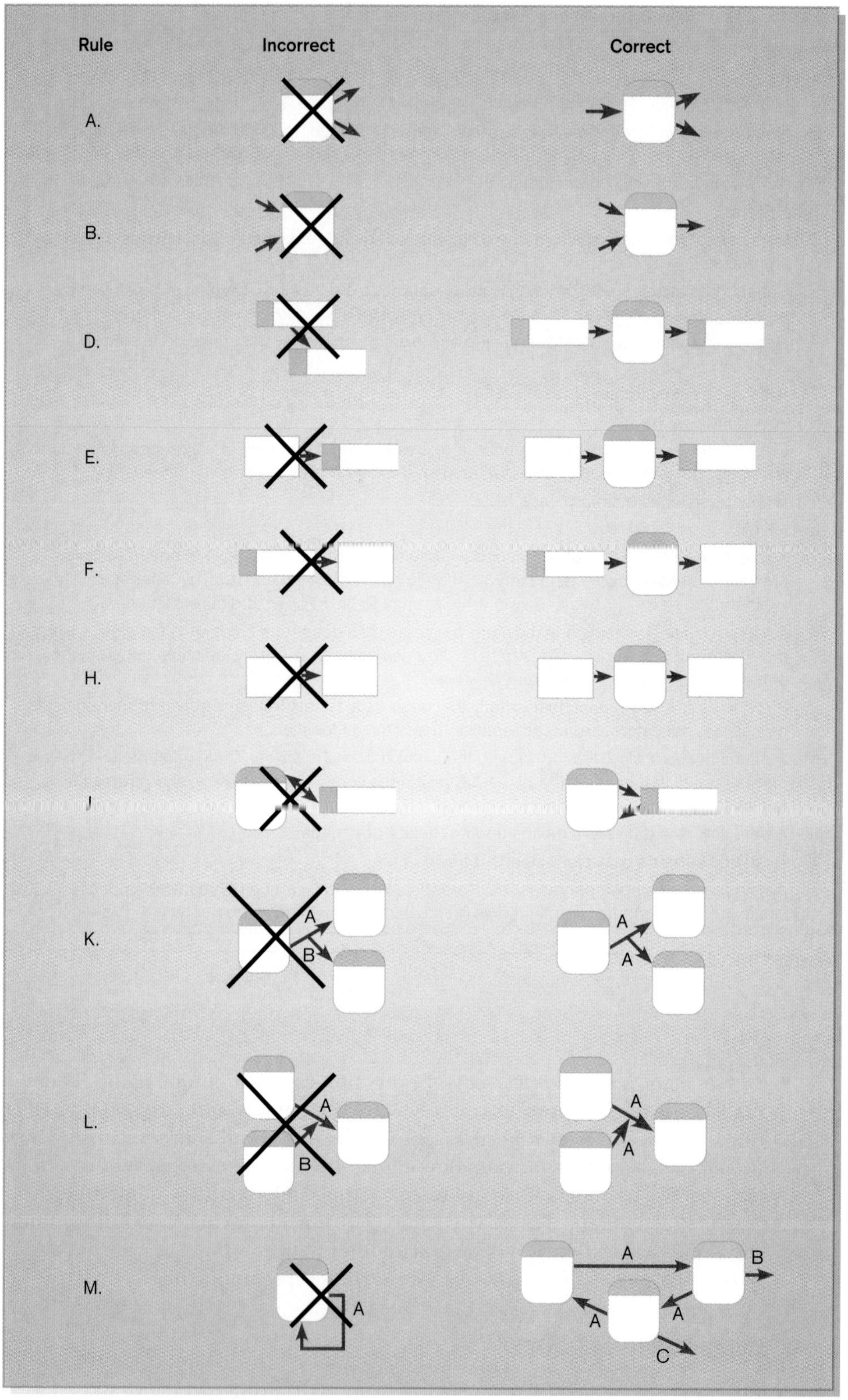

charts in which one process on a given chart is explained in greater detail on another chart. For the Hoosier Burger system, we broke down or decomposed the larger system into four processes. Each of those processes (or subsystems) are also candidates for decomposition. Each process may consist of several subprocesses. Each subprocess may also be broken down into smaller units. Decomposition continues until you have reached the point where no subprocess can logically be broken down any further. The lowest level of DFDs is called a *primitive DFD*, which we define later in this chapter.

Let's continue with Hoosier Burger's food ordering system to see how a level-0 DFD can be further decomposed. The first process in Figure 8-5, called Receive and Transform Customer Food Order, transforms a customer's verbal food order (for example, "Give me two cheeseburgers, one small order of fries, and one regular orange soda.") into four different outputs. Process 1.0 is a good candidate process for decomposition. Think about all of the different tasks that Process 1.0 has to perform: (1) receive a customer order, (2) transform the entered order into a form meaningful for the kitchen's system, (3) transform the order into a printed receipt for the customer, (4) transform the order into goods sold data, (5) transform the order into inventory data. There are at least these five logically separate functions that occur in Process 1.0. We can represent the decomposition of Process 1.0 as another DFD, as shown in Figure 8-7.

Note that each of the five processes in Figure 8-7 are labeled as subprocesses of Process 1.0: Process 1.1, Process 1.2, and so on. Also note that, just as with the other data flow diagrams we have looked at, each of the processes and data flows are named. You will also notice that there are no sources or sinks represented. Although you may include sources and sinks, the context and level-0 diagrams show the sources and sinks. The data flow diagram in Figure 8-7 is called a level-1 diagram. If we should decide to decompose Processes 2.0, 3.0, or 4.0 in a similar manner, the DFDs we create would also be called level-1 diagrams. In general, a **level-*n* diagram** is a DFD that is generated from *n* nested decompositions from a level-0 diagram.

Processes 2.0 and 3.0 perform similar functions in that they both use data input to update data stores. Since updating a data store is a singular logical function, neither of these processes need to be decomposed further. We can, on the other hand,

Level-*n* diagram: A DFD that is the result of *n* nested decomposition of a series of subprocesses from a process on a level-0 diagram.

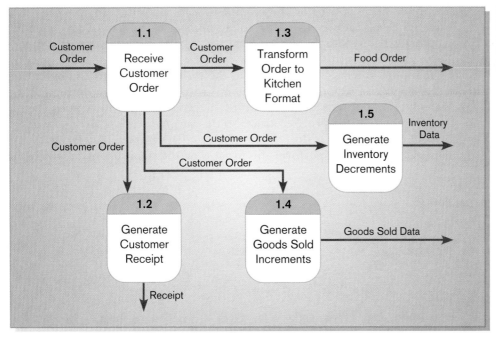

Figure 8-7
Level-1 diagram showing the decomposition of Process 1.0 from the level-0 diagram for Hoosier Burger's food ordering system

Figure 8-8
Level-1 diagram showing the decomposition of Process 4.0 from the level-0 diagram for Hoosier Burger's food ordering system

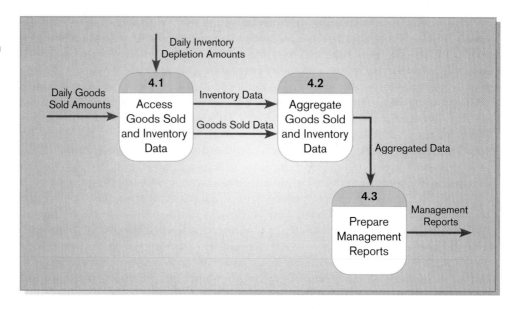

decompose Process 4.0, Produce Management Reports, into at least three subprocesses: Access Goods Sold and Inventory Data, Aggregate Goods Sold and Inventory Data, and Prepare Management Reports. The decomposition of Process 4.0 is shown in the level-1 diagram of Figure 8-8.

Each level-1, -2, or -*n* DFD represents one process on a level-*n*-1 DFD; each DFD should be on a separate page. As a rule of thumb, no DFD should have more than about seven processes in it, as too many processes will make the diagram too crowded and more difficult to understand.

To continue with the decomposition of Hoosier Burger's food ordering system, we examine each of the subprocesses identified in the two level-1 diagrams we have produced, one for Process 1.0 and one for Process 4.0. Should we decide that any of these subprocesses should be further decomposed, we would create a level-2 diagram showing that decomposition. For example, if we decided that Process 4.3 in Figure 8-8 should be further decomposed, we would create a diagram that looks something like Figure 8-9. Again, notice how the subprocesses are labeled.

Just as the labels for processes must follow numbering rules for clear communication, process names should also be clear yet concise. Typically, process names begin with an action verb, such as receive, calculate, transform, generate, or produce. Often process names are the same as the verbs used in many computer programming languages. Examples include merge, sort, read, write, and print. Process names should capture the essential action of the process in just a few words, yet be descriptive enough of the process' action so that anyone reading the name gets a good idea of what the process does. Many times, students just learning DFDs will use the names of people who perform the process or the department in which the

Figure 8-9
Level-2 diagram showing the decomposition of Process 4.3 from the level-1 diagram for Process 4.0 for Hoosier Burger's food ordering system

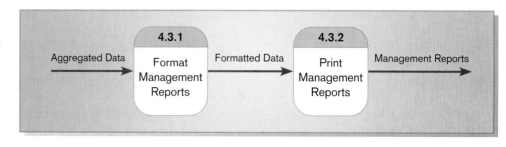

process is performed as the process name. This practice is not very useful, as we are more interested in the action the process represents than the person performing it or the place where it occurs.

Balancing DFDs

When you decompose a DFD from one level to the next, there is a conservation principle at work. You must conserve inputs and outputs to a process at the next level of decomposition. In other words, Process 1.0, which appears in a level-0 diagram, must have the same inputs and outputs when decomposed into a level-1 diagram. This conservation of inputs and outputs is called **balancing**.

Let's look at an example of balancing a set of DFDs. Look back at Figure 8-4. This is the context diagram for Hoosier Burger's food ordering system. Notice that there is one input to the system, the customer order, which originates with the customer. Notice also that there are three outputs: the customer receipt, the food order intended for the kitchen, and management reports. Now look at Figure 8-5. This is the level-0 diagram for the food ordering system. Remember that all data stores and flows to or from them are internal to the system. Notice that the same single input to the system and the same three outputs represented in the context diagram also appear at level-0. Further, no new inputs to or outputs from the system have been introduced. Therefore, we can say that the context diagram and level-0 DFDs are balanced.

Now look at Figure 8-7, where Process 1.0 from the level-0 DFD has been decomposed. As we have seen before, Process 1.0 has one input and four outputs. The single input and multiple outputs all appear on the level-1 diagram in Figure 8-7. No new inputs or outputs have been added. Compare Process 4.0 in Figure 8-5 to its decomposition in Figure 8-8. You see the same conservation of inputs and outputs.

Figure 8-10 shows you one example of what an unbalanced DFD could look like. Here, in the context diagram, there is one input to the system, A, and one output, B.

Balancing: The conservation of inputs and outputs to a data flow diagram process when that process is decomposed to a lower level.

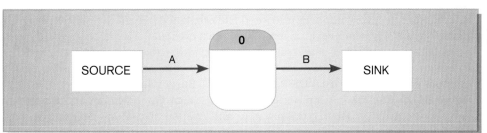

Figure 8-10
An unbalanced set of data flow diagrams
(a) Context diagram

(b) Level-0 diagram

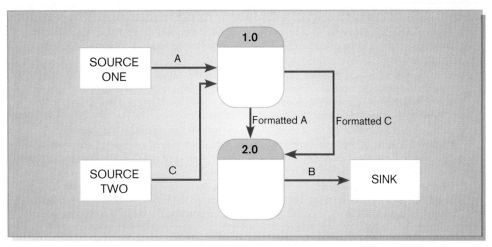

Figure 8-11
Example of data flow splitting
(a) Composite data flow

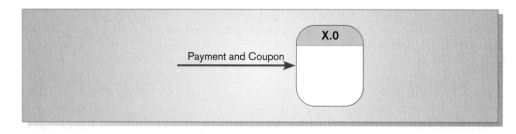

(b) Disaggregated data flows

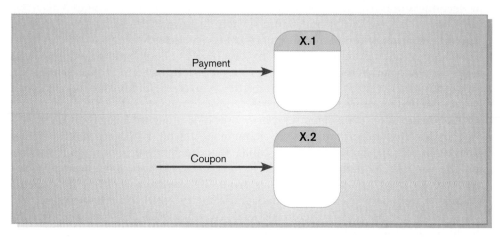

Yet, in the level-0 diagram, there is an additional input, C, and flows A and C come from different sources. These two DFDs are not balanced. If an input appears on a level-0 diagram, it must also appear on the context diagram. What happened in this example? Perhaps, when drawing the level-0 DFD, the analyst realized that the system also needed C in order to compute B. A and C were both drawn in the level-0 DFD, but the analyst forgot to update the context diagram. In making corrections, the analyst should also include "SOURCE ONE" and "SOURCE TWO" on the context diagram. It is very important to keep DFDs balanced, from the context diagram all the way through each level diagram you must create.

A data flow consisting of several subflows on a level-n diagram can be split apart on a level-$n + 1$ diagram for a process that accepts this composite data flow as input. For example, consider the partial DFDs from Hoosier Burger illustrated in Figure 8-11. In Figure 8-11a we see that a composite, or package, data flow, Payment and Coupon, is input to the process. That is, the payment and coupon always flow together and are

TABLE 8-3   Advanced Rules Governing Data Flow Diagramming

Q. A composite data flow on one level can be split into component data flows at the next level, but no new data can be added and all data in the composite must be accounted for in one or more subflows.

R. The inputs to a process must be sufficient to produce the outputs (including data placed in data stores) from the process. Thus, all outputs can be produced, and all data in inputs move somewhere, either to another process or to a data store outside the process or on a more detailed DFD showing a decomposition of that process.

S. At the lowest level of DFDs, new data flows may be added to represent data that are transmitted under exceptional conditions; these data flows typically represent error messages (e.g., "Customer not known; do you want to create a new customer") or confirmation notices (e.g., "Do you want to delete this record").

T. To avoid having data flow lines cross each other, you may repeat data stores or sources/sinks on a DFD. Use an additional symbol, like a double line on the middle vertical line of a data store symbol, or a diagonal line in a corner of a sink/source square, to indicate a repeated symbol.

Adapted from Celko, 1987

input to the process at the same time. In Figure 8-11b the process is decomposed (sometimes called exploded or nested) into two subprocesses, and each subprocess receives one of the components of the composite data flow from the higher-level DFD. These diagrams are still balanced since exactly the same data are included in each diagram.

The principle of balancing and the goal of keeping a DFD as simple as possible lead to four additional, advanced rules for drawing DFDs. These advanced rules are summarized in Table 8-3. Rule Q covers the situation illustrated in Figure 8-11. Rule R covers a conservation principle about process inputs and outputs. Rule S addresses one exception to balancing. Rule T tells you how you can minimize clutter on a DFD.

FOUR DIFFERENT TYPES OF DFDS

There are actually four different types of data flow diagrams used in the systems development process: (1) current physical, (2) current logical, (3) new logical, (4) new physical. When structured analysis and design was first introduced in the late 1970s, it was argued that analysts should prepare all four types of DFDs, in this particular order.

In a current physical DFD, process labels include the names of people or their positions or the names of computer systems that might provide some of the overall system's processing. That is, the label includes an identification of the "technology" used to process the data. Similarly, data flows and data stores are often labeled with the names of the actual physical media on which data flow or in which data are stored, such as file folders, computer files, business forms, or computer tapes. For the current logical model, the physical aspects of the system are removed as much as possible so that the current system is reduced to its essence, to the data and the processes that transform them, regardless of actual physical form. The new logical model would be exactly like the current logical model if the user were completely happy with the functionality of the current system but had problems with how it was implemented. Typically, though, the new logical model will differ from the current logical model by having additional functions, obsolete functions removed, and inefficient flows reorganized. Finally, the DFDs for the new physical system represent the physical implementation of the new system. The DFDs for the new physical system will reflect the decision of the analysts about which system functions, including those added in the new logical model, will be automated and which will be manual.

To illustrate the differences among the different types of DFDs, we will look at another example from Hoosier Burger. We saw that the food ordering system generates two types of usage data, for goods sold and for inventory. At the end of each day, the manager, Bob Mellankamp, generates the inventory report that tells him how much inventory should have been used for each item associated with sales. The amounts shown on the inventory report are just one input to a largely manual inventory control system Bob uses every day. Figure 8-12 lists the steps involved in Bob's inventory control system.

The data flow diagrams that model the current physical system are shown in Figure 8-13. The context diagram (Figure 8-13a) shows three sources of data outside the system: suppliers, the food ordering system inventory report, and stock-on-hand. Suppliers provide invoices as input, and the system returns payments and orders as outputs to the suppliers. Both the inventory report and the stock-on-hand provide inventory counts as system inputs. The level-0 DFD for Hoosier Burger's inventory system (Figure 8-13b) shows six different processes, most of which also appear in the list of inventory activities in Figure 8-12. You can see from the diagram that when Bob receives invoices from suppliers, he records their receipt on an invoice log sheet and files the actual invoices in his accordion file. Using the invoices, Bob records the amount of stock delivered on the stock logs,

Figure 8-12
List of activities involved in Bob Mellankamp's inventory control system for Hoosier Burger

1. Meet delivery trucks before opening restaurant.
2. Unload and store deliveries.
3. Log invoices and file in accordion file.
4. Manually add amounts received to stock logs.
5. After closing, print inventory report.
6. Count physical inventory amounts.
7. Compare inventory report totals to physical count totals.
8. Compare physical count totals to minimum order quantities; if the amount is less, make order; if not, do nothing.
9. Pay bills that are due and record them as paid.

Figure 8-13
Hoosier Burger's current physical inventory control system
(a) Context diagram

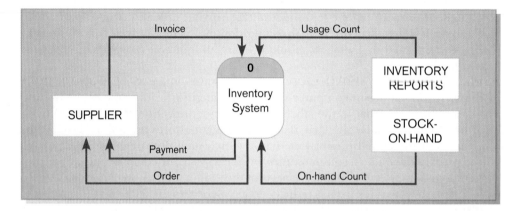

(b) Level-0 data flow diagram

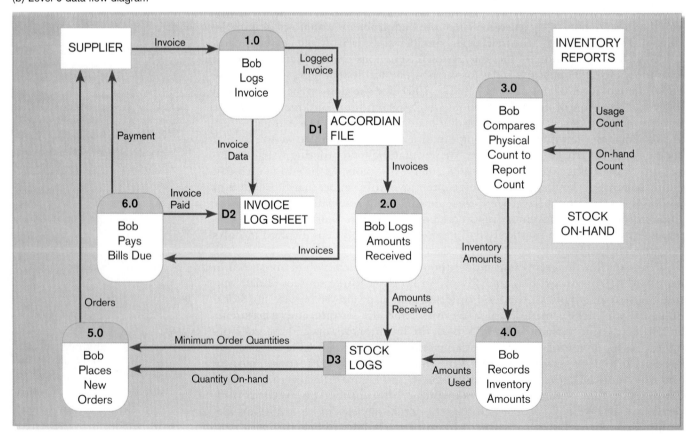

which are paper forms posted near the point of storage for each inventory item. Figure 8-13 also illustrates that a physical DFD shows only data movement, not the movement of materials or other physical items (e.g., energy). Data about physical items such as an Invoice, which is data about a shipment to Hoosier Burger, are shown on a physical DFD; the data about a physical item may or may not move with the actual physical item.

Figure 8-14 gives a partial example of Hoosier Burger's stock log. Notice that the minimum order quantities—the stock level at which orders must be placed in order to avoid running out of an item—appear on the log form. There are also spaces for entering the starting amount, amount delivered, and the amount used for each item. Amounts delivered are entered on the sheet when Bob logs stock deliveries; amounts used are entered after Bob has compared the amounts of stock used according to a physical count and according to the numbers on the inventory report generated by the food ordering system. We should note that Hoosier Burger has standing daily delivery orders for some perishable items that are used every day, like burger buns, meats, and vegetables.

As the DFD in Figure 8-13b shows, Bob uses the minimum order quantities and the amount of stock on hand to determine which orders need to be placed. He uses the invoices to determine which bills need to be paid, and he carefully records each payment.

To continue to the next step, creating the current logical model using DFDs, we need to identify the essence of the inventory system Bob has established. What are the key data necessary to keep track of inventory and to pay bills? What are the key processes involved? We also need to remove all elements of the physical system, such as Bob and the physical file folders he uses. There are at least four key processes that make up the Hoosier Burger's inventory system: (1) account for anything added to inventory, (2) account for anything taken from inventory, (3) place orders, (4) pay bills. Key data used by the system include inventories and stock-on-hand counts, however determined. Major outputs from the system continue to be orders and payments. Focusing on the essential elements of the system results in the DFD for the current logical system shown in Figure 8-15.

The purpose of the new logical DFD is (1) to show any additional functionality necessary in the new system, (2) to indicate which, if any, obsolete components have been eliminated, and (3) to describe any changes in the logical flow of data between system components, including different data stores. For Hoosier Burger's inventory system, Bob Mellankamp would like to add three additional functions. First, Bob would like data on new shipments to be entered into an automated system, thus

Figure 8-14
Hoosier Burger's stock log form

Stock Log						
Date:			Jan 1			Jan 2
	Reorder	Starting	Amount	Amount	Starting	
Item	Quantity	Amount	Delivered	Used	Amount	
Hamburger buns	50 dozen	5	50	43	12	
Hot dog buns	25 dozen	0	25	22	3	
English muffins	10 dozen	6	10	12	4	
Napkins	2 cases	10	0	2	8	
Straws	1 case	1	0	1	0	

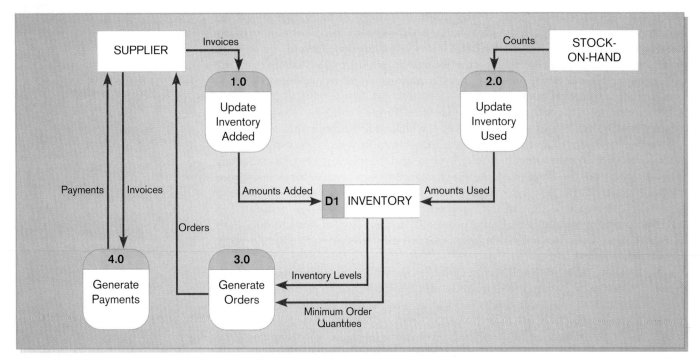

Figure 8-15
Level-0 data flow diagram for Hoosier Burger's current logical inventory control system

doing away with paper stock log sheets. Bob would like shipment data to be as current as possible by being entered into the system as soon as the new stock arrives at the restaurant. Second, Bob would like the system to determine automatically whether a new order should be placed. Automatic ordering would relieve Bob of worrying about whether Hoosier Burger has enough of everything in stock at all times. Finally, Bob would like to be able to know, at any time, the approximate inventory levels for all goods in stock. For some goods, such as hamburger buns, Bob can visually inspect the amount in stock and determine approximately how much is left and how much more is needed before closing time. For other items, however, Bob may need a quick, rough estimate of what is in stock more quickly than it would take him to do a visual inspection.

The new logical data flow diagram for Hoosier Burger's inventory system is shown in Figure 8-16. Notice how the DFD is almost identical to the DFD for current logical, in Figure 8-15. The only difference is a new Process 5.0, which allows for querying the inventory data to get an estimate of how much of an item is in stock. Bob's two other requests for change can both be handled within the existing logical view of the inventory system. Process 1.0, Update Inventory Added, does not indicate whether the updates are in real-time or batched or whether the updates occur on paper or as part of an automated system. Therefore, immediately entering shipment data into an automated system is encompassed by Process 1.0. Similarly, Process 2.0, Generate Orders, does not indicate whether Bob or a computer generates orders or whether the orders are generated on a real-time or batch basis, so Bob's request that orders be generated automatically by the system is already represented by Process 3.0. The next step would be to create a data flow diagram for the new physical system to represent the requests Bob has made. That step will be deferred, however, until several alternative solutions to Bob's inventory problems have been considered in Chapter 11.

You may be asking if it is really necessary for an analyst to construct not just one but four complete sets of data flow diagrams for each system on which he or she works. Many experts today say no, it is not necessary, that analysts should begin as

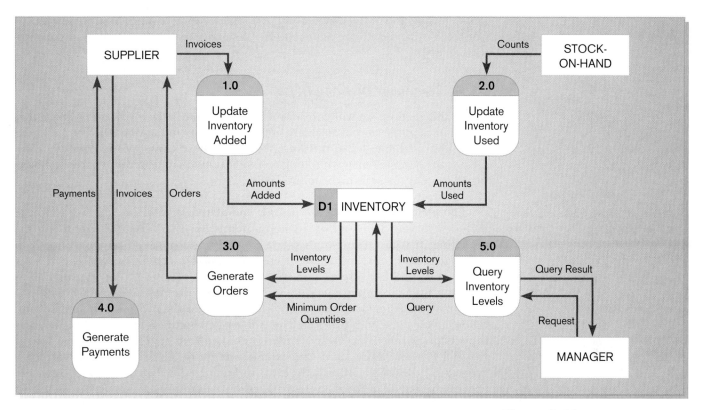

Figure 8-16
Level-0 data flow diagram for Hoosier Burger's new logical inventory control system

quickly as possible with the new logical DFD. Experts used to recommend that all four levels of DFDs be constructed because of three assumptions:

1. Analysts knew little about the user's business and needed to develop a detailed current physical DFD in order to understand the business.

2. Users were not able to work with a new logical DFD right away.

3. There is not much work in turning current logical DFDs into new logical DFDs.

These assumptions proved to be correct (Yourdon, 1989) but overlooked a greater danger: Analysts tended to devote a great deal of time to creating and refining a detailed set of DFDs for the current physical system, most of which was thrown away in the transition to the current logical DFDs. We recommend that you create a set of DFDs for the current physical system but that those DFDs only be detailed enough so that you come away with a good overview of the current system. We agree with the experts that the focus should be on the new logical system.

USING DATA FLOW DIAGRAMMING IN THE ANALYSIS PROCESS

Learning the mechanics of drawing data flow diagrams is important to you, as data flow diagrams have proven to be essential tools for the structured analysis process. Beyond the issues of drawing mechanically correct DFDs, there are other issues related to process modeling with which you as an analyst must be concerned. Such issues, including whether the DFDs are complete and consistent across levels, are dealt with in the next section on guidelines for drawing DFDs. Another issue to consider is how you can use data flow diagrams as a useful tool for analysis, discussed in the final section of the chapter. In these final sections, we also illustrate features of CASE tools that aid in compliance with rules and guidelines, as well as in good systems analysis.

Guidelines for Drawing DFDs

In this section, we will consider additional guidelines for drawing DFDs that extend beyond the simple mechanics of drawing diagrams and making sure that the rules listed in Tables 8-2 and 8-3 are followed. These guidelines include (1) completeness, (2) consistency, (3) timing considerations, (4) the iterative nature of drawing DFDs, and (5) drawing primitive DFDs.

DFD completeness: The extent to which all necessary components of a data flow diagram have been included and fully described.

Completeness   The concept of **DFD completeness** refers to whether you have included in your DFDs all of the components necessary for the system you are modeling. If your DFD contains data flows that do not lead anywhere, or data stores, processes, or external entities that are not connected to anything else, your DFD is not complete. Most CASE tools have built-in facilities that you can run to help you find incompleteness in your DFDs. For example, Figure 8-17 shows a draft of a DFD from Oracle's Designer CASE tool for the process Bob Mellankamp follows to hire new employees for Hoosier Burger. Figure 8-18 shows three analysis reports prepared using Oracle's Designer. The first report, in Figure 8-18a, is called a Process Source and Consequences Report. The report was created for Process 1.1, RECREV_APP, in the DFD in Figure 8-17. Process 1.1 is a complete process, in that all of the data flows in and out of the process are accounted for in the DFD. The report in Figure 8-18a shows this completeness by listing in tabular form all of the flows coming into and going out of Process 1.1. A careful look at these tables reveals the names of the flows and either where they originated or where they are going. The sources and the sinks, referred to in this report as Consequences, are also labeled with their types, whether data store, function, or

Figure 8-17
Hoosier Burger's hiring procedures in a data flow diagram

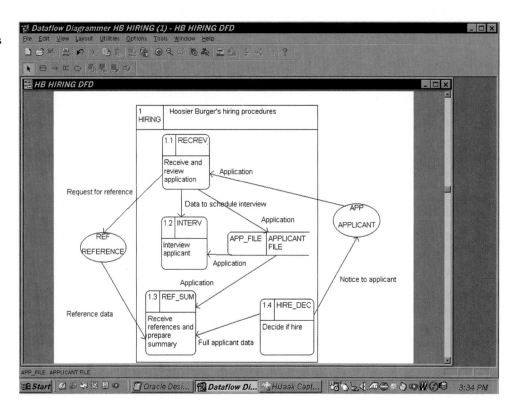

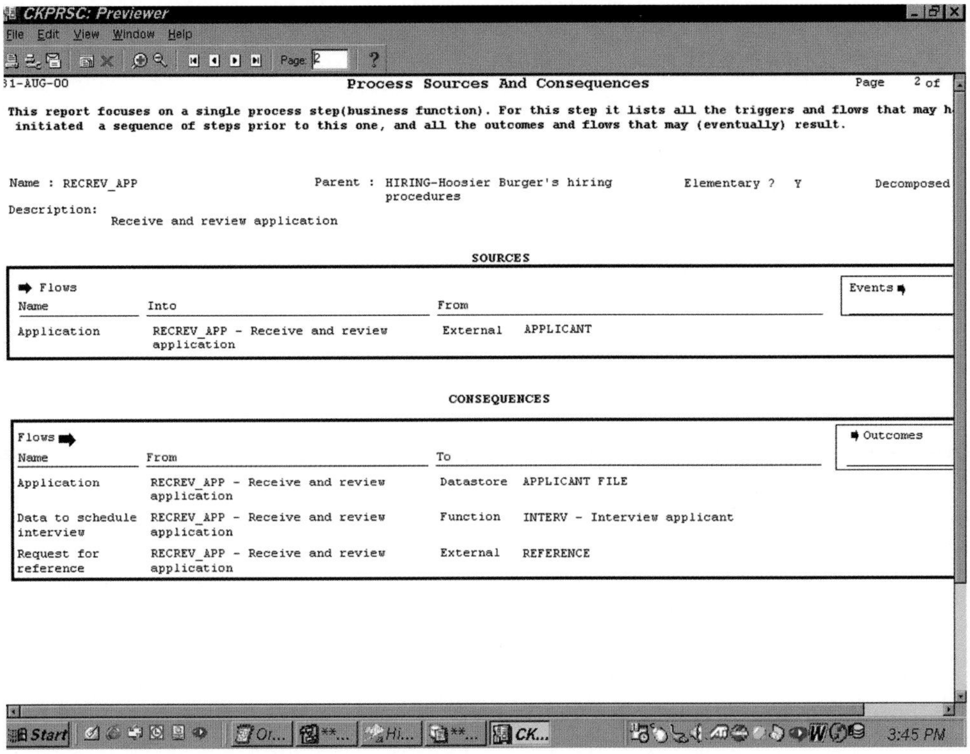

Figure 8-18
Oracle Designer Repository Reports
(a) Process Source and Consequences Report for a complete process

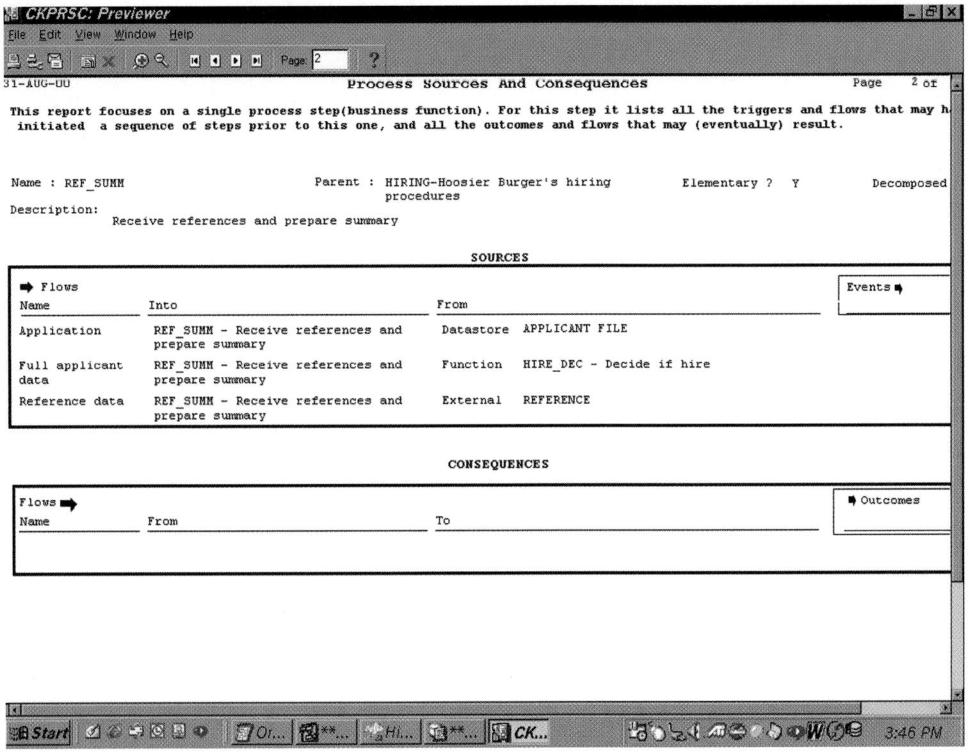

(b) Process Source and Consequences Report for an incomplete process

Figure 8-18
Oracle Designer Repository
Reports (continued)
(c) Dataflow Definition Report for an
incomplete process

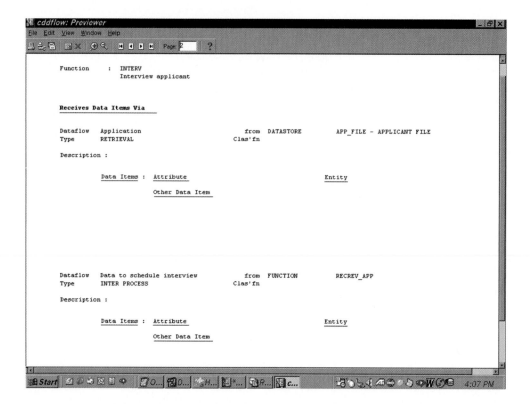

(d) Dataflow Definition Report for an
incomplete process, continued.

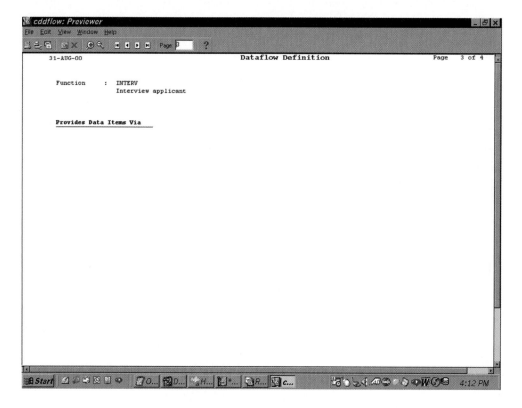

external. The second report, in Figure 8-18b, is also a Process Source and Consequences Report, but it was created for Process 1.3 in the DFD in Figure 8-17. Process 1.3 is called REF_SUMM. The report reveals that Process 1.3 has data flows coming in but none going on. A quick visual check of Figure 8-17 will confirm this. Figures 8-18c and 8-18d represent another type of repository report that is helpful in determining the completeness of data flow diagrams. Each figure represents a different page in the report, called a Dataflow Definition Report. The focus of this report is Process 1.2, INTERV, in Figure 8-17. Figure 8-18c shows that part of the report dealing with the data flows coming into Process 1.2. You can see that both of the flows coming into Process 1.3 are listed on the report. Figure 8-18d shows the second page of the report, focusing on data flows coming out of Process 1.2, and you can see that this page is blank. Like Process 1.3, Process 1.2 only receives data flows and does not send any data flows out. We do not include a repository report for Process 1.4 in Figure 8-17, but this process also has some problems. Can you determine what they are? Did you find all of the errors in Figure 8-17 before reading the repository reports? To which of the rules in Tables 8-2 and 8-3 are these errors associated? When you draw many DFDs for a system, it is not uncommon to make errors like those in Figure 8-17; either CASE tool analysis functions or walkthroughs with other analysts can help you identify such problems.

Not only must all necessary elements of a DFD be present, each of the components must be fully described in the project dictionary. With most CASE tools, the project dictionary is linked to the diagram. That is, when you define a process, data flow, source/sink, or data store on a DFD, an entry is automatically created in the repository for that element. You must then enter the repository and complete the element's description. Different descriptive information can be kept about each of the four types of elements on a DFD, and each CASE tool or project dictionary standard an organization adopts has different entry information. Figure 8-19 shows a display of the contents of the Oracle Designer repository entry for

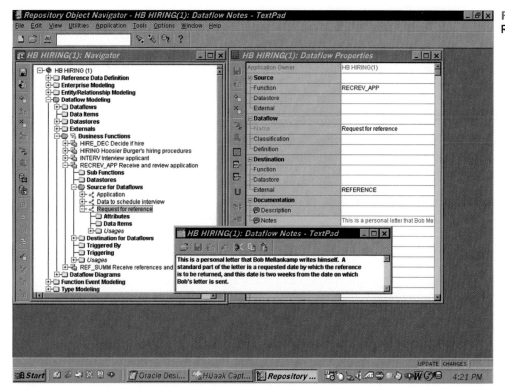

Figure 8-19
Repository entry for a data flow

the Request for Reference data flow on Figure 8-17. This data flow repository entry includes

- The label or name (e.g., Request for Reference) for the data flow as entered on DFDs (*Note:* Case and punctuation of the label matter, but if exactly the same label is used on multiple DFDs, whether nested or not, then the same repository entry applies to each reference.)
- A short description defining the data flow
- A list of other repository objects grouped into categories by type of object
- The composition or list of data elements contained in the data flow
- Notes supplementing the limited space for the description that go beyond defining the data flow to explaining the context and nature of this repository object
- A list of locations (the names of the DFDs) on which this data flow appears (e.g., the DFD was called "applicant") and the names of the sources and destinations on each of these DFDs for the data flow

By the way, it is this tight linkage between diagrams and the CASE repository that creates much of the value of a CASE tool. Although very sophisticated drawing tools as well as forms and word-processing systems exist, these stand-alone tools do not integrate graphical objects with their textual descriptions as CASE tools do.

DFD consistency: The extent to which information contained on one level of a set of nested data flow diagrams is also included on other levels.

Consistency The concept of **DFD consistency** refers to whether or not the depiction of the system shown at one level of a nested set of DFDs is compatible with the depictions of the system shown at other levels. A gross violation of consistency would be a level-1 diagram with no level-0 diagram. Another example of inconsistency would be a data flow that appears on a higher-level DFD but not on lower levels (a violation of balancing). Yet another example of inconsistency is a data flow attached to one object on a lower-level diagram but attached to another object at a higher level. For example, a data flow named "Payment," which serves as input to Process 1 on a level-0 DFD, appears as input to Process 2.1 on a level-1 diagram for Process 2.

CASE tools also have analysis facilities you can use to detect such inconsistencies across nested data flow diagrams. For example, to help you avoid making DFD consistency errors when you draw a DFD using a CASE tool, most tools will automatically place the inflows and outflows of a process on the DFD you create when you inform the tool to decompose that process. In manipulating the lower-level diagram, you could accidentally delete or change a data flow that would cause the diagrams to be out of balance; thus, a consistency check facility with a CASE tool is quite helpful.

Timing You may have noticed in some of the DFD examples we have presented that DFDs do not do a very good job of representing time. On a given DFD, there is no indication of whether a data flow occurs constantly in real-time, once per week, or once per year. There is also no indication of when a system would run. For example, many large transaction-based systems may run several large, computing-intensive jobs in batch mode at night, when demands on the computer system are lighter. A DFD has no way of indicating such overnight batch processing. When you draw DFDs, then, draw them as if the system you are modeling has never started and will never stop. You will learn in Chapter 20 that another type of diagram, a state-transition diagram, is used to show the timing of processes associated with a system.

Iterative Development The first DFD you draw will rarely capture perfectly the system you are modeling. You should count on drawing the same diagram over and over again, in an iterative fashion. With each attempt, you will come closer to a good approximation of the system or aspect of the system you are modeling. Iterative DFD development recognizes that requirements determination and requirements structuring are interacting, not sequential subphases of the analysis phase of the SDLC.

One rule of thumb is that it should take you about three revisions for each DFD you draw. Fortunately, CASE tools make revising drawings a lot easier than it would be if you had to draw each revision with pencil and template.

Primitive DFDs One of the more difficult decisions you need to make when drawing DFDs is when to stop decomposing processes. One rule is to stop drawing when you have reached the lowest logical level; however, it is not always easy to know what the lowest logical level is. Other more concrete rules for when to stop decomposing are

- When you have reduced each process to a single decision or calculation or to a single database operation, such as retrieve, update, create, delete, or read

- When each data store represents data about a single entity, such as a customer, employee, product, or order

- When the system user does not care to see any more detail, or when you and other analysts have documented sufficient detail to do subsequent systems development tasks

- When every data flow does not need to be split further to show that different data are handled in different ways

- When you believe that you have shown each business form or transaction, computer on-line display, and report as a single data flow (this often means, for example, that each system display and report title corresponds to the name of an individual data flow)

- When you believe there is a separate process for each choice on all lowest-level menu options for the system

Obviously, the iteration guideline discussed earlier and the various feedback loops in the SDLC (see Figure 8-1) suggest that when you think you have met the above rules for stopping, you may later discover nuances to the system that require you to further decompose a set of DFDs.

By the time you stop decomposing DFDs, a DFD can become quite detailed. Seemingly simple actions, such as generating an invoice, may pull information from several entities and may also return different results depending on the specific situation. For example, the final form of an invoice may be based on the type of customer (which would determine such things as discount rate), where the customer lives (which would determine such things as sales tax), and how the goods are shipped (which would determine such things as the shipping and handling charges). At the lowest-level DFD, called a **primitive DFD**, all of these conditions would have to be met. Given the amount of detail called for in a primitive DFD, perhaps you can see why many experts believe analysts should not spend their time completely diagramming the current physical information system as much of the detail will be discarded when the current logical DFD is created.

Primitive DFD: The lowest level of decomposition for a data flow diagram.

Using these guidelines will help you create DFDs that are more than just mechanically correct. Your data flow diagrams will also be robust and accurate representations of the information system you are modeling. Such primitive DFDs also facilitate consistency checks with the documentation produced from other requirements structuring techniques as well as make it easy for you to transition to system design steps. Having mastered the skills of drawing good DFDs, you can now use them to support the analysis process, the subject of the next section.

Using DFDs as Analysis Tools

We have seen that data flow diagrams are versatile tools for process modeling and that they can be used to model systems that are either physical or logical, current or new. Data flow diagrams can also be used in analysis for a process called **gap analysis**. In gap analysis, the role of the analyst is to discover discrepancies between two or more sets of data flow diagrams, representing two or more states of an information system, or discrepancies within a single DFD.

Gap analysis: The process of discovering discrepancies between two or more sets of DFDs or discrepancies within a single DFD.

Once the DFDs are complete, you can examine the details of individual DFDs for such problems as redundant data flows, data that are captured but are not used by the system, and data that are updated identically in more than one location. These problems may not have been evident to members of the analysis team or to other participants in the analysis process when the DFDs were created. For example, redundant data flows may have been labeled with different names when the DFDs were created. Now that the analysis team knows more about the system they are modeling, analysts can detect such redundancies. Such redundancies can be seen most easily from various CASE tool repository reports. For example, many CASE tools can generate a report listing all the processes that accept a given data element as input (remember, a list of data elements is likely part of the description of each data flow). From the label of these processes you can determine whether or not it appears as if the data are captured redundantly or if more than one process is maintaining the same data stores. In such cases, the DFDs may well accurately mirror the activities occurring in the organization. As the business processes being modeled took many years to develop, sometimes with participants in one part of the organization adapting procedures in isolation from other participants, redundancies and overlapping responsibilities may well have resulted. The careful study of the DFDs created as part of analysis can reveal these procedural redundancies and allow them to be corrected as part of system design.

Inefficiencies can also be identified by studying DFDs, and there are a wide variety of inefficiencies that might exist. Some inefficiencies relate to violations of DFD drawing rules. For example, a violation of rule R from Table 8-3 could occur because obsolete data are captured but never used within a system. Other inefficiencies are due to excessive processing steps. For example, consider the correct DFD in item M of Figure 8-6. Although this flow is mechanically correct, such a loop may indicate potential delays in processing data or unnecessary approval operations.

Similarly, a set of DFDs that models the current logical system can be compared to DFDs that model the new logical system to better determine which processes systems developers need to add or revise while building the new system. Processes for which inputs, outputs, and internal steps have not changed can possibly be reused in the construction of the new system. You can compare alternative logical DFDs to identify those few elements that must be discussed in evaluating competing opinions on system requirements. The logical DFDs for the new system can also serve as the basis for developing alternative design strategies for the new physical system. As we saw with the Hoosier Burger example, a process on a new logical DFD can be implemented in several different physical ways.

Using DFDs in Business Process Reengineering

Data flow diagrams also make a useful tool for modeling processes in business process reengineering (BPR), which you read about in Chapter 7. To illustrate their usefulness, let's look at an example from Hammer and Champy (1993). Hammer and Champy use IBM Credit Corporation as an example of a firm that successfully reengineered its primary business process. IBM Credit Corporation provides financing for customers making large purchases of IBM computer equipment. Its job is to analyze deals proposed by salespeople and write the final contracts governing those deals.

According to Hammer and Champy, IBM Credit Corporation typically took six business days to process each financing deal. The process worked like this: First, the salesperson called in with a proposed deal. The call was taken by one of a half-dozen people sitting around a conference table. Whoever received the call logged it and wrote the details on a piece of paper. A clerk then carried the paper to a second person who initiated the next step in the process by entering the data in a computer system and checking the client's creditworthiness. This person then wrote the details on a piece of paper and carried the paper, along with the original documentation, to a loan officer. Step three, the loan officer modified the standard IBM loan agreement for the customer. This step involved a separate computer system from the one used in step two.

Details of the modified loan agreement, along with the other documentation, were then sent on to the next station in the process, where a different clerk determined the appropriate interest rate for the loan. Step four also involved its own information system. In step five, the resulting interest rate, and all of the paper generated up to this point, were then carried to the next stop, where the quote letter was created. Once complete, the quote letter was sent via overnight mail back to the salesperson.

Only reading about this process makes it seem complicated. We can use data flow diagrams to illustrate how the overall process worked (see Figure 8-20). DFDs help us see that the process is not as complicated as it is tedious and wasteful, especially when you consider that so many different people and computer systems were used to support the work at each step.

According to Hammer and Champy, two IBM managers decided one day to see if they could improve the overall process at IBM Credit Corporation. They took a call from a salesperson and walked it through the system. These managers found the actual work being done on a contract only took 90 minutes. For much of the rest of the six days it took to process the deal, the various bits of documentation were sitting in someone's in-basket waiting to be processed.

IBM Credit Corporation management decided to reengineer its entire process. The five sets of task specialists were replaced with generalists. Now each call from the field comes to a single clerk, who does all the work necessary to process the contract. Now instead of having different people check for creditworthiness, modify the basic loan agreement, and determine the appropriate interest rate, one person does it all. IBM Credit Corporation still has specialists for the few cases that are significantly different from what the firm routinely encounters. There is also a single supporting computer system. The new process is modeled by the DFD in Figure 8-21. The most striking difference between the DFD in Figure 8-20 and the DFD in Figure 8-21, other than the number of process boxes in each one, is the lack of documentation flow in Figure 8-21. The resulting process is much simpler and cuts down dramatically on any chance of documentation getting lost between steps. Redesigning the process from beginning to end allowed IBM Credit Corporation to increase the number of contracts it can handle by one-hundred-fold—not 100 percent, which would only be doubling the amount of work. BPR allowed IBM Credit Corporation to handle 100 times more work, in the same amount of time, and with fewer people!

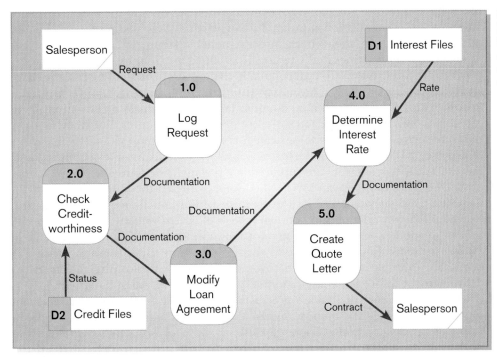

Figure 8-20
IBM Credit Corporation's primary work process before BPR
(Copyright © 1993 Harper Business, an imprint of HarperCollins Publishers. Adapted with permission.)

Figure 8-21
IBM Credit Corporation's primary work process after BPR
(Copyright © 1993 Harper Business, an imprint of HarperCollins Publishers. Adapted with permission.)

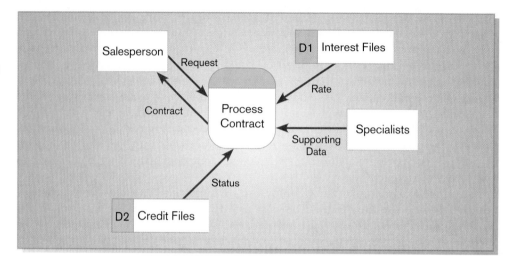

ORACLE'S PROCESS MODELER AND FUNCTIONAL HIERARCHY DIAGRAMS

Although data flow diagrams are widely known and used in systems analysis, they are not the only methods by which a system's process can be represented. We have used data flow diagrams in this chapter to illustrate the mechanics of process modeling and also to show the utility that process modeling has during analysis. Many of the same lessons you have learned about process modeling apply to two other modeling methods we will briefly introduce you to here. The first of these methods, process modeler, is unique to Oracle's Designer CASE toolset. The second method, functional hierarchy modeling, can be found in many different CASE toolsets, including Oracle's Designer and Computer Associate's COOL products.

A diagram created with Oracle's Process Modeler is shown in Figure 8-22. The processes represented in the diagram all come from Figure 8-12, which lists the activities involved in Hoosier Burger's inventory control system. Figure 8-12 lists the inventory control system activities in near temporal sequence, that is, the order in which the owner of Hoosier Burger completes the activities in a typical day. There are, however, other relationships among the activities than just time, and some of these relationships are shown in the process model in Figure 8-22.

Note that the activities in the figure are grouped into four categories, the names of which are listed in the left-most column in the process diagram. The categories are labeled "Loading Dock," "Inventory Control," "Accounting," and "Inventory Reconciliation." Typically, in a medium- to large-sized business organization, these category names would represent different organizational units, but since Hoosier Burger is such a small operation, the category names here represent different classes of activities Bob Mellankamp performs as part of his daily inventory control tasks. Each of Bob's daily activities are captured in a rectangle, organized according to the category of activities they can be classified into. Note that the lower right-hand corner of each task rectangle includes an estimate for how much time the task can take. Note also that any relationship between tasks is indicated with an arrow linking one process to another. Although they are not shown in Figure 8-22, it is also possible in a process model diagram to include triggers or events that prompt a process to begin. For example, the trigger that would prompt the start of the activity "Meet delivery trucks before opening restaurant" would be the arrival of the trucks.

An added feature of Oracle's process modeler is that it allows you to represent processes graphically and even to animate the processes to further illustrate the flow from one process to another. Figure 8-23 shows the same process model diagram as in Figure 8-22, but the processes are represented as icons, along with the text descrip-

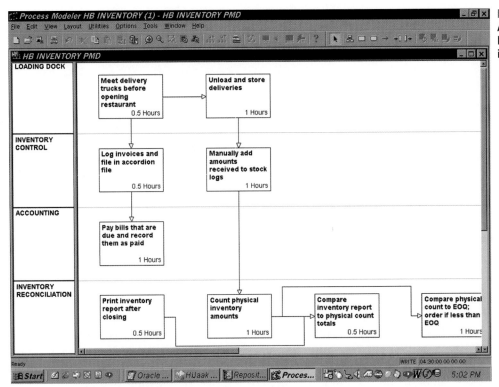

Figure 8-22
An Oracle process model based on
Figure 8-12, Hoosier Burger's
inventory control system

tions of the processes. Oracle offers several different icons for you to use to represent processes, and we have used just a few of them here. Note for example how the process dealing with meeting the trucks is represented by a truck, and the process dealing with unloading deliveries is represented by a forklift.

Oracle's process model diagrams are in many ways similar to data flow diagrams, in that they model some of the same elements, but the two methods and their out-

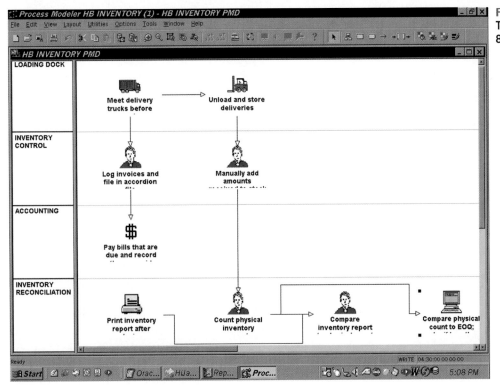

Figure 8-23
The same process model as Figure
8-22, but in graphical format

TABLE 8-4 Oracle's Process Models versus Data Flow Diagrams

Process Models	Data Flow Diagrams
Processes, flows, organizational units, but not external entities	Processes, flows, external entities, but no organizational units
Unit ownership of processes, data flows, and data stores	No ownership concept
No detail of data in flow or store	Show attributes of flows and stores
No numerical hierarchy	Numerical process hierarchy
External triggers	No external triggers
Can be animated with time parameters, run programs	No animation, time, external calls
Most useful in Strategy and Pre-Analysis phases of the life cyle	Most useful in requirements structuring subphase of Analysis

puts differ in important ways. These differences, which are summarized in Table 8-4, are the key to determining when you should use data flow diagrams and when you should use process models.

Yet another way to model processes and their relationships to each other is to create **functional hierarchy diagrams**. Functional hierarchy diagrams, or FHDs, focus on the functions performed on a business and how they are related to each other. FHDs allow the decomposition of functions and processes, just like DFDs, but unlike DFDs, all of the decomposition in an FHD is shown on the same diagram.

Figure 8-24 shows a functional hierarchy diagram based on the same information as contained in Figures 8-12, 8-22, and 8-23, the Hoosier Burger inventory control system. The nine basic functions that appear in the third level of the FHD had already been created when the process model diagrams were created and stored in the Designer repository. They only had to be referenced to create the FHD. The func-

Functional hierarchy diagram: A picture of the various tasks performed in a business and how they are related to each other. The tasks are broken down into their various parts, and all the parts are shown in the same representation.

Figure 8-24
A functional hierarchy diagram representing the same information as Figures 8-12, 8-22, and 8-23

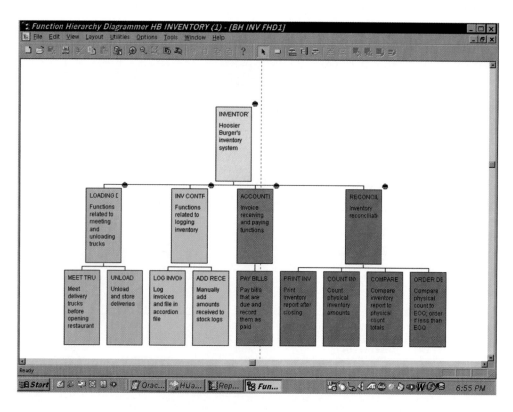

tions in the middle tier were then created as a way to group the lower-level functions by business activity, and also to show how decomposition of functions appears on an FHD. Note that each of the four areas is represented in a different color, for clarity. Also note that the functions having to do with the loading dock appear in gray. We decided to color these processes gray (this is not something Designer does automatically) in order to show that they would not be candidates for computer automation. Both of these functions deal with physical activities and so cannot be automated in a business-oriented computer system. In the FHD, the red circles by the root function and by each of the second-tier functions are switches that allow you to expand or contract the display. Right now, each of the circles has a black minus sign in it, meaning that the display can only be contracted or collapsed. A black plus sign in a red circle means the display can be expanded. This capability is convenient for complicated diagrams that include many functions, as contracting a display saves space and makes diagrams simpler and easier to understand.

You'll notice that there are not any flows in the FHD in Figure 8-24. Yet FHDs do show which data entities and elements are used by which functions. They do this internally. If Figure 8-24 were a computer screen, you could double-click on a function and its property box would open. Among the tabs in the property box would be one that showed entity usage by the function and another that showed attribute usage. You could then see if the function created, retrieved, updated, or deleted entity instances or particular attributes. This way of showing the relationships between data and process is more complicated than what a data flow diagram can do, and it is important, so we will come back to it in later chapters.

INTERNET DEVELOPMENT: PROCESS MODELING

Process modeling for an Internet-based electronic commerce application is no different than the process followed for other applications. In the last chapter, you read how Pine Valley Furniture determined the system requirements for their WebStore project, a project to sell furniture products over the Internet. In this section, we analyze the WebStore's high-level system structure and develop a level-0 DFD for those requirements.

TABLE 8-5   System Structure of the WebStore and Corresponding Level-0 Processes

WebStore System	Processes
❏ Main Page	Information Display (minor/no processes)
• Product Line (Catalog)	1.0 Browse Catalog
✓ Desks	2.0 Select Item for Purchase
✓ Chairs	
✓ Tables	
✓ File Cabinets	
• Shopping Cart	3.0 Display Shopping Cart
• Checkout	4.0 Check Out Process Order
• Account Profie	5.0 Add/Modify Account Profile
• Order Status/History	6.0 Order Status Request
• Customer Comments	Information Display (minor/no processes)
❏ Company Info	
❏ Feedback	
❏ Contact Information	

Process Modeling for Pine Valley Furniture's WebStore

After completing the JAD session, senior systems analyst, Jim Woo, went to work on translating the WebStore system structure into a data flow diagram. His first step was to identify the level-0—major system—processes. To begin, he carefully examined the outcomes of the JAD session that focused on defining the system structure of WebStore. From this analysis, he identified six high-level processes that would become the foundation of the level-0 DFD. These processes, listed in Table 8-5 were the "work" or "action" parts of the Website. Note that each of these processes correspond to the major processing items listed in the system structure.

Next, Jim determined that it would be most efficient if the WebStore system exchanged information with existing PVF systems rather than capturing and storing

Figure 8-25
Level-0 data flow diagram for the WebStore

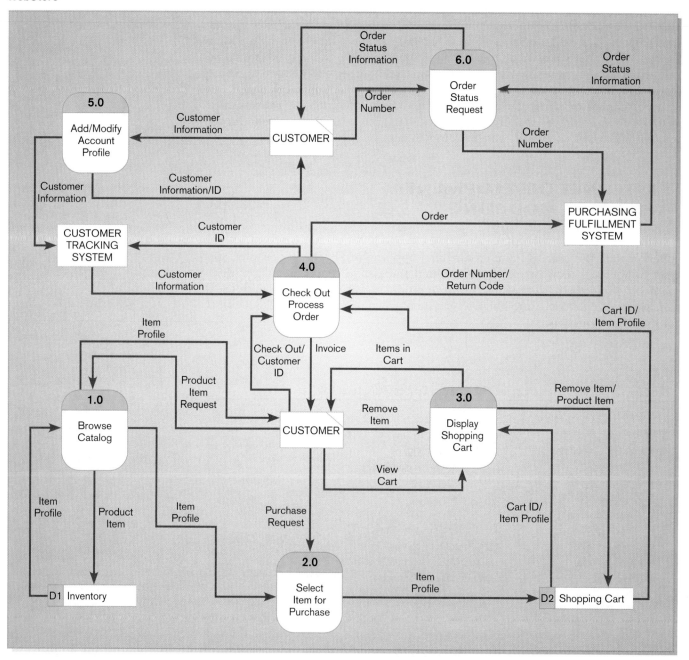

redundant information. This analysis concluded that the WebStore should exchange information with the Purchasing Fulfillment System—a system for tracking orders (see Chapter 2)—and the Customer Tracking System—a system for managing customer information. These two existing systems will be "sources" (provider) and "sinks" (receivers) of information for the WebStore system. When a customer opens an account, his/her information will be passed from the WebStore system to the Customer Tracking System. When an order is placed (or when a customer requests status information on a prior order), information will be stored (retrieved) in (from) the Purchasing Fulfillment System.

Finally, Jim found that the system would need to access two additional data sources. First, in order to produce an on-line product catalog, the system would need to access the inventory database. Second, to store the items a customer wanted to purchase in the WebStore's shopping cart, a temporary database would need to be created. Once the transaction was completed, the shopping cart data could be deleted. With this information, Jim was now able to develop the level-0 DFD for the WebStore system that is shown in Figure 8-25. He now felt he gained a good understanding of how information would flow through the WebStore, how a customer would interact with the system, and how the WebStore would share information with existing PVF systems. Each of these high-level processes would eventually need to be further decomposed before system design could proceed. Yet, before doing that, he wanted to get a clear picture of exactly what data was needed throughout the entire system. We will discover the outcomes of this analysis activity—conceptual data modeling—in Chapter 10, but first, in Chapter 9, we will look at the role of logic modeling in Internet application analysis.

Summary

Modeling processes can be done in many ways, including Oracle process models, functional hierarchy diagrams, and data flow diagrams, or DFDs. DFDs are very useful for representing the overall data flows into, through, and out of an information system. Data flow diagrams rely on only four symbols to represent the four conceptual components of a process model: data flows, data stores, processes, and sources/sinks. Data flow diagrams are hierarchical in nature and each level of a DFD can be decomposed into smaller, simpler units on a lower-level diagram. You begin with a context diagram, which shows the entire system as a single process. The next step is to generate a level-0 diagram, which shows the most important high-level processes in the system. You then decompose each process in the level-0 diagram, as warranted, until it makes no logical sense to go any further. When decomposing DFDs from one level to the next, it is important that the diagrams be balanced; that is, inputs and outputs on one level must be conserved on the next level.

There are four sets of data flow diagrams. The first set you create models the current physical system. The next set models the current logical system. The third set models the new logical system, which may be different from the current logical system to the extent that it shows desired functionality, even though the new system may radically reengineer the flow of data between system components. The fourth set models the new physical information system. Due to the time it takes to create a complete set of DFDs, many experts suggest that you begin work on the new logical DFDs as soon as possible.

Data flow diagrams should be mechanically correct, but they should also accurately reflect the information system being modeled. To that end, you need to check DFDs for completeness and consistency and draw them as if the system being modeled were timeless. You should be willing to revise DFDs several times. Complete sets of DFDs should extend to the primitive level where every component reflects certain irreducible properties; for example, a process represents a single database operation and every data store represents data about a single entity. Following these guidelines, you can produce DFDs to aid the analysis process by analyzing the gaps between existing procedures and desired procedures and between current and new systems.

Although the modeling of processes for information systems development is over 20 years old, dating back at least to the beginnings of the philosophy of structured analysis and design, it is just as important for Internet applications as it is for more traditional systems. Future chapters will show, as this one did, how traditional tools and techniques developed for structured analysis and design provide powerful assistance for the development of Internet applications.

Key Terms

1. Balancing
2. Context diagram
3. Data flow diagram
4. DFD completeness
5. DFD consistency
6. Data store
7. Functional decomposition
8. Functional hierarchy diagram
9. Gap analysis
10. Level-0 diagram
11. Level-*n* diagram
12. Primitive DFD
13. Process
14. Source/sink

Match each of the key terms above with the definition that best fits it.

_____ A picture of the movement of data between external entities and the processes and data stores within a system.

_____ The conservation of inputs and outputs to a data flow diagram process when that process is decomposed to a lower level.

_____ A data flow diagram that represents a system's major processes, data flows, and data stores at a high level of detail.

_____ The origin and/or destination of data; sometimes referred to as external entities.

_____ An overview of an organizational system that shows the system boundary, external entities that interact with the system, and the major information flows between the entities and the system.

_____ The lowest level of decomposition for a data flow diagram.

_____ The extent to which all necessary components of a data flow diagram have been included and fully described.

_____ The extent to which information contained on one level of a set of nested data flow diagrams is also included on other levels.

_____ A DFD that is the result of *n* nested decompositions of a series of subprocesses from a process on a level-0 diagram.

_____ The work or actions performed on data so that they are transformed, stored, or distributed.

_____ Data at rest, which may take the form of many different physical representations.

_____ The process of discovering discrepancies between two or more sets of data flow diagrams or discrepancies within a single DFD.

_____ An iterative process of breaking the description of a system down into finer and finer detail, which creates a set of charts in which one process on a given chart is explained in greater detail on another chart.

_____ A picture of the various tasks performed in a business and how they are related to each other. The tasks are broken down into their various parts, and all the parts are shown in the same representation.

Review Questions

1. What is a data flow diagram? Why do systems analysts use data flow diagrams?

2. Explain the rules for drawing good data flow diagrams.

3. What is decomposition? What is balancing? How can you determine if DFDs are not balanced?

4. Explain the convention for naming different levels of data flow diagrams.

5. What are the primary differences between current physical and current logical data flow diagrams?

6. Why don't analysts usually draw four complete sets of DFDs?

7. How can data flow diagrams be used as analysis tools?

8. Explain the guidelines for deciding when to stop decomposing DFDs.

9. How do you decide if a system component should be represented as a source/sink or as a process?

10. What unique rules apply to drawing context diagrams?

11. Compare data flow diagrams to Oracle's process model diagrams.

12. Compare data flow diagrams to functional hierarchy diagrams.

Problems and Exercises

1. Using the example of a retail clothing store in a mall, list relevant data flows, data stores, processes, and sources/sinks. Observe several sales transactions. Draw a context diagram and a level-0 diagram that represent the selling system at the store. Explain why you chose certain elements as processes versus sources/sinks.

2. Choose a transaction that you are likely to encounter, perhaps ordering a cap and gown for graduation, and develop a high-level DFD, or context diagram. Decompose this to a level-0 diagram.

3. Evaluate your level-0 DFD from the previous question using the rules for drawing DFDs in this chapter. Edit your DFD so that it does not break any of these rules.

4. Choose an example like that in the second question and draw a context diagram. Decompose this diagram until it doesn't make sense to continue. Be sure that your diagrams are balanced as discussed in this chapter.

5. Refer to Figure 8-26, which contains drafts of a context and level-0 DFD for a university class registration system. Identify and explain potential violations of rules and guidelines on these diagrams.

6. Why should you develop both logical and physical DFDs for systems? What advantage is there for drawing a logical DFD before a physical DFD for a new information system?

7. What is the relationship between DFDs and entries in the project dictionary or CASE repository?

8. This chapter has shown you how to model, or structure, just one aspect, or view, of an information system, namely the process view. Why do you think analysts have different types of diagrams and other documentation to depict different views (for example, process, logic, and data) of an information system?

9. Consider the DFD in Figure 8-27. List three errors (rule violations) on this DFD.

10. Consider the three DFDs in Figure 8-28. List three errors (rule violations) on these DFDs.

11. Starting with a context diagram, draw as many nested DFDs as you consider necessary to represent all the details of the employee hiring system described in the following narrative. You must draw at least a context and a level-0 diagram. In drawing these diagrams, if you discover that the narrative is incomplete, make up reasonable explanations to complete the story. Supply these extra explanations along with the diagrams. Here is the narrative. Projects, Inc. is an engineering firm with approximately 500 engineers of different types. The company keeps records on all employees, their skills,

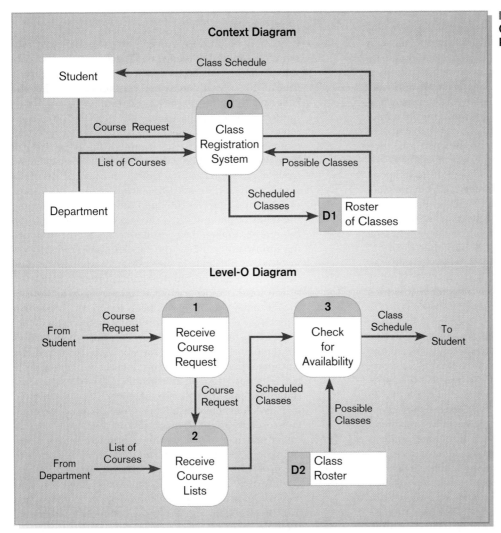

Figure 8-26
Class registration system for Problem and Exercise 5

Figure 8-27
DFD for Problem and Exercise 9

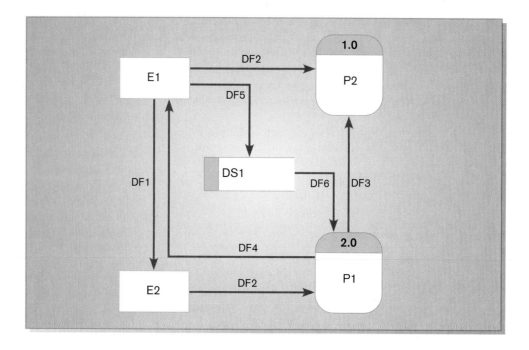

projects assigned, and departments worked in. New employees are hired by the personnel manager based on data in an application form and evaluations collected from other managers who interview the job candidates. Prospective employees may apply at any time. Engineering managers notify the personnel manager when a job opens and list the characteristics necessary to be eligible for the job. The personnel manager compares the qualifications of the available pool of applicants with the characteristics of an open job, then schedules interviews between the manager in charge of the open position and the three best candidates from the pool. After receiving evaluations on each interview from the manager, the personnel manager makes the hiring decision based upon the evaluations and applications of the candidates and the characteristics of the job, and then notifies the interviewees and the manager about the decision. Applications of rejected applicants are retained for one year, after which time the application is purged. When hired, a new engineer completes a nondisclosure agreement, which is filed with other information about the employee.

12. a. Starting with a context diagram, draw as many nested DFDs as you consider necessary to represent all the details of the system described in the following narrative. In drawing these diagrams, if you discover that the narrative is incomplete, make up reasonable explanations to make the story complete. Supply these extra explanations along with the diagrams. Here is the narrative. Maximum Software is a developer and supplier of software products to individuals and businesses. As part of their operations, Maximum provides an 800 telephone number help desk for clients who have questions about software purchased from Maximum. When a call comes in, an operator inquires about the nature of the call. For calls that are not truly help desk functions, the operator redirects the

call to another unit of the company (such as Order Processing or Billing). Since many customer questions require in-depth knowledge of a product, help desk consultants are organized by product. The operator directs the call to a consultant skilled on the software that the caller needs help with. Since a consultant is not always immediately available, some calls must be put into a queue for the next available consultant. Once a consultant answers the call, he determines if this is the first call from this customer about this problem. If so, he creates a new call report to keep track of all information about the problem. If not, he asks the customer for a call report number, and retrieves the open call report to determine the status of the inquiry. If the caller does not know the call report number, the consultant collects other identifying information such as the caller's name, the software involved, or the name of the consultant who has handled the previous calls on the problem in order to conduct a search for the appropriate call report. If a resolution of the customer's problem has been found, the consultant informs the client what that resolution is, indicates on the report that the customer has been notified, and closes out the report. If resolution has not been discovered, the consultant finds out if the consultant handling this problem is on duty. If so, he transfers the call to the other consultant (or puts the call into the queue of calls waiting to be handled by that consultant). Once the proper consultant receives the call, he records any new details the customer may have. For continuing problems and for new call reports, the consultant tries to discover an answer to the problem by using the relevant software and looking up information in reference manuals. If he can now resolve the problem, he tells the customer how to deal with the problem, and closes the call report. Otherwise, the consultant files the report for continued research

Figure 8-28
DFD for Problem and Exercise 10

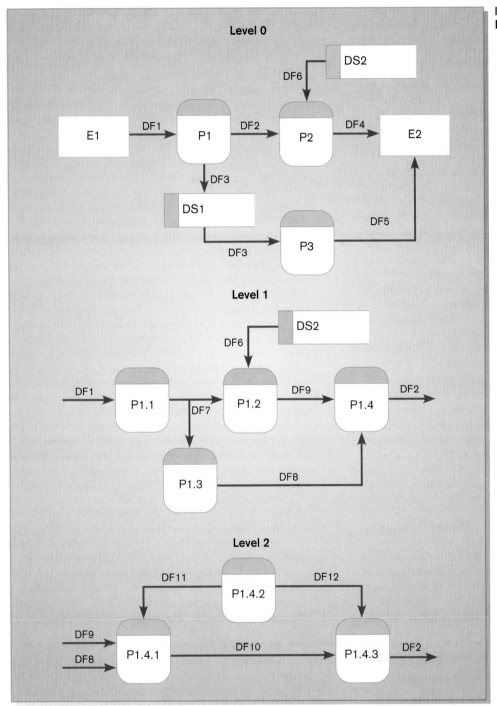

and tells the customer that someone at Maximum will get back to him, or if the customer discovers new information about the problem, to call back identifying the problem with a specified call report number.

b. Analyze the DFDs you created in part (a) of this question. What recommendations for improvements in the help desk system at Maximum can you make based upon this analysis? Draw new logical DFDs that represent the requirements you would suggest for an improved help

desk system. Remember, these are to be logical DFDs, so consider improvements independent of technology that can be used to support the help desk.

13. Develop a context diagram and level-0 diagram for the hospital pharmacy system described in the following narrative. If you discover that the narrative is incomplete, make up reasonable explanations to complete the story. Supply these extra explanations along with the diagrams. Here is

the narrative. The pharmacy at Mercy Hospital fills medical prescriptions for all patients and distributes these medications to the nurse stations responsible for the patient's care. Medical prescriptions are written by doctors and sent to the pharmacy. A pharmacy technician reviews the prescriptions and send them to the appropriate pharmacy station. Prescriptions for drugs that must be formulated (made on-site) are sent to the lab station, prescriptions for off-the-shelf drugs are sent to the shelving station, and prescriptions for narcotics are sent to the secure station. At each station, a pharmacist reviews the order, checks the patient's file to determine the appropriateness of the prescriptions, and fills the order if the dosage is at a safe level and it will not negatively interact with the other medications or allergies indicated in the patient's file. If the pharmacist does not fill the order, the prescribing doctor is contacted to discuss the situation. In this case, the order may ultimately be filled or the doctor may write another prescription depending on the outcome of the discussion. Once filled, a prescription label is generated listing the patient's name, the drug type and dosage, an expiration date, and any special instructions. The label is placed on the drug container and the orders are sent to the appropriate nurse stations. The patient's admission number, the drug type and amount dispensed, and the cost of the prescription are then sent to the billing department.

14. Develop a context diagram and a level-0 diagram for the contracting system described in the following narrative. If you discover that the narrative is incomplete, make up reasonable explanations to complete the story. Supply these extra explanations along with the diagrams. Here is the narrative. Government Solutions Company (GSC) sells computer equipment to federal government agencies. Whenever a federal agency needs to purchase equipment from GSC, it issues a purchase order against a standing contract previously negotiated with the company. GSC holds several standing contracts with various federal agencies. When a purchase order is received by GSC's contracting officer, the contract number referenced on the purchase order is entered into the contract database. Using information from the database, the contracting officer reviews the terms and conditions of the contract and determines whether the purchase order is valid. The purchase order is valid if the contract has not expired, the type of equipment ordered is listed on the original contract, and the total cost of the equipment does not exceed a predetermined limit. If the purchase order is not valid, the contracting officer sends the purchase order back to the requesting agency with a letter stating why the purchase order cannot be filled, and a copy of the letter is filed. If the purchase order is valid, the contracting officer enters the purchase order number into the contract database and flags the order as outstanding. The purchase order is then sent to the order fulfillment department. Here the inventory is checked for each item ordered. If any items are not in stock, the order fulfillment department creates a report listing the items not in stock and attaches it to the purchase order. All purchase orders are forwarded to the warehouse where the items in stock are pulled from the shelves and shipped to the customer. The warehouse then attaches to the purchase order a copy of the shipping bill listing the items shipped and sends it to the contracting officer. If all items were shipped, the contracting officer closes the outstanding purchase order record in the database. The purchase order, shipping bill, and exception report (if attached) are then filed in the contracts office.

15. Develop a context diagram and as many nested DFDs as you consider necessary to represent all the details of the training logistics system described in the following narrative. If you discover that the narrative is incomplete, make up reasonable explanations to complete the story. Supply these extra explanations along with the diagrams. Here is the narrative. Training, Inc., conducts training seminars in major U.S. cities. For each seminar, the logistics department must make arrangements for the meeting facilities, the training consultant's travel, and the shipment of any seminar materials. For each scheduled seminar, the bookings department notifies the logistics coordinator of the type of seminar, the dates and city location, and the name of the consultant who will conduct the training. To arrange for meeting facilities, the logistics coordinator gathers information on possible meeting sites in the scheduled city. The meeting site location decision is made based on date availability, cost, type of meeting space available, and convenience of the location. Once the site decision is made, the coordinator speaks with the sales manager of the meeting facility to reserve the meeting room(s), plan the seating arrangement(s), and reserve any necessary audiovisual equipment. The coordinator estimates the number and size of meeting rooms, the type of seating arrangements, and the audiovisual equipment needed for each seminar from the information kept in a logistics database on each type of seminar offered and the number of anticipated registrants for a particular booking. After negotiations are conducted by the logistics coordinator and the sales manager of the meeting facility, the sales manager creates a contract agreement specifying the negotiated arrangements and sends two copies of it to the logistics coordinator. The coordinator reviews the agreement and approves it if no changes are needed. One copy of the agreement is filed and the other copy is sent back to the sales manager. If changes are needed, the agreement copies are changed and returned to the sales manager for approval. This approval process continues until both parties have approved the agreement. The coordinator must also contact the training consultant to make travel arrangements. First, the coordinator reviews the consultant's travel information in the logistics database and researches flight schedules. Then the consultant is contacted to discuss possible travel arrangements; subsequently, the coordinator books a flight for the consultant with a travel agency. Once the consultant's travel arrangements have been completed, a written confirmation and itinerary are sent to the consultant. Two weeks before the date of the seminar, the coordinator determines what, if any, seminar materials (e.g., transparencies, training guides, pamphlets, etc.) need to be sent to the meeting facility. Each type of seminar has a specific set of materials assigned to it. For some materials, the coordinator must know how many participants have registered for the seminar in order to determine how many to send. A request for materials is sent to the material-handling department where the materials are gathered, boxed, and sent to the

meeting address listed on the request. Once the requested materials have been shipped, a notification is sent to the logistics coordinator.

Field Exercises

1. Talk to systems analysts who work at an organization. Ask the analyst to show you a complete set of DFDs from a current project. Interview the analyst about his or her views about DFDs and their usefulness for analysis.

2. Interview several people in an organization about a particular system. What is the system like now and what would they like to see changed? Create a complete set of DFDs for the current physical, current logical, and new logical system. Show some of the people you interviewed in your DFDs and ask for their reactions. What kinds of comments do they make? What kinds of suggestions?

3. Talk to systems analysts who use a CASE tool. Investigate what capabilities that CASE tool has for automatically checking for rule violations in DFDs. What reports can the CASE tool produce with error and warning messages to help analysts correct and improve DFDs?

4. Find out which, if any, drawing packages, word processors, forms design, and database management systems your uni-

16. Redraw each of the data flow diagrams you drew in Problems and Exercises 11 through 15, but this time, draw them as process models and as functional hierarchy diagrams.

versity or company supports. Research these packages to determine how they might be used in the production of a project dictionary. For example, do the drawing packages include either set of standard DFD symbols in their graphic symbol palette?

5. At an organization with which you have contact, ask one or more employees to draw a "picture" of the business process they interact with at that organization. Ask them to draw the process using whatever format suits them. Ask them to depict in their diagram each of the components of the process and the flow of information among these components at the highest level of detail possible. What type of diagram have they drawn? Why? In what ways does it resemble (and not resemble) a data flow diagram? Why? When they have finished, help them to convert their diagram to a standard data flow diagram as described in this chapter. In what ways is the data flow diagram stronger and/or weaker than the original diagram?

References

Celko, J. 1987. "I. Data Flow Diagrams." *Computer Language* 4(January): 41–43.

DeMarco, T. 1979. *Structured Analysis and System Specification.* Englewood Cliffs, NJ: Prentice-Hall.

Gane, C. and T. Sarson. 1979. *Structured Systems Analysis.* Englewood Cliffs, NJ: Prentice-Hall.

Gibbs, W. W. 1994. "Software's Chronic Crisis." *Scientific American* 271(Sept.): 86–95.

Hammer, M., and J. Champy. 1993. *Reengineering the Corporation.* New York, NY: Harper Business.

Yourdon, E. 1989. *Managing the Structured Techniques,* 4th ed. Englewood Cliffs, NJ: Prentice-Hall.

Yourdon, E. and L. L. Constantine. 1979. *Structured Design.* Englewood Cliffs, NJ: Prentice-Hall.

BROADWAY ENTERTAINMENT COMPANY, INC.

Structuring System Requirements: Process Modeling for the Web-Based Customer Relationship Management System

CASE INTRODUCTION

The BEC student team of Tracey Wesley, John Whitman, Missi Davies, and Aaron Sharp were feeling somewhat overwhelmed by all the system requirements they had captured during the requirements determination activities of their project to develop a Web-based customer relationship management system. Before they had begun requirements determination, they had structured what they had already learned. Based on the System Service Request and the initial meetings with Carrie, the team from Stillwater State University developed a context diagram for the system (see BEC Figure 8-1). The context diagram shows the system in the middle, the major external entities (Customer, Employee, and the Entertainment Tracker BEC in-store information system) that interact with the system on the outside, and the data flows between the system and the external entities.

Not too surprisingly, most of the data flows are between the system and customers. For this reason, the team decided to repeat the Customer, using one copy of Customer as a source and another copy as a sink of data flows.

The context diagram helped the team to organize for requirements determination. This data collection part of the analysis phase was used to verify this overview model of the customer relationship management system and to gather details for each data flow, processing activity, and data storage component inside the system.

The team needed one more result before beginning the detailed work of analysis and design—a catchy name for the system it was designing. "The BEC Customer Relationship Management System" was too long and dull. With the cooperation of Carrie Douglass, team members ran a contest among the other teams in their class to give each member of the team with the best name suggestion (as selected by Carrie) a free movie rental at the Centerville BEC store. Some teams tried to create acronyms using the words and acronyms *BEC, Broadway Entertainment Company,* and *customer relationship management,* but most of these were not pronounceable nor very meaningful. Other teams created phrases that conveyed

the Web technology to be used to build the system (e.g., one team suggested VideosByBEC, similar to AutoByTel for automobile sales and information on the Web). But, Carrie wanted a name that conveyed the personal relationship the system created with the customer. Thus, one suggested name stood out from the rest. The winner was MyBroadway.

STRUCTURING THE HIGH-LEVEL PROCESS FINDINGS FROM REQUIREMENTS DETERMINATION

The BEC student team used various methods to understand the requirements for MyBroadway. The following sections explain how they approached studying each data flow on the context diagram and what they discovered from their analysis.

Inventory and Rental Extension Request

The team studied documentation of the Entertainment Tracker system provided to store employees and the manager. From this documentation the team understood the data about products and product sales and rentals maintained in store records. This was a necessary step to determine what data could be in the Inventory data flow. It was clear that MyBroadway would not be the system of record to operate the store; Entertainment Tracker was this official record. For example, the official record of when a rented product was due to be returned would be recorded in the Entertainment Tracker database. Thus, product inventory, sales, and rental data needed by MyBroadway would be periodically extracted from Entertainment Tracker to be stored in MyBroadway for faster access and to keep the two systems as decoupled as possible. Because of the role of Entertainment Tracker, any activity in MyBroadway that changed data in Entertainment Tracker would have to submit a transaction to Entertainment Tracker that Entertainment Tracker understood. The only instance of this the team discovered related to the Rental Extension Request data flow. The process handling this inflow would find the due

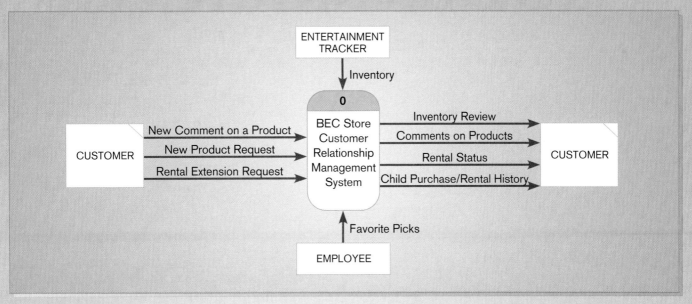

BEC Figure 8-1
Context diagram for the BEC customer relationship management system

date in the MyBroadway database and then interact with Entertainment Tracker to request the extension and to inform the customer whether the extension was accepted. Entertainment Tracker, however, would make the decision, based on its own rules, whether to accept the extension. Fortunately, requesting an extension is a transaction in Entertainment Tracker handled from a point-of-sale terminal in the store, so MyBroadway would simply need to simulate this transaction.

Favorite Picks

The team also learned from employees and customers what would be useful related to the employee Favorite Picks data flow. Both employees and customers agreed that there were only two broad groups of items for favorite picks: new releases and classics. Each week a different store employee would select one or two new release or classic products in a given product category. For example, each week one employee would select one or two new release children's videos, another employee would select one or two new release jazz and new age CDs, and yet another employee would select one or two classic romance DVDs. It is not possible to cover every category of videotape, DVD, and CD each week, but over time most categories would be selected. Selections would be retained for two years. Each week five store employees would make selections each in a different product category. An employee would be given a list of those 10 product categories for which favorite picks had not been made for the longest time. Each employee would be matched with the category with which he or she is most familiar

and given a list of those new release and classic products in that category. A classic product is one that continues to be rented or sold at least 10 years after its initial release. An employee selects one or two products on this list and provides a quality grade for each (A, A-, B, F), a description of its contents relevant to language and sexually explicit references, and a few sentences of personal comments about the product that a parent might want to know. The date of the entry would be recorded with the rest of the data.

New Product Request

The team used interviews with customers, Carrie Douglass, and the assistant store manager as well as observation of people using similar Web-based systems from major on-line bookstores and other shopping enterprises to determine the nature of the other six data flows on the context diagram. For New Product Request, MyBroadway will collect all the requests and at the request of Carrie print a list, in decreasing order of frequency of request, of each requested product. Carrie will then use this report to send a letter to the BEC purchasing department requesting the acquisition of these items. New Product Requests will be kept for two months and then purged.

New Comment on a Product

For New Comment on a Product, MyBroadway will show a parent or child basic information about the product (such as title, publisher, artist, and date released) and then allow him or her to enter an unstructured comment about that product. There is no limit on the length of the comment. Each comment is stored separately, and the same person may comment on the same item many times. The date and time of the comment are stored with the comment. An issue that required some discussion with Carrie was whether the person had to identify himself or herself for the comment to be recorded. Carrie was unsure what to do. So, the team convened a focus group of a few parents to explore this issue. The team discovered that the parents would not consider a comment valid unless it were attributed and that the parents thought they and their children would enter a more helpful comment if it were attributed. Carrie, however, saw no need to retain data about customers in MyBroadway, but how would bogus customer names be identified? Entertainment Tracker maintains data about each customer, including each child with a membership card. Thus, it was decided that the customer would have to enter his or her membership number along with the comment. This number would be sent to Entertainment Tracker for matching with its record of customer numbers after the comment was entered but before it was available to be reported to other customers. Whether the comment is entered by a parent or a child is also recorded with the comment. If the number does not match a membership number for a BEC customer, the comment will be dropped. When the comment is displayed, the name of the member entering the comment as well as whether that person is a parent or child will be shown with the comment.

Inventory Review

The Inventory Review data flow consolidates several data flows. A customer can ask to see product data by specific title, or to see data for all the products by artist, category (e.g., new age or jazz CD), publisher, release month, or any combination of these factors. In each case, for each product identified by the search criteria, the product title, artist, publisher, release date, media, description, and sale and rental price are shown.

Comments on Products

Comments on Products is produced when a customer enters the name of the product (and possibly searching through a set of products with approximately that name until the exact product is found). Once the exact product the customer is interested in is identified, then all the comments previously entered by customers are available for display. For the purpose of this data flow, Favorite Picks records are also considered comments. The customer may ask to see only those comments entered since some date they specify and may ask to see comments only by parents, only by children, only by employees (i.e., only Favorite Picks), or all comments. Comments are shown in reverse chronology entry order.

Rental Status

For this data flow, the customer enters her or his membership number and then MyBroadway displays a list of all the product titles and return due dates for all outstanding rented items. Often, customers obtain a Rental Status before they submit a Rental Extension Request, but the team decided to consider these separate data flows.

Child Purchase/Rental History

The team discovered that this is arguably the most complex of the data flows on the context diagram. The team decided to model this data flow at a fairly high level first and then decompose the process producing Child Purchase/Rental History later. At a high level, to produce this data flow MyBroadway needs access to sales and rental history data, including what products have been bought and rented by whom. Customers indicated that a simple history would not be sufficient. They also wanted to see the customer comments and favorite picks ratings for each item. So, an instance of the Child Purchase/Rental History data flow is a report that shows for a given child the title of each item he or she has bought or rented in the past six months, and for each item the rating entered by each employee who has rated that product, and the five most recent parent comments recorded about that item.

CASE SUMMARY

Accurately and thoroughly documenting business processes can be tedious and time-consuming, but very insightful. The student team working on the analysis and design of MyBroadway quickly discovered how extensive a system Carrie and the store employees and customers wanted for this customer relationship management system. The team was unsure whether it could do a thorough analysis and design for all the desired features. Starting with a context diagram and successively decomposing processes, however, allowed the team to show the total scope of the system as desired by the project sponsor and system users and yet focus attention on one piece of the system at a time. If only parts of the system could be built during the course project, at least the team would be able to show how those pieces fit into the complete system. The team members also recognized that structuring processes and data flows were only part of the systems analysis. They would also need to identify all the data stored inside MyBroadway (in data stores) and then structure these data into a database specification. Each primitive process on the lowest-level DFDs would have to be specified in sufficient detail for a programmer to build that functionality into the information system.

The BEC student team had made the decision to use automated tools to draw DFDs (and other system diagrams) and to record project dictionary data about system objects, such as external entities, data flows, data stores, and processes. (Because you will use whatever tools your instructor recommends, we do not refer to any specific tools by name in this or subsequent cases.) These automated tools are critical for making it easy to change diagrams, to produce clean documentation about the system requirements, and to make each aspect of the documentation consistent with each other. Drawing the initial diagrams and recording all the dictionary entries is very time-consuming. This automated data, however, can be changed by any team member, and team members can prepare new diagrams and dictionary reports at any time with minimal effort.

CASE QUESTIONS

1. Does the context diagram in BEC Figure 8-1 represent an accurate and complete overview of the system as described in this case for requirements collected during the analysis phase? If not, what is wrong or missing? If necessary, draw a new context diagram in light of what is explained in this case. Why might a context diagram initially drawn at the end of project initiation and planning need to be redrawn during the analysis phase?

2. In the context diagram of BEC Figure 8-1, why is the Rental Extension Request data flow shown as an inflow to the system? Why is the Rental Status data flow shown as an outflow from the system? Do you agree with these designations of the two data flows? Why or why not?

3. The store manager is not shown in the context diagram in BEC Figure 8-1, except implicitly as an Employee who enters Favorite Picks. Based on the descriptions in this case, does it make sense that store manager does not appear on the context diagram? If not on the context diagram, where might store manager appear? As an external entity on a lower-level diagram? As a process or data store on a lower-level diagram? Based on the description in this case, are there any external entities missing on the context diagram of BEC Figure 8-1?

4. Based on the descriptions in this case of each data flow from the context diagram, draw a level-0 data flow diagram for MyBroadway. Be sure it is balanced with the context diagram you might have drawn in answer to Case Question 1.

5. Write project dictionary entries (using standards given to you by your instructor) for all the data stores shown in the level-0 diagram in your answer for Case Question 4. Are there other data stores hidden inside processes for your level-0 diagram? If so, what kinds of data do you anticipate are retained in these hidden data stores? Why are these data stores hidden inside processes rather than appearing on the level-0 diagram?

6. Write project dictionary entries (using standards given to you by your instructor) for all the data flows shown in the level-0 diagram in your answer to Case Question 4. How detailed are these entries at this point? How detailed must these entries be for primitive DFDs?

7. Explain how you modeled in your answer to Case Question 4 the process that receives the New Product Request data flow. Was this a difficult process to model on the DFD? Did you consider several alternative ways to show this process? If so, explain the alternatives and why you chose the representation you drew in the level-0 diagram for Case Question 4.

8. Look at your answer to Case Question 4 and focus attention on the process for the Rental Extension Request data flow. Draw a level-1 diagram for this process based on the description of this data flow in the case and in the following explanation. A customer provides his or her customer number or name and a product number or title and then MyBroadway finds in its records the rental information for this customer's outstanding rental of this product, including the due date. Then the customer may decide that he or she can return the item by the due date, in which case no request for extension is made. If the customer decides to extend the due date, the customer can request a one-day or two-day extension, each with a different fee, which will be due when the product is returned. MyBroadway will then send a Rental Extension Request transaction to Entertainment Tracker as if it were a point-of-sale terminal from which the same request was being made. Entertainment Tracker may reject the request if the customer has delinquent fees. Once Entertainment Tracker makes its decision, it returns a code to MyBroadway indicating a yes or the reason for a no to the request. If the decision is no, the customer is given a message to explain rejection. If yes, MyBroadway rental data are updated to reflect the extension, and the user is given a confirmation message.

9. Does your answer to Case Question 7 necessitate any changes to your answer to Case Question 4? If so, what are these changes? Prepare a new level-0 diagram for MyBroadway.

Chapter 9

Structuring System Requirements:

Logic Modeling

LEARNING OBJECTIVES

After studying this chapter, you should be able to:

- Use Structured English as a tool for representing steps in logical processes in data flow diagrams.

- Use decision tables and decision trees to represent the logic of choice in conditional statements.

- Select among Structured English, decision tables, and decision trees for representing processing logic.

- Understand how logic modeling techniques apply to development for Internet applications.

INTRODUCTION

In Chapter 8 you learned how the processes that convert data to information are key parts of information systems. As good as data flow diagrams are for identifying processes, they are not very good at showing the logic inside the processes. Even the processes on the primitive-level data flow diagrams do not show the most fundamental processing steps. Just what occurs within a process? How are the input data converted to the output information? Since data flow diagrams are not really designed to show the detailed logic of processes, you must model process logic using other techniques. This chapter is about the techniques you use for modeling process decision logic.

First you will be introduced to Structured English, a modified version of the English language that is useful for representing the logic in information system processes.

You can use Structured English to represent all three of the fundamental statements necessary for structured programming: choice, repetition, and sequence.

Second, you will learn about decision tables. Decision tables allow you to represent a set of conditions and the actions that follow from them in a tabular format. When there are several conditions and several possible actions that can occur, decision tables can help you keep track of the possibilities in a clear and concise manner.

Third, you will also learn how to model the logic of choice statements using decision trees. Decision trees model the same elements as a decision table but in a more graphical manner.

Fourth, you will have to decide when to use Structured English, decision tables, and decision trees.

In this chapter, you will learn about the criteria you can use to make a choice among these three logic modeling techniques.

Fifth, you will see how logic modeling techniques can also be used for Internet applications.

LOGIC MODELING

In Chapter 7, you learned how the requirements for an information system are collected. Analysts structure the requirements information into data flow diagrams that model the flow of data into and through the information system. Data flow diagrams, though versatile and powerful techniques, are not adequate for modeling all of the complexity of an information system. Although decomposition allows you to represent a data flow diagram's processes at finer and finer levels of detail, the process names themselves cannot adequately represent what a process does and how it does it. For that reason, you must represent the logic contained in the process symbols on DFDs with other modeling techniques.

Logic modeling involves representing the internal structure and functionality of the processes represented on data flow diagrams. These processes appear on DFDs as little more than black boxes, in that we cannot tell from only their names or CASE repository descriptions precisely what they do and how they do it. Yet the structure and functionality of a system's processes are a key element of any information system. Processes must be clearly described before they can be translated into a programming language. In this chapter, we will focus on techniques you can use during the analysis phase to model the logic within processes; that is, data-to-information transformations and decisions. In the analysis phase, logic modeling will be complete and reasonably detailed, but it will also be generic in that it will not reflect the structure or syntax of a particular programming language. You will focus on more precise, language-based logic modeling in the design phase of the life cycle.

Modeling a System's Logic

The three subphases to systems analysis are requirements determination, requirements structuring, and generating alternative systems and selecting the best one (Figure 9-1). Modeling a system's logic is part of requirements structuring, just as was representing the system with data flow diagrams. Here our focus is on the processes pictured on the data flow diagrams and the logic contained within each. You can also use logic modeling to indicate when processes on a DFD occur (for example, when a process extracts a certain data flow from a given data store). Just as we use logic modeling to represent the logic contained in a data flow diagram's processes, we will use data modeling to represent the contents and structure of a data flow diagram's data flows and data stores.

Deliverables and Outcomes

In structured analysis, the primary deliverables from logic modeling are structured descriptions and diagrams that outline the logic contained within each DFD process as well as diagrams that show the temporal dimension of systems—when processes or events occur and how these events change the state of the system. Table 9-1 provides a list of deliverables that result from documenting the logic of a system's processes. Note that the analyst decides if a process requires more than one representation of its logic. Deliverables can also take the form of new entries into the project dictionary or CASE repository. These entries may update process descriptions or, if possible, store the new diagrams from logic and event-response modeling along with associated repository entries.

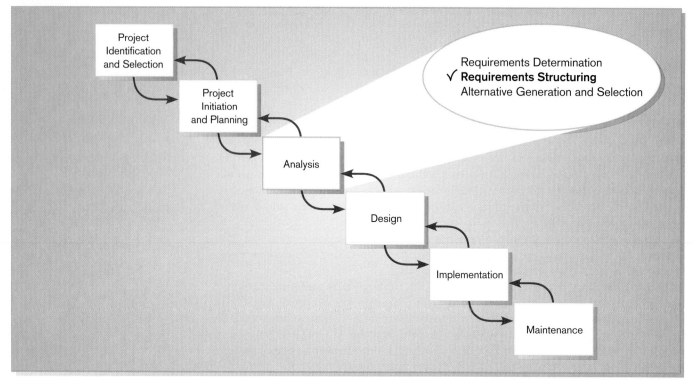

Figure 9-1
**Systems development life cycle
with the analysis phase highlighted**

Creating diagrams and descriptions of process logic is not an end in itself. Rather, these diagrams and descriptions are created ultimately to serve as part of an unambiguous and thorough explanation of the system's specifications. These specifications are used to explain the system requirements to developers, whether people or automated code generators. Users, analysts, and programmers use logic diagrams and descriptions throughout analysis to incrementally specify a shared understanding of requirements, without regard for programming languages or development environments. Such diagrams may be discussed during JAD sessions or project review meetings. Alternatively, system prototypes generated from such diagrams may be reviewed, and requested changes to a prototype will be implemented by changing logic diagrams and generating a new prototype from a CASE tool or other code generator.

As we have seen with other tools and techniques used in systems analysis, there are many ways to model logic in addition to the ones we focus on in this chapter. Structured English, decision tables, and decision trees are all methods of logic modeling that originated in the structured analysis approach. Yet they are not the only logic modeling techniques that originated in structured analysis. There are at least six different major approaches to structured systems development, each with its own particular tools and techniques (Wieringa, 1998). In addition to these toolsets,

TABLE 9-1   Deliverables for Logic Modeling

Where appropriate, each process on the lowest- (primitive-) level data flow diagrams will be represented with one or more of the following:
- Structured English representation of process logic
- Decision table representation
- Decision tree representation
- State-transition diagram or table (see Chapter 20)
- Sequence diagram (see Chapter 20)
- Activity diagram (see Chapter 20)

increasingly analysts use tools and techniques borrowed from the object-oriented approach to systems analysis and design. There are at least 19 different approaches to object-oriented analysis and design (OOAD), and each one of these approaches features a unique set of tools, methods, and techniques (Wieringa, 1998). Chapter 20, which provides an extensive overview of object-oriented analysis and design, features three tools for logic modeling—state transition diagrams, sequence diagrams, and activity diagrams. All three techniques are taken from the Unified Modeling Language approach to OOAD. All three of these tools are covered in detail in Chapter 20, and since they are an essential part of OOAD, it is important for you to learn about them in the OOAD context in which they were developed. You are encouraged, however, to study state transition, sequence, and activity diagrams in order to determine how they might fit in your analysis toolkit.

NET SEARCH
State transition diagrams are another way of structuring a type of process logic in systems analysis. Visit http://www.prenhall.com/hoffer to complete an exercise related to this topic.

MODELING LOGIC WITH STRUCTURED ENGLISH

You must understand more than just the flow of data into, through, and out of an information system. You must also understand what each identified process does and how it accomplishes its task. Starting with the processes depicted in the various sets of data flow diagrams you and others on the analysis team have produced, you must now begin to study and document the logic of each process. *Structured English* is one method used to illustrate process logic.

Structured English is a modified form of English that is used to specify the contents of process boxes in a DFD. It differs from regular English in that it uses a subset of English vocabulary to express information system process procedures. The same action verbs we listed in Chapter 8 for naming processes are also used in Structured English. These include such verbs as read, write, print, sort, move, merge, add, subtract, multiply, and divide. Structured English also uses noun phrases to describe data structures, such as patron-name and patron-address. Unlike regular English, Structured English does not use adjectives or adverbs. The whole point of using Structured English is to represent processes in a shorthand manner that is relatively easy for users and programmers to read and understand. As there is no standard version, each analyst will have his or her own particular dialect of Structured English.

It is possible to use Structured English to represent all three processes typical to structured programming: sequence, conditional statements, and repetition. Sequence requires no special structure but can be represented with one sequential statement following another. Conditional statements can be represented with a structure like the following:

```
BEGIN IF
    IF Quantity-in-stock is less than Minimum-order-quantity
    THEN GENERATE new order
    ELSE DO nothing
END IF
```

Another type of conditional statement is a case statement where there are many different actions a program can follow, but only one is chosen. A case statement might be represented as:

```
READ Quantity-in-stock
SELECT CASE
    CASE 1 (Quantity-in-stock greater than Minimum-order-quantity)
        DO nothing
    CASE 2 (Quantity-in-stock equals Minimum-order-quantity)
        DO nothing
```

Structured English: Modified form of the English language used to specify the logic of information system processes. Although there is no single standard, Structured English typically relies on action verbs and noun phrases and contains no adjectives or adverbs.

CASE 3 (Quantity-in-stock is less than Minimum-order-quantity)
   GENERATE new order
CASE 4 (Stock out)
   INITIATE emergency reorder routine
END CASE

Repetition can take the form of Do-Until loops or Do-While loops. A Do-Until loop might be represented as follows:

DO
   READ Inventory records
   BEGIN IF
      IF Quantity-in-stock is less than Minimum-order-quantity
      THEN GENERATE new order
      ELSE DO nothing
   END IF
UNTIL End-of-file

A Do-While loop might be represented as follows:

READ Inventory records
WHILE NOT End-of-File DO
   BEGIN IF
      IF Quantity-in-stock is less than Minimum-order-quantity
      THEN GENERATE new order
      ELSE DO nothing
   END IF
END DO

Let's look at an example of how Structured English would represent the logic of some of the processes identified in Hoosier Burger's current logical inventory control system. Figure 8-15 is reproduced below as Figure 9-2.

Figure 9-2
Current logical DFD for Hoosier Burger's inventory control system

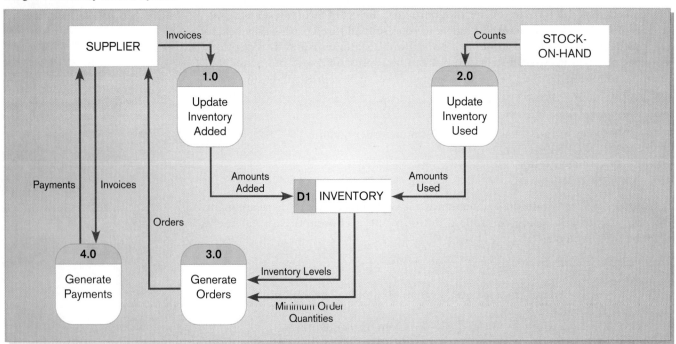

There are four processes depicted in Figure 9-2: Update Inventory Added, Update Inventory Used, Generate Orders, and Generate Payments. Structured English representations of each process are shown in Figure 9-3. Notice that in this version of Structured English the file names are connected with hyphens and file names and variable names are capitalized. Terms that signify logical comparisons, such as greater than and less than, are spelled out rather than represented by their arithmetic symbols. Also notice how short the Structured English specifications are, considering that these specifications all describe level-0 processes. The final specifications would model the logic in the lowest-level DFDs only. From reading the process descriptions in Figure 9-3, it should be obvious to you that much more detail would be required to actually perform the processes described. In fact, creating Structured English representations of processes in higher-level DFDs is one method you can use to help you decide if a particular DFD needs further decomposition.

Notice how the format of the Structured English process description mimics the format usually used in programming languages, especially the practice of indentation. This is the "structured part" of Structured English. Notice also that the language used is similar to spoken English, using verbs and noun phrases. The language

```
Process 1.0: Update Inventory Added
DO
    READ next Invoice-item-record
    FIND matching Inventory-record
    ADD Quantity-added from Invoice-item-record to Quantity-in-stock on
        Inventory-record
UNTIL End-of-file
```

```
Process 2.0: Update Inventory Used
DO
    READ next Stock-item-record
    FIND matching Inventory-record
    SUBTRACT Quantity-used on Stock-item-record from Quantity-in-stock on
        Inventory-record
UNTIL End-of-file
```

```
Process 3.0: Generate Orders
DO
    READ next Inventory-record
    BEGIN IF
        If Quantity-in-stock is less than Minimum-order-quantity
        THEN GENERATE Order
    END IF
UNTIL End-of-file
```

```
Process 4.0: Generate Payments
READ Today's-date
DO
    SORT Invoice-records by Date
    READ next Invoice-record
    BEGIN IF
        IF Date is 30 days or greater than Today's-date
        THEN GENERATE Payments
    END IF
UNTIL End-of-file
```

Figure 9-3
Structured English representations of the four processes depicted in Figure 9-2

is simple enough for a user who knows nothing about computer programming to understand the steps involved in performing the various processes, yet the structure of the descriptions makes it easy to eventually convert to a programming language. Using Structured English also means not having to worry about initializing variables, opening and closing files, or finding related records in separate files. These more technical details are left for later in the design process.

Structured English is intended to be used as a communication technique for analysts and users. Analysts and programmers have their own communication technique, *pseudocode*, which we will study later in this book. Whereas Structured English resembles spoken English, pseudocode resembles a programming language.

MODELING LOGIC WITH DECISION TABLES

Structured English can be used to represent the logic contained in an information system process, but sometimes a process' logic can become quite complex. If several different conditions are involved, and combinations of these conditions dictate which of several actions should be taken, then Structured English may not be adequate for representing the logic behind such a complicated choice. Not that Structured English cannot represent complicated logic; rather, Structured English becomes more difficult to understand and verify as logic becomes more complicated. Research has shown, for example, that people become confused in trying to interpret more than three nested IF statements (Yourdon, 1989). Where the logic is complicated, then a diagram may be much clearer than a Structured English statement. A **decision table** is a diagram of process logic where the logic is reasonably complicated. All of the possible choices and the conditions the choices depend on are represented in tabular form, as illustrated in the decision table in Figure 9-4.

The decision table in Figure 9-4 models the logic of a generic payroll system. There are three parts to the table: the **condition stubs**, the **action stubs**, and the **rules**. The condition stubs contain the various conditions that apply in the situation the table is modeling. In Figure 9-4, there are two condition stubs for employee type and hours worked. Employee type has two values: "S," which stands for salaried, and "H," which stands for hourly. Hours worked has three values: less than 40, exactly 40, and more than 40. The action stubs contain all the possible courses of action that result from combining values of the condition stubs. There are four possible courses of action in this table: Pay base salary, Calculate hourly wage, Calculate overtime, and Produce Absence Report. You can see that not all actions are triggered by all combinations of conditions. Instead, specific combinations trigger specific actions. The part of the table that links conditions to actions is the section that contains the rules.

N E T S E A R C H
Structured English has received considerable academic attention. Visit http://www.prenhall.com/hoffer to complete an exercise related to this topic.

Decision table: A matrix representation of the logic of a decision, which specifies the possible conditions for the decision and the resulting actions.

Condition stubs: That part of a decision table that lists the conditions relevant to the decision.

Action stubs: That part of a decision table that lists the actions that result for a given set of conditions.

Rules: That part of a decision table that specifies which actions are to be followed for a given set of conditions.

Figure 9-4
Complete decision table for payroll system example

	Conditions/ Courses of Action	Rules					
		1	2	3	4	5	6
Condition Stubs	Employee type	S	H	S	H	S	H
	Hours worked	<40	<40	40	40	>40	>40
Action Stubs	Pay base salary	X		X		X	
	Calculate hourly wage		X		X		X
	Calculate overtime						X
	Produce Absence Report		X				

To read the rules, start by reading the values of the conditions as specified in the first column: Employee type is "S," or salaried, and hours worked are less than 40. When both of these conditions occur, the payroll system is to pay the base salary. In the next column, the values are "H" and "<40," meaning an hourly worker who worked less than 40 hours. In such a situation, the payroll system calculates the hourly wage and makes an entry in the Absence Report. Rule 3 addresses the situation when a salaried employee works exactly 40 hours. The system pays the base salary, as was the case for Rule 1. For an hourly worker who has worked exactly 40 hours, Rule 4 calculates the hourly wage. Rule 5 pays the base salary for salaried employees who work more than 40 hours. Rule 5 has the same action as Rules 1 and 3, and governs behavior with regard to salaried employees. The number of hours worked does not affect the outcome for Rules 1, 3, or 5. For these rules, hours worked is an **indifferent condition** in that its value does not affect the action taken. Rule 6 calculates hourly pay and overtime for an hourly worker who has worked more than 40 hours.

Because of the indifferent condition for Rules 1, 3, and 5, we can reduce the number of rules by condensing Rules 1, 3, and 5 into one rule, as shown in Figure 9-5. The indifferent condition is represented with a dash. Whereas we started with a decision table with six rules, we now have a simpler table that conveys the same information with only four rules.

In constructing these decision tables, we have actually followed a set of basic procedures, as follows:

1. *Name the conditions and the values each condition can assume.* Determine all of the conditions that are relevant to your problem, and then determine all of the values each condition can take. For some conditions, the values will be simply "yes" or "no" (called a limited entry). For others, such as the conditions in Figures 9-4 and 9-5, the conditions may have more values (called an extended entry).

2. *Name all possible actions that can occur.* The purpose of creating decision tables is to determine the proper course of action given a particular set of conditions.

3. *List all possible rules.* When you first create a decision table, you have to create an exhaustive set of rules. Every possible combination of conditions must be represented. It may turn out that some of the resulting rules are redundant or make no sense, but these determinations should be made only *after* you have listed every rule so that no possibility is overlooked. To determine the number of rules, multiply the number of values for each condition by the number of values for every other condition. In Figure 9-4, we have two conditions, one with two values and one with three, so we need 2×3, or 6, rules. If we added a third condition with three values, we would need $2 \times 3 \times 3$, or 18, rules.

Indifferent condition: In a decision table, a condition whose value does not affect which actions are taken for two or more rules.

Conditions/ Courses of Action	Rules			
	1	2	3	4
Employee type	S	H	H	H
Hours worked	–	<40	40	>40
Pay base salary	X			
Calculate hourly wage		X	X	X
Calculate overtime				X
Produce Absence Report		X		

Figure 9-5
Reduced decision table for payroll system example

When creating the table, alternate the values for the first condition, as we did in Figure 9-4 for type of employee. For the second condition, alternate the values but repeat the first value for all values of the first condition, then repeat the second value for all values of the first condition, and so on. You essentially follow this procedure for all subsequent conditions. Notice how we alternated the values of hours worked in Figure 9-4. We repeated "<40" for both values of type of employee, "S" and "H." Then we repeated "40," and then ">40."

4. *Define the actions for each rule.* Now that all possible rules have been identified, provide an action for each rule. In our example, we were able to figure out what each action should be and whether all of the actions made sense. If an action doesn't make sense, you may want to create an "impossible" row in the action stubs in the table to keep track of impossible actions. If you can't tell what the system ought to do in that situation, place question marks in the action stub spaces for that particular rule.

5. *Simplify the decision table.* Make the decision table as simple as possible by removing any rules with impossible actions. Consult users on the rules where system actions aren't clear and either decide on an action or remove the rule. Look for patterns in the rules, especially for indifferent conditions. We were able to reduce the number of rules in the payroll example from six to four, but often greater reductions are possible.

Let's look at an example from Hoosier Burger. The Mellankamps are trying to determine how they reorder food and other items they use in the restaurant. If they are going to automate the inventory control functions at Hoosier Burger, they need to articulate their reordering process. In thinking through the problem, the Mellankamps realize that how they reorder depends on whether the item is perishable. If an item is perishable, such as meat, vegetables, or bread, the Mellankamps have a standing order with a local supplier stating that a prespecified amount of food is delivered each weekday for that day's use and each Saturday for weekend use. If the item is not perishable, such as straws, cups, and napkins, an order is placed when the stock on hand reaches a certain predetermined minimum reorder quantity. The Mellankamps also realize the importance of the seasonality of their work. Hoosier Burger's business is not as good during the summer months when the students are off campus as it is during the academic year. They also note that business falls off during Christmas and spring breaks. Their standing orders with all their suppliers are reduced by specific amounts during the summer and holiday breaks. Given this set of conditions and actions, the Mellankamps put together an initial decision table (see Figure 9-6).

Notice three things about Figure 9-6. First, notice how the values for the third condition have been repeated, providing a distinctive pattern for relating the values for all three conditions to each other. Every possible rule is clearly provided in this table. Second, notice we have 12 rules. Two values for the first condition (type of item) times two values for the second condition (time of week) times three values for the third condition (season of year) equals 12 possible rules. Third, notice how the action for nonperishable items is the same, regardless of day of week or time of year. For nonperishable goods, both time-related conditions are indifferent. Collapsing the decision table accordingly gives us the decision table in Figure 9-7. Now there are only seven rules instead of 12.

You have now learned how to draw and simplify decision tables. You can also use decision tables to specify additional decision-related information. For example, if the actions that should be taken for a specific rule are more complicated than one or two lines of text can convey, or if some conditions need to be checked only when other conditions are met (nested conditions), you may want to use separate, linked decision tables. In your original decision table, you can specify an action in the action stub that says "Perform Table B." Table B could contain an action stub that returns to the original table and the return would be the action for one or more rules in Table

Conditions/ Courses of Action	Rules											
	1	2	3	4	5	6	7	8	9	10	11	12
Type of item	P	N	P	N	P	N	P	N	P	N	P	N
Time of week	D	D	W	W	D	D	W	W	D	D	W	W
Season of year	A	A	A	A	S	S	S	S	H	H	H	H
Standing daily order	X				X				X			
Standing weekend order			X				X				X	
Minimum order quantity		X		X		X		X		X		X
Holiday reduction									X		X	
Summer reduction					X		X					

Type of item:	Time of week:	Season of year:
P = perishable	D = weekday	A = academic year
N = nonperishable	W = weekend	S = summer
		H = holiday

Figure 9-6
Complete decision table for Hoosier Burger's inventory reordering

B. Another way to convey more information in a decision table is to use numbers that indicate sequence rather than Xs where rules and action stubs intersect. For example, for Rules 3 and 4 in Figure 9-7, it would be important for the Mellankamps to account for the summer reduction to modify the existing standing order for supplies. "Summer reduction" would be marked with a "1" for Rules 3 and 4 while "Standing daily order" would be marked with a "2" for Rule 3, and "Standing weekend order" would be marked with a "2" for Rule 4.

You have seen how decision tables can model the relatively complicated logic of a process. Decision tables are more useful than Structured English for complicated logic in that they convey information in a tabular rather than a linear, sequential format. As such, decision tables are compact; you can pack a lot of information into a small table. Decision tables also allow you to check for the extent to which your logic is complete, consistent, and not redundant. Despite the usefulness of Structured English and decision tables, there are still other techniques available for modeling process logic. The next such technique you will learn about is decision trees.

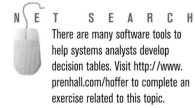

N ET S E A R C H
There are many software tools to help systems analysts develop decision tables. Visit http://www. prenhall.com/hoffer to complete an exercise related to this topic.

Conditions/ Courses of Action	Rules						
	1	2	3	4	5	6	7
Type of item	P	P	P	P	P	P	N
Time of week	D	W	D	W	D	W	–
Season of year	A	A	S	S	H	H	–
Standing daily order	X		X		X		
Standing weekend order		X		X		X	
Minimum order quantity							X
Holiday reduction					X	X	
Summer reduction			X	X			

Figure 9-7
Reduced decision table for Hoosier Burger's inventory reordering

Figure 9-8
Generic decision tree

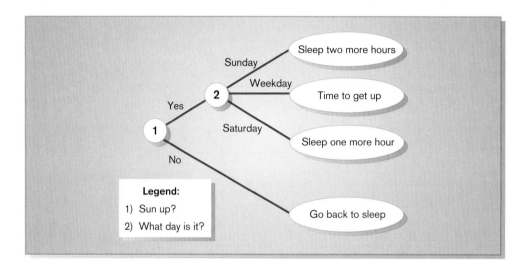

MODELING LOGIC WITH DECISION TREES

Decision tree. A graphical
representation of a decision
situation in which decision situation
points (nodes) are connected
together by arcs (one for each
alternative on a decision) and
terminate in ovals (the action that is
the result of all of the decisions
made on the path leading to that
oval).

A **decision tree** is a graphical technique that depicts a decision or choice situation as a connected series of nodes and branches. Decision trees were first devised as a management science technique to simplify a choice where some of the needed information is not known for certain. By relying on the probabilities of certain events, a management scientist can use a decision tree to choose the best course of action. Although this type of decision tree is beyond the scope of our text, we can use modified decision trees (without the probabilities) to diagram the same sorts of situations for which we used decision tables. Why introduce yet another diagramming technique to do what a decision table does? Both decision tables and decision trees are communication tools designed to make it easier for analysts to communicate with users. Deciding exactly which technique to use depends on various factors, which we discuss in detail in the next section, after you have an understanding of decision trees.

As used in requirements structuring, decision trees have two main components: decision points, which are represented by nodes, and actions, which are represented by ovals. Figure 9-8 shows a generic decision tree. To read a decision tree, you begin at the root node on the far left. Each node is numbered, and each number corresponds to a choice; the choices are spelled out in a legend for the diagram. Each path leaving a node corresponds to one of the options for that choice. From each node, there are at least two paths that lead to the next step, which is either another decision point or an action. Finally, all possible actions are listed on the far right of the diagram in leaf nodes. Each rule is represented by tracing a series of paths from the root node, down a path to the next node, and so on, until an action oval is reached.

Figure 9-9
**Decision tree representation of the
decision logic in the decision
tables in Figures 9-4 and 9-5, with
only two choices per decision point**

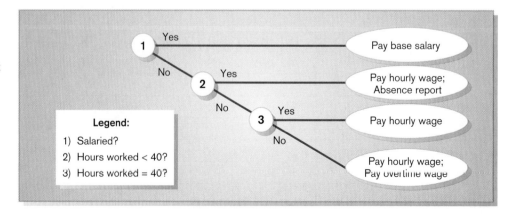

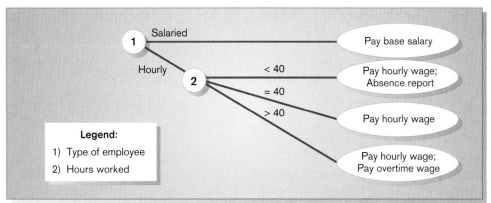

Figure 9-10
Decision tree representation of the
decision logic in the decision
tables in Figure 9-4 and 9-5, with
multiple choices per decision point

Look back at the decision tables we created for the payroll system logic (Figures 9-4 and 9-5). There are at least two ways to represent this same information as a decision tree. The first is shown in Figure 9-9. Here all of the choices are limited to two outcomes, either yes or no. However, looking at how the conditions are phrased in the decision tables, you remember that hours worked has three values, not two. You might argue that forcing a condition with three values into a set of conditions that have only yes or no available as values is somewhat artificial. To preserve the original logic of the decision situation, you can draw your decision tree as depicted in Figure 9-10. Here, there are only two conditions; the first condition has two values, and the second has three values, as is true in the decision table. We leave it as an exercise for the reader to diagram the logic of the situation depicted in Figures 9-6 and 9-7.

We have waited until now to make two important points about decision tables and decision trees. Once you have spent some time creating logic modeling aids such as these, be ready to refine them by drawing the diagrams again and again. As was the case with data flow diagrams, decision tables and decision trees benefit greatly from iteration. The second point is that you should always share your work with other team members and users to get feedback on the mechanical and content accuracy of your work. Other team members and users will often provide insight into issues you might have overlooked in describing the logic. For that reason, it is not uncommon for the analysis team leader to schedule a walkthrough at some point during the requirements structuring process.

DECIDING AMONG STRUCTURED ENGLISH, DECISION TABLES, AND DECISION TREES

How do you decide whether to use Structured English, decision tables, or decision trees when modeling process logic? On one level, the answer is to use whichever method you prefer and understand best. For example, some analysts and users prefer to see the logic of a complicated decision situation laid out in tabular form, as in a decision table; others will prefer the more graphical structure of a decision tree. Yet the issue actually extends beyond mere preferences. Just because you are very adept at using a hammer doesn't mean a hammer is the best tool for all home repairs—sometimes a screwdriver or a drill is the best tool. The same is true of logic modeling techniques: You have to consider the task you are performing and the purpose of the techniques in order to decide which technique is best. The relative advantages and disadvantages of Structured English, decision tables, and decision trees for different situations are presented in Tables 9-2 and 9-3.

Table 9-2 summarizes the research findings for comparisons of all three techniques. One study summarized in the table compared Structured English to decision

TABLE 9-2 Criteria for Deciding Among Structured English, Decision Tables, and Decision Trees

Criteria	Structured English	Decision Tables	Decision Trees
Determining conditions & actions	Second Best	Third Best	Best
Transforming conditions & actions into sequence	Best	Third Best	Best
Checking consistency & completeness	Third Best	Best	Best

tables and decision trees and analyzed the techniques for two different tasks (Vessey and Weber, 1986). The first task was determining the correct conditions and actions from a description of the problem, much the same situation analysts face when defining conditions and actions after an interview with a user. The study found that decision trees were the best technique to support this process as they naturally separate conditions and actions, making the logic of the decision rules more apparent. Even though Structured English does not separate conditions and actions, it was considered the second best technique for this task. Decision tables were the worst technique. The second task was converting conditions and actions to sequential statements, similar to what an analyst does when converting the stated conditions and actions to the sequence of pseudocode or a programming language. Structured English was the best technique for this task, as it is already written sequentially, but researchers found decision trees to be just as good. Decision tables were last again.

Both decision trees and decision tables do have at least one advantage over Structured English, however. Both decision tables and trees can be checked for completeness, consistency, and degree of redundancy. We checked all of our examples of decision tables for completeness when we made sure that each initial table included all possible rules. We knew the tables were complete when we multiplied the number of values for each condition to get the total number of possible rules. Following the other specific steps outlined earlier in the chapter will also help you check for a decision table's consistency and degree of redundancy. The same procedures can be easily adopted for decision trees. However, there are no such easy means to validate Structured English statements, giving decision tables and trees at least one advantage over Structured English.

Researchers have also compared decision tables to decision trees (Table 9-3). The pioneers of structured analysis and design thought decision tables were best for portraying complex logic while decision trees were better for simpler problems (Gane and Sarson, 1979). Others have found decision trees to be better for guiding decision making in practice (Subramanian et al., 1992), but decision tables have the advantage of being more compact than decision trees and easier to manipulate (Vanthienen, 1994). If more conditions are added to a situation, a decision table can easily accommodate more conditions, actions, and rules. If the table becomes too large, it can easily be divided into subtables, without the inconvenience of using flowchart-like tree connections used with decision trees. Creating and maintaining complex decision tables can be made easier with computer support (e.g., Prologa mentioned in Vanthienen, 1994).

TABLE 9-3 Criteria for Deciding Between Decision Tables and Decision Trees

Criteria	Decision Tables	Decision Trees
Portraying complex logic	Best	Worst
Portraying simple problems	Worst	Best
Making decisions	Worst	Best
More compact	Best	Worst
Easier to manipulate	Best	Worst

ELECTRONIC COMMERCE APPLICATION: LOGIC MODELING

Like many other analysis and design activities, logic modeling techniques for an Internet-based electronic commerce application are no different than the techniques used when depicting logic within processes for other types of applications. In the last chapter, you read how Jim developed a level-0 data flow diagram for the WebStore system. In this section, we examine how he represented the logic inside the DFD processes.

Logic Modeling for Pine Valley Furniture's WebStore

After defining the level-0 DFD for the WebStore system (see Figure 9-11), the Pine Valley Furniture development methodology dictated that Jim needed to represent the logic within each of the unique processes. Since the logic within each process was relatively straightforward, Jim decided to represent each using Structured English. For Processes 1.0 and 2.0, the logic was very straightforward (see Table 9-4). However, for Process 3.0, two distinct activities were being performed: (1) displaying the contents of the shopping cart and (2) removing items from the shopping cart. Therefore, Jim concluded that it would be best to first diagram the subprocesses using a DFD (see Figure 9-12), then write the logic for each separate process (see Table 9-5). The logic for the remaining processes is shown in Table 9-6.

Now that the high-level logic of the main processes of the WebStore has been defined, Jim needed to get a clear picture of exactly what data were needed throughout the entire system. We will learn how Jim and the PVF development group address this next analysis activity—conceptual data modeling—in the next chapter.

TABLE 9-4 Structured English Representations of Processes 1.0 and 2.0 from Figure 9-11

Process 1.0: Browse Catalog
READ Product-Item-Request
FIND Matching Product-Item from Inventory
DISPLAY Item-Profile
Process 2.0: Select Item for Purchase
READ Purchase-Request
READ Inventory-Item-Profile
ADD Item-Profile to Shopping-Cart

TABLE 9-5 Structured English Representations of Processes 3.1 and 3.2 from Figure 9-12

Process 3.1: Display Shopping Cart Details
READ View-Cart
DO
    READ Item-Profile
    DISPLAY Item-Profile
UNTIL Shopping-Cart is Empty
Process 3.2: Remove Item
READ Remove-Item
SUBTRACT Product-Item from Shopping-Cart

TABLE 9-6 Structured English Representations of Processes 4.0, 5.0, and 6.0 from Figure 9-11

Process 4.0: Check Out Process Orders
READ Customer-ID
READ Check-Out
FIND Customer-Information from Customer-Tracking-System
DO
    READ Item-Profile
    ADD Item-Profile to Order
UNTIL Shopping-Cart is Empty
ADD Order to Purchasing-Fulfillment-System
READ Order-Number from Purchasing-Fulfillment-System
READ Return-Code from Purchasing-Fulfillment-System
DISPLAY Invoice
Process 5.0: Add/Modify Account Profile
READ Customer-Information
ADD Customer-Information to Customer-Tracking-System
DISPLAY Customer-Information
Process 6.0: Order Status Request
READ Order-Number
FIND Order-Status-Information from Purchasing-Fulfillment-System
DISPLAY Order-Status-Information

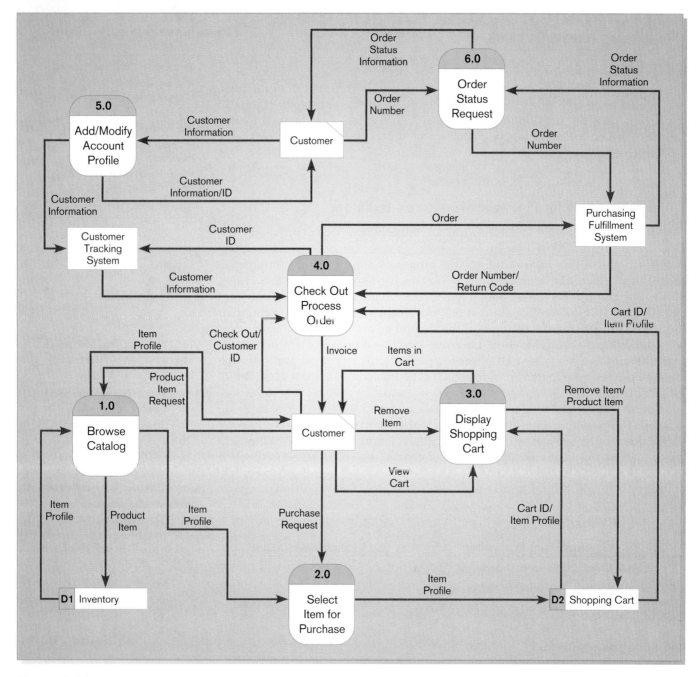

Figure 9-11
Level-0 DFD for the WebStore system

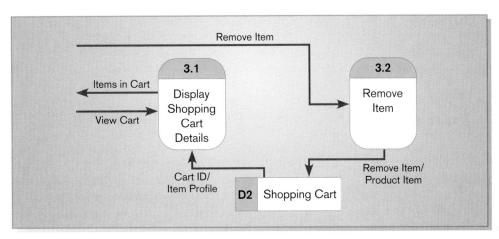

Figure 9-12
Level-1 DFD for the WebStore system for Process 3.0

Summary

Logic modeling is one of the three key activities in the requirements structuring phase of systems analysis. There are various techniques available for modeling the decision logic in information system processes. One method, Structured English, is a special form of spoken English analysts use to illustrate the logic of processes depicted in data flow diagrams. Structured English is primarily a communication technique for analysts and users. Structured English dialect varies with analysts, but it must represent sequence, conditional statements, and repetition in information system processes.

Decision tables and decision trees are graphical methods for representing process logic. In decision tables, conditions are listed in the condition stubs, possible actions are listed in the action stubs, and rules link combinations of conditions with the actions that should result. A first pass at a decision table involves listing all possible rules that result from all values of the conditions listed. Second and subsequent passes reduce the complexity by eliminating rules that don't make sense and by combining rules with indifferent conditions. The same logic portrayed in a decision table can be portrayed in a decision tree but, in a decision tree, the conditions are represented by decision points and the values are represented by paths between decision points and ovals that contain actions.

Several research studies have compared Structured English, decision tables, and decision trees as techniques for representing decision logic. Most studies show that decision trees are the best for many criteria while decision tables are the worst on some criteria and the best on other criteria. Because there is no best technique to use for structuring requirements, an analyst must be proficient at all three techniques, as required by different situations.

Key Terms

1. Action stubs
2. Condition stubs
3. Decision table
4. Decision tree
5. Indifferent condition
6. Rules
7. Structured English

Match each of the key terms above with the definition that best fits it.

_____ In a decision table, a condition whose value does not affect which actions are taken for two or more rules.

_____ A matrix representation of the logic of a decision, which specifies the possible conditions for the decision and the resulting actions.

_____ Modified form of the English language used to specify the logic of information system processes.

_____ That part of a decision table that specifies which actions are to be followed for a given set of conditions.

_____ That part of a decision table that lists the actions that result for a given set of conditions.

_____ That part of a decision table that lists the conditions relevant to the decision.

_____ A graphical representation of a decision situation in which decision points (nodes) are connected together by arcs (one for each alternative on a decision) and terminate in ovals (the action that is the result of all of the decisions made on the path leading to that oval).

Review Questions

1. What is the purpose of logic modeling? What techniques are used to model decision logic and what techniques are used to model temporal logic?

2. What is Structured English? How can Structured English be used to represent sequence, conditional statements, and repetition in an information systems process?

3. What is the difference between Structured English and pseudocode?

4. What are the steps in creating a decision table? How do you reduce the size and complexity of a decision table?

5. Explain the structure of a decision tree.

6. How do you know when to use Structured English, decision tables, or decision trees? Which is best for what situation?

7. What verbs are used in Structured English? What type of words are not used in Structured English?

8. What does the term "limited entry" mean in a decision table?

9. What is the formula that is used to calculate the number of rules a decision table must cover?

Problems and Exercises

1. Represent the decision logic in the decision table of Figure 9-5 in Structured English.

2. Look at the set of data flow diagrams created in Chapter 8 for Hoosier Burger's food-ordering system.
   a. Write Structured English to represent the logic in each process, at each level data flow diagram.
   b. Represent the decision logic of one or more of the processes as decision tables.

3. Diagram the decision logic illustrated in the decision tables in Figures 9-6 and 9-7 as decision trees.

4. What types of questions need to be asked during requirements determination in order to gather the information needed for logic modeling? Give examples.

5. In one company the rules for buying personal computers are such that if the purchase is over $15,000.00, it has to go out for bid and the Request for Proposals must be approved by the Purchasing Department. If the purchase is under $15,000.00, the personal computers can simply be bought from any approved vendor; however, the Purchase Order must still be approved by the Purchasing Department. If the purchase goes out for bid, there must be at least three proposals received for the bid. If not, the RFP must go out again. If there still are not enough proposals, then the process can continue with the one or two vendors that have submitted proposals. The winner of the bid must be on an approved list of vendors for the company and, in addition, must not have any violations against them for affirmative action or environmental matters. At this point, if the proposal is complete, the Purchasing Department can issue a Purchase Order. Use Structured English to represent the logic in this process. Notice the similarities between the text in this question and format of your answer.

6. If you were going to develop a computer-based tool to help an analyst interview users and quickly and easily create and edit Structured English logic models, what type of tool would you build? What features would it have, and how would it work? Why?

7. Present the logic of the business process described in Problem and Exercise 5 in a decision table and then in a decision tree. For this example, how do these two techniques compare with one another? Why? How do they compare with the Structured English technique? Why? Do either of these techniques seem better suited to helping us diagnose problems and/or inefficiencies in the business process? Why or why not?

8. If you were going to develop a computer-based tool to help an analyst interview users and quickly and easily create and edit decision table and decision tree logic models, what type of tool would you build? What features would it have, and how would it work? Why? Would this type of tool be more or less useful than a tool built for this purpose based on Structured English? Why?

9. In a relatively small company that sells thin, electronic keypads and switches, the rules for selling products are such that sales representatives are assigned to unique regions of the country. Sales come either from cold calling, referrals, or current customers with new orders. A sizable portion of their business comes from referrals from larger competitors, who send their excess and/or "difficult" projects to this company. The company tracks these and, similarly, returns the favors to these competitors by sending business their way. The sales reps receive a 10% commission on purchases, not on orders, in their region. They can collaborate on a sale with reps in other regions and share the commissions, with 8% going to the "home" rep, and 2% going to the "vis-

iting" rep. For any sales beyond the rep's previously stated and approved individual annual sales goals, he or she receives an additional 5% commission, an additional end-of-the-year bonus determined by management, and a special vacation for his or her family. Customers receive a 10% discount for any purchases over $100,000.00 per year, which are factored into the rep's commissions. In addition, the company focuses on customer satisfaction with the product and service, so there is an annual survey of customers in which they rate the sales rep. These ratings are factored into the bonuses such that a high rating increases the bonus amount, a moderate rating does nothing, and a low rating can lower the bonus amount. The company also wants to ensure that the reps close all sales. Any differences between the amount of orders and actual purchases is also factored into the rep's bonus amount. As best you can, present the logic of this business process first using Structured English, then using a decision table, and then using a decision tree. Write down any assumptions you have to make. Which of these techniques is most helpful for this problem? Why?

10. The following is an example that demonstrates the rules of the tenure process for faculty at many universities. Present the logic of this business process first using Structured English, then using a decision table, and then using a decision tree. Write down any assumptions you have to make. Which of these techniques is most helpful for this problem? Why?

A faculty member applies for tenure in his or her sixth year by submitting a portfolio summarizing his or her work. In rare circumstances he or she can come up for tenure earlier than the sixth year, but only if the faculty member has permission of the department chair and college dean. New professors, who have worked at other universities before taking their current jobs, rarely, if ever, come in with tenure. They are usually asked to undergo one "probationary" year during which they are evaluated and only can then be granted tenure. Top administrators coming in to a new university job, however, can often negotiate for retreat rights that enable them to become a tenured faculty member should their administrative post end. These retreat arrangements generally have to be approved by faculty. The tenure review process begins with an evaluation of the candidate's portfolio by a committee of faculty within the candidate's department. The committee then writes a recommendation on tenure and sends it to the department's chairperson who then makes a recommendation, and passes the portfolio and recommendation on to the next level, a college-wide faculty committee. This committee does the same as the department committee and passes its recommendation, the department's recommendation, and the portfolio on to the next level, a university-wide faculty committee. This committee does the same as the other two committees and passes everything on to the provost (or sometimes the academic vice president). The provost then writes his or her own recommendation and passes everything to the president, the final decision maker. This process, from the time the candidate creates his or her portfolio until the time the president makes a decision, can take an entire academic year. The focus of the evaluation is on research, which could be grants, presentations, and publications, though preference is given for empirical research that has been published in top-ranked, refereed journals and where the publication makes a contribution to the field. The candidate must also do well in teaching and service (i.e., to the university, the community, or to the discipline) but the primary emphasis is on research.

11. An organization is in the process of upgrading microcomputer hardware and software for all employees. Hardware will be allocated to each employee in one of three packages. The first hardware package includes a standard microcomputer with a color monitor of moderate resolution and moderate storage capabilities. The second package includes a high-end microcomputer with high-resolution color monitor and a great deal of RAM and ROM. The third package is a high-end notebook-sized microcomputer. Each computer comes with a network interface card so that it can be connected to the network for printing and e-mail. The notebook computers come with a modem for the same purpose. All new and existing employees will be evaluated in terms of their computing needs (e.g., the types of tasks they perform, how much and in what ways they can use the computer). Light users receive the first hardware package. Heavy users receive the second package. Some moderate users will receive the first package and some will receive the second package, depending on their needs. Any employee who is deemed to be primarily mobile (e.g., most of the sales force) will receive the third package. Each employee will also be considered for additional hardware. For example, those who need scanners will receive them and those needing their own printers will receive them. A determination will be made regarding whether or not the user receives a color or black-and-white scanner, and whether or not they receive a slow or fast, or color or black-and-white printer. In addition, each employee will receive a suite of software, including a word processor, spreadsheet, and presentation maker. All employees will be evaluated for their additional software needs. Depending on their needs, some will receive a desktop publishing package, some will receive a database management system (and some will also receive a developer's kit for the DBMS), and some will receive a programming language. Every 18 months those employees with the high-end systems will receive new hardware and then their old systems will be passed on to those who previously had the standard systems. All those employees with the portable systems will receive new notebook computers. Present the logic of this business process first using Structured English, then using a decision table, and then using a decision tree. Write down any assumptions you have to make. Which of these techniques is most helpful for this problem? Why?

12. Read the narratives below and follow the directions for each. If you discover that the narrative is incomplete, make up reasonable explanations to complete the story. Supply these extra explanations along with your answers.
    a. Samantha must decide which courses to register for this semester. She has a part-time job and she is waiting to find out how many hours per week she will be working during the semester. If she works 10 hours or less per week she will register for three classes, but if she works more than 10 hours per week she will register for only two classes. If she registers for two classes, she will take one class in her major area and one elective. If she regis-

ters for three classes, she will take two classes in her major area and one elective. Use both Structured English and a decision tree to represent this logic.

b. Jerry plans on registering for five classes this semester: English composition, physics, physics lab, COBOL, and music appreciation. However, he is not sure if these classes are being offered this semester or if there will be timing conflicts. Also, two of the classes, physics and physics lab, must be taken together during the same semester. Therefore, if he can only register for one of them he will not take either class. If, for any reason, he cannot register for a class, he will identify and register for a different class to take its place that fits in his time schedule. Use a decision tree and a decision table that show all rules to represent this logic.

13. Mary is trying to decide what graduate programs she will apply to. She wants to stay in the southeast region of the United States, but if a program is considered one of the top 10 in the country she is willing to move to another part of the United States. Mary is interested in both the MBA and Master of MIS programs. An MBA program must have at least one well-known faculty member and meet her location requirements before she will consider applying to it. Additionally, any program she applies to must offer financial aid unless she is awarded a scholarship. Use Structured English, a decision tree, and a decision table to represent this logic.

14. At a local bank, loan officers must evaluate loan applications before approving or denying them. During this evaluation process, many factors regarding the loan request and the applicant's background are considered. If the loan is for less than

$2,000, the loan officer checks the applicant's credit report. If the credit report is rated good or excellent, the loan officer approves the loan. If the credit report is rated fair, the officer checks to see if the applicant has an account at the bank. If the applicant holds an account, the application is approved; otherwise, the application is denied. If the credit report is rated poor, the application is denied. Loan applications for amounts between $2,000 and $200,000 are divided into four categories: car, mortgage, education, and other. For car, mortgage, and other loan requests, the applicant's credit report is reviewed and an employment check is made to verify the applicant's reported salary income. If the credit report rating is poor, the loan is denied. If the credit report rating is fair, good, or excellent and the salary income is verified, the loan is approved. If the salary income is not verifiable, the applicant is contacted and additional information is requested. In this case, the loan application along with the additional information is sent to the vice president for review and a final loan decision. For educational loans, the educational institution the applicant will attend is contacted to determine the estimated cost of attendance. This amount is then compared to the amount of the loan requested in the application. If the requested amount exceeds the cost of attendance, the loan is denied. Otherwise, education loan requests for amounts between $2,000 and $34,999 are approved if the applicant's credit rating is fair, good, or excellent. Education loan applications requesting amounts from $35,000 to $200,000 are approved only if the credit rating is good or excellent. All loan applications for amounts greater than $200,000 are sent to the vice president for review and approval. Use Structured English, a decision tree, and a decision table to represent this logic.

Field Exercises

1. Choose a transaction that you are likely to encounter, perhaps ordering a cap and gown for graduation or ordering clothing from a mail order catalog. Present the processes in this transaction using Structured English. Based on these examples, what appear to you to be the relative strengths and weaknesses of this technique versus decision tables or decision trees?

2. Why can't process logic simply be represented with regular written English? Research this issue in the library and write a position paper that explains the problems with representing procedure and action with the English language.

3. Explain the history of structured programming. How did it start? What was the basis for it? Why is it considered so

important? How does structured programming fit within fourth-generation programming languages? To what extent is structured programming used in organizations with which you have come in contact?

4. Visit a shopping mall near you and observe the many transactions taking place in the various stores. What techniques presented in this chapter for modeling the logic of processes are most helpful for each of these various processes? Why? Are any of the modeling techniques robust enough to model all of these processes well? Why or why not?

References

Gane, C., and T. Sarson. 1979. *Structured Systems Analysis.* Englewood Cliffs, NJ: Prentice-Hall.

Subramanian, G. H., J. Nosek, S. P. Raghunathan, and S. S. Kanitkar. 1992. "A Comparison of the Decision Table and Tree." *Communications of the ACM* 35 (January): 89–94.

Vanthienen, J. 1994. "Technical Correspondence." *Communications of the ACM* 37 (February): 109–111.

Vessey, I., and R. Weber. 1986. "Structured Tools and Conditional Logic." *Communications of the ACM* 29 (January): 48–57.

Wieringa, R. 1998. "A Survey of Structured and Object-Oriented Software Specification Methods and Techniques." *ACM Computing Surveys* 30(4), Dec., 459–527.

Yourdon, E. 1989. *Managing the Structured Techniques.* Englewood Cliffs, NJ: Prentice-Hall.

BROADWAY ENTERTAINMENT COMPANY, INC.

Structuring System Requirements: Logic Modeling for the Web-Based Customer Relationship Management System

CASE INTRODUCTION

The BEC student team of Tracey Wesley, John Whitman, Missi Davies, and Aaron Sharp were making good progress on structuring the requirements for MyBroadway, the Web-based customer relationship management for a Broadway Entertainment Company store. All of the information processes of system modules have been laid out in a level-0 data flow diagram (see BEC Figure 9-1) and each process/module was exploded into more detailed DFDs, down to system primitives. Now the team was ready to make sure that they accurately understood the rules by which each process was to operate.

THE LOGIC OF INFORMATION PROCESSES

To verify their understanding of system logic, team members met with Carrie Douglass, manager of the Centerville BEC store, and other store employees to hear them talk about how they thought each system function would or should operate. The BEC student team developed written descriptions for all eleven of the major system modules. Three of the key system modules are: Process 5: Enter New Comment; Process 6: Display Comment on Product; Process 7: Process New Product Request. The following sections contain the descriptions the team developed for these three modules.

Process 5: Process New Comment

There are two major steps to processing a new comment: entering and recording the comment. Recall from prior cases that it was decided that MyBroadway was not going to store customer data; also recall that comments are not to be recorded unless submitted by a valid and attributed customer. So, a comment is first accepted by MyBroadway as a tentative entry (not yet accessible to other customers), and then once Entertainment Tracker verifies that the customer exists, a comment is changed to active status and is available to be viewed by other customers. A customer starts to enter a comment by entering a customer ID and a piece of personal data (the cus-

tomer's birthday). Then the customer selects the type of product (DVD, CD, game). Then a customer can identify the associated product for the comment to be entered in one of two ways: either by selecting the product (of the designated type) from a drop-down list of products in inventory, or by a text search on the approximate title of the product. In the case of the search function, the customer verifies on which of possibly several products identified from the search he or she will enter a comment. If no product is found matching the entered title, or if none of those found in the one the customer seeks, he or she can enter a new text or exit. Once a product is selected, the system displays additional information about the product (full title, publisher, artist, date released) for further verification that the customer has referenced the desired product. At any point, the customer can stop the entry process. On the same screen with the product information, the customer can enter the desired comment. Once the comment is accepted by MyBroadway as coming from a valid customer, additional data about the customer is added to the record: whether the comment comes from a parent or a child, and the date and time of the comment. If the customer number and birthdate do not match Entertainment Tracker records, the comment is dropped.

Process 6: Display Comment on Product

Displaying comments seems more straightforward to the student team. From their discussions with store employees, the steps are as follows. A customer will find a product of a given type on which to display comments using the same alternatives described above for entering a comment on a product. Once the customer selects the product, all of the comments for that product are shown, along with the date and time the comment was entered, the name of the person entering the comment, and whether the entry was from a parent or a child. The process needs to handle the situation that no comments exist for the selected product. Comments are shown in reverse chronological order.

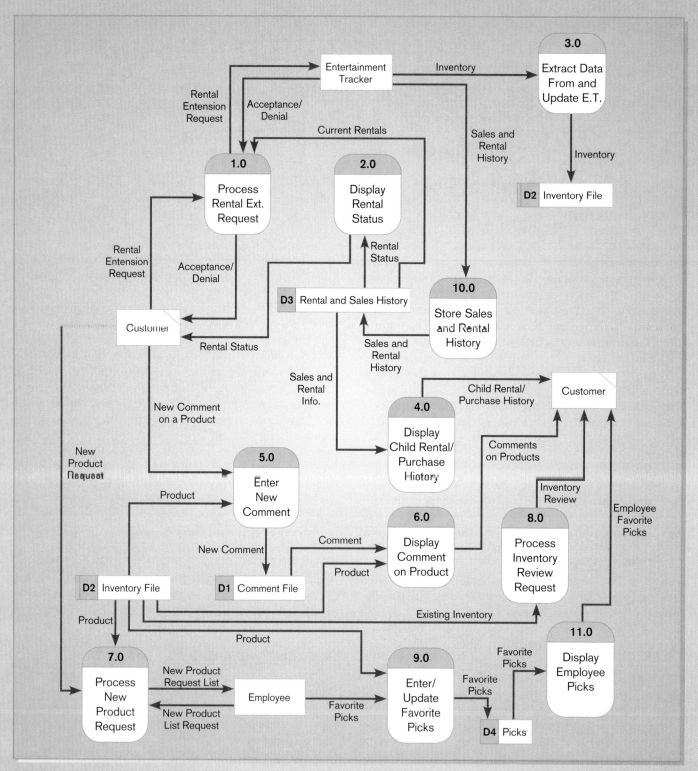

BEC Figure 9-1
Level-0 data flow diagram for MyBroadway

Process 7: Process New Product Request

For this process, the customer enters the type and approximate title, artist, or other information for the desired product. A search for the product is then conducted against the inventory records, similarly to the title search process of Processes 5 and 6, except that all entered parameters are used for the search. The system then displays any products that approximately match the search. If the customer sees a product carried by the BEC store that might match the one being requested, he or she can select that product for a display of the full description and other product information. If this is the desired product, the customer will exit the system or enter a new search for another desired product. If the customer is convinced that none of the found products from the search are the one he or she is requesting, he or she checks a box to place the request. The process then asks for contact information from the customer, so he or she can be notified when the product is stocked. The customer can enter as many requests as desired using this process before exiting the module.

CASE SUMMARY

The logic of system modules is a critical design element. Each logic explanation in a Structured English procedure, a decision table, a decision tree, or another alternative is one step removed from actual computer programming. In many organizations, those describing the processing rules are not the same individuals coding programs. In the case of MyBroadway, the students from Stillwater State University are both documenting the logic and writing the code. But, getting the logic right before coding means that they can concentrate on writing efficient and correct code when they are programming, no longer spending time determining if the logic of their programs meets business requirements.

CASE QUESTIONS

1. Refer to your answer to Case Question 4 from Chapter 8. What differences do you find between your answer to this question and BEC Figure 9-1? Make any changes you deem necessary and produce a new level-0 DFD for MyBroadway if neither your answer to the question in Chapter 8 or BEC Figure 9-1 is sufficient.

2. Process 5 is not necessarily a primitive process. Even though a logic model is developed only for a primitive process, prepare a Structured English description, a decision table, and a decision tree for the logic of this process as outlined in this case. Which type of logic model do you prefer to describe this process? Why?

3. Process 6 is not necessarily a primitive process. Even though a logic model is developed only for a primitive process, prepare a Structured English description, a decision table, and a decision tree for the logic of this process as outlined in this case. Which type of logic model do you prefer to describe this process? Why?

4. Process 7 is not necessarily a primitive process. Even though a logic model is developed only for a primitive process, prepare a Structured English description, a decision table, and a decision tree for the logic of this process as outlined in this case. Which type of logic model do you prefer to describe this process? Why?

5. If, in your opinion, Process 5, 6, or 7 is not a primitive process, develop DFDs to explode one of these nonprimitive processes into a lowest-level explosion. For this one process you chose, redo the associated Structured English from your previous answer into separate Structured English answers for each primitive process on the lowest-level diagrams for that level-0 process.

6. Case Question 8 in Chapter 8 described the Rental Extension Request process, Process 1.0 in Figure 9-1. Now answer that question from Chapter 8 not with a level-1 DFD but with a decision table.

7. Once the Stillwater students develop their process logic models for all the system modules (using any of the techniques developed in this chapter), what should the team do next?

10

Structuring System Requirements:

Conceptual Data Modeling

LEARNING OBJECTIVES

After studying this chapter, you should be able to:

- Concisely define each of the following key data-modeling terms: *entity type*, *attribute*, *multivalued attribute*, *relationship*, *degree*, *cardinality*, *business rule*, *associative entity*, *trigger*, *supertype*, and *subtype*.

- Draw an entity-relationship (E-R) diagram to represent common business situations.

- Explain the role of conceptual data modeling in the overall analysis and design of an information system.

- Distinguish between unary, binary, and ternary relationships, and give an example of each.

- Define four basic types of business rules in an E-R diagram.

- Explain the role of CASE technology in the analysis and documentation of data required in an information system.

- Relate data modeling to process and logic modeling as different views of describing an information system.

INTRODUCTION

In Chapters 8 and 9 you learned how to model and analyze two important views of an information system: (1) the flow of data between manual or automated steps and (2) the decision logic of processing data. None of the techniques discussed so far, however, has concentrated on the data that must be retained in order to support the data flows and processing described. For example, you learned how to show data stores, or *data at rest*, in a data flow diagram. The natural *structure* of data, however, was not shown. Data flow diagrams and various processing logic techniques show *how*, *where*, and *when* data are used or changed in an information system, but these techniques do not show the *definition*, *structure*, and *relationships* within the data. Data modeling develops this missing, and crucial, piece of the description of an information system.

In fact, some systems developers believe that a data model is the most important part of the statement of information system requirements. This belief is based on the following reasons. First, the characteristics of data captured during data modeling are crucial in the design of databases, programs, computer screens, and printed reports. For example, facts such as these—a data element is numeric, a product can be in only one product line at a time, a line item on a customer order can never be moved to another customer order, customer region name is limited to a specified set of values—are all essential pieces of information in ensuring data integrity in an information system.

Second, data rather than processes are the most complex aspects of many modern information systems and hence require a central role in structuring system requirements. Transaction processing systems can have considerable process complexity in validating data, reconciling errors, and coordinating the movement of data to various databases. These types of systems have been in place for years in most organizations; current systems development focuses more on management information systems (such as sales tracking), decision support systems (such as short-term cash investment), and executive support systems (such as product planning). MIS, DSS, and ESS are more data-intensive and require extracting data from various data sources. The exact nature of processing is also more ad hoc than with transaction processing systems, so the details of processing steps cannot be anticipated. Thus, the goal is to provide a rich data resource that might support any type of information inquiry, analysis, and summarization.

Third, the characteristics about data (such as length, format, and relationships with other data) are reasonably permanent. In contrast, the paths of data flow are quite dynamic. Who receives which data, the format of reports, and what reports are used change considerably and constantly over time. A data model explains the inherent nature of the organization, not its transient form. So, an information system design based on a data orientation, rather than a process or logic orientation, should have a longer useful life. Finally, structural information about data is essential for automatic generation of programs. For example, the fact that a customer order has many line items on it instead of just one line item affects the automatic design of a computer screen for entry of customer orders. Thus, although a data model specifically documents the file and database requirements for an information system, the business meaning, or semantics, of data included in the data model have broader impact on the design and construction of a system.

The most common format used for data modeling is entity-relationship (E-R) diagramming. A similar format used with object-oriented analysis and design methods is reviewed in Chapter 20. Data modeling using the E-R notation explains the characteristics and structure of data independent of how the data may be stored in computer memories. A data model using E-R notation is usually developed iteratively. Often IS planners use E-R notation to develop an enterprise-wide data model with very broad categories of data and little detail. Next, during the definition of a project, a specific E-R model is built to help explain the scope of a particular systems analysis and design effort. During requirements structuring, an E-R model represents conceptual data requirements for a particular system. Then, after system inputs and outputs are fully described during logical design, the conceptual E-R data model is refined before it is translated into a logical format (typically a relational data model) from which database definition and physical database design are done. The E-R notation represents certain types of business rules that govern the properties of data. Business rules are important statements of business policies that ideally will be enforced through the database and database management system ultimately used for the application you are designing. Thus, you will use E-R diagramming in many systems development project steps, and most IS project members need to develop and read E-R diagrams. Therefore, mastery of the requirements structuring methods and techniques addressed in this chapter is critical to your success on a systems development project team.

CONCEPTUAL DATA MODELING

Conceptual data model: A detailed model that captures the overall structure of organizational data while being independent of any database management system or other implementation considerations.

A **conceptual data model** is a representation of organizational data. The purpose of a conceptual data model is to show as many rules about the meaning and inter-relationships among data as are possible.

You typically do conceptual data modeling in parallel with other requirements analysis and structuring steps during systems analysis (see Figure 10-1). You collect the explanations of the business necessary for conceptual data modeling from infor-mation-gathering methods like interviewing, questionnaires, and JAD sessions. On larger systems development teams, a subset of the project team concentrates on data modeling while other team members focus attention on process or logic modeling. You develop (or use from prior systems development) a conceptual data model for the current system and build a conceptual data model that supports the scope and requirements for the proposed or enhanced system.

The work of all team members is coordinated and shared through the project dic-tionary or repository. As discussed in Chapter 4, this repository is often maintained by a common CASE tool, but some organizations still use manual documentation. Whether automated or manual, it is essential that the process, logic, and data model descriptions of a system be consistent and complete since each describes different but complemen-tary views of the same information system. For example, the names of data stores on the primitive-level DFDs often correspond to the names of data entities in entity-relation-ship diagrams, and the data elements associated with data flows on DFDs must be accounted for by attributes of entities and relationships in entity-relationship diagrams.

The Process of Conceptual Data Modeling

You typically begin the process of conceptual data modeling by developing a con-ceptual data model for the system being replaced, if a system exists. This is essential for planning the conversion of the current files or database into the database of the

Figure 10-1
Systems development life cycle with analysis phase highlighted

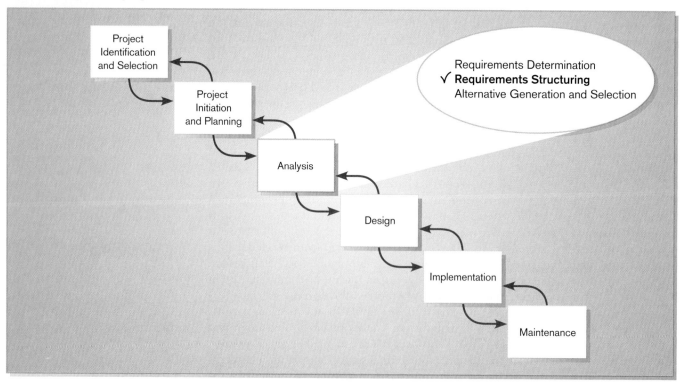

new system. Further, this is a good, but not a perfect, starting point for your understanding of the data requirements of the new system. Then, you build a new conceptual data model that includes all of the data requirements for the new system. You discovered these requirements from the fact-finding methods employed during requirements determination. Today, given the popularity of prototyping and other rapid development methodologies (see Chapter 19), these requirements often evolve through various iterations of a prototype, so the data model is constantly changing.

Conceptual data modeling is one kind of data modeling and database design carried out throughout the systems development process. Figure 10-2 shows the different kinds of data modeling and database design that go on during the whole systems development life cycle. The conceptual data modeling methods we discuss in this chapter are suitable for project identification and selection (including information systems planning that may lead to project identification), project initiation and planning, and analysis phases. These phases of the SDLC address issues of system scope, general requirements, and content—all independent of technical implementation. E-R diagramming is suited for this since E-R diagrams can be translated into a wide variety of technical architectures for data, such as relational, network, and hierarchical. An E-R data model evolves from project identification and selection through analysis as it becomes more specific and is validated by more detailed analysis of system needs.

In the design phase, the final E-R model developed in analysis is matched with designs for systems inputs and outputs and is translated into a format from which physical data storage decisions can be made. After specific data storage architectures are selected, then, in implementation, files and databases are defined as the system is coded. Through the use of the project repository, a field in a physical data record can, for example, be traced back to the conceptual data attribute that represents it on an E-R diagram. Thus, the data modeling and design steps in each of the SDLC phases are linked through the project repository.

Figure 10-2
Relationship between data modeling and the systems development life cycle

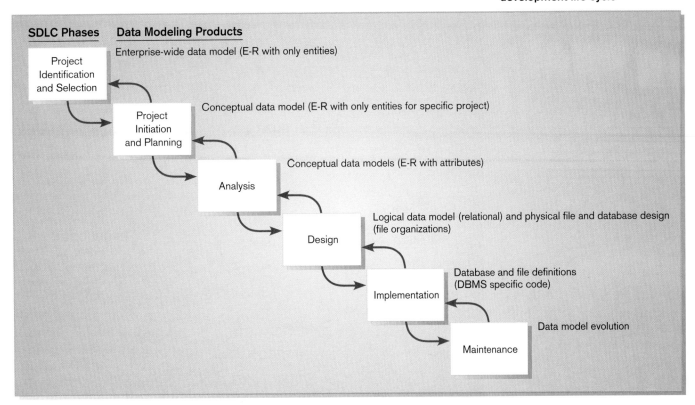

Deliverables and Outcomes

Most organizations today do conceptual data modeling using entity-relationship modeling, which uses a special notation to represent as much meaning about data as possible. Thus, the primary deliverable from the conceptual data modeling step within the analysis phase is an E-R diagram, similar to Figure 10-3. This figure shows the major categories of data (rectangles on the diagram) and the business relationships between them (lines connecting rectangles). For example, Figure 10-3 describes that, for the business represented by this diagram, a SUPPLIER *sometimes* Supplies ITEMs to the company, and an ITEM is *always* Supplied by one to four SUPPLIERS. The fact that a supplier only sometimes supplies items implies that the business wants to keep track of some suppliers without designating what they can supply. This diagram includes two names on each line giving you explicit language to read a relationship in each direction. For simplicity, we will not typically include two names on lines in E-R diagrams in this book; however, this is a standard used in many organizations.

There may be as many as four E-R diagrams produced and analyzed during conceptual data modeling:

1. An E-R diagram that covers just the data needed in the project's application. (This allows you to concentrate on the data requirements of the project's application without being constrained or confused by unnecessary details.)

2. An E-R diagram for the application system being replaced. (Differences between this diagram and the first show what changes you have to make to convert databases to the new application.) This is, of course, not produced if the proposed system supports a completely new business function.

Figure 10-3
Sample conceptual data model diagram

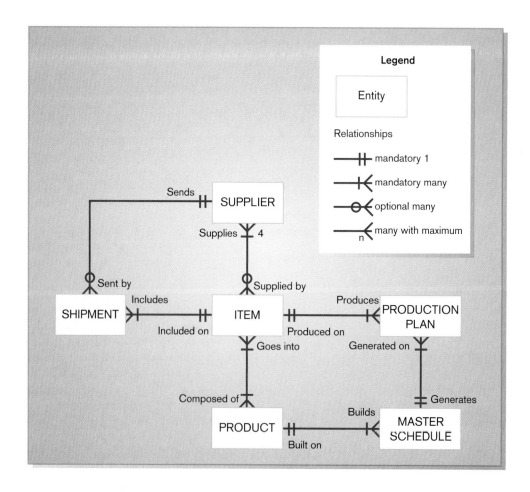

3. An E-R diagram that documents the whole database from which the new application's data is extracted. (Since many applications possibly share the same database or even several databases, this and the first diagram show how the new application shares the contents of more widely used databases.)

4. An E-R diagram for the whole database from which data for the application system being replaced is drawn. (Again, differences between this diagram and the third show what global database changes you have to make to implement the new application.) Even if there is no system being replaced, an understanding of the existing data systems is necessary to see where the new data will fit in or if existing data structures must change to accommodate new data.

The other deliverable from conceptual data modeling is a full set of entries about data objects to be stored in the project dictionary or repository. The repository is the mechanism to link data, process, and logic models of an information system. For example, there are explicit links between a data model and a data flow diagram. Some important links are briefly explained here.

- Data elements included in data flows also appear in the data model and vice versa. You must include in the data model any raw data captured and retained in a data store and a data model can include only data that has been captured or is computed from captured data. Since a data model is a general business picture of data, both manual and automated data stores will be included.

- Each data store in a process model must relate to business objects (what we will call data entities) represented in the data model. For example, in Figure 8-5, the Inventory File data store must correspond to one or several data objects on a data model.

Similar to what was shown in Figure 8-17, you can use an automated repository to verify these linkages.

GATHERING INFORMATION FOR CONCEPTUAL DATA MODELING

Requirements determination methods must include questions and investigations that take a data, not only a process and logic, focus. For example, during interviews with potential system users—during JAD sessions or within requirements questionnaires—you must ask specific questions in order to gain the perspective on data needed to develop a data model. In later sections of this chapter, we will introduce some specific terminology and constructs used in data modeling. Even without this specific data modeling language, you can begin to understand the kinds of questions that must be answered during requirements determination. These questions relate to understanding the rules and policies by which the area supported by the new information system operates. That is, a data model explains what the organization does and what rules govern how work is performed in the organization. You do not, however, need to know how or when data is processed or used to do data modeling.

You typically do data modeling from a combination of perspectives. The first perspective is generally called the *top-down approach*. This perspective derives the business rules for a data model from an intimate understanding of the nature of the business, rather than from any specific information requirements in computer displays, reports, or business forms. There are several very useful sources of typical questions that elicit the business rules needed for data modeling (see Aranow, 1989; Gottesdiener, 1999; and Sandifer and von Halle, 1991a and 1991b). Table 10-1 summarizes a few key questions you should ask system users and business managers so that you can develop an accurate and complete data model. The questions in this

TABLE 10-1   Requirements Determination Questions for Data Modeling

1. *What are the subjects/objects of the business?* What types of people, places, things, materials, events, etc., are used or interact in this business, about which data must be maintained? How many instances of each object might exist?—**data entities and their descriptions**

2. *What unique characteristic (or characteristics) distinguishes each object from other objects of the same type?* Might this distinguishing feature change over time or is it permanent? Might this characteristic of an object be missing even though we know the object exists?—**primary key**

3. *What characteristics describe each object?* On what basis are objects referenced, selected, qualified, sorted, and categorized? What must we know about each object in order to run the business?—**attributes and secondary keys**

4. *How do you use this data?* That is, are you the source of the data for the organization, do you refer to the data, do you modify it, and do you destroy it? Who is not permitted to use this data? Who is responsible for establishing legitimate values for this data?—**security controls and understanding who really knows the meaning of data**

5. *Over what period of time are you interested in this data?* Do you need historical trends, current "snapshot" values, and/or estimates or projections? If a characteristic of an object changes over time, must you know the obsolete values?—**cardinality and time dimensions of data**

6. *Are all instances of each object the same?* That is, are there special kinds of each object that are described or handled differently by the organization? Are some objects summaries or combinations of more detailed objects?—**supertypes, subtypes, and aggregations**

7. *What events occur that imply associations between various objects?* What natural activities or transactions of the business involve handling data about several objects of the same or different type?—**relationships, and their cardinality and degree**

8. *Is each activity or event always handled the same way or are there special circumstances?* Can an event occur with only some of the associated objects, or must all objects be involved? Can the associations between objects change over time (for example, employees change departments)? Are values for data characteristics limited in any way?—**integrity rules, minimum and maximum cardinality, time dimensions of data**

---

table are purposely posed in business terms. In this chapter you will learn the more technical terms included in bold at the end of each set of questions. Of course, these technical terms do not mean much to a business manager, so you must learn how to frame your questions in business terms for your investigation.

You can also gather the information you need for data modeling by reviewing specific business documents—computer displays, reports, and business forms—handled within the system. This process of gaining an understanding of data is often called a *bottom-up approach*. These items will appear as data flows on DFDs and will show the data processed by the system and, hence, probably the data that must be maintained in the system's database. Consider, for example, Figure 10-4, which shows a customer order form used at Pine Valley Furniture. From this form, we determine that the following data must be kept in the database:

PINE VALLEY FURNITURE

ORDER NO	CUSTOMER NO
ORDER DATE	NAME
PROMISED DATE	ADDRESS
PRODUCT NO	CITY-STATE-ZIP
DESCRIPTION	
QUANTITY ORDERED	
UNIT PRICE	

We also see that each order is from one customer, and an order can have multiple line items, each for one product. We will use this kind of understanding of an organization's operation to develop data models.

Figure 10-4
Sample customer order

PVF CUSTOMER ORDER

ORDER NO: 61384   CUSTOMER NO: 1273

NAME: Contemporary Designs
ADDRESS: 123 Oak St.
CITY-STATE-ZIP: Austin, TX 28384

ORDER DATE: 11/04/2001   PROMISED DATE: 11/21/2001

PRODUCT NO	DESCRIPTION	QUANTITY ORDERED	UNIT PRICE
M128	Bookcase	4	200.00
B381	Cabinet	2	150.00
R210	Table	1	500.00

INTRODUCTION TO E-R MODELING

The basic entity-relationship modeling notation (Chen, 1976) uses three main constructs: data entities, relationships, and their associated attributes. The E-R model notation has subsequently been extended to include additional constructs by Chen and others; for example, see Teorey et al. (1986) and Storey (1991). Several different E-R notations exist, and many CASE tools support multiple notations. For simplicity, we have adopted one common notation for this book, the so-called crow's foot notation. If you use another notation in courses or work, you should be able to easily translate between notations.

An **entity-relationship data model** (or E-R model) is a detailed, logical representation of the data for an organization or for a business area. The E-R model is expressed in terms of entities in the business environment, the relationships or associations among those entities, and the attributes or properties of both the entities and their relationships. An E-R model is normally expressed as an **entity-relationship diagram** (or E-R diagram), which is a graphical representation of an E-R model. The notation we will use for E-R diagrams appears in Figure 10-5, and subsequent sections explain this notation.

Entities

An entity (see the first question in Table 10-1) is a person, place, object, event, or concept in the user environment about which the organization wishes to maintain data. An entity has its own identity that distinguishes it from each other entity. Some examples of entities follow:

- Person: EMPLOYEE, STUDENT, PATIENT
- Place: STORE, WAREHOUSE, STATE
- Object: MACHINE, BUILDING, AUTOMOBILE, PRODUCT
- Event: SALE, REGISTRATION, RENEWAL
- Concept: ACCOUNT, COURSE, WORK CENTER

There is an important distinction between entity *types* and entity *instances*. An **entity type** (sometimes called an *entity class*) is a collection of entities that share common properties or characteristics. Each entity type in an E-R model is given a name. Since the name represents a class or set, it is singular. Also, since an entity is an

N E T   S E A R C H
There are several variations on the entity-relationship notation. Visit http://www.prenhall.com/hoffer to complete an exercise related to this topic.

Entity-relationship data model (E-R model): A detailed, logical representation of the entities, associations, and data elements for an organization or business area.

Entity-relationship diagram (E-R diagram): A graphical representation of an E-R model.

Entity type: A collection of entities that share common properties or characteristics.

Figure 10-5
Entity-relationship notation

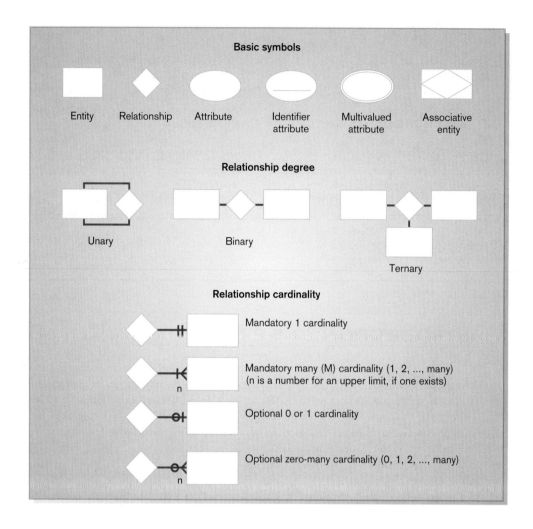

object, we use a simple noun to name an entity type. We use capital letters in naming an entity type and, in an E-R diagram, the name is placed inside a rectangle representing the entity:

Entity instance (instance): A single occurrence of an entity type.

An **entity instance** (or **instance**) is a single occurrence of an entity type. An entity type is described just once in a data model while many instances of that entity type may be represented by data stored in the database. For example, there is one EMPLOYEE entity type in most organizations, but there may be hundreds (or even thousands) of instances of this entity type stored in the database.

A common mistake made when you are just learning to draw E-R diagrams, especially if you already know how to do data flow diagramming, is to confuse data entities with sources/sinks or system outputs and relationships with data flows. A simple rule to avoid such confusion is that a true data entity will have many possible instances, each with a distinguishing characteristic, as well as one or more other descriptive pieces of data. Consider the entity types at the top of page 313 that might be associated with a sorority expense system.

In this situation, the sorority treasurer manages accounts, and records expense transactions against each account. However, do we need to keep track of data about the Treasurer and her supervision of accounts as part of this accounting system? The Treasurer is the person entering data about accounts and expenses and making inquiries about account balances and expense transactions by category. Since there is only one Treasurer, TREASURER data does not need to be kept. On the other hand, if each account has an account manager (for example, a sorority officer) who is responsible for assigned accounts, then we may wish to have an ACCOUNT MANAGER entity type, with pertinent attributes as well as relationships to other entity types.

In this same situation, is an expense report an entity type? Since an expense report is computed from expense transactions and account balances, it is a data flow, not an entity type. Even though there will be multiple instances of expense reports over time, the report contents are already represented by the ACCOUNT and EXPENSE entity types.

Often when we refer to entity types in subsequent sections we will simply say entity. This is common among data modelers. We will clarify that we mean an entity by using the term entity instance.

Naming and Defining Entity Types   Clearly naming and defining data, such as entity types, are important tasks during requirements determination and structuring. When naming and defining entity types, you should use the following guidelines:

- An entity type name is a *singular noun* (such as CUSTOMER, STUDENT, or AUTOMOBILE).
- An entity type name should be *descriptive and specific to the organization*. For example, a PURCHASE ORDER for orders placed with suppliers is distinct from CUSTOMER ORDER for orders placed with us by our customers. Both of these entity types cannot be named ORDER.
- An entity type name should be *concise*; for example, in a university database, using REGISTRATION for the event of a student registering for a class rather than STUDENT REGISTRATION FOR CLASS.
- *Event entity types* should be named for the *result of the event*, not the activity or process of the event. For example, the event of a project manager assigning an employee to work on a project results in an ASSIGNMENT.

There are also some specific guidelines for defining entity types, which follow:

- An entity type definition should include a statement of *what the unique characteristic(s) is (are) for each instance* of the entity type.
- An entity type definition should make it clear *what entity instances are included and not included* in the entity type. For example, "A customer is a person or organization that has placed an order for a product from us or that we have contacted to advertise or promote our products. A customer does not include persons or organizations that buy our products only through our customers, distributors, or agents."
- An entity type definition often includes a description of *when an instance of the entity type is created and deleted.*

- For some entity types, the definition must specify *when an instance might change into an instance of another entity type*; for example, a bid for a construction company becomes a contract once it is accepted.

- For some entity types, the definition must specify *what history is to be kept about entity instances*. Statements about keeping history may have ramifications about how we represent the entity type on an E-R diagram and eventually how we store data for the entity instances.

Attributes

Attribute: A named property or characteristic of an entity that is of interest to the organization.

Each entity type has a set of attributes (see the third question in Table 10-1) associated with it. An **attribute** is a property or characteristic of an entity that is of interest to the organization (relationships may also have attributes, as we will see in the section on Relationships). Following are some typical entity types and associated attributes:

> STUDENT: Student_ID, Student_Name, Home_Address, Phone_Number, Major
>
> AUTOMOBILE: Vehicle_ID, Color, Weight, Horsepower
>
> EMPLOYEE: Employee_ID, Employee_Name, Payroll_Address, Skill

We use an initial capital letter, followed by lowercase letters, and nouns in naming an attribute. In E-R diagrams, we can visually represent an attribute by placing its name in an ellipse with a line connecting it to the associated entity. Many CASE tools, to avoid placing a large number of symbols on a diagram, do not include attributes on an E-R diagram. Rather, the attributes of an entity (sometimes called its composition) are listed in the repository entry for the entity; then, each attribute may be separately defined as another object in the repository. This is similar to how the composition of a data flow is handled in Oracle's Designer (see Figure 8-19).

Naming and Defining Attributes

Often several attributes have approximately the same name and meaning. Thus, it is important to carefully name attributes using guidelines, which follow:

- An attribute name is a *noun* (such as Customer_ID, Age, or Product_Minimum_Price).

- An attribute name should be *unique*. No two attributes of the same entity type may have the same name, and it is desirable, for clarity purposes, that no two attributes across all entity types have the same name.

- To make an attribute name unique and for clarity purposes, *each attribute name should follow a standard format*. For example, your university may establish Student_GPA, as opposed to GPA_of_Student, as an example of the standard format for attribute naming.

- Similar attributes of different entity types should use the similar but distinguishing names; for example, the city of residence for faculty and students should be, respectively, Faculty_Residence_City_Name and Student_Residence_City_Name.

There are also some specific guidelines for defining attributes, which follow:

- An attribute definition states *what the attribute is and possibly why it is important*.

- An attribute definition should make it clear *what is included and not included* in the attribute's value; for example, "Employee_Monthly_Salary_Amount is the amount of money paid each month in the currency of the country of residence of the employee exclusive of any benefits, bonuses, reimbursements, or special payments."

- Any *aliases*, or alternative names, for the attribute can be specified in the definition.

- It may also be desirable to state in the definition *the source of values for the attribute*. Stating the source may make the meaning of the data clearer.

- An attribute definition should indicate *if a value for the attribute is required or optional*. This business rule about an attribute is important for maintaining data integrity.

- An attribute definition may indicate *if a value for the attribute may change* once a value is provided and before the entity instance is deleted. This business rule also controls data integrity.

- An attribute definition may also indicate any *relationships that attribute has with other attributes*; for example, "Employee_Vacation_Days_Number is the number of days of paid vacation for the employee. If the employee has a value of 'Exempt' for Employee_Type, then the maximum value for Employee_Vacation_Days_Number is determined by a formula involving the number of years of service for the employee."

Candidate Keys and Identifiers

Every entity type must have an attribute or set of attributes that distinguishes one instance from other instances of the same type (see the second question in Table 10-1). A **candidate key** is an attribute (or combination of attributes) that uniquely identifies each instance of an entity type. A candidate key for a STUDENT entity type might be Student_ID.

Candidate key: An attribute (or combination of attributes) that uniquely identifies each instance of an entity type.

Sometimes more than one attribute is required to identify a unique entity. For example, consider the entity type GAME for a basketball league. The attribute Team_Name is clearly not a candidate key, since each team plays several games. If each team plays exactly one home game against each other team, then the combination of the attributes Home_Team and Visiting_Team is a candidate key for GAME.

Some entities may have more than one candidate key. One candidate key for EMPLOYEE is Employee_ID; a second is the combination of Employee_Name and Address (assuming that no two employees with the same name live at the same address). If there is more than one candidate key, the designer must choose one of the candidate keys as the identifier. An **identifier** is a candidate key that has been selected to be used as the unique characteristic for an entity type. Bruce (1992) suggests the following criteria for selecting identifiers:

Identifier: A candidate key that has been selected as the unique, identifying characteristic for an entity type.

1. Choose a candidate key that will not change its value over the life of each instance of the entity type. For example, the combination of Employee_Name and Payroll_Address would probably be a poor choice as an identifier for EMPLOYEE because the values of Payroll_Address and Employee_Name could easily change during an employee's term of employment.

2. Choose a candidate key such that, for each instance of the entity, the attribute is guaranteed to have valid values and not be null. To insure valid values, you may have to include special controls in data entry and mainte- nance routines to eliminate the possibility of errors. If the candidate key is a combination of two or more attributes, make sure that all parts of the key will have valid values.

3. Avoid the use of so-called intelligent identifiers, whose structure indicates classifications, locations, and so on. For example, the first two digits of a key for a PART entity may indicate the warehouse location. Such codes are often modified as conditions change, which renders the primary key values invalid.

4. Consider substituting single-attribute surrogate keys for large composite keys. For example, an attribute called Game_ID could be used for the entity GAME instead of the combination of Home_Team and Visiting_Team.

For each entity, the name of the identifier is underlined on an E-R diagram. The following diagram shows the representation for a STUDENT entity type using E-R notation:

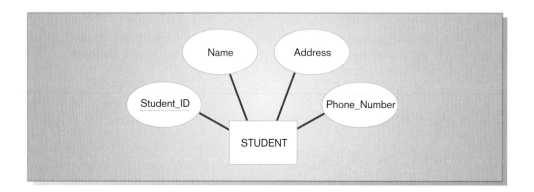

Multivalued Attributes

Multivalued attribute: An attribute that may take on more than one value for each entity instance.

A **multivalued attribute** may take on more than one value *for each entity instance.* Suppose that Skill is one of the attributes of EMPLOYEE. If each employee can have more than one skill, Skill is a multivalued attribute. During conceptual design, it is common to use a special symbol or notation to highlight multivalued attributes. Two ways of showing multivalued attributes are common. The first is to use a double-lined ellipse, so that the EMPLOYEE entity with its attributes is diagrammed as follows:

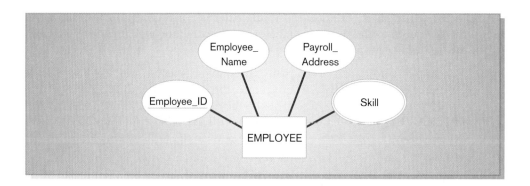

Repeating group: A set of two or more multivalued attributes that are logically related.

The second approach is to separate the repeating data into another entity, called a *weak* (or *attributive*) entity, and then using a relationship (relationships are discussed in the next section), link the weak entity to its associated regular entity. The approach also easily handles several attributes that repeat together, called a **repeating group.** For example, consider again an employee entity with multivalued attributes for data about each employee's dependents. In this situation, data such as dependent name, age, and relation to employee (spouse, child, parent, etc.) are multivalued attributes about an employee, and these attributes repeat together. We could show these multivalued attributes using the ellipse notation as

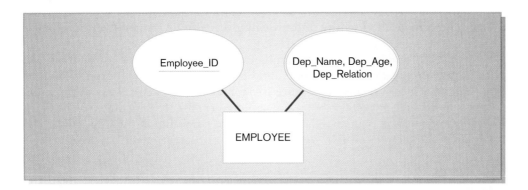

We can show this using a weak entity, DEPENDENT, and a relationship, shown here simply by a line between DEPENDENT and EMPLOYEE. The crow's foot next to DEPENDENT means that there may be many DEPENDENTs for the same EMPLOYEE. Some E-R notations and CASE tools use a special symbol to signify a weak entity. Common notations are a double-line border on the entity box or a mark on the relationship line.

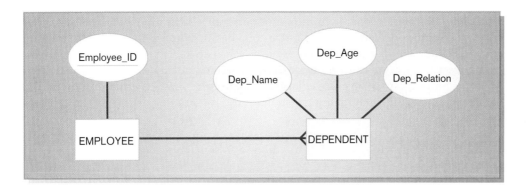

Relationships

Relationships are the glue that hold together the various components of an E-R model (see the fifth, seventh, and eighth questions in Table 10-1). A **relationship** is an association between the instances of one or more entity types that is of interest to the organization. An association usually means that event has occurred or that there exists some natural linkage between entity instances. For this reason, relationships are labeled with verb phrases. For example, a training department in a company is interested in tracking which training courses each of its employees has completed. This leads to a relationship (called Completes) between the EMPLOYEE and COURSE entity types that we diagram as follows:

Relationship: An association between the instances of one or more entity types that is of interest to the organization.

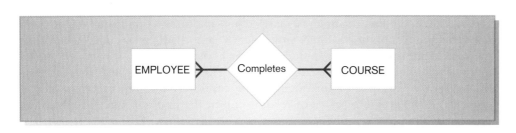

As indicated by the arrows, this is a many-to-many relationship: Each employee may complete more than one course, and each course may be completed by more than one employee. More significantly, we can use the Completes relationship to determine the specific courses that a given employee has completed. Conversely, we can determine the identity of each employee who has completed a particular course.

Many CASE tools, to avoid cluttering an E-R diagram with excess symbols, will not include the relationship diamond and simply place the verb phrase for the relationship name near the line (see Figure 10-3 for an example). We use both representations interchangeably in this book and, as noted earlier, we sometimes use two verb phrases so that there is an explicit name for the relationship in each direction. The standards you will follow will be determined by your organization.

CONCEPTUAL DATA MODELING AND THE E-R MODEL

The last section introduced the fundamentals of the E-R data modeling notation—entities, attributes, and relationships. The goal of conceptual data modeling is to capture as much of the meaning of data as is possible. The more details (business rules) about data that we can model, the better the system we can design and build. Further, if we can include all these details in a CASE repository, and if a CASE tool can generate code for data definitions and programs, then the more we know about data, the more code can be generated automatically. This will make system building more accurate and faster. More importantly, if we can keep a thorough repository of data descriptions, we can regenerate the system as needed as the business rules change. Since maintenance is the largest expense with any information system, the efficiencies gained by maintaining systems at the rule, rather than code, level drastically reduce the cost.

In this section, we explore more advanced concepts needed to more thoroughly model data and learn how the E-R notation represents these concepts.

Degree of a Relationship

Degree: The number of entity types that participate in a relationship.

The **degree** of a relationship (see question 7 in Table 10-1) is the number of entity types that participate in that relationship. Thus, the relationship Completes illustrated on the bottom of page 317 is of degree two, since there are two entity types: EMPLOYEE and COURSE. The three most common relationships in E-R models are *unary* (degree one), *binary* (degree two), and *ternary* (degree three). Higher-degree relationships are possible, but they are rarely encountered in practice, so we restrict our discussion to these three cases. Examples of unary, binary, and ternary relationships appear in Figure 10-6.

Unary relationship (recursive relationship): A relationship between the instances of one entity type.

Unary Relationship   Also called a *recursive relationship*, a **unary relationship** is a relationship between the instances of one entity type. Two examples are shown in Figure 10-6. In the first example, Is_married_to is shown as a one-to-one relationship between instances of the PERSON entity type. That is, each person may be currently married to one other person. In the second example, Manages is shown as a one-to-many relationship between instances of the EMPLOYEE entity type. Using this relationship, we could identify (for example) the employees who report to a particular manager, or reading the Manages relationship in the opposite direction, who the manager is for a given employee.

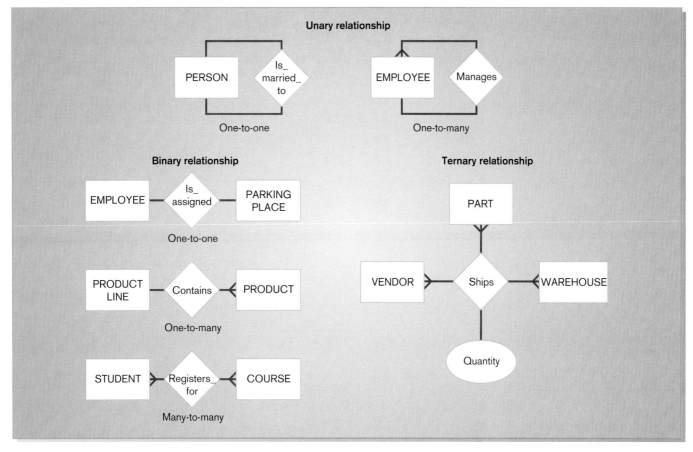

Figure 10-6
Example relationships of different degrees

Figure 10-7 shows an example of another common unary relationship, called a *bill-of-materials structure.* Many manufactured products are made of subassemblies, which in turn are composed of other subassemblies and parts, and so on. As shown in Figure 10-7a, we can represent this structure as a many-to-many unary relationship. In this figure, we use Has_components for the relationship name. The attribute Quantity, which is a property of the relationship, indicates the number of each component that is contained in a given assembly.

Two occurrences of this structure are shown in Figure 10-7b. Each diagram shows the immediate components of each item as well as the quantities of that component. For example, item X consists of item U (quantity 3) and item V (quantity 2). You can easily verify that the associations are in fact many-to-many. Several of the items have more than one component type (for example, item A has three immediate component types: V, X, and Y). Also, some of the components are used in several higher-level assemblies. For example, item X is used in both item A and item B. The many-to-many relationship guarantees, for example, that the same subassembly structure of X is used each time item X goes into making some other item.

Binary Relationship   A **binary relationship** is a relationship between instances of two entity types and is the most common type of relationship encountered in data modeling. Figure 10-6 shows three examples. The first (one-to-one) indicates that an employee is assigned one parking place, and each parking place is assigned to one employee. The second (one-to-many) indicates that a product line may contain sev-

Binary relationship: A relationship between instances of two entity types. This is the most common type of relationship encountered in data modeling.

Figure 10-7
Bill-of-materials unary relationship
(a) Many-to-many relationship

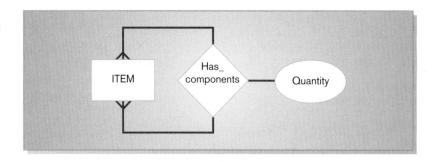

(b) Two instances

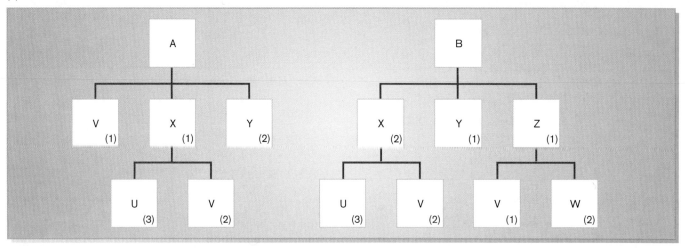

eral products, and each product belongs to only one product line. The third (many-to-many) shows that a student may register for more than one course, and that each course may have many student registrants.

Ternary relationship: A simultaneous relationship among instances of three entity types.

Ternary Relationship   A **ternary relationship** is a *simultaneous* relationship among instances of three entity types. In the example shown in Figure 10-6, the relationship Ships tracks the quantity of a given part that is shipped by a particular vendor to a selected warehouse. Each entity may be a one or a many participant in a ternary relationship (in Figure 10-6, all three entities are many participants).

Note that a ternary relationship is not the same as three binary relationships. For example, Quantity is an attribute of the Ships relationship in Figure 10-6. Quantity cannot be properly associated with any of the three possible binary relationships among the three entity types (such as that between PART and VENDOR) because Quantity is the amount of a particular PART shipped from a particular VENDOR to a particular WAREHOUSE. We strongly recommend that all ternary (and higher) relationships be represented as associative entities (described below). It is easier to draw an associative entity with most CASE tools and easier to understand cardinalities, which we define next.

Cardinalities in Relationships

Cardinality: The number of instances of entity B that can (or must) be associated with each instance of entity A.

Suppose that there are two entity types, A and B, that are connected by a relationship. The **cardinality** of a relationship (see the fifth, seventh, and eighth questions in Table 10-1) is the number of instances of entity B that can (or must) be associated with each instance of entity A. For example, consider the following relationship for video movies:

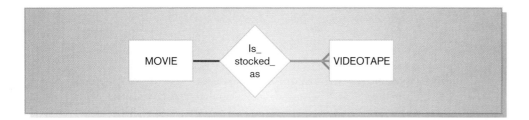

Clearly, a video store may stock more than one videotape of a given movie. In the terminology we have used so far, this example is intuitively a "many" relationship. Yet it is also true that the store may not have a single tape of a particular movie in stock. We need a more precise notation to indicate the *range* of cardinalities for a relationship. This notation was introduced in Figure 10-5, which you may want to review at this point.

Minimum and Maximum Cardinalities   The *minimum* cardinality of a relationship is the minimum number of instances of entity B that may be associated with each instance of entity A. In the preceding example, the minimum number of videotapes available for a movie is zero, in which case we say that VIDEOTAPE is an *optional participant* in the Is_stocked_as relationship. When the minimum cardinality of a relationship is one, then we say entity B is a *mandatory participant* in the relationship. The *maximum* cardinality is the maximum number of instances. For our example, this maximum is "many" (an unspecified number greater than one). Using the notation from Figure 10-5, we diagram this relationship as follows:

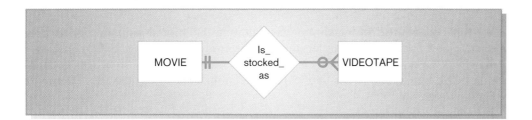

The zero through the line near the VIDEOTAPE entity means a minimum cardinality of zero, while the crow's foot notation means a "many" maximum cardinality.

Examples of three relationships that show all possible combinations of minimum and maximum cardinalities appear in Figure 10-8. A brief description of each relationship follows.

1. PATIENT Has PATIENT HISTORY (Figure 10-8a). Each patient has one or more patient histories (we assume that the initial patient visit is always recorded as an instance of PATIENT HISTORY). Each instance of PATIENT HISTORY "belongs to" exactly one PATIENT.

2. EMPLOYEE Is_assigned_to PROJECT (Figure 10-8b). Each PROJECT has at least one assigned EMPLOYEE (some projects have more than one). Each EMPLOYEE may or (optionally) may not be assigned to any existing PROJECT, or may be assigned to several PROJECTs.

3. PERSON Is_married_to PERSON (Figure 10-8c). This is an optional 0 or 1 cardinality in both directions, since a person may or may not be married.

N E T   S E A R C H
Business rules have become a very important concept in structuring system requirements. Visit http://www.prenhall.com/hoffer to complete an exercise related to this topic.

Figure 10-8
Examples of cardinalities in relationships
(a) Mandatory cardinalities

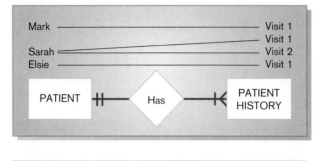

(b) One optional, one mandatory cardinality

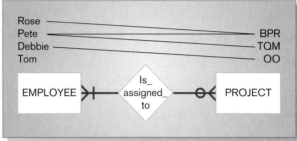

(c) Optional cardinalities

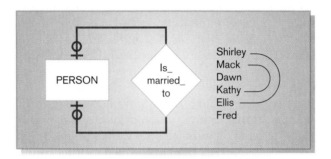

It is possible for the maximum cardinality to be a fixed number, not an arbitrary "many" value. For example, suppose corporate policy states that an employee may work on at most five projects at the same time. We could show this business rule by placing a 5 above or below the crow's foot next to the **PROJECT** entity in Figure 10-8b.

Naming and Defining Relationships

Relationships may be the most difficult component of an E-R diagram to understand. Thus, you should use a few special guidelines for naming relationships, which follow:

- A relationship name is a *verb phrase* (such as Assigned_to, Supplies, or Teaches). Relationships represent actions, usually in the present tense. A relationship name states the action taken, not the result of the action (e.g., use Assigned_to, not Assignment).

- You should *avoid vague names*, such as Has or Is_related_to. Use descriptive verb phrases, often taken from the action verbs found in the definition of the relationship.

There are also some specific guidelines for defining relationships, which follow:

- A relationship definition explains *what action is being taken and possibly why it is important*. It may be important to state who or what does the action, but it is not important to explain how the action is taken.

- It may be important to *give examples to clarify the action.* For example, for a relationship Registered_for between student and course, it may be useful to explain that this covers both on-site and on-line registration, and registrations made during the drop/add period.

- The definition should explain any *optional participation.* You should explain what conditions lead to zero associated instances, whether this can happen only when an entity instance is first created, or whether this can happen at any time.

- A relationship definition should also *explain the reason for any explicit maximum cardinality* other than many.

- A relationship definition should *explain any restrictions on participation in the relationship.* For example, "Supervised_by links an employee with the other employees he or she supervises and links an employee with the other employee who supervises him or her. An employee cannot supervise him- or herself, and an employee cannot supervise other employees if his or her job classification level is below 4."

- A relationship definition should *explain the extent of history that is kept in the relationship.*

- A relationship definition should *explain whether an entity instance* involved in a relationship instance *can transfer participation to another relationship instance.* For example, "Places links a customer with the orders they have placed with our company. An order is not transferable to another customer."

Associative Entities

As seen in the examples of the Ships relationship in Figure 10-6 and the Has_components relationship of Figure 10-7, attributes may be associated with a many-to-many relationship as well as with an entity. For example, suppose that the organization wishes to record the date (month and year) when an employee completes each course. Some sample data follows:

Employee_ID	Course_Name	Date_Completed
549-23-1948	Basic Algebra	March 2000
629-16-8407	Software Quality	June 2000
816-30-0458	Software Quality	Feb 2000
549-23-1948	C Programming	May 2000

From this limited data you can conclude that the attribute Date_Completed is not a property of the entity EMPLOYEE (since a given employee such as 549-23-1948 has completed courses on different dates). Nor is Date_Completed a property of COURSE, since a particular course (such as Software Quality) may be completed on different dates. Instead, Date_Completed is a property of the relationship between EMPLOYEE and COURSE. The attribute is associated with the relationship and diagrammed as follows:

Figure 10-9
Example associative entity

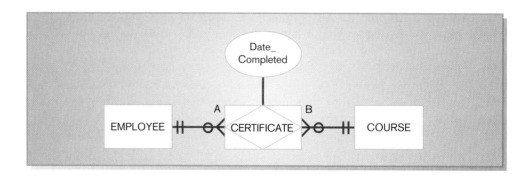

Associative entity: An entity type that associates the instances of one or more entity types and contains attributes that are peculiar to the relationship between those entity instances; also called a gerund.

Since many-to-many and one-to-one relationships may have associated attributes, the E-R data model poses an interesting dilemma: Is a many-to-many relationship actually an entity in disguise? Often the distinction between entity and relationship is simply a matter of how you view the data. An **associative entity** (sometimes called a *gerund*) is a relationship that the data modeler chooses to model as an entity type. Figure 10-9 shows the E-R notation for representing the Completes relationship as an associative entity. The diamond symbol is included within the entity rectangle as a reminder that the entity was derived from a relationship. The lines from CERTIFI-CATE to the two entities are not two separate binary relationships, so they do not have labels. Note that EMPLOYEE and COURSE have mandatory one cardinality since an instance of Completes must have an associated EMPLOYEE and COURSE. The labels A and B show where the cardinalities from the Completes relation now appear. The implicit identifier of Completes is the combination of the identifiers of EMPLOYEE and COURSE, Employee_ID and Course_Name, respectively.

An example of the use of an associative entity for a ternary relationship appears in Figure 10-10. This figure shows an alternative (and equally correct) representation of the ternary Ships relationship shown in Figure 10-6. In Figure 10-10, the entity type (associative entity) SHIPMENT replaces the Ships relationship from Figure 10-6. Each instance of SHIPMENT represents a real-world shipment by a given vendor of a particular part to a selected warehouse. The Quantity of that shipment is an attribute of SHIPMENT. A Shipment_ID is assigned to each shipment and is the identifier of SHIPMENT, as shown in Figure 10-10.

One situation in which a relationship *must* be turned into an associative entity is when the associative entity has other relationships with entities besides the relation-

Figure 10-10
SHIPMENT entity type (an associative entity)

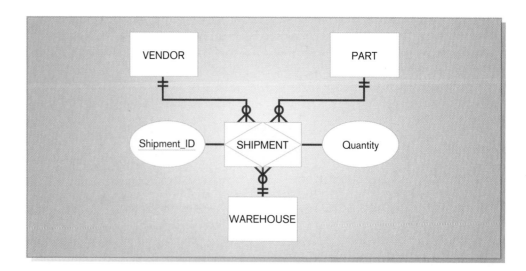

ship that caused its creation. For example, consider the following E-R model that represents price quotes from different vendors for purchased parts stocked by Pine Valley Furniture:

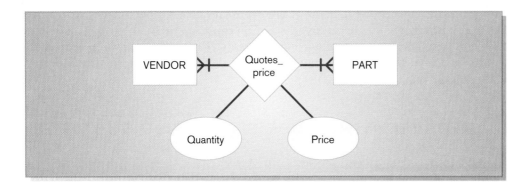

Now, suppose that we also need to know which price quote is in effect for each part shipment received. This additional data requirement *necessitates* that the Quotes_price relationship be transformed into an associative entity as follows:

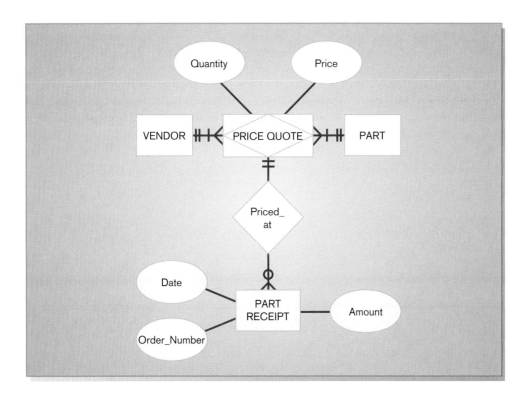

In this case, PRICE QUOTE is not a ternary relationship. Rather, PRICE QUOTE is a binary many-to-many relationship (associative entity) between VENDOR and PART. In addition, each PART RECEIPT, based on Amount, has an applicable, negotiated Price. Each PART RECEIPT is for a given PART from a specific VENDOR, and the Amount of the receipt dictates the purchase price in effect by matching with the Quantity attribute. Since the PRICE QUOTE pertains to a given PART and given VENDOR, PART RECEIPT does not need direct relationships with these entities.

Summary of Conceptual Data Modeling with E-R Diagrams

The purpose of E-R diagramming is to capture the richest possible understanding of the meaning of data necessary for an information system or organization. Besides the aspects shown in this chapter, there are many other semantics about data that E-R diagramming can represent. Some of these more advanced capabilities are explained in Hoffer, Prescott, and McFadden (2002). You can also find some general guidelines for effective conceptual data modeling in Moody (1996). The following section presents one final aspect of conceptual data modeling: capturing the relationship between similar entity types.

REPRESENTING SUPERTYPES AND SUBTYPES

Often two or more entity types seem very similar (maybe they have almost the same name), but there are a few differences. That is, these entity types share common properties but also have one or more distinct attributes or relationships. To address this situation, the E-R model has been extended to include supertype/subtype relationships. A **subtype** is a subgrouping of the entities in an entity type that is meaningful to the organization. For example, STUDENT is an entity type in a university. Two subtypes of STUDENT are GRADUATE STUDENT and UNDERGRADUATE STUDENT. A **supertype** is a generic entity type that has a relationship with one or more subtypes.

An example illustrating the basic notation used for supertype/subtype relationships appears in Figure 10-11. The supertype PATIENT is connected with a line to a circle, which in turn is connected by a line to each of the two subtypes, OUTPATIENT and RESIDENT PATIENT. The U-shaped symbol indicates that the subtype is a subset of the supertype. Attributes that are shared by all patients (including the identifier) are associated with the supertype; attributes that are unique to a particular subtype (for example, Checkback_Date for OUTPATIENT) are associated with that subtype. Relationships in which all types of patients participate (Is_cared_for) is associated with the supertype; relationships in which only a subtype participates (Is_assigned for RESIDENT PATIENTs) is associated only with the relevant subtype.

Subtype: A subgrouping of the entities in an entity type that is meaningful to the organization and that shares common attributes or relationships distinct from other subgroupings.

Supertype: A generic entity type that has a relationship with one or more subtypes.

Figure 10-11
Notation for supertype/subtype relationship

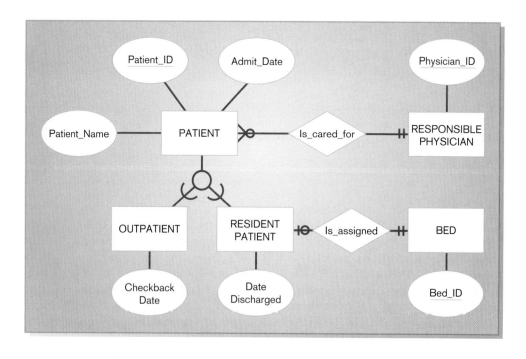

There are several important business rules for supertype/subtype relationships. The **total specialization rule** specifies that each entity instance of the supertype *must* be a member of some subtype in the relationship. The **partial specialization rule** specifies that an entity instance of the supertype is allowed not to belong to any subtype. Total specialization is shown on an E-R diagram by a double line from the supertype to the circle, and partial specialization is shown by a single line. The **disjoint rule** specifies that if an entity instance of the supertype is a member of one subtype, it cannot simultaneously be a member of any other subtype. The **overlap rule** specifies that an entity instance can simultaneously be a member of two (or more) subtypes. Disjoint versus overlap is shown by a "d" or an "o" in the circle.

Figure 10-12 illustrates several combinations of these rules for a hierarchy of supertypes and subtypes in a university database. In this example,

- a PERSON must be (total specialization) an EMPLOYEE, an ALUMNUS, or a STUDENT or any combination of these subtypes (overlap)
- an EMPLOYEE must be FACULTY or a STAFF (disjoint) or may be just an EMPLOYEE (partial specialization)
- a STUDENT can be only a GRADUATE STUDENT or an UNDERGRADUATE STUDENT, and nothing else (total specialization and disjoint)

Total specialization rule: Specifies that each entity instance of the supertype *must* be a member of some subtype in the relationship.

Partial specialization rule: Specifies that an entity instance of the supertype is allowed not to belong to any subtype.

Disjoint rule: Specifies that if an entity instance of the supertype is a member of one subtype, it cannot simultaneously be a member of any other subtype.

Overlap rule: Specifies that an entity instance can simultaneously be a member of two (or more) subtypes.

Figure 10-12
Example of supertype/subtype hierarchy

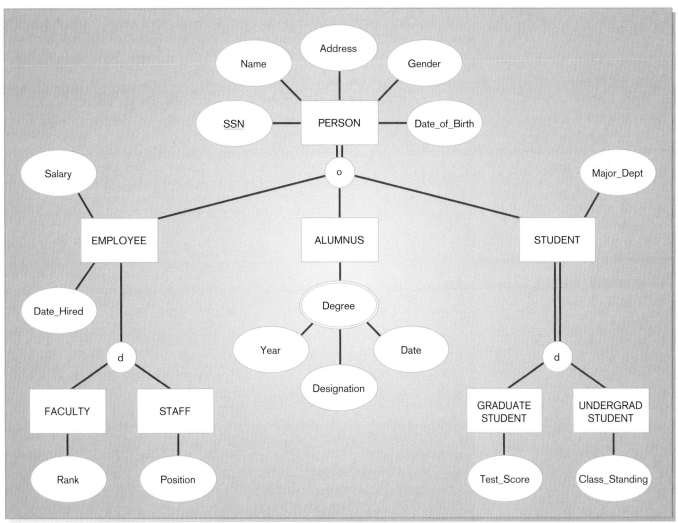

BUSINESS RULES

Business rules: Specifications that preserve the integrity of a conceptual or logical data model.

Conceptual data modeling is a step-by-step process for documenting information requirements that is concerned with both the structure of data and with rules about the integrity of that data (see the eighth question in Table 10-1). **Business rules** are specifications that preserve the integrity of the logical data model. Four basic types of business rules are:

1. *Entity integrity.* Each instance of an entity type must have a unique identifier that is not null.

2. *Referential integrity constraints.* Rules concerning the relationships between entity types.

3. *Domains.* Constraints on valid values for attributes.

4. *Triggering operations.* Other business rules that protect the validity of attribute values.

The entity-relationship model that we have described in this chapter is concerned primarily with the structure of data rather than with expressing business rules (although some elementary rules are implied in the E-R model). Generally the business rules are captured during requirements determination and stored in the CASE repository (described in Chapter 4) as they are documented. Entity integrity was described earlier in this chapter and referential integrity is described in Chapter 12 since it applies to database design. In this section, we briefly describe two types of rules: domains and triggering operations. These rules are illustrated with a simple example from a banking environment, shown in Figure 10-13a. In this example, an ACCOUNT entity has a relationship (Is_for) with a WITHDRAWAL entity.

Figure 10-13
Examples of business rules
(a) Simple banking relationship

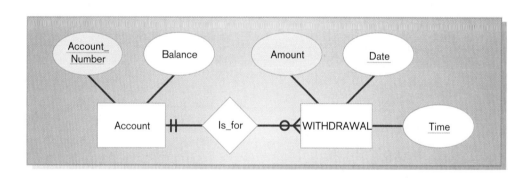

(b) Typical domain definitions

Name: Account_Number
Meaning: Customer account number in bank
Data type: Character
Format: nnn-nnnn
Uniqueness: Must be unique
Null support: Non-null

Name: Amount
Meaning: Dollar amount of transaction
Data type: Numeric
Format: 2 decimal places
Range: 0-10,000
Uniqueness: Nonunique
Null support: Non-null

(c) Typical triggering operation

User rule: WITHDRAWAL Amount may not exceed ACCOUNT Balance
Event: Insert
Entity Name: WITHDRAWAL
Condition: WITHDRAWAL Amount > ACCOUNT Balance
Action: Reject the insert transaction

Domains

A **domain** is the set of all data types and ranges of values that attributes may assume (Fleming and von Halle, 1990). Domain definitions typically specify some (or all) of the following characteristics of attributes: data type, length, format, range, allowable values, meaning, uniqueness, and null support (whether an attribute value may or may not be null).

Domain: The set of all data types and values that an attribute can assume.

Figure 10-13b shows two domain definitions for the banking example. The first definition is for Account_Number. Since Account_Number is an identifier attribute, the definition specifies that Account_Number must be unique and also must not be null (these specifications are true of all identifiers). The definition specifies that the attribute data type is character and that the format is nnn-nnnn. Thus any attempt to enter a value for this attribute that does not conform to its character type or format will be rejected, and an error message will be displayed.

The domain definition for the Amount attribute (dollar amount of the requested withdrawal) also may not be null, but is not unique. The format allows for two decimal places to accommodate a currency field. The range of values has a lower limit of zero (to prevent negative values) and an upper limit of 10,000. The latter is an arbitrary upper limit for a single withdrawal transaction.

The use of domains offers several advantages:

1. Domains verify that the values for an attribute (stored by insert or update operations) are valid.

2. Domains ensure that various data manipulation operations (such as joins or unions in a relational database system) are logical.

3. Domains help conserve effort in describing attribute characteristics.

Domains can conserve effort because we can define domains and then associate each attribute in the data model with an appropriate domain. To illustrate, suppose that a bank has three types of accounts, with the following identifiers:

Account Type	Identifier
CHECKING	Checking_Account_Number
SAVINGS	Savings_Account_Number
LOAN	Loan_Account_Number

If domains are not used, the characteristics for each of the three identifier attributes must be described separately. Suppose, however, that the characteristics for all three of the attributes are identical. Having defined the domain Account_Number once (as shown in Figure 10-13b), we simply associate each of these three attributes with Account_Number. Other common domains such as Date, Social_Security_Number, and Telephone_Number also need to be defined just once in the model.

Triggering Operations

A **triggering operation** (or **trigger**) is an assertion or rule that governs the validity of data manipulation operations such as insert, update, and delete. The scope of triggering operations may be limited to attributes within one entity, or it may extend to attributes in two or more entities. Complex business rules may often be stated as triggering operations.

Triggering operation (trigger): An assertion or rule that governs the validity of data manipulation operations such as insert, update, and delete.

A triggering operation normally includes the following components:

1. *User rule:* a concise statement of the business rule to be enforced by the triggering operation

2. *Event:* the data manipulation operation (insert, delete, or update) that initiates the operation

3. *Entity name:* the name of the entity being accessed and/or modified
4. *Condition:* condition that causes the operation to be triggered
5. *Action:* action taken when the operation is triggered

Figure 10-13c shows an example of a triggering operation for the banking situation. The business rule is a simple (and familiar) one: The amount of an attempted withdrawal may not exceed the current account balance. The event of interest is an attempted insert of an instance of the WITHDRAWAL entity type (perhaps from an automated teller machine). The condition is

Amount (of the withdrawal) > ACCOUNT Balance

When this condition is triggered, the action taken is to reject the transaction. You should note two things about this triggering operation: First, it spans two entity types; second, the business rule could not be enforced through the use of domains.

The use of triggering operations is an increasingly important component of database strategy. With triggering operations, the responsibility for data integrity lies within the scope of the database management system rather than with application programs or human operators. In the banking example, tellers could conceivably check the account balance before processing each withdrawal. Human operators would be subject to human error and, in any event, manual processing would not work with automated teller machines. Alternatively, the logic of integrity checks could be built into the appropriate application programs, but integrity checks would require duplicating the logic in each program. There is no assurance that the logic would be consistent (since the application programs may have been developed at different times by different people) or that the application programs will be kept up to date as conditions change.

As stated earlier, business rules should be documented in the CASE repository. Ideally, these rules will then be checked automatically by database software. Removing business rules from application programs and incorporating them in the repository (in the form of domains, referential integrity constraints, and triggering operations) has several important advantages:

1. Provides faster application development with fewer errors, since these rules can be generated into programs or enforced by the DBMS
2. Reduces maintenance effort and expenditures
3. Provides faster response to business changes
4. Facilitates end-user involvement in developing new systems and manipulating data
5. Provides for consistent application of integrity constraints
6. Reduces time and effort required to train application programmers
7. Promotes ease of use of a database

For a more thorough treatment of business rules, see Hoffer, Prescott, and McFadden (2002).

THE ROLE OF CASE IN CONCEPTUAL DATA MODELING

CASE tools provide two important functions in conceptual data modeling: (1) maintaining E-R diagrams as a visual depiction of structured data requirements and (2) linking objects on E-R diagrams to corresponding descriptions in a repository. Most CASE tools support one or more of several standard E-R diagramming notations, such as the crow's foot notation used in this text. Many tools do not support drawing ternary or higher relationships, so you may have to model these higher-

Figure 10-14
Typical conceptual data model
elements in a project dictionary

Entity (major category of data)

Name	A short and a long name that uniquely label the entity
Description	Explanation so that it is clear what objects are covered by this entity
Alias	Alternative names used for this entity (that is, synonyms)
Primary key	Name(s) attribute(s) that form the unique identifier for each instance of this entity
Attributes and repetition	List of attributes associated with this entity and the number of instances of each attribute for each entity instance
Abstraction	Indication of any superclasses or subclasses or composition of entity types involving this entity

Attribute (entity characteristic)

Name	A short and a long name that uniquely label the attribute
Description	Explanation of the attribute so that its meaning is clearly different from all other attributes
Alias	Alternative names used for this attribute (that is, synonyms)
Domain	The permitted values that this attribute may assume
Computation	If this is not raw data, the formula or method to calculate the attribute's value
Aggregation	Indication of any groupings of attributes involving this attribute (e.g., a month attribute as part of a date attribute)

Relationship (association between entity instances)

Name	A short and a long name that uniquely label the relationship
Description	Explanation of the relationship so that its meaning is clearly different from all other relationships
Degree	Names of entities involved in the relationship
Cardinality	The potential number of instances of each entity involved in the relationship
Insertion rules	Business rules that control the inclusion of entity instances in this relationship
Deletion rules	Business rules that control the elimination of entity instances from this relationship

degree relationships as several binary relationships, even though this is not semantically correct.

Figure 10-14 lists the typical data model elements that are placed in the project dictionary or CASE repository during conceptual data modeling. A CASE tool will typically allow you to move directly to the repository entry for an object once you select it on an E-R diagram. The precise list of object characteristics will vary by CASE tool or the standard you use for a project dictionary. Figure 10-14, however, provides a basic set of repository contents used in almost any circumstance. Later data modeling and database design steps develop additional data model elements for the repository, as we will show in subsequent chapters.

AN EXAMPLE OF CONCEPTUAL DATA MODELING AT HOOSIER BURGER

Chapters 8 and 9 structured the process and logic requirements for a new inventory control system for Hoosier Burger. The data flow diagram and decision table (repeated here as Figures 10-15 and 10-16) describe requirements for this new system. The purpose of this system is to monitor and report changes in raw material inventory levels, and to issue material orders and payments to suppliers. Thus, the central data entity for this system will be an INVENTORY ITEM, corresponding to data store D1 in Figure 10-15.

Changes in inventory levels are due to two types of transactions: receipt of new items from suppliers and consumption of items from sales of products. Inventory is added upon receipt of new raw materials, for which Hoosier Burger receives a supplier INVOICE (see Process 1.0 in Figure 10-15). Each INVOICE indicates that the supplier has sent a specific quantity of one or more INVOICE ITEMs, which correspond to Hoosier's INVENTORY ITEMs. Inventory is used when customers order

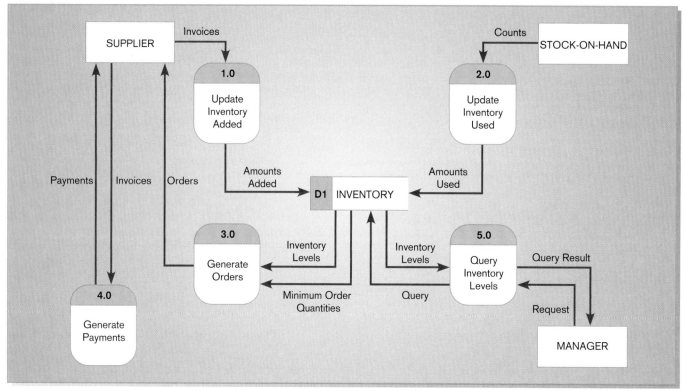

Figure 10-15
Level-0 data flow diagram for Hoosier Burger's new logical inventory control system (same as Figure 8-16)

and pay for PRODUCTs. That is, Hoosier makes a SALE for one or more ITEM SALEs, each of which corresponds to a food PRODUCT. Since the real-time customer order processing system is separate from the inventory control system, a source, STOCK ON-HAND on Figure 10-15, represents how data flows from the order processing to the inventory control system. Finally, since food PRODUCTs are made up of various INVENTORY ITEMs (and vice versa), Hoosier maintains a RECIPE to indicate how much of each INVENTORY ITEM goes into making one PRODUCT. From this discussion, we have identified the data entities required in a data model for the new Hoosier Burger inventory control system: INVENTORY ITEM, INVOICE, INVOICE ITEM, PRODUCT, SALE, ITEM SALE, and RECIPE. To

Figure 10-16
Reduced decision table for Hoosier Burger's inventory reordering (Same as Figure 9-7)

Conditions/ Courses of Action	Rules						
	1	2	3	4	5	6	7
Type of item	P	P	P	P	P	P	N
Time of week	D	W	D	W	D	W	–
Season of year	A	A	S	S	H	H	–
Standing daily order	X		X		X		
Standing weekend order		X		X		X	
Minimum order quantity							X
Holiday reduction					X	X	
Summer reduction			X	X			

complete the data model, we must determine a necessary relationship between these entities as well as attributes for each entity.

The wording in the previous description tells us much of what we need to know to determine relationships:

- An INVOICE includes one or more INVOICE ITEMs, each of which corresponds to an INVENTORY ITEM. Obviously, an INVOICE ITEM cannot exist without an associated INVOICE, and over time there will be zero-to-many receipts, or INVOICE ITEMs, for an INVENTORY ITEM.

- Each PRODUCT has a RECIPE of INVENTORY ITEMs. Thus, RECIPE is an associative entity supporting a bill-of-materials type relationship between PRODUCT and INVENTORY ITEM.

- A SALE indicates that Hoosier sells one or more ITEM SALEs, each of which corresponds to a PRODUCT. An ITEM SALE cannot exist without an associated SALE, and over time there will be zero-to-many ITEM SALEs for a PRODUCT.

Figure 10-17 shows an E-R diagram with the entities and relationships described above. We include on this diagram two labels for each relationship, one to be read in either relationship direction (e.g., an INVOICE Includes one-to-many INVOICE ITEMs, and an INVOICE ITEM Is_included_on exactly one INVOICE). RECIPE, since it is an associative entity, also serves as the label for the many-to-many relationship between PRODUCT and INVENTORY ITEM. Now that we understand the entities and relationships, we must decide which data elements are associated with the entities and gerunds in this diagram.

You may wonder at this point why only the INVENTORY data store is shown in Figure 10-15 when there are seven entities and associative entities on the E-R diagram. The INVENTORY data store corresponds to the INVENTORY ITEM entity in Figure 10-17. The other entities are hidden inside other processes for which we have not shown lower-level diagrams. In actual requirements structuring steps, you would have to match all entities with data stores: Each data store represents some subset of an E-R diagram, and each entity is included in one or more data stores. Ideally, each data store on a primitive DFD will be an individual entity.

To determine data elements for an entity, we investigate data flows in and out of data stores that correspond to the data entity, and supplement this with a study of decision logic and temporal logic that use or change data about the entity. There are six data flows associated with the INVENTORY data store in Figure 10-15. The descrip-

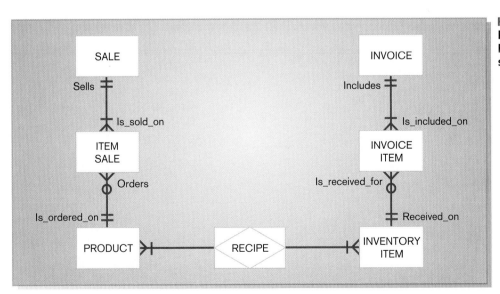

Figure 10-17
Preliminary E-R diagram for Hoosier Burger's inventory control system

tion of each data flow in the project dictionary or CASE repository would include the data flow's composition, which then tells us what data are flowing in or out of the data store. For example, the Amounts Used data flow coming from Process 2.0 indicates how much to decrement an attribute Quantity_in_Stock due to use of the INVENTORY ITEM to fulfill a customer sale. Thus, the Amounts Used data flow implies that Process 2.0 will first read the relevant INVENTORY ITEM record, then update its Quantity_in_Stock attribute, and finally store the updated value in the record. Structured English for Process 2.0 would depict this logic. Each data flow would be analyzed similarly (space does not permit us to show the analysis for each data flow).

The analysis of data flows for data elements is supplemented by a study of decision logic. For example, consider the decision table of Figure 10-16. One condition used to determine the process of reordering an INVENTORY ITEM involves the Type_of_Item. Thus, Process 3.0 in Figure 10-15 (to which this decision table relates) needs to know this characteristic of each INVENTORY ITEM, so this identifies another attribute of this entity.

Although we do not illustrate a state-transition diagram for this system, the analysis of such a chart could also reveal additional data requirements. For example, a state-transition diagram on an inventory item might show states of below reorder point, above reorder point, and projected above reorder point. This last state occurs when an invoice is received for a new shipment but before the shipment's quantity is verified. Such a state could imply the need for a new attribute on an INVENTORY ITEM to specify whether the Quantity_in_Stock is an actual value or an estimate.

After having considered all data flows in and out of data stores related to data entities, plus all decision and temporal logic related to inventory control, we derive the full E-R data model, with attributes, shown in Figure 10-18.

Figure 10-18
Final E-R diagram for Hoosier Burger's inventory control system

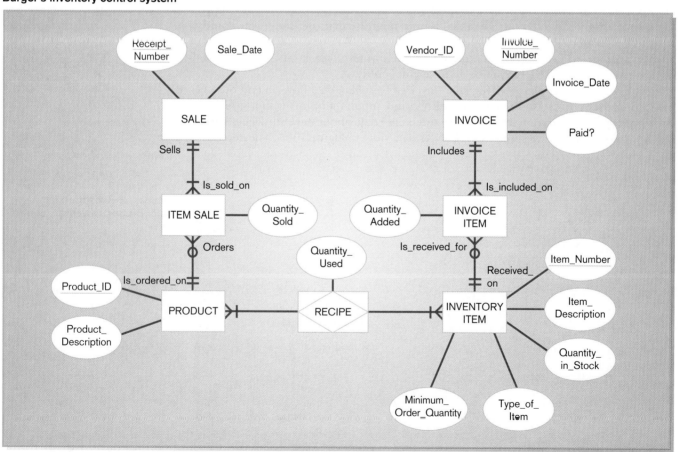

INTERNET DEVELOPMENT: CONCEPTUAL DATA MODELING

Conceptual data modeling for an Internet-based electronic commerce application is no different than the process followed when analyzing the data needs for other types of applications. In the preceding chapters, you read how Jim analyzed the flow of information within the WebStore and developed a data flow diagram, and you studied decision-making logic within the WebStore and developed Structured English. In this section, we examine the process he followed when developing the WebStore's conceptual data model.

Conceptual Data Modeling for Pine Valley Furniture's WebStore

To better understand what data would be needed within the WebStore, Jim Woo carefully reviewed the information from the JAD session and his previously developed data flow diagram. Table 10-2 shows a summary of the customer and inventory information identified during the JAD session. He wasn't sure if this information was complete, but he knew that it was a good starting place for identifying what information the WebStore needed to capture, store, and process. To identify additional information, he carefully studied the DFD shown in Figure 10-19. In this diagram, two data stores—Inventory and Shopping Cart—are clearly identified; both were strong candidates to become entities within the conceptual data model. Finally, Jim examined the data flows from the DFD as additional possible sources for entities. This analysis resulted in the identification of five general categories of information that he needed to consider:

- Customer
- Inventory
- Order
- Shopping Cart
- Temporary User/System Messages

After identifying these multiple categories of data, his next step was to carefully define each item. To do this, he again examined all data flows within the DFD and recorded the source and destination of all data flows. By carefully listing these flows, he could more easily move through the DFD and more thoroughly understand what information needed to move from point-to-point. This activity resulted in the creation of two tables that documented his growing understanding of the WebStore's

TABLE 10-2   Customer and Inventory Information for WebStore

Corporate Customer	Home Office Customer	Student Customer	Inventory Information
Company name	Name	Name	SKU
Company address	Doing business as	School	Name
Company phone	(company's name)	Address	Description
Company fax	Address	Phone	Finished product size
Company preferred	Phone	E-mail	Finished product weight
shipping method	Fax		Available materials
Buyer name	E-mail		Available colors
Buyer phone			Price
Buyer e-mail			Lead time

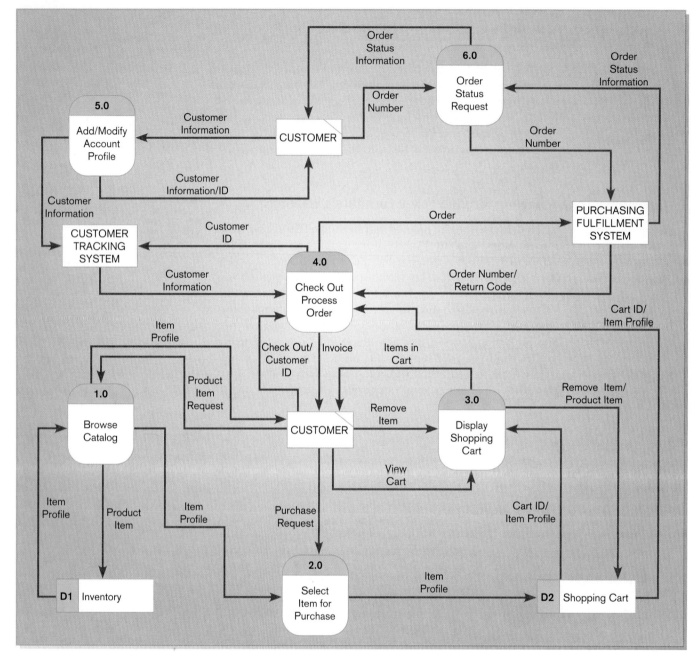

Figure 10-19
Level-0 data flow diagram
for the Webstore

requirements. The first, Table 10-3, lists each of the data flows within each data category and its corresponding description. The second, Table 10-4, lists each of the unique data flows within each data category. He now felt ready to construct an entity-relationship diagram for the WebStore.

He concluded that Customer, Inventory, and Order were each a unique entity and would be part of his E-R diagram. Recall that an entity is a person, place, or object; all three of these items meet this criteria. Because the Temporary User/System Messages data were not permanently stored items—nor were they a person, place, or object—he concluded that this should not be an entity in the conceptual data model. Alternatively, although the shopping cart was also a temporarily stored item, its contents needed to be stored for at least the duration of a customer's visit to the WebStore and should be considered an object. As shown in Figure 10-19,

TABLE 10-3   Data Category, Data Flow, and Data Flow Descriptions for the WebStore DFD

Data Category Data Flow	Description
Customer Related	
Customer ID	Unique identifier for each customer (generated by Customer Tracking System)
Customer Information	Detailed customer information (stored in Customer Tracking System)
Inventory Related	
Product Item	Unique identifier for each product item (stored in Inventory Database)
Item Profile	Detailed product information (stored in Inventory Database)
Order Related	
Order Number	Unique identifier for an order (generated by Purchasing Fulfillment System)
Order	Detailed order information (stored in Purchasing Fulfillment System)
Return Code	Unique code for processing customer returns (generated by/ stored in Purchasing Fulfillment System)
Invoice	Detailed order summary statement (generated from order information stored in Purchasing Fulfillment System)
Order Status Information	Detailed summary information on order status (stored/ generated by)
Shopping Cart	
Cart ID	Unique identifier for shopping cart
Temporary User/System Messages	
Product Item Request	Request to view information on a catalog item
Purchase Request	Request to move an item into the shopping cart
View Cart	Request to view the contents of the shopping cart
Items in Cart	Summary report of all shopping cart items
Remove Item	Request to remove item from shopping cart
Check Out	Request to check out and process order

Process 4.0, Check Out Process Order, moves the Shopping Cart contents to the Purchasing Fulfillment System, where the order details are stored. Thus, he concluded that Shopping Cart—along with Customer, Inventory, and Order—would be entities in his E-R diagram.

The final step was to identify the interrelationships between these four entities. After carefully studying all the related information, he concluded the following:

1. Each Customer *owns* zero-or-one Shopping Cart instances; each Shopping Cart instance *is owned by* one-and-only-one Customer.

2. Each Shopping Cart instance *contains* one-and-only-one Inventory item; each Inventory item *is contained in* zero-or-many Shopping Cart instances.

3. Each Customer *places* zero-to-many Orders; each Order *is placed by* one-and-only-one Customer.

4. Each Order *contains* one-to-many Shopping Cart instances; each Shopping Cart instance *is contained in* one-and-only-one Order.

With these relationships defined, Jim drew the E-R diagram shown in Figure 10-20. He now had a very good understanding of the requirements, the flow of information within the WebStore, the flow of information between the WebStore and existing PVF systems, and now the conceptual data model. Over the next few hours, Jim planned to further refine his understanding by listing the specific attributes for each

TABLE 10-4   Data Category, Data Flow, and the Source/Destination of Data Flows Within the WebStore DFD

Data Flow	From/To
Customer Related	
Customer ID	From Customer to Process 4.0
	From Process 4.0 to Customer Tracking System
	From Process 5.0 to Customer
Customer Information	From Customer to Process 5.0
	From Process 5.0 to Customer
	From Process 5.0 to Customer Tracking System
	From Customer Tracking System to Process 4.0
Inventory Related	
Product Item	From Process 1.0 to Data Store D1
	From Process 3.0 to Data Store D2
Item Profile	From Data Store D1 to Process 1.0
	From Process 1.0 to Customer
	From Process 1.0 to Process 2.0
	From Process 2.0 to Data Store D2
	From Data Store D2 to Process 3.0
	From Data Store D2 to Process 4.0
Order Related	
Order Number	From Purchasing Fulfillment System to Process 4.0
	From Customer to Process 6.0
	From Process 6.0 to Purchasing Fulfillment System
Order	From Process 4.0 to Purchasing Fulfillment System
Return Code	From Purchasing Fulfillment System to Process 4.0
Invoice	From Process 4.0 to Customer
Order Status	From Process 6.0 to Customer
Information	From Purchasing Fulfillment System to Process 6.0
Shopping Cart	
Cart ID	From Data Store D2 to Process 3.0
	From Data Store D2 to Process 4.0
Temporary User/System Messages	
Product Item Request	From Customer to Process 1.0
Purchase Request	From Customer to Process 2.0
View Cart	From Customer to Process 3.0
Items in Cart	From Process 3.0 to Customer
Remove Item	From Customer to Process 3.0
	From Process 3.0 to Data Store D2
Check Out	From Customer to Process 4.0

Figure 10-20
Entity-relationship diagram for the WebStore system

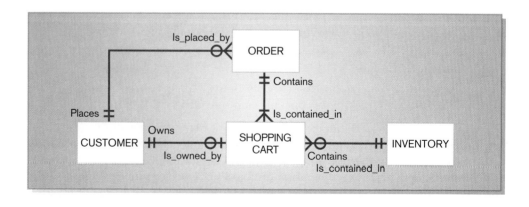

entity and then compare these lists with the existing inventory, customer, and order database tables. Making sure that all attributes were accounted for would be the final conceptual data modeling activity before beginning the process of selecting a final design strategy.

Summary

We have presented the process and basic notation used to model the data requirements of an information system. We outlined the structuring of conceptual data models using the entity-relationship notation and also discussed how the components of a conceptual data model relate to data flows and data stores as well as states on a state-transition diagram.

Conceptual data modeling is based on certain constructs about the structure, not use, of data. These constructs include: entity, relationship, degree, and cardinality. A data model shows the relatively permanent business rules that define the nature of an organization. Rules define such characteristics of data as the legitimate domain of values for data attributes, the unique characteristics (identifier) of entities, the relationships between different entities, and the triggering operations that protect the validity of attributes during data maintenance.

A data model shows major categories of data, called entities, the associations or relationships between entities, and the attributes of both entities and relationships. A special type of entity called an associative entity is often necessary to represent a many-to-many relationship

between entities. Entity types are distinct from entity instances. Each entity instance is distinguished from other instances of the same type by an identifier attribute (or attributes).

Relationships are the glue that hold a data model together. Three common relationship types are unary, binary, and ternary. The minimum and maximum number of entity instances that participate in a relationship represent important rules about the nature of the organization, as captured during requirements determination. Supertype/subtype relationships can be used to show a hierarchy of more general to more specific, related entity types that share common attributes and relationships. Rules for total and partial specialization between the supertype and subtypes and disjoint and overlap among the subtypes clarify the meaning of the related entity types.

This chapter completes our coverage of the traditional techniques used to structure information system requirements. The next chapter, the last in the section on the analysis phase of the life cycle, addresses the selection among alternative directions for designing the new or replacement system.

Key Terms

1. Associative entity
2. Attribute
3. Binary relationship
4. Business rules
5. Candidate key
6. Cardinality
7. Conceptual data model
8. Degree
9. Disjoint rule
10. Domain
11. Entity-relationship data model (E-R model)
12. Entity-relationship diagram (E-R diagram)
13. Entity instance (instance)
14. Entity type
15. Identifier
16. Multivalued attribute
17. Overlap rule
18. Partial specialization rule
19. Relationship
20. Repeating group
21. Subtype
22. Supertype
23. Ternary relationship
24. Total specialization rule
25. Triggering operation (trigger)
26. Unary relationship (recursive relationship)

Match each of the key terms above with the definition that best fits it.

_____ A detailed model that captures the overall structure of organizational data while being independent of any database management system or other implementation considerations.

_____ A detailed, logical representation of the entities, associations, and data elements for an organization or business area.

_____ A graphical representation of an E-R model.

_____ A collection of entities that share common properties or characteristics.

_____ A single occurrence of an entity type.

_____ A named property or characteristic of an entity that is of interest to the organization.

_____ An attribute (or combination of attributes) that uniquely identifies each instance of an entity type.

_____ A candidate key that has been selected as the unique, identifying characteristic for an entity type.

_____ An attribute that may take on more than one value for each entity instance.

_____ A set of two or more multivalued attributes that are logically related.

_____ An association between the instances of one or more entity types that is of interest to the organization.

_____ The number of entity types that participate in a relationship.

_____ A relationship between the instances of one entity type.

_____ A relationship between instances of two entity types.

_____ A simultaneous relationship among instances of three entity types.

_____ The number of instances of entity B that can (or must) be associated with each instance of entity A.

_____ An entity type that associates the instances of one or more entity types and contains attributes that are peculiar to the relationship between those entity instances.

_____ Specifications that preserve the integrity of a conceptual or logical data model.

_____ A subgrouping of the entities in an entity type that is meaningful to the organization.

_____ A generic entity type that has a relationship with one or more subtypes.

_____ Specifies that each entity instance of the supertype must be a member of some subtype in the relationship.

_____ Specifies that an entity instance of the supertype is allowed not to belong to any subtype.

_____ Specifies that if an entity instance of the supertype is a member of one subtype, it cannot simultaneously be a member of any other subtype.

_____ Specifies that an entity instance can simultaneously be a member of two (or more) subtypes.

_____ The set of all data types and values that an attribute can assume.

_____ An assertion or rule that governs the validity of data manipulation operations such as insert, update, and delete.

Review Questions

1. Discuss why some systems developers believe that a data model is one of the most important parts of the statement of information system requirements.

2. Distinguish between the data modeling done during information systems planning, project initiation and planning, and analysis phases of the systems development life cycle.

3. What elements of a data flow diagram should be analyzed as part of data modeling?

4. Explain why a ternary relationship is not the same as three binary relationships.

5. When must a many-to-many relationship be modeled as an associative entity?

6. What is the significance of triggering operations business rules in the analysis and design of an information system?

7. Which of the following types of relationships can have attributes associated with them: one-to-one, one-to-many, many-to-many?

8. What are the linkages between data flow diagrams, decision tables, state-transition diagrams, and entity-relationship diagrams?

9. What is the degree of a relationship? Give an example of each of the relationship degrees illustrated in this chapter.

10. Give an example of a ternary relationship (different from any example in this chapter).

11. List the deliverables from the conceptual data modeling part of the analysis phase of the systems development process.

12. Explain the relationship between minimum cardinality and optional and mandatory participation.

13. List the ideal characteristics of an entity identifier attribute.

14. List the four types of E-R diagrams produced and analyzed during conceptual data modeling.

15. Contrast the following terms:
    a. subtype; supertype
    b. total specialization rule; partial specialization rule
    c. disjoint rule; overlap rule

Problems and Exercises

1. Assume that at Pine Valley Furniture each product (described by Product No., Description, and Cost) is comprised of at least three components (described by Component No., Description, and Unit of Measure) and components are used to make one or many products (that is, must be used in at least one product). In addition,

assume that components are used to make other components and that raw materials are also considered to be components. In both cases of components being used to make products and components being used to make other components, we need to keep track of how many components go into making something else. Draw an E-R diagram for this situation and place minimum and maximum cardinalities on the diagram.

2. Much like Pine Valley Furniture's sale of products, stock brokerages sell stocks and the prices are continually changing. Draw an E-R diagram that takes into account the changing nature of stock price.

3. If you were going to develop a computer-based tool to help an analyst interview users and quickly and easily create and edit entity-relationship diagrams, what type of tool would you build? What features would it have, and how would it work? Why?

4. A software training program is divided into training modules and each module is described by module name and the approximate practice time. Each module sometimes has prerequisite modules. Model this situation of training programs and modules with an E-R diagram.

5. Each semester, each student must be assigned an advisor who counsels students about degree requirements and helps students register for classes. Students must register for classes with the help of an advisor, but if their assigned advisor is not available, they may register with any advisor. We must keep track of students, their assigned advisor, and with whom the student registered for the current term. Represent this situation of students and advisors with an E-R diagram.

6. Assume that entity PART has attributes Part_Number, Drawing_Number, Weight, Description, Storage_Location, and Cost. Which attributes are candidate keys? Why? Which attribute would you select for the identifier of PART? Why? Or do you think that you should create another attribute to be the identifier? Why or why not?

7. Consider the E-R diagram in Figure 10-9.
   a. What is the identifier for the COMPLETES associative entity?
   b. Now, assume that the same employee may take the same course multiple times, on different dates. Does this change your answer to part a? Why or why not?

8. Study the E-R diagram of Figure 10-21. Based on this E-R diagram, answer the following questions:
   a. How many PROJECTs can an employee work on?
   b. What is the degree of the Includes relationship?
   c. Are there any associative entities on this diagram? If so, name them.
   d. How else could the attribute Skill be modeled?
   e. Is it possible to attach any attributes to the Includes relationship?
   f. Could TASK be modeled as an associative entity?

9. For the entity-relationship diagram provided in Figure 10-22, draw in the relationship cardinalities and describe them. Describe any assumptions you must make about relevant

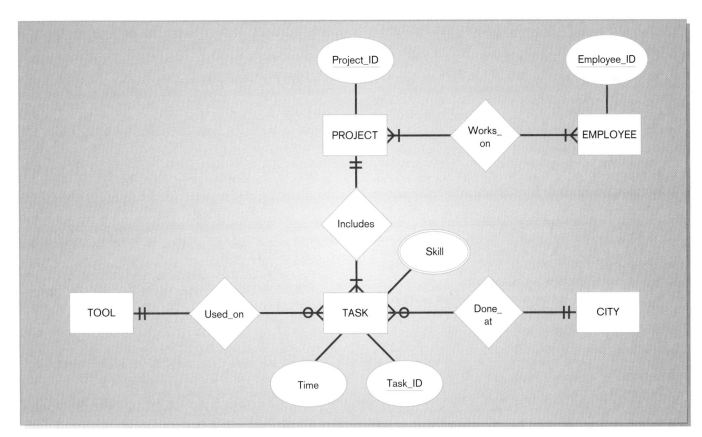

Figure 10-21
E-R diagram for Problem and Exercise 8

Figure 10-22
E-R diagram for Problem
and Exercises 9 and 10

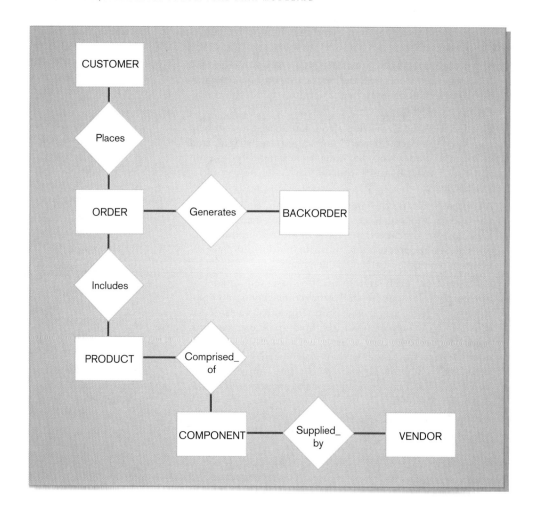

business rules. Are there any changes or additions you would make to this diagram to make it better? Why or why not?

10. For the entity-relationship diagram provided in Figure 10-22, assume that this company decided to assign each sales representative to a small, unique set of customers; some customers can now become "members" and receive unique benefits; small manufacturing teams will be formed and each will be assigned to the production of a small, unique set of products; and each purchasing agent will be assigned to a small, unique set of vendors. Make the necessary changes to the entity-relationship diagram and draw and describe the new relationship cardinalities.

11. Obtain a copy of an invoice, order form, or bill used in one of your recent business transactions. Create an E-R diagram to describe your sample document.

12. Using Table 10-1 as a guide, develop the complete script (questions and possible answers) of an interview between analysts and users within the order entry function at Pine Valley Furniture.

13. An airline reservation is an association between a passenger, a flight, and a seat. Select a few pertinent attributes for each of these entity types and represent a reservation in an E-R diagram.

14. Choose from your own experiences with organizations and draw an E-R diagram for a situation that has a ternary relationship.

15. Consider the E-R diagram in Figure 10-23. Are all three relationships—Holds, Goes_on, and Transports—necessary (i.e., can one of these be deduced from the other two)? Are there reasonable assumptions that make all three relationships necessary? If so, what are these assumptions?

16. Draw an E-R diagram to represent the sample customer order in Figure 10-4.

17. In a real estate database, there is an entity called PROPERTY, which is a property for sale by the agency. Each time a potential property buyer makes a purchase offer on a property, the agency records the date, offering price, and name of the person making the offer.
   a. Represent the PROPERTY entity and its purchase offer attributes using the notation for multivalued attributes.
   b. Represent the PROPERTY entity and its purchase offer attributes using two entity types.
   c. Finally, assume the agency decides to also keep data about buyers and potential buyers, including their name, phone number, and address. Augment your answer to part b above to accommodate this new entity type.

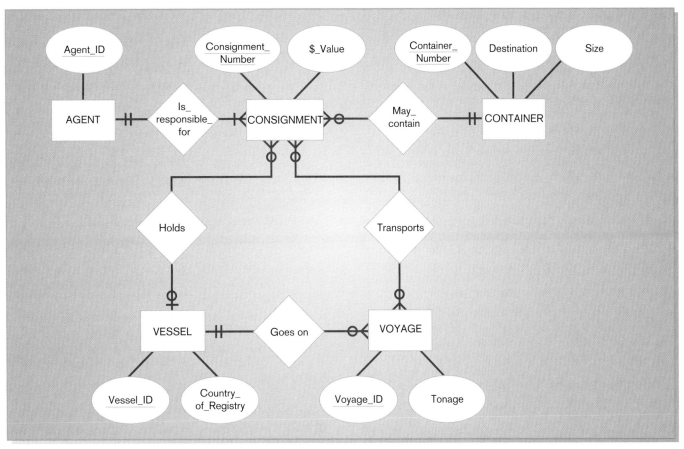

Figure 10-23
E-R diagram for Problem and Exercise 15

18. Consider the Is_married_to unary relationship in Figure 10-8c.
    a. Assume we wanted to know the date on which a marriage occurred. Augment this E-R diagram to include a Date_married attribute.
    b. Since sometimes persons remarry after the death of a spouse or divorce, redraw this E-R diagram to show the whole history of marriages (not just the current marriage) for persons. Show the Date_married attribute on this diagram.

19. Consider Figure 10-13.
    a. Write a domain integrity rule for Balance.
    b. Write a triggering operation business rule for the Balance attribute for the event of inserting a new ACCOUNT.

20. As best you can, describe the process logic and business rules underlying the entity-relationship diagrams from Problems and Exercises 9 and 10. How are entity-relationship diagrams similar to and different from the logic modeling techniques from the previous chapter (e.g., Structured English, decision tables, decision trees)? In what ways are these data and logic modeling techniques complimentary? What problems might be encountered if either data or logic modeling techniques were not performed well or not performed at all as part of the systems development process?

21. In the Purchasing Department at one company, each purchase request is assigned to a "case worker" within the Purchasing Department. This case worker follows the purchase request through the entire purchasing process and acts as the sole contact person with the person or unit buying the goods or services. The Purchasing Department refers to its fellow employees buying goods and services as "customers." The purchasing process is such that purchase requests over $1,500.00 must go out for bid to vendors, and the associated Request for Bids for these large requests must be approved by the Purchasing Department. If the purchase is under $1,500.00, the product or service can simply be bought from any approved vendor, but the purchase request must still be approved by the Purchasing Department and they must issue a Purchase Order. For large purchases, once the winning bid is accepted, the Purchasing Department can issue a Purchase Order. List the relevant entities and attributes and draw an entity-relationship diagram for this business process. List whatever assumptions you must make to define identifiers, assess cardinality, and so on.

Field Exercises

1. Interview a friend or family member to elicit from them each of the entities, attributes, relationships, and relevant business rules they come into contact with at work. Use this information to construct and present to this person an entity-relationship diagram. Revise the diagram until it seems appropriate to you and to your friend or family member.

2. Visit an organization that provides primarily a service, such as a dry cleaners, and a company that manufactures a more tangible product. Interview employees from these organizations to elicit from them each of the entities, attributes, relationships, and relevant business rules that are commonly encountered by these companies. Use this information to construct entity-relationship diagrams. What differences and similarities are there between the diagrams for the service- and the product-oriented companies? Does the entity-relationship diagramming technique handle both situations equally well? Why or why not? What differences, if any, might there be in the use of this technique for a public agency?

3. Discuss with a systems analyst the role of conceptual data modeling in the overall systems analysis and design of information systems at his or her company. How, and by whom, is conceptual data modeling performed? What training in this technique is given? At what point(s) is this done in the development process? Why?

4. Ask a systems analyst to give you examples of unary, binary, and ternary relationships that they have heard of or dealt with personally at their company. Ask them which is the most common. Why?

5. Talk to MIS professionals at a variety of organizations and determine the extent to which CASE tools are used in the creation and editing of entity-relationship diagrams. Try to determine whether or not they use CASE tools for this purpose, what CASE tools are used, and why, when, and how they use CASE tools for this. In companies that do not use CASE tools for this purpose, determine why not and what would have to change to have them use CASE tools.

6. Ask a systems analyst to give you a copy of the standard notation he or she uses to draw E-R diagrams. In what ways is this notation different from notation in this text? Which notation do you prefer and why? What is the meaning of any additional notation?

7. Ask a systems analyst in a manufacturing company to show you an E-R diagram for a database in that organization that contains bill-of-materials data. Compare that E-R diagram to the one in Figure 10-7. What are the differences between these diagrams and why?

References

Aranow, E. B. 1989. "Developing Good Data Definitions." *Database Programming & Design* 2 (8) (August): 36–39.

Bruce, T. A. 1992. *Designing Quality Databases with IDEF1X Information Models.* New York, NY: Dorset House Publications.

Chen, P. P-S. 1976. "The Entity-Relationship Model—Toward a Unified View of Data." *ACM Transactions on Database Systems* 1 (March): 9–36.

Fleming, C. C., and B. von Halle. 1990. "An Overview of Logical Data Modeling." *Data Resource Management* 1 (1) (Winter): 5–15.

Gottesdiener, E. 1999. "Turning Rules into Requirements." *Application Development Trends* 6(7): 37–50.

Hoffer, J. A., M. B. Prescott, and F. R. McFadden, 2002. *Modern Database Management.* 6th ed. Upper Saddle River, NJ: Prentice Hall.

Moody, D. 1996. "The Seven Habits of Highly Effective Data Modelers." *Database Programming & Design* 9 (October): 57, 58, 60–62, 64.

Sandifer, A., and B. von Halle. 1991a. "A Rule by Any Other Name." *Database Programming & Design* 4 (2) (February): 11–13.

Sandifer, A., and B. von Halle. 1991b. "Linking Rules to Models." *Database Programming & Design* 4 (3) (March): 13–16.

Storey, V. C. 1991. "Relational Database Design Based on the Entity-Relationship Model." *Data and Knowledge Engineering* 7 (1991): 47–83.

Teorey, T. J., D. Yang, and J. P. Fry. 1986. "A Logical Design Methodology for Relational Databases Using the Extended Entity-Relationship Model." *Computing Surveys* 18 (2) (June): 197–221.

BROADWAY ENTERTAINMENT COMPANY, INC.

Structuring System Requirements: Conceptual Data Modeling for the Web-based Customer Relationship Management System

CASE INTRODUCTION

Requirements determination activities for the MyBroadway project yielded what at times seemed to the student team to be an overwhelming amount of data. The team of students from Stillwater State University has several hundred pages of notes from various data collection activities including twelve interviews with employees and customers, six hours of observation of employees using on-line shopping services, a one-hour focus group session with customers, and investigations of Broadway Entertainment documents. Structuring these requirements for the analysis of the MyBroadway information system is a much bigger effort than any class exercise the team members had ever encountered.

Also adding to the complexity of requirements structuring activities in the analysis phase of the project is that work is not easily compartmentalized. It seems to the team members that while they are documenting data movement and processing requirements, they also have to find ways to understand the meaning of data the system will handle. Conceptual data modeling techniques, primarily entity-relationship diagramming, help but changes are frequent. The steps are very repetitive. As the team decomposes a business process, members need to redesign the E-R diagram for MyBroadway. When they change the E-R model, they gain new insights about the data and suggest issues of data handling processes for validation, special cases, and capturing relationships.

STRUCTURING THE HIGH-LEVEL DATA MODELING FINDINGS FROM REQUIREMENTS DETERMINATION

The various BEC student team members have taken responsibility for different requirements collection activities and for developing the explosions of each process on the level-0 DFD. So, no one team member has a complete picture of all the data needs. This is not uncommon on real development projects. The team has yet to appoint someone to be the data administrator for the project. To gain a shared understanding of the database needs for MyBroadway, the team members read all of their notes carefully in preparation for a team meeting.

At the team meeting each member suggests the data entities he or she thinks are needed in his or her part of the system. After some discussion the team concludes that six entity types are referenced repeatedly in data flows and data stores across all business processes. See BEC Figure 10-1 for the initial entity-relationship diagram the team draws. The entity types identified by the team follow:

- *Product:* An item made available for sale or rent by BEC to customers. Each product is a CD, DVD, or videocassette title. For example, a product is the movie *Star Wars Episode I: The Phantom Menace* on videocassette. Although the Entertainment Tracker operational system must keep track of each copy of a movie available for rent at a store, MyBroadway simply needs to track titles, not individual copies. For items for sale, a product is the generic title, not an individual copy, of the product for sale.

- *Request:* An inquiry by a customer asking BEC to stock a product. BEC may choose never to stock that product. Yet if enough requests for the same product are submitted, it is likely that the item will eventually appear on the store shelves.

- *Sale:* A record that a particular product (by title) was sold to a specified customer on what date. Entertainment Tracker keeps the official record of each sale transaction, including which items are sold in the same transaction. MyBroadway, however, does not need this information but rather needs only when and to whom each product was ever sold.

- *Rental:* A record that a particular product (by title) was rented by a specified customer on what date. As with the SALE entity type, MyBroadway does not need to track the rental transaction in detail.

- *Comment:* An unstructured statement by a customer about a specified product (by title).

- *Pick:* An unstructured comment and rating by an employee about a specified product (by title).

BEC Figure 10-1
Initial E-R diagram for MyBroadway

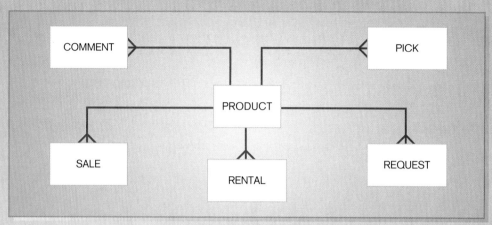

The MyBroadway team finds it very interesting that although both customers and employees are very prominent actors in the system and data about each are needed in data flows and are retained in data stores, neither of these appear to be useful entities themselves. The team concluded this because data about these objects of the business seem to have no usefulness on their own (in MyBroadway), but only when associated with other data. For example, MyBroadway needs to know what members bought which products when, but customer data such as name, address, standard credit card number, and so forth—prominent in Entertainment Tracker—never appear in any data flow.

The team also concluded that attributes about PRODUCT, SALE, and RENTAL would not be captured within MyBroadway; rather, data for these entities would come from the Entertainment Tracker database. The team's initial thought was that MyBroadway probably needs only a minor subset of the data from Entertainment Tracker on these data entities. For example, the team had not identified any data flow that needed product price, cost, location in store, or a host of other product attributes in the Entertainment Tracker database useful for transaction processing and management reporting. In MyBroadway, the transactions of interest are entries of comments, favorite picks, and product requests. Each comment, pick, or request is considered an independent data item, whereas product sales and rentals frequently appear together all in one transaction (e.g., someone rents three movies and buys one CD all in the same point-of-sale transaction). These observations suggest to the team that the structure of the MyBroadway database may be simpler than for most operational databases.

CASE SUMMARY

Of course, whether these six entities are all the team needs still remains to be finalized. The team must carefully match this list of data entities with the data stores

and data flows from data flow diagrams they are developing. For example, every attribute of a data flow going into a data store must be an attribute of some entity type. Also, there must be an attribute either in a data store or directly passed through the system to generate all the attributes of each data flow leaving MyBroadway to some external entity. The team has many more questions to answer before it can produce an E-R diagram for the MyBroadway system.

CASE QUESTIONS

1. Review the data flow diagrams you developed for questions in the BEC case at the end of Chapter 8 (or diagrams given to you by your instructor). Study the data flows and data stores on these diagrams and decide if you agree with the team's conclusion that there are only the six entity types listed in this case and in BEC Figure 10-1. If you disagree, define additional entity types, explain why they are necessary, and modify BEC Figure 10-1.

2. Again, review the DFDs you developed for the MyBroadway system (or those given to you by your instructor). Use these DFDs to identify the attributes of each of the six entities listed in this case plus any additional entities identified in your answer to Case Question 1. Write an unambiguous definition for each attribute. Then, redraw BEC Figure 10-1 by placing the six (and additional) entities in this case on the diagram along with their associated attributes.

3. Using your answer to Case Question 2, designate which attribute or attributes form the identifier for each entity type. Explain why you chose each identifier.

4. Using your answer to Case Question 3, draw the relationships between entity types needed by the system. Remember, a relationship is needed only if the system wants data about associated entity instances. Give a meaningful name to each relationship. Specify cardi-

nalities for each relationship and explain how you decided on each minimum and maximum cardinality on each end of each relationship. State any assumptions you made if the BEC cases you have read so far and the answers to questions in these cases do not provide the evidence to justify the cardinalities you chose.

5. Now that you have developed in your answer to Case Question 4 a complete E-R diagram for the MyBroadway database, what are the consequences of not having customer or employee entity types on this diagram? Assuming only the attributes you show on the E-R diagram, would any attribute be moved from the entity it is currently associated with to a customer or employee entity type if such entity types were on the diagram? Why or why not?

6. Write project dictionary entries (using standards given to you by your instructor) for all the entities, attributes, and relationships shown in the E-R diagram in your answer to Case Question 4. How detailed are these entries at this point? What other details still must be filled in? Are any of the entities on the E-R diagram in your answer to Case Question 4 weak entities? Why? In particular, is the REQUEST entity type a weak entity? If so, why? If not, why not?

7. What date-related attributes did you identify in each of the entity types in your answer to Case Question 4? Why are each of these needed? Can you make some general observations about why date attributes must be kept in a database based on your analysis of this database?

11

Selecting the Best Alternative Design Strategy

LEARNING OBJECTIVES

After studying this chapter, you should be able to:

- Describe the different sources of software.
- Assemble the various pieces of an alternative design strategy.
- Generate at least three alternative design strategies for an information system.
- Select the best design strategy using both qualitative and quantitative methods.
- Update a Baseline Project Plan based on the results of the analysis phase.
- Understand how selecting the best design strategy applies to development for Internet applications.

INTRODUCTION

You have now reached the point in the analysis phase where you are ready to transform all of the information you have gathered and structured into some concrete ideas about the nature of the design for the new or replacement information system. This is called the *design strategy*. From requirements determination, you know what the current system does and you know what the users would like the replacement system to do. From requirements structuring, you know what forms the replacement system's process flow, process logic, and data should take, at a logical level independent of any physical implementation. For example, every data flow from a system process to the external environment represents an on-line display, printed report, or business form or document that the system can produce, often for use by a human. Every data flow from a data store to a process implies a retrieval capability of the system's files and database. Most processes represent a capability of the system to transform inputs into outputs. And data flows from the environment to system processes indicate capabilities to capture data on on-line displays, or some batch method, to validate the data and protect access to the system from the environment and to route raw data to the appropriate processing points.

Thus, at this point in the systems development process you have a preliminary specification of what the new information system should do and you understand why a replacement system is necessary to fix problems in the current system and to respond to new needs and opportunities to use information. Actually, there still may be some uncertainty about the capabilities of a new system. This uncertainty is due to competing ideas from different users and stakeholders on what they would like the system to do and to existing alternatives for an implementation environment for the new system. To bring analysis to a conclusion, your job is to take these structured requirements and transform them into several competing design strategies, one of which will be pursued in the design phase of the life cycle.

Part of generating a design strategy is determining how you want to acquire the replacement system using a combination of sources inside and outside the organization. If you decide to proceed with development in-house, you will have to answer general questions about software, such as whether all of the software should be built in-house or whether some software components should be bought off-the-shelf or contracted to software development companies. You will have to answer general questions about hardware and system software, such as whether the new system will run on a mainframe platform, stand-alone personal computers, or on a client/server platform, and whether the system can run on existing hardware. It is also not too early to begin thinking about data conversion issues, which must be addressed as you move from your current system to the new one. You even have to start thinking about how much training will be required for users, and how easy or difficult the system will be to implement. You have to determine whether you can build and implement the system you desire given the funding and management support you can count on. And you have to address these concerns for each alternative you generate. These issues need to be addressed so that you can update the Baseline Project Plan with detailed activities and resource requirements for the next life-cycle phase—logical design—and probably for the physical design phase as well. That is, in this step of the analysis phase you bring the current phase to a close, prepare a report and presentation to management concerning continuation of the project, and get ready to move the project into the design phases.

In this chapter, you will learn why you need to come up with alternative design strategies and about guidelines for generating alternatives. You will then learn about the different issues that must be addressed for each alternative. Once you have generated your alternatives, you will have to choose the best design strategy to pursue. We include a discussion of one technique that analysts and users often use to decide among system alternatives and to help them agree on the best approach for the new information system.

Throughout this chapter we emphasize the need for sound project management. Now that you have seen the various techniques and steps of the analysis phase, we outline what a typical analysis phase project schedule might look like and discuss the execution of the analysis phase and the transition from analysis to design.

SELECTING THE BEST ALTERNATIVE DESIGN STRATEGY

Selecting the best alternative system involves at least two basic steps: (1) generating a comprehensive set of alternative design strategies and (2) selecting the one that is most likely to result in the desired information system, given all of the organizational, economic, and technical constraints that limit what can be done. In a sense then, the most likely strategy is the best one. A system **design strategy** is an approach to developing the system. The strategy includes the system's functionality, hardware and system software platform, and method for acquisition. We use the term design

Design strategy: A high-level statement about the approach to developing an information system. It includes statements on the system's functionality, hardware and system software platform, and method for acquisition.

strategy in this chapter rather than the term alternative system because, at the end of analysis, we are still quite a long way from specifying an actual system. This delay is purposeful since we do not want to invest in design efforts until there is agreement on which direction to take the project and the new system. The best we can do at this point is to outline rather broadly the approach we can take in moving from logical system specifications to a working physical system. The overall process of selecting the best system strategy and the deliverables from this step in the analysis process are discussed next.

The Process of Selecting the Best Alternative Design Strategy

As Figure 11-1 shows, there are three subphases to systems analysis: requirements determination, requirements structuring, and generating alternative system design strategies and selecting the best one. After the system requirements have been structured in terms of process flow, process logic (decision and temporal), and data, analysts again work with users to package the requirements into different system configurations. Shaping alternative system design strategies involves the following processes:

- Dividing requirements into different sets of capabilities, ranging from the bare minimum that users would accept (the required features) to the most elaborate and advanced system the company can afford to develop (which includes all the features desired across all users). Alternatively, different sets of capabilities may represent the position of different organizational units with conflicting notions about what the system should do.

- Enumerating different potential implementation environments (hardware, system software, and network platforms) that could be used to deliver the different sets of capabilities. (Choices on the implementation environment may place technical limitations on the subsequent design phase activities.)

- Proposing different ways to source or acquire the various sets of capabilities for the different implementation environments.

Figure 11-1
Systems development life cycle with the analysis phase highlighted

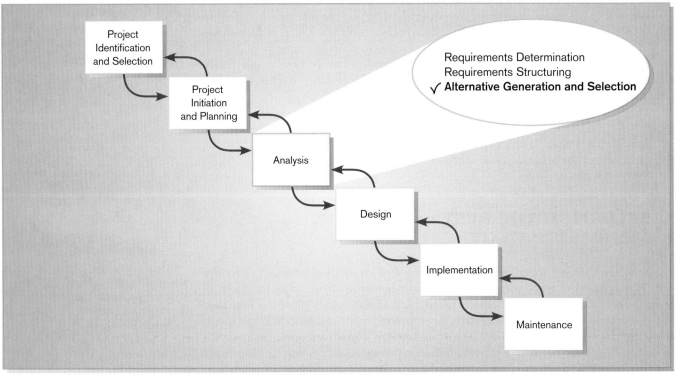

TABLE 11-1   Deliverables for Generating Alternatives and Selecting the Best One

1. At least three substantively different system design strategies for building the replacement information system
2. A design strategy judged most likely to lead to the most desirable information system
3. A Baseline Project Plan for turning the most likely design strategy into a working information system

In theory, if there are three sets of requirements, two implementation environments, and four sources of application software, there would be 24 possible design strategies. In practice, some combinations are usually infeasible or uninteresting. Further, usually only a small number—typically three—can be easily considered. Selecting the best alternative is usually done with the help of a quantitative procedure. Analysts will recommend what they believe to be the best alternative, but management (a combination of the steering committee and those who will fund the rest of the project) will make the ultimate decision about which system design strategy to follow. At this point in the life cycle, it is also certainly possible for management to end a project before the more expensive phases of design and implementation are begun if the costs or risks seem to outweigh the benefits, if the needs of the organization have changed since the project began, or if other competing projects appear to be of greater worth and development resources are limited.

Deliverables and Outcomes

The primary deliverables from generating alternative design strategies and selecting the best one are outlined in Table 11-1. The primary deliverable that is carried forward into design is an updated Baseline Project Plan detailing the work necessary to turn the selected design strategy into the desired replacement information system. Of course, that plan cannot be assembled until a strategy has been selected, and no strategy can be selected until alternative strategies have been generated and compared. Therefore, all three objects—the alternatives, the selected alternative, and the plan—are listed as deliverables in Table 11-1. Further, these three deliverables plus the supporting deliverables from requirements determination and structuring steps are necessary information to conduct systems design, so all of this information is carried in the project dictionary and CASE repository for reference in subsequent phases.

GENERATING ALTERNATIVE DESIGN STRATEGIES

In many cases, it may seem to an analyst that the solution to an organizational problem is obvious. Typically, the analyst is very familiar with the problem, having conducted an extensive analysis of it and how it has been solved in the past, or the analyst is very familiar with a particular solution that he or she attempts to apply to all organizational problems encountered. For example, if an analyst is an expert at using advanced database technology to solve problems, then there is a tendency for the analyst to recommend advanced database technology as a solution to every possible problem. Or if the analyst designed a similar system for another customer or business unit, the "natural" design strategy would be the one used before. Given the role of experience in the solutions analysts suggest, analysis teams typically generate at least two alternative solutions for every problem they work on.

A good number of alternatives for analysts to generate is three. Why three? Three alternatives can neatly represent both ends and the middle of a continuum or spectrum of potential solutions. One alternative represents the low end of the spectrum. Low-end solutions are the most conservative in terms of the effort, cost, and technology involved in developing a new system. In fact, low-end solutions may not involve computer technology at all, focusing instead on making paper flows more efficient or reducing the redundancies in current processes. A low-end strategy provides all the required functionality users demand with a system that is minimally different from the current system.

Another alternative represents the high end of the spectrum. High-end alternatives go beyond simply solving the problem in question and focus instead on systems that contain many extra features users may desire. Functionality, not cost, is the primary focus of high-end alternatives. A high-end alternative will provide all desired features using advanced technologies, which often allow the system to expand to meet future requirements. Finally, the third alternative lies between the extremes of the low-end and high-end systems. Such alternatives combine the frugality of low-end alternatives with the focus on functionality of high-end alternatives. Midrange alternatives represent compromise solutions. There are certainly other possible solutions that exist outside of these three alternatives. Defining the bottom, middle, and top possibilities allows the analyst to draw bounds around what can be reasonably done.

How do you know where to draw bounds around the potential solution space? The analysis team has already gathered the information it needs to identify the solution space, but first that information must be systematically organized. There are two major considerations. The first is determining what the minimum requirements are for the new system. These are the mandatory features, any of which, if missing, make the design strategy useless. Mandatory features are those that everyone agrees are necessary to solve the problem or meet the opportunity. Which features are mandatory can be determined from a survey of users and other stakeholders who have been involved in requirements determination. You would conduct this survey near the end of the analysis phase after all requirements have been structured and analyzed. In this survey, stakeholders rate features discovered during requirements determination or categorize features on some desirable-mandatory scale, and an arbitrary breakpoint is used to divide mandatory from desired features. Some organizations will break the features into three categories: mandatory, essential, and desired. Whereas mandatory features screen out possible solutions, essential features are the important capabilities of a system that will serve as the primary basis for comparison of different design strategies. Desired features are those that users could live without but are used to select between design strategies of almost equal value in terms of essential features. Features can take many different forms. Features might include

- Data kept in system files (for example, multiple customer addresses so that bills can be sent to addresses different from where we ship goods)
- System outputs (printed reports, on-line displays, transaction documents—for example, a paycheck or sales summary graph)
- Analyses to generate the information in system outputs (for example, a sales forecasting module or an installment billing routine)
- Expectations on accessibility, response time, or turnaround time for system functions (for example, on-line, real-time updating of inventory files)

The second consideration in drawing bounds around alternative design strategies is determining the constraints on system development. Constraints may exist on such factors as

- A date when the replacement system is needed
- Available financial and human resources
- Elements of the current system that cannot change
- Legal and contractual restrictions

- The importance or dynamics of the problem that may limit how the system can be acquired (for example, a strategically important system that uses highly proprietary data probably cannot be outsourced or purchased)

Remember, be impertinent and question whether stated constraints are firm; you may want to consider some design alternatives that violate constraints you consider to be flexible.

Both requirements and constraints must be identified and ranked in order of importance. The reason behind such a ranking should be clear. Whereas you can design a high-end alternative to fulfill every wish users have for a new system, you design low-end alternatives to fulfill only the most important wishes. The same is true of constraints. Low-end alternatives will meet every constraint; high-end alternatives will ignore all but the most daunting constraints.

ISSUES TO CONSIDER IN GENERATING ALTERNATIVES

The required functionality of the replacement system and the constraints that limit that functionality form the basis for the many issues that must be considered in putting together all of the pieces that comprise alternative design strategies. That is, most of the substantive debate about alternative design strategies hinges on the relative importance of system features. Issues of functionality lead, however, to other associated issues such as whether the system should be developed and run in-house, software and hardware selection, implementation, and organizational limitations such as available funding levels. This list is not comprehensive, but it does remind you that an information system is more than just software. Each issue must be considered when framing alternatives. We will discuss each consideration in turn, beginning with the outsourcing decision.

Outsourcing

If another organization develops or runs a computer application for your organization, that practice is called **outsourcing**. Outsourcing includes a spectrum of working arrangements. At one extreme is having a firm develop and run your application on their computers—all you do is supply input and take output. A common example of such an arrangement is a company that runs payroll applications for clients so that clients don't have to develop an independent in-house payroll system. Instead they simply provide employee payroll information to the company and, for a fee, the company returns completed paychecks, payroll accounting reports, and tax and other statements for employees. For many organizations, payroll is a very cost-effective operation when outsourced in this way. In another example of outsourcing arrangements, you hire a company to run your applications at your site on your computers. In some cases, an organization employing such an arrangement will dissolve some or all of its information systems unit and fire all of its information systems employees. Many times the company brought in to run the organization's computing will hire many of the information systems unit employees.

Outsourcing is a large and growing segment of the IS industry, with a global market of $76 billion in 1995 and a projected global market of over $121 billion in the year 2000 (Lacity and Willcocks, 1998). Outsourcing provides a way for firms to leapfrog their current position in IS and to turn over development and operations to staff with skills not found internally. Outsourcing continues to grow in popularity. According to a 1998 Corbett Group study, 97% of more than 200 executives polled said they increased spending on outsourcing in 1998 over 1997. These executives also expected to increase spending in 1999 over 1998 levels. A total of 60% were satisfied with their outsourcing initiatives (Merrill, 1999).

Outsourcing: The practice of turning over responsibility of some to all of an organization's information systems applications and operations to an outside firm.

Why would an organization outsource its information systems operations? As we saw in the payroll example, outsourcing may be cost-effective. If a company specializes in running payroll for other companies, it can leverage the economies of scale it achieves from running one very stable computer application for many organizations into very low prices. Other reasons include access to increased knowledge and expertise, which may not be available internally, and the availability and quality of vendors who provide outsourcing services (Ketler and Willems, 1999). But why would an organization dissolve its entire information processing unit and bring in an outside firm to manage its computer applications? One reason may be to overcome operating problems the organization faces in its information systems unit. For example, the city of Grand Rapids, Michigan, hired an outside firm to run its computing center 30 years ago in order to better manage its computing center employees. Union contracts and civil service constraints then in force made it difficult to fire people, so the city brought in a facilities management organization to run its computing operations, and it was able to get rid of problem employees at the same time. Another reason for total outsourcing is that an organization's management may feel its core mission does not involve managing an information systems unit and that it might achieve more effective computing by turning over all of its operations to a more experienced, computer-oriented company. Kodak decided in the late 1980s that it was not in the computer applications business and turned over management of its mainframes to IBM and management of its personal computers to Businessland (Applegate and Montealagre, 1991).

Outsourcing is an alternative analysts need to be aware of. When generating alternative system development strategies for a system, as an analyst you should consult organizations in your area that provide outsourcing services. It may well be that at least one such organization has already developed and is running an application very close to what your users are asking for. Perhaps outsourcing the replacement system should be one of your alternatives. Knowing what your system requirements are before you consider outsourcing means that you can carefully assess how well the suppliers of outsourcing services can respond to your needs. However, should you decide not to consider outsourcing, you need to consider whether some software components of your replacement system should be purchased and not built.

N E T   S E A R C H

Outsourcing has become very popular for a wide variety of organizations. Visit http://www.prenhall.com/hoffer to complete an exercise related to this topic.

Sources of Software

We can group organizations that produce software into six major categories: hardware manufacturers, packaged software producers, custom software producers, enterprise-wide solutions, application service providers, and in-house developers.

Hardware Manufacturers   At first it may seem counterintuitive that hardware manufacturers would develop information systems or software. Yet hardware manufacturers are among the largest producers of software; for example, IBM is a leader in software development. Table 11-2 ranks the top 10 global software companies and their total revenues in 1999 (from all sales). However, IBM actually develops relatively little application software. Rather, IBM's leadership comes from its operating systems and utilities (like sort routines or database management systems) for the hardware it manufactures, as well as its middleware, the software that links one set of software services to another. IBM has also become a leader in developing software for supporting electronic-business applications.

Packaged Software Producers   The growth of the software industry has been phenomenal since its beginnings in the mid-1960s. Now, some of the largest computer companies in the world, as measured by Software Magazine, are companies that produce software exclusively. Table 11-3 lists the top 10 global software companies in rank order according to their revenues from software licensing. By comparing

TABLE 11-2   The Top 10 Global Software Companies

Rank Among Software Companies	Company	1999 Revenues (in millions)	Software Specialization
1	IBM	44,900	Operating systems, Middleware, Databases
2	Microsoft	21,591	Operating systems, Applications
3	PriceWaterhouse-Coopers	17,300	IT services/consulting
4	Oracle	9,328	Databases, ERP
5	Andersen Consulting	8,941	IT services/consulting
6	Hewlett Packard	8,734	Infrastructure & systems management
7	Compaq	7,779	IT services, Operating systems
8	Computer Associates	6,268	Infrastructure & systems management
9	Hitachi	5,900	System integration services
10	SAP	5,071	ERP, ASP

(From Software Magazine's Website, www.softwaremag.com)

Tables 11-2 and 11-3, you can see the percentage that software revenue is of total revenue for those firms on both lists; other revenues come from hardware, consulting services, and other items. As Table 11-3 shows, for the top 10 companies alone the total revenues for software licensing exceeded $50 billion in 1999.

Software companies develop what are sometimes called prepackaged or off-the-shelf systems. Microsoft's Project and Intuit's Quicken™, QuickPay™, and QuickBooks™ are popular examples of such software. The packaged software development industry serves many market segments. Their software offerings range from general, broad-based packages, such as general ledger, to very narrow, niche packages, such as software to help manage a day care center. Software companies develop software to run on many different computer platforms, from microcomputers to large mainframes. The companies range in size from just a few people to thousands of employees. Software companies consult with system users after the initial software design has been completed and an early version of the system has been built. The systems are then tested in actual organizations to determine whether there are any problems or if any improvements can be made. Until testing is completed, the system is not offered for sale to the public.

TABLE 11-3   The Top 10 Global Software Companies by Software License Revenues

Rank by Software Licenses	Company	1999 Revenues (in millions)
1	Microsoft	21,591
2	IBM	12,700
3	Computer Associates	4,962
4	Oracle	3,873
5	Hewlett Packard	2,542
6	SAP	1,946
7	Sun	1,302
8	Unisys	1,207
9	Compaq	1,156
10	Novell	1,092

(From Software Magazine's Website, www.softwaremag.com)

NET SEARCH
Changes are constantly occurring in the software industry. Visit http://www.prenhall.com/hoffer to complete an exercise related to this topic.

Some off-the-shelf software systems cannot be modified to meet the specific, individual needs of a particular organization. Such application systems are sometimes called turnkey systems. The producer of a turnkey system will only make changes to the software when a substantial number of users ask for a specific change. Other off-the-shelf application software can be modified or extended, however, by the producer or by the user, to more closely fit the needs of the organization. Even though many organizations perform similar functions, no two organizations do the same thing in quite the same way. A turnkey system may be good enough for a certain level of performance but it will never perfectly match the way a given organization does business. A reasonable estimate is that off-the-shelf software can at best meet 70% of an organization's needs. Thus, even in the best case, 30% of the software system doesn't match the organization's specifications.

Custom Software Producers   If a company needs an information system but does not have the expertise or the personnel to develop the system in-house and a suitable off-the-shelf system is not available, the company will likely consult a custom software company. Consulting firms, such as PriceWaterhouseCoopers or Electronic Data Systems, will help a firm develop custom information systems for internal use. These firms employ people with expertise in the development of information systems. Their consultants may also have expertise in a given business area. For example, consultants who work with banks understand financial institutions as well as information systems. Consultants use many of the same methodologies, techniques, and tools that companies use to develop systems in-house. The 10 largest global computer-services firms (based on revenues not only from custom software development but also outsourcing and other services) are listed in Table 11-4. Note that while three large and well-known consulting firms are on the list, the top service company is IBM. Other firms in this market segment include ADP, Cap Gemini, and Unisys.

Enterprise Solutions Software   As mentioned in Chapter 1, more and more organizations are choosing complete software solutions, called enterprise solutions or **enterprise resource planning systems (ERP)**, to support their operations and business processes. These ERP software solutions consist of a series of integrated modules. Each module supports an individual, traditional business function, such as accounting, distribution, manufacturing, and human resources. The difference between the modules and traditional approaches is that the modules are integrated to focus on business processes rather than on business functional areas. For example, a series of modules will support the entire order entry process, from receiving an

Enterprise resource planning (ERP) systems: A system that integrates individual traditional business functions into a series of modules so that a single transaction occurs seamlessly within a single information system rather than several separate systems.

TABLE 11-4   The Top 10 Global Software Companies by Service Revenue

Rank by Service Revenues	Company	1999 Revenues (in millions)
1	IBM	32,200
2	PriceWaterhouseCoopers	17,300
3	Andersen Consulting	8,941
4	Compaq	6,623
5	Hewlett Packard	6,192
6	Oracle	5,455
7	SAP AG	3,125
8	Bull Worldwide Information Systems	2,790
9	Ernst & Young	2,000
10	Sun	1,035

(From Software Magazine's Website, www.softwaremag.com)

order to adjusting inventory to shipping to billing to after-the-sale service. The traditional approach would use different systems in different functional areas of the business, such as a billing system in accounting and an inventory system in the warehouse. Using enterprise software solutions, a firm can integrate all parts of a business process in a unified information system. All aspects of a single transaction occur seamlessly within a single information system, rather than in a series of disjointed, separate systems focused on business functional areas.

The benefits of the enterprise solutions approach include a single repository of data for all aspects of a business process and the flexibility of the modules. A single repository ensures more consistent and accurate data, as well as less maintenance. The modules are very flexible because additional modules can be added as needed, once the basic system is in place. Added modules are immediately integrated into the existing system.

There are also disadvantages to enterprise solutions software. The systems are very complex, so implementation can take a long time to complete. Organizations typically do not have the necessary expertise in-house to implement the systems, so they must rely on consultants or employees of the software vendor, which can be very expensive. In some cases, organizations must change how they do business in order to benefit from a migration to enterprise solutions.

There are several major vendors of enterprise solution software. The best known is probably SAP AG, a German firm, known for its flagship product R/3. SAP stands for Systems, Applications, and Products in Data Processing. SAP AG was founded in 1972, but most of its growth has occurred since 1992. In 1999, SAP AG was the tenth largest software company in the world and sixth in terms of revenues from software licensing (see Tables 11-2 and 11-3). Another well-known supplier of enterprise solution software is Oracle, a company more often associated with its database products. Other market leaders include PeopleSoft, Inc., a U.S. firm founded in 1987, best known for its human resources management systems; J. D. Edwards, a leader in manufacturing and procurement systems; and The Baan Company, NV, a Dutch firm founded in 1978, which has focused on manufacturing, finance, distribution and transportation, and service. The 1999 market for enterprise-wide systems was estimated to have been $16.9 billion, with SAP AG holding about one-third of the market. SAP's 1998 revenues of $5 billion represented over 19,000 installations of its key product R/3 in over 90 countries. The total ERP market is projected to grow to $21.4 billion by 2004.

Although the general trend with regards to ERP systems has been for a firm to deal exclusively with a single vendor for a single enterprise-wide implementation of the vendor's ERP products, some firms have instead followed a best-of-breed strategy. Such a strategy typically entails using different products from different ERP vendors, as well as specialized products from non-ERP vendors, and developing other software in-house to fill in the gaps and ease cross-product integration. The key advantage of a best-of-breed strategy is that it capitalizes on the strengths of individual vendor products. For example, a firm might use SAP's order entry modules, Oracle's financial systems, and PeopleSoft's human resources products. For the increased system functionality offered in such a scheme, however, the firm gives up the single architecture, single interface, and single vendor connection that comes with adopting one ERP vendor's products to support all functions throughout the firm.

Application Service Providers

Another method for organizations to obtain computerized applications is to rent them or license them from third-party providers who run the applications at remote sites. Users have access to the applications through the Internet or through virtual private networks. The application provider buys, installs, maintains, and upgrades the applications. Users pay on a per use basis, or they license the software, typically month to month. The companies that host the applications are called **Application Service Providers**, or ASPs.

N E T   S E A R C H
The enterprise resource planning software market is very competitive. Visit http://www.prenhall.com/hoffer to complete an exercise related to this topic.

Application Service Provider:
Organizations that host and run computer applications for other companies, typically on a per use or license basis.

The market for ASPs was relatively small in 1999, estimated to be about $296 million in the United States. However, the market was forecasted to grow to $7.8 billion by 2004. Some analysts have predicted that by 2010, 80% of all corporate applications will be hosted (Harney, 2000). Although ASPs sometimes develop and rent or license their own applications to customers, for the most part, ASPs purchase or license applications from other software vendors. For example, Oracle and Microsoft make their applications available through ASPs (Wilcox and Farmer, 2000; Holohan and Hall, 2000). Microsoft offers its Windows operating system and Office software to ASPs, while Oracle offers its ERP applications. Another example is EDS, which has made available its Wholesale ASP Services offering, which provides a set of Internet-based applications ASPs can host and sell to customers (Ferranti, 2000).

As the forecasts for growth indicate, taking the ASP route has its advantages. One of the most important advantages for many companies, especially medium to small ones, is less need for internal information technology staff. Companies that get applications from ASPs also save money on internal infrastructure and initial capital outlay. Companies going through ASPs can also gain access to big and complex systems without having to go through the expensive and time-consuming process of implementing the systems themselves in-house. Using an ASP also makes it easier to walk away from an unsatisfactory systems solution. Finally, going through an ASP for an application means that a company can be quicker to market with whatever service or product the application supports.

There are disadvantages to using an ASP, however. Since the application runs at some remote site, the company leasing the application has less control over the application, such as when it will be upgraded or how access to it is facilitated. Applications made available through ASPs also tend to address routine problems, where there is not too much difference from company to company in how a particular problem is addressed. ASP-provided solutions, therefore, tend to be rather generic, allowing only about 20% customization for any given company.

In-House Development   We have talked about three different types of external organizations that serve as sources of software, but in-house development remains an option. In-house development has become a progressively smaller piece of all systems development work that takes place in and for organizations. Internal corporate IS shops now spend a smaller and smaller proportion of their time and effort on developing systems from scratch. In 1998, corporate IS groups reported spending 33% less time and money on traditional software development and maintenance than they did in 1997 (King and Cole-Gomolski, 1999). Instead, they increased work on packaged applications by a factor of three, and they increased outsourcing by 42%. Where in-house development occurred, it was related to Internet technology. In-house development can lead to a larger maintenance burden than other development methods, such as packaged applications, according to a recent study (Banker, Davis and Slaughter, 1998). The study found that using a code generator as the basis for in-house development was related to an increase in maintenance hours, while using packaged applications was associated with a decrease in maintenance effort.

Of course, in-house development need not entail development of all of the software that will comprise the total system. Hybrid solutions involving some purchased and in-house software components are common. Table 11-5 compares six different software sources.

If you choose to acquire software from outside sources, this choice is made at the end of the analysis phase. Choosing between a package or an external supplier will be determined by your needs, not by what the supplier has to sell. As we will discuss, the results of your analysis study will define the type of product you want to buy and will make working with an external supplier much easier, productive, and worthwhile.

TABLE 11-5   Comparison of Six Different Sources of Software Components

Producers	Source of Application Software?	When to Go to This Type of Organization for Software	Internal Staffing Requirements
Hardware manufacturers	Generally not	For system software and utilities	Varies
Packaged software producers	Yes	When supported task is generic	Some IS and user staff to define requirements and evaluate packages
Custom software producers	Yes	When task requires custom support and system can't be built internally	Internal staff may be needed, depending on application
Application service providers	Yes	When supported task is generic, or buying and installing the system locally would be too expensive, or for instant access to an application	Ideally, none
Enterprise-wide solutions	Yes	For complete systems that cross functional boundaries	Some internal staff necessary but mostly need consultants
In-house developers	Yes	When resources and staff are available and system must be built from scratch	Internal staff necessary though staff size may vary

Choosing Off-the-Shelf Software

Once you have decided to purchase off-the-shelf software rather than write some or all of the software for your new system, how do you decide what to buy? There are several criteria to consider, and special criteria may arise with each potential software purchase. For each criterion, an explicit comparison should be made between the software package and the process of developing the same application in-house. The most common criteria are as follows:

- Cost
- Functionality
- Vendor support
- Viability of vendor
- Flexibility
- Documentation
- Response time
- Ease of installation

These criteria are presented in no particular order. The relative importance of the criteria will vary from project to project and from organization to organization. If you had to choose two criteria that would always be among the most important, those two would probably be vendor viability and vendor support. You don't want to get involved with a vendor that might not be in business tomorrow. Similarly, you don't want to license software from a vendor with a reputation for poor support. How you rank the importance of the remaining criteria will very much depend on the specific situation in which you find yourself.

Cost involves comparing the cost of developing the same system in-house to the cost of purchasing or licensing the software package. You should include a comparison of the cost of purchasing vendor upgrades or annual license fees with the costs you would incur to maintain your own software. Costs for purchasing and developing in-house can be compared based on the economic feasibility measures outlined in Chapter 6 (for example, a present value can be calculated for the cash flow associated with each alternative). Functionality refers to the tasks the software can perform and the mandatory, essential, and desired system features. Can the software package perform all or just some of the tasks your users need? If some, can it perform the necessary core tasks? Note that meeting user requirements occurs at the end of the analysis phase because you cannot evaluate packaged software until user requirements have been gathered and structured. Purchasing application software is not a substitute for conducting the systems analysis phase; rather, purchasing software is part of one design strategy for acquiring the system identified during analysis.

As we said earlier, vendor support refers to whether and how much support the vendor can provide. Support occurs in the form of assistance to install the software, to train user and systems staff on the software, and to provide help as problems arise after installation. Recently, many software companies have significantly reduced the amount of free support they will provide customers, so the cost to use telephone, on-site, fax, or computer bulletin board support facilities should be considered. Related to support is the vendor's viability. You don't want to get stuck with software developed by a vendor that might go out of business soon. This latter point should not be minimized. The software industry is quite dynamic, and innovative application software is created by entrepreneurs working from home offices—the classic cottage industry. Such organizations, even with outstanding software, often do not have the resources or business management ability to stay in business very long. Further, competitive moves by major software firms can render the products of smaller firms outdated or incompatible with operating systems. One software firm we talked to while developing this book was struggling to survive just trying to make their software work on any supposedly IBM-compatible PC (given the infinite combination of video boards, monitors, BIOS chips, and other components). Keeping up with hardware and system software change may be more than a small firm can handle, and good off-the-shelf application software is lost.

Flexibility refers to how easy it is for you, or the vendor, to customize the software. If the software is not very flexible, your users may have to adapt the way they work to fit the software. Are they likely to adapt in this manner? Purchased software can be modified in several ways. Sometimes, the vendor will be willing to make custom changes for you, if you are willing to pay for the redesign and programming. Some vendors design the software for customization. For example, the software may include several different ways of processing data and, at installation time, the customer chooses which to initiate. Also, displays and reports may be easily redesigned if these modules are written in a fourth-generation language. Reports, forms, and displays may be easily customized using a process whereby your company name and chosen titles for reports, displays, forms, column headings, and so forth are selected from a table of parameters you provide. You may want to employ some of these same customization techniques for in-house developed systems so that the software can be easily adapted for different business units, product lines, or departments.

Documentation includes the user's manual as well as technical documentation. How understandable and up-to-date is the documentation? What is the cost for multiple copies, if required? Response time refers to how long it takes the software package to respond to the user's requests in an interactive session. Another measure of time would be how long it takes the software to complete running a job. Finally, ease of installation is a measure of the difficulty of loading the software and making it operational.

Validating Purchased Software Information   One way to get all of the information you want about a software package is to collect it from the vendor. Some of this information may be contained in the software documentation and technical marketing literature. Other information can be provided upon request. For example, you can send prospective vendors a questionnaire asking specific questions about their packages. This may be part of a **request for proposal (RFP)** or request for quote (RFQ) process your organization requires when major purchases are made. Space does not permit us to discuss the topic of RFPs and RFQs here; you may wish to refer to purchasing and marketing texts if you are unfamiliar with such processes (additional references about RFPs and RFQs are found at the end of the chapter).

There is, of course, no replacement for actually using the software yourself and running it through a series of tests based on the criteria for selecting software. Remember to test not only the software but also the documentation, training materials, and even the technical support facilities. One requirement you can place on prospective software vendors as part of the bidding process is that they install (free or at an agreed-upon cost) their software for a limited amount of time on your computers. This way you can determine how their software works in your environment, not in some optimized environment they have.

One of the most reliable and insightful sources is other users of the software. Vendors will usually provide a list of customers (remember, they will naturally tell you about satisfied customers, so you may have to probe for a cross section of customers) and people who are willing to be contacted by prospective customers. And here is where your personal network of contacts, developed through professional groups, college friends, trade associations, or local business clubs, can be a resource; do not hesitate to find some contacts on your own. Such current or former customers can provide a depth of insight on use of a package at their organizations.

To gain a range of opinion about possible packages, you can use independent software testing and abstracting services that periodically evaluate software and collect user opinions. Such surveys are available for a fee either as subscription services or on demand (two popular services are Auerbach Publishers and DataPro); occasionally, unbiased surveys appear in trade publications. Often, however, articles in trade publications, even software reviews, are actually seeded by the software manufacturer and are not unbiased.

If you are comparing several software packages, you can assign scores for each package on each criterion and compare the scores using the quantitative method we demonstrate at the end of the chapter for comparing alternative system design strategies.

Hardware and Systems Software Issues

The first question you need to ask yourself about hardware and system software is whether the new system that follows a particular design strategy can be run on your firm's existing hardware and software platform. System software refers to such key components as operating systems, database management systems, programming languages, code generators, and network software. To determine if current hardware and system software is sufficient, you should consider such factors as the age and capacity of the current hardware and system software, the fit between the hardware and software and your new application's goals and proposed functionality and, if some of your system components are off-the-shelf software, whether the software can run on the existing hardware and system software. The advantages to running your new system on the existing platform are persuasive:

1. Lower costs as little, if any, new hardware and system software has to be purchased and installed.

Request for proposal (RFP): An RFP is a document provided to vendors to ask them to propose hardware and system software that will meet the requirements of your new system.

N E T   S E A R C H
A Request for Proposal has a rather standard format. Visit http://www. prenhall.com/hoffer to complete an exercise related to this topic.

2. Your information systems staff is quite familiar with the existing platform and how to operate and maintain it.

3. The odds of integrating your new application system with existing applications are enhanced.

4. No added costs of converting old systems to a new platform, if necessary, or of translating existing data between current technology and the new hardware and system software you have to acquire for your system.

On the other hand, there are also very persuasive reasons for acquiring new hardware or system software:

1. Some software components of your new system will only run on particular platforms with particular operating systems.

2. Developing your system for a new platform gives your organization the opportunity to upgrade or expand its current technology holdings.

3. New platform requirements may allow your organization to radically change its computing operations, as in moving from mainframe-centered processing to a database machine or a client-server architecture.

As the determination of whether or not to acquire new hardware and system software is so context-dependent, providing platform options as part of your design strategy alternatives is an essential practice.

At this point, if you decide that new hardware or system software is a strong possibility, you may want to issue a request for proposal (RFP) to vendors. The RFP will ask the vendors to propose hardware and system software that will meet the requirements of your new system. Issuing an RFP gives you the opportunity to have vendors carry out the research you need in order to decide among various options. You can request that each bid submitted by a vendor contain certain information essential for you to decide on what best fits your needs. For example, you can ask for performance information related to speed and number of operations per second. You can ask about machine reliability and service availability and whether there is an installation nearby that you can visit for more information. You can ask to take part in a demonstration of the hardware. And of course the bid will include information on cost. You can then use the information you have collected in generating your alternative design strategies.

Implementation Issues

As you will see in Chapter 17, implementing a new information system is just as much an organizational change process as it is a technical process. Implementation involves more than installing a piece of software, turning it on, and moving on to the next software project. New systems often entail new ways of performing the same work, new working relationships, and new skills. Users have to be trained. Disruptions in work procedures have to be found and addressed. In addition, system implementation may be phased in over many weeks or even months. You must address the technical and social aspects of implementation as part of any alternative design strategy. Management and users will want to know how long the implementation will take, how much training will be required, and how disruptive the process will be.

Organizational Issues

One reason management is so interested in the outlook for implementation is that all of the concerns we listed previously cost management money. The longer the implementation process, the more training required; the more disruption expected, the more it will cost to implement the system. Implementation costs are just one cost management has to consider for the new system. Management must also consider the costs of the design process that precedes implementation and the cost of maintaining the system once implementation is over. Overall cost and the availability of funding is just one of the organizational issues to consider in developing alternative design strategies.

A second organizational issue is determining what management will support. Even if adequate funding is available, your organization's management may not be willing or able to support one or another of your alternatives. For example, if your new system differs dramatically from what the corporate office has determined is adequate or from the corporate standard computing environment, your local management may not be willing to support a system that will irritate the corporate office. Most organizations like to have a manageably small number of basic technologies since only a few choices on hardware and software platforms can be supported well. If your new system calls for high levels of cooperation across departments when current operations involve very little of such cooperation, your management may not be willing to support such a system. Your management may also have politically inspired reasons not to support your new system. For example, if your information systems unit reports to the organization's chief financial officer and your new system strengthens a rival department such as manufacturing, management may not support such a system because they prefer to continue the status quo whereby finance is stronger than manufacturing.

A third organizational issue is the extent to which users will accept the new system and use it as designed. A high-end, high-technology solution that represents a radical break in what users are familiar with may have less of a chance of acceptance than a system that is closer to what users know and use. On the other hand, acceptance of a high-end alternative will depend on the users. Some users may demand nothing less than what leading-edge technology can deliver.

These three organizational issues are not the only organizational issues you could, or should, consider. Your assessment of operational and political feasibility (see Chapter 6), which is updated during analysis, may identify other organizational issues. Most likely, such organizational issues will affect all of the alternative design strategies you will develop, but it may be possible to offer a range of options in your alternatives that will allow management to make trade-offs between different approaches to various organizational issues. For example, management may be willing to support a system that requires making a little more funding available if it means a little less of a change in the status quo than you might believe is necessary.

DEVELOPING DESIGN STRATEGIES FOR HOOSIER BURGER'S NEW INVENTORY CONTROL SYSTEM

As an example of alternative generation and selection, let's look again at Hoosier Burger's inventory control system. Figure 11-2 lists ranked requirements and constraints for the enhanced information system being considered by Hoosier Burger. The requirements represent a sample of those developed from the requirements determination and structuring carried out in prior analysis steps. The system in question is an upgrade to the company's existing inventory system, which was used as an example in Chapters 8, 9, and 10. As you remember, there were several steps to Bob Mellankamp's largely manual inventory control system (Figure 11-3).

Remember that when Bob receives invoices from suppliers, he records their receipt on an invoice log sheet, and he puts the actual invoices in his accordion file. Using the invoices, Bob records the amount of stock delivered on the stock logs, paper forms posted near the point of storage for each inventory item. The stock logs

Figure 11-2
Ranked system requirements and constraints for Hoosier Burger's inventory system

SYSTEM REQUIREMENTS
(in descending priority)
1. Must be able to easily enter shipments into system as soon as they are received.
2. System must automatically determine whether and when a new order should be placed.
3. Management should be able to determine at any time approximately what inventory levels are for any given item in stock.

SYSTEM CONSTRAINTS
(in descending order)
1. System development can cost no more than $50,000.
2. New hardware can cost no more than $25,000.
3. The new system must cost no more than $5,000 per year to operate.
4. Training needs must be minimal, that is, the new system must be very easy to use.

include minimum order quantities as well as spaces for posting the starting amount, amount delivered, and the amount used for each item. Amounts delivered are entered on the sheet when Bob logs stock deliveries; amounts used are entered after Bob has compared the amounts of stock used, according to a physical count, and according to the numbers on the inventory report generated by the food ordering system. You remember too that some Hoosier Burger items, especially perishable goods, have standing orders for daily (or Saturday) delivery.

The Mellankamps want to improve their inventory system so that new orders are immediately accounted for, so that the system can determine when new orders should be placed, and so that management can obtain accurate inventory levels at any time of the day, as we saw in Chapter 8. All three of these system requirements have been ranked in order of descending priority in Figure 11-2.

The constraints on developing an enhanced inventory system at Hoosier Burger are also listed in Figure 11-2, again in order of descending priority. The first two constraints cover costs for systems development and for new computer hardware. Development can cost no more than $50,000. New hardware can cost no more than $25,000. The third constraint involves operating costs, which must be less than $5,000 per year. Finally, Hoosier Burger would prefer that training for the system be simple; the new system must be designed so that it is easy to use. However, as this is the fourth most important constraint, the demands it makes are more flexible than those contained in the other three.

Any set of alternative solutions to Hoosier Burger's inventory system problems must be developed with the company's prioritized requirements and constraints in mind. Figure 11-4 illustrates how each of three possible alternatives meets (or exceeds) the criteria implied in Hoosier Burger's requirements and constraints. Alternative A is a low-end solution, proposed by one of the restaurant's best cus-

Figure 11-3
The steps in Hoosier Burger's inventory control system

1. Meet delivery trucks before opening restaurant
2. Unload and store deliveries
3. Log invoices and file in accordion file
4. Manually add amounts received to stock logs
5. After closing, print inventory report
6. Count physical inventory amounts
7. Compare inventory reports totals to physical count totals
8. Compare physical count totals to minimum order quantities; if the amount is less, make order; if not, do nothing
9. Pay bills that are due and record them as paid

CRITERIA	ALTERNATIVE A	ALTERNATIVE B	ALTERNATIVE C
Requirements			
1. Easy real-time entry of new shipment data	Yes	Yes	Yes
2. Automatic reorder decisions	No	Yes	Yes
3. Real-time data on inventory levels	Not available	Yes	Yes
Constraints			
1. Cost to develop	$15,000	None	$50,000
2. Cost of hardware	$10,000	$10,000	$25,000
3. Operating costs	$1,000/year	$6,000 + $1,000 for support per year + $250 per year for ISP connection	$1,000/year
4. Ease of training	One day	One week	One week

Figure 11-4
Description of three alternative systems that could be developed for Hoosier Burger's inventory system

tomers, a faculty member who works as a systems development consultant on the side. Alternative A involves automating Hoosier Burger's current inventory files using a PC-based relational database program. It only meets the first requirement and adds nothing for the second and third requirements. However, Alternative A is relatively inexpensive to develop and operate, and it requires hardware that is less expensive than the largest amount Hoosier Burger is willing to pay. Alternative A also meets the requirement for the fourth constraint: Users will require only one day of training. Alternative C is the high-end solution. The Mellankamps went to a local consulting firm to get the bid for Alternative C. It meets all of the requirements criteria. On the other hand, Alternative C is close to violating two of the four constraints: It will cost $50,000 to develop and will require $25,000 in new hardware. Alternative B is in the middle. It represents a new inventory and purchasing system made available over the Web by an Application Service Provider that specializes in the restaurant industry. The system costs only $25 per month per module (there are two different modules, one for inventory and one for purchasing). Support is available at $100 per incident requiring support. The only other operating cost is for the Internet Service Provider connection so that Hoosier Burger can access the system over the Internet. The system is designed to issue purchase orders and to enter the amount of stock received when the goods ordered are delivered. The inventory module also allows for re-entry points to be set up for each inventory item. All of these same features are available in Alternative C, but unlike Alternative C, there are no development costs for Alternative B. This alternative solution meets all of the requirements and violates only one constraint—the annual operating costs of $7,250 are higher than the limit of $5,000.

Now that three plausible alternative solutions have been generated for Hoosier Burger, the analyst hired to study the problem, the same analyst that generated the proposal for Alternative C, has to decide which one to recommend to management for development. Management will then decide whether to continue with the development project (incremental commitment) and whether the system recommended by the analyst is the system that should be developed.

SELECTING THE MOST LIKELY ALTERNATIVE

One method we can use to decide among the alternative solutions to Hoosier Burger's inventory system problem is illustrated in Figure 11-5. On the left, you see that we have listed all three system requirements and all four constraints from Figure 11-2. These are our decision criteria. We have weighted requirements and constraints equally; that is, we believe that requirements are just as important as constraints. We do not have to weigh requirements and constraints equally; it is certainly possible to make requirements more or less important than constraints. Weights are arrived at in discussions among the analysis team, users, and sometimes managers. Weights tend to be fairly subjective and, for that reason, should be determined through a process of open discussion to reveal underlying assumptions, followed by an attempt to reach consensus among stakeholders. We have also assigned weights to each individual requirement and constraint. Notice that the total of the weights for both requirements and constraints is 50. Our weights correspond with our prioritization of the requirements and constraints.

The next step we have taken is to rate each requirement and constraint for each alternative, on a scale of 1 to 5. A rating of one indicates that the alternative does not meet the requirement very well or that the alternative violates the constraint. A rating of five indicates that the alternative meets or exceeds the requirement or clearly abides by the constraint. Ratings are even more subjective than weights and should also be determined through open discussion among users, analysts, and managers. The next step we have taken is to multiply the rating for each requirement and each constraint by its weight, and we have followed this procedure for each alternative. The final step is to add up the weighted scores for each alternative. Notice that we have included three sets of totals: for requirements, for constraints, and overall totals. If you look at the totals for requirements, Alternative C is the best choice, as it meets or exceeds all requirements. However, if you look only at constraints, Alternative A is the best choice, as it does not violate any constraints. When we combine the totals for requirements and constraints, we see that the best choice is Alternative C, even though it had the lowest score for constraints, as it has the highest overall score.

Alternative C, then, appears to be the best choice for Hoosier Burger. Whether Alternative C is actually chosen for development is another issue. The

Figure 11-5
Weighted approach for comparing the three alternative systems for Hoosier Burger's inventory system

Criteria	Weight	Alternative A		Alternative B		Alternative C	
		Rating	Score	Rating	Score	Rating	Score
Requirements							
Real-time data entry	18	5	90	5	90	5	90
Automatic reorder	18	1	18	5	90	5	90
Real-time data query	14	1	14	5	70	5	70
	50		122		250		250
Constraints							
Developer costs	15	4	60	5	75	3	45
Hardware costs	15	4	60	4	60	3	45
Operating costs	15	5	75	1	15	5	75
Ease of training	5	5	25	3	15	3	15
	50		220		165		180
Total	100		342		415		430

Mellankamps may be concerned that Alternative C comes close to violating two constraints, including the most important one, development costs. On the other hand, the owners (and chief users) at Hoosier Burger may so desire the full functionality Alternative C offers that they are willing to ignore the constraints. Or Hoosier Burger's management may be so interested in cutting costs they prefer Alternative A, even though its functionality is severely limited. What may appear to be the best choice for a systems development project may not always be the one that ends up being developed.

UPDATING THE BASELINE PROJECT PLAN

You will recall that the Baseline Project Plan was developed during project initiation and planning (see Chapter 6) to explain the nature of the requested system and the project to develop it. The plan includes (we presented this originally in Figure 6-10 and reproduce it here as Figure 11-6) a preliminary description of the system as requested, an assessment of the feasibility or justification for the system (the business case), and an overview of management issues for the system and project. It was this plan that was presented to a steering committee or other body who approved the commitment of funds to conduct the analysis phase just completed. Thus, it is time to report back (in written and oral form) to this group on the project's progress and to update the group on the findings from analysis. This group will make the final decision on the design strategy to be followed and approve the commitment of resources outlined from the logical (and possibly physical) design steps. Of course, this group could determine that the business case has not developed as originally thought and either stop or drastically redirect the project.

The outline of the Baseline Project Plan can still be used for the analysis phase status report. The updated plan will typically be longer as more is known on each topic. Further, the various process, logic, and data models are often included to make the system description more specific. Usually only high-level versions of the diagrams are included within section 2.0, and more detailed versions are provided as appendices.

Every section of the Baseline Project Plan Report is updated at this point. For example, section 1.0.B will now contain the recommendation for the design strategy chosen by the analysis team. Section 2.0.A provides the descriptions of the competing strategies studied during alternative generation and selection, often including the types of comparison charts shown earlier in this chapter. Section 3.0 is typically significantly changed since you now know much better than you did during project initiation and planning what the needs of the organization are. For example, economic benefits that were intangible before may now be tangible. Risks, especially operational ones, are likely better understood.

Section 3.0.F will now show the actual activities and their durations during the analysis phase, as well as include a detailed schedule for the activities in the design phases and whatever additional details can be anticipated for later phases. Many Gantt charting packages can show actual progress versus planned activities. It is important to show in this section how well the actual conduct of the analysis phase matched the planned activities. This helps you and management understand how well the project is understood and how likely it is that the stated future schedule will occur. Those activities whose actual durations differed significantly from planned durations may be very useful to you in estimating future activity durations. For example, a longer than expected task to analyze a certain process on a DFD may suggest that the design of system features to support this process may take longer than originally anticipated.

Figure 11-6
Outline of Baseline Project Plan

BASELINE PROJECT PLAN REPORT

1.0 Introduction
  A. Project Overview–Provides an executive summary that specifies the project's scope, feasibility, justification, resource requirements, and schedules. Additionally, a brief statement of the problem, the environment in which the system is to be implemented, and constraints that affect the project are provided.
  B. Recommendation–Provides a summary of important findings from the planning process and recommendations for subsequent activities.

2.0 System Description
  A. Alternatives–Provides a brief presentation of alternative system configurations.
  B. System Description–Provides a description of the selected configuration and a narrative of input information, tasks performed, and resultant information.

3.0 Feasibility Assessment
  A. Economic Analysis–Provides an economic justification for the system using cost-benefit analysis.
  B. Technical Analysis–Provides a discussion of relevant technical risk factors and an overall risk rating of the project.
  C. Operational Analysis–Provides an analysis of how the proposed system solves business problems or takes advantage of business opportunities in addition to an assessment of how current day-to-day activities will be changed by the system.
  D. Legal and Contractual Analysis–Provides a description of any legal or contractual risks related to the project (e.g., copyright or nondisclosure issues, data capture or transferring, and so on).
  E. Political Analysis–Provides a description of how key stakeholders within the organization view the proposed system.
  F. Schedules, Timeline, and Resource Analysis–Provides a description of potential timeframe and completion date scenarios using various resource allocation schemes.

4.0 Management Issues
  A. Team Configuration and Management–Provides a description of the team member roles and reporting relationships.
  B. Communication Plan–Provides a description of the communication procedures to be followed by management, team members, and the customer.
  C. Project Standards and Procedures–Provides a description of how deliverables will be evaluated and accepted by the customer.
  D. Other Project-Specific Topics–Provides a description of any other relevant issues related to the project uncovered during planning.

Often the design phase activities will be driven by the capabilities chosen for the recommended design strategy. For example, you will place specific design activities on the schedule for such design deliverables as the following:

- Layout of each report and data input and display screens (the DFDs include data flows for each of these)

- Structuring of data into logical tables or files (the E-R diagrams identify what data entities are involved in this)

- Programs and program modules that need to be described (processes on DFD and process and temporal logic models explain how complicated these tasks will be)

- Training on new technologies to be used in implementing the system

Many design phase activities result in developing design specifications for one or more examples of the types of design deliverables listed previously.

Section 4.0 is also updated. It is likely that the project team needs to change as new skills are needed in the next and subsequent project phases. Also, since project team members are often evaluated after each phase, the project leader may request the reassignment of a team member who has not performed as required. The communication plan needs to be reassessed to see if other communication methods need to be employed (see Table 3-2 for a list of common communication methods). New standards and procedures will be necessary as the team discovers that some current procedures are inadequate for the new tasks. Section 4.0.D is often used to outline issues for management that have been discovered during analysis. Recall, for example, that we discussed in Chapter 7 how you might find redundancies and inconsistencies in job descriptions and the way people actually do their jobs. Since these issues must be resolved by management, and must be addressed before you can progress into detailed system design, now is the last time to call these issues to the attention of management.

As the project leader, you and other analysts also must ensure that the project workbook and CASE repository are completely up-to-date as you finalize the analysis phase. Since the project team composition will likely change and, as time passes, you forget facts learned in earlier stages, the workbook and repository are necessary to transfer information between phases. This is also a good time for the project leader to do a final check that all elements of project execution have been properly handled (Figure 3-13).

Besides the written Baseline Project Plan Report update, an oral presentation is typically made, and it may be at this meeting that a decision to approve your recommendations, redirect your recommendations, or kill the project is made. It is not uncommon for the analysis team to follow this project review meeting with a suitable celebration for reaching an important project milestone.

Before and After Baseline Project Plans for Hoosier Burger

Even though their Inventory Control System was relatively small, Hoosier Burger developed a Baseline Project Plan for the project. The plan included information for each area listed in the outline in Figure 11-6, parts of which are reproduced below. Now that the analysis phase of the life cycle has ended, the plan must be updated, and those updated sections are also reproduced below. The first item we will consider is the cost-benefit analysis prepared as part of Section 3.0.A, on economic analysis. Hoosier Burger's initial cost-benefit analysis for the inventory project is shown in Figure 11-7. The format is the same as you saw in Chapter 6.

The numbers in the spreadsheet are based in part on the constraints listed in Figure 11-2. Part of the worksheet the Mellankamps used to determine the values in the spreadsheet in Figure 11-7 is shown in Table 11-6. The Mellankamps estimated that the benefits of the new system could be quantified in two ways: first, instant entry of shipment data would lead to more accurate inventory data; second, Hoosier Burger would be less likely to run out of stock with an automatic order determination as part of the system. Savings from more accurate data amount to $1,250 per month or $15,000 per year; savings from fewer stockouts amount to $1,000 per month or $12,000 per year. As you can see from Figure 11-7, although a new inventory control system for Hoosier Burger would result in a positive return, it is not a very good investment, with only a 5% return, given a 12% discount rate.

Hoosier Burger
Economic Feasibility Analysis
Inventory Control System

		Year of Project					
	Year 0	1	2	3	4	5	TOTALS
Net economic benefit	$0	$27,000	$27,000	$27,000	$27,000	$27,000	
Discount rate (12%)	1	0.8928571	0.7971939	0.7117802	0.6355181	0.5674269	
PV of benefits	$0	$24,107	$21,524	$19,218	$17,159	$15,321	
NPV of all BENEFITS	$0	$24,107	$45,631	$64,849	$82,008	$97,329	$97,329
One-time COSTS	($75,000)						
Recurring costs	$0	($5,000)	($5,000)	($5,000)	($5,000)	($5,000)	
Discount rate (12%)	1	0.8928571	0.7971939	0.7117802	0.6355181	0.5674269	
PV of Recurring Costs	$0	($4,464)	($3,986)	($3,559)	($3,178)	($2,837)	
NPV of all COSTS	($75,000)	($79,464)	($83,450)	($87,009)	($90,187)	($93,024)	($93,024)
Overall NPV							$4,305
Overall ROI (Overall NPV/ NPV of all COSTS)							0.05

Figure 11-7
Hoosier Burger's initial cost-benefit analysis for its Inventory Control System project

Figure 11-8 shows the cost-benefit analysis after the analysis phase has ended, which appears in the updated Baseline Project Plan. Notice that developing the new system, represented by Alternative C, is now a much better investment, with a 58% return. What happened?

Much of Figure 11-8 is the same as Figure 11-7. The discount rate is the same. What has changed, in addition to lower maintenance costs, is that net benefits are now estimated to be larger than in Figure 11-7. Details on these new estimates are

TABLE 11-6 Hoosier Burger's Initial Economic Analysis Worksheet

One-time costs: Development	$50,000
One-time costs: Hardware	$25,000
Recurring costs: Maintenance	$5,000 per year
Savings: Fewer stockouts due to automatic reordering	$12,000 per year
Savings: More accurate data from shipment logging	$15,000 per year
Intangible benefit: Better management information	

Hoosier Burger
Economic Feasibility Analysis
Inventory Control System

		Year of Project					
	Year 0	1	2	3	4	5	TOTALS
Net economic benefit	$0	$36,000	$36,000	$36,000	$36,000	$36,000	
Discount rate (12%)	1	0.8928571	0.7971939	0.7117802	0.6355181	0.5674269	
PV of benefits	$0	$32,143	$28,699	$25,624	$22,879	$20,427	
NPV of all BENEFITS	$0	$32,143	$60,842	$86,466	$109,345	$129,772	$129,772
One-time COSTS	($75,000)						
Recurring costs	$0	($2,000)	($2,000)	($2,000)	($2,000)	($2,000)	
Discount rate (12%)	1	0.8928571	0.7971939	0.7117802	0.6355181	0.5674269	
PV of Recurring Costs	$0	($1,786)	($1,594)	($1,424)	($1,271)	($1,135)	
NPV of all COSTS	($75,000)	($76,786)	($78,380)	($79,804)	($81,075)	($82,210)	($82,210)
Overall NPV							$47,562
Overall ROI (Overall NPV/ NPV of all COSTS)							0.58

Figure 11-8
Hoosier Burger's initial cost-benefit analysis for its Inventory Control System project

shown in Table 11-7. The Mellankamps reestimated the savings from the new system, using more accurate data, and found they had been too optimistic. The expected savings from instantly logging new shipments of supplies were reduced from $15,000 per year to $12,000 per year. But the Mellankamps also found a new benefit: They realized that better management information, and its ready availability through a new query capability, could be quantified at about $1,000 per month. The Mellankamps did not just dream up the $1,000 per month savings estimate. They

TABLE 11-7   Hoosier Burger's Updated Economic Analysis Worksheet

One-time costs: Development	$50,000
One-time costs: Hardware	$25,000
Recurring costs: Maintenance	$2,000 per year
Savings: Fewer stockouts due to automatic reordering	$12,000 per year
Savings: More accurate data from shipment logging	$12,000 per year
Savings: Better management information and availability	$12,000 per year

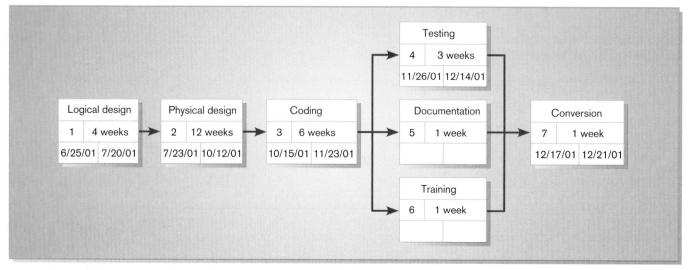

Figure 11-9
Hoosier Burger's initial schedule for its Inventory Control System project

developed it from thinking about how timely, accurate management information would affect their ability to prepare reports as well as their ability to improve their operations through better inventory control. So even though Alternative C was more expensive to develop than the other alternatives, it actually resulted in a higher level of tangible benefits.

Figure 11-9 shows the part of the project schedule from the initial version of the Baseline Project Plan that applies to subsequent steps. Notice that the schedule covers only the design and implementation phases of the life cycle and that the schedule is very general. The physical design subphase is not broken down into its constituent parts. The task times in the schedule are also driven by two of the constraints listed in Figure 11-2. The entire schedule spans exactly six months of activity and training takes only one week. The estimates of how long each task should take to complete are all very rough.

Compare Figure 11-9 to Figure 11-10, the revised schedule that goes in the updated Baseline Project Plan. The schedule is more detailed and it more closely

Figure 11-10
Hoosier Burger's revised schedule for its Inventory Control System project

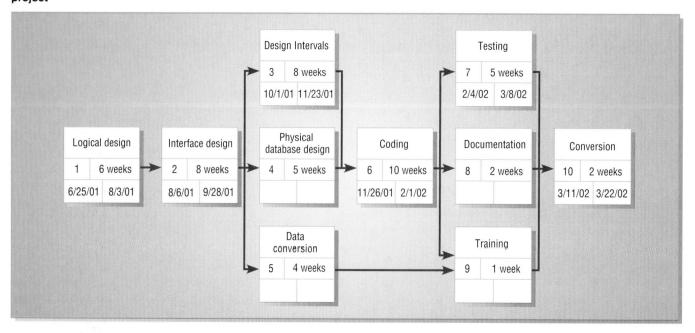

reflects the development time necessary for Alternative C. Training still takes only one week, but now that estimate is based on a clear understanding of the requirements of a particular system rather than based primarily on positive thinking. Also, the entire schedule now spans nine months, the time necessary to fully develop and implement Alternative C. Some design tasks in Figure 11-10 have been decomposed into four different subtasks, many of which can be worked on concurrently: interface design, designing the internals, physical database design, and data conversion. Note that even this schedule presents the project at a very high level. It would be typical in actual projects to show not only the major steps but also the individual activities needed to complete each step. For example, Interface design might be broken down into many steps for each display or report and for different activities, such as meetings with users to walk through the tentative designs.

We have shown you only two parts to the Hoosier Burger Baseline Project Plan for their Inventory Control System. Even for a project this small, a complete Baseline Project Plan would be too much to include in this book. From these examples, though, you should get a good general idea both of what an initial Baseline Project Plan contains and how it changes when a major life-cycle phase, like analysis, ends.

INTERNET DEVELOPMENT: SELECTING THE BEST ALTERNATIVE DESIGN STRATEGY

Like many other analysis and design activities, selecting the best design strategy for an Internet-based electronic commerce application is no different than the process followed when selecting the optimal design strategy for other types of applications. In the last chapter, you read how Jim modeled the data requirements for the WebStore system. In this section, we examine the process he followed when assessing and selecting a design strategy for the WebStore.

Selecting the Best Alternative Design Strategy for Pine Valley Furniture's WebStore

PINE VALLEY FURNITURE

As Jim began to evaluate the possible design options for the WebStore, he quickly realized that he and PVF's technical group had limited understanding of Internet application development. Consequently, he recommended to PVF management that a consulting firm be hired to assist in setting the WebStore design options. Management quickly approved this recommendation and Jim retained a small consulting organization that had a very strong reputation for designing and developing very high-quality electronic commerce solutions. Once on contract, Jim worked with the consulting firm to solidify the system requirements and constraints. During this process, they organized the requirements into three categories: minimum system requirements, essential system requirements, and desired system requirements. The list of system requirements is summarized in Table 11-8. In addition to the system requirements, they also identified four significant constraints that any design must address. The list of system constraints is also summarized in Table 11-8.

Next, Jim and the consultants defined three alternative system designs, with advantages and disadvantages for each. PVF management requested that three alternative designs be defined so that clear comparisons could be made between a low-end (low cost and limited features), high-end (high cost and extensive features), and mid-level designs (moderate cost and features). Table 11-9 summarizes the results of this analysis. Now that both the system requirements and constraints were defined as well as the alternative system designs, a meeting was held with PVF management to

TABLE 11-8   WebStore System Requirements and Constraints

Requirements	Constraints
Minimum System Requirements: • Full integration with current inventory, sales, and customer tracking systems • 99.9% uptime and availability *Essential System Requirements*: • Flexibility and scalability for future systems integration • Efficient and cost-effective system management *Desired System Requirements*: • Available support and/or emergency response • Documentation	• Christmas season roll-out • Small development/ support staff • Transaction-style interaction with current systems • Limited external consultation budget

Scalable: The ability to seamlessly upgrade the capabilities of the system through either hardware upgrades, software upgrades, or both.

Web server: A computer that is connected to the Internet and that stores files written in HTML—Hypertext Markup Language—that are publicly available through an Internet connection.

Application server: A "middle-tier" software and hardware combination that lies between the Web Server and the corporate network and systems.

select a design strategy for the WebStore. At this meeting, it was unanimously agreed upon that option 3 in Table 11-9, "Application Server/Object Framework," best suited both PVF's current needs and future growth initiatives.

The proposed system would incorporate a scalable three-tier architecture to integrate the WebStore with the current systems. A **scalable** system has the ability to seamlessly upgrade the capabilities of the system through hardware upgrades, software upgrades, or both. As Figure 11-11 shows, Tier 1, the Web Server layer, processes incoming Internet requests. For example, a *scalable* electronic commerce system would be one that could effectively handle six requests per second with one server and by adding a second server, twelve requests per second could be effectively handled. The **Web Server** is a computer that is connected to the Internet and stores files written in HTML—Hypertext Markup Language—that are publicly available through an Internet connection. As shown in Figure 11-11, the Web Server layer communicates with Tier 2, the Application Server layer. The **Application Server** is a "middle-tier" software and hardware combination that lies between the Web Server and the corporate network and systems such as the Customer Tracking System, Inventory System, and the Order Fulfillment System. In other words, the Web Server will manage the client interaction and broker requests to the middle-tier Application

TABLE 11-9   Three Alternative Systems and Their Advantages and Disadvantages

1. Outsource Application Service Provider (Low-end)

Advantages:	*Disadvantages*:
• All hardware is located off-site • Application is developed and professionally managed off-site • Excellent emergency response	• Inflexible • Difficult to integrate with current systems • Shared resources with other clients

2. Enterprise Resource Planning System (High-end)

Advantages:	*Disadvantages*:
• Stability • Available documentation	• Requires skilled internal staff • Expensive hardware and software • Big learning curve

3. Application Server/Object Framework (Moderate)

Advantages:	*Disadvantages*:
• Excellent integration with current system • Scalability • Flexible	• Requires internal development (and/or a professional consultation) • Proprietary • Documentation must be created during planning and development

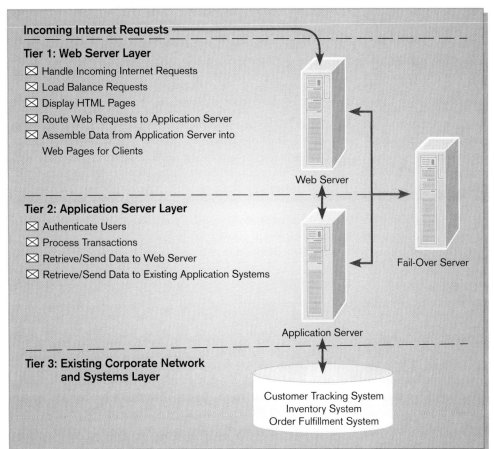

Figure 11-11
WebStore multi-tier system architecture

Incoming Internet Requests

Tier 1: Web Server Layer
- ☒ Handle Incoming Internet Requests
- ☒ Load Balance Requests
- ☒ Display HTML Pages
- ☒ Route Web Requests to Application Server
- ☒ Assemble Data from Application Server into Web Pages for Clients

Web Server

Tier 2: Application Server Layer
- ☒ Authenticate Users
- ☒ Process Transactions
- ☒ Retrieve/Send Data to Web Server
- ☒ Retrieve/Send Data to Existing Application Systems

Fail-Over Server

Application Server

Tier 3: Existing Corporate Network and Systems Layer

Customer Tracking System
Inventory System
Order Fulfillment System

Server. The Application Server will manage the data specific to running the WebStore (shopping carts, promotions, site logs, etc.) and will also manage all interactions with existing PVF systems for managing customers, inventory, and orders. A third server, the Fail-Over Server, will be an emergency backup system that will be on standby, ready to take the place of either server should one fail. Each of these separate components—Web Server, Application Server, Fail-Over Server—can be thought of as an *object* (see Chapter 20), each with a well-defined role that can be easily defined, designed, implemented, and modified. For this reason, option 3 is referred to as an *Application Server/ Object Framework* architecture.

Now that the basic architecture of the WebStore has been defined, Jim worked with the consultants to further refine the specifications of the system. The PVF development staff will use these specifications as a blueprint in their development efforts. With this detailed specification, PVF developers will now be able to implement all six systems requirements and comply with each of the four constraints listed in Table 11-8.

Summary

In the alternative generation and choice phase of systems analysis, you develop alternative solutions, or design strategies, to the organization's information system problem. A design strategy is a combination of system features, hardware and system software platform, and acquisition method that characterize the nature of the system and how it will be developed. A good number of alternative design strategies to develop is three, because three alternatives can represent the high-end, middle, and low-end of the spectrum of possible systems that can be built.

When developing design strategy alternatives, you must become aware of all of the possible options available. You must be aware of where you can obtain software that meets some or all of an organization's needs. You can obtain application (and system) software from hardware vendors, packaged software vendors, and custom software developers, as well as from internal systems development resources. You can even hire an organization to handle all of your systems development needs, called outsourcing. You must also know which criteria to use to be able to choose among off-the-shelf software products. These criteria include cost, functionality, vendor support, vendor viability, flexibility, documentation, response time, and ease of installation. You must also determine whether new hardware and system software are needed. Requests for proposals are one way you can collect more information about hardware and system software, their performance, and costs. In addition to software and hardware issues, you must consider implementation issues and broader organizational concerns, such as the availability of funding and management support.

Alternative design strategies are developed after a system's requirements and constraints have been identified and prioritized. Once developed, alternatives can be compared to each other through quantitative methods, but the actual decision may depend on other criteria, such as organizational politics.

Since generating and selecting alternative design strategies completes the analysis phase of the SDLC, the systems development project has reached a major milestone. Once an analysis of alternative design strategies is completed, you and other members of the analysis team will present your findings to a management steering committee and/or the client requesting the system change. In this presentation (both written and oral), you will summarize the requirements discovered, evaluate alternative design strategies and justify the recommended alternative as well as present an updated Baseline Project Plan for the project to follow if the committee decides to fund the next life-cycle phase.

Most of the same techniques used for selecting the best design strategy for traditional systems can also be fruitfully applied to the development of Internet applications. In the PVF example, you saw how analysts generated three alternative designs for the WebStore Internet application, based on system requirements and constraints, just as was done in the Hoosier Burger example. Although Internet applications are very different in some ways from traditional systems, many of the same analysis tools and techniques can be used profitably in each.

Key Terms

1. Application server
2. Application service provider
3. Design strategy
4. Enterprise resource planning (ERP) systems
5. Outsourcing
6. Request for proposal (RFP)
7. Scalable
8. Web server

Match each of the key terms above with the definition that best fits it.

_____ A computer that is connected to the Internet and stores files written in HTML—hypertext markup language—which are publicly available through an Internet connection.

_____ A particular approach to developing an information system. It includes statements on the system's functionality, hardware and system software platform, and method for acquisition.

_____ The ability to seamlessly upgrade the capabilities of the system through either hardware upgrades, software upgrades, or both.

_____ A system that integrates individual traditional business functions into a series of modules so that a single transaction occurs seamlessly within a single information system rather than several separate systems.

_____ The practice of turning over responsibility of some to all of an organization's information systems applications and operations to an outside firm.

_____ A document provided to vendors to ask them to propose hardware and system software that will meet the requirements of your new system.

_____ A "middle-tier" software and hardware combination that lies between the Web Server and the corporate network and systems.

_____ An organization that hosts and runs computer applications for other companies, typically on a per use or license basis.

Review Questions

1. What are the deliverables from generating alternatives and selecting the best one?

2. Why generate at least three alternatives?

3. Describe the four sources of software.

4. How do you decide among various off-the-shelf software options? What criteria do you use?

5. What issues are considered when analysts try to determine whether new hardware or system software is necessary? What is an RFP and how do analysts use one to gather information on hardware and system software?

6. What issues other than hardware and software must analysts consider in preparing alternative system design strategies?

7. How do analysts generate alternative solutions to information systems problems?

8. How do managers decide which alternative design strategy to develop?

9. Which elements of a Baseline Project Plan might be updated during the alternative generation and selection step of the analysis phase of the SDLC?

10. What methods can a systems analyst employ to verify vendor claims about a software package?

11. What are enterprise resource planning systems? What are the benefits and disadvantages of such systems as a design strategy?

Problems and Exercises

1. Find the most current issue of *Datamation* that includes the most recent *Datamation* 100 listings. How much has the rank order of the top software companies changed compared to the list printed here? Read the issue and determine why your list is different from the list in this chapter. What changes are occurring in the computer industry that might affect this list?

2. Research how to prepare a Request for Proposal. Prepare an outline of an RFP for Hoosier Burger to use in collecting information on its new inventory system hardware.

3. Recreate the spreadsheet in Figure 11-6 in your spreadsheet package. Change the weights and compare the outcome to Figure 11-6. Change the rankings. Add criteria. What additional information does this "what if" analysis provide for you as a decision maker? What insight do you gain into the decision-making process involved in choosing the best alternative system design?

4. Prepare a list for evaluating computer hardware and system software that is comparable to the list of criteria for selecting off-the-shelf application software presented earlier.

5. The method for evaluating alternatives used in Figure 11-6 is called weighting and scoring. This method implies that the total utility of an alternative is the product of the weights of each criterion times the weight of the criterion for the alternative. What assumptions are characteristic of this method for evaluating alternatives? That is, what conditions must be true for this to be a valid method of evaluation alternatives?

6. Weighting and scoring (see Problem and Exercise 5) is only one method for comparing alternative solutions to a problem. Go to the library and find a book or articles on qualitative and quantitative decision making and voting methods and outline two other methods for evaluating alternative solutions to a problem. What are the pros and cons of these methods compared to the weighting and scoring method? Under weighting and scoring and the other alternatives you find, how would you incorporate the opinions of multiple decision makers?

7. Prepare an agenda for a meeting at which you would present the findings of the analysis phase of the SDLC to Bob Mellankamp concerning his request for a new inventory control system. Use information provided in Chapters 8 through 10 as background in preparing this agenda. Concentrate on which topics to cover, not the content of each topic.

8. Review the criteria for selecting off-the-shelf software presented in this chapter. Use your experience and imagination and describe other criteria that are or might be used to select off-the-shelf software in the "real world." For each new criterion, explain how use of this criterion might be functional (i.e., it is useful to use this criterion), dysfunctional, or both.

9. The owner of two pizza parlors located in adjacent towns wants to computerize and integrate sales transactions and inventory management within and between both stores. The point-of-sale component must be very easy to use and flexible enough to accommodate a variety of pricing strategies and coupons. The inventory management, which will be linked to the point-of-sale component, must also be easy to use and fast. The systems at each store must be linked so that sales and inventory levels can be determined instantly for each store and for both stores combined. The owner can allocate $40,000 for hardware and $20,000 for software and must have the new system operational in three months. Training must be very short and easy. Briefly describe three alternative systems for this situation and explain how each would meet the requirements and constraints. Are the requirements and constraints realistic? Why or why not?

10. Compare the alternative systems from Problem and Exercise 9 using the weighted approach demonstrated in Figure 11-6. Which system would you recommend? Why? Was the approach taken in this and Problem and Exercise 9 useful even for this relatively small system? Why or why not?

11. Suppose that an analysis team did not generate alternative design strategies for consideration by a project steering committee or client. What might the consequences be of having only one design strategy? What might happen during the oral presentation of project progress if only one design strategy is offered?

12. In the section on Choosing Off-the-Shelf Software there are eight criteria proposed for evaluating alternative packages.

Suppose the choice was between alternative custom software developers rather than prewritten packages. What criteria would be appropriate to select and compare among competing bidders for custom development of an application? Define each of these criteria.

13. How might the project team recommending an enterprise resource planning design strategy justify its recommendation as compared to other types of design strategies?

Field Exercises

1. Consider the purchase of a new PC to be used by you at your work (or by you at a job that you would like to have). Describe in detail three alternatives for this new PC that represent the low, mid-, and high points of a continuum of potential solutions. Be sure that the low-end PC meets at least your minimum requirements and the high end PC is at least within a reasonable budget. At this point, without quantitative analysis, which alternative would you choose?

2. For the new PC described above, develop ranked lists of your requirements and constraints as displayed in Figure 11-6. Display the requirements and constraints, along with the three alternatives, in a diagram like the one displayed in Figure 11-6, and note how each alternative is rated on each requirement and constraint. Calculate scores for each alternative on each criterion and compute total scores. Which alternative has the highest score? Why? Does this choice fit with your selection in the previous question? Why or why not?

3. One of the most competitive software markets today is electronic spreadsheets. Pick three packages (for example, Microsoft Excel, Lotus 1-2-3, and Quattro Pro—but any three spreadsheet packages would do). Study how you use spreadsheet packages for school, work, and personal financial management. Develop a list of criteria important to you on which to compare alternative packages. Then contact each vendor and ask for the information you need to evaluate their package and company. Request a demonstration copy or trial use of their software. If they cannot provide a sample copy, then try to find a computer software dealer or club where you can

test the software and documentation. Based on the information you receive and the software you use, rate each package using your chosen criteria. Which package is best for you? Why? Talk to other students and find out which package they rated as best. Why are there differences between what different students determined as best?

4. Interview businesspeople who participate in the purchase of off-the-shelf software in their organizations. Review with them the criteria for selecting off-the-shelf software presented in this chapter. Have them prioritize the list of criteria as they are used in their organization and provide an explanation of the rationale for the ranking of each criterion. Ask them to list and describe any other criteria that are used in their organization.

5. Obtain copies of actual Requests for Proposals used for information systems developments and/or purchases. If possible, obtain RFPs from public and private organizations. Find out how they are used. What are the major components of these proposals? Do these proposals seem to be useful? Why or why not? How and why are the RFPs different for the public versus the private organizations?

6. Contact an organization that has or is implementing an enterprise resource planning integrated application. Why did it choose this design strategy? How has it managed this development project differently from prior large projects? What organizational changes have occurred due to this design strategy? How long did the implementation last and why?

References

Applegate, L. M., and R. Montealegre. 1991. "Eastman Kodak Company: Managing Information Systems Through Strategic Alliances." Harvard Business School case 9-192-030. Cambridge, MA: President and Fellows of Harvard College.

Banker, R. D., Davis, G. B., and Slaughter, S. A. 1998. "Software Development Practices, Software Complexity, and Software Maintenance Performance: A Field Study." *Management Science* 44(4): 433–450.

Ferranti, M. 2000. "EDS Packages Services for Business of All Sizes." *Infoworld.com* (*www.infoworld.com*), March 2.

Harney, J. 2000. "Lost Among the ASPs." *Intelligent Enterprise* (Feb. 9): 27, 30–31, 34.

Holohan, M., and Hall, M. 2000. "Update: Oracle Embraces ASPs." *ComputerWorld* (*www.computerworld.com*), July 31.

Ketler, K., and Willems, J.R. 1999. "A Study of the Outsourcing Decision: Preliminary Results." *Proceedings of SIGCPR '99*, New Orleans, LA, 182–189.

King, J., and Cole-Gomolski, B. 1999. "IT Doing Less Development, More Installation, Outsourcing." *Computerworld*, 1/25/99, 4+.

Lacity, M., and Willcocks, L. 1998. "An Empirical Investigation of Information Technology Sourcing Practices: Lessons From Experience." *MIS Quarterly* 22(3): 363–408.

Merrill, K. 1999. "Poll: IT Outsourcing Shows No Signs of Slowing." *TechWeb* (www.techweb.com), March 31, 1999.

Wilcox, J., and Farmer, M. A. 2000. "Microsoft to Unveil Software-for-Rent Strategy." *CNET News.com* (news.cnet.com), July 14.

BROADWAY ENTERTAINMENT COMPANY, INC.

Formulating a Design Strategy for the Web-Based Customer Relationship Management System

CASE INTRODUCTION

Defining a design strategy for a systems development project is typically a crucial step for a project team. The team must select the scope of functionality, the implementation platform, and the method of acquisition for the system. There may be many alternatives, and different sets of users may support different design strategies. The design strategy, chosen near the end of the analysis phase, is also crucial because it sets the direction for the rest of the project. In contrast, a development approach that uses a prototype to prove the concept for the value of a system need not be perfect. Further, the prototype does not need to take the exact form of the final system. For these reasons, some choices can be made to meet shorter-term rather than longer-term objectives. This is the case with MyBroadway, the customer relationship information system being developed by a team of students from Stillwater State University for Carrie Douglass, manager of the Broadway Entertainment Company store in Centerville, Ohio.

DETERMINING THE SYSTEM FUNCTIONALITY

The activities of systems analysis for the MyBroadway project have identified possible system requirements. Usually a prototyping methodology follows a repetitive development process of building, testing, and rebuilding a system until the user agrees that the system functions as desired. In the case of MyBroadway, the prototype will be used more during the design (rather than analysis) phase to refine system requirements, screens, and report layouts, and to test customer acceptance (see BEC Figure 11-1). So far, a structured approach has been taken to determine initial requirements. The team took this combined approach of initially a structured and then a prototyping methodology because data flow and storage issues needed to be addressed. For example, structured techniques allowed the team to study connections between MyBroadway and the Entertainment Tracker store management system. Prototyping works best once the data requirements are in place, but the use of the data still

needs to be decided. Thus, the team fully expects changes to be made to the functional requirements during design.

Because prototyping will be used in subsequent steps, the student team decided not to propose several alternative (or extreme) functional scopes to Carrie Douglass. The team members suggest that alternatives should be considered in an evolutionary fashion during design. The only functional scope requirement the team raised with Carrie was a big one and needed to be clarified before the prototyping began. The issue related to one finding from the observation of customers using other, noncompeting on-line shopping services. Several customers voiced their interest in being able to preview products before buying or renting. Their suggestion was to include a video or sound clip with the other data about products and then be able to play these as streaming video or audio via the Web.

The team explained to Carrie that although this capability is definitely technically feasible, it would present several problems. First, the data storage requirements for MyBroadway would greatly increase. Second, some Web browsers might have difficulty playing the video or audio streams, which would cause user frustration. Third, although client tools for playing multimedia streams are inexpensive, the server-side software is rather costly (in terms of Carrie's budget), so platform cost would increase. Finally, there would be considerable effort to create the video or audio streams. The team suggested that once the system being developed for Carrie is proven without this feature and BEC management has a chance to see the system, they could decide to include this capability in a company-wide rollout of the product. Carrie also pointed out that for a company-wide rollout, there would be multiple-language capabilities required of streaming multimedia, because BEC operates stores in both Spanish- and French-speaking locations. Carrie agreed to keep this feature out of the prototype. She added, however, that it had been her experience in using Amazon.com and other services that once she actually used the system, her concept of what features she wanted available to her changed. Thus, she wanted the team to

BEC Figure 11-1
Design process for MyBroadway system

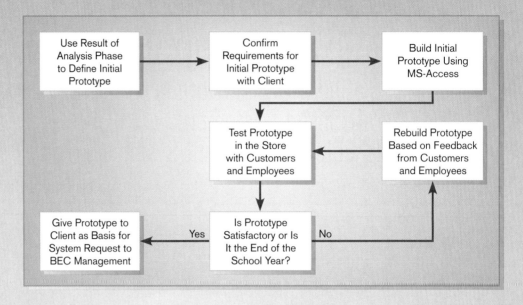

know that although its analysis so far had been good, the functional capabilities should be able to change as easily as possible during the design prototyping.

DETERMINING THE SYSTEM PLATFORM

The platform chosen by the students was somewhat a surprise to Carrie—the team suggested using Microsoft Access. Carrie thought the team would choose a specialized Web development tool, such as Microsoft FrontPage, DreamWeaver, or ColdFusion. The Stillwater student team explained that it felt that Access was a better prototyping platform than these Web development environments. The final system would likely be built using such a tool, using active server pages accessing a backend database. However, the team members suggested that the prototype system be tested in the store on a stand-alone PC, and not be placed on the Web. They felt that a Website, although it would allow use from home, was too risky for the prototype. The prototype would likely change frequently, and the changing user interface could be confusing to customers. Testing could be more controlled and monitored by allowing customers to test the prototype only in the store. Also, and very importantly, Carrie would not have to rent space from an Internet service provider (ISP) for the prototype. Instead of contracting with an ISP, the prototype could be placed on a Web server at Stillwater State University, although customers might be confused by the Stillwater URL being involved in the site name.

Microsoft Access screens could be designed to look very similar to browser screens. Both customers and employees could use the system. The students are quite familiar with Access, so there would be minimal, if any, training time. Access should provide fast response during

prototype testing. Also, the Management Information Systems department at Stillwater has a laptop computer and extra 17-inch monitor the team had arranged to use for the project. The monitor could be left at the store, and the team could develop the prototype on the laptop and bring it to the store for customer and employee use. The laptop would have sufficient power for one user at a time. The laptop already had the latest version of Access loaded on it, part of the campus computing requirement instituted two years ago, through which each student received a license for all Microsoft Office and Visual Studio tools.

Carrie was pleased with the prototype platform recommendation, primarily because it meant no additional money to be spent on the project. She had already bought more than enough drinks, pizza, and wings at a nearby restaurant for the various project status report sessions.

DETERMINING THE SYSTEM ACQUISITION METHODOLOGY

The Stillwater student team felt that it had sufficient skills and time to build the prototype. Thus, no third party would be needed during design and implementation. Also, because prototyping would be used to fine-tune requirements and to develop experience with usage of the system, it did not seem feasible to use a preprogrammed package. Actually, the team never considered a package, and Carrie never asked. Carrie was already committed to a custom-developed system, one that she could say would be hers.

The team did raise one acquisition question with Carrie. The question was whether Carrie knew if any other BEC store had tried to build such a system, or if the BEC corporate IT organization was developing a prototype of a Web-based system. Carrie did not know the answer to either question and was still reluctant to let other BEC stores or corporate employees know about the project she was sponsoring. Carrie and the student team spent about half an hour using Web search engines to try to find a similar service from any video and audio media store but were unable to find such a system. Not that Carrie would have supported buying a system from another store, but it would have at least been interesting to see what others might be doing.

CASE SUMMARY

The Stillwater student team had accomplished a lot since the first meeting with Carrie Douglass. Carrie is pleased with the progress. Carrie gave the team the go-ahead to start to build a prototype, using the design strategy the team suggested. But, Carrie was still concerned about the rest of the project. What would happen once customers and employees saw a real working system? Would they start asking for many additional features? How could their expectations be tempered? How would the team know when the project was done, if people kept suggesting new features? Even though the student team felt confident that a laptop computer would work well to simulate both a client and server for a Web-based customer relationship management system, would this assessment be right? When should Carrie expose BEC corporate IS people to the project? So far, the team members had been able to find out everything they needed to know about Entertainment Tracker from the user documentation provided to the store, but would this continue as more technical design issues arose?

CASE QUESTIONS

1. Was the Stillwater student team wise to not suggest several alternative sets of functional capabilities for MyBroadway? What are the risks of considering only one starting point for prototyping?

2. Although the Stillwater student team and Carrie could find no other entertainment media store that had a Website, certainly companies in competition with BEC have Websites. Visit the Websites for Blockbuster and at least one other BEC competitor. From reviewing those sites, what other features would you anticipate might be suggested during customer and employee use of the initial MyBroadway prototype? Given your answers to questions in previous BEC cases in this text, will the architecture of the prototype (including the database model and process flow models for acquiring and reporting data) be changed much by some of the Website features you find? Why or why not?

3. What is your assessment of the recommendation to use Microsoft Access as the prototyping platform? Would you suggest an alternative? Why? Under what circumstances would an alternative platform be better?

4. Carrie is concerned with user expectation management as the project moves forward. Given the additional information you know now about the system being developed, MyBroadway, what would you recommend the team do to limit user expectations or to handle frustrations if users are underwhelmed by what they see in the prototypes?

5. This case states that the student team thought that including video and audio clips would be a problem, at least for the prototype. Do you agree? Why or why not? One might also argue that because a BEC store carries over 4,000 different products, there is already a database problem even without including multimedia product clips data. What prototype implementation issues does this large product inventory have on the development of the prototypes? What suggestions would you make for changes (or refinements) to the design strategy to handle this extensive set of products? Justify your suggestions.

6. The acceptance of a design strategy concludes the analysis phase of a project. Hence, this is a natural time to reassess how to conduct the rest of the project. The statement of how to progress is done by updating the Baseline Project Plan. In several questions for the BEC case at the end of Chapter 6, you developed components of the BPP. Given your answers to those questions and what you now know about the MyBroadway system, answer the following questions related to the components of a BPP.

   a. List tangible and intangible benefits and costs for this project. Be sure to quantify tangible costs and benefits.

   b. What are the remaining risks of the project? How would you suggest the project team deal with these risks?

   c. How would you continue to use the concept of incremental commitment for the rest of the project?

   d. What would be the next steps in the project? If possible, develop a PERT or Gantt chart for the remaining steps of the project. Is such a project schedule chart possible for the design strategy recommended by the student team?

   e. What contact, if any, do you think the student team should have during the design phase with BEC corporate IT staff, especially those responsible for the Entertainment Tracker system? Under what circumstance would such contact not be necessary? Can the team organize design activities so that it does not need to know anything about the technical architecture of Entertainment Tracker?

Part

FOUR

Design

An Overview of Part FOUR

Design

The focus of Part IV is system design, which is often the first phase of the systems development life cycle in which you and the user develop a concrete understanding of how the system will operate. The activities within design are not necessarily sequential. For example, the design of data, system inputs and outputs, and interfaces interact to identify flaws and missing elements. This means that the project dictionary or CASE repository becomes an active and evolving component of system development management during design. It is only when each design element is consistent with others and satisfactory to the end user that you know that design is complete.

Data make up a core system element that is studied in all systems development methodologies. You have seen how data flow diagrams (DFDs) and E-R diagrams are used to depict the data requirements of a system. Both of these diagrams are flexible and allow considerable latitude in how you represent data. For example, you can use one or many data stores with a process in a DFD. E-R diagrams provide more structure, but an entity can still be either very detailed or rather aggregate. When designing databases, you define data in its most fundamental form, called normalized data. *Normalization* is a well-defined method of identifying relationships between each data attribute and representing all the data so that they cannot logically be broken down into more detail. The goal is to rid the data design of unwanted anomalies that would make a database susceptible to errors and inefficiencies. This is the topic of Chapter 12.

In Chapter 13, you will learn the principles and guidelines for usable system inputs and outputs. Your overall goal in formatting the presentation of data to users should be usability: helping users of all types to use the system efficiently, accurately, and with satisfaction. The achievement of these goals can be greatly improved if you follow certain guidelines to present data on business forms, visual display screen, printed documents, and other kinds of media. Fortunately, there has been considerable research on how to present data to users, and this chapter summarizes and illustrates the most useful of these guidelines. Closely related, Chapter 14 addresses principles you should follow in tying all the system inputs and outputs together into an overall pattern of interaction between users and the system. System interfaces and dialogues form a conversation that provides user access to and navigation between each system function. This chapter focuses on providing specifications for designing effective system interfaces and dialogues, and a technique for representing these designs called dialogue diagramming.

Next, Chapter 15 addresses techniques and principles you should follow when finalizing design specifications. The design specifications include functional descriptions for each part of the system, as well as information about input received and output generated for each program and its component parts. It does not matter what form the specifications take, but all relevant aspects of the system must be included. Structure charts are often used when finalizing a system design to show all of the different software modules in the system and their relationships to each other, including the data and messages passed among modules. Alternatively, specifications can be captured and represented using a working prototype. In some cases, these prototypes become the basis for the production system.

Before the design specifications can be handed over to the developers to begin the implementation process, questions about multiple users, multiple platforms, and program and data distribution have to be considered. Additionally, the rise in the use of Internet-based systems has resulted in numerous new design issues. The focus of

Chapter 16 is on the intricacies of designing distributed and Internet systems.

The deliverables of design include detailed, functional specifications for system inputs, outputs, interfaces, dialogues, and databases. Often these elements are represented in prototypes, or working versions. The project dictionary or CASE repository is updated to include each form, report, interface, dialogue, and relation design. Due to considerable user involvement in reviewing prototypes and specifications during design, and due to the

fact that activities within design can be scheduled with considerable overlap in the project baseline plan, there often is not a formal review milestone or walkthrough after each activity. If prototyping is not done, however, then you should conduct a formal walkthrough at the completion of system design.

All of the chapters in Part IV conclude with a BEC case. These cases illustrate numerous relevant design activities for an ongoing systems development project within the company.

Chapter 12

Designing

Databases

LEARNING OBJECTIVES

After studying this chapter, you should be able to:

● Concisely define each of the following key database design terms: *relation*, *primary key*, *normalization*, *functional dependency*, *foreign key*, *referential integrity*, *field*, *data type*, *null value*, *denormalization*, *file organization*, *index*, and *secondary key*.

● Explain the role of designing databases in the analysis and design of an information system.

● Transform an entity-relationship (E-R) diagram into an equivalent set of well-structured (normalized) relations.

● Merge normalized relations from separate user views into a consolidated set of well-structured relations.

● Choose storage formats for fields in database tables.

● Translate well-structured relations into efficient database tables.

● Explain when to use different types of file organizations to store computer files.

● Describe the purpose of indexes and the important considerations in selecting attributes to be indexed.

INTRODUCTION

In Chapter 10 you learned how to represent an organization's data graphically using an entity-relationship (E-R) diagram. In this chapter, you learn guidelines for well-structured and efficient database files and about logical and physical database design. It is likely that the human interface and database design steps will happen in parallel, as illustrated in the SDLC in Figure 12-1.

Database design has five purposes:

1. Structure the data in stable structures, called normalized tables, that are not likely to change over time and that have minimal redundancy.

2. Develop a logical database design that reflects the actual data requirements that exist in the

forms (hard copy and computer displays) and reports of an information system. This is why database design is often done in parallel with the design of the human interface of an information system.

3. Develop a logical database design from which we can do physical database design. Because most information systems today use relational database management systems, logical database design usually uses a relational database model, which represents data in simple tables with common columns to link related tables.

4. Translate a relational database model into a technical file and database design that balances several performance factors.

5. Choose data storage technologies (such as floppy disk, CD-ROM, or optical disk) that will efficiently, accurately, and securely process database activities.

The implementation of a database (i.e., creating and loading data into files and databases) is done during the next phase of the systems development life cycle. Because implementation is very technology specific, we address implementation issues only at a general level in Chapter 17.

DATABASE DESIGN

File and database design occurs in two steps. You begin by developing a logical database model, which describes data using a notation that corresponds to a data organization used by a database management system. This is the system software responsible for storing, retrieving, and protecting data (such as Microsoft Access, Oracle, or SQL Server). The most common style for a logical database model is the relational database model. Once you develop a clear and precise logical database

Figure 12-1
Systems development life cycle with logical design phase highlighted

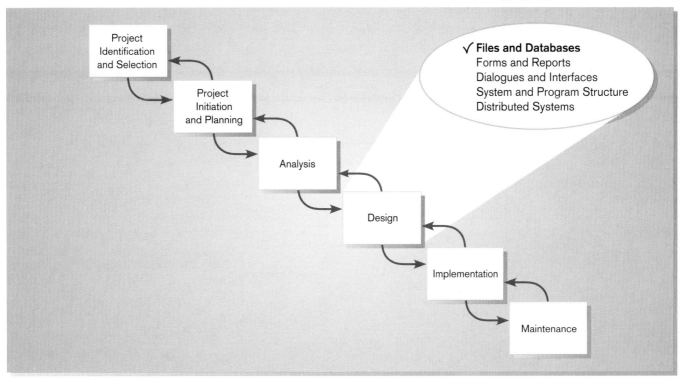

model, you are ready to prescribe the technical specifications for computer files and databases in which to store the data ultimately. A physical database design provides these specifications.

You typically do logical and physical database design in parallel with other systems design steps. Thus, you collect the detailed specifications of data necessary for logical database design as you design system inputs and outputs. Logical database design is driven not only from the previously developed E-R data model for the application but also from form and report layouts. You study data elements on these system inputs and outputs and identify interrelationships among the data. As with conceptual data modeling, the work of all systems development team members is coordinated and shared through the project dictionary or repository. The designs for logical databases and system inputs and outputs are then used in physical design activities to specify to computer programmers, database administrators, network managers, and others how to implement the new information system. We assume for this text that the design of computer programs and distributed information processing and data networks are topics of other courses, so we concentrate on the aspect of physical design most often undertaken by a systems analyst—physical file and database design.

The Process of Database Design

Figure 12-2 shows that database modeling and design activities occur in all phases of the systems development process. In this chapter we discuss methods that help you finalize logical and physical database designs during the design phase. In logical database design you use a process called normalization, which is a way to build a data model that has the properties of simplicity, nonredundancy, and minimal maintenance.

Figure 12-2
Relationship between data modeling and the systems development life cycle

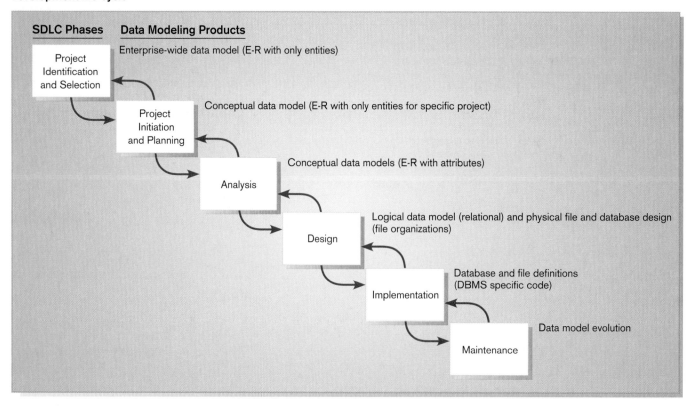

In most situations, many physical database design decisions are implicit or eliminated when you choose the data management technologies to use with the application. We concentrate on those decisions you will make most frequently and use Oracle to illustrate the range of physical database design parameters you must manage. The interested reader is referred to Hoffer, Prescott, and McFadden (2002) for a more thorough treatment of techniques for logical and physical database design.

There are four key steps in logical database modeling and design:

1. Develop a logical data model for each known user interface (form and report) for the application using normalization principles.

2. Combine normalized data requirements from all user interfaces into one consolidated logical database model; this step is called view integration.

3. Translate the conceptual E-R data model for the application, developed without explicit consideration of specific user interfaces, into normalized data requirements.

4. Compare the consolidated logical database design with the translated E-R model and produce, through view integration, one final logical database model for the application.

During physical database design, you use the results of these four key logical database design steps. You also consider definitions of each attribute; descriptions of where and when data are entered, retrieved, deleted, and updated; expectations for response time and data integrity; and descriptions of the file and database technologies to be used. These inputs allow you to make key physical database design decisions, including the following:

1. Choosing the storage format (called data type) for each attribute from the logical database model; the format is chosen to minimize storage space and to maximize data quality. Data type involves choosing length, coding scheme, number of decimal places, minimum and maximum values, and potentially many other parameters for each attribute.

2. Grouping attributes from the logical database model into physical records (in general, this is called selecting a stored record, or data, structure).

3. Arranging related records in secondary memory (hard disks and magnetic tapes) so that individual and groups of records can be stored, retrieved, and updated rapidly (called file organizations). You should also consider protecting data and recovering data after errors are found.

4. Selecting media and structures for storing data to make access more efficient. The choice of media affects the utility of different file organizations. The primary structure used today to make access to data more rapid is key indexes, on unique and nonunique keys.

We show how to do each of the logical database design steps and discuss factors to consider in making each physical file and database design decision in this chapter.

Deliverables and Outcomes

During logical database design, you must account for every data element on a system input or output—form or report—and on the E-R model. Each data element (like customer name, product description, or purchase price) must be a piece of raw data kept in the system's database, or in the case of a data element on a system output, the element can be derived from data in the database. Figure 12-3 illustrates the outcomes from the four-step logical database design process listed above. Figure 12-3(a) and 12-3(b) (step 1) contain two sample system outputs for a customer order

PINE VALLEY FURNITURE

Figure 12-3
Simple example of logical data modeling
(a) Highest-volume customer query screen

HIGHEST VOLUME CUSTOMER

ENTER PRODUCT ID.:   M128
START DATE: 11/01/2001
END DATE: 12/31/2001
- - - - - - - - - - - - - - - - - - -
CUSTOMER ID.:   1256
NAME: Commonwealth Builder
VOLUME: 30

This inquiry screen shows the customer with the largest volume total sales of a specified product during an indicated time period.

Relations:
    CUSTOMER(Customer_ID,Name)
    ORDER(Order_Number,Customer_ID,Order_Date)
    PRODUCT(Product_ID)
    LINE ITEM(Order_Number,Product_ID,Order_Quantity)

(b) Backlog summary report

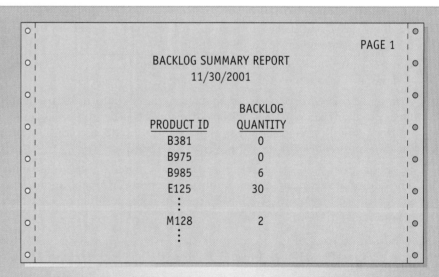

PAGE 1

BACKLOG SUMMARY REPORT
11/30/2001

PRODUCT ID	BACKLOG QUANTITY
B381	0
B975	0
B985	6
E125	30
⋮	
M128	2
⋮	

This report shows the unit volume of each product that has been ordered less that amount shipped through the specified date.

Relations:
    PRODUCT(Product_ID)
    LINE ITEM(Product_ID,Order_Number,Order_Quantity)
    ORDER(Order_Number,Order_Date)
    SHIPMENT(Product_ID,Invoice_Number,Ship_Quantity)
    INVOICE(Invoice_Number,Invoice_Date,Order_Number)

Figure 12-3 (continued)
(c) Integrated set of relations

```
CUSTOMER(Customer_ID,Name)
PRODUCT(Product_ID)
INVOICE(Invoice_Number,Invoice_Date,Order_Number)
ORDER(Order_Number,Customer_ID,Order_Date)
LINE ITEM(Order_Number,Product_ID,Order_Quantity)
SHIPMENT(Product_ID,Invoice_Number,Ship_Quantity)
```

(d) Conceptual data model and transformed relations

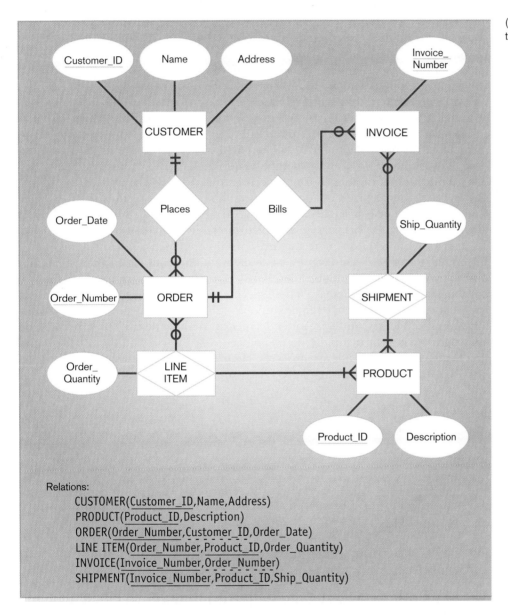

Relations:
```
CUSTOMER(Customer_ID,Name,Address)
PRODUCT(Product_ID,Description)
ORDER(Order_Number,Customer_ID,Order_Date)
LINE ITEM(Order_Number,Product_ID,Order_Quantity)
INVOICE(Invoice_Number,Order_Number)
SHIPMENT(Invoice_Number,Product_ID,Ship_Quantity)
```

(e) Final set of normalized relations

```
CUSTOMER(Customer_ID,Name,Address)
PRODUCT(Product_ID,Description)
ORDER(Order_Number,Customer_ID,Order_Date)
LINE ITEM(Order_Number,Product_ID,Order_Quantity)
INVOICE(Invoice_Number,Order_Number,Invoice_Date)
SHIPMENT(Invoice_Number,Product_ID,Ship_Quantity)
```

Primary key: An attribute whose value is unique across all occurrences of a relation.

processing system at Pine Valley Furniture. A description of the associated database requirements, in the form of what we call normalized relations, is listed below each output diagram. Each relation (think of a relation as a table with rows and columns) is named and its attributes (columns) are listed within parentheses. The **primary key** attribute—that attribute whose value is unique across all occurrences of the relation—is indicated by an underline, and an attribute of a relation that is the primary key of another relation is indicated by a dashed underline.

In Figure 12-3(a), data are shown about customers, products, and the customer orders and associated line items for products. Each of the attributes of each relation either appears in the display or is needed to link related relations. For example, because an order is for some customer, an attribute of ORDER is the associated Customer_ID. The data for the display in Figure 12-3(b) are more complex. A backlogged product on an order occurs when the amount ordered (Order_Quantity) is less than the amount shipped (Ship_Quantity) for invoices associated with an order. The query refers to only a specified time period, so the Order_Date is needed. The INVOICE Order_Number links invoices with the associated order.

Figure 12-3(c) (step 2) shows the result of integrating these two separate sets of normalized relations. Figure 12-3(d) (step 3) shows an E-R diagram for a customer order processing application that might be developed during conceptual data modeling along with equivalent normalized relations. Finally, Figure 12-3(e) (step 4) shows a set of normalized relations that would result from reconciling the logical database designs of Figures 12-3(c) and 12-3(d). Normalized relations like those in Figure 12-3(e) are the primary deliverable from logical database design.

It is important to remember that relations do not correspond to computer files. In physical database design, you translate the relations from logical database design into specifications for computer files. For most information systems, these files will be tables in a relational database. These specifications are sufficient for programmers and database analysts to code the definitions of the database. The coding, done during systems implementation, is written in special database definition and processing languages, such as Structured Query Language (SQL), or by filling in table definition forms, such as with Microsoft Access. Figure 12-4 shows a possible definition for the SHIPMENT relation from Figure 12-3(e) using Microsoft Access. This display of

Figure 12-4
Definition of SHIPMENT table in Microsoft Access

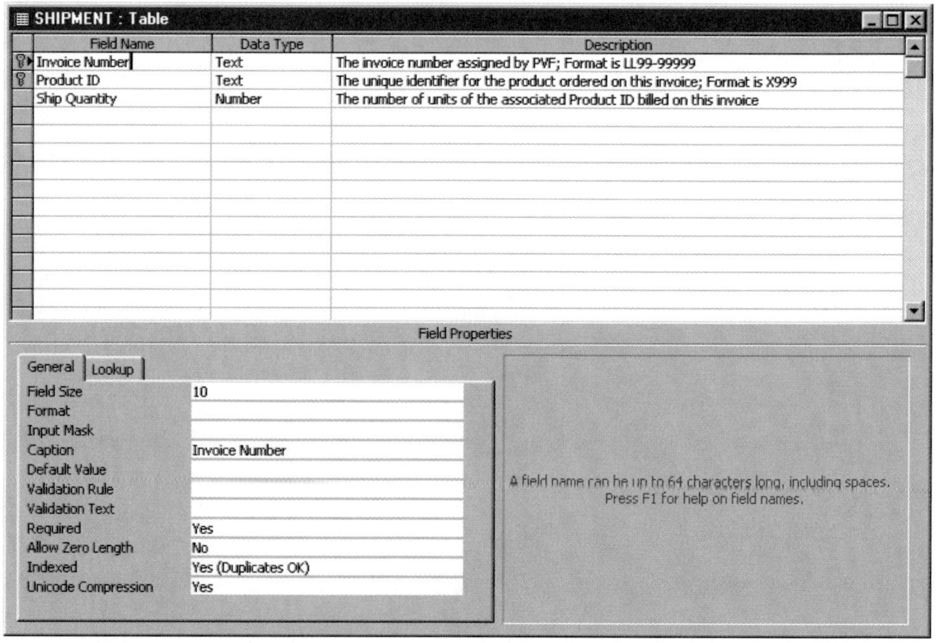

the SHIPMENT table definition illustrates choices made for several physical database design decisions.

- All three attributes from the SHIPMENT relation, and no attributes from other relations, have been grouped together to form the fields of the SHIPMENT table.

- The Invoice Number field has been given a data type of Text, with a maximum length of 10 characters.

- The Invoice Number field is required because it is part of the primary key for the SHIPMENT table (the value that makes every row of the SHIPMENT table unique is a combination of Invoice Number and Product ID).

- An index is defined for the Invoice Number field, but because there may be several rows in the SHIPMENT table for the same invoice (different products on the same invoice), duplicate index values are allowed (so Invoice Number is what we will call a secondary key).

Many other physical database design decisions were made for the SHIPMENT table, but they are not apparent on the display in Figure 12-4. Further, this table is only one table in the PVF Order Entry database, and other tables and structures for this database are not illustrated in this figure.

RELATIONAL DATABASE MODEL

Many different database models are in use and are the basis for database technologies. Although hierarchical and network models have been popular in the past, these are not used very often today for new information systems. Object-oriented database models are emerging, but are still not common. The vast majority of information systems today use the relational database model. The **relational database model** (Codd, 1970) represents data in the form of related tables or relations. A **relation** is a named, two-dimensional table of data. Each relation (or table) consists of a set of named columns and an arbitrary number of unnamed rows. Each column in a relation corresponds to an attribute of that relation. Each row of a relation corresponds to a record that contains data values for an entity.

Figure 12-5 shows an example of a relation named EMPLOYEE1. This relation contains the following attributes describing employees: Emp_ID, Name, Dept, and Salary. There are five sample rows in the table, corresponding to five employees.

You can express the structure of a relation by a shorthand notation in which the name of the relation is followed (in parentheses) by the names of the attributes in the relation. The identifier attribute (called the primary key of the relation) is underlined. For example, you would express EMPLOYEE1 as follows:

EMPLOYEE1(Emp_ID,Name,Dept,Salary)

N E T   S E A R C H
Investigate the origins of the relational database model. Visit http://www.prenhall.com/hoffer to complete an exercise related to this topic.

Relational database model: Data represented as a set of related tables or relations.

Relation: A named, two-dimensional table of data. Each relation consists of a set of named columns and an arbitrary number of unnamed rows.

EMPLOYEE1

Emp_ID	Name	Dept	Salary
100	Margaret Simpson	Marketing	42,000
140	Allen Beeton	Accounting	39,000
110	Chris Lucero	Info Systems	41,500
190	Lorenzo Davis	Finance	38,000
150	Susan Martin	Marketing	38,500

Figure 12-5
EMPLOYEE1 relation with sample data

Not all tables are relations. Relations have several properties that distinguish them from nonrelational tables:

1. Entries in cells are simple. An entry at the intersection of each row and column has a single value.

2. Entries in a given column are from the same set of values.

3. Each row is unique. Uniqueness is guaranteed because the relation has a non-empty primary key value.

4. The sequence of columns can be interchanged without changing the meaning or use of the relation.

5. The rows may be interchanged or stored in any sequence.

Well-Structured Relations

Well-structured relation (or table): A relation that contains a minimum amount of redundancy and allows users to insert, modify, and delete the rows without errors or inconsistencies.

What constitutes a **well-structured relation** (or **table**)? Intuitively, a well-structured relation contains a minimum amount of redundancy and allows users to insert, modify, and delete the rows in a table without errors or inconsistencies. EMPLOYEE1 (Figure 12-5) is such a relation. Each row of the table contains data describing one employee, and any modification to an employee's data (such as a change in salary) is confined to one row of the table.

In contrast, EMPLOYEE2 (Figure 12-6) contains data about employees and the courses they have completed. Each row in this table is unique for the combination of Emp_ID and Course, which becomes the primary key for the table. This is not a well-structured relation, however. If you examine the sample data in the table, you notice a considerable amount of redundancy. For example, the Emp_ID, Name, Dept, and Salary values appear in two separate rows for employees 100, 110, and 150. Consequently, if the salary for employee 100 changes, we must record this fact in two rows (or more, for some employees).

The problem with this relation is that it contains data about two entities: EMPLOYEE and COURSE. You will learn to use principles of normalization to divide EMPLOYEE2 into two relations. One of the resulting relations is EMPLOYEE1 (Figure 12-5). The other we will call EMP COURSE, which appears with sample data in Figure 12-7. The primary key of this relation is the combination of Emp_ID and Course (we emphasize this by underlining the column names for these attributes).

NORMALIZATION

Normalization: The process of converting complex data structures into simple, stable data structures.

We have presented an intuitive discussion of well-structured relations; however, we need rules and a process for designing them. **Normalization** is a process for converting complex data structures into simple, stable data structures. For

Figure 12-6
Relation with redundancy

EMPLOYEE2

Emp_ID	Name	Dept	Salary	Course	Date_Completed
100	Margaret Simpson	Marketing	42,000	SPSS	6/19/2002
100	Margaret Simpson	Marketing	42,000	Surveys	10/7/2002
140	Alan Beeton	Accounting	39,000	Tax Acc	12/8/2002
110	Chris Lucero	Info Systems	41,500	SPSS	1/22/2002
110	Chris Lucero	Info Systems	41,500	C++	4/22/2002
190	Lorenzo Davis	Finance	38,000	Investments	5/7/2002
150	Susan Martin	Marketing	38,500	SPSS	6/19/2002
150	Susan Martin	Marketing	38,500	TQM	8/12/2002

EMP COURSE

Emp_ID	Course	Date_Completed
100	SPSS	6/19/2002
100	Surveys	10/7/2002
140	Tax Acc	12/8/2002
110	SPSS	1/22/2002
110	C++	4/22/2002
190	Investments	5/7/2002
150	SPSS	6/19/2002
150	TQM	8/12/2002

Figure 12-7
EMP COURSE relation

example, we used the principles of normalization to convert the EMPLOYEE2 table with its redundancy to EMPLOYEE1 (Figure 12-5) and EMP COURSE (Figure 12-7).

Rules of Normalization

Normalization is based on well-accepted principles and rules. There are many normalization rules, more than can be covered in this text (see Hoffer, Prescott, and McFadden (2002), for a more complete coverage). Besides the five properties of relations outlined above, there are two other frequently used rules:

1. *Second normal form (2NF).* Each nonprimary key attribute is identified by the whole key (what we call full functional dependency). For example, in Figure 12-7, both Emp_ID and Course identify a value of Date_Completed because the same Emp_ID can be associated with more than one Date_Completed and the same for Course.

2. *Third normal form (3NF).* Nonprimary key attributes do not depend on each other (what we call no transitive dependencies). For example, in Figure 12-5, neither Name, Dept, nor Salary can be guaranteed to be unique for each other.

The result of normalization is that every nonprimary key attribute depends upon the whole primary key and nothing but the primary key. We discuss second and third normal form in more detail next.

Functional Dependence and Primary Keys

Normalization is based on the analysis of functional dependence. A **functional dependency** is a particular relationship between two attributes. In a given relation, attribute B is functionally dependent on attribute A if, for every valid value of A, that value of A uniquely determines the value of B (Dutka and Hanson, 1989). The functional dependence of B on A is represented by an arrow, as follows: $A \rightarrow B$ (e.g., $Emp\_ID \rightarrow Name$ in the relation of Figure 12-5). Functional dependence does not imply mathematical dependence—that the value of one attribute may be computed from the value of another attribute; rather, functional dependence of B on A means that there can be only one value of B for each value of A. Thus, for a given Emp_ID value, there can be only one Name value associated with it; the value of Name, however, cannot be derived from the value of Emp_ID. Other examples of functional dependencies from Figure 12-3(b) are in ORDER, $Order\_Number \rightarrow Order\_Date$, and in INVOICE, $Invoice\_Number \rightarrow Invoice\_Date$ and $Order\_Number$.

Functional dependency: A particular relationship between two attributes. For a given relation, attribute B is functionally dependent on attribute A if, for every valid value of A, that value of A uniquely determines the value of B. The functional dependence of B on A is represented by $A \rightarrow B$.

Figure 12-8
EXAMPLE relation

EXAMPLE

A	B	C	D
X	U	X	Y
Ⓨ	X	Z	X
Z	Y	Y	Y
Ⓨ	Z	W	Z

An attribute may be functionally dependent on two (or more) attributes, rather than on a single attribute. For example, consider the relation EMP COURSE (Emp_ID,Course,Date_Completed) shown in Figure 12-7. We represent the functional dependency in this relation as follows: Emp_ID,Course→Date_Completed. In this case, Date_Completed cannot be determined by either Emp_ID or Course alone, because Date_Completed is a characteristic of an employee taking a course.

You should be aware that the instances (or sample data) in a relation do not prove that a functional dependency exists. Only knowledge of the problem domain, obtained from a thorough requirements analysis, is a reliable method for identifying a functional dependency. However, you can use sample data to demonstrate that a functional dependency does not exist between two or more attributes. For example, consider the sample data in the relation EXAMPLE(A,B,C,D) shown in Figure 12-8. The sample data in this relation prove that attribute B is not functionally dependent on attribute A because A does not uniquely determine B (two rows with the same value of A have different values of B).

Second Normal Form

Second normal form (2NF): A relation is in second normal form if every nonprimary key attribute is functionally dependent on the whole primary key.

A relation is in **second normal form (2NF)** if every nonprimary key attribute is functionally dependent on the whole primary key. Thus no nonprimary key attribute is functionally dependent on part, but not all, of the primary key. Second normal form is satisfied if any one of the following conditions apply:

1. The primary key consists of only one attribute (such as the attribute Emp_ID in relation EMPLOYEE1).

2. No nonprimary key attributes exist in the relation.

3. Every nonprimary key attribute is functionally dependent on the full set of primary key attributes.

EMPLOYEE2 (Figure 12-6) is an example of a relation that is not in second normal form. The shorthand notation for this relation is

EMPLOYEE2(Emp_ID,Name,Dept,Salary,Course,Date_Completed)

The functional dependencies in this relation are the following:

Emp_ID→Name,Dept,Salary

Emp_ID,Course→Date_Completed

The primary key for this relation is the composite key Emp_ID,Course. Therefore, the nonprimary key attributes Name, Dept, and Salary are functionally dependent on only Emp_ID but not on Course. EMPLOYEE2 has redundancy, which results in problems when the table is updated.

To convert a relation to second normal form, you decompose the relation into new relations using the attributes, called *determinants*, that determine other attributes; the determinants are the primary keys of these relations. EMPLOYEE2 is decomposed into the following two relations:

1. EMPLOYEE(Emp_ID,Name,Dept,Salary): This relation satisfies the first second normal form condition (sample data shown in Figure 12-5).

2. EMP COURSE(Emp_ID,Course,Date_Completed): This relation satisfies second normal form condition three (sample data appear in Figure 12-7).

Third Normal Form

A relation is in **third normal form (3NF)** if it is in second normal form and there are no functional dependencies between two (or more) nonprimary key attributes (a functional dependency between nonprimary key attributes is also called a *transitive dependency*). For example, consider the relation SALES (Customer_ID,Customer_Name,Salesperson,Region) (sample data shown in Figure 12-9[a]).

The following functional dependencies exist in the SALES relation:

1. Customer_ID→Customer_Name,Salesperson,Region (Customer_ID is the primary key.)

2. Salesperson→Region (Each salesperson is assigned to a unique region.)

Notice that SALES is in second normal form because the primary key consists of a single attribute (Customer_ID). However, Region is functionally dependent on Salesperson, and Salesperson is functionally dependent on Customer_ID. As a result, there are data maintenance problems in SALES.

1. A new salesperson (Robinson) assigned to the North region cannot be entered until a customer has been assigned to that salesperson (because a value for Customer_ID must be provided to insert a row in the table).

2. If customer number 6837 is deleted from the table, we lose the information that salesperson Hernandez is assigned to the East region.

3. If salesperson Smith is reassigned to the East region, several rows must be changed to reflect that fact (two rows are shown in Figure 12-9[a]).

Third normal form (3NF): A relation is in third normal form (3NF) if it is in second normal form and there are no functional (transitive) dependencies between two (or more) nonprimary key attributes.

SALES

Customer_ID	Customer_Name	Salesperson	Region
8023	Anderson	Smith	South
9167	Bancroft	Hicks	West
7924	Hobbs	Smith	South
6837	Tucker	Hernandez	East
8596	Eckersley	Hicks	West
7018	Arnold	Faulb	North

Figure 12-9
Removing transitive dependencies
(a) Relation with transitive dependency

(b) Relations in 3NF

SALES1

Customer_ID	Customer_Name	Salesperson
8023	Anderson	Smith
9167	Bancroft	Hicks
7924	Hobbs	Smith
6837	Tucker	Hernandez
8596	Eckersley	Hicks
7018	Arnold	Faulb

SPERSON

Salesperson	Region
Smith	South
Hicks	West
Hernandez	East
Faulb	North

These problems can be avoided by decomposing SALES into the two relations, based on the two determinants, shown in Figure 12-9(b). These relations are the following:

SALES1(Customer_ID,Customer_Name,Salesperson)
SPERSON(Salesperson,Region)

Note that Salesperson is the primary key in SPERSON. Salesperson is also a foreign key in SALES1. A **foreign key** is an attribute that appears as a nonprimary key attribute in one relation (such as SALES1) and as a primary key attribute (or part of a primary key) in another relation. You designate a foreign key by using a dashed underline.

A foreign key must satisfy **referential integrity**, which specifies that the value of an attribute in one relation depends on the value of the same attribute in another relation. Thus, in Figure 12-9(b), the value of Salesperson in each row of table SALES1 is limited to only the current values of Salesperson in the SPERSON table. Referential integrity is one of the most important principles of the relational model.

Foreign key: An attribute that appears as a nonprimary key attribute in one relation and as a primary key attribute (or part of a primary key) in another relation.

Referential integrity: An integrity constraint specifying that the value (or existence) of an attribute in one relation depends on the value (or existence) of the same attribute in another relation.

TRANSFORMING E-R DIAGRAMS INTO RELATIONS

Normalization produces a set of well-structured relations that contains all of the data mentioned in system inputs and outputs developed in human interface design. Because these specific information requirements may not represent all future information needs, the E-R diagram you developed in conceptual data modeling is another source of insight into possible data requirements for a new application system. To compare the conceptual data model and the normalized relations developed so far, your E-R diagram must be transformed into relational notation, normalized, and then merged with the existing normalized relations.

Transforming an E-R diagram into normalized relations and then merging all the relations into one final, consolidated set of relations can be accomplished in four steps. These steps are summarized briefly here, and then steps 1, 2, and 4 are discussed in detail in the remainder of this chapter.

1. *Represent entities.* Each entity type in the E-R diagram becomes a relation. The identifier of the entity type becomes the primary key of the relation, and other attributes of the entity type become nonprimary key attributes of the relation.

2. *Represent relationships.* Each relationship in an E-R diagram must be represented in the relational database design. How we represent a relationship depends on its nature. For example, in some cases we represent a relationship by making the primary key of one relation a foreign key of another relation. In other cases, we create a separate relation to represent a relationship.

3. *Normalize the relations.* The relations created in steps 1 and 2 may have unnecessary redundancy. So, we need to normalize these relations to make them well structured.

4. *Merge the relations.* So far in database design we have created various relations from both a bottom-up normalization of user views and from transforming one or more E-R diagrams into sets of relations. Across these different sets of relations, there may be redundant relations (two or more relations that describe the same entity type) that must be merged and renormalized to remove the redundancy.

Represent Entities

Each regular entity type in an E-R diagram is transformed into a relation. The identifier of the entity type becomes the primary key of the corresponding relation. Each nonkey attribute of the entity type becomes a nonkey attribute of the relation. You should check to make sure that the primary key satisfies the following two properties:

1. The value of the key must uniquely identify every row in the relation.
2. The key should be nonredundant; that is, no attribute in the key can be deleted without destroying its unique identification.

Some entities may have keys that include the primary keys of other entities. For example, an EMPLOYEE DEPENDENT may have a Name for each dependent, but, to form the primary key for this entity, you must include the Employee_ID attribute from the associated EMPLOYEE entity. Such an entity whose primary key depends upon the primary key of another entity is called a weak entity.

Representation of an entity as a relation is straightforward. Figure 12-10(a) shows the CUSTOMER entity type for Pine Valley Furniture Company. The corresponding CUSTOMER relation is represented as follows:

CUSTOMER(Customer_ID,Name,Address,City_State_ZIP,Discount)

In this notation, the entity type label is translated into a relation name. The identifier of the entity type is listed first and underlined. All nonkey attributes are listed after the primary key. This relation is shown as a table with sample data in Figure 12-10(b).

Represent Relationships

The procedure for representing relationships depends on both the degree of the relationship—unary, binary, ternary—and the cardinalities of the relationship.

Binary 1:N and 1:1 Relationships   A binary one-to-many (1:N) relationship in an E-R diagram is represented by adding the primary key attribute (or attributes) of the entity on the one side of the relationship as a foreign key in the relation that is on the many side of the relationship.

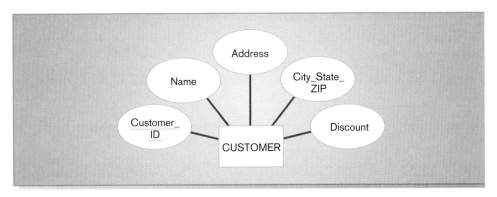

Figure 12-10
Transforming an entity type to a relation
(a) E-R diagram

(b) Relation

CUSTOMER

Customer_ID	Name	Address	City_State_ZIP	Discount
1273	Contemporary Designs	123 Oak St.	Austin, TX 28384	5%
6390	Casual Corner	18 Hoosier Dr.	Bloomington, IN 45821	3%

Figure 12-11(a), an example of this rule, shows the Places relationship (1:*N*) linking CUSTOMER and ORDER at Pine Valley Furniture Company. Two relations, CUSTOMER and ORDER, were formed from the respective entity types (see Figure 12-11[b]). Customer_ID, which is the primary key of CUSTOMER (on the one side of the relationship) is added as a foreign key to ORDER (on the many side of the relationship).

One special case under this rule was mentioned in the previous section. If the entity on the many side needs the key of the entity on the one side as part of its primary key (this is a so-called weak entity), then this attribute is added not as a nonkey but as part of the primary key.

For a binary or unary one-to-one (1:1) relationship between two entities A and B (for a unary relationship, A and B would be the same entity type), the relationship can be represented by any of the following choices:

1. Adding the primary key of A as a foreign key of B
2. Adding the primary key of B as a foreign key of A
3. Both of the above

Figure 12-11
Representing a 1:*N* relationship
(a) E-R Diagram

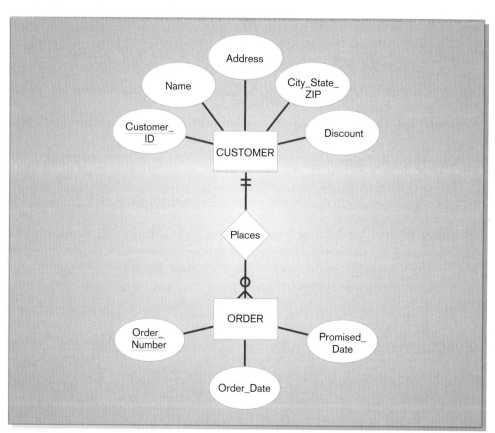

(b) Relations

CUSTOMER

Customer_ID	Name	Address	City_State_ZIP	Discount
1273	Contemporary Designs	123 Oak St.	Austin, TX 28384	5%
6390	Casual Corner	18 Hoosier Dr.	Bloomington, IN 45821	3%

ORDER

Order_Number	Order_Date	Promised_Date	Customer_ID
57194	3/15/0X	3/28/0X	6390
63725	3/17/0X	4/01/0X	1273
80149	3/14/0X	3/24/0X	6390

Binary and Higher-Degree M:N Relationships   Suppose that there is a binary many-to-many (*M:N*) relationship (or associative entity) between two entity types A and B. For such a relationship, we create a separate relation C. The primary key of this relation is a composite key consisting of the primary key for each of the two entities in the relationship. Any nonkey attributes that are associated with the *M:N* relationship are included with the relation C.

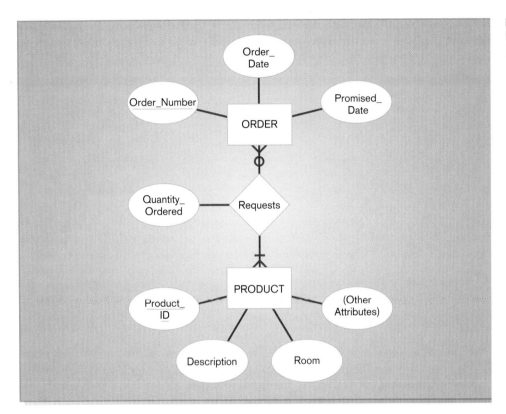

Figure 12-12
Representing an *M:N* relationship
(a) E-R Diagram

(b) Relations

ORDER

Order_Number	Order_Date	Promised_Date
61384	2/17/2002	3/01/2002
62009	2/13/2002	2/27/2002
62807	2/15/2002	3/01/2002

ORDER LINE

Order_Number	Product_ID	Quantity_ Ordered
61384	M128	2
61384	A261	1

PRODUCT

Product_ID	Description	Room	(Other Attributes)
M128	Bookcase	Study	–
A261	Wall unit	Family	–
R149	Cabinet	Study	–

Figure 12-12(a), an example of this rule, shows the Requests relationship (*M:N*) between the entity types ORDER and PRODUCT for Pine Valley Furniture Company. Figure 12-12(b) shows the three relations (ORDER, ORDER LINE, and PRODUCT) that are formed from the entity types and the Requests relationship. A relation (called ORDER LINE in Figure 12-12[b]) is created for the Requests relationship. The primary key of ORDER LINE is the combination (Order_Number, Product_ID), which is the respective primary keys of ORDER and PRODUCT. The nonkey attribute Quantity_Ordered also appears in ORDER LINE.

Occasionally, the relation created from an *M:N* relationship requires a primary key that includes more than just the primary keys from the two related relations. Consider, for example, the following situation:

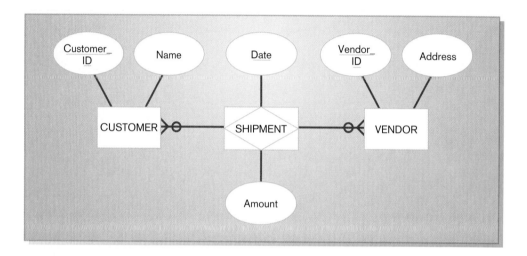

In this case, Date must be part of the key for the SHIPMENT relation to uniquely distinguish each row of the SHIPMENT table, as follows:

SHIPMENT(<u>Customer_ID</u>,<u>Vendor_ID</u>,<u>Date</u>,Amount)

If each shipment has a separate nonintelligent key, say a shipment number, then Date becomes a nonkey and Customer_ID and Vendor_ID become foreign keys, as follows:

SHIPMENT(<u>Shipment_Number</u>,Customer_ID,Vendor_ID,Date,Amount)

In some cases, there may be a relationship among three or more entities. In such cases, we create a separate relation that has as a primary key the composite of the primary keys of each of the participating entities (plus any necessary additional key elements). This rule is a simple generalization of the rule for a binary *M:N* relationship.

Unary Relationships   To review, a unary relationship is a relationship between the instances of a single entity type, which are also called *recursive relationships*. Figure 12-13 shows two common examples. Figure 12-13(a) shows a one-to-many relationship named Manages that associates employees with another employee who is their manager. Figure 12-13(b) shows a many-to-many relationship that associates certain items with their component items. This relationship is called a *bill-of-materials structure*.

For a unary 1:*N* relationship, the entity type (such as EMPLOYEE) is modeled as a relation. The primary key of that relation is the same as for the entity type. Then a foreign key is added to the relation that references the primary key values. A **recursive foreign key** is a foreign key in a relation that references the primary key values of that same relation. We can represent the relationship in Figure 12-13(a) as follows:

Recursive foreign key: A foreign key in a relation that references the primary key values of that same relation.

EMPLOYEE(<u>Emp_ID</u>,Name,Birthdate,Manager_ID)

In this relation, Manager_ID is a recursive foreign key that takes its values from the same set of worker identification numbers as Emp_ID.

For a unary $M:N$ relationship, we model the entity type as one relation. Then we create a separate relation to represent the $M:N$ relationship. The primary key of this new relation is a composite key that consists of two attributes (which need not have the same name) that both take their values from the same primary key. Any attribute associated with the relationship (such as Quantity in Figure 12-13[b]) is included as a nonkey attribute in this new relation. We can express the result for Figure 12-13[b] as follows:

ITEM(Item_Number,Name,Cost)

ITEM-BILL(Item_Number,Component_Number,Quantity)

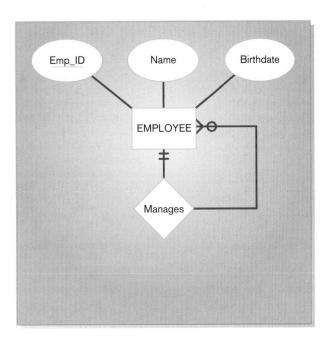

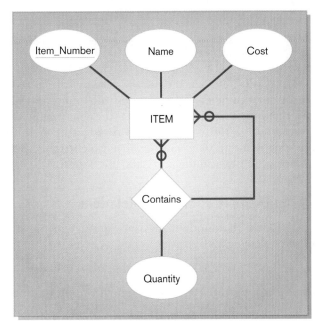

Figure 12-13
Two unary relations
(a) EMPLOYEE with Manages relationship (1:N)

(b) Bill-of-materials structure ($M:N$)

TABLE 12-1 E-R to Relational Transformation

E-R Structure	Relational Representation
Regular entity	Create a relation with primary key and nonkey attributes.
Weak entity	Create a relation with a composite primary key (which includes the primary key of the entity on which this weak entity depends) and nonkey attributes.
Binary or unary 1:1 relationship	Place the primary key of either entity in the relation for the other entity or do this for both entities.
Binary 1:N relationship	Place the primary key of the entity on the one side of the relationship as a foreign key in the relation for the entity on the many side.
Binary or unary M:N relationship or associative entity	Create a relation with a composite primary key using the primary keys of the related entities, plus any nonkey attributes of the relationship or associative entity.
Binary or unary M:N relationship or associative entity with additional key(s)	Create a relation with a composite primary key using the primary keys of the related entities and additional primary key attributes associated with the relationship or associative entity, plus any nonkey attributes of the relationship or associative entity.
Binary or unary M:N relationship or associative entity with its own key	Create a relation with the primary key associated with the relationship or associative entity, plus any nonkey attributes of the relationship or associative entity and the primary keys of the related entities (as foreign key attributes).
Supertype-subtype relationship	Create a relation for the superclass, which contains the primary key and all nonkey attributes in common with all subclasses, plus create a separate relation for each subclass with the same primary key (with the same or local name) but with only the nonkey attributes related to that subclass.

Summary of Transforming E-R Diagrams to Relations

We have now described how to transform E-R diagrams to relations. Table 12-1 lists the rules discussed in this section for transforming entity-relationship diagrams into equivalent relations. After this transformation, you should check the resulting relations to determine whether they are in third normal form and, if necessary, perform normalization as described earlier in the chapter.

MERGING RELATIONS

As part of the logical database design, normalized relations likely have been created from a number of separate E-R diagrams and various user interfaces. Some of the relations may be redundant—they may refer to the same entities. If so, you should merge those relations to remove the redundancy. This section describes merging relations or *view integration*, which is the last step in logical database design and prior to physical file and database design.

An Example of Merging Relations

Suppose that modeling a user interface or transforming an E-R diagram results in the following 3NF relation.

EMPLOYEE1(Emp_ID,Name,Address,Phone)

Modeling a second user interface might result in the following relation:

EMPLOYEE2(Emp_ID,Name, Address,Jobcode,Number_of_Years)

Because these two relations have the same primary key (Emp_ID) and describe the same entity, they should be merged into one relation. The result of merging the relations is the following relation:

EMPLOYEE(Emp_ID,Name,Address,Phone,Jobcode,Number_of_Years)

Notice that an attribute that appears in both relations (such as Name in this example) appears only once in the merged relation.

View Integration Problems

When integrating relations, you must understand the meaning of the data and must be prepared to resolve any problems that may arise in that process. In this section, we describe and illustrate four problems that arise in view integration: synonyms, homonyms, dependencies between nonkeys, and class/subclass relationships.

Synonyms   In some situations, two or more attributes may have different names but the same meaning, as when they describe the same characteristic of an entity. Such attributes are called **synonyms**. For example, Emp_ID and Employee_Number may be synonyms.

Synonyms: Two different names that are used for the same attribute.

When merging the relations that contain synonyms, you should obtain, if possible, agreement from users on a single standardized name for the attribute and eliminate the other synonym. Another alternative is to choose a third name to replace the synonyms. For example, consider the following relations:

STUDENT1(Student_ID,Name)

STUDENT2(Matriculation_Number,Name,Address)

In this case, the analyst recognizes that both the Student_ID and the Matriculation_Number are synonyms for a person's social security number and are identical attributes. One possible resolution would be to standardize on one of the two attribute names, such as Student_ID. Another option is to use a new attribute name, such as SSN, to replace both synonyms. Assuming the latter approach, merging the two relations would produce the following result:

STUDENT(SSN,Name,Address)

Homonyms   In other situations, a single attribute name, called a **homonym**, may have more than one meaning or describe more than one characteristic. For example, the term *account* might refer to a bank's checking account, savings account, loan account, or other type of account; therefore, *account* refers to different data, depending on how it is used.

Homonym: A single attribute name that is used for two or more different attributes.

You should be on the lookout for homonyms when merging relations. Consider the following example:

STUDENT1(Student_ID,Name,Address)

STUDENT2(Student_ID,Name,Phone_Number,Address)

In discussions with users, the systems analyst may discover that the attribute Address in STUDENT1 refers to a student's campus address, whereas in STUDENT2 the same attribute refers to a student's home address. To resolve this conflict, we would probably need to create new attribute names and the merged relation would become

STUDENT(Student_ID,Name,Phone_Number,Campus_Address,
Permanent_Address)

Dependencies Between Nonkeys   When two 3NF relations are merged to form a single relation, dependencies between nonkeys may result. For example, consider the following two relations:

>STUDENT1(Student_ID,Major)
>STUDENT2(Student_ID,Adviser)

Because STUDENT1 and STUDENT2 have the same primary key, the two relations may be merged:

>STUDENT(Student_ID,Major,Adviser)

However, suppose that each major has exactly one adviser. In this case, Adviser is functionally dependent on Major:

>Major→Adviser

If the above dependency exists, then STUDENT is in 2NF but not 3NF, because it contains a functional dependency between nonkeys. The analyst can create 3NF relations by creating two relations with Major as a foreign key in STUDENT:

>STUDENT(Student_ID,Major)
>MAJOR ADVISER(Major,Adviser)

Class/Subclass   Class/subclass relationships may be hidden in user views or relations. Suppose that we have the following two hospital relations:

>PATIENT1(Patient_ID,Name,Address,Date_Treated)
>PATIENT2(Patient_ID,Room_Number)

Initially, it appears that these two relations can be merged into a single PATIENT relation. However, suppose that there are two different types of patients: inpatients and outpatients. PATIENT1 actually contains attributes common to *all* patients. PATIENT2 contains an attribute (Room_Number) that is a characteristic only of inpatients. In this situation, you should create *class/subclass* relationships for these entities:

>PATIENT(Patient_ID,Name,Address)
>INPATIENT(Patient_ID,Room_Number)
>OUTPATIENT(Patient_ID,Date_Treated)

LOGICAL DATABASE DESIGN FOR HOOSIER BURGER

In Chapter 10 we developed an E-R diagram for a new Inventory Control System at Hoosier Burger (Figure 12-14 repeats the diagram from Chapter 10). In this section we show how this E-R model is translated into normalized relations and how to normalize and then merge the relations for a new report with the relations from the E-R model.

In this E-R model, four entities exist independently of other entities: SALE, PRODUCT, INVOICE, and INVENTORY ITEM. Given the attributes shown in Figure 12-14, we can represent these entities in the following four relations:

>SALE(Receipt_Number,Sale_Date)
>PRODUCT(Product_ID,Product_Description)
>INVOICE(Vendor_ID,Invoice_Number,Invoice_Date,Paid?)
>INVENTORY ITEM(Item_Number,Item_Description,Quantity_in_Stock, Minimum_Order_Quantity,Type_of_Item)

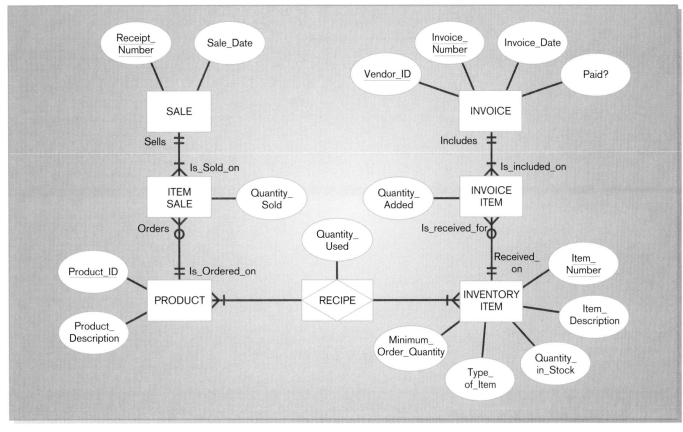

Figure 12-14
Final E-R diagram for Hoosier Burger's Inventory Control System

The entities ITEM SALE and INVOICE ITEM as well as the associative entity RECIPE each have composite primary keys taken from the entities to which they relate, so we can represent these three entities in the following three relations:

ITEM SALE(Receipt_Number, Product_ID,Quantity_Sold)
INVOICE ITEM(Vendor_ID,Invoice_Number,Item_Number,Quantity_Added)
RECIPE(Product_ID,Item_Number,Quantity_Used)

Because there are no many-to-many, one-to-one, or unary relationships, we have now represented all the entities and relationships from the E-R model. Also, each of the above relations is in 3NF because all attributes are simple, all nonkeys are fully dependent on the whole key, and there are no dependencies between nonkeys in the INVOICE and INVENTORY ITEM relations.

Now suppose that Bob Mellankamp wanted an additional report that was not previously known by the analyst who designed the Inventory Control System for Hoosier Burger. A rough sketch of this new report, listing volume of purchases from each vendor by type of item in a given month, appears in Figure 12-15. In this report, the same type of item may appear many times if multiple vendors supply the same type of item.

This report contains data about several relations already known to the analyst, including

INVOICE(Vendor_ID,Invoice_Number,Invoice_Date): primary keys, and the date is needed to select invoices in the specified month of the report
INVENTORY ITEM(Item_Number,Type_of_Item): primary key and a nonkey in the report
INVOICE ITEM(Vendor_ID,Invoice_Number,Item_Number,Quantity_Added): primary keys and the raw quantity of items invoiced that are subtotaled by vendor and type of item in the report

Figure 12-15
Hoosier Burger Monthly Vendor Load Report

Vendor			
ID	Name	Type of Item	Total Quantity Added
V1	V1name	aaa	nnn1
		bbb	nnn2
		ccc	nnn3
V2	V2name	bbb	nnn4
		mmm	nnn5
x			
x			
x			

Monthly Vendor Load Report
for Month: xxxxx
Page x of n

In addition, the report includes a new attribute—Vendor name. After some investigation, an analyst determines that Vendor_ID→Vendor_Name. Because the whole primary key of the INVOICE relation is Vendor_ID and Invoice_Number, if Vendor_Name were part of the INVOICE relation, this relation would violate the 3NF rule. So, a new VENDOR relation must be created as follows:

VENDOR(Vendor_ID,Vendor_Name)

Figure 12-16
E-R diagram corresponding to normalized relations of Hoosier Burger's Inventory Control System

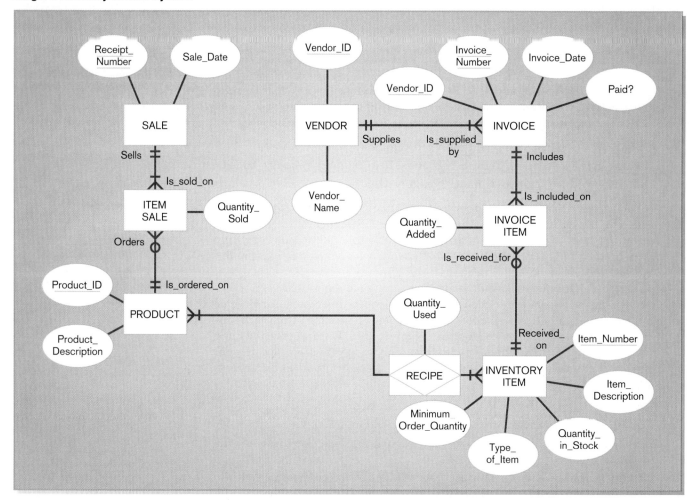

Now, Vendor_ID not only is part of the primary key of INVOICE but also is a foreign key referencing the VENDOR relation. Hence, there must be a one-to-many relationship from VENDOR to INVOICE. The systems analyst determines that an invoice must come from a vendor, and there is no need to keep data about a vendor unless the vendor invoices Hoosier Burger. An updated E-R diagram, reflecting these enhancements for new data needed in the monthly vendor load report, appears in Figure 12-16. The normalized relations for this database are:

SALE(Receipt_Number,Sale_Date)
PRODUCT(Product_ID,Product_Description)
INVOICE(Vendor_ID,Invoice_Number,Invoice_Date,Paid?,Vendor_ID)
INVENTORY ITEM(Item_Number,Item_Description,Quantity_in_Stock,
    Minimum_Order_Quantity, Type_of_Item)
ITEM SALE(Receipt_Number,Product_ID,Quantity_Sold)
INVOICE ITEM(Vendor_ID,Invoice_Number,Item_Number,Quantity_Added)
RECIPE(Product_ID,Item_Number,Quantity_Used)
VENDOR(Vendor_ID,Vendor_Name)

PHYSICAL FILE AND DATABASE DESIGN

Designing physical files and databases requires certain information that should have been collected and produced during prior SDLC phases. This information includes:

- Normalized relations, including volume estimates
- Definitions of each attribute
- Descriptions of where and when data are used: entered, retrieved, deleted, and updated (including frequencies)
- Expectations or requirements for response time and data integrity
- Descriptions of the technologies used for implementing the files and database so that the range of required decisions and choices for each is known

Normalized relations are, of course, the result of logical database design. Statistics on the number of rows in each table as well as the other information listed above may have been collected during requirements determination in systems analysis. If not, these items need to be discovered to proceed with database design.

We take a bottom-up approach to reviewing physical file and database design. Thus, we begin the physical design phase by addressing the design of physical fields for each attribute in a logical data model.

DESIGNING FIELDS

A **field** is the smallest unit of application data recognized by system software, such as a programming language or database management system. An attribute from a logical database model may be represented by several fields. For example, a student name attribute in a normalized student relation might be represented as three fields: last name, first name, and middle initial. In general, you will represent each attribute from each normalized relation as one or more fields. The basic decisions you must make in specifying each field concern the type of data (or storage type) used to represent the field and data integrity controls for the field.

Field: The smallest unit of named application data recognized by system software.

Choosing Data Types

Data type: A coding scheme recognized by system software for representing organizational data.

A **data type** is a coding scheme recognized by system software for representing organizational data. The bit pattern of the coding scheme is usually immaterial to you, but the space to store data and the speed required to access data are of consequence in the physical file and database design. The specific file or database management software you use with your system will dictate which choices are available to you. For example, Table 12-2 lists the most commonly used data types available in Oracle 8i:

Selecting a data type balances four objectives that will vary in degree of importance dependent on the application:

1. Minimize storage space
2. Represent all possible values of the field
3. Improve data integrity for the field
4. Support all data manipulations desired on the field

You want to choose a data type for a field that minimizes space, represents every possible legitimate value for the associated attribute, and allows the data to be manipulated as needed. For example, suppose a quantity sold field can be represented by a Number data type. You would select a length for this field that would handle the maximum value, plus some room for growth of the business. Further, the Number data type will restrict users from entering inappropriate values (text), but it does allow negative numbers (if this is a problem, application code or form design may be required to restrict the values to positive).

Be careful—the data type must be suitable for the life of the application; otherwise, maintenance will be required. Choose data types for future needs by anticipating growth. Also, be careful that date arithmetic can be done so that dates can be subtracted or time periods can be added to or subtracted from a date.

Several other capabilities of data types may be available with some database technologies. We discuss a few of the most common of these features next: calculated fields and coding and compression techniques.

TABLE 12-2   Oracle 8i Data Types

Data Type	Description
VARCHAR2	Variable-length character data with a maximum length of 4000 characters; you must enter a maximum field length (e.g., VARCHAR2(30) for a field with a maximum length of 30 characters). A value less than 30 characters will consume only the required space.
CHAR	Fixed-length character data with a maximum length of 255 characters; default length is 1 character (e.g., CHAR(5) for a field with a fixed length of five characters, capable of holding a value from 0 to 5 characters long).
LONG	Capable of storing up to two gigabytes of one variable-length character data field (e.g., to hold a medical instruction or a customer comment).
NUMBER	Positive and negative numbers in the range 10^{-130} to 10^{126}; can specify the precision (total number of digits to the left and right of the decimal point) and the scale (the number of digits to the right of the decimal point) (e.g., NUMBER(5) specifies an integer field with a maximum of 5 digits and NUMBER(5, 2) specifies a field with no more than five digits and exactly two digits to the right of the decimal point).
DATE	Any date from January 1, 4712 B.c to December 31, 4712 AD; date stores the century, year, month, day, hour, minute, and second.
BLOB	Binary large object, capable of storing up to four gigabytes of binary data (e.g., a photograph or sound clip).

Calculated Fields   It is common that an attribute is mathematically related to other data. For example, an invoice may include a total due field, which represents the sum of the amount due on each item on the invoice. A field that can be derived from other database fields is called a **calculated** (or **computed** or **derived**) **field** (recall that a functional dependency between attributes does not imply a calculated field). Some database technologies allow you to explicitly define calculated fields along with other raw data fields. If you specify a field as calculated, you would then usually be prompted to enter the formula for the calculation; the formula can involve other fields from the same record and possibly fields from records in related files. The database technology will either store the calculated value or compute it when requested.

Calculated (or **computed** or **derived**) **field:** A field that can be derived from other database fields.

Coding and Compression Techniques   Some attributes have very few values from a large range of possible values. For example, suppose in Pine Valley Furniture each product has a finish attribute, with possible values Birch, Walnut, Oak, and so forth. To store this attribute as Text might require 12, 15, or even 20 bytes to represent the longest finish value. Suppose that even a liberal estimate is that Pine Valley Furniture will never have more than 25 finishes. Thus, a single alphabetic or alphanumeric character would be more than sufficient. We not only reduce storage space but also increase integrity (by restricting input to only a few values), which helps to achieve two of the physical file and database design goals. Codes also have disadvantages. If used in system inputs and outputs, they can be more difficult for users to remember, and programs must be written to decode fields if codes will not be displayed.

N E T   S E A R C H
Investigate the capabilities of several data compression programs. Visit http://www.prenhall.com/hoffer to complete an exercise related to this topic.

Controlling Data Integrity

We have already explained that data typing helps control data integrity by limiting the possible range of values for a field. There are additional physical file and database design options you might use to ensure higher-quality data. Although these controls can be imposed within application programs, it is better to include these as part of the file and database definitions so that the controls are guaranteed to be applied all the time as well as uniformly for all programs. There are four popular data integrity control methods: default value, range control, referential integrity, and null value control.

- *Default value.* A **default value** is the value a field will assume unless an explicit value is entered for the field. For example, the city and state of most customers for a particular retail store will likely be the same as the store's city and state. Assigning a default value to a field can reduce data entry time (the field can simply be skipped during data entry) and data entry errors, such as typing *IM* instead of *IN* for *Indiana*.

Default value: A value a field will assume unless an explicit value is entered for that field.

- *Range control.* Both numeric and alphabetic data may have a limited set of permissible values. For example, a field for the number of product units sold may have a lower bound of 0, and a field that represents the month of a product sale may be limited to the values JAN, FEB, and so forth.

- *Referential integrity.* As noted earlier in this chapter, the most common example of referential integrity is cross-referencing between relations. For example, consider the pair of relations in Figure 12-17(a). In this case, the values for the foreign key Customer_ID field within a customer order must be limited to the set of Customer_ID values from the customer relation; we would not want to accept an order for a nonexisting or unknown customer. Referential integrity may be useful in other instances. Consider the employee relation example in Figure 12-17(b). In this example, the employee relation has a field of Supervisor_ID. This field refers to the Employee_ID of the employee's supervisor and should have referential integrity on the Employee_ID field within the same relation. Note in this case that because some employees do not have supervisors, this is a weak referential integrity constraint because the value of a Supervisor_ID field may be empty.

Figure 12-17
Examples of referential integrity field controls
(a) Referential integrity between relations

CUSTOMER(**Customer_ID**,Cust_Name,Cust_Address, . . .)

CUST_ORDER(Order_ID, **Customer_ID**,Order_Date, . . .)

and Customer_ID may not be null because every order must be for some existing customer

(b) Referential integrity within a relation

EMPLOYEE(**Employee_ID,Supervisor_ID**,Empl_Name, . . .)
and Superviosr_ID may be null because not all employees have supervisors

Null value: A special field value, distinct from 0, blank, or any other value, that indicates that the value for the field is missing or otherwise unknown.

- *Null value control.* A **null value** is a special field value, distinct from 0, blank, or any other value, that indicates that the value for the field is missing or otherwise unknown. It is not uncommon that when it is time to enter data—for example, a new customer—you might not know the customer's phone number. The question is whether a customer, to be valid, must have a value for this field. The answer for this field is probably initially no, because most data processing can continue without knowing the customer's phone number. Later a null value may not be allowed when you are ready to ship product to the customer. On the other hand, you must always know a value for the Customer_ID field. Due to referential integrity, you cannot enter any customer orders for this new customer without knowing an existing Customer_ID value, and customer name is essential for visual verification of correct data entry. Besides using a special null value when a field is missing its value, you can also estimate the value, produce a report indicating rows of tables with critical missing values, or determine whether the missing value matters in computing needed information.

DESIGNING PHYSICAL TABLES

A relational database is a set of related tables (tables are related by foreign keys referencing primary keys). In logical database design you grouped into a relation those attributes that concern some unifying, normalized business concept, such as a customer, product, or employee. In contrast, a **physical table** is a named set of rows and columns that specifies the fields in each row of the table. A physical table may or may not correspond to one relation. Whereas normalized relations possess properties of well-structured relations, the design of a physical table has two goals different from those of normalization: efficient use of secondary storage and data processing speed.

The efficient use of secondary storage (disk space) relates to how data are loaded on disks. Disks are physically divided into units (called pages) that can be read or written in one machine operation. Space is used efficiently when the physical length of a table row divides close to evenly into the length of the storage unit. For many information systems, this even division is very difficult to achieve because it depends on factors, such as operating system parameters, outside the control of each database. Consequently, we do not discuss this factor of physical table design in this text.

A second and often more important consideration when selecting a physical table design is efficient data processing. Data are most efficiently processed when they are stored close to one another in secondary memory, thus minimizing the number of input/output (I/O) operations that must be performed. Typically, the data in one

Physical table: A named set of rows and columns that specifies the fields in each row of the table.

physical table (all the rows and fields in those rows) are stored close together on disk. **Denormalization** is the process of splitting or combining normalized relations into physical tables based on affinity of use of rows and fields. Consider Figure 12-18. In Figure 12-18(a), a normalized product relation is split into separate physical tables with each containing only engineering, accounting, or marketing product data; the primary key must be included in each table. Note, the Description and Color attributes are repeated in both the engineering and marketing tables because these attrib-

Denormalization: The process of splitting or combining normalized relations into physical tables based on affinity of use of rows and fields.

Figure 12-18
Examples of denormalization
(a) Denormalization by columns

Normalized Product Relation
    Product(Product_ID,Description,Drawing_Number,Weight,Color,Unit_Cost,
        Burden_Rate,Price,Product_Manager)

Denormalized Functional Area Product Relations for Tables
    Engineering: E_Product(Product_ID,Description,Drawing_Number,Weight,Color)
    Accounting: A_Product(Product_ID,Unit_Cost,Burden_Rate)
    Marketing: M_Product(Product_ID,Description,Color, Price,Product_Manager)

(b) Denormalization by rows

Normalized Customer Table
CUSTOMER

Customer_ID	Name	Region	Annual_Sales
1256	Rogers	Atlantic	10,000
1323	Temple	Pacific	20,000
1455	Gates	South	15,000
1626	Hope	Pacific	22,000
2433	Bates	South	14,000
2566	Bailey	Atlantic	12,000

Denormalized Regional Customer Tables
A_CUSTOMER

Customer_ID	Name	Region	Annual_Sales
1256	Rogers	Atlantic	10,000
2566	Bailey	Atlantic	12,000

P_CUSTOMER

Customer_ID	Name	Region	Annual_Sales
1323	Temple	Pacific	20,000
1626	Hope	Pacific	22,000

S_CUSTOMER

Customer_ID	Name	Region	Annual_Sales
1455	Gates	South	15,000
2433	Bates	South	14,000

utes relate to both kinds of data. In Figure 12-18(b), a customer relation is denormalized by putting rows from different geographic regions into separate tables. In both cases, the goal is to create tables that contain only the data used together in programs. By placing data used together close to one another on disk, the number of disk I/O operations needed to retrieve all the data needed in a program is minimized.

The capability to split a table into separate sections, often called partitioning, is possible with most relational database products. With Oracle 8i, there are three types of table partitioning:

- *Range partitioning.* Partitions are defined by nonoverlapping ranges of values for a specified attribute (so, separate tables are formed of the rows whose specified attribute values fall in indicated ranges)
- *Hash partitioning.* A table row is assigned to a partition by an algorithm and then maps the specified attribute value to a partition.
- *Composite partitioning.* Combines range and hash partitioning by first segregating data by ranges on the designated attribute and then within each of these partition it further partitions by hashing on the designated attribute.

Each partition is stored in a separate contiguous section of disk space, which Oracle calls a tablespace.

Denormalization can increase the chance of errors and inconsistencies that normalization avoided. Further, denormalization optimizes certain data processing at the expense of others, so if the frequencies of different processing activities change, the benefits of denormalization may no longer exist (Finkelstein, 1988).

Various forms of denormalization, which involves combining data from several normalized tables, can be done, but there are no hard-and-fast rules for deciding when to denormalize data. Here are three common situations (Rodgers, 1989) in which denormalization across tables often makes sense (see Figure 12-19 for illustrations):

1. *Two entities with a one-to-one relationship.* Figure 12-19(a) shows student data with optional data from a standard scholarship application a student may complete. In this case, one record could be formed with four fields from the STUDENT and SCHOLARSHIP APPLICATION FORM normalized relations. (*Note:* In this case, fields from the optional entity must have null values allowed.)

Figure 12-19
Possible denormalization situations
(a) Two entities with a one-to-one relationship

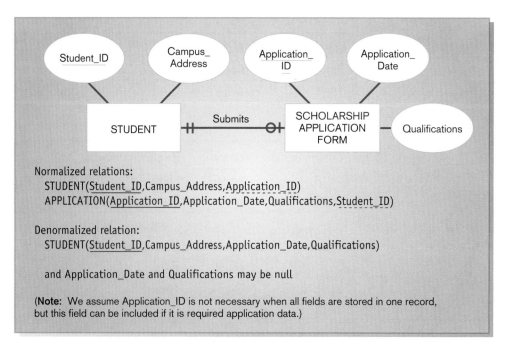

Normalized relations:
    STUDENT(Student_ID,Campus_Address,Application_ID)
    APPLICATION(Application_ID,Application_Date,Qualifications,Student_ID)

Denormalized relation:
    STUDENT(Student_ID,Campus_Address,Application_Date,Qualifications)

and Application_Date and Qualifications may be null

(**Note:** We assume Application_ID is not necessary when all fields are stored in one record, but this field can be included if it is required application data.)

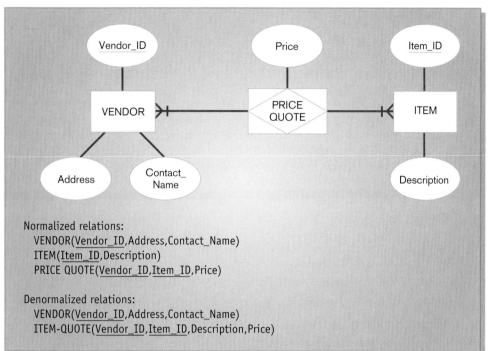

Normalized relations:
  VENDOR(Vendor_ID,Address,Contact_Name)
  ITEM(Item_ID,Description)
  PRICE QUOTE(Vendor_ID,Item_ID,Price)

Denormalized relations:
  VENDOR(Vendor_ID,Address,Contact_Name)
  ITEM-QUOTE(Vendor_ID,Item_ID,Description,Price)

(c) Reference data

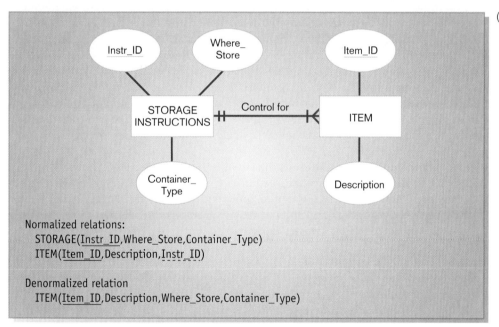

Normalized relations:
  STORAGE(Instr_ID,Where_Store,Container_Type)
  ITEM(Item_ID,Description,Instr_ID)

Denormalized relation
  ITEM(Item_ID,Description,Where_Store,Container_Type)

2. *A many-to-many relationship (associative entity) with nonkey attributes.* Figure
   12-19(b) shows price quotes for different items from different vendors. In this
   case, fields from ITEM and PRICE QUOTE relations might be combined into
   one physical table to avoid having to combine all three tables together. (*Note*:
   This may create considerable duplication of data—in the example, the ITEM
   fields, such as Description, would repeat for each price quote—and excessive
   updating if duplicated data changes.)

3. *Reference data.* Figure 12-19(c) shows that several ITEMs have the same STOR-
   AGE INSTRUCTIONS and STORAGE INSTRUCTIONS relate only to ITEMs.
   In this case, the storage instruction data could be stored in the ITEM table,
   thus reducing the number of tables to access but also creating redundancy
   and the potential for extra data maintenance.

Arranging Table Rows

Physical file: A named set of table rows stored in a contiguous section of secondary memory.

The result of denormalization is the definition of one or more physical files. A computer operating system stores data in a **physical file**, which is a named set of table rows stored in a contiguous section of secondary memory. A file contains rows and columns from one or more tables, as produced from denormalization. To the operating system—like Windows, Linux, or UNIX—each table may be one file or the whole database may be in one file, depending on how the database technology and database designer organize data. The way the operating system arranges table rows in a file is called a **file organization**. With some database technologies, the systems designer can choose among several organizations for a file.

File organization: A technique for physically arranging the records of a file.

If the database designer has a choice, he or she chooses a file organization for a specific file to provide:

1. Fast data retrieval
2. High throughput for processing transactions
3. Efficient use of storage space
4. Protection from failures or data loss
5. Minimal need for reorganization
6. Accommodation of growth
7. Security from unauthorized use

Often these objectives conflict, and you must select an organization for each file that provides a reasonable balance among the criteria within the resources available.

Pointer: A field of data that can be used to locate a related field or row of data.

To achieve these objectives, many file organizations utilize the concept of a pointer. A **pointer** is a field of data that can be used to locate a related field or row of data. In most cases, a pointer contains the address of the associated data, which has no business meaning. Pointers are used in file organizations when it is not possible to store related data next to each other. Because this is often the case, pointers are common. In most cases, fortunately, pointers are hidden from a programmer. Yet, because a database designer may need to decide if and how to use pointers, we introduce the concept here.

Literally hundreds of different file organizations and variations have been created, but we outline the basics of three families of file organizations used in most file management environments: sequential, indexed, and hashed, as illustrated in Figure 12-20. You need to understand the particular variations of each method available in the environment for which you are designing files.

Sequential file organization: The rows in the file are stored in sequence according to a primary key value.

Sequential File Organizations

In a **sequential file organization**, the rows in the file are stored in sequence according to a primary key value (see Figure 12-20[a]). To locate a particular row, a program must normally scan the file from the beginning until the desired row is located. A common example of a sequential file is the alphabetic list of persons in the white pages of a phone directory (ignoring any index that may be included with the directory). Sequential files are very fast if you want to process rows sequentially, but they are essentially impractical for random row retrievals. Deleting rows can cause wasted space or the need to compress the file. Adding rows requires rewriting the file, at least from the point of insertion. Updating a row may also require rewriting the file, unless the file organization supports rewriting over the updated row only. Moreover, only one sequence can be maintained without duplicating the rows.

Indexed file organization: The rows are stored either sequentially or nonsequentially, and an index is created that allows software to locate individual rows.

Index: A table used to determine the location of rows in a file that satisfy some condition.

Indexed File Organizations

In an **indexed file organization**, the rows are stored either sequentially or nonsequentially, and an index is created that allows the application software to locate individual rows (see Figure 12-20[b]). Like a card catalog in a library, an **index** is a structure that is used to determine the rows in a file that satisfy

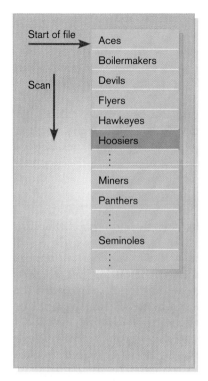

(a) Sequential

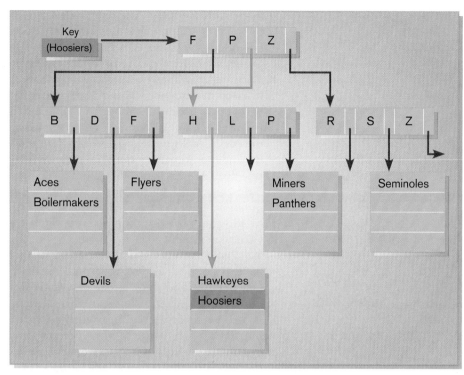

(b) Indexed

Figure 12-20
Comparison of file organizations

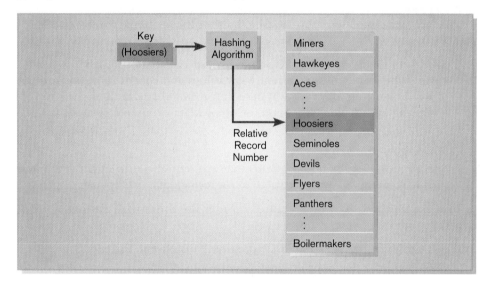

(c) Hashed

some condition. Each entry matches a key value with one or more rows. An index can point to unique rows (a primary key index, such as on the Product_ID field of a product table) or to potentially more than one row. An index that allows each entry to point to more than one record is called a **secondary key** index. Secondary key indexes are important for supporting many reporting requirements and for providing rapid ad hoc data retrieval. An example would be an index on the Finish field of a product table.

The example in Figure 12-20(b), typical of many index structures, illustrates that indexes can be built on top of indexes, creating a hierarchical set of indexes, and the data are stored sequentially in many contiguous segments. For example, to find the record with key "Hoosiers," the file organization would start at the top index and take the pointer after the entry *P*, which points to another index for all keys that

Secondary key: One or a combination of fields for which more than one row may have the same combination of values.

begin with the letters *G* through *P* in the alphabet. Then the software would follow the pointer after the *H* in this index, which represents all those records with keys that begin with the letters *G* through *H*. Eventually, the search through the indexes either locates the desired record or indicates that no such record exists. The reason for storing the data in many contiguous segments is to allow room for some new data to be inserted in sequence without rearranging all the data.

The main disadvantages to indexed file organizations are the extra space required to store the indexes and the extra time necessary to access and maintain indexes. Usually these disadvantages are more than offset by the advantages. Because the index is kept in sequential order, both random and sequential processing are practical. Also, because the index is separate from the data, you can build multiple index structures on the same data file (just as in the library where there are multiple indexes on author, title, subject, and so forth). With multiple indexes, software may rapidly find records that have compound conditions, such as finding books by Tom Clancy on espionage.

The decision of which indexes to create is probably the most important physical database design task for relational database technology, such as Microsoft Access, Oracle, DB2, and similar systems. Indexes can be created for both primary and secondary keys. When using indexes, there is a trade-off between improved performance for retrievals and degrading performance for inserting, deleting, and updating the rows in a file. Thus, indexes should be used generously for databases intended primarily to support data retrievals, such as for decision support applications. Because they impose additional overhead, indexes should be used judiciously for databases that support transaction processing and other applications with heavy updating requirements.

Here are some guidelines for choosing indexes for relational databases (Gibson et al., 1989):

1. Specify a unique index for the primary key of each table (file). This selection ensures the uniqueness of primary key values and speeds retrieval based on those values. Random retrieval based on primary key value is common for answering multitable queries and for simple data maintenance tasks.

2. Specify an index for foreign keys. As in the first guideline, this speeds processing multitable queries.

3. Specify an index for nonkey fields that are referenced in qualification and sorting commands for the purpose of retrieving data.

PINE VALLEY FURNITURE

To illustrate the use of these rules, consider the following relations for Pine Valley Furniture Company:

PRODUCT(<u>Product_Number</u>,Description,Finish,Room,Price)
ORDER(<u>Order_Number</u>,<u>Product_Number</u>,Quantity)

You would normally specify a unique index for each primary key: Product_Number in PRODUCT and Order_Number in ORDER. Other indexes would be assigned based on how the data are used. For example, suppose that there is a system module that requires PRODUCT and PRODUCT_ORDER data for products with a price below $500, ordered by Product_Number. To speed up this retrieval, you could consider specifying indexes on the following nonkey attributes:

1. Price in PRODUCT because it satisfies rule 3

2. Product_Number in ORDER because it satisfies rule 2

Because users may direct a potentially large number of different queries against the database, especially for a system with a lot of ad hoc queries, you will probably have to be selective in specifying indexes to support the most common or frequently used queries.

TABLE 12-3 Comparative Features of Sequential, Indexed, and Hashed File Organizations

| | File Organization | | |
Factor	Sequential	Indexed	Hashed
Storage space	No wasted space	No wasted space for data, but extra space for index	Extra space may be needed to allow for addition and deletion of records.
Sequential retrieval on primary key	Very fast	Moderately fast	Impractical
Random retrieval on primary key	Impractical	Moderately fast	Very fast
Multiple key retrieval	Possible, but requires scanning whole file	Very fast with multiple indexes	Not possile
Deleting rows	Can create wasted space or require reorganizing	If space can be dynamically allocated, this is easy, but requires maintenance of indexes.	Very easy
Adding rows	Requires rewriting file	If space can be dynamically allocated, this is easy, but requires maintenance of indexes.	Very easy, except multiple keys with same address require extra work
Updating rows	Usually requires rewriting file	Easy, but requires maintenance of indexes	Very easy

Hashed File Organizations   In a **hashed file organization**, the location of each row is determined using an algorithm (see Figure 12-20[c]) that converts a primary key value into a row address. Although there are several variations of hashed files, in most cases the rows are located nonsequentially as dictated by the hashing algorithm. Thus, sequential data processing is impractical. On the other hand, retrieval of random rows is very fast. There are issues in the design of hashing file organizations, such as how to handle two primary keys that translate into the same address, but again, these issues are beyond our scope (see Hoffer, Prescott, and McFadden [2002] for a thorough discussion).

Hashed file organization: The address for each row is determined using an algorithm.

Summary of File Organizations   The three families of file organizations—sequential, indexed, and hashed—cover most of the file organizations you will have at your disposal as you design physical files and databases. Table 12-3 summarizes the comparative features of these file organizations. You can use this table to help choose a file organization by matching the file characteristics and file processing requirements with the features of the file organization.

Designing Controls for Files

Two of the goals of physical table design mentioned earlier are protection from failures or data loss and security from unauthorized use. These goals are achieved primarily by implementing controls on each file. Data integrity controls, a primary type of control, was mentioned earlier in the chapter. Two other important types of controls address file backup and security.

It is almost inevitable that a file will be damaged or lost, due to either software or human errors. When a file is damaged, it must be restored to an accurate and reasonably current condition. A file and database designer has several techniques for file restoration, including:

- Periodically making a backup copy of a file
- Storing a copy of each change to a file in a transaction log or audit trail
- Storing a copy of each row before or after it is changed

For example, a backup copy of a file and a log of rows after they were changed can be used to reconstruct a file from a previous state (the backup copy) to its current values. This process would be necessary if the current file were so damaged that it could not be used. If the current file is operational but inaccurate, then a log of before images of rows can be used in reverse order to restore a file to an accurate but previous condition. Then a log of the transactions can be reapplied to the restored file to bring it up to current values. It is important that the information system designer make provisions for backup, audit trail, and row image files so that data files can be rebuilt when errors and damage occur.

An information system designer can build data security into a file by several means, including:

N E T   S E A R C H
Investigate further data security methods and techniques. Visit http://www.prenhall.com/hoffer to complete an exercise related to this topic.

- Coding, or encrypting, the data in the file so that they cannot be read unless the reader knows how to decrypt the stored values

- Requiring data file users to identify themselves by entering user names and passwords, and then possibly allowing only certain file activities (read, add, delete, change) for selected users to selected data in the file

- Prohibiting users from directly manipulating any data in the file, but rather force programs and users to work with a copy (real or virtual) of the data they need; the copy contains only the data that users or programs are allowed to manipulate, and the original version of the data will change only after changes to the copy are thoroughly checked for validity

Security procedures such as these all add overhead to an information system, so only necessary controls should be included.

PHYSICAL DATABASE DESIGN FOR HOOSIER BURGER

A set of normalized relations and an associated E-R diagram for Hoosier Burger (Figure 12-16) were presented in the section Logical Database Design for Hoosier Burger earlier in this chapter. The display of a complete design of this database would require more documentation than space permits in this text, so we illustrate in this section only a few key decisions from the complete physical database.

As outlined in this chapter, to translate a logical database design into a physical database design, you need to make the following decisions:

- Create one or more fields for each attribute and determine a data type for each field.
- For each field, decide if it is calculated, needs to be coded or compressed, must have a default value or picture, or must have range, referential integrity, or null value controls.
- For each relation, decide if it should be denormalized to achieve desired processing efficiencies.
- Choose a file organization for each physical file.
- Select suitable controls for each file and the database.

Remember, the specifications for these decisions are made in physical database design, and then the specifications are coded in the implementation phase using the capabilities of the chosen database technology. These database technology capabilities determine what physical database design decisions you need to make. For example, for Oracle, which we assume is the implementation environment for this illustration, the only choice for file organization is indexed, so the file organization decision becomes on which primary and secondary key attributes to build indexes.

An index is information about the primary or secondary keys of a file. Each index entry contains the key value and a pointer to the row that contains that key value. An index facilitates rapid retrieval to rows for queries that involve AND, OR, and NOT qualifications of keys (e.g., all products with a maple finish and unit cost greater than $500 or all products in the office furniture product line). When using indexes, there is a trade-off between improved performance for retrievals and degrading performance for inserting, deleting, and updating the rows in a file. Thus, indexes should be used generously for databases intended primarily to support data retrievals, such as for decision support applications. Because they impose additional overhead, indexes should be used judiciously for databases that support transaction processing and other applications with heavy updating requirements. Typically, you create indexes on a file for its primary key, foreign keys, and other attributes used in qualification and sorting clauses in queries, forms, reports, and other system interfaces.

Key Terms

1. Calculated (or computed or derived) field
2. Data type
3. Default value
4. Denormalization
5. Field
6. File organization
7. Foreign key
8. Functional dependency
9. Hashed file organization
10. Homonym
11. Index
12. Indexed file organization
13. Normalization
14. Null value
15. Physical file
16. Physical table
17. Pointer
18. Primary key
19. Recursive foreign key
20. Referential integrity
21. Relation
22. Relational database model
23. Second normal form (2NF)
24. Secondary key
25. Sequential file organization
26. Synonyms
27. Third normal form (3NF)
28. Well-structured relation (or table)

Match each of the key terms above to the definition that best fits it.

_____ A named, two-dimensional table of data. Each relation consists of a set of named columns and an arbitrary number of unnamed rows.

_____ A relation that contains a minimum amount of redundancy and allows users to insert, modify, and delete the rows without errors or inconsistencies.

_____ The process of converting complex data structures into simple, stable data structures.

_____ A particular relationship between two attributes.

_____ A relation for which every nonprimary key attribute is functionally dependent on the whole primary key.

_____ A relation that is in second normal form and that has no functional (transitive) dependencies between two (or more) nonprimary key attributes.

_____ An attribute that appears as a nonprimary key attribute in one relation and as a primary key attribute (or part of a primary key) in another relation.

_____ An integrity constraint specifying that the value (or existence) of an attribute in one relation depends on the value (or existence) of the same attribute in another relation.

_____ A foreign key in a relation that references the primary key values of that same relation.

_____ Two different names that are used for the same attribute.

_____ A single attribute name that is used for two or more different attributes.

_____ The smallest unit of named application data recognized by system software.

_____ A coding scheme recognized by system software for representing organizational data.

_____ A field that can be derived from other database fields.

_____ A value a field will assume unless an explicit value is entered for that field.

_____ A special field value, distinct from 0, blank, or any other value, that indicates that the value for the field is missing or otherwise unknown.

_____ A named set of rows and columns that specify the fields in each row of the table.

_____ The process of splitting or combining normalized relations into physical tables based on affinity of use of rows and fields.

_____ A named set of table rows stored in a contiguous section of secondary memory.

_____ A technique for physically arranging the records of a file.

_____ A field of data that can be used to locate a related field or row of data.

_____ The rows in the file are stored in sequence according to a primary key value.

_____ The rows are stored either sequentially or nonsequentially, and an index is created that allows software to locate individual rows.

_____ A table used to determine the location of rows in a file that satisfy some condition.

_____ One or a combination of fields for which more than one row may have the same combination of values.

_____ The address for each row is determined using an algorithm.

_____ An attribute whose value is unique across all occurrences of a relation.

_____ Data represented as a set of related tables or relations.

Review Questions

1. What is the purpose of normalization?
2. List five properties of relations.
3. What problems can arise during view integration or merging relations?
4. How are relationships between entities represented in the relational data model?
5. What is the relationship between the primary key of a relation and the functional dependencies among all attributes within that relation?
6. How is a foreign key represented in relational notation?
7. Can instances of a relation (sample data) prove the existence of a functional dependency? Why or why not?
8. In what way does the choice of a data type for a field help to control the integrity of that field?
9. What is the difference between how a range control statement and a referential integrity control statement are handled by a file management system?
10. What is the purpose of denormalization? Why might you not want to create one physical table or file for each relation in a logical data model?
11. What factors influence the decision to create an index on a field?
12. Explain the purpose of data compression techniques.
13. What are the goals of designing physical tables?
14. What are the seven factors that should be considered in selecting a file organization?

Problems and Exercises

1. Assume that at Pine Valley Furniture products are comprised of components, products are assigned to salespersons, and components are produced by vendors. Also assume that in the relation PRODUCT(Prodname, Salesperson,Compname,Vendor)Vendor is functionally dependent on Compname, and Compname is functionally dependent on Prodname. Eliminate the transitive dependency in this relation and form 3NF relations.

2. Transform the E-R diagram of Figure 10-3 into a set of 3NF relations. Make up a primary key and one or more nonkeys for each entity.

3. Transform the E-R diagram of Figure 12-21 into a set of 3NF relations.

4. Consider the list of individual 3NF relations below. These relations were developed from several separate normalization activities.

   PATIENT(Patient_ID,Room_Number,Admit_Date, Address)
   ROOM(Room_Number,Phone,Daily_Rate)
   PATIENT(Patient_Number,Treatment_Description,Address)
   TREATMENT(Treatment_ID,Description,Cost)
   PHYSICIAN(Physician_ID,Name,Department)
   PHYSICIAN(Physician_ID,Name,Supervisor_ID)

   a. Merge these relations into a consolidated set of 3NF relations. Make and state whatever assumptions you consider necessary to resolve any potential problems you identify in the merging process.
   b. Draw an E-R diagram for your answer to part a.

5. Consider the following 3NF relations about a sorority or fraternity:

   MEMBER(Member_ID,Name,Address,Dues_Owed)
   OFFICE(Office_Name,Officer_ID,Term_Start_Date,Budget)
   EXPENSE(Ledger_Number,Office_Name,Expense_Date, Amt_Owed)
   PAYMENT(Check_Number,Expense_Ledger_Number, Amt_Paid)
   RECEIPT(Member_ID,Receipt_Date,Dues_Received)
   COMMITTEE(Committee_ID,Officer_in_Charge)
   WORKERS(Committee_ID,Member_ID)

   a. Foreign keys are not indicated in these relations. Decide which attributes are foreign keys and justify your decisions.
   b. Draw an E-R diagram for these relations, using your answer to part a.
   c. Explain the assumptions you made about cardinalities in your answer to part b. Explain why it is said that the E-R data model is more expressive or more semantically rich than the relational data model.

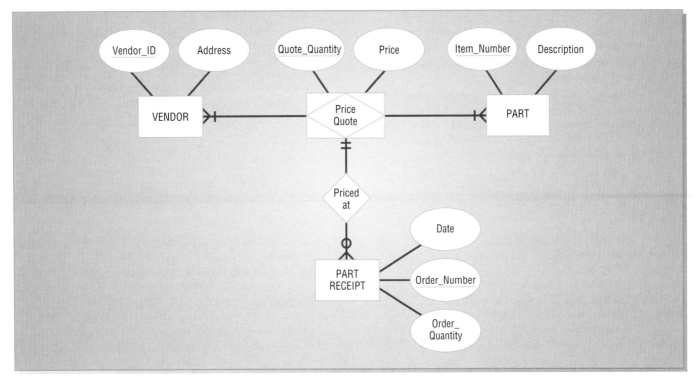

Figure 12-21
E-R diagram for Problem and Exercise 3

6. Consider the following functional dependencies:

Applicant_ID→Applicant_Name
Applicant_ID→Applicant_Address
Position_ID→Position_Title
Position_ID→Date_Position_Opens
Position_ID→Department
Applicant_ID + Position_ID→Date_Applied
Applicant_ID + Position_ID + Date_Interviewed→

   a. Represent these attributes with 3NF relations. Provide meaningful relation names.
   b. Represent these attributes using an E-R diagram. Provide meaningful entity and relationship names.

7. Suppose you were designing a file of student records for your university's placement office. One of the fields that would likely be in this file is the student's major. Develop a coding scheme for this field that achieves the objectives outlined in this chapter for field coding.

8. In Problem and Exercise 3, you developed integrated normalized relations. Choose primary keys for the files that would hold the data for these relations. Did you use attributes from the relations for primary keys or did you design new fields? Why or why not?

9. Suppose you created a file for each relation in your answer to Problem and Exercise 3. If the following queries represented the complete set of accesses to this database, suggest and justify what primary and secondary key indexes you would build.

   a. For each PART in Item_Number order list in Vendor_ID, sequence all the vendors and their associated prices for that part.
   b. List all PART RECEIPTs including related PART fields for all the parts received on a particular day.
   c. For a particular VENDOR, list all the PARTs and their associated prices that VENDOR can supply.

10. Suppose you were designing a default value for the age field in a student record at your university. What possible values would you consider and why? How might the default vary by other characteristics about the student, such as school within the university or degree sought?

11. Consider Figure 12-19(b). Explain a query that would likely be processed more quickly using the denormalized relations rather than the normalized relations.

Field Exercises

1. Find in the library books or articles with discussions of additional normal forms other than second and third normal forms. Describe each of these additional normal forms and give examples of each. How are these additional normal forms different from those presented in this chapter? What additional benefit does their use provide?

2. Describe the deliverables from file and database design.

3. Discuss what additional information should be collected during requirements analysis that is needed for file and database design and that is not very useful for earlier phases of systems development.

4. Find out what database management systems are available at your university for student use. Investigate which data types these DBMSs support. Compare these DBMSs based upon data types supported, and suggest which types of applications each DBMS is best suited for based on this comparison.

5. Find out what database management systems are available at your university for student use. Investigate what physical file and database design decisions need to be made. Compare this list of decisions to those discussed in this chapter. For physical database and design decisions (or options) not discussed in the chapter, investigate what choices you have and how you should choose among these choices. Submit a report to your instructor with your findings.

References

Codd, E. F. 1970, "A Relational Model of Data for Large Relational Databases." *Communications of the ACM* 13 (June): 77–87.

Dutka, A. F., and H. H. Hanson, 1989. *Fundamentals of Data Normalization*. Reading, MA: Addison-Wesley.

Hoffer, J. A., M. B. Prescott, and F. R. McFadden 2002. *Modern Database Management*, 6th edition. Upper Saddle River, NJ: Prentice Hall.

Finkelstein, R. 1988. "Breaking the Rules Has a Price." *Database Programming & Design* 1 (June): 11–14.

Gibson, M., C. Hughes, and W. Remington, 1989. "Tracking the Trade-Offs with Inverted Lists." *Database Programming & Design* 2 (Jan): 28–34.

Rodgers, U. 1989. "Denormalization: Why, What, and How?" *Database Programming & Design* 2(12) (Dec.): 46–53.

BROADWAY ENTERTAINMENT COMPANY, INC.

Designing the Relational Database for the Customer Relationship Management System

CASE INTRODUCTION

The students from Stillwater State University are making good progress in the design of MyBroadway, the Web-based customer relationship management system for Broadway Entertainment Company stores. They recently completed the design for all human interfaces and received tentative approval for these from Carrie Douglass, their client at the Centerville, Ohio, BEC store. This was an important step not only for making progress on the human interfaces but also because this approval validated all the data needed in system inputs and outputs. The entity-relationship diagram the team developed earlier in the project (see the BEC case at the end of Chapter 10) was not grounded in actual system inputs and outputs, but rather from a general understanding of system requirements. The approved inputs and outputs allow the student team to check that the entity types in the E-R diagram can hold all the input data that must be kept and can be used to produce all the outputs from MyBroadway.

IDENTIFYING RELATIONS

BEC Figure 10-1 at the end of Chapter 10 identified six entities the students decided were required for the MyBroadway database. As the students discuss the translation of this diagram into normalized relations, they conclude that the task is fairly straightforward. The translation looks easy because all the relationships are one-to-many. Based on procedures they have been taught in courses at Stillwater, each data entity becomes a rela-

tional table. The identifier of each entity can be used at the primary key of the associated relation, and the other attributes of an entity become the nonkey attributes of the associated relation. The relationships are represented as foreign keys, in which the primary key of the entity on the one side of a relationship becomes a foreign key in the relation for the entity on the many side of the relationship. BEC Figure 12-1 shows the team's initial relational data model, based on these rules.

The students are fairly confident that the relations in BEC Figure 12-1 are accurate (both complete and in third normal form), but they see some implementation issucs with these relations. First, the primary key of the PRODUCT relation, which is a foreign key in every other relation, is awkward because it has three components. The students are concerned that linking the tables based on three attributes will be inefficient in terms of both storage space and producing pages. The second issue they identify is with the PRODUCT Description, the COMMENT Member_Comment, and the PICK Employee_Comment attributes. These attributes are highly variable in length and may be quite long. These two traits also mean that retrieving and storing data can be time-consuming. The students will have to address these issues before the database is defined.

DESIGNING THE PHYSICAL DATABASE

The students decide to address the issue of the compound primary key by creating what is called a nonintelligent key. A nonintelligent key is a system-assigned value that has no

BEC Figure 12-1
Initial relations for MyBroadway

PRODUCT(<u>Title,Artist,Type</u>,Publisher,Category,Media,Description,Release_Date,
      Sale_Price,Rental_Price)
COMMENT(<u>Membership_ID,Comment_Time_Stamp</u>,Title,Artist,Type,Parent/Child?,
      Member_Comment)
REQUEST(<u>Membership_ID,Request_Time_Stamp</u>,Title,Artist,Type)
SALE(<u>Membership_ID,Sale_Time_Stamp</u>,Title,Artist,Type)
RENTAL(<u>Membership_ID,Rental_Time_Stamp</u>,Title,Artist,Type,Due_Date,Refund?)
PICK(<u>Employee_Name,Pick_Time_Stamp</u>,Title,Artist,Type,Rating,
     Employee_Comment)

PRODUCT(Product_ID,Title,Artist,Type,Publisher,Category,Media,Description,
Release_Date,Sale_Price,Rental_Price)
COMMENT(Membership_ID,Comment_Time_Stamp,Product_ID,Parent/Child?,
Member_Comment)
REQUEST(Membership_ID,Request_Time_Stamp,Product_ID)
SALE(Membership_ID,Sale_Time_Stamp,Product_ID)
RENTAL(Membership_ID,Rental_Time_Stamp,Product_ID,Due_Date,Refund?)
PICK(Employee_Name,Pick_Time_Stamp,Product_ID,Rating,Employee_Comment)

BEC Figure 12-2
Relations with nonintelligent primary key for Products

business meaning. It simply is an artificial attribute that will have a unique value for each row in the PRODUCT table. The three attributes of Title, Artist, and Type then become nonkey attributes, with the nonintelligent key of Product_ID becoming the primary key of PRODUCT and the foreign key in the other relations. BEC Figure 12-2 shows the relations with this modification.

The students address the second issue about the long and variable length of the Description, Member_Comment and Employee_Comment attributes by defining them each as a memo field (similar to a long data type in Oracle). Microsoft Access stores memo fields separately from the other attributes of a relation, which overcomes the problems with long and variable length data.

Because the team members have chosen to use Microsoft Access for the prototype, very few specific physical database design decisions must be made. They will have to select a data type for each attribute. For numeric data, such as PRODUCT Sale_Price, the students will have to decide on a format with number of decimal places, and for text fields, like PRODUCT Category, they will have to establish a maximum length. Access does not allow a designer to choose between file organizations for each table, but the students will need to decide on which attributes to build indexes. They immediately decide to create a primary key index for each table, but they are not sure which secondary key indexes will be best. Other decisions to be made are (1) whether to save storage space by coding fields, like PRODUCT media; (2) whether to define any data integrity controls on each field, such as a default value, input mask, validation rule; and (3) whether a value for the field is required for a new row of the table to be stored (e.g., must there be a specified Release_Date for each PRODUCT?).

CASE SUMMARY

The student team has many specific decisions to make in order to finalize the design of the database for MyBroadway. For a prototype, the students remember that some developers will not take the time to make intelligent physical design decisions. However, because the students plan for actual BEC customers to use the

MyBroadway prototype, they want the system to be reasonably efficient. Thus, they plan on taking the time to use all the power of Microsoft Access to create an efficient and reliable database. The team decides to analyze further each data input and information output page to understand better how those pages use the database. So, each team member is assigned several pages, and they agree to meet in two days with suggestions for all the physical database design decisions before they begin implementation of the initial prototype.

CASE QUESTIONS

1. In the questions associated with the BEC case at the end of Chapter 10, you were asked to modify the entity-relationship diagram drawn by the Stillwater student team to include any other entities and the attributes you identified from the BEC cases. Review your answers to these questions, and modify the relations in BEC Figure 12-2 to include your changes.

2. Study your answer to Question 1. Verify that the relations you say represent the MyBroadway database are in third normal form. If they are not, change them so that they are.

3. The E-R diagram you developed in questions in the BEC case at the end of Chapter 10 should have shown minimum cardinalities on both ends of each relationship. Are minimum cardinalities represented in some way in the relations in your answer to Question 2? If not, how are minimum cardinalities enforced in the database?

4. You have probably noticed that the Stillwater students chose to include a time stamp field as part of the primary key for all of the relations except PRODUCT. Explain why you think they decided to include this field in each relation and why it is part of the primary key. Are there other alternatives to a time stamp field for creating the primary key of these relations?

5. This BEC case indicated the data types chosen for only a few of the fields of the database. Using your answer to Question 2, select data types, and lengths for each attribute of each relation. Use the data

types and formats supported by Microsoft Access. What data type should be used for the nonintelligent primary keys? Do you agree with the use of Memo for Description and Member_Comment attributes?

6. This BEC case also mentioned that the students will consider if any fields should be coded. Are any fields good candidates for coding? If so, suggest a coding scheme for each coding candidate field. How would you implement field coding through Microsoft Access?

7. Complete all table and field definitions for the MyBroadway database using Microsoft Access. Besides the decisions you have made in answers to the other case questions, fill in all other field definition parameters for each field of each table.

8. The one decision for a relational database that usually influences efficiency the most is index definition. Besides the primary key indexes the students have chosen, what other secondary key indexes do you recommend for this database? Justify your selection of each secondary key index.

chapter 13

Designing Forms and Reports

LEARNING OBJECTIVES

After studying this chapter, you should be able to:

- Explain the process of designing forms and reports and the deliverables for their creation.

- Apply the general guidelines for formatting forms and reports.

- Use color and know when color improves the usability of information.

- Format text, tables, and lists effectively.

- Explain how to assess usability and describe how variations in users, tasks, technology, and environmental characteristics influence the usability of forms and reports.

- Discuss guidelines for the design of forms and reports for Internet-based electronic commerce systems.

INTRODUCTION

In this chapter, you will learn guidelines to follow when designing forms and reports. In general, forms are used to present or collect information on a single item such as a customer, product, or event. Forms can be used for both input and output. Reports, on the other hand, are used to convey information on a collection of items. Form and report design is a key ingredient for successful systems. As users often equate the quality of a system to the quality of its input and output methods, you can see that the design process for forms and reports is an especially important activity. And since information can be collected and formatted in many ways, gaining an understanding of the dos and don'ts and the trade-offs between various formatting options is a useful skill for all systems analysts.

In the next section, the process of designing forms and reports is briefly described and we also provide guidance on the deliverables produced during this process. Guidelines for formatting information are then provided that serve as the building blocks for designing all forms and reports. The next section describes methods for assessing the usability of form and report designs. The chapter concludes by examining how to design forms and reports for Internet-based electronic commerce applications.

DESIGNING FORMS AND REPORTS

This is the second chapter focusing on system design within the systems development life cycle (see Figure 13-1). In this chapter, we describe issues related to the design of system inputs and outputs—forms and reports. In Chapter 14, we focus on the design of dialogues and interfaces, which are how users interact with systems. Due to the highly related topics and guidelines in these two chapters, they form one conceptual body of guidelines and illustrations that jointly guide the design of all aspects of system inputs and outputs. In each of these chapters, your objective is to gain an understanding of how you can transform information gathered during analysis into a coherent design. Although all system design issues are related, topics discussed in this chapter on designing forms and reports are especially relative to those in the following chapter—the design of dialogues and interfaces.

System inputs and outputs—forms and reports—were identified during requirements structuring. The kinds of forms and reports the system will handle were established as part of the design strategy formed at the end of the analysis phase of the systems development process. During analysis, however, you may not have been concerned with the precise appearance of forms and reports, only with which ones needed to exist and what their contents were. You may have distributed prototypes of forms and reports that emerged during analysis as a way to confirm requirements

Figure 13-1
Systems development life cycle with logical design phase highlighted

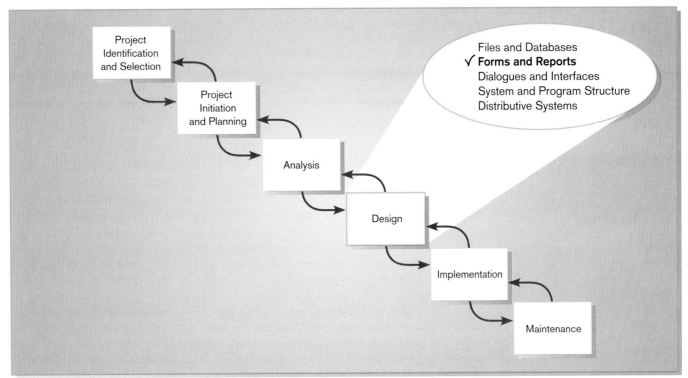

with users. Forms and reports are integrally related to various diagrams developed during requirements structuring. For example, every input form will be associated with a data flow entering a process on a DFD, and every output form or report will be a data flow produced by a process on a DFD. This means that the contents of a form or report correspond to the data elements contained in the associated data flow. Further, the data on all forms and reports must consist of data elements in data stores and on the E-R data model for the application, or must be computed from these data elements. (In rare instances, data simply go from system input to system output without being stored within the system.) It is common that, as you design forms and reports, you will discover flaws in DFDs and E-R diagrams; the project dictionary or CASE tool repository, therefore, continues to be the central and constantly updated source of all project information.

If you are unfamiliar with computer-based information systems, it will be helpful to clarify exactly what we mean by a form or report. A **form** is a business document containing some predefined data and often includes some areas where additional data are to be filled in. Most forms have a stylized format and are usually not in a simple row and column format. Examples of business forms are product order forms, employment applications, and class registration sheets. Traditionally, forms have been displayed on a paper medium, but today video display technology allows us to duplicate the layout of almost any printed form, including an organizational logo or any graphic, on a video display terminal. Forms displayed on a video display may be used for data display or data entry. Additional examples of forms are an electronic spreadsheet, computer sign-on or menu, and an ATM transaction layout. On the Internet, form interaction is the standard method of gathering and displaying information when consumers order products, request product information, or query account status.

A report is a business document containing only predefined data; it is a passive document used solely for reading or viewing. Examples of reports are invoices, weekly sales summaries by region and salesperson, and a pie chart of population by age categories. We usually think of a report as printed on paper, but it may be printed to a computer file, a visual display screen, or some other medium such as microfilm. Often a report has rows and columns of data, but a report may consist of any format—for example, mailing labels. Frequently, the differences between a form and a report are subtle. A report is only for reading and often contains data about multiple unrelated records in a computer file. On the other hand, a form typically contains data from only one record or is, at least, based on one record, such as data about one customer, one order, or one student. The guidelines for the design of forms and reports are very similar.

Form: A business document that contains some predefined data and may include some areas where additional data are to be filled in. An instance of a form is typically based on one database record.

Report: A business document that contains only predefined data; it is a passive document used solely for reading or viewing. A report typically contains data from many unrelated records or transactions.

The Process of Designing Forms and Reports

Designing forms and reports is a user-focused activity that typically follows a prototyping approach (see Figure 1-10). First, you must gain an understanding of the intended user and task objectives by collecting initial requirements during requirements determination. During this process, several questions must be answered. These questions attempt to answer the "who, what, when, where, and how" related to the creation of all forms or reports (see Table 13-1). Gaining an understanding of these questions is a required first step in the creation of any form or report.

For example, understanding who the users are—their skills and abilities—will greatly enhance your ability to create an effective design. In other words, are your users experienced computer users or novices? What is their educational level, business background, and task-relevant knowledge? Answers to these questions will provide guidance for both the format and content of your designs. Also, what is the pur-

TABLE 13-1   Fundamental Questions when Designing Forms and Reports

1. Who will use the form or report?
2. What is the purpose of the form or report?
3. When is the form or report needed and used?
4. Where does the form or report need to be delivered and used?
5. How many people need to use or view the form or report?

pose of the form or report? What task will users be performing and what information is needed to complete this task? Other questions are also important to consider. Where will the users be when performing this task? Will users have access to on-line systems or will they be in the field? Also, how many people will need to use this form or report? If, for example, a report is being produced for a single user, the design requirements and usability assessment will be relatively simple. A design for a larger audience, however, may need to go through a more extensive requirements collection and usability assessment process.

After collecting the initial requirements, you structure and refine this information into an initial prototype. Structuring and refining the requirements are completed independently of the users, although you may need to occasionally contact users in order to clarify some issue overlooked during analysis. Finally, you ask users to review and evaluate the prototype. After reviewing the prototype, users may accept the design or request that changes be made. If changes are needed, you will repeat the construction-evaluate-refinement cycle until the design is accepted. Usually, several iterations of this cycle occur during the design of a single form or report. As with any prototyping process, you should make sure that these iterations occur rapidly in order to gain the greatest benefits from this design approach.

The initial prototype may be constructed in numerous environments, including DOS, Unix, Windows, Linux, or Apple. The obvious choice is to use a CASE tool or the standard development tools used within your organization. Often, initial prototypes are simply mock screens that are not working modules or systems. Mock screens can be produced from a word processor, computer graphics design package, or electronic spreadsheet. It is important to remember that the focus of this phase within the SDLC is on the *design*—content and layout. How specific forms or reports are implemented (for example, the programming language or screen painter code) is left to later phases. Nonetheless, tools for designing forms and reports are rapidly evolving. In the past, inputs and outputs of all types were typically designed by hand on a coding or layout sheet. For example, Figure 13-2 shows the layout of a data input form using a coding sheet.

Although coding sheets are still used, their importance has diminished due to significant changes in system operating environments and the evolution of automated design tools. Prior to the creation of graphical operating environments, for example, analysts designed many inputs and outputs that were 80 columns (characters) by 25 rows, the standard dimensions for most video displays. These limits in screen dimensions are radically different in graphical operating environments such as Microsoft's Windows® where font sizes and screen dimensions can often be changed from user to user. Consequently, the creation of new tools and development environments was needed to help analysts and programmers develop these graphical and flexible designs. Figure 13-3 shows an example of the same data input form as designed in Microsoft's Visual Basic®. Note the variety of fonts, sizes, and highlighting that was used. On-line graphical tools for designing forms and reports are rapidly becoming the de facto standard in most professional development organizations.

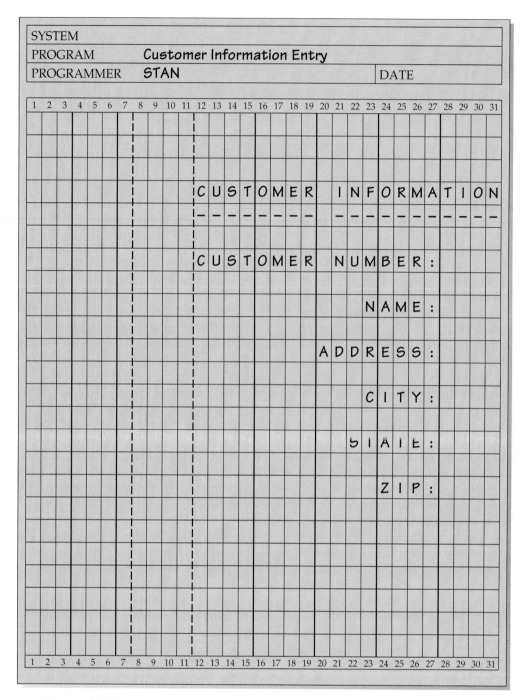

Figure 13-2
The layout of a data input form using a coding sheet

Deliverables and Outcomes

Each SDLC phase helps you to construct a system. In order to move from phase to phase, each phase produces some type of deliverable that is used in a later phase or activity. For example, within the project initiation and planning phase of the SDLC, the Baseline Project Plan serves as input to many subsequent SDLC activities. In the case of designing forms and reports, design specifications are the major deliv-

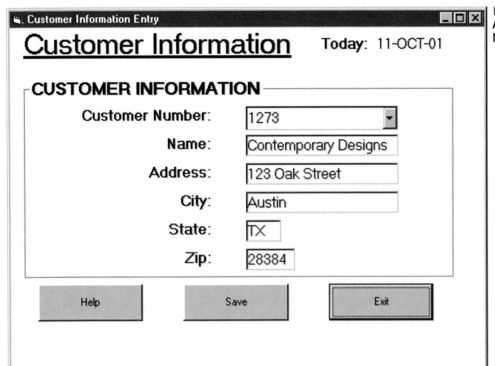

Figure 13-3
A data input screen designed in
Microsoft's Visual Basic

erables and are inputs to the system implementation phase. Design specifications
have three sections:

1. Narrative overview
2. Sample design
3. Testing and usability assessment

The first section of a design specification contains a general overview of the characteristics of the target users, tasks, system, and environmental factors in which the
form or report will be used. The purpose is to explain to those who will actually
develop the final form why this form exists and how it will be used so that they can
make the appropriate implementation decisions. In this section, you list general information and the assumptions that helped shape the design. For example, Figure 13-4
shows an excerpt of a design specification for a Customer Account Status form for Pine
Valley Furniture. The first section of the specification, Figure 13-4a, provides a narrative overview containing the relevant information to developing and using the form
within PVF. The overview explains the tasks supported by the form, where and when
the form is used, characteristics of the people using the form, the technology delivering the form, and other pertinent information. For example, if the form is delivered on
a visual display terminal, this section would describe the capabilities of this device, such
as whether it has a touch screen and whether color and a mouse are available.

In the second section of the specification, Figure 13-4b, a sample design of the form is
shown. This design may be hand-drawn using a coding sheet although, in most instances,
it is developed using CASE or standard development tools. Using actual development
tools allows the design to be more thoroughly tested and assessed. The final section of the
specification, Figure 13-4c, provides all testing and usability assessment information.
Procedures for assessing designs are described later in the chapter. Some specification
information may be irrelevant when designing some forms and reports. For example, the
design of a simple Yes/No selection form may be so straightforward that no usability
assessment is needed. Also, much of the narrative overview may be unnecessary unless
intended to highlight some exception that must be considered during implementation.

Figure 13-4
Design specification for the design of forms and reports
(a) Narrative overview

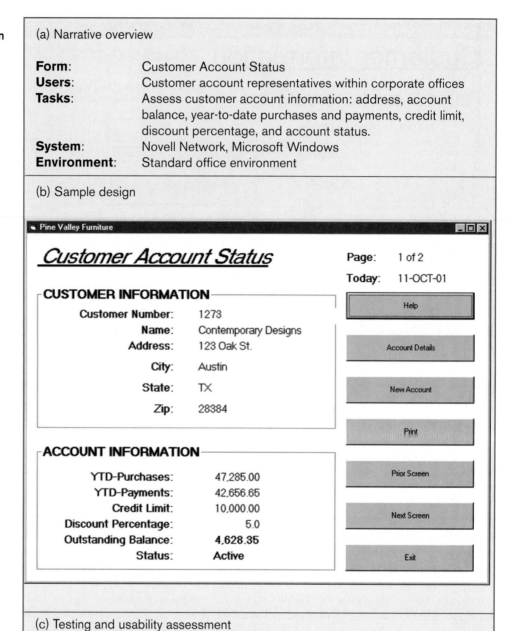

(a) Narrative overview

Form:	Customer Account Status
Users:	Customer account representatives within corporate offices
Tasks:	Assess customer account information: address, account balance, year-to-date purchases and payments, credit limit, discount percentage, and account status.
System:	Novell Network, Microsoft Windows
Environment:	Standard office environment

(b) Sample design

(b) Sample design

(c) Testing and usability assessment

(c) Testing and usability assessment

User Rated Perceptions (average 14 users):
consistency [1 = consistent to 7 = inconsistent]:   1.52
sufficiency [1 = sufficient to 7 = insufficiency]:   1.43
accuracy [1 = accurate to 7 = inaccurate]:   1.67
. . .

FORMATTING FORMS AND REPORTS

A wide variety of information can be provided to users of information systems and, as technology continues to evolve, a greater variety of data types will be used. Unfortunately, a definitive set of rules for delivering every type of information to users has yet to be defined and these rules are continuously evolving along with the rapid changes in technology. Nonetheless, a large body of human-computer interaction research has provided numerous general guidelines for formatting information. Many of these guidelines will undoubtedly apply to the formatting of all evolving information types on yet-to-be-determined devices. Keep in mind that the mainstay of designing usable forms and reports requires your active interaction with users. If this single and fundamental activity occurs, it is likely that you will create effective designs.

For example, "palm" computing devices like the Palm Pilot and those based on Microsoft's Windows CE operating system are becoming increasingly popular. Palm computers are used to manage personal schedules, send and receive electronic mail, and browse the Web. One of the greatest challenges of palm computing is in the design of the human-computer interface because the video display is significantly smaller than full-size displays and many devices do not use a color display. For example, surfing the Web on a palm computer is very difficult because most Internet sites assume that users will have a full-sized, color display. To address this problem, the Web browser in Windows CE is "smart" and automatically shrinks images so that user's viewing experience is good. Alternatively, a growing number of Websites are designed with the palm computer user in mind. For example, these sites provide a vast array of information preformatted for smaller screens. As these and other mobile computing devices like WAP cellular phones (WAP stands for Wireless Application Protocol, the set of specifications for developing Web-like applications that run over wireless networks) evolve and gain popularity, standard guidelines will emerge to make the process of designing interfaces for them much less challenging.

General Formatting Guidelines

Over the past several years, industry and academic researchers have spent considerable effort investigating how information formatting influences individual task performance and perceptions of usability. Through this work, several guidelines for formatting information have emerged (see Table 13-2). These guidelines reflect some of the general truths that apply to the formatting of most types of information (for more information, the interested reader should see the books by Shneiderman, 1997; Johnson, 2000; Nielson, 1999; and Flanders and Will, 1998).

The differences between a well-designed form or report and one that is poorly designed will often be obvious to you. For example, Figure 13-5a shows a poorly designed form for viewing a current account balance for a PVF customer. Figure 13-5b (page 2 of 2) is a better design incorporating several general guidelines from Table 13-2.

TABLE 13-2   General Guidelines for the Design of Forms and Reports

Meaningful Titles:

Clear and specific titles describing content and use of form or report

Revision date or code to distinguish a form or report from prior versions

Current date which identifies when the form or report was generated

Valid date which identifies on what date (or time) the data in the form or report were accurate

Meaningful Information:

Only needed information should be displayed

Information should be provided in a manner that is usable without modification

Balance the Layout:

Information should be balanced on the screen or page

Adequate spacing and margins should be used

All data and entry fields should be clearly labeled

Design an Easy Navigation System:

Clearly show how to move forward and backward

Clearly show where you are (e.g., page 1 of 3)

Notify user when on the last page of a multipaged sequence

The first major difference between the two forms has to do with the title. The title on Figure 13-5a is ambiguous whereas the title on Figure 13-5b clearly and specifically describes the contents of the form. The form in Figure 13-5b also includes the date on which the form was generated so that, if printed, it will be clear to the reader when this occurred. Figure 13-5a displays information that is extraneous to the intent of the form—viewing the current account balance—and provides information that is not in the most useful format for the user. For example, Figure 13-5a provides all customer data as well as account transactions and a summary of year-to-date purchases and payments. The form does not, however, provide the current outstanding balance of the account without making a manual calculation. The layout of information between the two forms also varies in balance and information density. Gaining an understanding of the skills of the intended system users and the tasks they will be performing is invaluable when constructing a form or report. By following these general guidelines, your chances of creating effective forms and reports will be enhanced. In the next sections we will discuss specific guidelines for highlighting information, using color, displaying text, and presenting numeric tables and lists.

Highlighting Information

As display technologies continue to improve, there will be a greater variety of methods available to you for highlighting information. Table 13-3 provides a list of the most commonly used methods for highlighting information. Given this vast array of options, it is more important than ever to consider how highlighting can be used to enhance an output and not prove a distraction. In general, highlighting should be

Figure 13-5
Contrasting customer information forms (Pine Valley Furniture)
(a) Poorly designed form

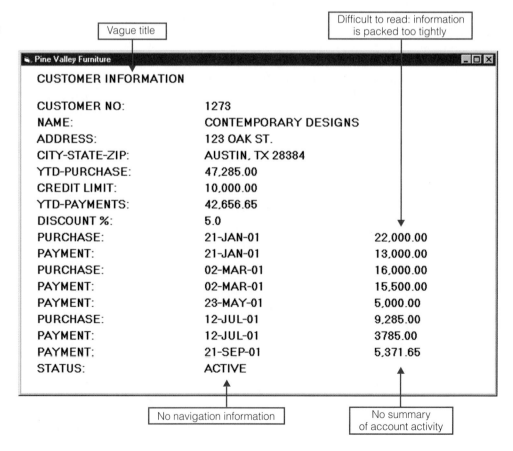

used sparingly to draw the user to or away from certain information and to group together related information. There are several situations when highlighting can be a valuable technique for conveying special information:

- Notifying users of errors in data entry or processing
- Providing warnings to users regarding possible problems such as unusual data values or an unavailable device
- Drawing attention to keywords, commands, high priority messages, and data that have changed or gone outside normal operating ranges

Additionally, many highlighting techniques can be used singularly or in tandem, depending upon the level of emphasis desired by the designer. Figure 13-6 shows a form where several types of highlighting are used. In this example, boxes clarify different categories of data; capital letters and different fonts distinguish labels from actual data; and intensity is used to draw attention to important data.

Much research has focused on the effects of varying highlighting techniques on task performance and user perceptions. A general guideline resulting from this research is that highlighting should be used conservatively. For example, blinking and audible tones should only be used to highlight critical information requiring an immediate response from the user. Once a response is made, these highlights should

TABLE 13-3   **Methods of Highlighting**
Blinking and audible tones
Color differences
Intensity differences
Size differences
Font differences
Reverse video
Boxing
Underlining
All capital letters
Offsetting the position of nonstandard information

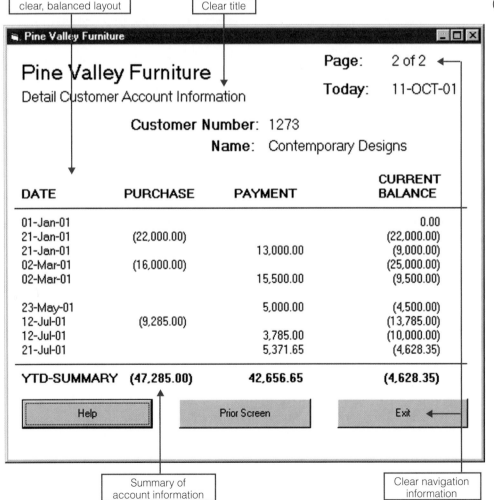

Figure 13-5 (continued)
(b) Improved design for form

Figure 13-6
Customer account status display using various highlighting techniques (Pine Valley Furniture)

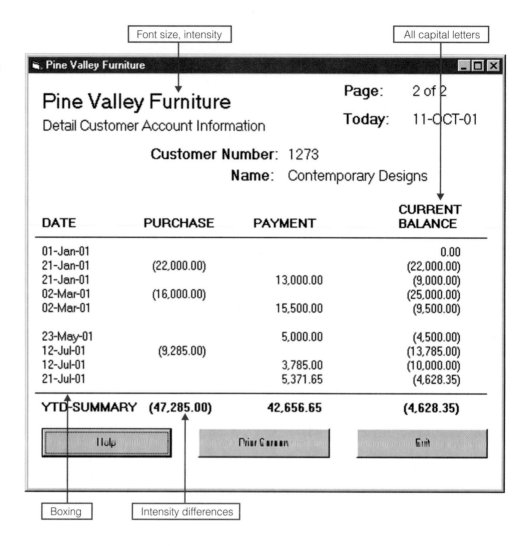

be turned off. Additionally, highlighting methods should be consistently used and selected based upon the level of importance of the emphasized information. It is also important to examine how a particular highlighting method appears on all possible output devices that could be used with the system. For example, some color combinations may convey appropriate information on one display configuration but wash out and reduce legibility on another.

Recent advances in the development of graphical operating environments have provided designers with some standard highlighting guidelines. However, these guidelines are often quite vague and are continuously evolving, leaving a great deal of control in the hands of the systems developer. Therefore, in order for organizations to realize the benefits of using standard graphical operating environments—such as reduced user training time and interoperability among systems—you must be disciplined in how you use highlighting.

Color versus No-Color

Color is a powerful tool for the designer in influencing the usability of a system. When applied appropriately, color provides many potential benefits to forms and reports, summarized in Table 13-4. As the use of color displays became widely available during the 1980s, a substantial amount of color versus no-color research was conducted. The objective of this research was to gain a better understanding of the effects of color on human task performance (see, for example, Benbasat, Dexter, and Todd, 1986).

The general findings from this research were that the use of color had positive effects on user task performance and perceptions when the user was under time constraints for the completion of a task. Color was also beneficial for gaining greater understanding from a display or chart. An important conclusion from this research was that color was *not* universally better than no-color. *The benefits of color only seem to apply if the information is first provided to the user in the most appropriate presentation format.* That is, if information is most effectively displayed in a bar chart, color can be used to enhance or supplement the display. If information is displayed in an inappropriate format, color has little or no effect on improving understanding or task performance.

There are also several problems associated with using color, also summarized in Table 13-4. Most of these dangers are related more to the technical capabilities of the display and hard copy devices than misuse. However, color blindness is a particular user issue that is often overlooked in the design of systems. Shneiderman (1997) reports that approximately 8 percent of the males in the European and North American communities have some form of color blindness. He suggests that you first design video displays for monochrome and allow color (or better yet, a flexible palette of colors) to be a user-activated option. He also suggests that you limit the number of colors and where they are applied, using color primarily as a tool to assist in the highlighting and formatting of information.

Displaying Text

In business-related systems, textual output is becoming increasingly important as text-based applications such as electronic mail, bulletin boards, and information services (e.g., Dow Jones) are more widely used. The display and formatting of system help information, which often contains lengthy textual descriptions and examples, is one example of textual data which can benefit from following a few simple guidelines that have emerged from past research (see Table 13-5). The first guideline is simple: you should display text using common writing conventions such as mixed upper and lower case and appropriate punctuation. If space permits, text should be double-spaced and, minimally, a blank line should be placed between each paragraph. You should also left-justify text with a ragged right margin—research shows that a ragged right margin makes it easier to find the next line of text when reading than when text is both left- and right-justified.

When displaying textual information, you should also be careful not to hyphenate words between lines or use obscure abbreviations and acronyms. Users may not know whether the hyphen is a significant character if it is used to continue words across lines. Information and terminology that are not widely understood by the intended users may significantly influence the usability of the system. Thus, you should use abbreviations and acronyms only if they are significantly shorter than the full text and are commonly known by the intended system users. Figure 13-7 shows two versions of a help screen from an application systems at PVF. Figure 13-7a shows many violations of the general guidelines for displaying text whereas 13-7b shows the same information but follows the general guidelines for displaying text. Formatting

TABLE 13-4 Benefits and Problems from Using Color

Benefits from Using Color:

Soothes or strikes the eye.

Accents an uninteresting display.

Facilitates subtle discriminations in complex displays.

Emphasizes the logical organization of information.

Draws attention to warnings.

Evokes more emotional reactions.

Problems from Using Color:

Color pairings may wash out or cause problems for some users (e.g., color blindness).

Resolution may degrade with different displays.

Color fidelity may degrade on different displays.

Printing or conversion to other media may not easily translate.

Adapted from Shneiderman, 1997.

N E T S E A R C H

There are millions of commercial Websites that have emerged over the past few years, some having a highly usable interface whereas others continue to have relatively weak interface. Visit http://www.prenhall.com/hoffer to complete an exercise related to this topic.

TABLE 13-5 Guidelines for Displaying Text

Case	Display text in mixed upper and lower case and use conventional punctuation.
Spacing	Use double spacing if space permits. If not, place a blank line between paragraphs.
Justification	Left-justify text and leave a ragged right margin.
Hyphenation	Do not hyphenate words between lines.
Abbreviations	Use abbreviations and acronyms only when they are widely understood by users and are significantly shorter than the full text.

Figure 13-7
Contrasting the display of textual help information
(a) Poorly designed form

(b) Improved design for form

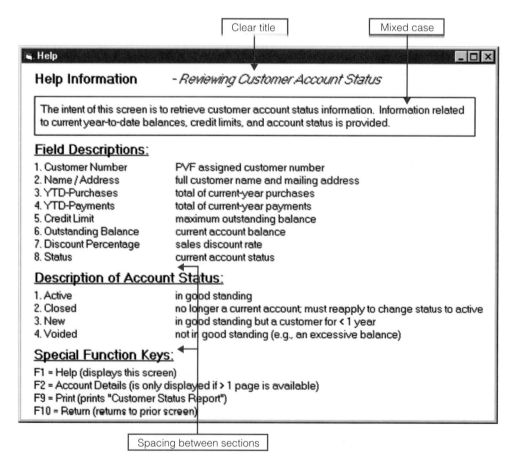

guidelines for the *entry* of text and alphanumeric data is also a very important topic. These guidelines are presented in Chapter 14, *Designing Interfaces and Dialogues*, where we focus on issues of human-computer interaction.

Designing Tables and Lists

Unlike textual information, where context and meaning are significantly derived through reading, the context and meaning of tables and lists are significantly derived from the format of the information. Consequently, the usability of information displayed in tables and alphanumeric lists is likely to be much more influenced by effective layout than most other types of information display. As with the display of textual information, tables and lists can also be greatly enhanced by following a few simple guidelines. These are summarized in Table 13-6.

Figure 13-8 displays two versions of a form design from a Pine Valley Furniture application system that displays customer year-to-date transaction information in a table format. Figure 13-8a displays the information without consideration of the guidelines presented in Table 13-6 and Figure 13-8b (only page 2 of 2 is shown) displays this information after consideration of these guidelines.

One key distinction between these two display forms relates to labeling. The information reported in Figure 13-8b has meaningful labels that more clearly stand out as labels compared to the display in Figure 13-8a. Transactions are sorted by date and numeric data are right-justified and aligned by decimal point in Figure 13-8b, which helps to facilitate scanning. Adequate space is left between columns, and blank lines are inserted after every five rows in Figure 13-8b to help ease the finding and reading of information. Such spacing also provides room for users to annotate data that catch their attention. Taken together, using a few simple rules significantly enhanced the layout of the information for the user.

Most of the guidelines in Table 13-6 are rather obvious, but this and other tables serve as a quick reference to validate that your form and report designs will be usable. It is beyond our scope here to discuss each of these guidelines, but you should read each carefully and think about why each is appropriate. For example, why should labels be repeated on subsequent screens and pages (the third guideline in Table 13-6)?

TABLE 13-6   General Guidelines for Displaying Tables and Lists

Use meaningful labels:

All columns and rows should have meaningful labels.

Labels should be separated from other information by using highlighting.

Re-display labels when the data extend beyond a single screen or page.

Formatting columns, rows, and text:

Sort in a meaningful order (e.g., ascending, descending, or alphabetic).

Place a blank line between every five rows in long columns.

Similar information displayed in multiple columns should be sorted vertically (that is, read from top to bottom, not left to right).

Columns should have at least two spaces between them.

Allow white space on printed reports for user to write notes.

Use a single typeface, except for emphasis.

Use same family of typefaces within and across displays and reports.

Avoid overly fancy fonts.

Formatting numeric, textual, and alphanumeric data:

Right-justify *numeric data* and align columns by decimal points or other delimiter.

Left-justify *textual data*. Use short line length, usually 30–40 characters per line (this is what newspapers use, and it is easier to speed read).

Break long sequences of *alphanumeric data* into small groups of three to four characters each.

Figure 13-8
**Contrasting the display of tables
and lists (Pine Valley Furniture)**
(a) Poorly designed form

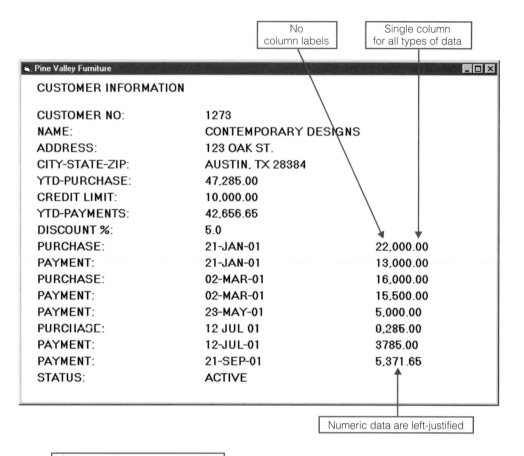

(b) Improved design for form

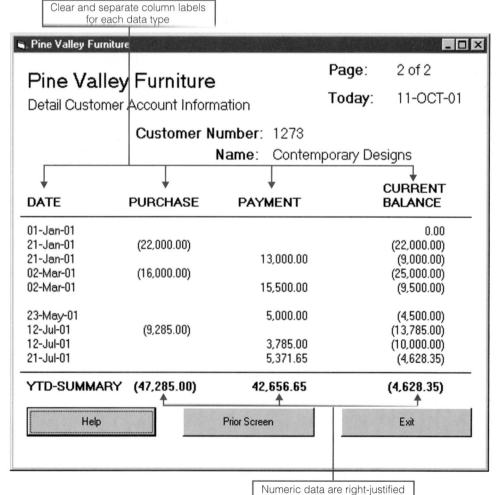

One explanation is that pages may be separated or copied and the original labels will no longer be readily accessible to the reader of the data. Why should long alphanumeric data (see the last guideline) be broken into small groups? (If you have a credit card or bank check, look at how your account number is displayed.) Two reasons are that the characters will be easier to remember as you read and type them, and there will be a natural and consistent place to pause when you speak them over the phone; for example, when you are placing a phone order for products in a catalog.

When you design the display of numeric information, you must determine whether a table or a graph should be used. A considerable amount of research focusing on this topic has been conducted (see, for example, Jarvenpaa and Dickson, 1988, for very specific guidelines on the use of tables and graphs). In general, this research has found that tables are best when the user's task is related to finding an individual data value from a larger data set whereas line and bar graphs are more appropriate for gaining an understanding of data changes over time (see Table 13-7). For example, if the marketing manager for Pine Valley Furniture needed to review the actual sales of a particular salesperson for a particular quarter, a tabular report like the one shown in Figure 13-9 would be most useful. This report has been annotated to emphasize good report design practices. The report has both a printed date as well as a clear indication, as part of the report title, of the period over which the data apply. There is also sufficient white space to provide some room for users to

TABLE 13-7   Guidelines for Selecting Tables vs. Graphs

Use Tables for
Reading individual data values
Use Graphs for
Providing a quick summary of data
Detecting trends over time
Comparing points and patterns of
    different variables
Forecasting activities
Reporting vast amounts of
    information when relatively simple
    impressions are to be drawn

Adapted from Jarvenpaa and Dickson, 1988.

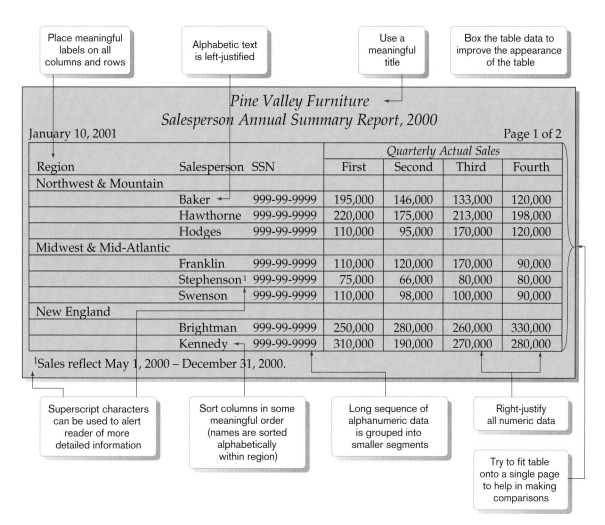

Figure 13-9
Tabular report illustrating numerous design guidelines (Pine Valley Furniture)

Figure 13-10
Graphs for comparison
(a) Line graph

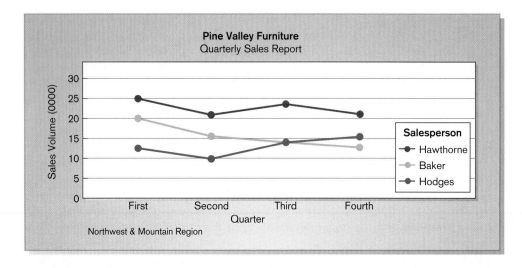

(b) Bar graph

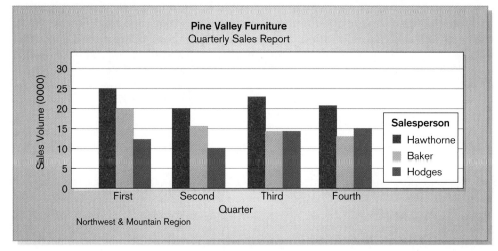

add personal comments and observations. Often, to provide such white space, a report must be printed in landscape, rather than portrait, orientation. Alternatively, if the marketing manager wished to compare the overall sales performance of each sales region, a line or bar graph would be more appropriate (see Figure 13-10). As with other formatting considerations, the key determination as to when you should select a table or a graph is the task being performed by the user.

Paper versus Electronic Reports

When a report is produced on paper rather than on a computer display, there are some additional things that you need to consider. For example, laser printers (especially color laser printers) and ink jet printers allow you to produce a report that looks exactly as it does on the display screen. Thus, when using these types of printers, you can follow our general design guidelines to create a report with high usability. However, other types of printers are not able to closely reproduce the display screen image onto paper. For example, many business reports are produced using high-speed impact printers that produce characters and a limited range of graphics by printing a fine pattern of dots. The advantages of impact printers are that they are very fast, very reliable, and relatively inexpensive. Their drawbacks are that they have a limited ability to produce graphics and have a somewhat lower print quality. In other words, they are good at rapidly producing reports that contain primarily alphanumeric information, but cannot exactly replicate a screen report onto paper. Because of this, impact printers are mostly used for producing large batches of reports, like a batch of phone bills for your telephone company, on a wide range of paper widths and types. When designing reports for impact printers, you use a coding sheet like that displayed in Figure 13-2, although coding sheets for designing printer reports typically can have up to 132 columns. Like the process for designing all forms and reports, you follow a prototyping process and carefully control the spacing of characters in order to produce a high-quality report. However, unlike other form and report designs, you may be limited in the range of formatting, text types, and highlighting options. Nonetheless, you can easily produce a highly usable report of any type if you carefully and creatively use the formatting options that are available.

ASSESSING USABILITY

There are many factors to consider when you design forms and reports. The objective for designing forms, reports, and all human-computer interactions is **usability**. Usability typically refers to the following three characteristics:

1. *Speed.* Can you complete a task efficiently?
2. *Accuracy.* Does the output provide what you expect?
3. *Satisfaction.* Do you like using the output?

In other words, usability means that your designs assist, not hinder, user performance. In the remainder of this section, we describe numerous factors that influence usability and several techniques for assessing the usability of a design.

Usability: An overall evaluation of how a system performs in supporting a particular user for a particular task.

Usability Success Factors

Research and practical experience have found that design consistency is the key ingredient in designing usable systems (Nielsen, 2000). Consistency significantly influences users' ability to gain proficiency when interacting with a system. Consistency means, for example, that titles, error messages, menu options, and other

TABLE 13-8   General Design Guidelines for Usability of Forms and Reports

Usability Factor	Guidelines for Achievement of Usability
Consistency	Consistent use of terminology, abbreviations, formatting, titles, and navigation within and across outputs. Consistent response time each time a function is performed.
Efficiency	Formatting should be designed with an understanding of the task being performed and the intended user. Text and data should be aligned and sorted for efficient navigation and entry. Entry of data should be avoided where possible (e.g., computing rather than entering totals).
Ease	Outputs should be self-explanatory and not require users to remember information from prior outputs in order to complete tasks. Labels should be extensively used, and all scales and units of measure should be clearly indicated.
Format	Information format should be consistent between entry and display. Format should distinguish each piece of data and highlight, not bury, important data. Special symbols, such as decimal places, dollar signs, and +/– signs should be used as appropriate.
Flexibility	Information should be viewed and retrieved in a manner most convenient to the user. For example, users should be given options for the sequence in which to enter or view data and for use of shortcut keystrokes, and the system should remember where the user stopped during the last use of the system.

TABLE 13-9   **Characteristics for Consideration When Designing Forms and Reports**

Characteristic	Consideration for Form and Report Design
User	Issues related to experience, skills, motivation, education, and personality should be considered.
Task	Tasks differ in amount of information that must be obtained from or provided to the user. Task demands such as time pressure, cost of errors, and work duration (fatigue) will influence usability.
System	The platform on which the system is constructed will influence interaction styles and devices.
Environment	Social issues such as the users' status and role should be considered in addition to environmental concerns such as lighting, sound, task interruptions, temperature, and humidity. The creation of usable forms and reports may necessitate changes in the users' physical work facilities.

Adapted from Norman, 1991.

design elements appear in the same place and look the same on all forms and reports. Consistency also means that the same form of highlighting has the same meaning each time it is used and that the system will respond in roughly the same amount of time each time a particular operation is performed. Other important factors found to be important include efficiency, ease (or understandability), format, and flexibility. Each of these usability factors, with associated guidelines, is described in more detail in Table 13-8.

When designing outputs, you must also consider the context in which the screens, forms, and reports will be used. As mentioned, numerous characteristics play an important role in shaping a system's usability. These characteristics are related to the intended users and task being performed in addition to the technological, social, and physical environment in which the system and outputs are used. Table 13-9 lists several factors in which variations in any item may influence the usability of a design. Your role is to gain a keen awareness of these factors so that your chances of creating highly usable designs are increased.

Measures of Usability

NET   SEARCH
For Internet electronic commerce sites, high system usability is a necessity because switching costs for customers are so low. Visit http://www.prenhall.com/hoffer to complete an exercise related to this topic.

User friendliness is a term often used, and misused, to describe system usability. Although the term is widely used, it is too vague from a design standpoint to provide adequate information because it means different things to different people. Consequently, most development groups use several methods for assessing usability, including these considerations (Shneiderman, 1997):

- Time to learn
- Speed of performance
- Rate of errors
- Retention over time
- Subjective satisfaction

In assessing usability, you can collect information by observation, interviews, keystroke capturing, and questionnaires. Time to learn simply reflects how long it takes the average system user to become proficient using the system. Equally important is the extent to which users remember how to use inputs and outputs over time. The manner in which the processing steps are sequenced and the selection of one set of keystrokes over others can greatly influence learning time, the user's task performance, and error rates. For example, the most commonly used functions should be quickly accessed in the fewest number of steps (for example, pressing one key to save your work). Additionally, the layout of information should be consistent, both *within and across* applications, whether the information is delivered on a screen display or hard copy report.

ELECTRONIC COMMERCE APPLICATION: DESIGNING FORMS AND REPORTS FOR PINE VALLEY FURNITURE'S WEBSTORE

PINE VALLEY FURNITURE

Designing the forms and reports for an Internet-based electronic commerce application is a central and critical design activity. Since this is where a customer will interact with a company, much care must be put into its design. Like the process followed when designing the forms and reports for other types of systems, a prototyping design process is most appropriate. Although the techniques and technology for building Internet sites is rapidly evolving, several general design guidelines have emerged. In this section, we examine some of these as they apply to design of Pine Valley Furniture's WebStore.

General Guidelines

The rapid deployment of Internet Websites has resulted in having countless people design sites who, arguably, have limited ability to do so. To put this into perspective, consider the following quote from Web design guru, Jakob Nielsen (1999; pp. 65–66):

> If the [Web's] growth rate does not slow down, the Web will reach 200 million sites sometime during 2003. . . . The world has about 20,000 user interface professionals. If all sites were to be professionally designed by a single UI professional, we can conclude that every UI professional in the world would need to design one Web site every working hour from now on to meet demand. This is obviously not going to happen. . . .
> There are three possible solutions to the problem:
>
> - Make it possible to design reasonably usable sites without having UI expertise;
> - Train more people in good Web design; and
> - Live with poorly designed sites that are hard to use.

When designing forms and reports, there are several errors that are specific to Website design. It is unfortunately beyond the scope of this book to critically examine all possible design problems with contemporary Websites. Here, we will simply summarize those errors that commonly occur and are particularly detrimental to the user's experience (see Table 13-10). Fortunately, there are numerous excellent sources for learning more about designing useful Websites (Johnson, 2000; Flanders and Willis, 1998; Nielson, 1999; Nielson, 2000; www.useit.com; www.webpagesthatsuck.com).

Designing Forms and Reports at Pine Valley Furniture

When Jim and the PVF development team focused on designing the forms and reports, that is the "pages" for the WebStore, they first reviewed many popular electronic commerce Websites. From this review they established the following design guidelines.

- Lightweight Graphics
- Forms and Data Integrity Rules
- Template-Based HTML

N E T   S E A R C H
A well-designed corporate Website can help attract potential customers. Visit http://www.prenhall.com/hoffer to complete an exercise related to this topic.

TABLE 13-10   Common Errors when Designing the Layout of Web Pages

Error	Recommendation
Nonstandard Use of GUI Widgets	Make sure that when using standard design items, that they behave in accordance to major interface design standards. For example, the rules for radio buttons state that they are used to select one item among a set of items, that is, not confirmed until "OK'ed" by a user. In many Websites, selecting radio buttons are used as both *selection* and *action*.
Anything that Looks Like Advertising	Since research on Web traffic has shown that many users have learned to stop paying attention to Web advertisement, make sure that you avoid designing any legitimate information in a manner that resembles advertising (e.g., banners, animations, pop-ups).
Bleeding-edge Technology	Make sure that users don't need the latest browsers or plug-ins to view your site.
Scrolling Test and Looping Animations	Avoid scrolling text and animations since they are both hard to read and users often equate such content as advertising.
Nonstandard Link Colors	Avoid using nonstandard colors to show links and for showing links that users have already used; nonstandard colors will confuse the user and reduce ease of use.
Outdated Information	Make sure your site is continuously updated so that users "feel" that the site is regularly maintained and updated. Outdated content is a sure way to lose credibility.
Slow Download Times	Avoid using large images, lots of images, unnecessary animations, or other time-consuming content that will slow the downloading time of a page.
Fixed-Formatted Text	Avoid fixed-formatted text that requires users to scroll horizontally to view content or links
Displaying Long Lists as Long Pages	Avoid requiring users to scroll down a page to view information, especially navigational controls. Manage information by showing only *N* items at a time, using multiple pages, or by using a scrolling container within the window.

In order to assure that all team members understood what was meant by each guideline, Jim organized a design briefing to explain how each would be incorporated into the WebStore interface design.

Lightweight Graphics

Lightweight graphics: The use of small simple images to allow a Web page to more quickly be displayed.

In addition to easy menu and page navigation, the PVF development team wants a system where Web pages load quickly. A technique to assist in making pages load quickly is through the use of **lightweight graphics.** Lightweight graphics is the use of small simple images that allow a page to load as quickly as possible. "Using lightweight graphics allows pages to load quickly and helps users to reach their final location in the site—hopefully the point of purchase area—as quickly as possible. Large color images will only be used for displaying detailed product pictures that customers explicitly request to view," explained Jim. Experienced Web designers have found that customers are not willing to wait at each hop of navigation for a page to load, just so they have to click and wait again. The quick feedback that a Website with lightweight graphics can provide will help to keep customers at the WebStore longer.

Forms and Data Integrity Rules

Because the goal of the WebStore is to have users place orders for products, all forms that request information should be clearly labeled and provide adequate room for input. If a specific field requires a specific input format such as a date of birth or phone number, it must provide a clear example for the user so that data errors can be reduced. Additionally, the site must clearly designate which fields are optional, which are required, and which fields have a range of values.

Jim emphasized, "all of this seems a bit overkill, but it makes processing the data much simpler. Our site will check all data before submitting it to the server for processing. This will allow us to provide quicker feedback to the user on any data entry error and eliminate the possibility of writing erroneous data into the permanent database. Additionally, we want to provide a disclaimer to reassure our customers that the data will only be used for processing orders, that it will never be sold to marketers, and that it will be kept strictly confidential."

Template-Based HTML

When Jim talked with the consultants about the WebStore during the analysis phase, they emphasized the advantages of using **template-based HTML.** He was told that when displaying individual products, it would be very advantageous to try and have a few "templates" that could be used to display the entire product line. In other words, not every product needs its own page, the development time for that would be far too great. Jim explained, "we need to look for ways to write a module once and reuse it. This way, a change requires modifying one page, not 700. Using HTML templates will help us create an interface that is very easy to maintain. For example, a desk and a filing cabinet are two completely different products. Yet, both have an array of finishes to choose from. Logically, each item requires the same function—namely, 'display all finishes.' If designed correctly, this function can be applied to all products in the store. On the other hand, if we write a separate module for each product, it would require us to change each and every module every time we make a product change, like adding a new finish. But, a function such as 'display all finishes,' written once and associated with all appropriate products, will require the modification of one generic or 'abstract' function, not hundreds."

Template-based HTML: Templates to display and process common attributes of higher-level, more abstract items.

Summary

This chapter focuses on a primary product of information systems: designing forms and reports. As organizations move into more complex and competitive business environments with greater diversity in the workforce, the quality of the business processes will determine success. One key to designing quality business processes is the delivery of the right information to the right people, in the right format, at the right time. The design of forms and reports concentrates on this goal. A major difficulty of this process comes from the great variety of information-formatting options available to designers.

Specific guidelines should be followed when designing forms and reports. These guidelines, proven over years of experience with human-computer interaction, help you to create professional, usable systems. The chapter presented a variety of guidelines covering use of titles, layout of fields, navigation between pages or screens, highlighting data, use of color, format of text and numeric data, appropriate use and layout of tables and graphs, avoiding bias in information display, and achieving usable forms and reports.

Form and report designs are created through a prototyping process. Once created, designs may be stand-alone or integrated into actual working systems. The purpose, however, is to show users what a form or report will look like when the system is implemented. The outcome of this activity is the creation of a specification document where characteristics of the users, tasks, system, and environment are outlined along with each form and report design. Performance testing and usability assessments may also be included in the design specification.

The goal of form and report design is usability. Usability means that users can use a form or report quickly, accurately, and with high satisfaction. To be usable, designs must be consistent, efficient, self-explanatory, well-formatted, and flexible. These objectives are achieved by applying a wide variety of guidelines concerning such aspects as navigation, use of highlighting and color, display of text, tables, and lists.

Key Terms

1. Form
2. Lightweight graphics
3. Report
4. Template-Based HTML
5. Usability

Match each of the key terms above with the definition that best fits it.

_____ Templates to display and process common attributes of high-level, more abstract items.

_____ An overall evaluation of how a system performs in supporting a particular user for a particular task.

_____ A business document that contains only predefined data; it is a passive document used only for reading or viewing. A form typically contains data from many unrelated records or transactions.

_____ A business document that contains some predefined data and may include some areas where additional data are to be filled in. An instance of a form is typically based on one database record.

_____ The use of small simple images to allow a Web page to more quickly be displayed.

Review Questions

1. Describe the prototyping process of designing forms and reports. What deliverables are produced from this process? Are these deliverables the same for all types of system projects? Why or why not?

2. To which initial questions must the analyst gain answers in order to build an initial prototype of a system output?

3. When can highlighting be used to convey special information to users?

4. Discuss the benefits, problems, and general design process for the use of color when designing system output.

5. How should textual information be formatted on a help screen?

6. What type of labeling can you use in a table or list to improve its usability?

7. What column, row, and text formatting issues are important when designing tables and lists?

8. Describe how numeric, textual, and alphanumeric data should be formatted in a table or list.

9. What is meant by usability and what characteristics of an interface are used to assess a system's usability?

10. What measures do many development groups use to assess a system's usability?

11. List and describe the common errors when designing the layouts of Websites.

12. Provide some examples where variations in user, task, system, and environmental characteristics might impact the design of system forms and reports.

Problems and Exercises

1. Imagine that you are to design a budget report for a colleague at work using a spreadsheet package. Following the prototyping discussed in the chapter (see also Figure 1-10), describe the steps you would take to design a prototype of this report.

2. Consider a system that produces budget reports for your department at work. Alternatively, consider a registration system that produces enrollment reports for a department at a university. For whichever system you choose, answer the following design questions. Who will use the output? What is the purpose of the output? When is the output needed and when is the information that will be used within the output available? Where does the output need to be delivered? How many people need to view the output?

3. Imagine the worst possible reports from a system. What is wrong with them? List as many problems as you can. What are the consequences of such reports? What could go wrong as a result? How does the prototyping process help guard against each problem?

4. Imagine an output display form for a hotel registration system. Using a software package for drawing such as Visio, follow the design suggestions in this chapter and design this form entirely in black and white. Save the file and then, following the color design suggestions in this chapter, redesign the form using color. Based on this exercise, discuss the relative strengths and weaknesses of each output form.

5. Consider reports you might receive at work (e.g., budgets or inventory reports) or at a university (e.g., grade reports or transcripts). Evaluate the usability of these reports in terms of speed, accuracy, and satisfaction. What could be done to improve the usability of these outputs?

6. List the PC-based software packages you like to use. Describe each package in terms of the following usability characteristics: time to learn, speed of performance, rate of errors by users, retention over time, and subjective satisfaction. Which of these characteristics has led to your wanting to continue to use this package?

7. Given the guidelines presented in this chapter, identify flaws in the design of the Report of Customers that follows. What assumptions about users and tasks did you make in order to assess this design? Redesign this report to correct these flaws.

Report of Customers - 26-Oct-01

Cust-ID	Organization
AC-4	A.C. Nielson Co.
ADTRA-20799	Adran
ALEXA-15812	Alexander & Alexander, Inc.
AMERI-1277	American Family Insurance
AMERI-28157	American Residential Mortgage
ANTAL-28215	Antalys
ATT-234	AT&T Residential Services
ATT-534	AT&T Consumer Services
. . .	
DOLE-89453	Dole United, Inc.
DOME-5621	
DO-67	Doodle Dandies
. . .	
ZNDS-22267	Zenith Data System

8. Review the guidelines for attaining usability of forms and reports in Table 13-8. Consider an on-line form you might use to register a guest at a hotel. For each usability factor, list two examples of how this form could be designed to achieve that dimension of usability. Use examples other than those mentioned in Table 13-8.

9. How can differences in user, task, system, or the environment influence the design of a form or report? Provide an example that contrasts characteristics for each difference.

10. Go out to the Internet and find commercial Websites that demonstrate each of the common errors listed in Table 13-10.

Field Exercises

1. Find your last grade report. Given the guidelines presented in this chapter, identify flaws in the design of this grade report. Redesign this report to correct these flaws.

2. As stated in the chapter, most forms and reports are designed for contemporary information systems by using software to prototype output. Packages like Visual Basic, PowerBuilder, and Paradox for Windows have very sophisticated output design modules. Gain access to such a tool at your university or where you work and study all the features the software provides for the design of printed output. Write a report that lists and explains all the features for layout, highlighting, summarizing data, etc.

3. Investigate the display uses in another field (e.g., aviation). What types of forms and reports are used in this field? What standards, if any, are used to govern the use of these outputs?

4. Interview a variety of people you know about the different types of forms and reports they use in their jobs. Ask to examine a few of these documents and answer the following questions for each one:

a. What types of tasks does each support and how is it used?

b. What types of technologies and devices are used to deliver each one?

c. Assess the usability of each form or report. Is each usable? Why or why not? How could each be improved?

5. Scan the annual reports from a dozen or so companies for the past year. These reports can usually be obtained in a university library. Describe the types of information and the ways that information has been presented in these reports. How have color and graphics been used to improve the usability of information? Describe any instances where formatting has been used to hide or enhance the understanding of information.

6. Choose a PC-based package you like to use and choose one that you don't like to use. Interview other users to determine their evaluations of these two packages. Ask each individual to evaluate each package in terms of the usability characteristics described in the previous question. Is there a consensus among these evaluations, or do the respondents' evaluations differ from each other or from your own evaluations? Why?

References

Benbasat, I., A. S. Dexter, and P. Todd. 1986. "The Influence of Color and Graphical Information Presentation in a Managerial Decision Simulation." *Human-Computer Interaction* 2: 65–92.

Flanders, V., and M. Willis. 1998. *Web Pages That Suck: Learn Good Design by Looking at Bad Design*. Alameda, CA: Sybex Publishing.

Jarvenpaa, S. L., and G. W. Dickson. 1988. "Graphics and Managerial Decision Making: Research Based Guidelines." *Communications of the ACM* 31 (6): 764–74.

Johnson, J. 2000. *GUI Bloopers: Don'ts and Do's for Software Developers and Web Designers*. San Diego, CA: Academic Press.

Nielsen, J. 1999. "User Interface Directions for the Web." *Communications of the ACM* 42(1): 65–71.

Nielsen, J. 2000. *Designing Web Usability: The Practice of Simplicity*. Indianapolis, IN: New Riders Publishing.

Norman, K. L. 1991. *The Psychology of Menu Selection*. Norwood, NJ: Ablex.

Shneiderman, B. 1997. *Designing the User Interface: Strategies for Effective Human-Computer Interaction*. 3rd ed. Reading, MA: Addison-Wesley.

BROADWAY ENTERTAINMENT COMPANY, INC.

Designing Forms and Reports for the Customer Relationship Management System

CASE INTRODUCTION

The students from Stillwater State University are eager to begin building a prototype of MyBroadway, the Web-based customer relationship management system for Carrie Douglass, manager of the Centerville, Ohio, Broadway Entertainment Company (BEC) store. Prototyping seems like an ideal design approach for this system, because the final project product is not intended to be a production system. Rather, the student team is producing a proof-of-concept, initial system version, to be used to justify full development by BEC. Before building the prototype in Microsoft Access, the team is ready to plan the structure for the human interface of the system. For a Web-based system, the human interface is, to the customer, the system. Although the MyBroadway prototype system is not meant to be extensive, the prototype will be effective only if the human interface, including on-screen forms and reports, delights BEC customers. The students first decide to do a pencil-and-paper prototype before development in Access. This initial prototype will be used primarily for discussion among the team and for sharing with other teams in their information systems projects class at Stillwater. Professor Tann, their instructor, encourages collaborative learning, and the members of other teams will be valued, impartial evaluators of the usability of the system's human interface design.

IDENTIFYING THE FORMS AND REPORTS

Many of the forms and reports for MyBroadway are clearly visible from the system's context diagram (see BEC Figure 8-1 at the end of Chapter 8). The main forms and reports are data flows from and to each human external entity—customers and employees. The team decides to concentrate on the customer interfaces for the purpose of the pencil-and-paper prototype. BEC Table 13-1 summarizes the seven customer-related and one employee-related data flows and categorizes them as either system input or system output oriented.

The team quickly realizes that these data flows are not the only human interfaces. Any Web page that is needed

in a navigation path leading to these data flows is also a form or report. For example, to produce the Inventory Review output page (O1), the customer must enter criteria for selecting which inventory items to display. The content and design for the pages leading to the main forms and reports can be addressed later after the sequence and structure of these intermediate pages are organized into a dialogue design (see the BEC case after Chapter 14).

DESIGNING FORMS AND REPORTS FOR MYBROADWAY

Each of the eight pages identified in BEC Table 13-1 needs to be designed for customer usability. The team realizes that usability means that the page is easy to understand, helps the user perform a given task, and is efficient for the user to use. From their courses at Stillwater State University, the students are familiar with many usability guidelines for human-computer interfaces. Also from their education the students know that usability is improved if there are frequent reviews of the proposed human interfaces. This is another reason why prototyping will be an effective development strategy for MyBroadway. Initial designs for each page will be reviewed by the team's classmates from Stillwater, and then working prototype iterations will be evaluated by actual customers in the Centerville store.

Two of the pencil-and-paper page prototypes appear in BEC Figures 13-1 and 13-2. The page in BEC Figure 13-1 is an intermediate page that helps the user formulate the criteria to specify which items from inventory to include in the Inventory Review, output 1 from BEC Table 13-1. This

BEC TABLE 13-1 Forms and Reports for MyBroadway

Inputs from Users	Outputs to Users
I1 New Comment on a Product	O1 Inventory Review
I2 New Product Request	O2 Comments on Products
I3 Rental Extension Request	O3 Rental Status
I4 Favorite Picks	O4 Child Purchase/Rental History

PRODUCT SELECTION

You can see what products we carry in our BEC store by entering values below. For each value you want to search on, click in the check box next to that criterion. If you know the name of the video, music, or game, enter the title—we'll scan our inventory to find the products with a title that comes closest to what you enter. You can also look for products by category (for example, R&B, hard rock, jazz), publisher, or release date. For these criteria, select a value from the available options from the drop-down menu. If you enter or select values from more than one criterion, we'll look for products that satisfy all of the selections you make.

☐ Title: [   ]

☐ Category: [▼]

☐ Publisher: [▼]

               Year   Month

☐ Release Date: [▼] [▼]

[Submit]　[Reset]　[Return to Welcome Page]　[Back]

BEC Figure 13-1
Design for Product Selection

selection page has a title and explanation of its purpose. Because the page is not too full of data, the team decides to include an explanation of its purpose and content. Alternatively the team considered excluding this explanation, instead making it accessible by the user via a help button. The user clicks in a check box to indicate that a value will be entered or selected for each type of product selection criteria. Because there are thousands of titles in a BEC store's inventory, the team decides not to use a drop-down menu for entering a title, but rather the user enters the approximate title of the product. The team realizes that in many cases the title entered will not exactly match the title stored in the database. The logic that processes a

title will have to search for the best match. All other criteria, if checked by the user, have a more limited set of options, so drop-down selection boxes are used. The team decides that there are four ways for the user to exit the page. The first option is to submit the product selection, which will take the user to the page in BEC Figure 13-2, the Inventory Review. The second option is to return to the welcome page. The third (possible) option is to go back to a page, yet to be designed, from which the user navigated to this page. A fourth option is to clear the values in the page to consider another selection. This might occur because the user changes his or her mind after indicating some selections, or upon returning from the Inventory Review page, the user may want to enter another selection starting with no selection values.

BEC Figure 13-2 shows a design for the Inventory Review system output page. The team decides to show the selection criteria in a top frame to remind the user of the selections. Because many items may satisfy a selection, the team allows for scrolling through all results in the bottom frame. A user may exit the page in four ways—(1) go back to the selection page, (2) return to the Product Information page (at this point in the design, only a generic page dealing with production information), (3) return to the Welcome Page, and (4) print the query results. The students had not even considered the need for printing until designing this screen. The team will ask Carrie about this at their next meeting. The cost of the printer will not be that great, but the team is more concerned about paper costs, keeping a supply of paper in the printer, and the extra effort to keep the printer free of paper jams, with full ink cartridges, and otherwise in working order. The use of a printer requires more active involvement of store staff than the team had previously explained to Carrie would be necessary.

BEC Figure 13-2
Design for Inventory Review

INVENTORY REVIEW

| These are the criteria you selected: | Title:
Category:
Publisher:
Release Date: | | [Return to Product Information]　[Return to Welcome Page]
[Back]　[Print] |

Title:　　　　　　　　　　　　　Artist:

Category:　　　　　Publisher:　　　　　Release Date:

Description:

Price—Sale:　　　Rental:

CASE SUMMARY

The student team feels that the design of the product selection and inventory review pages represents a suitable project deliverable that should be reviewed by someone outside the team. Fortunately, Professor Tann has scheduled project status report presentations for the next session of the team's project class. The MyBroadway student team will present the page designs to obtain an initial reaction to the style for pages they have adopted. Because the team has not invested a great deal of time in the initial design, the team members believe they can be open to, not defensive in response to, constructive suggestions. Because other teams will also likely walk through form and report designs, the BEC team will see some other creative designs, which will give additional ideas for improvement.

CASE QUESTIONS

1. Using guidelines from this chapter and other sources, evaluate the usability of the system input design depicted in BEC Figure 13-1. Specifically evaluate the use of drop-down selection lists for some fields and not others, the inclusion of the instructions for the page on the page itself, and the exit options. If you would recommend changes to this design, provide an alternative screen layout with a justification of why your page has a better design.

2. Using guidelines from this chapter and other sources, evaluate the usability of the system output design depicted in BEC Figure 13-2. Specifically evaluate the layout of fields, use of highlighting, the way multiple results are handled, and the exit options. If you would recommend changes to this design, provide an alternative screen layout with a justification of why your page has a better design.

3. One of the exit options for the page in BEC Figure 13-2 is to print the information on this page. Design the printed version of this page. Justify your design by explaining how it satisfies the design guidelines found in this chapter or in other sources you reference.

4. Using your answers to Case Questions 1 through 3 as reference, design a page to accept input I1 from BEC Table 13-1, the New Comment on a Product page. Also design any page (similar to the page in BEC Figure 13-1) that might immediately precede the New Comment on a Product page to select the prod-uct on which to comment, if you do not design product selection into the New Comment on a Product page. Justify your design by explaining how it satisfies the design guidelines found in this chapter or in other sources you reference.

5. Using your answers to Case Questions 1 through 4 as reference, design a page to produce output O2 from BEC Table 13-1, the Comments on Products page. Also design any page (similar to the page in BEC Figure 13-1) that might immediately precede the Comments on Products page to select the product on which to display comments, if you do not design product selection into the Comments on Products page. Justify your design by explaining how it satisfies the design guidelines found in this chapter or in other sources you reference.

6. Given your answer to Case Question 5, design the printed version of the Comments on Products page. Justify your design by explaining how it satisfies the design guidelines found in this chapter or in other sources you reference.

7. A future enhancement that Carrie once mentioned to the Stillwater student team is to allow wireless, palm computing device access to MyBroadway. Given your answers to Case Questions 1 and 2, how would your proposed designs for these forms need to change to accommodate the display technology of palm devices? Justify your answer.

8. The pencil-and-paper designs in BEC Figures 13-1 and 13-2 are monochrome. Using your answers to Case Questions 1 through 2 as reference, in what ways would you recommend the use of color to improve the usability of these form designs? Justify your answer by explaining how it satisfies the design guidelines found in this chapter or in other sources you reference.

9. You may have noticed by now that the context diagram in BEC Figure 8-1 and BEC Table 13-1 includes an input for employee favorite picks (I4 in BEC Table 13-1), but there is no explicit associated system output for customers. Did you include employee favorite pick information in your answer to Case Question 5? If not, return to this case question and redesign the output forms to allow for the inclusion of favorite pick information about products. Justify your answer by explaining how it satisfies the design guidelines found in this chapter or in other sources you reference.

Chapter 14

Designing Interfaces
and Dialogues

LEARNING OBJECTIVES

After studying this chapter, you should be able to:

● Explain the process of designing interfaces and dialogues and the deliverables for their creation.

● Contrast and apply several methods for interacting with a system.

● List and describe various input devices and discuss usability issues for each in relation to performing different tasks.

● Describe and apply the general guidelines for designing interfaces and specific guidelines for layout design, structuring data entry fields, providing feedback, and system help.

● Design human-computer dialogues, including the use of dialogue diagramming.

● Design graphical user interfaces.

● Discuss guidelines for the design of interfaces and dialogues for Internet-based electronic commerce systems.

INTRODUCTION

In this chapter, you learn about system interface and dialogue design. Interface design focuses on how information is provided to and captured from users; dialogue design focuses on the sequencing of interface displays. Dialogues are analogous to a conversation between two people. The grammatical rules followed by each person during a conversation are analogous to the interface. Thus, the design of interfaces and dialogues is the process of defining the manner in which humans and computers exchange infor-mation. A good human-computer interface provides a uniform structure for finding, viewing, and invoking the different components of a system. This chapter complements Chapter 13, which addressed design guidelines for the content of forms and reports. Here, you will learn about navigation between forms, alternative ways for users to cause forms and reports to appear, and how to supplement the content of forms and reports with user help and error messages, among other topics.

In the next section, the process of designing interfaces and dialogues and the deliverables produced during this activity are described. This is followed by a section that describes interaction methods and devices. Next, interface design is described. This discussion focuses on layout design, data entry, providing feedback, and designing help. The next section examines techniques for designing human-computer dialogues. Finally, we examine the design of interfaces and dialogues within electronic commerce applications.

DESIGNING INTERFACES AND DIALOGUES

This is the third chapter that focuses on design within the systems development life cycle (see Figure 14-1). In Chapter 13, you learned about the design of forms and reports. As you will see, the guidelines for designing forms and reports also apply to the design of human-computer interfaces.

The Process of Designing Interfaces and Dialogues

Similar to designing forms and reports, the process of designing interfaces and dialogues is a user-focused activity. This means that you follow a prototyping methodology of iteratively collecting information, constructing a prototype, assessing usability, and making refinements. To design usable interfaces and dialogues, you must answer the same who, what, when, where, and how questions used to guide the design of forms and reports (see Table 13-1). Thus, this process parallels that of designing forms and reports.

Figure 14-1
Systems development life cycle with design phase highlighted

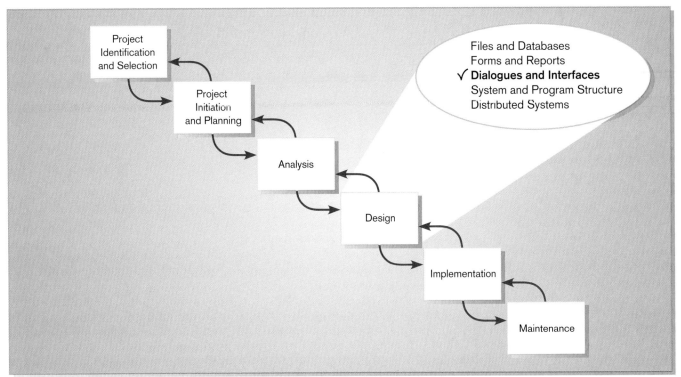

Deliverables and Outcomes

The deliverable and outcome from system interface and dialogue design is the creation of a design specification. This specification is also similar to the specification produced for form and report designs—with one exception. Recall that design specifications in Chapter 13 had three sections (see Figure 13-4):

1. Narrative overview
2. Sample design
3. Testing and usability assessment

For interface and dialogue designs, one additional subsection is included: a section outlining the dialogue sequence—the ways a user can move from one display to another. Later in the chapter you will learn how to design a dialogue sequence by using dialogue diagramming and state-transition diagramming. An outline for a design specification for interfaces and dialogues is shown in Figure 14-2.

INTERACTION METHODS AND DEVICES

Interface: A method by which users interact with information systems.

The human-computer **interface** defines the ways in which users interact with an information system. All human-computer interfaces must have an interaction style and use some hardware device(s) for supporting this interaction. In this section we therefore describe various interaction methods and guidelines for designing usable interfaces.

Methods of Interacting

When designing the user interface, the most fundamental decision you make relates to the methods used to interact with the system. Given that there are numerous approaches for designing the interaction, we briefly provide a review of those

Figure 14-2
Specification outline for the design of interfaces and dialogues

Design Specification
1. Narrative overview a. Interface/Dialogue Name b. User Characteristics c. Task Characteristics d. System Characteristics e. Environmental Characteristics
2. Interface/Dialogue Designs a. Form/Report Designs b. Dialogue Sequence Diagram(s) and Narrative Description

3. Testing and Usability Assessment
 a. Testing Objectives
 b. Testing Procedures
 c. Testing Results
   i) Time to Learn
   ii) Speed of Performance
   iii) Rate of Errors
   iv) Retention Over Time
   v) User Satisfaction and Other Perceptions

most commonly used.[1] Our review will examine the basics of five widely used styles: command language, menu, form, object, and natural language. We will also describe several devices for interacting, focusing primarily on their usability for various interaction activities.

Command Language Interaction   In **command language interaction**, the user enters explicit statements to invoke operations within a system. This type of interaction requires users to remember command syntax and semantics. For example, to copy a file named PAPER.DOC from one storage location (C:) to another (A:) using Microsoft's disk operating system (DOS), a user would type

COPY C:PAPER.DOC A:PAPER.DOC

Command language interaction places a substantial burden on the user to remember names, syntax, and operations. Most newer or large-scale systems no longer rely entirely on a command language interface. Yet, command languages are good for experienced users, for systems with a limited command set, and for rapid interaction with the system.

A relatively simple application such as a word processor may have hundreds of commands for such operations as saving a file, deleting words, canceling the current action, finding a specific piece of data, or switching between windows. Some of the burden of assigning keys to actions has been taken off your shoulders through the development of user interface standards such as those for the Macintosh, Microsoft Windows, or Java (Apple Computer, 1993; McKay, 1999; Sun Microsystems, 1999). For example, Figure 14-3a shows the help screen from Microsoft Word describing the function key commands, and Figure 14-3b shows the same screen for Microsoft Excel. Note how many of the same keys have been assigned the same function. Also note that designers still have great flexibility in how they interpret and implement these standards. This means that you still need to pay attention to usability factors and conduct formal assessments of designs. Any application may adopt the CUA function key standard, as appropriate for the functionality of that application.

Menu Interaction   A significant amount of interface design research has stressed the importance of a system's ease of use and understandability. **Menu interaction** is a means by which many designers have accomplished this goal. A menu is simply a list of options; when an option is selected by the user, a specific command is invoked, or another menu is activated. Menus have become the most widely used interface method because the user only needs to understand simple signposts and route options to effectively navigate through a system.

Menus can differ significantly in their design and complexity. The variation of their design is most often related to the capabilities of the development environment, the skills of the developer, and the size and complexity of the system. For smaller and less complex systems with limited system options, you may use a single menu or a linear sequence of menus. A single menu has obvious advantages over a command language but may provide little guidance beyond invoking the command. A single menu for a DOS shell program is shown in Figure 14-4.

For large and more complex systems, you can use menu hierarchies to provide navigation between menus. These hierarchies can be simple tree structures or variations wherein children menus have multiple parent menus or that allow multilevel traversal. Variations as to how menus are arranged can greatly influence the usability

Command language interaction: A human-computer interaction method where users enter explicit statements into a system to invoke operations.

Menu interaction: A human-computer interaction method where a list of system options is provided and a specific command is invoked by user selection of a menu option.

---

[1] Readers interested in learning more about interaction methods are encouraged to see the books by Johnson (2000) or Shneiderman (1997).

Figure 14-3
Function key assignments in Microsoft Office 97
(a) Help screen from Microsoft Word describing function key commands

(b) Help screen from Microsoft Excel describing function key commands

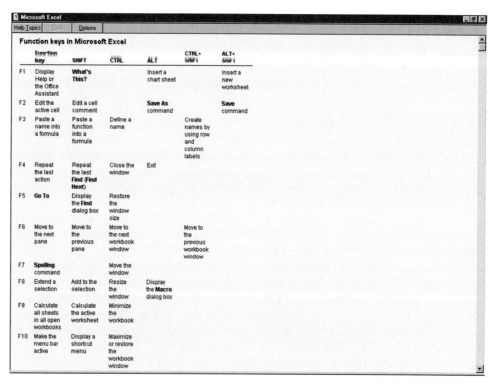

of a system. Figure 14-5 shows a variety of ways in which menus can be structured and traversed. An arc on this diagram signifies the ability to move from one menu to another. Although more complex menu structures provide greater user flexibility, they may also confuse users about exactly where they are in the system. Structures with multiple parent menus also require the application to remember which path has been followed so that users can correctly backtrack.

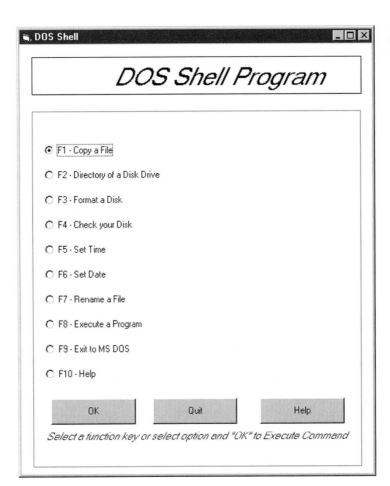

Figure 14-4
Single-level menu for disk
operating system application

There are two common methods for positioning menus. In a **pop-up menu** (also called a dialogue box), menus are displayed near the current cursor position so users don't have to move the position or their eyes to view system options (Figure 14-6a). A pop-up menu has a variety of potential uses. One is to show a list of commands relevant to the current cursor position (for example, delete, clear, copy, or validate current field). Another is to provide a list of possible values (from a look-up table) to fill in for the current field. For example, in a customer order form, a list of current customers could pop up next to the customer number field so the user can select the correct customer without having to know the customer's identifier. In a **drop-down menu**, menus drop down from the top line of the display (Figure 14-6b). Drop-down menus have become very popular in recent years because they provide consistency in menu location and operation among applications and efficiently use display space. Most advanced operating environments such as Microsoft Windows, Macintosh, and the Unix graphical operating environment called X-Windows provide a combination of both pop-up and drop-down menus.

Pop-up menu: A menu positioning method that places a menu near the current cursor position.

Drop-down menu: A menu positioning method that places the access point of the menu near the top line of the display; when accessed, menus open by dropping down onto the display.

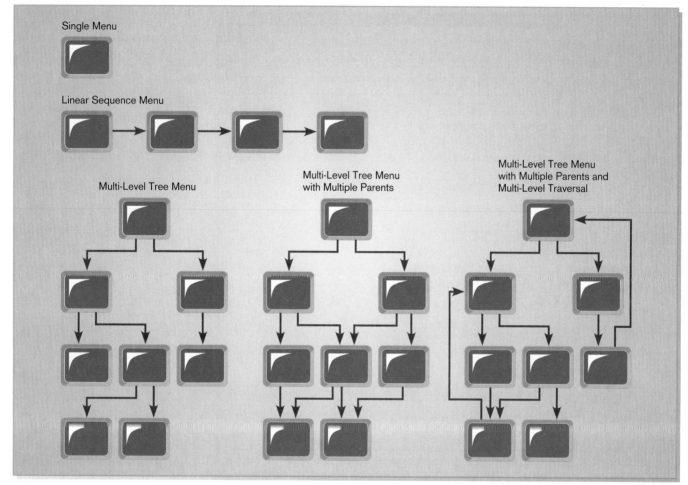

Figure 14-5
Various types of menu configurations (Adapted from Shneiderman, 1997)

When designing menus, there are several general rules that should be followed and these are summarized in Table 14-1. For example, each menu should have a meaningful title and be presented in a meaningful manner to users. A menu option of Quit, for instance, is ambiguous—does it mean return to the previous screen, exit the program, or exit to DOS? To more easily see how to apply these guidelines, Figure 14-7 contrasts a poorly designed menu with a menu following the menu

TABLE 14-1   Guidelines for Menu Design

Wording	• Each menu should have a meaningful title
	• Command verbs should clearly and specifically describe operations
	• Menu items should be displayed in mixed upper- and lower-case letters and have a clear, unambiguous interpretation
Organization	• A consistent organizing principle should be used that relates to the tasks the intended users perform; for example, related options should be grouped together and the same option should have the same wording and codes each time it appears
Length	• The number of menu choices should not exceed the length of the screen
	• Submenus should be used to break up exceedingly long menus
Selection	• Selection and entry methods should be consistent and reflect the size of the application and sophistication of the users
	• How the user is to select each option and the consequences of each option should be clear (e.g., whether another menu will appear)
Highlighting	• Highlighting should be minimized and used only to convey selected options (e.g., a check mark) or unavailable options (e.g., dimmed text)

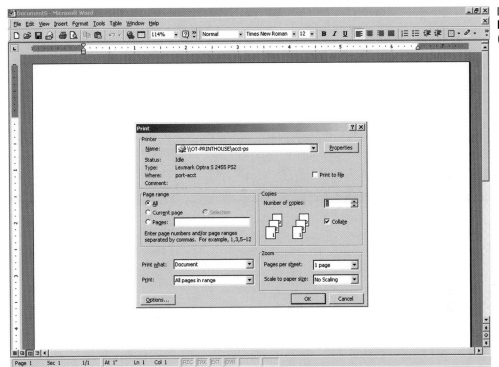

Figure 14-6
Menus from Microsoft Word 2000
(a) Pop-up menu

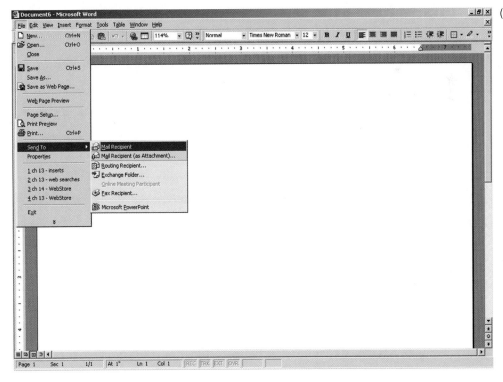

(b) Drop-down menu

design guidelines. Annotations on the two parts of this figure highlight poor and improved menu interface design features.

Many advanced programming environments provide powerful tools for designing menus. For example, Microsoft's Visual Basic allows you to quickly design a menu structure using a menu design facility. Figure 14-8a shows the design form, Menu Design Window, in which a menu structure is defined. Figure 14-8b shows how the menu looks to a user within the actual information system. To build a menu you only

Figure 14-7
Contrasting menu designs
(a) Poor menu design

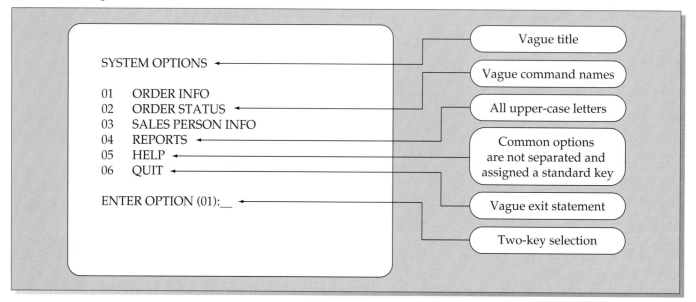

(b) Improved menu design

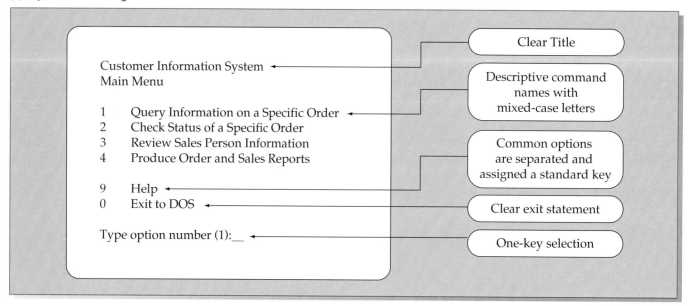

need to type the words that will represent each item on the menu. With the use of a few easily invoked options, you can also assign shortcut keys to menu items, connect help screens to individual menu items, define submenus, and set usage properties. Usage properties, for example, include the ability to dim the color of a menu item while a program is running to indicate that a function is currently unavailable. Menu building tools allow a designer to quickly and easily prototype a design that will look exactly as it will in the final system.

Form interaction: A highly intuitive human-computer interaction method whereby data fields are formatted in a manner similar to paper-based forms.

Form Interaction   The premise of **form interaction** is to allow users to fill in the blanks when working with a system. Form interaction is effective for both the input and presentation of information. An effectively designed form includes a self-explanatory title and field headings, has fields organized into logical groupings with

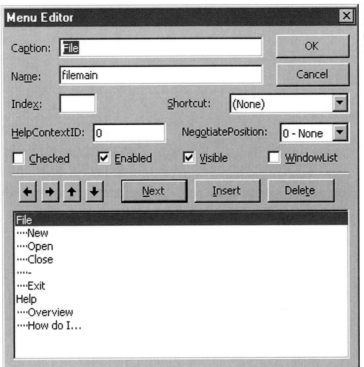

Figure 14-8
Menu building within a graphical user interface environment
(a) Menu design window

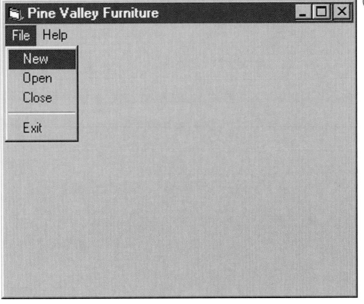

(b) Menu design as viewed by user

distinctive boundaries, provides default values when practical, displays data in appropriate field lengths, and minimizes the need to scroll windows (Shneiderman, 1997). You saw many other design guidelines for forms in Chapter 13. Form interaction is the most commonly used method for data entry and retrieval in business-based systems. Figure 14-9 shows a form from the Google Advanced Search engine. Using interactive forms, organizations can easily provide all types of information to Web surfers.

Object-based interaction: A human-computer interaction method where symbols are used to represent commands or functions.

Icon: Graphical pictures that represent specific functions within a system.

Object-Based Interaction   The most common method for implementing **object-based interaction** is through the use of icons. **Icons** are graphic symbols that look like the processing option they are meant to represent. Users select operations by pointing to the appropriate icon with some type of pointing device. The primary advantages to icons are that they take up little screen space and can be quickly understood by most users. An icon may also look like a button that, when selected or depressed, causes the system to take an action relevant to that form, such as cancel, save, edit a record, or ask for help. For example, Figure 14-10 illustrates an icon-based interface when entering Microsoft Visual Basic.

Natural language interaction: A human-computer interaction method whereby inputs to and outputs from a computer-based application are in a conventional speaking language such as English.

Natural Language Interaction   A branch of artificial intelligence research studies techniques for allowing systems to accept inputs and produce outputs in a conventional language like English. This method of interaction is referred to as **natural language interaction**. Presently, natural language interaction is not as viable an interaction style as other methods presented. Current implementations can be tedious, frustrating, and time-consuming for the user and are often built to accept input in narrowly constrained domains (e.g., database queries). Natural language interaction is being applied within both keyboard and voice entry systems.

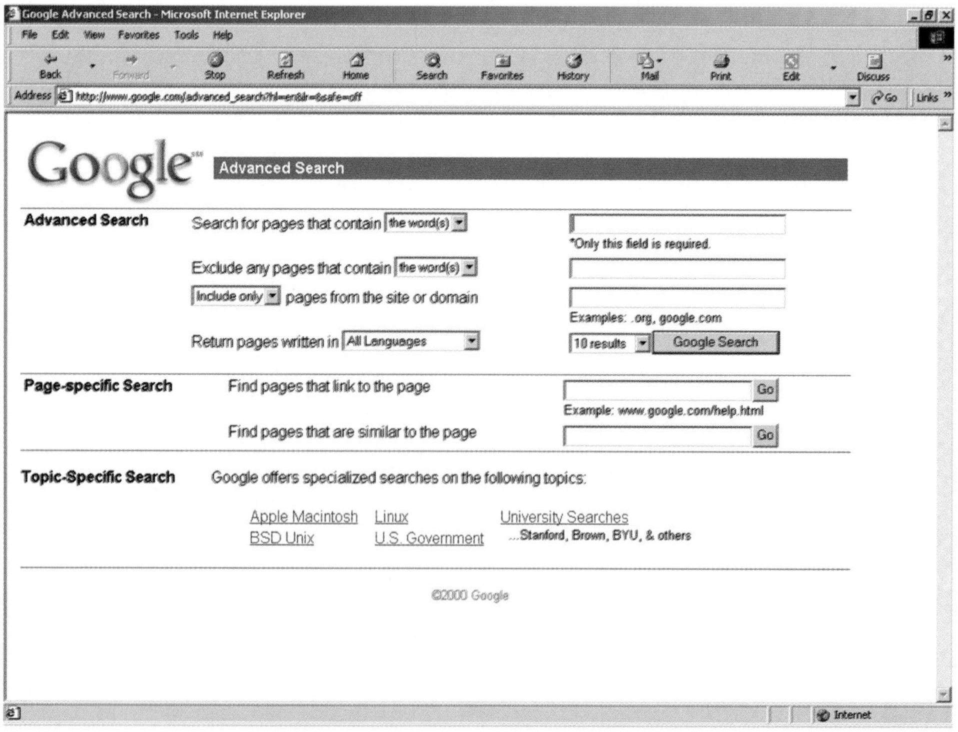

Figure 14-9
Example of form interaction from the Google Advanced Search engine

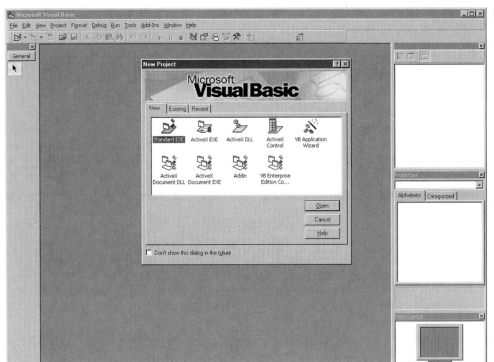

Figure 14-10
Icon-based interface for beginning a new session in Microsoft Visual Basic

Hardware Options for System Interaction

In addition to the variety of methods used for interacting with a system, there is also a growing number of hardware devices employed to support this interaction (see Table 14-2 for a list of interaction devices along with brief descriptions of the typical usage of each). The most fundamental and widely used is the keyboard, which is the mainstay of most computer-based applications for the entry of alphanumeric information. Keyboards vary, from the typewriter kind of keyboards used with personal computers to special-function keyboards on point-of-sale or shop floor devices. The growth in graphical user environments, however, has spurred the broader use of pointing devices such as mice, joysticks, trackballs, and graphics tablets. The creation of notebook and pen-based computers with trackballs, joysticks, or pens attached directly to the computer has also brought renewed interest to the usability of these various devices.

Research has found that each device has its strengths and weaknesses that must guide your selection of the appropriate devices to aid users in their interaction with an application. The selection of devices users will use for interaction must be made during logical design since different interfaces require different devices. Table 14-3 summarizes much of the usability assessment research by relating each device to various types of human-computer interaction problems. For example, for many applications keyboards do not give users a precise feel for cursor movement, do not provide direct feedback on each operation, and can be a slow way to enter data (depending on the typing skill of the user). Another means to gain an understanding of device usability is to highlight which devices have been found most useful for completing specific tasks. The results of this research are summarized in Table 14-4. The rows of this table list common user-computer interaction tasks, and the columns show three criteria for evaluating the usability of the different devices. After reviewing these three tables, it should be evident that no device is perfect and that some are more appropriate for performing some tasks than others. To design the most effective interfaces for a given application, you should understand the capabilities of various interaction methods and devices.

N E T   S E A R C H

Many have felt that voice interaction would become a dominant interaction method. Visit http://www.prenhall.com/hoffer to complete an exercise related to this topic.

TABLE 14-2 Common Devices for Interacting with an Information System

Device	Description and Primary Characteristics or Usage
Keyboard	Users push an array of small buttons that represent symbols which are then translated into words and commands. Keyboards are widely understood and provide considerable flexibility for interaction.
Mouse	A small plastic box that users push across a flat surface and whose movements are translated into cursor movement on a computer display. Buttons on the mouse tell the system when an item is selected. A mouse works well on flat desks but may not be practical in dirty or busy environments, such as a shop floor or check-out area in a retail store. Newer pen-based mice provide the user with more of the feel of a writing implement.
Joystick	A small vertical lever mounted on a base that steers the cursor on a computer display. Provides similar functionality to a mouse.
Trackball	A sphere mounted on a fixed base that steers the cursor on a computer display. A suitable replacement for a mouse when work space for a mouse is not available.
Touch Screen	Selections are made by touching a computer display. This works well in dirty environments or for users with limited dexterity or expertise.
Light Pen	Selections are made by pressing a pen-like device against the screen. A light pen works well when the user needs to have a more direct interaction with the contents of the screen.
Graphics Tablet	Moving a pen-like device across a flat tablet steers the cursor on a computer display. Selections are made by pressing a button or by pressing the pen against the tablet. This device works well for drawing and graphical applications.
Voice	Spoken words are captured and translated by the computer into text and commands. This is most appropriate for users with physical challenges or when hands need to be free to do other tasks while interacting with the application.

TABLE 14-3 Summary of Interaction Device Usability Problems

Device	Problem						
	Visual Blocking	User Fatigue	Movement Scaling	Durability	Adequate Feedback	Speed	Pointing Accuracy
Keyboard	□	□	■	□	■	■	□
Mouse	□	□	■	□	■	□	□
Joystick	□	□	■	□	■	□	■
Trackball	□	□	■	■	■	□	□
Touch Screen	■	■	□	■	□	□	■
Light Pen	■	■	□	□	□	□	■
Graphics Tablet	□	□	■	□	■	□	□
Voice	□	□	■	□	■	□	■

Adapted from Blattner & Schultz, 1988.
Key:

□ = little or no usability problems
■ = potentially high usability problems for some applications
Visual Blocking = extent to which device blocks display when using
User Fatigue = potential for fatigue over long use
Movement Scaling = extent to which device movement translates to equivalent screen movement
Durability = lack of durability or need for maintenance (e.g., cleaning) over extended use
Adequate Feedback = extent to which device provides adequate feedback for each operation
Speed = cursor movement speed
Pointing Accuracy = ability to precisely direct cursor

TABLE 14-4 Summary of General Conclusions from Experimental Comparisons of Input Devices in Relation to Specific Task Activities

Task	Most Accurate	Shortest Positioning	Most Preferred
Target Selection	trackball, graphics tablet, mouse, joystick	touch screen, light pen, mouse, graphics tablet, trackball	touch screen, light pen
Text Selection	mouse	mouse	—
Data Entry	light pen	light pen	—
Cursor Positioning	—	light pen	—
Text Correction	light pen, cursor keys	light pen	light pen
Menu Selection	touch screen	—	keyboard, touch screen

(Adapted from Blattner & Schultz, 1988)
Key:

Target Selection = moving the cursor to select a figure or item
Text Selection = moving the cursor to select a block of text
Data Entry = entering information of any type into a system
Cursor Positioning = moving the cursor to a specific position
Text Correction = moving the cursor to a location to make a text correction
Menu Selection = activating a menu item
— = no clear conclusion from the research

DESIGNING INTERFACES

Building on the information provided in Chapter 13 on the design of content for forms and reports, in this section we discuss issues related to the design of interface layouts. This discussion provides guidelines for structuring and controlling data entry fields, providing feedback, and designing on-line help. Effective interface design requires that you gain a thorough understanding of each of these concepts.

Designing Layouts

To ease user training and data recording, you should use standard formats for computer-based forms and reports similar to paper-based forms and reports for recording or reporting information. A typical paper-based form for reporting customer sales activity is shown in Figure 14-11. This form has several general areas common to most forms:

PINE VALLEY FURNITURE

- Header information
- Sequence and time-related information
- Instruction or formatting information
- Body or data details
- Totals or data summary
- Authorization or signatures
- Comments

In many organizations, data is often first recorded on paper-based forms and then later recorded within application systems. When designing layouts to record or display information on paper-based forms, you should try to make both as similar as possible. Additionally, data entry displays should be consistently formatted across applications to speed data entry and reduce errors. Figure 14-12 shows an equivalent computer-based form to the paper-based form shown in Figure 14-11.

Figure 14-11
Paper-based form for reporting customer sales activity (Pine Valley Furniture)

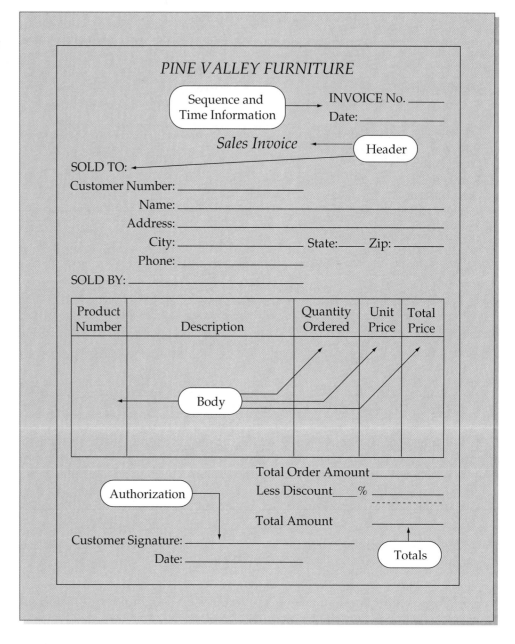

Another concern when designing the layout of computer-based forms is the design of between-field navigation. Since you can control the sequence for users to move between fields, standard screen navigation should flow from left to right and top to bottom just as when you work on paper-based forms. For example, Figure 14-13 contrasts the flow between fields on a form used to record business contacts. Figure 14-13a uses a consistent left to right, top to bottom flow. Figure 14-13b uses a flow that is nonintuitive. When appropriate, you should also group data fields into logical categories with labels describing the contents of the category. Areas of the screen not used for data entry or commands should be inaccessible to the user.

When designing the navigation procedures within your system, flexibility and consistency are primary concerns. Users should be able to freely move forward and backward or to any desired data entry fields. Users should be able to navigate each form in the same way or in as similar a manner as possible. Additionally, data should not *usually* be permanently saved by the system until the user makes an explicit

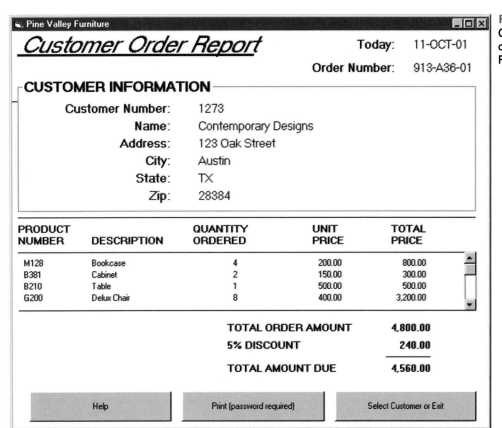

Figure 14-12
Computer-based form reporting
customer sales activity (Pine Valley
Furniture)

request to do so. This allows the user to abandon a data entry screen, back up, or move forward without adversely impacting the contents of the permanent data.

Consistency extends to the selection of keys and commands. Each key or command should have only one function and this function should be consistent throughout the entire system and across systems, if possible. Depending upon the application, various types of functional capabilities will be required to provide smooth navigation and data entry. Table 14-5 provides a list of the functional requirements for providing smooth and easy navigation within a form. For example, a functional and consistent interface will provide common ways for users to move the cursor to different places on the form, editing characters and fields, moving among form displays, and obtaining help. These functions may be provided by keystrokes, mouse or other pointing device operations, or menu selection or button activation. It is possible that, for a single application, all functional capabilities listed in Table 14-5 may not be needed in order to create a flexible and consistent user interface. Yet, the capabilities that are used should be consistently applied to provide an optimal user environment. As with other tables in Chapters 13 and 14, Table 14-5 can serve as a checklist for you to validate the usability of user interface designs.

Structuring Data Entry

Several rules should be considered when structuring data entry fields on a form (see Table 14-6). The first is simple, but often violated by designers. To minimize data entry errors and user frustration, *never* require the user to enter information that is already available within the system or information that can be easily computed by the system. For example, never require the user to enter the current date and time, since each of these values can be easily retrieved from the computer system's internal cal-

Figure 14-13
Contrasting the navigation flow within a data entry form
(a) Proper flow between data entry fields

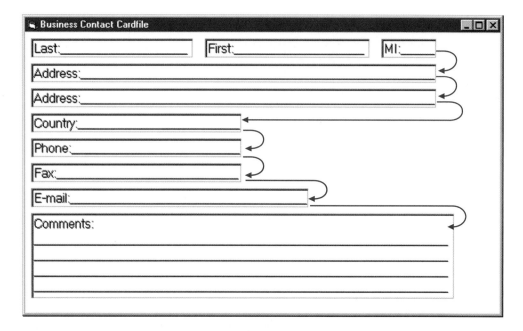

(b) Poor flow between data entry fields

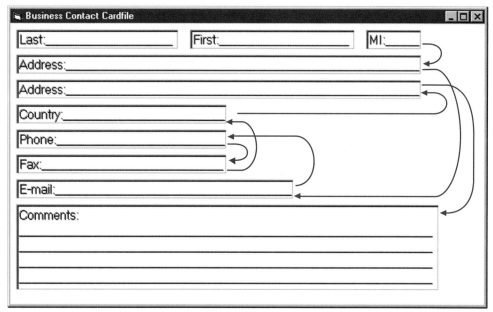

endar and clock. By allowing the system to do this, the user simply confirms that the calendar and clock are working properly.

Other rules are equally important. For example, suppose that a bank customer is repaying a loan on a fixed schedule with equal monthly payments. Each month when a payment is sent to the bank, a clerk needs to record that the payment has been received into a loan processing system. Within such a system, default values for fields should be provided whenever appropriate. This means that *only* in the instances where the customer pays *more or less* than the scheduled amount should the clerk have to enter data into the system. In all other cases, the clerk would simply verify that the check is for the default amount provided by the system and press a single key to confirm the receipt of payment.

When entering data, the user should also not be required to specify the dimensional units of a particular value. For example, a user should not be required to spec-

TABLE 14-5 Data Entry Screen Functional Capabilities

Cursor Control Capabilities:

Move the cursor forward to the next data field

Move the cursor backward to the previous data field

Move the cursor to the first, last, or some other designated data field

Move the cursor forward one character in a field

Move the cursor backward one character in a field

Editing Capabilities:

Delete the character to the left of the cursor

Delete the character under the cursor

Delete the whole field

Delete data from the whole form (empty the form)

Exit Capabilities:

Transmit the screen to the application program

Move to another screen/form

Confirm the saving of edits or go to another screen/form

Help Capabilities:

Get help on a data field

Get help on a full screen/form

ify that an amount is in dollars or that a weight is in tons. Field formatting and the data entry prompt should make clear the type of data being requested. In other words, a caption describing the data to be entered should be adjacent to each data field. Within this caption, it should be clear to the user what type of data is being requested. As with the display of information, all data entered onto a form should automatically justify in a standard format (e.g., date, time, money). Table 14-7 illustrates a few options appropriate for printed forms. For data entry on video display terminals, you should highlight the area in which text is entered so that the exact number of characters per line and number of lines are clearly shown. You can also use check boxes or radio buttons to allow users to choose standard textual responses. And, you can use data entry controls to ensure that the proper type of data (alphabetic or numeric, as required) are entered. Data entry controls are discussed next.

TABLE 14-6 Guidelines for Structuring Data Entry Fields

Entry	Never require data that are already on-line or that can be computed; for example, do not enter customer data on an order form if that data can be retrieved from the database, and do not enter extended prices which can be computed from quantity sold and unit prices.
Defaults	Always provide default values when appropriate; for example, assume today's date for a new sales invoice, or use the standard product price unless overridden
Units	Make clear the type of data units requested for entry; for example, indicate quantity in tons, dozens, pounds, etc.
Replacement	Use character replacement when appropriate; for example, allow the user to look up the value in a table or automatically fill in the value once the user enters enough significant characters.
Captioning	Always place a caption adjacent to fields; see Table 14-7 for caption options.
Format	Provide formatting examples when appropriate; for example, automatically show standard embedded symbols, decimal points, credit symbol, or dollar sign.
Justify	Automatically justify data entries; numbers should be right-justified and aligned on decimal points, and text should be left-justified.
Help	Provide context-sensitive help when appropriate; for example, provide a hot key, such as the F1 key, that opens the help system on an entry that is most closely related to where the cursor is on the display.

TABLE 14-7 Options for Entering Text

Options	Example
Line caption	Phone Number ( ) - _____
Drop caption	( ) - _____ Phone Number
Boxed caption	Phone Number
Delimited characters	|(| | |)| | | |-| | | | Phone Number
Check-off boxes	Method of payment (check one) ☐ Check ☐ Cash ☐ Credit card: Type

Controlling Data Input

One objective of interface design is to reduce data entry errors. As data are entered into an information system, steps must be taken to ensure that the input is valid. As a systems analyst, you must anticipate the types of errors users may make and design features into the system's interfaces to avoid, detect, and correct data entry mistakes. Several types of data errors are summarized in Table 14-8. In essence, data errors can occur from appending extra data onto a field, truncating characters off a field, transcripting the wrong characters into a field, or transposing one or more characters within a field. Systems designers have developed numerous tests and techniques for catching invalid data before saving or transmission, thus improving the likelihood that data will be valid (see Table 14-9 which summarizes these techniques). These tests and techniques are often incorporated into both data entry screens and intercomputer data transfer programs.

TABLE 14-8 Sources of Data Errors

Data Error	Description
Appending	Adding additional characters to a field
Truncating	Losing characters from a field
Transcripting	Entering invalid data into a field
Transposing	Reversing the sequence of one or more characters in a field

TABLE 14-9 Validation Tests and Techniques to Enhance the Validity of Data Input

Validation Test	Description
Class or Composition	Test to assure that data are of proper type (e.g., all numeric, all alphabetic, alphanumeric)
Combinations	Test to see if the value combinations of two or more data fields are appropriate or make sense (e.g., does the quantity sold make sense given the type of product?)
Expected Values	Test to see if data is what is expected (e.g., match with existing customer names, payment amount, etc.)
Missing Data	Test for existence of data items in all fields of a record (e.g., is there a quantity field on each line item of a customer order?)
Pictures/Templates	Test to assure that data conform to a standard format (e.g., are hyphens in the right places for a student ID number?)
Range	Test to assure data are within proper range of values (e.g., is a student's grade point average between 0 and 4.0?)
Reasonableness	Test to assure data are reasonable for situation (e.g., pay rate for a specific type of employee)
Self-Checking Digits	Test where an extra digit is added to a numeric field in which its value is derived using a standard formula (see Figure 14-14)
Size	Test for too few or too many characters (e.g., is social security number exactly nine digits?)
Values	Test to make sure values come from set of standard values (e.g., two-letter state codes)

Practical experience has also found that correcting erroneous data is much easier to accomplish before it is permanently stored in a system. On-line systems can notify a user of input problems as data are being entered. When data are processed on-line as events occur, it is much less likely that data validity errors will occur and not be caught. In an on-line system, most problems can be easily identified and resolved before permanently saving data to a storage device using many of the techniques described in Table 14-9. However, in systems where inputs are stored and entered (or transferred) in batch, the identification and notification of errors is more difficult. Batch processing systems can, however, reject invalid inputs and store them in a log file for later resolution.

Most of the tests and techniques shown in Table 14-9 are widely used and straightforward. Some of these tests can be handled by data management technologies, such as a database management system (DBMS), to ensure that they are applied for all data maintenance operations. If a DBMS cannot perform these tests, then you must design the tests into program modules. An example of one item that is a bit sophisticated, self-checking digits, is shown in Figure 14-14. The figure provides a description and an outline of how to apply the technique as well as a short example. The example shows how a check digit is added to a field before data entry or transfer. Once entered or transferred, the check digit algorithm is again applied to the field to "check" whether the check digit received obeys the calculation. If it does, it is likely (but not guaranteed, since two different values could yield the same check digit) that no data transmission or entry error occurred. If not equal, then some type of error occurred.

In addition to validating the data values entered into a system, controls must be established to verify that all input records are correctly entered and that they are only processed once. A common method used to enhance the validity of entering batches

Description	Techniques where extra digits are added to a field to assist in verifying its accuracy
Method	1. Multiply each digit of a numeric field by a weighting factor (e.g., 1,2,1,2, . . .). 2. Sum the results of weighted digits. 3. Divide sum by modulus number (e.g., 10). 4. Subtract remainder of division from modulus number to determine check digit. 5. Append check digits to field.
Example	Assume a numeric part number of: 12473 1-2. Multiply each digit of part number by weighting factor from right to left and sum the results of weighted digits: $$\begin{array}{ccccc} 1 & 2 & 4 & 7 & 3 \\ \times 1 & \times 2 & \times 1 & \times 2 & \times 1 \\ \hline 1 + & 4 + & 4 + & 14 + & 3 = 26 \end{array}$$ 3. Divide sum by modulus number. $$26/10 = 2 \text{ remainder } 6$$ 4. Subtract remainder from modulus number to determine check digit. $$\text{check digit} = 10 - 6 = 4$$ 5. Append check digits to field. Field value with appended check digit = 124734

Figure 14-14
Using check digits to verify data correctness

of data records is to create an audit trail of the entire sequence of data entry, processing, and storage. In such an audit trail, the actual sequence, count, time, source location, human operator, and so on are recorded into a separate transaction log in the event of a data input or processing error. If an error occurs, corrections can be made by reviewing the contents of the log. Detailed logs of data inputs are not only useful for resolving batch data entry errors and system audits but also serve as a powerful method for performing backup and recovery operations in the case of a catastrophic system failure. These types of file and database controls are discussed further in Hoffer, Prescott, and McFadden (2002).

Providing Feedback

When talking with a friend, you would be concerned if he or she did not provide you with feedback by nodding and replying to your questions and comments. Without feedback, you would be concerned that he or she was not listening, likely resulting in a less than satisfactory experience. Similarly, when designing system interfaces, providing appropriate feedback is an easy method for making a user's interaction more enjoyable; not providing feedback is a sure way to frustrate and confuse. System feedback can consist of three types:

1. Status information
2. Prompting cues
3. Error or warning messages

N E T   S E A R C H
Users have limits on how long they will patiently wait for a system to respond. Visit http://www.prenhall. com/hoffer to complete an exercise related to this topic.

Status Information Providing status information is a simple technique for keeping users informed of what is going on within a system. For example, relevant status information such as displaying the current customer name or time, placing appropriate titles on a menu or screen, or identifying the number of screens following the current one (e.g., Screen 1 of 3) all provide needed feedback to the user. Providing status information during processing operations is especially important if the operation takes longer than a second or two. For example, when opening a file you might display "Please wait while I open the file" or, when performing a large calculation, flash the message "Working . . ." to the user. Further, it is important to tell the user that besides working, the system has accepted the user's input and the input was in the correct form. Sometimes it is important to give the user a chance to obtain more feedback. For example, a function key could toggle between showing a "Working . . ." message and giving more specific information as each intermediate step is accomplished. Providing status information will reassure users that nothing is wrong and make them feel in command of the system, not vice versa.

Prompting Cues A second feedback method is to display prompting cues. When prompting the user for information or action, it is useful to be specific in your request. For example, suppose a system prompted users with the following request:

READY FOR INPUT:_____

With such a prompt, the designer assumes that the user knows exactly what to enter. A better design would be specific in its request, possibly providing an example, default values, or formatting information. An improved prompting request might be as follows:

Enter the customer account number (123–456–7):____-____-__

Errors and Warning Messages A final method available to you for providing system feedback is using error and warning messages. Practical experience has found that a few simple guidelines can greatly improve their usefulness. First, mes-

TABLE 14-10   Examples of Poor and Improved Error Messages

Poor Error Messages	Improved Error Messages
ERROR 56 OPENING FILE	The file name you typed was not found. Press F2 to list valid file names.
WRONG CHOICE	Please enter an option from the menu.
DATA ENTRY ERROR	The prior entry contains a value outside the range of acceptable values. Press F9 for list of acceptable values.
FILE CREATION ERROR	The file name you entered already exists. Press F10 if you want to overwrite it. Press F2 if you want to save it to a new name.

N E T   S E A R C H
Standard error messages have emerged for Internet-related computing that some believe are very cryptic and difficult to understand. Visit http://www.prenhall.com/hoffer to complete an exercise related to this topic.

sages should be specific and free of error codes and jargon. Additionally, messages should never scold the user and should attempt to guide the user toward a resolution. For example, a message might say "No customer record found for that Customer ID. Please verify that digits were not transposed." Messages should be in user, not computer, terms. Hence, such terms as "end of file," "disk I/O error," or "write protected" may be too technical and not helpful for many users. Multiple messages can be useful so that a user can get more detailed explanations if wanted or needed. Also, error messages should appear in roughly the same format and placement each time so that they are recognized as error messages and not as some other information. Examples of good and bad messages are provided in Table 14-10. Using these guidelines, you will be able to provide useful feedback in your designs. A special type of feedback is answering help requests from users. This important topic is described next.

Providing Help

Designing how to provide help is one of the most important interface design issues you will face. When designing help, you need to put yourself in the user's place. When accessing help, the user likely does not know what to do next, does not understand what is being requested, or does not know how the requested information needs to be formatted. A user requesting help is much like a ship in distress, sending an SOS. In Table 14-11, we provide our SOS guidelines for the design of system help: *S*implicity, *O*rganize, and *S*how. Our first guideline, *simplicity*, suggests that help messages should be short, to the point, and use words that enable understanding. This leads to our second guideline, *organize*, which means that help messages should be written so that information can be easily absorbed by users. Practical experience has found that long paragraphs of text are often difficult for people to understand. A better design organizes lengthy information in a manner easier for users to digest through the use of bulleted and ordered lists. Finally, it is often useful to explicitly *show* users how to perform an operation and the outcome of procedural steps. Figure 14-15 contrasts the designs of two help screens, one employing our guidelines and one that does not.

TABLE 14-11   Guidelines for Designing Usable Help

Guideline	Explanation
Simplicity	Use short, simple wording, common spelling, and complete sentences. Give users only what they need to know, with ability to find additional information.
Organize	Use lists to break information into manageable pieces.
Show	Provide examples of proper use and the outcomes of such use.

Figure 14-15
Contrasting help screens
(a) Poorly designed help display

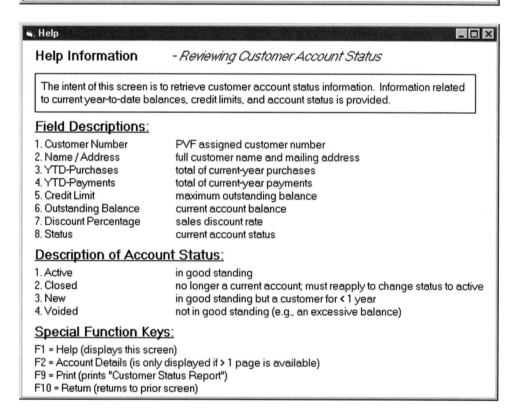

(b) Improved design for help display

Many commercially available systems provide extensive system help. For example, Table 14-12 lists the range of help available in a popular electronic spreadsheet. Many systems are also designed so that users can vary the level of detail provided. Help may be provided at the system level, screen or form level, and individual field level. The ability to provide field level help is often referred to as "context-sensitive" help. For some applications, providing context-sensitive help for all system options is a tremendous undertaking that is virtually a project in itself. If you do decide to design an extensive help system with many levels of detail, you must be sure that you know exactly what the user needs help with, or your efforts may confuse users more

TABLE 14-12   Types of Help

Type of Help	Example of Question
Help on Help	How do I get help?
Help on Concepts	What is a customer record?
Help on Procedures	How do I update a record?
Help on Messages	What does "Invalid File Name" mean?
Help on Menus	What does "Graphics" mean?
Help on Function Keys	What does each Function key do?
Help on Commands	How do I use the "Cut" and "Paste" commands?
Help on Words	What do "merge" and "sort" mean?

than help them. After leaving a help screen, users should always return back to where they were prior to requesting help. If you follow these simple guidelines, you will likely design a highly usable help system.

As with the construction of menus, many programming environments provide powerful tools for designing system help. For example, Microsoft's Help Compiler allows you to quickly construct hypertext-based help systems. In this environment, you use a text editor to construct help pages that can be easily linked to other pages containing related or more specific information. Linkages are created by embedding special characters into the text document that make words hypertext buttons—that is, direct linkages—to additional information. The Help Compiler transforms the text document into a hypertext document. For example, Figure 14-16 shows a hypertext-based help screen from Microsoft's Internet Explorer 5. Hypertext-based help systems have become the standard environment for most commercial Windows-based applications. This has occurred for two primary reasons. First, standardizing system help across applications eases user training. Second, hypertext allows users to selectively access the level of help they need, making it easier to provide effective help for both novice and experienced users within the same system.

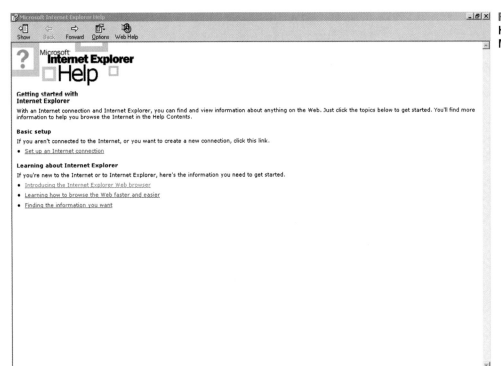

Figure 14-16
Hypertext-based help system from Microsoft's Internet Explorer 5

DESIGNING DIALOGUES

Dialogue: The sequence of
interaction between a user and a
system.

The process of designing the overall sequences that users follow to interact with
an information system is called dialogue design. A **dialogue** is the sequence in which
information is displayed to and obtained from a user. As the designer, your role is to
select the most appropriate interaction methods and devices (described above) and
to define the conditions under which information is displayed and obtained from
users. There are three major steps:

1. Designing the dialogue sequence
2. Building a prototype
3. Assessing usability

There are a few general rules that should be followed when designing a dialogue
and they are summarized in Table 14-13. For a dialogue to have high usability, it must
be consistent in form, function, and style. All other rules regarding dialogue design
are mitigated by the consistency guideline. For example, the effectiveness of how well
errors are handled or feedback is provided will be significantly influenced by consistency in design. If the system does not consistently handle errors, the user will often
be at a loss as to why certain things happen.

One example of these guidelines concerns removing data from a database or file
(see the Reversal entry in Table 14-13). It is good practice to display the information
that will be deleted before making a permanent change to the file. For example, if
the customer service representative wanted to remove a customer from the database,
the system should ask only for the customer ID in order to retrieve the correct customer account. Once found, and before allowing the confirmation of the deletion,

TABLE 14-13 Guidelines for the Design of Human-Computer Dialogues

Guideline	Explanation
Consistency	Dialogues should be consistent in sequence of actions, keystrokes, and terminology (e.g., the same labels should be used for the same operations on all screens, and the location of the same information should be the same on all displays).
Shortcuts and Sequence	Allow advanced users to take shortcuts using special keys (e.g., CTRL-C to copy highlighted text). A natural sequence of steps should be followed (e.g., enter first name before last name, if appropriate).
Feedback	Feedback should be provided for every user action (e.g., confirm that a record has been added, rather than simply putting another blank form on the screen).
Closure	Dialogues should be logically grouped and have a beginning, middle, and end (e.g., the last in the sequence of screens should indicate that there are no more screens).
Error Handling	All errors should be detected and reported; suggestions on how to proceed should be made (e.g., suggest why such errors occur and what user can do to correct the error). Synonyms for certain responses should be accepted (e.g., accept either "t," "T," or "TRUE").
Reversal	Dialogues should, when possible, allow the user to reverse actions (e.g., undo a deletion); data should not be destroyed without confirmation (e.g., display all the data for a record the user has indicated is to be deleted).
Control	Dialogues should make the user (especially an experienced user) feel in control of the system (e.g., provide a consistent response time at a pace acceptable to the user).
Ease	Dialogues should be simple for users to enter information and navigate between screens (e.g., provide means to move forward, backward, and to specific screens, such as first and last).

Adapted from Shneiderman, 1997.

the system should display the account information. For actions making permanent changes to system data files and when the action is not commonly performed, many system designers use the *double-confirmation* technique where the users must confirm their intention twice before being allowed to proceed.

Designing the Dialogue Sequence

Your first step in dialogue design is to define the sequence. In other words, you must first gain an understanding of how users might interact with the system. This means that you must have a clear understanding of user, task, technological, and environmental characteristics when designing dialogues. Suppose that the marketing manager at Pine Valley Furniture (PVF) wants sales and marketing personnel to be able to review the year-to-date transaction activity for any PVF customer. After talking with the manager, you both agree that a typical dialogue between a user and the Customer Information System for obtaining this information might proceed as follows:

1. Request to view individual customer information
2. Specify the customer of interest
3. Select the year-to-date transaction summary display
4. Review customer information
5. Leave system

As a designer, once you understand how a user wishes to use a system, you can then transform these activities into a formal dialogue specification.

A formal method for designing and representing dialogues is **dialogue diagramming**. Dialogue diagrams have only one symbol, a box with three sections; each box represents one display (which might be a full screen or a specific form or window) within a dialogue (see Figure 14-17). The three sections of the box are used as follows:

> **Dialogue diagramming:** A formal method for designing and representing human-computer dialogues using box and line diagrams.

1. *Top:* Contains a unique display reference number used by other displays for referencing it
2. *Middle:* Contains the name or description of the display
3. *Bottom:* Contains display reference numbers that can be accessed from the current display

All lines connecting the boxes within dialogue diagrams are assumed to be bi-directional and thus do not need arrowheads to indicate direction. This means that users are allowed to always move forward and backward between adjacent displays. If you desire only unidirectional flows within a dialogue, arrowheads should be placed on one end of the line. Within a dialogue diagram, you can easily represent the sequencing of displays, the selection of one display over another, or the repeated use

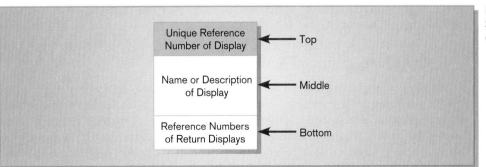

Figure 14-17
Sections of a dialogue diagramming box

Figure 14-18
Dialogue diagram illustrating
sequence, selection, and iteration

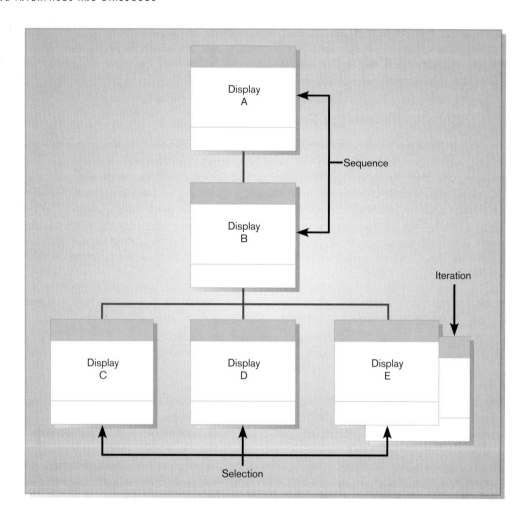

of a single display (e.g., a data entry display). These three concepts—sequence, selection, and iteration—are illustrated in Figure 14-18.

Continuing with our PVF example, Figure 14-19 shows a partial dialogue diagram for processing the marketing manager's request. In this diagram, the analyst placed the request to view year-to-date customer information within the context of the overall Customer Information System. The user must first gain access to the system through a log-on procedure (item 0). If log-on is successful, a main menu is displayed that has four items (item 1). Once the user selects the Individual Customer Information (item 2), control is transferred to the Select Customer display (item 2.1). After a customer is selected, the user is presented with an option to view customer information four different ways (item 2.1.1). Once the user views the customer's year-to-date transaction activity (item 2.1.1.2), the system will allow the user to back up to select a different customer (2.1), return to the main menu (1), or exit the system (see bottom of item 2.1.1.2).

Building Prototypes and Assessing Usability

Building dialogue prototypes and assessing usability are often optional activities. Some systems may be very simple and straightforward. Others may be more complex but are extensions to existing systems where dialogue and display standards have already been established. In either case, you may not be required to build prototypes and do a formal assessment. However, for many other systems, it is critical that you build prototype displays and then assess the dialogue; this can pay numerous divi-

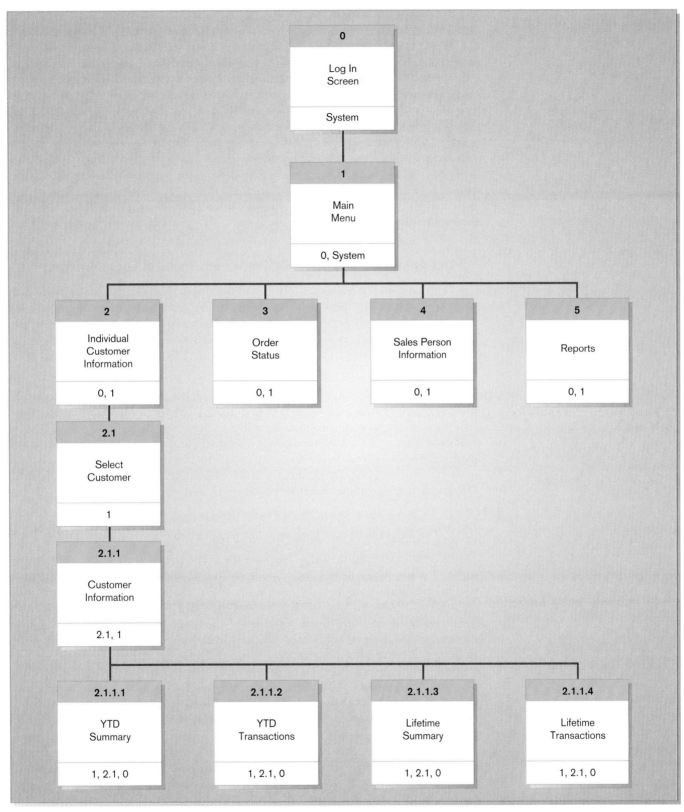

Figure 14-19
Dialogue diagram for the Customer Information System (Pine Valley Furniture)

dends later in the systems development life cycle (for example, it may be easier to implement a system or train users on a system they have already seen and used).

Building prototype displays is often a relatively easy activity for you to perform when using CASE tools or many of the graphical development environments such as Microsoft's Visual Basic. Some systems development environments include easy-to-use input and output (form, report, or window) design utilities. There are also several tools called "Prototypers" or "Demo Builders" that allow you to quickly design displays and show how an interface will work within a full system. These demo systems allow users to enter data and move through displays as if using the actual system. Such activities are not only useful for you to show how an interface will look and feel, they are also useful for assessing usability and for performing user training long before actual systems are completed. In the next section, we extend our discussion of interface and dialogue design to consider issues specific to graphical user interface environments.

Designing Interfaces and Dialogues in Graphical Environments

Graphical user interface (GUI) environments are rapidly growing in popularity. Although all of the interface and dialogue design guidelines presented previously apply to designing GUIs, there are additional issues that are unique to these environments that must be considered. Here, we briefly discuss some of these issues.

Graphical Interface Design Issues   When designing GUIs for an operating environment such as Microsoft Windows or for the Macintosh, there are numerous factors to consider. Some factors are common to all GUI environments while others are specific to a single environment. We will not, however, discuss the subtleties and details of any single environment. Instead, our discussion will focus on a few general truths that experienced designers mention as critical to the design of usable GUIs. In most discussions of GUI programming, two rules repeatedly emerge as comprising the first step to becoming an effective GUI designer:

1. *Become an expert user of the GUI environment.*
2. *Understand the available resources and how they can be used.*

The first step should be an obvious one. The greatest strength of designing within a standard operating environment is that *standards* for the behavior of most system operations have already been defined. For example, how you cut and paste, set up your default printer, design menus, or assign commands to functions have been standardized both within and across applications. This allows experienced users of one GUI-based application to easily learn a new application. Thus, in order to design effective interfaces in such environments, you must first understand how other applications have been designed so that you will adopt the established standards for "look and feel." Failure to adopt the standard conventions in a given environment will result in a system that will likely frustrate and confuse users.

The second rule—gaining an understanding of the available resources and how they can be used—is a much larger undertaking. For example, within Windows you can use menus, forms, and boxes in many ways. In fact, the flexibility with which these resources *can be used* versus the established standards for how most designers *actually use* these resources makes design especially challenging. For example, you have the ability to design menus using all upper-case text, putting multiple words on the top line of the menu, and other nonstandard conventions. Yet, the standards for menu design require that top-level menu items consist of one word and follow a specific ordering. Numerous other standards for menu design have also been established (see Figure 14-20 for illustrations of many of these standards). Failure to follow standard design conventions will likely prove very confusing to users.

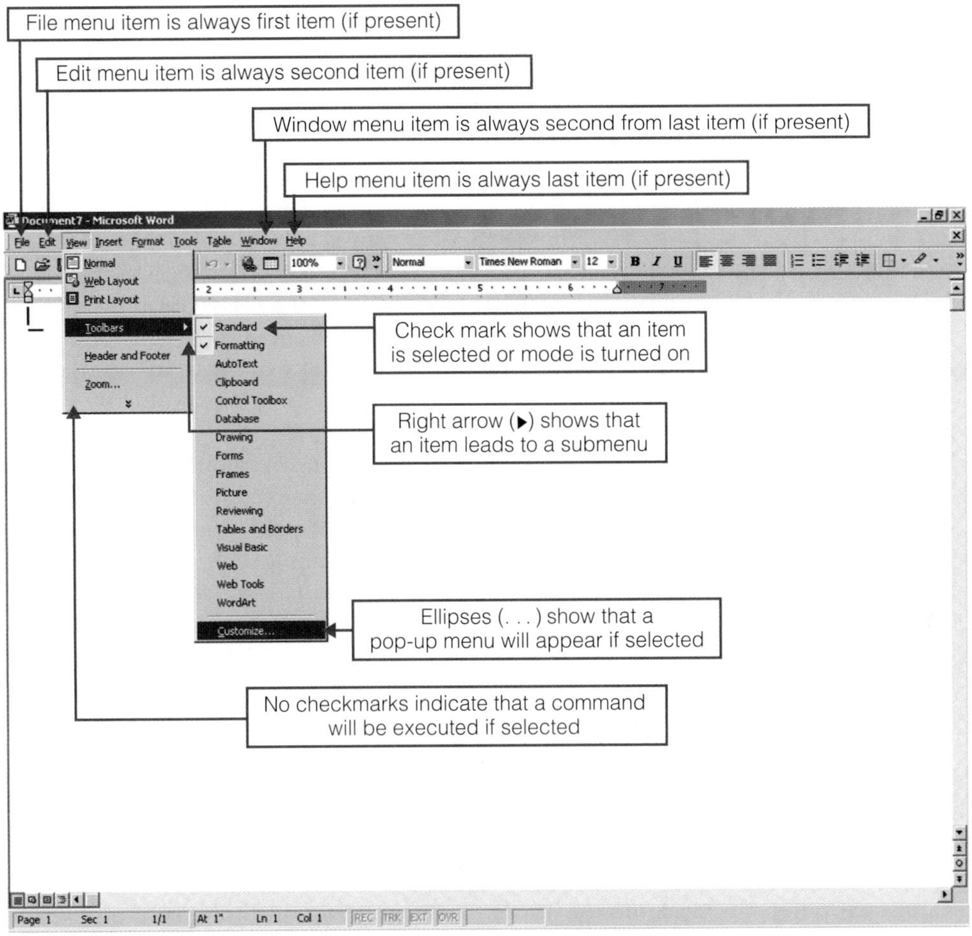

Figure 14-20
Highlighting graphical user
interface design standards

In GUIs, information is requested by placing a window (or form) on the visual display screen. Like menu design, forms can also have numerous properties that can be mixed and matched (see Table 14-14). For example, properties about a form determine whether a form is resizable or movable after being opened. Since properties define how users can actually work with a form, the effective application of properties is fundamental to gaining usability. This means that, in addition to designing the lay-

**TABLE 14-14   Common Properties of Windows and Forms in a Graphical User
Interface Environment that Can Be Active or Inactive**

Property	Explanation
Modality	Requires users to resolve the request for information before proceeding (e.g., need to cancel or save before closing a window).
Resizable	Allows users to resize a window or form (e.g., to make room to see other windows also on the screen).
Movable	Allows users to move a window or form (e.g., to allow another window to be seen).
Maximize	Allows users to expand a window or form to a full screen in size (e.g., to avoid distraction from other active windows or forms).
Minimize	Allows users to shrink a window or form to an icon (e.g., to get the window out of the way while working on other active windows).
System Menu	Allows a window or form to also have a system menu to directly access system-level functions (e.g., to save or copy data).

Adapted from Wagner, 1994.

out of a form, you must also define the "personality" of the form with its characteristic properties. Fortunately, numerous GUI design tools have been developed that allow you to "visually" design forms and interactively engage properties. Interactive GUI design tools have greatly facilitated the design and construction process.

In addition to the issues related to interface design, the sequencing of displays turns out to be a bit more challenging in graphical environments. This topic is discussed next.

Dialogue Design Issues in a Graphical Environment   When designing a dialogue, your goal is to establish the sequence of displays (full screens or windows) that users will encounter when working with the system. Within many GUI environments, this process can be a bit more challenging due to the GUI's ability to suspend activities (without resolving a request for information or exiting the application altogether) and switch to another application or task. For example, within Microsoft Word, the spell checker executes independently from the general word processor. This means that you can easily jump between the spell checker and word processor without exiting either one. Conversely, when selecting the print operation, you must either initiate printing or abort this request before returning to the word processor. This is an example of the concept of "modality" described in Table 14-14. Thus, Windows-type environments allow you to create forms that either *require* the user to resolve a request before proceeding (print example) or *selectively choose* to resolve a request before proceeding (the spell checker). Creating dialogues that allow the user to jump from application to application or from module to module within a given application requires that you carefully think through the design of dialogues.

One easy way to deal with the complexity of designing advanced graphical user interfaces is to require users to *always* resolve all requests for information before proceeding. For such designs, the dialogue diagramming technique is an adequate design tool. This, however, would make the system operate in a manner similar to a traditional non-GUI environment where the sequencing of displays is tightly controlled. The drawback to such an approach would be the failure to capitalize on the task-switching capabilities of these environments. Consequently, designing dialogues in environments where the sequence between displays cannot be predetermined offers significant challenges to the designer. Using tools like dialogue diagramming helps analysts to better manage the complexity of designing graphical interfaces.

PINE VALLEY FURNITURE

ELECTRONIC COMMERCE APPLICATION: DESIGNING INTERFACES AND DIALOGUES FOR PINE VALLEY FURNITURE'S WEBSTORE

Designing the human interface for an Internet-based electronic commerce application is a central and critical design activity. Since this is where a customer will interact with a company, much care must be put into its design. Like the process followed when designing the interface for other types of systems, a prototyping design process is most appropriate when designing the human interface for an Internet electronic commerce system. Although, the techniques and technology for building the human interface for Internet sites is rapidly evolving, several general design guidelines have emerged. In this section, we examine some of these as they apply to design of Pine Valley Furniture's WebStore.

General Guidelines

Over the years, interaction standards have emerged for virtually all of the commonly used desktop computing environments such as Windows or the Macintosh. However, some interface design experts believe that the growth of the Web has resulted in a big step backwards for interface design. One problem, as discussed in Chapter 13, is that countless nonprofessional developers are designing commercial Web applications. In addition to this, there are four other important contributing factors (Johnson, 2000):

- Web's single "click-to-act" method of loading static hypertext documents (i.e., most buttons on the Web do not provide click feedback);
- Limited capabilities of most Web-browsers to support finely grained user interactivity;
- Limited agreed-upon standards for encoding Web content and control mechanisms; and
- Lack of maturity of Web scripting and programming languages as well as limitations in commonly used Web GUI component libraries.

In addition to these contributing factors, designers of Web interfaces and dialogues are often guilty of many design errors. Although not inclusive of all possible errors, Table 14-15 summarizes those errors that are particularly troublesome. Fortunately,

TABLE 14-15   Common Errors when Designing the Interface and Dialogues of Websites

Error	Description
Opening New Browser Window	Avoid opening a new browser window when a user clicks on a link unless it is clearly marked that a new window will be opened; users may not see that a new window has been opened, which will complicate navigation, especially moving backwards.
Breaking or Slowing Down the Back Button	Make sure users can use the back button to return to prior pages. Avoid opening new browser windows, using an immediate redirect where, when a user clicks the back button, they are pushed forward to an undesired location, or prevent caching such that each click of the back button requires a new trip to the server.
Complex URLs	Avoid overly long and complex URLs since it makes it more difficult for users to understand where they are and can cause problems if users want to e-mail page locations to colleagues.
Orphan Pages	Avoid having pages with no "parent" that can be reached by using a back button; requires users to "hack" the end of the URL to get back to some other prior page.
Scrolling Navigation Pages	Avoid placing navigational links below where a page opens, since many users may miss these important options that are below the opening window.
Lack of Navigation Support	Make sure your pages conform to users expectation by providing commonly used icon links such as a site logo at the top or other major elements. Also place these elements on pages in a consistent manner.
Hidden Links	Make sure you leave a border around images that are links, don't change link colors from normal defaults, and avoid embedding links within long blocks of text.
Links that Don't Provide Enough Information	Avoid not turning off link marking borders so that links clearly show which links users have clicked and which they have not. Make sure users know which links are internal anchor points versus external links and indicate if a link brings up a separate browser window from those that do not. Finally, make sure link images and text provide enough information to users so that they understand the meaning of the link.
Buttons that Provide No Click Feedback	Avoid using image buttons that don't clearly change when being clicked; use Web GUI toolkit buttons, HTML form-submit buttons, or simple textual links.

there are numerous excellent sources on how to avoid these and other interface and design errors (Johnson, 2000; Flanders and Willis, 1998; Nielson, 1999; Nielson, 2000; www.useit.com; www.webpagesthatsuck.com).

Designing Interfaces and Dialogues at Pine Valley Furniture

To establish design guidelines for the human-computer interface, Jim and the PVF development team again reviewed many popular electronic commerce Websites. The key feature they wanted to incorporate into the design was an interface with "Menu Driven Navigation with Cookie Crumbs." In order to assure that all team members understood what was meant by this guideline, Jim organized a design briefing to explain how this feature would be incorporated into the WebStore interface.

Menu-Driven Navigation with Cookie Crumbs

After reviewing several sites, the team concluded that menus should stay in the exact same place throughout the entire site. They concluded that placing a menu in the same location on every page will help customers to more quickly become familiar with the site and therefore more rapidly navigate through the site. Experienced Web developers know that the quicker a customer can reach a specific destination at a site, the quicker they can purchase the product they are looking for, or get the information they set out to find. Jim emphasized this point by stating, "these details may seem silly, but the second a user finds themselves 'lost' in our site, they're gone. One mouse click and they're no longer shopping at Pine Valley, but at one of our competitor's sites."

Cookie crumbs: A technique for showing a user where they are in a Website by placing a series of "tabs" on a Web page that shows a user where they are and where they have been.

A second design feature, and one that is being used on many electronic commerce sites, are cookie crumbs. **Cookie crumbs** are a technique for showing a user where they are in the site by placing "tabs" on a Web page that remind the user where they are and where they have been. These tabs are hypertext links that can allow the user to quickly move backwards in the site. For example, suppose that a site is four levels deep, with the top level called "Entrance," the second called "Products," the third called "Options," and the fourth called "Order." As the user moves deeper into the site, a tab is displayed across the top of the page showing the user where she is and giving her the ability to quickly jump backwards one or more levels. In other words, when first entering the store, a tab will be displayed at the top (or some other standard place) of the screen with the word "Entrance." After moving down a level, two tabs will be displayed, "Entrance" and "Products." After selecting a product on the second level, a third level is displayed where a user can choose product options. When this level is displayed, a third tab is produced with the label "Options." Finally, if the customer decides to place an order and selects this option, a fourth level screen is displayed and a fourth tab is displayed with the label "Order." In summary:

Level 1: Entrance

Level 2: Entrance → Products

Level 3: Entrance → Products → Options

Level 4: Entrance → Products → Options → Order

"By using cookie crumbs, the user knows exactly how far they have wandered from 'home'. If each tab is a link, the user can quickly jump back to a broader part of the store should they not find exactly what they are looking for. Cookie crumbs serve two important purposes. First, they allow users to navigate to a point previously visited and will assure that they are not lost. Second, it clearly shows users where they have been and how far they have gone from home."

Summary

In this chapter, our focus was to acquaint you with the process of designing human-computer interfaces and dialogues. Understanding the characteristics of various interaction methods (command language, menu, form, object, natural language) and devices (keyboard, mouse, joystick, trackball, touch screen, light pen, graphics tablet, voice) is a fundamental skill you must master. No single interaction style or device is the most appropriate in all instances: each has its strengths and weaknesses. You must consider characteristics of the intended users, the tasks being performed, and various technical and environmental factors when making design decisions.

The chapter also reviewed design guidelines for computer-based forms. You learned that most forms have a header, sequence or time-related information, instructions, a body, summary data, authorization, and comments. Users must be able to move the cursor position, edit data, exit with different consequences, and obtain help. Techniques for structuring and controlling data entry were presented along with guidelines for providing feedback, prompting, and error messages. A simple, well-organized help function that shows examples of proper use of the system should be provided. A variety of help types were reviewed.

Next, guidelines for designing human-computer dialogues were presented including a description of dialogue diagramming. These guidelines are consistency, allowing for shortcuts, providing feedback and closure on tasks, handling errors, allowing for operations reversal, giving the user a sense of control, and ease of navigation. Designing dialogues and procedures for assessing their usability were also reviewed. Several interface and dialogue design issues were described within the context of designing graphical user interfaces. These included the need to follow standards to provide the capabilities of modality, resizing, moving, maximizing, and minimizing windows, and to offer a system menu choice. This discussion highlighted how concepts presented earlier in the chapter can be applied or augmented in these emerging environments. Finally, interface and dialogue design issues for Internet-based applications were discussed where several common design errors were highlighted.

Our goal was to provide you with a foundation for building highly usable human-computer interfaces. As more and more development environments provide rapid prototyping tools for the design of interfaces and dialogues, many complying to common interface standards, the difficulty of designing usable interfaces will be reduced. However, you still need a solid understanding of the concepts presented in this chapter in order to succeed. Learning to use a computer system is like learning to use a parachute—if a person fails on the first try, odds are he or she won't try again (Blattner and Schultz, 1988). If this analogy is true, it is important that a user's first experience with a system be a positive one. By following the design guidelines outlined in this chapter, your chances of providing a positive first experience to users will be greatly enhanced.

Key Terms

1. Command language interaction
2. Cookie crumbs
3. Dialogue
4. Dialogue diagramming
5. Drop-down menu
6. Form interaction
7. Icon
8. Interface
9. Menu interaction
10. Natural language interaction
11. Object-based interaction
12. Pop-up menu

Match each of the key terms above with the definition that best fits it.

_____ A method by which users interact with information systems.

_____ A human-computer interaction method where users enter explicit statements into a system to invoke operations.

_____ A formal method for designing and representing human-computer dialogues using box and line diagrams.

_____ A menu positioning method that places a menu near the current cursor position.

_____ A human-computer interaction method where a list of system options is provided and a specific command is invoked by user selection of a menu option.

_____ A technique for showing a user where they are in a Website by placing a series of "tabs" on a Web page that shows a user where they are and where they have been.

_____ A menu positioning method that places the access point of the menu near the top line of the display; when accessed, menus open by dropping down onto the display.

_____ A highly intuitive human-computer interaction method whereby data fields are formatted in a manner similar to paper-based forms.

—— A human-computer interaction method where symbols are used to represent commands or functions.

—— Graphical pictures that represent specific functions within a system.

—— A human-computer interaction method whereby inputs to and outputs from a computer-based application are in a conventional speaking language such as English.

—— The sequence of interaction between a user and a system.

Review Questions

1. Contrast the following terms:
   a. dialogue, interface
   b. command language interaction, form interaction, menu interaction, natural language interaction, object-based interaction
   c. drop-down menu, pop-up menu

2. Describe the process of designing interfaces and dialogues. What deliverables are produced from this process? Are these deliverables the same for all types of system projects? Why or why not?

3. Describe five methods of interacting with a system. Is one method better than all others? Why or why not?

4. Describe several input devices for interacting with a system. Is one device better than all others? Why or why not?

5. Describe the general guidelines for the design of menus. Can you think of any instances when it would be appropriate to violate these guidelines?

6. List and describe the general sections of a typical business form. Do computer-based and paper-based forms have the same components? Why or why not?

7. List and describe the functional capabilities needed in an interface for effective entry and navigation. Which capabili-

ties are most important? Why? Will this be the same for all systems? Why or why not?

8. Describe the general guidelines for structuring data entry fields. Can you think of any instances when it would be appropriate to violate these guidelines?

9. Describe four types of data errors.

10. Describe the methods used to enhance the validity of data input.

11. Describe the types of system feedback. Is any form of feedback more important than the others? Why or why not?

12. Describe the general guidelines for designing usable help. Can you think of any instances when it would be appropriate to violate these guidelines?

13. What steps do you need to follow when designing a dialogue? Of the guidelines for designing a dialogue, which is most important? Why?

14. Describe the properties of windows and forms in a graphical user interface environment. Which property do you feel is most important? Why?

15. List and describe the common interface and dialogue design errors found on Websites.

Problems and Exercises

1. Consider software applications that you regularly use that have menu interfaces, whether they be PC- or mainframe-based applications. Evaluate these applications in terms of the menu design guidelines outlined in Table 14-1.

2. Consider the design of a registration system for a hotel. Following design specification items in Figure 14-2, briefly describe the relevant users, tasks, and displays involved in such a system.

3. Imagine the design of a system used to register students at a university. Discuss the user, task, system, and environmental characteristics (see Table 13-9) that should be considered when designing the interface for such a system.

4. For each of the following interaction methods recall a software package that you have used recently: command language, menus, and objects. What did you like and dislike about each package? What were the strengths and weaknesses of each? Which type do you prefer for which circumstances? Which type do you believe will become most prevalent? Why?

5. Briefly describe several different business tasks that are good candidates for form-based interaction within an information system.

6. List the physical input devices described in this chapter that you have seen or used. For each device, briefly describe your experience and provide your personal evaluation. Do your personal evaluations parallel the evaluations provided in Tables 14-3 and 14-4?

7. Propose some specific settings where natural language interaction would be particularly useful and explain why.

8. Examine the help systems for some software applications that you use. Evaluate each using the general guidelines provided in Table 14-11.

9. Design one sample data entry screen for a hotel registration system using the data entry guidelines provided in this chapter (see Table 14-6). Support your design with arguments for each of the design choices you made.

10. Describe some typical dialogue scenarios between users and a hotel registration system. For hints, reread the section in this chapter that provides sample dialogue between users and the Customer Information System at Pine Valley Furniture.

11. Represent the dialogues from the previous question through the use of dialogue diagrams.

12. List four contributing factors that have acted to impede the design of high-quality interfaces and dialogues on Internet-based applications.

13. Go out to the Internet and find commercial Websites that demonstrate each of the common errors listed in Table 14-15.

Field Exercises

1. Research the topic "natural language" at your library. Determine the status of applications available with natural language interaction. Forecast how long it will be before natural language capabilities are prevalent in information systems use.

2. Examine two PC-based graphical user interfaces (e.g., Microsoft's Windows and Macintosh). If you do not own these interfaces, you are likely to find them at your university, workplace, or at a computer retail store. You may want to supplement your hands-on evaluation with recent formal evaluations published in magazines such as *PC Magazine*. In what ways are these two interfaces similar and different? Are these interfaces intuitive? Why or why not? Is one more intuitive than the other? Why or why not? Which interface seems easier to learn? Why? What types of system requirements does each interface have, and what are the costs of each interface? Which do you prefer? Why?

3. Interview a variety of people you know about the various ways they interact, in terms of inputs, with systems at their workplaces. What types of technologies and devices are used to deliver these inputs? Are the input methods and devices easy to use and do they help these people complete their tasks effectively and efficiently? Why or why not? How could these input methods and devices be improved?

4. Interview systems analysts and programmers in an organization where graphical user interfaces are used. Describe the ways that these interfaces are developed and used. How does the use of such interfaces enhance or complicate the design of interfaces and dialogues?

References

Apple Computer. 1993. *Macintosh Human Interface Guidelines.* Reading, MA: Addison-Wesley.

Blattner, M., and E. Schultz. 1988. "User Interface Tutorial." Presented at the 1988 Hawaii International Conference on System Sciences, Kona, Hawaii, January, 1988.

Flanders, V., and Willis, M. 1998. *Web Pages That Suck: Learn Good Design by Looking at Bad Design.* Alameda, CA: Sybex Publishing.

Hoffer, J. A., M. B. Prescott, and F. R. McFadden. 2002. *Modern Database Management.* 6th ed. Upper Saddle River, NJ: Prentice Hall.

Johnson, J. 2000. *GUI Bloopers: Don'ts and Do's for Software Developers and Web Designers.* San Diego, CA: Academic Press.

McKay, E. N. 1999. *Developing User Interfaces for Microsoft Windows.* Redmond, WA: Microsoft Press.

Nielsen, J. 1999. "User Interface Directions for the Web." *Communications of the ACM* 42(1): 65–71.

Nielsen, J. 2000. *Designing Web Usability: The Practice of Simplicity.* Indianapolis, IN: New Riders Publishing.

Shneiderman, B. 1997. *Designing the User Interface: Strategies for Effective Human-Computer Interaction.* 3rd ed. Reading, MA: Addison-Wesley.

Sun Microsystems. 1999. *Java Look and Feel Design Guidelines.* Reading, MA: Addison-Wesley.

Wagner, R. 1994. "A GUI Design Manifesto." *Paradox Informant* 5 (June): 36–42.

BROADWAY ENTERTAINMENT COMPANY, INC.

Designing the Human Interface for the Customer Relationship Management System

CASE INTRODUCTION

The students from Stillwater State University are almost ready to build a prototype of MyBroadway, the Web-based customer relationship management system for Carrie Douglass, manager of the Centerville, Ohio, Broadway Entertainment Company (BEC) store. The team has decided on the style and specific design for individual pages on the Website. Before building the prototype in Microsoft Access, the team is ready to plan the structure for the navigation between the pages of the system. For a Web-based system, the human interface is, to the customer, the system. As with the page designs, the students decide to do a pencil-and-paper prototype before development in Access. This initial prototype will be used primarily for discussion among the team and for sharing with other teams in their information systems projects class at Stillwater.

DESIGNING THE DIALOGUE BETWEEN MYBROADWAY AND USERS

The human interfaces for MyBroadway are clearly visible from the system's context diagram (see BEC Figure 8–1 at the end of Chapter 8). The main human interfaces are data flows from and to each human external entity—customers and employees. The team has already designed one or more pages for these data flows. The team quickly realizes that these data flows are not the only human interfaces. Any Web page that is needed in a navigation path leading to these data flows is also a human interface. For example, to produce the Inventory Review output page, the customer must enter criteria for selecting which inventory items to display and must get to these pages from the welcome page.

The student team decides that the home or welcome page should contain a catchy graphic and menu selections for accessing different parts of the system. One way to categorize system functions would logically be to group all inputs, or data entry pages, together and all system outputs, or form and report display pages,

together into a second group. The team decides, however, that this is a system-centric view, not a user-centric view of the system's functionality. After some brainstorming, the team decides that it would be more logical for users to understand and use the system if pages were grouped by the type of data the users want to use. There appear to be two natural data groupings: product and purchase/rental data.

BEC Figure 14-1 is a dialogue diagram that represents the relationships between system Web pages developed by the team using a data orientation for human interfaces. Page 0 is the welcome page. Besides information to introduce MyBroadway to customers, this page provides menu options or buttons for the user to indicate which data group he or she wants to use. If the user wants to work with product data, then page 1 provides the user a way to enter the request for a new product (page 1.3, which is input 2 from BEC Table 8-1) or to work with existing product data (pages 1.1 and 1.2). Page 1.1 guides the user either to enter a new comment on a product (page 1.1.1 for input 1) or to view existing product comments (page 1.1.2 for output 2). Page 1.1, thus, must provide a way for the user to select or enter data to identify the product for use in subordinate pages.

In general, each system input or output is a terminal (or leaf) node of the dialogue diagram. Each superior node above a leaf is a step for guiding the user to a system input or output. Sometimes a system output can be the basis for a customer to create a system input. For example, consider pages 2.1, 2.1.1, and 2.1.1.1. Page 2.1 is the Rental Status report, output 3. The team decides that users will want to see this report before requesting an extension to a particular rental, which is done on page 2.1.1, representing input 3. Thus, page 2.1 not only displays the Rental Status report but also provides a way for a user to select a particular outstanding rental for which to request an extension in page 2.1.1. Page 2.1.1.1 is a message page (possibly not a totally separate page but rather a message window to overlay on top of page 2.1.1) that will say whether the extension request is accepted.

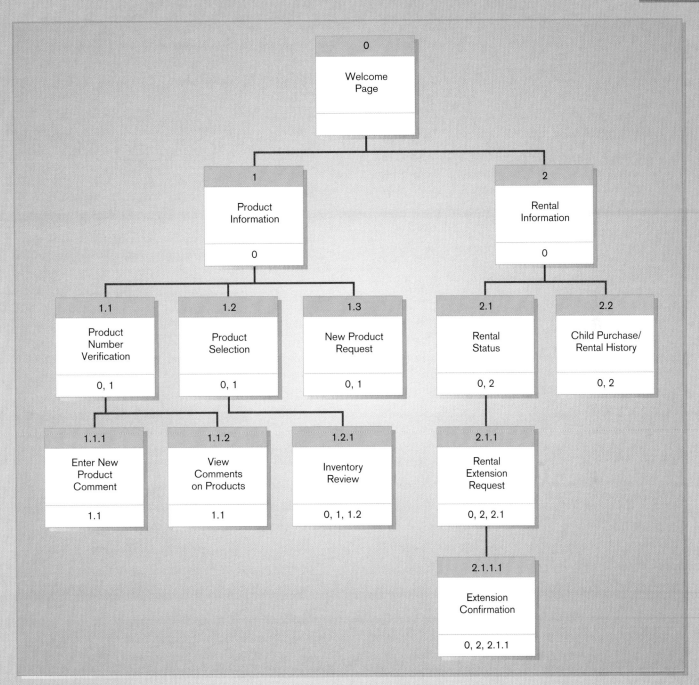

BEC Figure 14-1
Dialogue diagram for MyBroadway

CASE SUMMARY

The student team feels that the design of the user dialogue represents a suitable project deliverable that should be reviewed by someone outside the team. Thus, they schedule a structured walkthrough of their dialogue design along with the design of pages, which they did previously (see the BEC case at the end of Chapter 13).

Because the team has not invested a great deal of time in the initial design, the team members believe they can be open to, not defensive in response to, constructive suggestions. Because other teams will also likely walk through their dialogue designs, the BEC team will see some other creative designs, which will give additional ideas for improvement.

CASE QUESTIONS

1. Using guidelines from this chapter and other sources, evaluate the usability of the dialogue design depicted in BEC Figure 14-1. Specifically, consider the overall organization, grouping of pages, navigation paths between pages, and depth of the dialogue diagram and how this depth might affect user efficiency.

2. Are there any missing pages in BEC Figure 14-1? Can you anticipate the need for additional pages in the customer interface for MyBroadway? If so, where do these pages come from if not from the list of system inputs and outputs listed in BEC Table 14-1?

3. Chapter 14 encourages the design of a help system early in the design of the human interface. How would you incorporate help into the interface as designed by the Stillwater students?

4. The designs for pages 1.2 and 1.2.1 include a Back button. Is this button necessary or desirable?

5. Are there any other possible navigation paths exiting page 1.2 that are not shown on BEC Figure 14-1? Is page 1.2.1 the only possible result of searching on the selection criteria? If not, design pages for other results and the navigation paths to these pages.

6. BEC Figure 14-1 indicates a multi-level tree menu design for MyBroadway. Do you see opportunities for other menu configurations? Explain. Do you see opportunities for interaction methods other than menus? Explain.

7. Based on the guidelines in Chapter 14 on designing interfaces, return to the page designs from the BEC case in Chapter 13 and reevaluate those page layouts. Do you now see any additional issues or do you have any specific recommendations for those pages based on the material in this chapter?

8. What types of errors might users of MyBroadway make? Design error messages and a way to display those messages. Justify your design based on guidelines in Chapter 14.

9. Now that you have studied both the form and report design and the navigation design developed by the Stillwater students for MyBroadway, evaluate the overall usability of their design. Consider the definition of usability in Chapter 13 as well as the guidelines in Chapters 13 and 14. Suggest changes to make the system more usable.

Chapter 15

Finalizing Design Specifications

LEARNING OBJECTIVES

After studying this chapter, you should be able to:

- Discuss the need for system design specifications.
- Define quality requirements and write quality requirements statements.
- Read and understand a structure chart.
- Distinguish between evolutionary and throwaway prototyping.
- Explain the role of CASE tools in capturing design specifications.
- Discuss other methods for representing design specifications.
- Demonstrate how design specifications can be declared for Web-based applications.

INTRODUCTION

There is no question that, in today's fast-paced and dynamic business environment, speed has become a major part of doing business. That applies to systems development just as much as it does to other areas of business. Systems must be developed more and more quickly. Given the emphasis on speed, and the tools that are currently available to support analysts in the development process, the lines between analysis and design, and between design and implementation, are blurring.

The systems development life cycle we use in this book has clear breaks between design and implementation. According to this SDLC, coding and testing take place in the implementation phase. One reason for this is historical. Traditionally, detailed design specifications were completed by analysts at the end of the design phase and handed over to programmers for coding at the beginning of the implementation phase. The break between phases was pretty clearly defined. These specifications were paper-based and contained thorough descriptions of the different software modules that were to comprise the new system, along with detailed descriptions of their functions. Programmers then coded and tested the new system based on these design specifications. Today, however, with CASE tools, prototyping, and other approaches to systems development, the design specifications are delivered to the programmers in forms other than paper, such as a working prototype in executable code generated from a CASE tool. At least some of the coding that would traditionally occur in the implementation phase, then, has already taken place in the design phase. Many times these prototypes go on to become the basis for the final production system. The desire for rapid development has caused these two basic phases of the SDLC to overlap.

Whether design and implementation overlap or not on the projects you will be assigned to as a systems analyst, you will still have the responsibility for transforming everything you have learned about the new system up to this point into design specifications. You will take what you know about the system's input, output, interface, dialog, and database, as well as what you have determined it will take to make the interface operable, generate output, and access and update the organization's databases, and you will turn all of that into design specifications. The many choices you have for finalizing these design specifications is the subject of this chapter. In the pages that follow, you will read about traditional paper-based design specification documents, and about a graphical way of presenting basis system organization, called structure charts. You will learn about the increasingly common role prototypes play in representing design specifications, both as evolutionary prototypes and throwaway prototypes. To illustrate how CASE tools can be used to generate prototypes, we will show how Oracle's Designer can generate physical databases and the software modules that make up a working prototype. Finally, you will learn about development approaches that combine elements of design and implementation, called "eXtreme Programming" and Rapid Application Development (RAD), and how they represent design specifications.

FINALIZING DESIGN SPECIFICATIONS

Much of the work in analysis and in the early stages of design involves gathering and organizing an enormous amount of information about the target system. In earlier chapters, you have seen how a lot of this information has been structured as diagrams and models and as initial designs for forms, reports, interfaces, and dialogs. Before the task of detailed system construction can be passed on to programmers, however, many more aspects of the system must be specified. The result of all of this detailed work in the design specification document, also known as a software requirements specification. The specification document is one of the major deliverables from the design phase of the systems development life cycle (Figure 15-1).

The Process of Finalizing Design Specifications

Though time-consuming, the process of developing detailed design specifications is extremely important. It is much easier, and much cheaper, to detect and correct errors at this point than it is to detect and correct errors during implementation. Your job now is to take the logical design information and turn it into a blueprint for the physical information system. There are several guidelines you can follow to help you in your conversion from logical to physical design (Wiegers, 1999). The first set of guidelines deals with writing quality specification statements (Table 15-1). The second set of guidelines deals with the quality of the specifications themselves (Table 15-2).

As you can see from Table 15-1, there are six different characteristics of a quality specification statement. Notice that all six characteristics require considerable knowledge, not just about a particular requirement, but also about the system and its environment, on the part of the author. To know if a requirement is correct, the author of the requirement statement must understand the system, its users, and their needs. To determine if a requirement is feasible, the author of the statement must understand the operating environment into which the system will be implemented. To

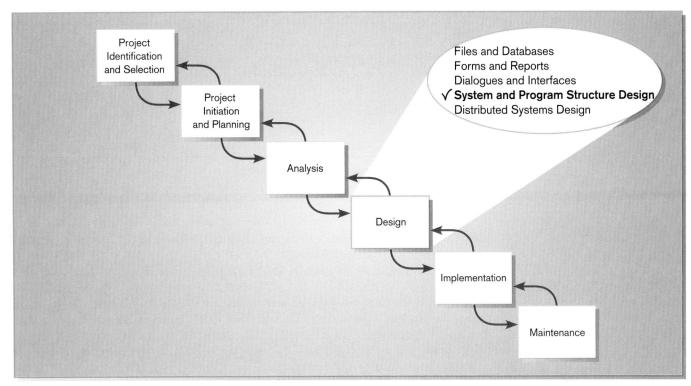

Figure 15-1
The systems development life cycle with design phase highlighted

know if a requirement is necessary, the author of the requirement statement must have a good understanding of the difference between what users want and what they really need. The same type of analysis and knowledge is also true for the other three requirements in Table 15-1. At this point in the design phase of the SDLC, however, the analyst writing requirements statements should have the high level of familiarity with the system to be able to write high-quality requirement statements.

Table 15-2 lists the characteristics of quality requirements, which is different from the characteristics of quality requirement statements. It's possible for a design specifications document to be filled with quality requirements that are poorly represented by the requirements statements that describe them. It is also possible to have extremely well-written requirements statements that describe poorly conceptualized requirements. In short, a quality requirement is completely described, does not con-

TABLE 15-1 **Characteristics of Quality Requirement Statements**

Characteristic	Definition
Correct	Each statement should accurately describe the functionality to be developed.
Feasible	Each requirement must be possible, given the capabilities and constraints of the system and the environment.
Necessary	Each requirement must be something the users really need.
Prioritized	Each requirement should be assigned a priority rating, which reflects how important it is to the final product.
Unambiguous	Each requirement should be clear to anyone who reads its description. There should be only one way to interpret each requirement statement.
Verifiable	It should be possible to determine if each requirement has been successfully implemented in the system.

(Source: From Wiegers, 1999, used with permission.)

TABLE 15-2   **Characteristics of Quality Requirements**

Characteristic	Definition
Complete	A quality requirement is not missing any key description information.
Consistent	A quality requirement does not conflict with any other requirement specified for the system.
Modifiable	A quality requirement can be altered, with a history kept of the changes that are made. Each quality requirement, then, must be unambiguously labeled and kept separate from other requirements.
Traceable	A quality requirement must be traceable to its original source.

(Source: From Wiegers, 1999, used with permission.)

TABLE 15-3   **Contents of a Design Specification Document**

Overall system description
- Functions
- Operating environment
- Types of users
- Constraints
- Assumptions and dependencies

Interface requirements
- User interfaces
- Hardware interfaces
- Software interfaces
- Communication interfaces

System features
- Feature 1 description
- Feature 2 description
- . . .
- Feature n description

Nonfunctional requirements
- Performance
- Safety
- Security
- Software quality
- Business rules

Other requirements

Supporting diagrams and models

(Source: Based on System Requirements Specifications found at www.processimpact.com/goodies.shtml; used with permission.)

flict with other requirements, can be easily changed without adversely affecting other requirements, and can be traced back to its origin.

Here is an example of a poorly written requirements statement for a low-quality requirement (Wiegers, 1999): "The product shall provide status messages at regular intervals not less than every 60 seconds." The statement is ambiguous. We can't tell what "the product" is, and the information about the timing of messages is not at all clear. In addition, the requirement itself is incomplete. The description is missing a lot of information that a developer would need to convert this specification into a working system feature. Here is a suggested solution:

1. Status Messages.

1.1. The Background Task Manager shall display status messages in a designated area of the user interface at intervals of 60 plus or minus 10 seconds.

1.2. If background processing is progressing normally, the percentage of the background task processing that has been completed shall be displayed.

1.3. A message shall be displayed when the background task is completed.

1.4. An error message shall be displayed if the background task has stalled (Wiegers, 1999).

This solution solves the problems with both ambiguity and lack of completeness. The product has been clearly identified, the information about the timing of the error messages is clear, and the information necessary to implement the feature has been provided.

Deliverables and Outcomes

There is really only one key deliverable for finalizing the design specifications: a set of physical design specifications for the entire system, which also includes detailed specifications for each separate part of the system. The design specifications include functional descriptions for each part of the system, as well as information about input received and output generated for each program and its component parts. It doesn't matter what form the specifications take: All of the different relevant aspects of the system must be included. Table 15-3 provides more detailed information on the contents of a design specification.

Table 15-3 provides at least a starting point for considering all of the information that must be included in a design specification document. The contents of the table are by no means exhaustive. Every organization that requires design specification documents will have a different list of items that must be included, and each list will be organized a little differently. The point to remember is that a complete design specification is comprehensive. In the rest of the chapter, we introduce some ways to represent design specifications, starting in the next section with traditional methods.

TRADITIONAL METHOD

Traditional methods for representing design specifications date back to pre-CASE days, when there were no other choices for representing specifications than to prepare comprehensive documents that described all of the design particulars for the system being developed. These documents were written in natural language and augmented with various graphical models (Wiegers, 2000). As you might imagine, generating the specifications took an enormous amount of time and energy, but it was just as necessary as generating all of the detailed documents needed to move from architectural designs for a building to the documents that describe its physical implementation. In the sections that follow, we will briefly discuss design specification documents and one particular method of representing part of the specifications in graphical format, called structure charts.

Specification Documents

Information systems are extremely complicated things, especially in today's distributed and Internet-oriented environment. There are many pieces that must work together, and many different parts of the organization that are affected by a single system. Providing detailed written design specifications, then, is quite a chore. To get an indication of just how much information needs to be provided in a specification document, examine the partial list of information required for such a document for the government of the state of Maryland (Figure 15-2). Given the time and effort necessary to collect and spell out all of this information, it should not be surprising that organizations and developers have been looking for ways to make the creation and maintenance of these documents faster and cheaper. One method that has been developed is the computer-based requirements management tool (Figure 15-3). These tools make it easier for analysts to keep requirements documents up-to-date, to add additional information about specifications, and to define links between different parts of the overall specifications package. Another method for getting control of specification documents is to manifest some of the requirements in digital form, as can be done with a CASE tool, a visual development environment, or a working prototype. One reason that the boundaries between design and implementation can blur is that many design specifications are increasingly being captured in computer-based form.

Given the volume of most design specifications documents, it's common for many aspects of them to be represented in graphical form. Although there are many graphical techniques available, a particularly useful tool for representing the overall hierarchical structure of an information system is a structure chart, the topic of the next section.

Structure Charts

Just as you can diagram how information flows through a system and how data entities are related to each other, you can also diagram how the various program parts of an information system are physically organized. The most common architecture for representing the physical structure of a system is hierarchical. A hierarchy looks like an inverted tree or organization chart with one root or main routine at the top and multiple levels of other modules nested underneath. A **structure chart** shows how an information system is organized in a hierarchy of components, called modules. The purpose of a structure chart is to show graphically how the parts of a system or program are related to each other, in terms of passing data and in terms of the basic components of structured programming: sequence, selection, and repetition. A structure chart redefines the flow and processing of data from data flow diagrams into a structure of system components that follow certain principles of good program design.

N E T   S E A R C H
Design specification is a process that requires much attention to detail. Visit http://www.prenhall.com/hoffer to complete an exercise related to this topic.

N E T   S E A R C H
There are many requirement management tools other than the one briefly mentioned in this chapter. Visit http://www.prenhall.com/hoffer to complete an exercise related to this topic.

Structure chart: Hierarchical diagram that shows how an information system is organized.

Figure 15-2
State of Maryland Structured Systems Development Life-Cycle Methodology

Planning Phase

The planning phase consists of formulating and formalizing the new system's functional, quality, and architecture requirements, and then designing the system to meet those requirements. The agency must identify and assess the critical system specification issues before actual system development begins. The exact degree of detail should be appropriate for the size, type, and scope of the project.

The planning phase includes the following steps.

[Note: This is a partial list of activities in the Planning Phase, which includes Analysis and Design activities. The activities listed here pertain to design specifications.]:

1. Technical design.
    1. Define application architecture.
        1. Review architectural model.
            1. Application style.
            2. Distribution.
            3. Class of service.
        2. Resolve architectural issues.
            1. Select program models.
            2. Select approach to reuse.
            3. Select external interface style.
            4. Select approach to recovery processing.
            5. Define data and process distribution guidelines.
            6. Define usability features.
        3. Map requirements to application architecture.
        4. Document application architecture.
    2. Distribute data and processes over system network.
        1. Study, in depth, the consequences of putting certain processes or data at certain locations.
    3. Define messages and processing flow.
        1. Identify all programs (batch, on-line, client, server, etc.).
        2. Identify processing sequence.
        3. Identify interfaces with other systems.
        4. Design program and file controls.
        5. Identify candidates for reuse.
        6. Document resulting process flow.
    4. Design logical database.
        1. Choose data management software for each entity.
        2. Design Relational Database Management System (RDBMS) structures.
        3. Design file structures.
        4. Review and update data models as required with Database Administrator, system developers, and users.
    5. Design automated processes.
        1. Design each program.
        2. Develop database access and Input/Output (I/O) operations.
    6. Design system interfaces.
        1. Design interfaces.
            1. Type (shared database, file, message).
            2. Record or message layout.
            3. Exchange protocol.
            4. Controls.
        2. Create change request for other systems.
    7. Design IS operations processes.

(continues)

      1. Design I/O control processes.
      2. Design job control for batch programs.
      3. Design backup, restart, and recovery procedures.
      4. Outline operations manual.
  8. Design physical database.
      1. Design physical RDBMS structures.
          1. Design grouping scheme (tablespace, databases).
          2. Design indexes.
          3. Denormalize where appropriate.
      2. Design database file structures.
      3. Review design with appropriate parties.
2. Migration design.
  1. Plan training curriculum.
      1. Determine training needs.
      2. Determine how to train.
      3. Develop training program and schedule.
  2. Design testing approach.
      1. Identify required testing stages (unit through acceptance).
      2. Determine testing approach in each stage.
          1. Test objectives
          2. Requirements to test.
          3. Who will execute.
          4. Environment.
          5. Regression testing.
          6. Test data.
          7. Testing tools.
      3. Develop test plan addressing the following areas.
          1. Specific test objectives and expected results.
          2. System requirements.
          3. Programs.
          4. Security controls.
          5. Performance and response times.
          6. Telecommunications.
          7. System output.
          8. System documentation.
          9. Backup and recovery procedures.
         10. Methods for evaluating results.
  3. Design implementation configuration.
      1. Identify and describe the stages in implementing the system.
          1. Structure.
          2. Interfaces.
          3. Environment changes.
  4. Design data conversion processes.
      1. Identify sources of data for each database.
      2. Identify integrity constraints.
      3. Identify interface requirements.
      4. Design data conversion work flow.
          1. Data validation.
          2. Data cleanup.
          3. Data loading.
      5. Design data conversion programs.
      6. Plan for data conversion contingencies.
      7. Obtain management sign-off for plan

Source: http://www.dbm.state.md.us/mdplan/mplan.htm#toc

Figure 15-3
A screen from DOORS Enterprise Requirements Suite (a product of Quality Systems and Software, Inc.).

Source: www.qssinc.com; *used by permission.*

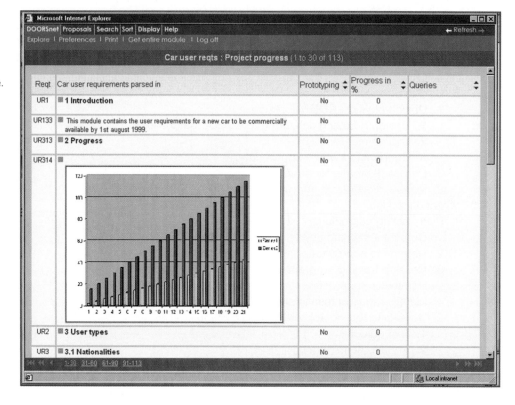

Structure charts are used to show the breakdown of a system into programs and the internal structure of programs written in third- and fourth-generation languages. The structure of programs written in newer object-oriented or event-driven programming languages is usually depicted by state-transition diagrams and Structured English, which we discuss in Chapters 20 and 9, respectively.

Module: A self-contained component of a system, defined by function.

A **module** is a relatively small unit of a system that is defined by its function. Modules are self-contained system components. As much as possible, all of the computer instructions contained in a module should contribute to the same function. Modules are executed as units and, in most instances, have a single point of entry and a single point of exit. Standard programming terms used to identify modules include COBOL sections, paragraphs, or subprograms, and subroutines in BASIC or FORTRAN. For object-oriented programming languages, a module would be roughly a method. A computer program is typically made up of several modules. Modules may also represent separately compiled programs, subprograms, or identifiable internal procedures.

In a structure chart, there is a single coordinating module at the root and on the next level are modules that the coordinating module calls. For on-line systems, you can think of the root as the main menu and each of the subordinate modules as the main menu options. Each of the modules that report to the coordinating module may call additional modules in the next row down. A module calls another, lower-level module when the subordinate module is needed to perform its function. Modules at the lowest levels do not call any other modules; instead they only perform specific tasks. Middle-level modules act as coordinating modules for those lower-level modules they control but may perform some processing as well. In a structure chart, each module is represented as a rectangle containing a descriptive name of its function (see Figure 15-4). Each module name should concisely and accurately reflect what that module does. When naming modules, you should avoid names that include conjunctions because conjunctions indicate that the module performs more than a single function. Modules that seem to need compound names would probably be better represented as multiple modules. Modules are called in order from left to right.

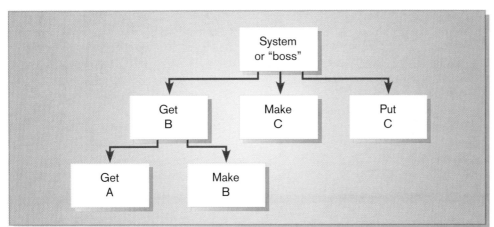

Figure 15-4
An illustration of the hierarchy of a structure chart

Modules in a structure chart communicate with each other through passing parameters. These parameters take the form of data, represented as data couples and flags. A **data couple** shows data being exchanged between two modules. Data couples are drawn as circles with arrows coming out of them (see Figure 15-5a). The arrow indicates the direction of movement of the data couple between modules. A data couple's circle is not filled in. A data couple is usually a single data element, although it can also be a data structure or even an entire record. A **flag** shows control data, or a message, being passed between modules. You represent a flag (also called a control flag) using the same symbol as a data couple except that you fill in the circle. A flag represents information the system needs for processing; a flag itself is never processed. For example, a flag may carry a message such as "End of file" or "Value out of range." A flag should never represent one module telling another module what to do, such as "Write EOF message."

You use special symbols with modules to indicate specific types of processing or special types of modules. A diamond shape at the bottom of a module means that only one of the subordinates attached to the diamond will be called; other modules, called each time, are not attached to the diamond. The diamond indicates there is a conditional statement in the module's code that determines which subordinate module to call (see Figure 15-5b). The diamond is how we show selection in structure charts. If a module's subordinates are called over and over again until some terminal condition, a curved line is drawn through the arrows connecting the module to these subordinates to indicate repetition (see Figure 15-5c). Again, other subordinates may be called only once and, hence, the curved line does not intersect the connection to these modules. If a module is predefined, meaning that its function is dictated by some preexisting part of the system, then you represent the module with a vertical bar drawn down each side (see Figure 15-5d). For example, a predefined module may already exist because it is part of the operating system or because its function is dictated by some piece of hardware necessary for the system to work. Predefined modules are most commonly found on the very bottom layer of a structure chart because the types of functions they perform deal with such things as reading raw data from the original source or printing to a specific type of printer.

In Figure 15-5e, the special symbol connecting the two modules is called a "hat." A hat means that the function in the subordinate module is important logically to the system, but so few lines of code are needed to perform the function that the code itself is actually contained in the superior module. For example, a university registrar's system will definitely contain code for calculating a student's grade point average (GPA). A module that calculates GPA will be a central transform to a registrar's system, so it will appear on a structure chart as a separate module. However, to calculate GPA takes only a few lines of code, not enough to really merit a separate module. The Calculate GPA

Data couple: A diagrammatic representation of the data exchanged between two modules in a structure chart.

Flag: A diagrammatic representation of a message passed between two modules.

Figure 15-5
Special symbols used in structure charts

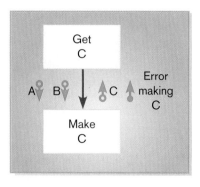

(a) Data couples and control flag

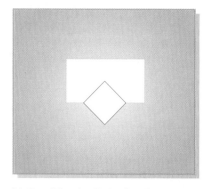

(b) Conditional call of subordinates

(c) Repetitive calls of subordinates

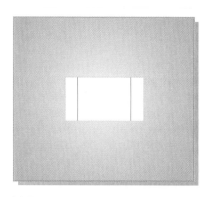

(d) Predefined module

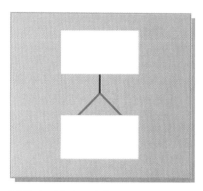

(e) Embedded module

module, then, appears on the structure chart as a separate module logically, but the actual code needed for the function will be contained in the boss module.

Structure charts are read from left to right, and the order in which modules are called is determined not so much by the placement of the modules as it is by the placement of the arrows connecting the modules. Look at Figure 15-6a. The arrows connecting the modules do not overlap. The system module first calls the Get Valid A module. Get Valid A first calls the module Read A, which executes its function and returns the data couple A to its coordinating module, often called the boss module. A is then passed to Validate A, which returns the data couple Valid A. Control is returned to Get Valid A, which then returns the data couple Valid A and control to the boss module. A similar set of steps occurs in connection with the data couple Valid B. Next, Valid A and Valid B are passed to Make C, which returns the data couple C and control to the system module. Finally, the system module calls Put C and passes C to the module.

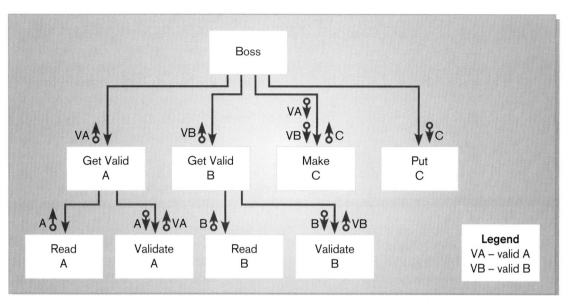

Figure 15-6
**How to read a
structure chart**
(a) Nonoverlapping
arrows

(b) Overlapping
arrows

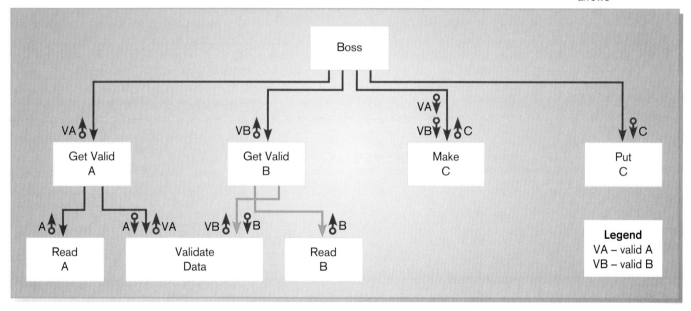

Now look at Figure 15-6b. Here the system designer has decided that the Validate B module is redundant and that the validation process implemented in the module Validate A works just as well for validating B. Therefore, there is only one validation module, called Validate data. Processing continues here as in the structure chart described previously. Notice, however, that the arrows leaving the module Get Valid B now cross over each other. The left-most arrow that leaves the module points to Read B, and the right-most arrow points to Validate Data. Read B is called first, as the arrow that points to it is the left-most arrow leaving Get Valid B. Validate Data is called next. You can also tell the order in which the modules are called by reading the labels—you can't validate B until you have read it into the system.

Once you understand all of the different symbols that make up structure charts, you can begin to use them to represent program designs. The overall program design is just one part of the design specification, however. You also need to describe what happens inside each module in enough detail to allow programmers to begin coding the process. You could use Structured English to represent the processing

Pseudocode: A method for representing the instructions in a module with language very similar to computer programming code.

logic inside each module, but Structured English is too cumbersome and inexact to serve as a reliable source of process function information for programmers. Instead, you should use **pseudocode**, a more exact representation of processing logic. Representing program code as text is the purpose of pseudocode. Think of pseudocode as "almost code," a personalized way to describe a computer program and all of its steps using a language that is not quite a programming language but not quite English either. Pseudocode serves two functions:

1. It helps an analyst think in a structured way about the task a module is designed to perform.

2. It acts as a communication tool between the analyst, who understands what the module needs to do, and the programmer, who must write the code to make the module perform appropriately.

Every analyst develops his or her own version of pseudocode, so there is no one right way to write it. However, some organizations have standardized on a version of pseudocode and require its use. Whether its origin is personal or standard, all pseudocode must have a way to represent at least sequence, conditional statements, and iteration. The actual form pseudocode takes will depend on the programming languages the analyst knows best.

Obviously, preparing detailed design specification documents can take a great deal of time and effort. There is a need for other ways to provide the same amount of information as is contained in specifications, but in a form that is easier to access and manage, and is faster to create and faster to change as needed. Prototyping is one of those ways, and you will read about different approaches to prototyping, as well as how CASE tools can be used to generate prototypes, in the next section.

PROTOTYPING

As we have seen in previous chapters, prototyping is the construction of a model of a system. The practice is borrowed from engineering, where it is common to construct scale models of objects to be built. Models can be physical, shaped from balsawood or clay, for example, or they can be digital, as in computer-aided design (CAD) systems. The models allow the engineers the opportunity to test certain aspects of a design. Prototyping in systems development works the same way. The prototype allows the developer, and the user, to test certain aspects of the overall design, to check for functionality as well as appearance and usability. Remember also that the prototyping process is iterative. Changes in the model are made based on the results of the design tests. As the prototype changes through iterations, more and more of the design specifications for the system are captured in the prototype. The prototype can then serve as the basis for the production system itself, in a process called evolutionary prototyping. Alternatively, the prototype can serves only as a model, which is used as a reference for the construction of the actual system. In this process, called throwaway prototyping, the prototype is discarded after it has been used. Both of these processes are described in the following two sections. The third section illustrates prototype creation using Oracle's Designer CASE tool.

Evolutionary Prototyping

In evolutionary prototyping, you begin by modeling parts of the target system, and if the prototyping process is successful, you evolve the rest of the system from those parts (McConnell, 1996). A life-cycle model of evolutionary prototyping illustrates the iterative nature of the process, and the tendency to refine the prototype until it is ready to release (Figure 15-7). One key aspect of this approach is that the

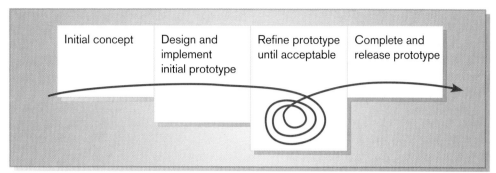

Figure 15-7
McConnell's evolutionary
prototyping model

prototype becomes the actual production system. Because of this, you often start with those parts of the system that are most difficult and uncertain. If the difficult parts of a system can be successfully modeled, then the rest of the system should follow without too much difficulty. If the difficult parts cannot be successfully modeled, then you may want to consider canceling the project. If the prototype is to become the basis for the production system, analysts must choose prototyping languages and tools that are consistent with the development environment for production system.

Although a prototype system may do a great job at representing easy-to-see aspects of a system, such as the user interface, the production system itself will perform many more functions, many of which are transparent or invisible to the users. Any given system must be designed to facilitate database access, database integrity, system security, and networking. Systems also must be designed to support scalability, multi-user support, and multi-platform support. Few of these design specifications will be coded into a prototype. Further, as much as 90 percent of a system's functioning is devoted to handling exceptional cases (McConnell, 1996). Prototypes are designed to handle only the typical cases, so exception handling must be added to the prototype as it is converted to the production system. Clearly, the prototype is only part of the final design specification.

Throwaway Prototyping

Unlike evolutionary prototyping, throwaway prototyping does not preserve the prototype that has been developed. In throwaway prototyping, there is never any intention to convert the prototype into a working system. Instead, the prototype is developed quickly to demonstrate some aspect of a system design that is unclear, or to help users decide between different features or interface characteristics. Once the uncertainty the prototype was created to address has been reduced, the prototype can be discarded, and the principles learned from its creation and testing can then become part of the design specifications documentation. To best support throwaway prototyping, you need to use a language or development environment that is fast and easy to use. The next section illustrates how prototypes can be built using Oracle's Designer CASE tool. The same approach can be followed, whether the prototype is kept as the basis for the production system or discarded.

An Example: Oracle

As we have stated many times previously, one of the key attractions of an integrated CASE tool is its ability to quickly generate working prototypes, moving from high-level logical diagrams through transformations to executable code. Although many products offer this capability, the example we present here uses Oracle's Designer CASE tool. We will use the Hoosier Burger inventory control system as the basis for our example (see Figure 10–18 for the original ERD and Figure 8-16 for the original DFD).

Transforming and Generating the Database   To begin the rapid prototyping process for Hoosier Burger's inventory control system, we go back to the logical diagrams created for modeling the system. We begin with the entity-relationship diagram, first created in Figure 10-18. This ERD has been entered into Designer's Entity Relationship Diagramming tool (Figure 15-8). If you compare Figure 10-18 to Figure 15-8, you'll notice that Figure 15-8 has only four entities instead of seven. You'll also notice that all of the relationships in Figure 15-8 are many-to-many. The reason we have done this is that Designer will create all of the associative entities and their primary keys, as well as all of the foreign keys, once we transform the ERD. We can include the attributes of the associative entities later.

After creating the ERD, we next transform it, using Designer's Database Design Transformer tool. Figure 15-9 shows the transformer tool window. You can see that we have decided to transform all of the entities in the ERD by looking at the radio button chosen under the choice "Run the Transformer in Default Mode." We have decided to transform all of the entities because none have been transformed yet. In the box under "Summary of run set," you see that we have four entities to transform and that none have been mapped to tables yet. Figure 15-10 shows the results of the transformation. You can see in the Output Window that all four of the entities in the ERD have been transformed, and you can also see that three new entities have been created in the process. These three new entities are the associate entities we knew would be created.

The next task we need to perform in Designer is to create and edit a Server Model Diagram, which shows us the transformed ERD (Figure 15-11). The server model looks a lot like a Designer ERD, except all of the foreign keys and associate entities have been included. We can now create any new attributes that belong to the associative entities. Each of the three associate entities in this set require a quantity attribute to be added. Once we are satisfied with the server model diagram and its contents, we are ready to generate the physical database tables that will reside on the database server. As was the case with transformation, generation is a straightforward process. The end result is the set of Database Definition Language (DDL) necessary to create the tables, indexes, constraints, and other objects contained in the server model diagram. Once the DDL gen-

Figure 15-8
Entity-relationship diagram for Hoosier Burger's inventory control system, in Oracle Designer

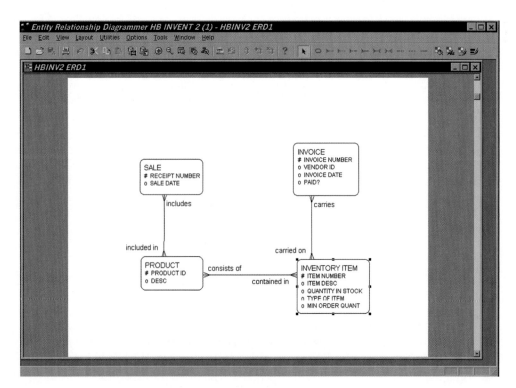

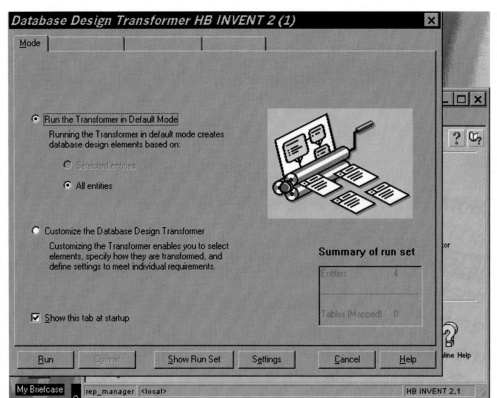

Figure 15-9
Opening screen for the database design transformer in Oracle Designer

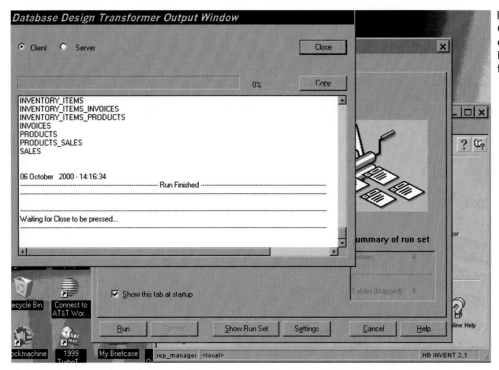

Figure 15-10
Oracle Designer database design output window, after the ERD for Hoosier Burger has been transformed

Figure 15-11
Server model diagram created for the transformed ERD for Hoosier Burger's inventory control system, in Oracle Designer's Design Editor

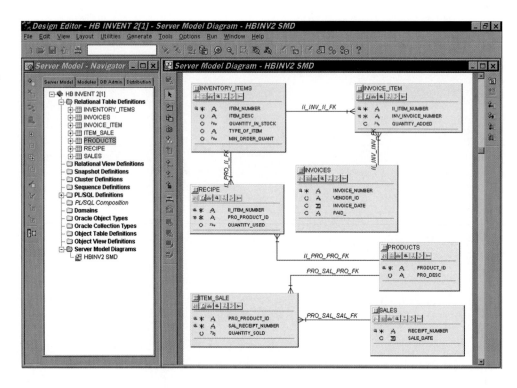

eration is complete, running the DDL scripts creates your database tables on the database server (Figure 15-12). At some point, you will need to populate your database. Some people prefer to do this using SQL scripts; others prefer creating simple forms in Oracle's Developer tool. Now that the database has been created, we can move to generating the forms and reports in the prototype that will use the database.

Transforming and Generating Software Modules The first step in generating software modules in Oracle's Designer is a data flow diagram or a functional hierarchy diagram. We start with the logical DFD for Hoosier Burger's inventory control

Figure 15-12
Oracle Designer's Design Editor, after the database has been successfully generated for Hoosier Burger's inventory control system, showing the generated database definition language (DDL) for creating tables

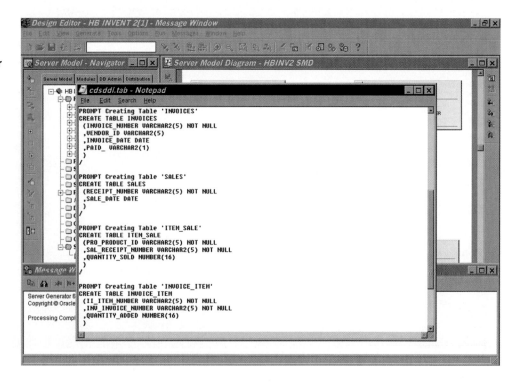

system (Figure 8–16), which we have transformed into a functional hierarchy diagram, or FHD (Figure 15-13). The system itself appears as the root function, and all five processes in the level-0 diagram appear as functions on the next level. Although it is not shown in the FHD, the relationships between the data entities and their attributes and each function have been designated in a CRUD (create, retrieve, update, delete) matrix for each function. Also, each function has been designated as either a form or a report by choosing immediate or overnight response, respectively, in the properties dialog for each function. Using Designer's Application Design Transformer (Figure 15-14), the FHD will be transformed into a set of software modules, from which forms and reports can be generated.

Going back to the Design Editor, we see the results of transforming the functional hierarchy diagram for Hoosier Burger's inventory control system (Figure 15-15). The five modules are listed in the Navigator window on the left. The first four are forms; the last one is a report. We have created a module diagram for each module, the names for which approximate the work each function performs, and these are listed in the Navigator window under "Module Diagrams." The module diagram for module HB INV0010, Update Inventory Added, is shown in the right-hand window in Figure 15-15. Although it is beyond the scope of this example to go through all of the details involved, the module diagram provides a visually oriented way to edit and enhance the specific design for a module. Using the module diagram, you can add navigation controls, for moving within or between modules. You can specify default values for data fields, and you can create non-bound data items, like totals. All of the details of the form or report would be addressed now, in the module diagrams, before the final forms or reports are generated.

Once you are satisfied with the design of your form or report, you generate the form or report in the Design Editor, the same place you generated the DDL for the physical database tables (Figure 15-16). Once the form or report has been generated, you are given the opportunity to save it. Forms and reports are saved to the client directory of your choice. From the Design Editor, you can compile and run the form or report you just generated and saved (Figure 15-17).

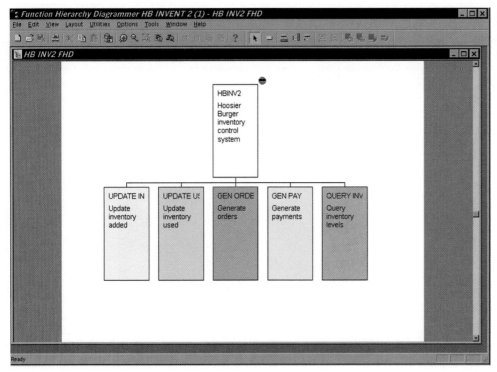

Figure 15-13
Functional hierarchy diagram for Hoosier Burger's inventory control system

Figure 15-14
Application Design Transformer dialog box, for initiating the transformation process for the functional hierarchy diagram for Hoosier Burger's inventory control system

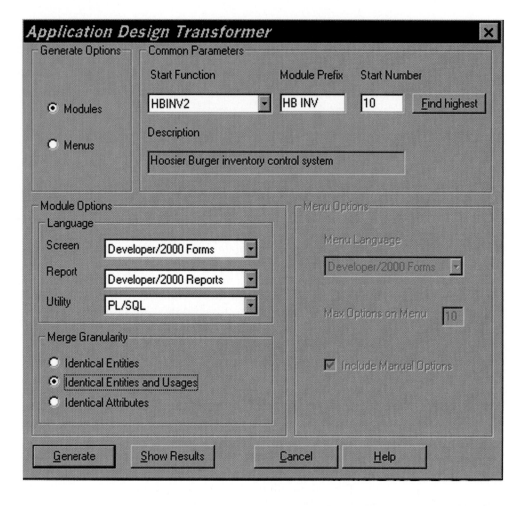

Figure 15-15
Module diagram for the Update Inventory Added function in the Design Editor in Oracle Designer

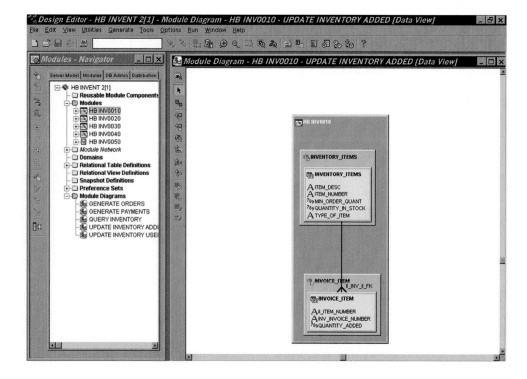

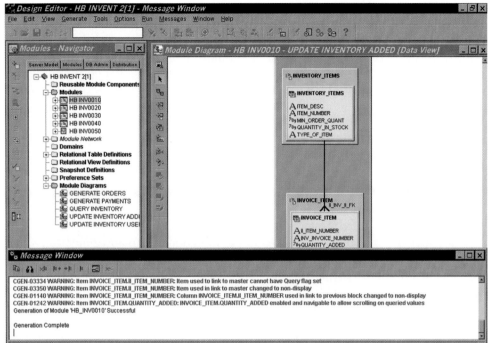

Figure 15-16
Oracle Designer Design Editor, featuring a message window indicating that the generation of the Update Inventory Added form has been successfully completed

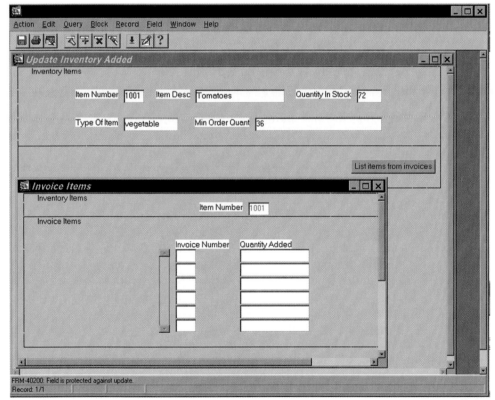

Figure 15-17
The form generated from the Update Inventory Added software module in Hoosier Burger's inventory control system

N E T   S E A R C H
There are many CASE tools that can generate prototypes, and there are many tools created specifically to support the prototyping process. Visit http://www.prenhall.com/hoffer to complete an exercise related to this topic.

If you are not completely happy with the form or report, for the purposes of the prototype you are creating, you can continue to edit the module diagram and regenerate the form or report from it, or you can edit the form or report in Oracle's Developer tool. Developer gives you much more control over all of the different aspects of your forms and reports, but it also requires more in-depth knowledge, including knowledge of PL/SQL for creating and refining GUI controls. Although these Oracle tools, and others like them from other vendors, allow you to quickly create a working prototype, it is worth repeating that prototyping is an iterative process. The prototype you create, using the process we have briefly explained above, will be only the first step in the prototyping process for capturing design specifications.

RADICAL METHODS

Producing comprehensive design specification documents and generating working prototypes are not the only methods available for specifying software design details. In the interest of saving time and effort, bulky design specification documents may not be created at all. The lines between design and implementation may even disappear as these two phases of the SDLC converge into a single phase. Two examples of such approaches are described next. The first approach is called eXtreme Programming (Beck, 2000). The second approach is Rapid Application Development or RAD. Since it is the subject of Chapter 19, RAD is only briefly described here.

eXtreme Programming

eXtreme programming is an approach to software development put together by Kent Beck (2000). It is distinguished by its short cycles, its incremental planning approach, its focus on automated tests written by programmers and customers to monitor the process of development, and its reliance on an evolutionary approach to development that lasts throughout the lifetime of the system. One of the key emphases of eXtreme programming is its use of two-person programming teams and having a customer on-site during the development process, both of which we will discuss in more detail in Chapter 17 on implementation. The relevant parts of eXtreme programming that relate to design specifications are (1) how planning, analysis, design, and construction are all fused together into a single phase of activity, and (2) its unique way of capturing and presenting system requirements and design specifications. All phases of the life cycle converge together into series of activities based on the basic processes of coding, testing, listening, and designing.

What is of interest here, however, is the way requirements and specifications are dealt with. Both of these activities take place in what Beck calls the "Planning Game." The Planning Game is really just a stylized approach to development that seeks to maximize fruitful interaction between those who need a new system and those who build it. The players in the Planning Game, then, are Business and Development. Business is the customer and is ideally represented by someone who knows the processes to be supported by the system being developed. Development is represented by those actually designing and constructing the system. The game pieces are what Beck calls "Story Cards." These cards are created by Business and contain a description of a procedure or feature to be included in the system. Each card is dated and numbered and has space on it for tracking its status throughout the development effort.

The Planning Game has three phases: exploration, commitment, and steering (Figure 15-18). In exploration, Business creates a Story Card for something it wants the new system to do. Development responds with an estimation of how long it would take to implement the procedure. At this point, it may make sense to split a Story Card

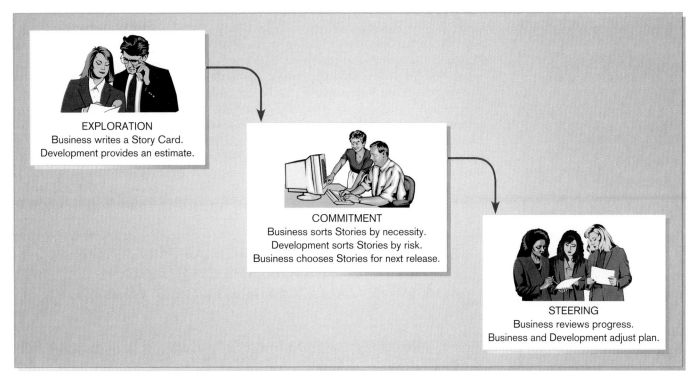

Figure 15-18
eXtreme Programming's planning game

into multiple Story Cards, as the scope of features and procedures becomes more clear during discussion. In the commitment phase, Business sorts Story Cards into three stacks, one for essential features, one for features that are not essential but would still add value, and one for features that would be nice to have. Development then sorts Story Cards according to risk, based on how well they can estimate the time needed to develop each feature. Business then selects the cards that will be included in the next release of the product. In the final phase, steering, Business has a chance to see how the development process is progressing and to work with Development to adjust the plan accordingly. Steering can take place as often as once every three weeks.

The Planning Game between Business and Development is followed by the Iteration Planning Game, played only by programmers. Instead of Story Cards, programmers write Task Cards, which are based on Story Cards. Typically, several Task Cards are generated for each Story Card. The Iteration Planning Game has the same three phases as the Planning Game, exploration, commitment, and steering. During exploration, programmers convert Story Cards into Task Cards. During commitment, they accept responsibility for tasks and balance their workloads. During steering, the programmers write the code for the feature, test it, and if it works, they integrate the feature into the product being developed. The Iteration Planning Game takes place during the time intervals between steering phase meetings in the Planning Game.

You can see how the approach embodied by eXtreme programming differs from that taken for generating design specification documents or prototypes. Yet in many ways, some of the core principles are the same. Customers, or Business, are still the source for what the system is supposed to do. Requirements are still captured and negotiated. The overall process is still documented, although much of the documentation is in the code itself in eXtreme programming. Given the way requirements are identified and recorded and broken down from stories to tasks, design specifications can easily incorporate the characteristics of quality requirements: completeness, consistency, modifiability, and traceability.

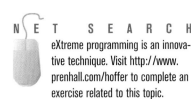

NET SEARCH
eXtreme programming is an innovative technique. Visit http://www.prenhall.com/hoffer to complete an exercise related to this topic.

RAD

In Rapid Application Development, there are only four life-cycle phases: planning, design, construction, and cutover (see Chapter 19 for more detail). There tends to be heavy iteration between design, where requirements are captured, and construction, where the system is designed and built. Both phases make heavy use of Joint Application Design (JAD) workshops, CASE tools, and prototyping. Developers and customers work together to define requirements during the design phase, and customers are also involved in the construction effort. Requirements tend not to be formally represented in RAD, where the emphasis is on speed. Rather, they are captured during JAD workshops and formalized in prototype construction using CASE tools. Depending on how much emphasis is placed on speed, many of the characteristics of quality requirements and quality requirement statements are sometimes ignored.

ELECTRONIC COMMERCE APPLICATION: FINALIZING DESIGN SPECIFICATIONS FOR PINE VALLEY FURNITURE'S WEBSTORE

Over the past few chapters, you have read how analysts design various system features. In this chapter, we discussed how analysts finalize design specifications using various techniques. Techniques that are particularly useful for designing Internet-based electronic commerce applications are throwaway and evolutionary prototyping. Analysts use throwaway prototyping many times as a test, as a proof of concept, or to simply show others what they have in mind. For example, when building a Website, you could use this technique to experiment with layout, decide where banners might go, experiment with navigation, or test whether or not to even use buttons with GUI controls. Many times, prototypes that were created as "throwaways" can become the basis of an evolutionary prototype, ultimately becoming pieces of the actual system. In this section, we examine how throwaway prototyping was used to help finalize some design issues for the Pine Valley Furniture's WebStore.

Finalizing Design Specifications for Pine Valley Furniture's WebStore

In the past two chapters you read how Jim defined specifications for the forms and reports as well as the interface and dialogs for Pine Valley Furniture's WebStore. In this design work, he and his development team concluded that they wanted the human-computer interface of their site to have four key features:

- Menu-driven navigation with cookie crumbs
- Lightweight graphics
- Form and data integrity rules
- Template-based HTML

In addition to these specifications, his team defined the required fields for each of the pages identified in the design process.

One thing that he and the team failed to define was the "look" of the page. Although the actual construction of the site had been outsourced (recall from Chapter 11 that PVF selected a consulting firm to actually construct the WebStore), Jim felt that some additional design issues needed to be resolved. In particular, he had some ideas on how the site would look in regard to color schemes and backgrounds.

To demonstrate to his group what he had in mind, Jim decided to build a throwaway prototype of a sample screen or two using Microsoft's Frontpage™. He felt that since most of the customers would be remodeling an existing area, or designing a completely new building or room in which to put PVF furnishings, he wanted to use a blueprint or construction metaphor throughout the site. He recalled when building his personal Web pages using Frontpage that it had numerous themes that could be used to quickly design a working prototype (see Figure 15-19).

Once he selected the blueprint theme for the prototype pages, Jim imported a small "lightweight" graphic logo image of a pine tree and placed it in the upper left corner of the root page of his prototype (see Figure 15-20a). Because the root page acts as a style sheet for the entire site, this image is propagated to all other pages (see Figure 15-20b). Also, using options within Frontpage, he could easily establish many of the navigational controls that he wanted to demonstrate. In particular, he wanted to show how navigational menus and cookie crumbs could be used to assist users in successfully navigating throughout the site. By setting the navigational properties in the root page, the properties were applied to all pages within the site. Notice in Figure 15-20a the "Home" navigational button below the "Pine Valley Furniture WebStore" banner is different than the "Products, Services, News, and Feedback" buttons. This *difference* indicates to users that they are on the "Home" page of the site. In Figure 15-20b, the "Services" page, note the associated change in the navigational buttons. Now, Services is *different* than the others. These changes in the buttons as a user moves through the site demonstrate the concept of cookie crumbs.

Being able to establish navigational controls as well as defining the overall look of the site from a single style sheet demonstrates the concept of "template-based HTML." Using a single template to control the look and behavior of pages throughout the site not only eases initial development, but also greatly simplifies subsequent system maintenance since one change can be propagated throughout the entire site.

Now that Jim completed his throwaway prototype, he called a meeting with his team and the consultants building the WebStore. At this meeting he demonstrated the prototype so that everyone could understand his vision for the site. Since the consultants had substantial experience in building such systems, numerous issues were clarified, discussed, and resolved. As a result, everyone now had a very clear understanding of exactly how the WebStore would look and how users would navigate throughout the site.

Figure 15-19
Establishing the Blueprint Theme for the WebStore throwaway prototype

Figure 15-20
(a) The Home page within the WebStore throwaway prototype

(b) The Services page within the WebStore throwaway prototype

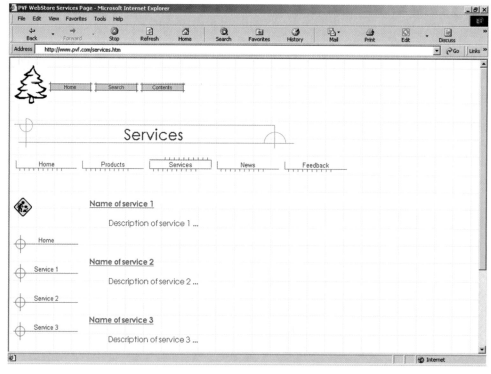

Summary

As the information systems design process comes to an end, designers must be concerned with the system's input, output, databases, and programs' physical specifications. Designers must consider what each component will look like and how it will be physically implemented in the new system. In this chapter, we have focused on finalizing system design specifications. You have learned about the different ways that design specifications can be represented. Specifications can be put together as a written document, augmented by various diagrams. The diagrams can include structure charts, which illustrate the hierarchical layout of a system. This layout shows all of the different software modules in the system and their relationships to each other, including the data and messages passed among modules. The specifications can also be input into and managed by a computer-based requirements management tool. Such tools can save time and effort in creating and maintaining requirements and their status. Specifications can also be captured in working prototypes. In some cases, these prototypes become the basis for the production system. In other cases, they are created to test an idea or to prove a concept and then are thrown away. You saw how Oracle's Designer tool can be used to rapidly create a prototype system. Design specification documents and prototypes are not the only way to capture requirements, however. In eXtreme programming, requirements are captured on Story Cards, refined on Task Cards, and then captured in code as part of the production system itself. In Rapid Application Development, as many of the design specifications as possible are captured in the CASE-generated prototype, which often becomes the basis for the production system. The same processes and techniques that have been used for traditional systems development can be used profitably for Internet-based systems. In the example here, throwaway prototyping was used to help analysts with design specifications for their WebStore application.

Before the design specifications can be handed over to the developers to begin the implementation process, questions about multiple users, multiple platforms, and program and data distribution have to be considered. Designing distributed systems is the topic of the next chapter.

Key Terms

1. Data couple
2. Flag
3. Module
4. Pseudocode
5. Structure chart

Match each of the key terms above with the definition that best fits it.

_____ Hierarchical diagram that shows how an information system is organized.

_____ A self-contained component of a system, defined by function.

_____ A diagrammatic representation of a message passed between two modules.

_____ A diagrammatic representation of the data exchanged between two modules in a structure chart.

_____ A method for representing the instructions in a module with language very similar to computer programming code.

Review Questions

1. What is a design specification? What role does it play in the design phase of the SDLC?

2. What are the characteristics of a quality requirement? Of a quality requirement statement? What is the difference between a requirement and a requirement statement?

3. What is a requirement management tool? What purpose does it serve?

4. What is a structure chart's role in physical information system design?

5. What is the difference between a data couple and a flag?

6. List and explain all of the special structure chart symbols in Figure 15-5.

7. What is pseudocode and what is its purpose?

8. What is the difference between evolutionary and throwaway prototyping?

9. In your own words, explain how Oracle's Designer tool can be used to rapidly create a working prototype.

10. How are design specifications captured in eXtreme programming?

11. How are design specifications captured in RAD?

12. Explain how design specifications can be captured for Web-based information systems.

Problems and Exercises

1. Using Figure 15-2 as a guide, create a detailed outline for a design specification document for the Hoosier Burger inventory control system. Append any diagrams that will be helpful to the developers who will eventually be given your specification document for implementation.

2. Convert the functional hierarchy diagram in Figure 15-13 into a structure chart.

3. Represent as pseudocode the processing logic for each module you create in Problem and Exercise 1.

4. A module of a payroll system calculates the overtime pay for employees. Employees who work over 40 hours in a one-week period are paid their base salary rate for hours 1 through 40 plus one and a half times their base salary rate for every additional hour. However, if the employee data indicate that an employee worked over 70 hours in a one-week period, the record is flagged, an overtime calculation is not made, and an error message is printed. Represent the processing logic of this module using pseudocode.

5. Choose an entity-relationship diagram from elsewhere in the book, and working in Oracle's Designer or another CASE tool, create an ERD and generate physical database tables from it.

6. Choose a data flow diagram or functional hierarchy diagram from elsewhere in the book, and working in Oracle's Designer or another CASE tool, create a DFD or FHD and generate working software modules from it.

Field Exercises

1. Interview a systems analyst about the process of finalizing design specifications where the analyst works. What specific design methodology does the analyst's organization follow? Does the methodology rely on structure charts? Prototyping? Some other technique? Report on how design specifications are created and updated in this organization.

2. Research the information system design process in Japan and in Europe. How do these processes compare to what you have learned here about North American practice? How are the processes similar? How are they different?

References

Beck, K. 2000. *eXtreme Programming eXplained.* Upper Saddle River, NJ: Addison-Wesley.

McConnell, S. 1996. *Rapid Development.* Redmond, WA: Microsoft Press.

Wiegers, K. E. 1999. "Writing Quality Requirements." *Software Development,* May (www.processimpact.com/pubs.shtml).

Wiegers, K. E. 2000. "When Telepathy Won't Do: Requirements Engineering Key Practices." *Cutter IT Journal,* May.

BROADWAY ENTERTAINMENT COMPANY, INC.

Finalizing Design Specifications for the Customer Relationship Management System

CASE INTRODUCTION

The students from Stillwater State University have gathered and organized an enormous amount of information about MyBroadway, the Web-based customer relationship management system for Carrie Douglass, manager of the Centerville, Ohio, Broadway Entertainment Company (BEC) store. The team has also developed prototypes for several Web pages and for the overall navigation of the site, as well as data, process, and logic models for the workings of the system. Through all of its previous work, the student team has tried to create quality requirements statements and quality requirements. In fact, by this point in the project, the team has produced a fairly thorough set of specification documents. Because the team has chosen a prototyping methodology, they have, however, avoided producing traditional program documentation. MyBroadway will likely be a throwaway prototype, so a final system will most likely be built by BEC using corporate Web development tools. Thus, the students believe that program documentation is much less important than business logic and requirements documentation, along with the working prototype. They feel that key to the final system specifications will be the final prototype, along with an explanation of how and why it came into being.

USING THROWAWAY PROTOTYPING FOR DOCUMENTING SYSTEM SPECIFICATIONS

BEC Figure 15-1 depicts the prototyping process the Stillwater students plan to use to finalize the system specifications. Each iteration of the prototype will add new functionality or improve on the functionality in a previous prototype. Each prototype iteration will be used by selected customers and/or employees. The assumption under which the team members are operating is that feedback at each iteration will be used to produce a better prototype that meets more requirements or meets those requirements in a better way. Although their initial discussions with Carrie Douglass and other BEC store employees produced system requirement statements, the prototyping process will be refining and expanding these require-

ments. From their classes at Stillwater, the team members know that quality requirements statements must be correct, feasible, necessary, prioritized, unambiguous, and verifiable (see Table 15-1, earlier in this chapter). However, the team has no procedures in place, yet, to help the prototyping process achieve these objectives.

CASE SUMMARY

Prototyping is a wonderful tool for requirements determination and producing a design specification. However, the evolution of the prototype can consume so much of the focus of a development team that the business reasons for the prototype are lost. The Stillwater student team working on MyBroadway is looking for a way to manage their prototyping process so that each iteration produces a better prototype and the justification for the final prototype is integrally produced along with the prototype. The team is especially sensitive to this concern because their client, Carrie Douglass is not the ultimate client. The students know that if their work is going to have real impact, what they produce must present a compelling case to BEC corporate management and IT professionals that MyBroadway, and in particular the students' design of MyBroadway, represents a high-quality system.

CASE QUESTIONS

1. Design a feedback form that the Stillwater students could use when a customer or employee uses some iteration of the MyBroadway prototype. Explain why each element is included on this form and what you are trying to accomplish by collecting each feedback element.

2. The feedback received at each prototyping iteration across multiple users may be inconsistent. For example, different users may have different reactions to an existing system feature, may have conflicting suggestions on how to improve the system, and may believe

BEC Figure 15-1
The MyBroadway prototyping process

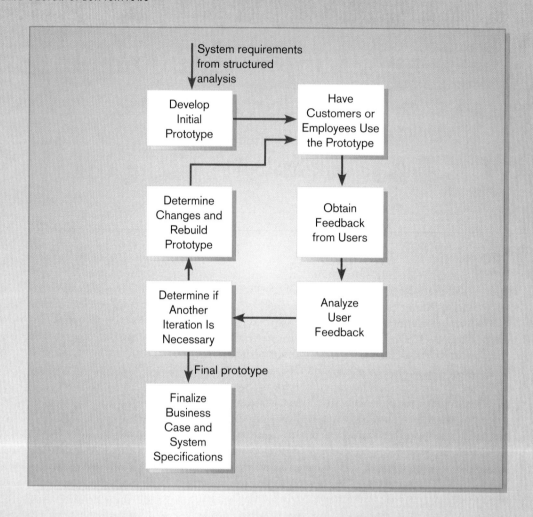

that different new features will be of greatest benefit. What procedures would you recommend to the Stillwater team to deal with possible inconsistencies in feedback so that each iteration can make the greatest progress?

3. How will the Stillwater students know when the prototyping process outlined in BEC Figure 15-1 is done? That is, when is it time to stop iterating?

4. Once the last prototype is developed, what further steps will be necessary to build the business case that MyBroadway meets the needs of the business (that is, delivers quality requirements)? How will you use the feedback collected from the forms you designed in Case Question 1 to produce this business case?

5. Some iterations of the prototype may invalidate the structured system documentation (data, process, logic models, etc.) developed previously by the Stillwater students. This invalidation can occur because prototyping is a discovery process, which can uncover flaws in previous analyses and deficiencies in requirements. Should the team continuously update the structured documentation, wait until the end to make the structured documentation consistent with the final prototype, or forget about updating the structured documentation? Justify your answer.

Chapter 16

Designing Distributed and Internet Systems

LEARNING OBJECTIVES

After studying this chapter, you should be able to:

- Define the key terms *client/server architecture*, *local area network*, *distributed database*, and *middleware*.

- Distinguish between file server and client/server environments, and contrast how each is used in a local area network.

- Describe alternative designs for distributed systems and their trade-offs.

- Describe how standards shape the design of Internet-based systems.

- Describe options for assuring Internet design consistency.

- Describe how site management issues can influence customer loyalty and trustworthiness as well as system security.

- Discuss issues related to managing on-line data including context development, on-line transaction processing, on-line analytical processing, and data warehousing.

INTRODUCTION

The advances in computing technology and the rapid evolution of graphical user interfaces, networking, and the Internet are changing the way today's computing systems are being used to meet ever more demanding business needs. In many organizations, previous stand-alone personal computers are being linked together to form networks that support workgroup computing (this process is sometimes called *upsizing*). At the same time, other organizations (or even the same organization) are migrating mainframe applications to personal computers, work-stations, and networks (this process is sometimes called *downsizing*) to take advantage of the greater cost-effectiveness of these environments. Organizations are also using the World Wide Web for delivering applications to internal and external customers.

A variety of new opportunities and competitive pressures are driving the trend toward these technologies. Corporate restructuring—mergers, acquisitions, consolidations—makes it necessary to connect or replace existing stand-alone applications. Similarly, corporate downsizing has caused individual managers to have a broader span of control, thus requiring access to a wider range of data, applications, and people. Applications are being downsized from expensive mainframes to networked microcomputers and workstations (possibly with a mainframe as a server) that are much more cost-effective and scalable, and can also be more user-friendly. The explosion of electronic commerce is today's biggest driver for developing new types of systems. How distributed and Internet systems are designed can significantly influence system performance, usability, and maintenance.

In this chapter we describe several technologies that are being used to upsize, downsize, and distribute information systems and data. These technologies are LAN-based DBMSs, client/server DBMSs, and the Internet. The capabilities and issues surrounding these technologies are the foundation for understanding how to migrate single processor applications and designs into a multiprocessor, distributed computing environment.

DESIGNING DISTRIBUTED AND INTERNET SYSTEMS

In this section we will briefly discuss the process and deliverables when designing distributed and Internet systems. Given the direction of organizational change and technological evolution, it is likely that most future systems development efforts will need to consider the issues surrounding the design of distributed systems.

The Process of Designing Distributed and Internet Systems

This is the last chapter in the text dealing with system design within the systems development life cycle (see Figure 16-1). In the previous chapters on system design, specific techniques for representing and refining data, screens, interfaces, and events were presented. In this chapter, however, no specific techniques will be presented on how to represent the design of distributed and Internet systems since no generally accepted techniques exist. Alternatively, we will focus on increasing your awareness of common environments for deploying these systems and the issues you will confront surrounding their design and implementation. To distinguish between distributed and Internet-focused system design, we will use *distributed* to reflect LAN-based file server and client/server architectures.

Designing distributed and Internet systems is much like designing single-location systems. The primary difference is that because a system will be deployed over two or more locations, numerous design issues must be considered due to their influence on the reliability, availability, and survivability of the system when it is implemented. Because these systems have more components than a single-location system—that is, more processors, networks, locations, data, and so on—there are more potential places for a failure to occur. Consequently, various strategies can be used when designing and implementing these systems.

Thus, when designing distributed and Internet systems, you will need to consider numerous trade-offs. To create effective designs, you need to understand the characteristics of the commonly used architectures for supporting these systems.

Deliverables and Outcomes

When designing distributed and Internet systems, the deliverable is a document that will consolidate the information that must be considered when implementing a system design. Figure 16-2 lists the types of information that must be considered

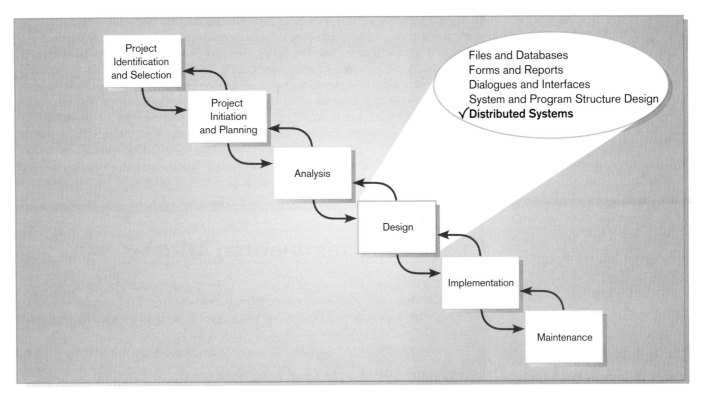

Figure 16-1
Systems development life cycle with design phase highlighted

1. Description of Site (for each site)
   a. geographical information
   b. physical location
   c. infrastructure information
   d. personnel characteristics (education, technical skills, etc.)
   e. ...

2. Description of Data Usage (for each site)
   a. data elements used
   b. data elements created
   c. data elements updated
   d. data elements deleted

3. Description of Business Process (for each site)
   a. list of processes
   b. description of processes

4. Contrasts of Alternative IS Architectures for Site, Data, and Process Needs (for each site)
   a. pros and cons of no technological support
   b. pros and cons of non-networked, local system
   c. pros and cons of various distributed configurations
   d. ...

Figure 16-2
Outcomes and deliverables from designing distributed systems

when implementing such a system. In general, the information that must be considered is the site, processing, and data information for *each* location (or processor) in a distributed environment. Specifically, information related to physical distances between locations, counts and usage patterns by users, building and location infrastructure issues, personnel capabilities, data usage (use, create, update, or destroy), and local organizational processes must be described. Additionally, the pros and cons of various implementation solutions for each location should be reviewed. The collection of this information, in conjunction with the physical design information already developed, will provide the basis for implementing the information system in the distributed environment. Note, however, that our discussion assumes that any required network infrastructure is already in place. In other words, we focus only on those issues in which you will likely have a choice.

DESIGNING DISTRIBUTED SYSTEMS

In this section we will focus on issues related to the design of distributed systems that use LAN-based file server or client/server architectures. The section begins by providing a high-level description of both architectures. This is followed by a brief description of various configurations for designing client/server systems.

Designing Systems for Local Area Networks

Personal computers and workstations can be used as stand-alone systems to support local applications. However, organizations have discovered that if data are valuable to one employee, they are probably also valuable to other employees in the same workgroup or in other workgroups. By interconnecting their computers, workers can exchange information electronically and can also share devices such as laser printers that are too expensive to be used by only a single user.

Local area network (LAN): The cabling, hardware, and software used to connect workstations, computers, and file servers located in a confined geographical area (typically within one building or campus).

A **local area network (LAN)** supports a network of personal computers, each with its own storage, that are able to share common devices and software attached to the LAN. Each PC and workstation on a LAN is typically within 100 feet of the others, with a total network cable length of under 1 mile. At least one computer (a microcomputer or larger) is designated as a file server, where shared databases and applications are stored. The LAN modules of a DBMS, for example, add concurrent access controls, possibly extra security features, and query or transaction queuing management to support concurrent access from multiple users of a shared database.

File server: A device that manages file operations and is shared by each client PC attached to a LAN.

File Servers In a basic LAN environment (see Figure 16-3), all data manipulation occurs at the workstations where data are requested. One or more file servers are attached to the LAN. A **file server** is a device that manages file operations and is shared by each client PC that is attached to the LAN. In a file server configuration, each file server acts as an additional hard disk for each client PC. For example, your PC might recognize a logical F: drive, which is actually a disk volume stored on a file server on the LAN. Programs on your PC refer to files on this drive by the typical path specification, involving this drive and any directories, as well as the file name.

When using a DBMS in a file server environment, each client PC is authorized to use the DBMS application program on that PC. Thus, there is one database on the file server and many concurrently running copies of the DBMS on each active PC. The primary characteristic of a client-based LAN is that all data manipulation is performed at the client PC, not at the file server. The file server acts simply as a shared data storage device and is an extension of a typical PC. It also provides additional resources (e.g., disk drives, shared printing), collaborative applications (e.g., electronic mail), in addition to the shared data. Software at the file server only queues

Figure 16-3
File server model

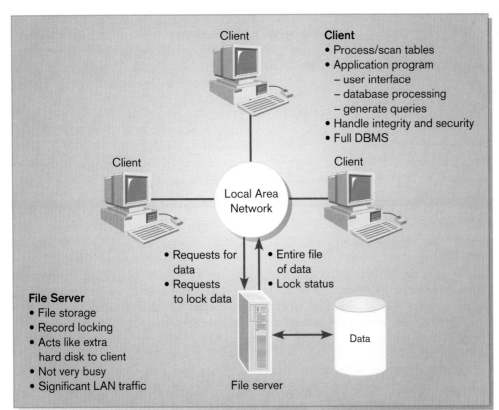

access requests; it is up to the application program at each client PC, working with the copy of the DBMS on that PC, to handle all data management functions. This means that in an application that wants to view a single customer account record in a database stored on the server, the file containing all customer account records will be sent over the network to the PC. Once at the PC, the file will be searched to find the desired record. Additionally, data security checks and file and record locking are done at the client PCs in this environment, making multi-user application development a relatively complex process.

Limitations of File Servers   There are three limitations when using file servers on local area networks:

1. Excessive data movement
2. Need for powerful client workstation
3. Decentralized data control

First, when using a file server architecture, considerable data movement is generated across the network. For example, when an application program running on a client PC in Pine Valley Furniture wants to access the Birch products, the whole Product table is transferred to the client PC; then, the table is scanned at the client to find the few desired records. Thus, the server does very little work, the client is busy with extensive data manipulation, and the network is transferring large blocks of data (see Figure 16-4). Consequently, a client-based LAN places a considerable burden on the client PC to carry out functions that have to be performed on all clients and creates a high network traffic load.

Second, since each client workstation must devote memory to a full version of the DBMS, there is less room on the client PC to rapidly manipulate data in high-speed random access memory (RAM). Often, data must be swapped between RAM

Figure 16-4
File servers transfer entire files when data are requested from a client

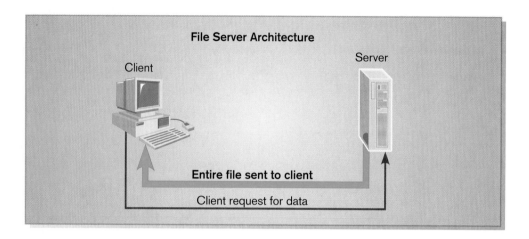

Figure 16-4
File servers transfer entire files when data are requested from a client

and a relatively slower hard disk when processing a particularly large database. Further, because the client workstation does most of the work, each client must be rather powerful to provide a suitable response time. File server-based architectures also benefit from having a very fast hard disk and cache memory in both clients and the server to enhance their ability to transfer files to and from the network, RAM, and hard disk.

Third, and possibly most important, the DBMS copy in each workstation must manage the shared database integrity. In addition, each application program must recognize, for example, locks on data and take care to initiate the proper locks. A lock is necessary to stop users from accessing data that are in the process of being updated. Thus, application programmers must be rather sophisticated to understand various subtle conditions that can arise in a multiple-user database environment. Programming is more complex since you have to program each application with the proper concurrency, recovery, and security controls.

Designing Systems for a Client Server Architecture

Client/server architecture: A LAN-based computing environment in which a central database server or engine performs all database commands sent to it from client workstations, and application programs on each client concentrate on user interface functions.

An improvement in LAN-based systems is the **client/server architecture** in which application processing is divided (not necessarily evenly) between client and server. The client workstation is most often responsible for managing the user interface, including presenting data, and the database server is responsible for database storage and access such as query processing. The typical client/server architecture is illustrated in Figure 16-5.

In the typical client/server architecture, all database recovery, security, and concurrent access management is centralized at the server, whereas this is the responsibility of each user workstation in a simple LAN. These central DBMS functions are often called a **database engine** in a client/server environment. Some people refer to the central DBMS functions as the back-end functions whereas client-based delivery of applications to users using PCs and workstations are called front-end applications. Further, in the client/server architecture, the server executes all requests for data so that only data that match the requested criteria are passed across the network to client stations. This is a significant advantage of client/server over simple file server-based designs. Since the server provides all shared database services, this leaves the **client** software to concentrate on user interface and data manipulation functions. The trade-off is that the server must be more powerful than the server in a file server environment.

Database engine: The (back-end) portion of the client/server database system running on the server and providing database processing and shared access functions.

Client: The (front-end) portion of the client/server database system that provides the user interface and data manipulation functions.

An application built using the client/server architecture is also different from a centralized database system on a mainframe. The primary difference is that each client is an *intelligent* part of the application processing system. In other words, the

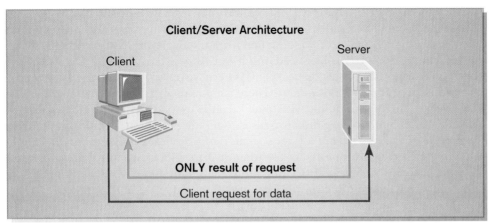

Figure 16-5
Client/server architecture transfers only the required data after a request from a client

application program executed by a user is running on the client, not on the server. The application program handles all interactions with the user and local devices (printer, keyboard, screen, etc.). Thus, there is a division of duties between the server (database engine) and the client: The database engine handles all database access and control functions and the client handles all user interaction and data manipulation functions. The client PC sends database commands to the database engine for processing. Alternatively, in a mainframe environment, all parts of the information system are managed and executed by the central computer.

Another advantage of client/server architectures is the ability to decouple the client environment from the server environment. Clients can consist of multiple types (for example, different computers, operating systems, and application programs), which means that the client can be running any application system that can generate the proper commands (often SQL) to request data from the server. For example, the application program might be written in Visual Basic, a report writer, a sophisticated screen painter, or any fourth-generation language that has an **application program interface**, or API, for the database engine. The database engine might be DB2 on an IBM mainframe or midrange computer, or Sybase or Oracle running on a variety of platforms. An API calls library routines that transparently route SQL commands from the front-end client application to the database server. An API might work with existing front-end software, like a third-generation language or custom report generator, and it might include its own facilities for building applications. When APIs exist for several program development tools, then you have considerable independence to develop client applications in the most convenient front-end programming environment, yet still draw from a common server database. With some APIs, it is possible to access data from both the client and server in one database operation, as if the data were in one location managed by one DBMS.

Application program interface (API): Software building blocks that are used to assure that common system capabilities like user interfaces and printing are standardized as well as modules to facilitate the data exchange between clients and servers.

Client/Server Advantages and Cautions There are several significant benefits that can be realized by adopting a client server architecture:

1. It allows companies to leverage the benefits of microcomputer technology. Today's workstations deliver impressive computing power at a fraction of the costs of mainframe.

2. It allows most processing to be performed close to the source of processed data, thereby improving response times and reducing network traffic.

3. It facilitates the use of graphical user interfaces (GUIs) and visual presentation techniques commonly available for workstations.

4. It allows for and encourages the acceptance of open systems.

Many vendors of relational DBMSs and other LAN-based technologies have or are attempting to migrate their products into the client/server environment. However, products that were not designed from the beginning under a client/server architecture may have problems adapting to this new environment (see Radding, 1992, for a discussion of such issues). This is because new issues and new spins on old issues arise in this new environment. These issues and areas include compatibility of data types, query optimization, distributed databases, data administration of distributed data, CASE tool code generators, cross operating system integration, and more. In general, there is a lack of tools for systems design and performance monitoring in a client/server environment. As versions of different front- and back-end tools change, problems may arise with compatibility, until the API evolves, and these problems must be handled directly by the programmer, not by development tools.

Now that you have an understanding of the general differences between file server and client/server architectures, we will next discuss how data can be managed within a distributed environment. After discussing data management options, we will present several design alternatives for distributed systems. All LAN-based distributed system designs are implemented using some configuration of the general file server or client/server architectures and data management options.

Alternative Designs for Distributed Systems

A clear trend in systems design is to move away from central mainframe systems and stand-alone PC applications to some form of system that distributes data and processing across multiple computers. In this section, we briefly review the major differences between file servers and database servers. In the following section, we discuss the trade-offs among various ways to separate processing between clients and servers.

Choosing Between File Server and Client/Server Architectures   Both file server and client/server architectures use personal computers and workstations and are interconnected using a LAN. Yet, a file server architecture is very different from a client/server architecture. A file server architecture supports only the distribution of data whereas the client/server architecture supports both the distribution of data and the distribution of processing. This is an important distinction that has ramifications for systems design.

Table 16-1 summarizes some of the key differences between file server and client/server architectures. Specifically, a file server architecture is the simplest method for interconnecting PCs and workstations. In this architecture, the file server simply acts as a shared storage device for all clients on the network. Entire programs and databases must be transferred to each client when accessed. This means that a file server architecture is most appropriate for applications that are relatively small in size with little or no concurrent data access by multiple users.

TABLE 16-1   Several Differences between File Server and Client/Server Architectures

Characteristic	File Server	Client/Server
Processing	Client only	Both client and server
Concurrent Data Access	Low—managed by each client	High—managed by server
Network Usage	Large file and data transfers	Efficient data transfers
Database Security and Integrity	Low—managed by each client	High—managed by server
Software Maintenance	Low—software changes just on server	Mixed—some new parts must be delivered to each client
Hardware and System Software Flexibility	Client and server decoupled and can be mixed	Need for greater coordination between client and server

Alternatively, a client/server architecture overcomes many of the limitations of the file server architecture since both client and server share the processing workload of a task and only transfer needed information. This characteristic has resulted in having many organizations migrate very large applications with extensive data sharing requirements to client/server environments. In fact, client/server computing has become the workhorse architecture for many organizations where multiple clients are likely to be concurrently working with the same data. Also, if the systems and databases are relatively large in size, the client/server architecture is preferred because of the client and server's ability to distribute the work and to transfer only needed information (e.g., only a record if that is all that is needed rather than an entire database as in a file server environment).

Advanced Forms of Client/Server Architectures   Client/server architectures represent the way different application system functions can be distributed between client and server computers. These variations are based on the concept that there are three general components to any information system:

1. *Data management*: These functions manage all interaction between software and files and databases, including data retrieval/querying, updating, security, concurrency control, and recovery.

2. *Data presentation*: These functions manage just the interface between system users and the software, including the display and printing of forms and reports and possibly validating system inputs.

3. *Data analysis*: These functions transform inputs into outputs, including simple summarization to complex mathematical modeling like regression analysis.

Different client/server architectures distribute, or partition, each of these functions to one or both of the client or server computers, and more increasingly, into a third computer referred to as the **application server**. In fact, it is becoming commonplace to use three or more distinct computers in many advanced client/server architectures (see Askren, 1996; Kara, 1997). This evolution in client/server computing has resulted in two new terms, **three-tiered client/server** and **middleware**, to represent this evolution. Three-tiered client/server combines three logical and distinct applications—data management, presentation, and analysis—into a single information system application. Middleware brings together distinct hardware, software, and communication technologies in order to create a three-tiered client/server environment.

A typical use of middleware is shown in Figure 16-6. This figure shows how client applications can access databases on database servers. Open DataBase Connectivity (ODBC) is a Microsoft standard for database middleware. ODBC drivers, residing on both client and server computers, allow, for example, an Access query to retrieve data stored in an Oracle database as if they were in an Access database. By referring to a different ODBC driver, the Access database could reference data in an Informix database using the same query.

There are three primary reasons for creating three-tiered client/server architectures (Stevens, 1996). First, applications can be partitioned in a way that best fits the organizational computing needs. For example, in a traditional two-tiered client/server system, the application (data analysis) resides on the client, which would access information like customer data from a database server. In a three-tiered architecture, data analysis can reside on a powerful application server, resulting in substantially faster response times for users. In addition, a three-tiered architecture provides greater flexibility by allowing the partitioning of applications in different ways for different users in order to optimize performance.

A second advantage of three-tiered architecture is that since most or all of the data analysis is contained in the application server, making global changes or cus-

Application server: A computing server where data analysis functions primarily reside.

Three-tiered client/server: Advanced client/server architectures in which there are three logical and distinct applications—data management, presentation, and analysis—which are combined to create a single information system.

Middleware: A combination of hardware, software, and communication technologies that bring together data management, presentation, and analysis into a three-tiered client/server environment.

Figure 16-6
ODBC middleware environment

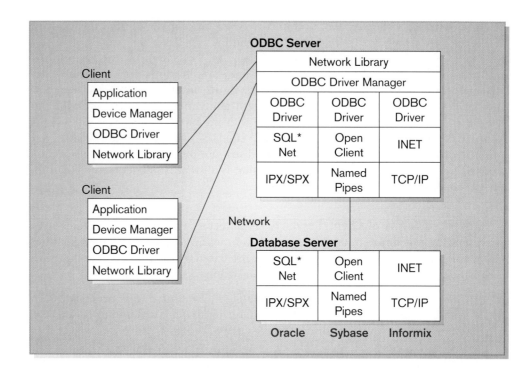

Figure 16-7
Types of client/server architectures
(a) Distributed presentation

FUNCTION	CLIENT	SERVER
Data management		All data management
Data analysis		All data analysis
Data presentation	Data for presentation on server are reformatted for presentation to user	Data delivered to client using server presentation technologies

(b) Remote presentation

FUNCTION	CLIENT	SERVER
Data management		All data management
Data analysis		All data analysis
Data presentation	Data from analysis on server are formatted for presentation to user	

(c) Remote data management

FUNCTION	CLIENT	SERVER
Data management		All data management
Data analysis	Raw data from server are retrieved and analyzed	
Data presentation	All data presentation	

(*continues*)

Figure 16-7 (continued)
Types of client/server architectures
(d) Distributed function

FUNCTION	CLIENT	SERVER
Data management		All data management
Data analysis	Selective data from server retrieved and analyzed	Selective data from server retrieved and analyzed, then transmitted to client
Data presentation	All data presentation, from analyses on both server and client	

(e) Distributed database

FUNCTION	CLIENT	SERVER
Data management	Local data management	Shared management of data on server
Data analysis	Data retrieved from both client and server for analysis	
Data presentation	All data presentation	

(f) Distributed processing

FUNCTION	CLIENT	SERVER
Data management	Local data management	Shared management of data on server
Data analysis	Data retrieved from both client and server for analysis	Data retrieved from server for analysis, then sent to client for further analysis and presentation
Data presentation	All data presentation	

tomizing processes for individual users is relatively easy. This allows developers to easily create custom versions of large-scale systems without creating a completely separate system. Finally, because the data analysis is separate from the user interface, it is a lot easier to change one or both without having a major maintenance effort. By separating the data analysis from the data presentation (the user interface), either can be changed independently without affecting the other—greatly simplifying system maintenance. The combinations of these three benefits—application partitioning, easier customization, and easier maintenance—are driving many organizations to adopt this powerful alternative to standard client/server computing.

Given the flexibility of placing data management, presentation, and analysis on two or more separate machines, there are countless possible architectures. In practice, however, only six possible configurations have emerged (see Figure 16-7). Technology exists to allow you to develop an application using any of these six architectures; although automated development tools do not yet have equal code-generation capabilities for each. A brief description of each is provided in Table 16-2.

As the designer of information systems, you have more choices available to you today than ever before. You must weigh the factors discussed above and outlined in Table 16-2 to determine a distributed system design that will be most beneficial to the

TABLE 16-2 Approaches to Designing Client/Server Architectures

Approach	Architecture Description
Distributed Presentation	This architecture is used to freshen up the delivery of existing server-based applications to distributed clients. Often the server is a mainframe, and the existing mainframe code is not changed. Here, technologies called "screen scrappers" work on the client machines to simply reformat mainframe screen data in a more appealing and easier-to-use interface.
Remote Presentation	This architecture places all data presentation functions on the client machine so that the client has total responsibility for formatting data. This architecture gives you greater flexibility compared to the distributed presentation style since the presentation on the client will not be constrained by having to be compatible with that on the server.
Remote Data Management	This architecture places all software on the client except for the data management functions. This form is the closest to what we have called client/server earlier in the chapter. A mixed client environment may be more difficult to support than in the previous architectures since you must learn multiple analysis programming environments, not just those for presentation tools.
Distributed Function	This architecture splits analysis functions between the client and server, leaving all presentation on the client and all data management on the server. This is a very difficult environment in which to develop, test, and maintain software due to the potential for considerable coordination between analysis functions on both client and server.
Distributed Database	This architecture places all functionality on the client, except data storage and management that is divided between client and server. Although possible today, this is a very unstable architecture since it requires considerable compatibility and communication between software on the client and server, which may never have been meant to be compatible.
Distributed Processing	This architecture combines the best features of distributed function and distributed database by splitting both of these across client and server, with presentation functions under the exclusive responsibility of the client machine. This permits even greater flexibility since analysis functions and data both can be located wherever it makes the most sense.

organization. As with other physical design decisions, organizational standards may limit your choices, and you will make such application design decisions in cooperation with other system professionals. You, however, are in the best position to understand user requirements and to be able to estimate the ramifications of distributed system design decisions on response time and other factors for the user.

DESIGNING INTERNET SYSTEMS

The vast majority of new systems development in organizations is focusing on Internet-based applications. The Internet can be used for delivering internal organizational systems, business-to-business systems, or business-to-consumer systems. The rapid migration to Internet-based systems should not be a surprise because it is motivated by the desire to take advantage of the global computing infrastructure of the Internet and the comprehensive set of tools and standards that have been developed. However, like the design of any other type of system, there are numerous choices that have to be made when designing an Internet application. The design choices you make can greatly influence the ease of development and the future maintainability of any system. In this section we focus on several fundamental issues that you must consider when designing Internet-based systems.

Internet Design Fundamentals

Standards play a major role when designing Internet-based systems. In this section, we examine many fundamental and emerging building blocks of the Internet and how each of these pieces influences system design.

Standards Drive the Internet   Designing Internet-based systems is much simpler than designing traditional client/server systems because of the use of standards. For example, information is located throughout the Internet via the use of the standard **domain naming system (BIND)** (the "B" in BIND refers to Berkley, California, where the standard was first developed at the University of California; for more information see www.isc.org/products/BIND/bind-history.html). BIND provides the ability to locate that information using common domain names that are translated into corresponding Internet Protocol (IP) addresses. For example, the domain name www.wsu.edu translates to 134.121.1.61.

> **Domain naming system (BIND):** A method for translating Internet domain names into Internet Protocol (IP) addresses.

Universal user access on a broad variety of clients is achieved through a standardized communication protocol: the Hypertext Transfer Protocol (HTTP). **HTTP** is the agreed upon format for exchanging information on the World Wide Web (see www.w3.org/Protocols/ for more information). The HTTP protocol defines how messages are formatted and transmitted as well as how Web servers and browsers respond to commands. For example, when you enter a URL in your browser, an HTTP command is sent to the appropriate Web server requesting the desired Web page.

> **HTTP (hypertext transfer protocol):** A communications protocol for exchanging information on the Internet.

Beyond the naming standards of BIND and the transfer mechanism of HTTP, an Internet-based system has another advantage over other types of systems: the Hypertext Markup Language (HTML). HTML is the standard language for representing content on the Web through the use of hundreds of command tags. Examples of command tags include those to bold text (<b> . . . </b>), to create tables (<table> . . . </table>), or to insert links onto a Web page (<A href=http:// www.wsu.edu /> Washington State University Home Page</A>).

Having standardized naming (BIND), translating (HTTP), and formatting (HTML) enable designers to quickly craft systems because much of the complexity of the design and implementation are removed. These standards also free the designer from much of the worry of delivering applications over a broad range of computing devices and platforms. Together BIND, HTTP, and HTML provide a standard for designers when developing Internet-based applications. In fact, without these standards, the Internet as we know it would not be possible.

Separating Content and Display   As a method to build first-generation electronic commerce applications, HTML has been a tremendous success. It is a very easy language to learn and there are countless tools to assist in authoring Web pages. In addition to its ease of use, it is also extremely tolerant to variations in usage such as the use of both upper or lower case letters for representing the same command or even the leniency to allow some commands to not *require* closing tags. HTML's simplicity also greatly limits its power (Castro, 2001). For example, most of HTML's tags are formatting-oriented, making it difficult for distinguishing data from formatting information. Additionally, because formatting information is inherently embedded into HTML documents, the migration of electronic commerce applications to emerging types of computing devices—like wireless handheld computers—is much more difficult. Some new devices, for example, like wireless Internet phones, cannot display HTML due to limited screen space and other limitations. To address this problem, new languages are being developed to separate content (data) from its display.

XML (extensible markup language): An Internet authoring language that allows designers to create customized tags, enabling the definition, transmission, validation, and interpretation of data between applications.

A language specifically designed to separate content from display is XML (see Castro, 2001; www.w3.org/XML/). **XML** (extensible markup language) is a lot like HTML, with tags, attributes, and values, but it also allows designers to create customized tags, enabling the definition, transmission, validation, and interpretation of data between applications. For electronic commerce applications, XML is rapidly growing in popularity. Whereas HTML has a fixed set of tags, designers can create custom languages—called vocabularies—for any type of application in XML. This ability to create customized languages is at the root of the power of XML; however, this power comes at a price. Whereas HTML is very forgiving on the formatting of tags, XML is very strict. Additionally, as mentioned above, XML documents do not contain any formatting information. XML tags simply define what the data mean. For this reason, many believe that HTML will remain a popular tool for developing personal Web pages, but that XML will become the tool of choice for commercial Internet applications.

Future Evolution The infrastructure currently supporting HTML-based data exchange is the same infrastructure that will support the wide spread use of XML and other emerging standards. As we move beyond desktop computers and standard Web browsers, the greatest driver of change and evolution of Internet standards will be the desire to support wireless, mobile computing devices. Wireless, mobile computing devices are often referred to as thin client technologies. **Thin clients** such as network PCs, handheld computers, and wireless phones are being designed to operate as clients in Internet-based environments (see Figure 16-8). Thin clients are most appropriate for doing a minimal amount of client-side processing, essentially displaying information sent to the client from the server. Alternatively, a workstation that can provide significant amounts of client-side storage and processing is referred to as a *fat* client. Current PC workstations connected to the Internet can be thought of as fat clients. For desktop PC workstations, Internet browsers render content marked up in HTML documents. However, as thin clients gain in popularity, designing applications in XML will enable content to be more effectively displayed on any client device, regardless of the screen size or resolution.

Thin client: A client device designed so that most processing and data storage occurs on the server.

Similar to the way that the HTTP and HTML standards support the delivery of Internet content to desktop PC, other standard architectures are needed for emerging devices. Delivering Internet applications to wireless mobile devices, for example, is being accomplished with the Wireless Application Protocol (WAP) and the Wireless Markup Language (WML). WAP is a variation of HTTP and WML is an XML-based markup language that was designed specifically to describe how WAP content is presented on a wireless terminal. Wireless devices have relatively small

Figure 16-8
Thin clients used to access the Internet

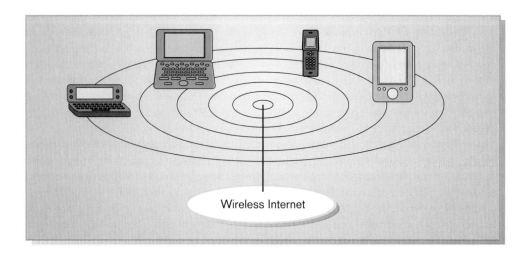

Wireless Internet

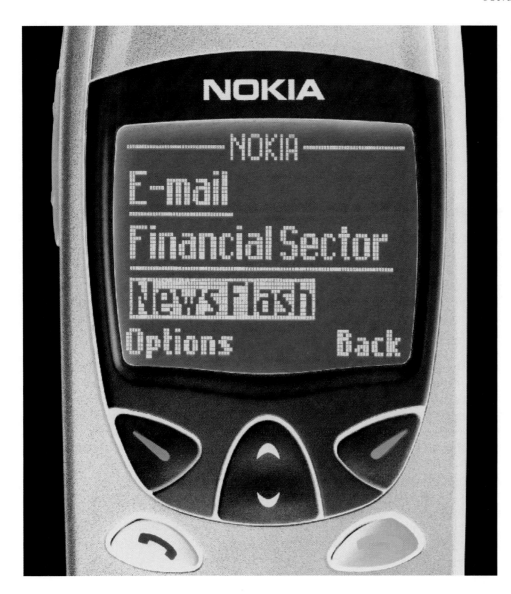

Figure 16-9
WAP Internet display has limited screen size
(Source: http://www.nokia.com, *Copyright © 2001 Nokia, Inc. Reprinted with permission.)*

screens compared to desktop computers; the average desktop runs at a resolution of 1024 pixels by 768 pixels while a typical Internet wireless phone has a resolution of approximately 100 by 60 pixels (see Figure 16-9).

Regardless of whether the device is an Internet wireless phone, a handheld computer, or a desktop PC, the use of standards will drive Internet-based system design. A well-designed system will isolate the content presentation from the business logic and data, allowing any Internet capable device to become a part of the overall distributed system. Techniques to assure the consistency of the site's appearance for any type of device are discussed next.

Site Consistency

A consistent "look and feel" is fundamental to conveying the image that the site is professionally designed. A site with high consistency is also much easier for users to navigate and is much more intuitive for users to anticipate the meaning of links. From a practical standpoint, it is a poor design decision to not enforce a standard look and feel to an entire site. There can also be a development and maintenance nightmare when implementing changes to colors, fonts, or other

NET SEARCH
Wireless mobile devices have relatively small display screens that create some unique challenges for system designers. Visit http://www.prenhall.com/hoffer to complete an exercise related to this topic.

elements across thousands of Web pages within a site. In this section we discuss ways to help you enforce design consistency across an entire site and to simplify page maintenance.

Cascading Style Sheets   One of the biggest difficulties in developing a large-scale Website is to assure consistency throughout the site in regard to color, background, fonts, and other page elements. Experienced Website designers have found that the use of **cascading style sheets (CSS)** can greatly simplify site maintenance and also assure that pages are consistent. CSS are simply a set of style rules that tell a Web browser how to present a document. Although there are various ways of linking these style rules to HTML documents (see www.w3.org/Style/CSS/), the simplest method is to use HTML's *STYLE* element, that is, to embed the style elements within each page. To do this, style elements can be placed in the document *HEAD*, containing the style rules for the page. This method, however, is not the best method for implementing CSS because each page will have to be changed if a single change is made to a site's style. The best way to implement CSS is through the use of *linked* style sheets. Using this method, through HTML's *LINK* element, only a single file needs to be updated when changing style elements across an entire site. The *LINK* element indicates some sort of a relationship between an HTML document and some other object or file (see Figure 16-10). CSS is the most basic way to implement a standard style design within a Website.

Cascading style sheets (CSS): A set of style rules that tells a Web browser how to present a document.

Extensible Style Language (XSL)   A second and more sophisticated method for implementing standard page styles throughout a site is via the **extensible style language (XSL)**. XSL is a specification for separating style from content when generating HTML documents (see www.w3.org/TR/xsl/ for more information). XSL allows designers to apply single style templates to multiple pages in a manner similar to that of Cascading Style Sheets. XSL allows designers to dictate how Web pages are displayed and whether the client device is a Web browser, handheld device, speech, or some other media. In other words, XSL provides designers with specifications that allow XML content to be seamlessly displayed on various client devices. This method of separating style from content is a significant departure from how normal HTML content is displayed.

Extensible style language (XSL): A specification for separating style from content when generating HTML documents.

In practical terms, XSL allows designers to separate presentation logic from site content. This separation standardizes "look and feel" without having to customize

Figure 16-10
Using HTML's link command for cascading style sheets

Sample Command:	
LINK HREF="style5.css" REL=StyleSheet TYPE="text/css" TITLE="Common Background Style" MEDIA="screen, print">	
Command Parameters:	
HREF="filename or URL"	Indicate the location of the linked object or document.
REL="relationship"	Specify the type of relationship between the document and linked object or document.
TITLE="object or document title"	Declare the title of the linked object or document.
TYPE="object to document type"	Declare the type of the linked object or document.
MEDIA="type of media"	Declare the type of medium or media to which the style sheet will be applied (e.g., screen, print, projection, aural, braille, tty, tv, all).

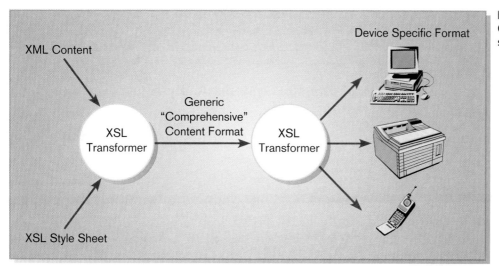

Figure 16-11
Combining XML data with XML stylesheet to format content

specifically to the capabilities of individual devices. Given the rapid evolution of devices (e.g., desktop computers, mobile computing devices, or televisions), XSL is a powerful method to assure that information is displayed in a consistent manner and utilizes the capabilities of the client device. XSL-based formatting consists of two parts:

1. Methods for *transforming* XML documents into a generic comprehensive form.
2. Methods for *formatting* the generic comprehensive form into a device specific form.

In other words, XML content, queried from a remote data source, is formatted based on rules within an associated XSL stylesheet (see Figure 16-11). This content is then translated to a device specific format and then displayed to the user. For example, if the user has made the request from a Web browser, the presentation layer will produce an HTML document. If the request has been made from a wireless mobile phone, the content will be delivered as a WML document.

Other Site Consistency Issues   In addition to using style sheets to enforce consistency in the design of a Website, it is also important that designers adopt standards for defining page and link titles. Every Web page should have a title that helps users better navigate through the site (see Nielsen, 1996). Page titles are used as the default description in bookmark lists, history lists, and within listings retrieved from search engines. Given this variety of use, page titles need to be clear and meaningful. However, care should be taken so that overly long titles—no more than 10 words— are not used. The selection of the actual words for the title are also extremely important, with two key issues to consider:

1. *Use unique titles.* Give each page a unique identity that represents its purpose and assists user navigation. If all pages have the same title, user will have difficulty returning to a prior page from the history list or distinguishing pages from the results of a search engine.
2. *Choose words carefully.* Given that titles are used for summarizing page content, chose words that assist users. Bookmark lists, history archives, and search engine results may be listed alphabetically; eliminate the use of articles like "An," "A," or "The" in the beginning of the title. Likewise, don't use a title such as "Welcome to My Company" but rather use "My Company—Home Page." The latter title will not only be much more meaningful to users, it will also provide a standard model for defining the titles of subsequent titles with your site (e.g., "My Company—Feedback" or "My Company—Products").

Figure 16-12
Using link titles to explain hyperlink

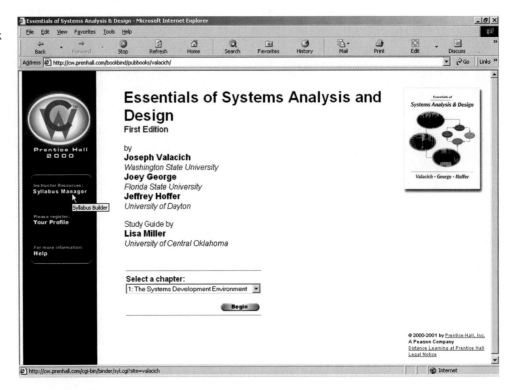

A major problem on the Internet is that many users do not know where they are going when they follow a hyperlink. In addition to taking great care when choosing link names, most browsers support the ability to pop up a brief explanation of the link before a user selects it (see Figure 16-12). Using link titles, users will be less likely to follow the wrong link and have more success when navigating your site. Some guidelines for defining link titles are summarized in Table 16-3.

This section highlighted issues that focused on the need for design consistency within an Internet Website. Experienced designers have learned that consistency not only makes the site easier for users, it also greatly simplifies site implementation and maintenance. It should be clear that careful attention to issues of design consistency would yield tremendous benefits.

Design Issues Related to Site Management

Maintenance is part of the ongoing management of the system. Many design issues will significantly influence the long-term successful operation of the system. Therefore we will discuss in this section those issues that are particularly important when designing an Internet-based system.

TABLE 16-3 Guidelines for Link Titles (Nielsen, 1998a)

Guideline	Description
Appropriate Information to Include	• Name of site or subsite link will lead to if different from current
	• Details about the type of information found on the destination page
	• Warnings about the selection of the link (e.g., "password required")
Length	Usually less than 80 characters—shorter is better
Limit Usage	Only add titles to links that are not obvious

Customer Loyalty and Trustworthiness   In order for your Website to become the preferred method for your customers to interact with you, they must feel that the site—and their data—is secure. There are many ways that the design of the site can convey trustworthiness to your users. Customers build trust from positive experiences interacting with a site. According to Web design guru Jakob Nielsen (1999), designers can convey trustworthiness in a Website in the following ways:

1. *Design quality.* A professional appearance and clear navigation conveys respect for customers and an implied promise of good service.
2. *Up-front disclosure.* Immediately inform users of all aspects of the customer relationship (e.g., shipping charges, data privacy policy) helps to convey an open and honest relationship.
3. *Comprehensive, correct, and current content.* Up-to-date content conveys a commitment to provide users with the most up-to-date information.
4. *Connected to the rest of the Web.* Linking to outside sites is a sign of confidence and credibility is gained; an isolated site feels like it may have something to hide.

In addition to these methods, protecting your customers' data will also be a significant factor for conveying trustworthiness. For example, many users are fearful of disclosing their e-mail address in fear of getting frequent unsolicited messages (*spam*). As a result, many users have learned to provide secondary e-mail address—using services like Hotmail or Yahoo mail—when trust has not yet been established. Consequently, if you need to gather a customer's e-mail address or other information, you should disclose why this is being done and how this information will be used in the future (e.g., information will only be used for order confirmation). Failing to consider how you convey trust to your customers may result in a system design that is not a success.

Another way to increase loyalty and to convey trustworthiness to customers is to provide useful, *personalized* content (see Nielsen, 1998b). **Personalization** refers to providing content to users based upon knowledge of that customer. For example, once you register and place orders on Amazon.com, each time you visit you are presented with a customized page based upon your prior purchase behavior (see Figure 16-13).

Personalization: Providing Internet content to users based upon knowledge of that customer.

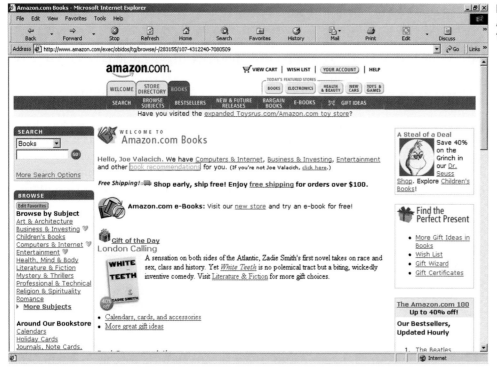

Figure 16-13
Amazon.com page showing links for purchasing recommendations

Figure 16-14
Microsoft MSN portal customized to a user's preferences

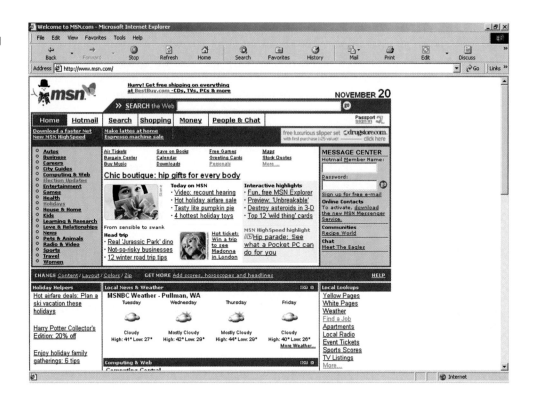

Customization: Internet sites that allow a user to customize information to their personal preferences.

N E T   S E A R C H

Many Internet portals provide users with the ability to customize what information they are provided. Visit http://www.prenhall.com/hoffer to complete an exercise related to this topic.

Personalization should not be confused with **customization** that refers to sites that allow a user to customize to their personal preferences and once defined have customized page content. For example, the popular Internet portals—Websites that offer a broad array of resources and services, such at Yahoo, MSN, and many of the popular search engines—allow users to customize the site to the user's preferences and interests (see Figure 16-14).

Because a personalized site *knows* you, each time you visit you are presented with new personalized content without having to enter any additional information. The site is able to personalize content because the system learns customer's buying preferences and builds a profile based upon this history. This method for personalizing site content is a success because users do not have to do anything to set it up. For example, users typically view the personalized data from Amazon.com favorably. To personalize each customer's content, Amazon compares a user's prior purchases with the purchasing behavior of millions of customers to reliably make purchase recommendations that may never have occurred to a customer. Amazon does a nice job of not making personalization recommendations too obtrusive so that if the system makes a bad guess at what the user might be interested in, users aren't annoyed by having the site trying to be smarter than it actually is. For example, many users visit Amazon to purchase books as gifts for friends; using this data to personalize the site may impede the user's experience when shopping for personal items.

Web Pages Must Live Forever   For commercial Internet sites, your pages must live forever and there are four primary reasons why professional developers have come to this conclusion (Nielsen, 1998c):

1. *Customer Bookmarks.* Because customers may bookmark any page on your site, you cannot ever remove a page without running the risk of losing a customer who would not find a working link if they encountered a *dead* link.

2. *Links from Other Sites.* Like your customers that bookmark pages, other sites may link directly to pages within your site; removing a page my result in losing customer referrals.

3. *Search Engines Referrals.* Because search engines are often slow to update their databases, this is another source for old and dead pages.

4. *Old Content Adds Value.* Some customers may find significant value from old content. The first three reasons are why links cannot die; but in addition to these practical issues, many users may actually find value from old content. Don't conclude from this discussion that Web content cannot change and evolve, but that the links themselves cannot die. In other words, when users bookmark a page and return to your site, this link should return something useful to the user, otherwise you run the risk of losing the customer.

Old content can remain useful to users because of historic interest, old product support, or background information for recent events. Additionally, the cost of keeping old content is relatively small, but it is important that old content is maintained so that links do not die and that obsolete or misleading information is corrected or removed. Finally, make sure that you explicitly date old content, provide disclaimers that point out what no longer applies or is accurate, and provide forward-pointing links to current pages. With a small amount of maintenance on your site's old content, you will provide a valuable resource to your customers. It should be obvious that customers who visit your site infrequently should easily find what they are looking for; otherwise they will become frustrated, leave, and not come back.

System Security   A paradox lies in the fact that, within a distributed system, security and ease of use are in conflict with each other. A secure system is often much less "user-friendly," whereas an easy-to-use system is often less secure. When designing an Internet-based system, successful sites strike an appropriate balance between security and ease of use. For example, many sites that require a password for site entry provide the functionality of "remember my password." This feature will make a user's experience at a site much more convenient and smooth, but this feature also results in a less secure environment. By remembering the password, anyone using this computer potentially has access the initial user's account and personal information.

In addition, if you must require customers to register to use your site and gain access via passwords, experienced designers have learned that it is best to delay customer registration and to not require registration to gain access to the top levels of the site. If you ask for registration too early before you have demonstrated value to a new customer, you run the risk of turning away the customer (Nielsen, 1997). Once a customer chooses to register on your site, make sure that the process is as simple as possible. Also, if possible, store user information in *cookies* rather than requiring users to reenter information each time they visit your site. Of course, if your site requires high-security requirement (e.g., a stock trading site), you may want to require users to enter an explicit password each visit. Security is clearly a double-edge sword. Too much and you might turn customers away; too little and you run the risk of losing customers because they do not trust the security of the site. Careful system design is needed so that the right balance is taken between security and ease of use.

Managing On-Line Data

Modern organizations are said to be drowning in data but starving for information. Despite the mixed metaphor, this statement seems to portray quite accurately the situation in many organizations. The advent of Internet-based electronic commerce has resulted in the collection of an enormous amount of customer and transactional data. How this data is collected, stored, and manipulated is a significant factor influencing the success of a commercial Internet Website. In this section we discuss system design issues for managing on-line data.

Figure 16-15
Context diagram comparing four distinct systems

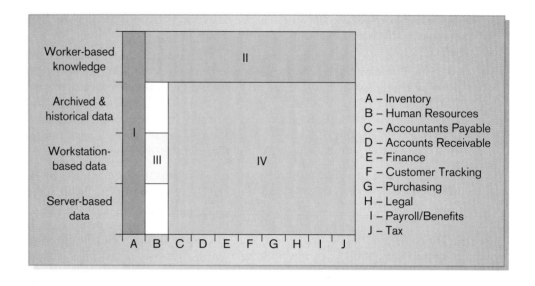

Context Development   Gaining an understanding of how a new system fits within the *context* of an organization's existing application portfolio is a fundamental part to effectively designing and managing system data. This understanding is necessary to assure that data can be effectively collected, stored, and managed. **Context development** is a method for helping you to gain an understanding of how a system fits within existing business activities and data. Two metrics—integration depth and organizational breadth—can be used to define a system's context. **Integration depth** reflects how far into the existing technology infrastructure a system penetrates. A "deep" integration both retrieves data from and sends data directly into existing systems. A "shallow" system will have minimal real-time coexistence with existing data sources. **Organizational breadth** tracks the core business functions that are affected by a system. A "wide" breadth reflects a situation when many distinct organizational areas have some type of interaction with the system; a "narrow" breadth reflects a situation when very few departments utilize or access the system.

Simply put, the context of an Internet-based system is an assessment of the integration depth and the organizational breadth of a system and can be represented using an *X-Y* graph. The *X*-axis measures the business functions affected by the system and the *Y*-axis measures how far a system penetrates into the existing technology infrastructure. For example, Figure 16-15 is a context graph for an organization like Pine Valley Furniture that has a broad range of organizational functions and a variety of information systems. Table 16-4 provides a relative comparison of these systems shown in Figure 16-15, each with varying depth and breadth. With this understanding, you are able to better understand how a new system would fit within an organization's existing application portfolio.

Context development: A method that helps you to better understand how a system fits within existing business activities and data.

Integration depth: A measurement of how far into the existing technology infrastructure a system penetrates.

Organizational breadth: A measurement that tracks the core business functions affected by a system.

PINE VALLEY FURNITURE

TABLE 16-4   Comparison of Relative System Context

System	Depth	Breadth	Context
I	Deep	Narrow	Inventory control
II	Shallow	Wide	Knowledge management across all business functions
III	Shallow	Narrow	System that gathers information residing on employee's workstations in HR department
IV	Deep	Wide	Enterprise Resource Planning (ERP) system

On-Line Transaction Processing (OLTP)   Figure 16-16 shows the level-0 data flow diagram from Pine Valley Furniture's WebStore application. Each of the processes defined in the DFD can be viewed as an autonomous transaction. For example, Process 5.0, Add/Modify Account Profile, shows one input from the Customer (Customer Information) and two outputs; one output to the Customer (Customer Information/ID) and one output to the Customer Tracking System (Customer Information). All of these operations are transactions. **On-line transaction processing** refers to immediate automated responses to the requests of users. OLTP systems are designed to specifically handle multiple concurrent transactions. Typically, these transactions have a fixed number of inputs and a specified output such as those represented in the DFD in Figure 16-16. Common transactions include receiving user

On-line transaction processing (OLTP): The immediate automated responses to the requests of users.

Figure 16-16
Dataflow diagram from Pine Valley Furniture's WebStore

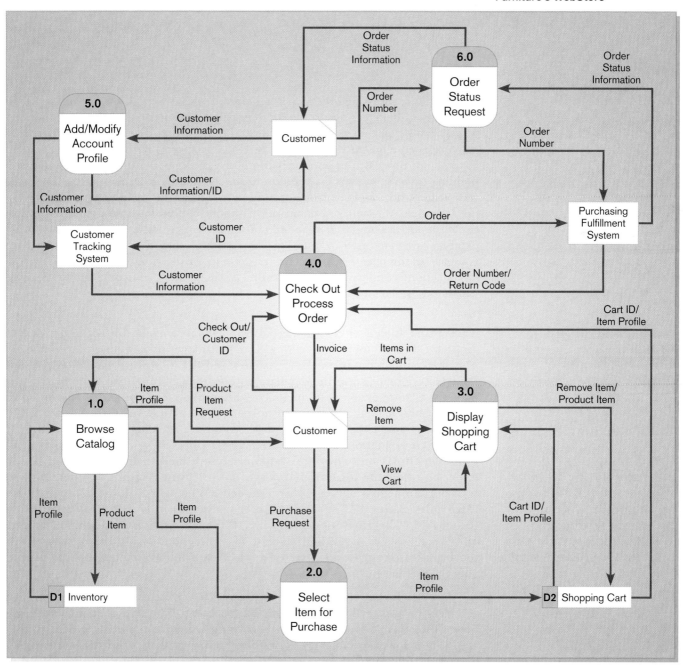

Figure 16-17
Global customers require that
on-line transactions be processed
efficiently

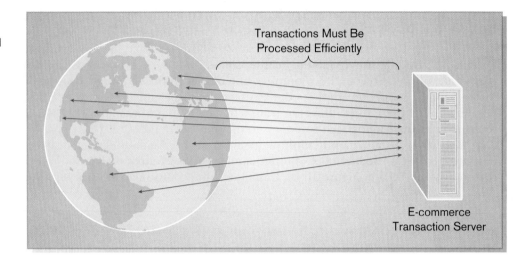

Transactions Must Be
Processed Efficiently

E-commerce
Transaction Server

information, processing orders, and generating sales receipt. Consequently, OLTP is a big part of interactive electronic commerce applications on the Internet. Since customers can be located virtually anywhere in the world, it is critical that transactions are processed efficiently (see Figure 16-17). The speed that database management systems can process transactions is therefore an important design decision when building Internet systems. In addition to which technology is chosen to process the transactions, how the data is organized is also a major factor in determining system performance. Although the database operations behind most transactions are relatively simple, designers often spend considerable time making adjustments to the database design in order to "tune" processing for optimal system performance. Once you have all this data, organizations must design ways to gain the greatest value from its collection; on-line analytical processing is one method being used to analyze these vast amounts of data.

**On-line analytical processing
(OLAP):** The use of graphical
software tools that provide complex
analysis of data stored in a database.

On-Line Analytical Processing (OLAP) **On-line analytical processing** refers to graphical software tools that provide complex analysis of data stored in a database. The chief component of an OLAP system is the "OLAP server," which understands how data is organized in the database and has special functions for analyzing the data. OLAP tools enable users to analyze different dimensions of data, beyond data summary and data aggradations of normal database queries. For example, OLAP can provide time series and trend analysis views of data, data drill-downs to deeper levels of consolidation, as well as the ability to answer "what if" and "why" questions. An OLAP query for Pine Valley Furniture might be: "What would be the effect on wholesale furniture costs if wood prices increased by 10 percent and transportation costs decreased by 5 percent?" Managers use the complex query capabilities of an OLAP system to answer questions within Executive Information Systems (EIS), Decision Support Systems (DSS), and Enterprise Resource Planning (ERP) systems. Given the high volume of transactions within Internet-based systems, analysts must provide extensive OLAP capabilities to managers in order to gain the greatest business value.

N E T   S E A R C H
As organizations collect more and
more data, using OLAP tools to
enhance decision making have
become very popular. Visit
http://www.prenhall.com/hoffer to
complete an exercise related to this
topic.

Merging Transaction and Analytical Processing The requirements for designing and supporting transactional and analytical systems are quite different. In a distributed on-line environment, performing real-time analytical processing will diminish the performance of transaction processing. For example, complex analytical

TABLE 16-5 Comparison of Operational and Informational Systems

Characteristic	Operational System	Informational System
Primary purpose	Run the business on a current basis	Support managerial decision making
Type of data	Current representation of state of the business	Historical or point-in-time (snapshot)
Primary users	On-line customers, clerks, salespersons, administrators	Managers, business analysts, customers (checking status, history)
Scope of usage	Narrow vs. simple updates and queries	Broad vs. complex queries and analysis
Design goal	Performance	Ease of access and use

queries from an OLAP system requires the locking of data resources for extended periods of execution time, whereas transactional events—data insertions and simple queries—are fast and can often occur simultaneously. Thus, a well-tuned and responsive transaction system may have uneven performance for customers while analytical processing occurs. As a result, many organizations replicate all transactions on a second server, so that analytical processing doesn't slow customer transaction processing performance. This replication typically occurs in batch during off-peak hours when site traffic volumes are at a minimum.

The systems that are used to interact with customers and run a business in real time are called the **operational systems**. Examples of operational systems are sales order processing and reservation systems. The systems designed to support decision making based on stable point-in-time or historical data are called **informational systems**. The key differences between operational and informational systems are shown in Table 16-5. Increasingly, data from informational systems are being consolidated with other organizational data into a comprehensive data warehouse, where OLAP tools can be used to extract the greatest and broadest understanding from the data.

Data Warehousing A **data warehouse** is a subject-oriented, integrated, time-variant, nonvolatile collection of data used in support of management decision making (Inmon and Hackathorn, 1994). The meaning of each of the key terms in this definition is the following:

1. *Subject-oriented.* A data warehouse is organized around the key subjects (or high-level entities) of the enterprise such as customers, patients, students, or products.

2. *Integrated.* Data housed in the data warehouse are defined using consistent naming conventions, formats, encoding structures, and related characteristics, collected from many operational systems within the organization and external data sources.

3. *Time-variant.* Data in the data warehouse contains a time dimension so that it may be used in historical record pertaining to the business.

4. *Nonvolatile.* Data in the data warehouse are loaded and refreshed from operational systems, but cannot be updated by end users.

In other words, data warehouses contain a broad range of data that, if analyzed appropriately, can provide a broad and coherent picture of business conditions at a single point in time. The basic architectures used for data warehouses are either a generic two-level architecture, or a more sophisticated three-level architecture. Of course, there are more complex architectures beyond these two basic models; however, this topic is beyond the scope of our discussion.

Operational systems: Systems that are used to interact with customers and run a business in real time.

Informational systems: Systems designed to support decision making based on stable point-in-time or historical data.

Data warehouse: A subject-oriented, integrated, time-variant, nonvolatile collection of data used in support of management decision making.

The generic two-level architecture is shown in Figure 16-18. Building a data warehouse using this architecture requires four basic steps.

1. Data are extracted from the various source systems files and databases. In a large organization there may be dozens or hundreds of such files and databases.

2. The data from the various source systems are transformed and integrated before being loaded into the data warehouse.

3. The data warehouse is a read-only database organized for decision support. It contains both detailed and summary data.

4. Users access the data warehouse by means of a variety of query languages and analytical tools.

The two-level architecture represents the earliest model but it is still widely used today. The two-level architecture works well in small- to medium-sized companies with a limited number of hardware and software platforms and a relatively homogeneous computing environment. However, for larger companies with a large number of data sources and a heterogeneous computing environment, this approach leads to problems in maintaining data quality and managing the data extraction processes (Devlin, 1997). These problems, together with the trend toward distributed computing, led to the expanded, three-level architecture shown in Figure 16-19. The three-level architecture has the following components:

1. Operational systems and data

2. Enterprise data warehouse

3. Data marts

Figure 16-18
Generic two-tier data warehouse architecture

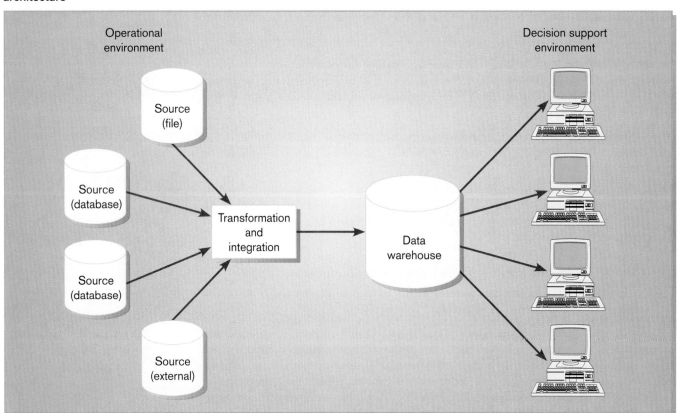

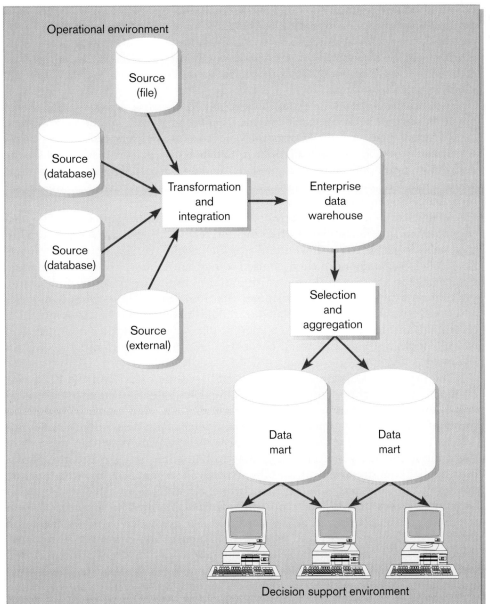

Figure 16-19
Three-layer data warehouse architecture

The first major difference between the two-level and the three-level architecture is the creation of the enterprise data warehouse. An **enterprise data warehouse (EDW)** is a centralized, integrated data repository that serves as the control point of all data made available to end users. This single data source drives decision-support applications throughout the entire organization. The EDW's purpose is twofold:

1. A centralized control point assures the quality and integrity of data, before it is made available to end-users.

2. The single data source provides an accurate, consolidated historical record of business, for time-sensitive data.

Although the enterprise data warehouse is the single source of all data for decision support, it is not typically accessed directly by end users. For most large organi-

Enterprise data warehouse (EDW): A centralized, integrated data warehouse that is the control point and single source of all data made available to end users for decision-support applications throughout the entire organization.

zations, the EDW is simply too large and too complex for users to navigate for most decision-support applications. This leads to the second major difference between the two- and three-level architecture. In a two-level architecture, users access data directly from the data warehouse via decision support tools. However, in a three-level architecture, users access the data that have been derived from the EDW that are stored in data marts.

A **data mart** is a data warehouse that is limited in scope. A data mart may be a physically separate subset of data extracted from the EDW, or it may be a customized, logical view of the data in the EDW of relevance to a class of users. Data marts contain selected information from the EDW such that each data mart is customized for the decision support applications of a particular end-user group. For example, an organization may have several data marts that are customized for a particular type of user such as a marketing data mart or a finance data mart.

A data warehouse for an Internet-based business can become huge. The data coming from Internet activities often include records of clickstream user actions, such as which links were clicked and in which sequence. Analysis of data warehouse clickstream data can then be used to customize and personalize a marketing message to a customer during a visit to the Website. For example, understanding that on a travel Website a customer first looks at airplane flight itineraries and then looks at books about related travel destinations would suggest including certain ads on that customer's Web page display whereas another customer who typically looks at rental cars when checking flight information would receive different ads.

The use of clickstream and other event-based data stored in the EDW (for example, purchase transactions, help desk inquiries, sales staff contacts) allows an organization to create an active data warehouse. For example, in a banking environment suppose a customer receives a large electronic direct deposit of funds into an account from, say, a Treasury note interest payment. Within minutes of this transaction the same customer may independently log on to the bank's Internet banking site to pay utility bills. Typically, the bank will be using separate operational applications to manage electronic, direct funds transfers and Internet banking. Without an EDW, the bank will have no way to link these transactions in real-time, and may miss a timely opportunity for generating new business and increasing customer loyalty and trust. With an active data enterprise warehouse, the transactional data from the separate operational applications are quickly fed to the EDW, which acts as a hub for sending messages to separate operational applications. In this environment, the bank can develop rules that would allow the Internet banking system to recognize the opportunity to attempt to automatically cross-sell the customer to a certificate of deposit or other investment account while the customer is using the Internet site.

Some Internet electronic commerce applications can receive and process millions of transactions per day. To gain the greatest understanding of customer behavior and to assure adequate system performance for customers, you must effectively manage on-line data. For example, Amazon.com is the world's largest bookstore with more than 2.5 million titles, and is open 24 hours a day, 365 days a year with customers all over the world ordering books and a broad range of other products. Amazon's servers log millions of transactions per day. Amazon is a vast departure from a more traditional physical bookstore. In fact, the largest physical bookstore carries "only" about 170,000 titles and it would not be economically feasible to build a physical bookstore the size of Amazon; a physical bookstore that carried Amazon's 2.5 million titles would need to be the size of nearly 25 football fields! The key to effectively designing an on-line electronic commerce business is clearly the effective management of on-line data. In this section we provided a very brief overview of this important topic (to learn more about managing on-line data, see Hoffer, Prescott, and McFadden, 2002).

Data mart: A data warehouse that is limited in scope; whose data are obtained by selecting and (where appropriate) summarizing data from the enterprise data warehouse.

ELECTRONIC COMMERCE APPLICATION: DESIGNING A DISTRIBUTED ADVERTISEMENT SERVER FOR PINE VALLEY FURNITURE'S WEBSTORE

In this chapter, we have examined numerous issues to consider when designing Internet-based systems. As we saw in the last chapter, prototyping can be useful in conceptualizing the look-and-feel of a Website. The look-and-feel is a function of the data presentation layer within an Internet-based application. What prototyping also provides is a view of the transactions and processes within the system. Transactions and processes managed by the middle layer, data analysis, of a three-tiered architecture. In this section, we will see how a distributed system, the advertisement rotation system, is integrated into Pine Valley Furniture's WebStore.

Advertising on Pine Valley Furniture's WebStore

PINE VALLEY FURNITURE

Having reviewed Jim Woo's throwaway prototype of the WebStore (see Chapter 15), Jackie Judson wanted to assess the feasibility of adding advertisements to the site. She came up with the following list of potential benefits for including advertising:

- Potential to increase revenue generated from the WebStore
- Potential to create cross-promotions and alliances with other on-line commerce systems
- Potential to provide customers improved service when looking for additional products that accessorize PVF's product line

Jim agreed with the principles of advertising on the site, and researched advertising examples on an array of different Internet sites. He compiled the following list of potential concerns that need to be addressed in the system design in order to implement a successful advertisement rotation system within the WebStore:

- Advertisements must be served quickly so that site performance is not affected.
- Advertisements must be uniform in size and resolution, as to not *disrupt* the site layout.
- Advertisement links must not redirect the user's browser away from the WebStore.

Designing the Advertising Component

To begin the process, Jim modified the style sheets of the prototype to include a space where the advertisement would appear. Since all advertisements would be approved by the marketing department before being included in the rotation, Jim could rely on the fact that they would be uniform in size and resolution. If the advertisement is clicked, a new, smaller window is opened and directed to the advertiser's site. The link is not direct, though. It is first directed to the advertising server within the WebStore system, the same server the advertisement came from. This "click-thru" transaction is logged and the user is sent to the appropriate destination.

Jim identified two distinct sets of data that would be generated by the advertisement rotation system: the number of advertisements served and the number of "click-thru's." This data being generated must be stored quickly and function transparently within the overall operation of the system. The transactional requirements of the advertisement system are:

1. Determine which advertisements apply, based on where the user is in the WebStore;

2. Personalize the advertisement if the identity of the user has been established and preferences are known;

3. Check for any seasonal or promotional advertisements; and

4. Log the transaction.

These requirements are part of the business rules that govern the rotation system. Jim and Jackie want these parameters to be flexible and scalable, so that future systems can incorporate these rules. To demonstrate how an advertisement might be placed on the WebStore, Jim modified the throwaway prototype to include a small banner ad (see Figure 16-20).

Designing the Management Reporting Component

Once the transactional requirements of the system were established, Jackie turned her attention to what reports she and other upper-level managers would like to see generated. Jim immediately began to scribble down all of the demographic information stored in the customer-tracking system, and cross-referenced it to the information stored when an advertisement was clicked. This led Jim and Jackie to identify numerous potential analytical queries that tied information from the customer-tracking system with the transactional data in the advertisement rotation system. A few of the queries they came up with were:

- "How many women, when shopping for desks, clicked on an advertisement for lamps?"

- "How many advertisements were served to shoppers looking at filing cabinets?"

- "How many people clicked on the first advertisement they saw?"

- "How many people clicked on an advertisement and then purchased something from the WebStore?"

Figure 16-20
Adding advertising to the WebStore throwaway prototype

Being able to analyze these and other results will provide critical feedback from targeted marketing campaigns, seasonal promotions, and product tie-ins. Using a distributed, transaction-based advertisement system in the WebStore will keep maintenance costs low and should increase the revenue generated from the site. Information derived from the analytical queries of advertisement transaction data increases the site's value even further.

Jackie and Jim reviewed the advertising model with the entire marketing staff. Many of the client account reps expressed interest in seeking a partnership with frequent customers to do advertising on the site. Junior sales staff members were eager to sell advertising space with the knowledge that they could provide purchasers with feedback on "click-thru" rates and overall advertisement views. One of the graphic designers even produced an advertisement on the spot for an upcoming product release. Everyone seemed to agree that the advertisement rotation system would increase the value of the WebStore to Pine Valley Furniture.

Summary

We have covered in this chapter various issues and technologies involved in the sharing of systems and data by multiple people across space and time in distributed and Internet systems. You learned about the client/server architecture, which is being used in both network personal computers and workstations (upsizing) and to replace older mainframe applications (downsizing). Components of the client/server architecture, including local area networks, database servers, application programming interfaces, and application development tools were described.

Two common types of local area network-based architectures—file server and client/server—were compared. It was shown that the newer client/server technologies have significant advantages over the older file servers. These advantages include less network traffic, greater flexibility to develop applications in convenient environments, and a more sensible distribution of queries to form a cooperative computing situation.

We also outlined in this chapter the evolution of distributed systems and three-tiered client/server technologies that are giving you more options for distributed system design. These client/server architectures—distributed presentation, remote presentation, remote data management, distributed function, distributed database, and distributed processing—are supported by some CASE tool code generators, but CASE tool support may be limited. Thus, the decision to use very sophisticated distributed processing environments must consider the potential for considerable human support for both programming and training as well as the availability of middleware to allow software and hardware on each tier to interact.

When designing Internet-based systems, standardized location naming, content translating, and document formatting enable designers to quickly craft systems because much of the complexity of the design and implementation are removed. These standards also free the designer from much of the worry of delivering applications over a broad range of computing devices and platforms. Many commercial Internet applications are vast and can contain thousands of distinct pages. A consistent "look and feel" is fundamental to conveying the image that the site is professionally designed. A site with high consistency is also much easier for users to navigate and is much more intuitive for users to anticipate the meaning of links. Two techniques for enforcing consistency when designing large-scale Web applications are through the use of Cascading Style Sheets (CSS) and the eXtensible Style Language (XSL). Given the desire to deliver the Internet onto a broader array of client devices, there is a trend to separate Web content from its delivery. Electronic commerce applications are particularly embracing this trend and are adopting standards like eXtensible Markup Language (XML) to author Web data and XSL to manage content formatting. In addition to using style sheets to enforce consistency in the design of a Website across client devices, it is also important that designers adopt standards for defining page and link titles. Finally, a successful design will make users feel that the site—and their data—is secure. Customers build trust from positive experiences interacting with a site. Taking steps to convey trustworthiness will help to attract and retain customers.

Gaining an understanding of how a new system fits within the context of an organization's existing application portfolio is a fundamental part to designing and managing system data. The major source of data within an Internet-based application is through the accumulation of customer transactions. On-line transaction processing (OLTP) refers to collection and immediate response to the requests of users interacting with a Web application. To improve decision making, organizations use on-line analytical processing (OLAP) to analyze the vast amounts of transaction data. OLAP refers to graphical software tools that provide complex analysis of data stored in a database or data warehouse. The purpose of a data warehouse is to consolidate and integrate data from a variety of sources, and to format those data in a context for making accurate business decisions. Most data warehouses today follow a three-layer architecture. The first layer consists of data distributed throughout the various operational systems. The second layer is an enterprise data warehouse, which is a centralized, integrated data warehouse that is the control point and single source of all data made available to end users for decision-support applications. The third layer is a series of data marts. A data mart is a data warehouse whose data are obtained by selecting and (where appropriate) summarizing data from the enterprise data warehouse. Each data mart is designed to support the decision-making needs of a department or other work group.

We did not have the space in this chapter to address several additional issues concerning distributed and Internet systems. Many of these issues are handled by other systems professionals, such as database administrators, telecommunications experts, and computer security specialists. Systems analysts must work closely with other professionals to build sound distributed systems.

Key Terms

1. Application program interface (API)
2. Application server
3. Cascading style sheets (CSS)
4. Client
5. Client/server architecture
6. Context development
7. Customization
8. Data mart
9. Data warehouse
10. Database engine
11. Domain naming system (BIND)
12. Enterprise data warehouse (EDW)
13. Extensible style language (XSL)
14. File server
15. Hypertext transfer protocol (HTTP)
16. Informational system
17. Integration depth
18. Local area network (LAN)
19. Middleware
20. On-line analytic processing (OLAP)
21. On-line transaction processing (OLTP)
22. Operational systems
23. Organizational breadth
24. Personalization
25. Thin client
26. Three-tiered client/server
27. eXtensible markup language (XML)

Match each of the key terms above with the definition that best fits it.

_____ The cabling, hardware, and software used to connect workstations, computers, and file servers located in a confined geographical area (typically within one building or campus).

_____ A device that manages file operations and is shared by each client PC attached to a LAN.

_____ A LAN-based computing environment in which a central database server or engine performs all database commands sent to it from client workstations, and application programs on each client concentrate on user interface functions.

_____ The (back-end) portion of the client/server database system running on the server and providing database processing and shared access functions.

_____ The (front-end) portion of the client/server database system that provides the user interface and data manipulation functions.

_____ Software building blocks that are used to assure that common system capabilities like user interfaces and printing are standardized as well as modules to facilitate the data exchange between clients and servers.

_____ A computing server where data analysis functions primarily reside.

_____ Advanced client/server architectures in which there are three logical and distinct applications—data management, presentation, and analysis—which are combined to create a single information system.

_____ A combination of hardware, software, and communication technologies that bring together data management, presentation, and analysis into a three-tiered client/server environment.

_____ A method for translating Internet domain names into Internet Protocol (IP) addresses.

_____ A communications protocol for exchanging information on the Internet.

_____ An Internet authoring language that allows designers to create customized tags, enabling the definition, transmission, validation, and interpretation of data between applications.

_____ A client device designed so that most processing and data storage occurs on the server.

_____ A set of style rules that tell a Web browser how to present a document.

_____ A specification for separating style from content when generating HTML documents.

_____ Providing Internet content to users based upon knowledge of that customer.

_____ Internet sites that allow a user to customize information to their personal preferences.

_____ A method that helps you to better understand how a system fits within existing business activities and data.

_____ A measurement of how far into the existing technology infrastructure a system penetrates.

_____ A measurement that tracks the core business functions affected by a system.

_____ The immediate automated responses to the requests of users.

_____ The use of graphical software tools that provide complex analysis of data stored in a database.

_____ Systems that are used to interact with customers and run a business in real time.

_____ Systems designed to support decision making based on stable point-in-time or historical data.

_____ A subject-oriented, integrated, time-variant, nonvolatile collection of data used in support of management decision making.

_____ A centralized, integrated data warehouse that is the control point and single source of all data made available to end users for decision-support applications throughout the entire organization.

_____ A data warehouse that is limited in scope; whose data are obtained by selecting and (where appropriate) summarizing data from the enterprise data warehouse.

Review Questions

1. Contrast the following terms.
   a. file server, client/server architecture, local area network
   b. hypertext markup language (HTML), hypertext transfer protocol (HTTP), domain naming system (BIND)
   c. Cascading style sheets (CSS), extensible style language (XSL)
   d. personalization, customization
   e. operational system, informational system
   f. integration depth, organizational depth
   g. on-line transaction processing (OLTP), on-line analytical processing (OLAP)
   h. data warehouse, enterprise data warehouse, data mart

2. Describe the limitations of a file server architecture.

3. Describe the advantages of a client/server architecture.

4. What are the major advantages and issues of the client/server architecture?

5. Summarize the six possible architectures for client/server systems.

6. Summarize the reasons for a three-tiered client/server architecture.

7. Explain the role of middleware in client/server computing.

8. In what ways do Internet standards like BIND, HTTP, and HTML assist designers when building Internet-based systems?

9. Why is it important to separate content from display when designing an Internet-based electronic commerce system?

10. How can cascading style sheets (CSS) and extensible style language (XSL) help to assure design consistency when designing an Internet-based electronic commerce system?

11. Discuss how you can instill customer loyalty and trustworthiness when designing an Internet-based electronic commerce system?

12. Why is it important that "Web pages live forever" when designing an Internet-based electronic commerce system?

13. Why do many commercial Websites have both operational and informational systems?

14. Briefly describe and contrast the components of a two-tier versus a three-tier data warehouse.

15. What is a data mart and why do some organizations use these to support organizational decision making?

Problems and Exercises

1. Under what circumstances would you recommend that a file server approach, as opposed to a client/server approach, be used for a distributed information system application? What warnings would you give the prospective user for this file server approach? What factors would have to change for you to recommend the move to a client/server approach?

2. Develop a table that summarizes the comparative capabilities of the six client/server architectures. You might start with Table 16-1 for some ideas.

3. Suppose you are responsible for the design of a new order entry and sales analysis system for a national chain of auto part stores. Each store has a PC that supports office functions. The company also has regional managers who travel from store to store working with the local managers to promote sales. There are four national offices for the regional managers, who each spend about one day a week in their office and four on the road. Stores place orders to replenish stock on a daily basis, based on sales history and inventory

levels. The company uses high-speed dial-in lines and modems to attach store PCs into the company's main computer. Each regional manager has a laptop computer with a modem and a network connection for times when the manager is in the office. Would you recommend a client/server distributed system for this company and, if so, which architecture would you recommend? Why?

4. The Internet is a network of networks. Using the terminology of this chapter, what type of distributed network architecture is used on the Internet?

5. Suppose you were designing applications for a standard file server environment. One issue discussed in the chapter for this distributed processing environment is that the application software on each client PC must share in the responsibilities for data management. One data management problem that can arise is that applications running concurrently on two clients may want to update the same data at the same time. What could you do to manage this potential conflict? Is there any way this conflict might result in both PCs making no progress (in other words, going into an infinite loop)? How might you avoid such problems?

6. An extension of the three-tiered client/server architecture is the *n*-tiered architecture, in which there are many specialized application servers. Extend the reasons for the three-tiered architecture to an *n*-tiered architecture.

7. You read in this chapter about the advantages of client/server architecture. What operational and management problems can client/server create? Considering both the advantages and disadvantages of client/server, suggest the characteristics of an application that could be implemented in a client/server architecture.

8. Obtain access to a typical PC database management system, such as Microsoft Access. What steps do you have to follow to link an Access database to a database on a server? Do any of these steps change depending on the database management system on the server?

9. There is a movement toward wireless mobile computing using thin client technology. Go to the Web and visit some of the major computer vendors who are producing thin client products like handheld computers, WAP phones, and personal digital assistants (PDAs). Investigate the features of each category of device and prepare a report that contrasts each type of device on at least the following criteria: screen size and color, networking options and speed, permanent memory, and embedded applications.

10. Building on the research conducted in Problem and Exercise 9 above, what challenges does each device present for designers when delivering an electronic commerce application? Are some devices more suitable for supporting some applications than for others?

11. Design consistency within an Internet site is an important way to build customer loyalty and trustworthiness. Visit one of your favorite Websites and analyze this site for design consistency. Your analysis should consider general layout, colors and fonts, labeling, links, and other such items.

12. Go to the Web and find a site that provides personalized content and a site that allows you to customize the site's content to your preferences. Prepare a report that demonstrates personalization and customization. Is one method better than the other? Why or why not?

13. Data warehousing is an important part of most large-scale commercial electronic commerce sites. Assume you are an executive with a leading company like Amazon.com; develop a list of questions that you would like to be answered by analyzing information within your company's data warehouse.

Field Exercises

1. Visit an organization that has installed a local area network (LAN). Explore the following questions.

   a. Inventory all application programs that are delivered to client PCs using a file server architecture. How many users use each application? What are their professional and technical skills? What business processes are supported by the application? What data are created, read, updated, or destroyed in each application? Could the same business processes be performed without using technology? If so, how? If not, why not?

   b. Inventory all application programs that are delivered to client PCs using a client/server architecture. How many users use each application? What are their professional and technical skills? What business processes are supported by the application? What data are created, read, updated, or destroyed in each application? Could the same business processes be performed without using technology? If so, how? If not, why not?

2. Scan the literature and determine the various local area network operating systems available. Describe the relative strengths and weaknesses of these systems. Do these systems seem to be adequate for distributed information system needs in organizations? Why or why not? Determine the current sales volume and approximate market shares for these systems. Why are they selling so well?

3. In this chapter file servers were described as one way of providing information to users of a distributed information system. What file servers are available and what are their relative strengths and weaknesses and costs? What other types of servers are available and/or for what other uses are file servers employed (e.g., print servers)?

4. Examine the capabilities of a client/server API environment. List and describe the types of client-based operations that you can perform with the API. List and describe the types of server-based operations that you can perform with the API. How are these operations the same/different?

5. The references in this chapter point to many sources of Website design guidelines. Visit these sites and summarize into a report guidelines not addressed in this chapter. Did you find inconsistencies or contradictions across the sites you studied? Why do these differences exist?

References

Askren, S. 1996. "Building Mult-Tier Apps Is About To Get Easier." *Client/Server Computing*, April: 61–64.

Castro, E. (2001). *XML for the World Wide Web*. Berkley, CA: Peachpit Press.

Devlin, B. (1997). *Data Warehouse: From Architecture to Implementation*. Reading, MA: Addison Wesley Longman.

Hoffer, J. A., M. B. Alexander, and F. R. McFadden. *2002. Modern Database Management*. 6th edition. Upper Saddle River, NJ: Prentice Hall.

Inmon, W. H, and Hackathorn, R. D. (1994). *Using the Data Warehouse*. New York: John Wiley & Sons.

Kara, D. 1997. "Why Partition?" *Application Development Trends*, May: 38–46.

Nielsen, J. (1996). "Marginalia of Web Design." November; www.useit.com/alertbox/9611.html.

Nielsen, J. (1997). "Loyalty on the Web." August 1; www.useit.com/alertbox/9708a.html.

Nielsen, J. (1998a). "Using Link Titles to Help Users Predict Where They Are Going." January 11; www.useit.com/alertbox/980111.html.

Nielsen, J. (1998b). "Personalization is Over-Rated." October 4; www.useit.com/alertbox/981004.html.

Nielsen, J. (1998c). "Web Pages Must Live Forever." November 29; www.useit.com/alertbox/981129.html.

Nielsen, J. (1999). "Trust or Bust: Communicating Trustworthiness in Web Design." March 7; www.useit.com/alertbox/990307.html.

Radding, A. 1992. "DBAs Find Tools Gap in C/S." *Software Magazine*. Client/Server Special Section 12 (16) (November), 33–38.

Stevens, L. 1996. "Consider Three-Tier Client/Server." *Datamation* (February 15): 61–64.

BROADWAY ENTERTAINMENT COMPANY, INC.

Designing Internet Features into the Customer Relationship Management System

CASE INTRODUCTION

The students from Stillwater State University are using Microsoft Access to build the prototype for MyBroadway, the Web-based customer relationship management system for Carrie Douglass, manager of the Centerville, Ohio, Broadway Entertainment Company (BEC) store. Although Access can be used to create passive, static Web pages for system output, it is not really an Internet application development tool. Nevertheless, the student team members want to follow good Website design principles so that the forms and reports in their final prototype can represent a design that the BEC corporate IS staff can, as directly as possible, migrate to a pure Web environment.

ESTABLISHING WEBSITE DESIGN PRINCIPLES FOR MYBROADWAY

The Stillwater students have already considered many human interface design guidelines (see the BEC cases in Chapters 13 and 14). However, before they develop all of the pages and finalize the navigation elements, the students decide to do some more research into Website design. From what the students learned in their MIS classes, they know that one well-recognized and impartial resource for Website design principles is the Website maintained by Jakob Nielsen (www.useit.com). This site is extensive, with many short articles of helpful hints for making Websites usable.

The BEC team decides to start to compile a set of guiding principles they believe will be important to the usability of the MyBroadway customer relationship management site. Based on scanning some of the articles on the Nielsen Website, the students develop an initial set of guidelines (see BEC Figure 16-1). The students recall from their classes that guidelines are just that, guidelines, not requirements. Exceptions can be made, with reason. Because Microsoft Access does not support style sheets, guidelines will be essential to gain consistency across the several team members who will be developing all the pages for the Website.

CASE SUMMARY

Sometimes the Stillwater students think their work will never end. There always seems to be one more page to design, one more requirement to include, one new perspective on system requirements, or one more factor to influence the usability of MyBroadway. This attraction to change is becoming a real hindrance to completing the project. Thus, the team members believe that they must finalize their design guidelines very soon, freeze the requirements from scope creep, and converge on the final prototype of all of MyBroadway. The team members are eager to implement the MyBroadway prototype. To insure that subsequent iterations of the prototype actually converge to a final prototype, they must firmly hold to well-developed design guidelines.

CASE QUESTIONS

1. The Stillwater students looked at articles on the Nielsen Website posted only through the end of year 2000 to form BEC Figure 16-1. Revisit this Website and update this figure based on guidelines and articles posted since the beginning of year 2001. Add only elements you believe are essential and relevant to the design of MyBroadway.

2. Review Chapters 13 and 14. Combine into your answer to Case Question 1 guidelines from these chapters. How unique do you consider the human interface design guidelines for a Website to be from general application design guidelines? Justify your answer.

3. Review the designs for MyBroadway pages presented in BEC Figures 13-1 and 13-2 in light of your answer to Case Question 2. Do you have specific recommendations for changes that need to be made to these pages?

4. Review your answers to Case Questions 4 and 5 in the BEC case in Chapter 13. Evaluate your designs based upon the guidelines you developed in your answer to Case Question 2 above.

Feature	Guideline
Interacting menus– avoid	An interacting menu changes when users select something in another menu on the same page. Users get very confused when options come and go, and it is often hard to make a desired option visible when it depends on a selection in a different widget.
Very long menus– avoid	Very long menus that require scrolling make it impossible for users to see all their choices in one glance. It's often better to present such long lists of options as a regular HTML list of traditional hypertext links.
Menus of abbreviations– avoid	It is usually faster for users to simply type the abbreviation (e.g., a two-character state code) than to select it from a drop-down menu. Free-form input requires validation by a code on the Web page or on the server.
Menus of well-known data– avoid	Selecting well-known data, such as month, city, or country often breaks the flow of typing for users, and creates other data entry problems.
Frames– use sparingly	Frames can be confusing when a user tries to print a page or when trying to link to another site. Frames can prevent a user from e-mailing a URL to other users and can be more clumsy for inexperienced users.
Moving page elements– use sparingly	Moving images have an overpowering effect on the human peripheral vision and can distract a user from productive use of other page content. Moving text may be difficult to read.
Scrolling– minimize	Some users will not scroll beyond the information that is visible on the screen. Thus, critical content and navigation elements should be obvious (on the top of the page, possibly in a frame on the top of the page so that these elements never leave the page).
Context– emphasize	Don't assume that users know as much about your site as you do. Users have difficulty finding information, so they need support in the form of a strong sense of structure and place. Start your design with a good understanding of the structure of the information from the user's perspective and communicate this structure explicitly to the user.
System status– make visible	The system should always keep users informed about what is going on, through appropriate feedback within reasonable time.

(continues)

BEC Figure 16-1 (continued)
Guidelines for Design of
MyBroadway Website

Feature	Guideline
Language– use user's terms	The system should consistently speak the users' language, with words, phrases, and concepts familiar to the user, rather than system-oriented terms. Follow real-world conventions, making information appear in a natural and logical order.
Fixing mistakes– make it easy	Users often choose system functions by mistake and will need a clearly marked "emergency exit" to leave the unwanted state without having to go through an extended dialog. Support undo and redo and default values. Even better than good error messages is a careful design that prevents a problem from occurring in the first place.
Actions– make them obvious	Make objects, actions, and options visible. The user should not have to remember information from one part of the dialog to another. Instructions for use of the system should be visible or easily retrievable whenever appropriate.
Customize– for flexibility and efficiency	Design the system for both novice and experienced users. Allow users to tailor the system to their frequent actions.
Content– make it relevant	Every extra unit of information in a dialog competes with the relevant units of information and diminishes their relative visibility.
Cancel button– use sparingly	The Web is a navigation environment where users move between pages of information. Since hypertext navigation is the dominant user behavior, users have learned to rely on the *Back* button for getting out of unpleasant situations. Offer a *Cancel* button when users may fear that they have committed to something they want to avoid. Having an explicit way to *Cancel* provides an extra feeling of safety that is not afforded by simply leaving.

Source: Adapted from the following sources: Jakob Nielsen Website (www.useit.com), specifically pages:
    http://www.useit.com/alertbox/20001112.html
    http://www.useit.com/alertbox/9605.html
    http://www.useit.com/papers/heuristic/heuristic_list.html
    http://www.useit.com/alertbox/20000416.html
    http://www.useit.com/alertbox/990502.html

5. Review your answer to Case Question 1 in the BEC case in Chapter 14. Evaluate your answer based upon the guidelines you developed in your answer to Case Question 2.

6. Search for other Web-based resources, besides the Nielsen Website, for Website design (Hint: look at the references at the end of this and prior chapters). In what ways do the design guidelines you find contra-

dict your answer to Case Question 2. Explain the differences.

7. Chapter 16 introduced the concepts of loyalty and trustworthiness as necessary for customers to interact with a Website. What elements, if any, are missing from the design of the MyBroadway Website that would improve the levels of loyalty and trustworthiness of BEC customers?

Part FIVE

Implementation and Maintenance

An Overview of Part FIVE

Implementation and Maintenance

Implementation and maintenance are the last two phases of the systems development life cycle. The purpose of implementation is to build a properly working system, install it in the organization, replace old systems and work methods, finalize system and user documentation, train users, and prepare support systems to assist users. Implementation also involves close-down of the project, including evaluating personnel, reassignment staff, assessing the success of the project, and turning all resources over to those who will support and maintain the system.

The purpose of maintenance is to fix and enhance the system to respond to problems and changing business conditions. Maintenance includes activities from all systems development phases. Maintenance also involves responding to requests to change the system, transforming requests into changes, designing the changes, and implementing changes.

We address the variety of work done during system implementation in Chapter 17. This chapter concentrates on the management of coding, in which you as an analyst will be directly involved. You will learn about testing systems and system components and methods to ensure and measure software quality. Your role as a systems analyst is to develop a plan for testing, which includes developing all the test data required to exercise every part of the system. You start developing the test plan early in the project, usually during analysis, since testing requirements are highly related to system functional requirements. You will learn how to document each test case and testing results. A testing plan usually follows a bottom-up approach, beginning with small modules, followed by extensive alpha testing by the programming group, beta testing with users, and final acceptance testing. Testing ensures the quality of software by using measures and methods such as structured walk-throughs. You will learn about a radical new technique, eXtreme programming, that combines coding and testing by using small programming teams.

Installing a new system involves more than making technical changes to computer systems. Managing installation includes managing organization changes as much as it does technical changes. We review several approaches to installation and several frameworks you can use to anticipate and control human and organizational resistance to change.

Documentation is extensive for any system. You have been developing most of the system documentation needed by system maintenance staff by keeping a thorough project workbook or CASE repository. You now need to finalize user documentation. In Chapter 17, you will see a generic outline for a user's guide as well as a wide range of guidelines you can use to develop high-quality user documentation. Remember, documentation must be tested for completeness, accuracy, and readability.

While documentation is being finalized, user support activities also need to be designed and implemented. Support includes training, whether through face-to-face classes, computer-based tutorials, or vendors. New electronic performance support systems deliver on-demand training. Increasingly, training is available from various sources over the Internet or corporate Intranets. Once trained, users will still encounter difficulties. Therefore, you must design ongoing support from help desks, newsletters, user groups, on-line bulletin boards, and other mechanisms, and these need to be tested and implemented. We conclude Chapter 17 with a brief review of project close-down activities, since the end of implementation means the end of the project, and with an example of implementation for the Webstore in Pine Valley Furniture.

But work on the system is just beginning. Today, as much as 80 percent of the life cycle cost of a system occurs after implementation. Maintenance handles updates to correct flaws and to accommodate new technologies as well as to meet new business conditions, regulations, and other requirements. In Chapter 18 you will learn about your role in systems maintenance.

There are four kinds of maintenance: corrective, adaptive, perfective, and preventive. You can help control the potentially monumental cost of a system by making systems maintainable. You can affect maintainability by reducing the number of defects, improving the skill of users, preparing high-quality documentation, and developing a sound system structure.

You may also be involved in establishing a maintenance group for a system. You will learn about different organizational structures for maintenance personnel and reasons for each. You will learn how to measure maintenance effectiveness. Configuration management and deciding how to handle change requests are important. You will learn how a systems librarian keeps track of baseline software modules, checks these out to maintenance staff, and then rebuilds systems. You will also learn about special issues for maintaining Web sites, and read about an example of a maintenance situation for the Webstore at Pine Valley Furniture.

Chapters 17 and 18 also include the final installments of the Broadway Entertainment Company (BEC) project case. These final two case segments help you to understand implementation and maintenance issues in an organizational context.

[17]

System Implementation

After studying this chapter, you should be able to:

● Describe the process of coding, testing, and converting an organizational information system and outline the deliverables and outcomes of the process.

● Prepare a test plan for an information system.

● Apply four installation strategies: direct, parallel, single location, and phased installation.

● List the deliverables for documenting the system and for training and supporting users.

● Distinguish between system and user documentation and determine which types of documentation are necessary for a given information system.

● Compare the many modes available for organizational information system training, including self-training and electronic performance support systems.

● Discuss the issues of providing support for end users.

● Explain why systems implementation sometimes fails.

● Compare the factor and political models of the implementation process.

● Show how traditional implementation issues apply to Internet-based systems.

INTRODUCTION

After maintenance, the implementation phase of the systems development life cycle is the most expensive and time-consuming phase of the entire life cycle. Implementation is expensive because so many people are involved in the process; it is time-consuming because of all the work that has to be completed during implementation. Physical design specifications must be turned into working computer code, the code must be tested until most of the errors have been detected and corrected, the system must be installed, user sites must be prepared for the new system, and users must come to rely on the new system rather than the existing one to get their work done.

Implementing a new information system into an organizational context is not a mechanical process. The organizational context has been shaped and reshaped by the people who work in the organization. The work habits, beliefs, interrelationships, and personal goals of an organization's members all affect the implementation process. Although factors important to successful implementation have been identified, there are no sure recipes you can follow. During implementation, you must be attuned to key aspects of the organizational context, such as history, politics, and environmental demands—aspects that can contribute to implementation failures if ignored.

Here you will learn about the many activities that the implementation phase comprises. In this chapter, we discuss coding, testing, installation, documentation, user training, support for a system after it is installed, and implementation success. Our intent is not to teach you how to program and test systems—most of you have already learned about writing and testing programs in the courses you took before this one. Rather, this chapter shows you where coding and testing fit in the overall scheme of implementation and stresses the view of implementation as an organizational change process that is not always successful.

In addition, you will learn about providing documentation about the new system for the information systems personnel who will maintain the system and for the system's users. These same users must be trained to use what you have developed and installed in their workplace. Once training has ended and the system has become institutionalized, users will have questions about the system's implementation and how to use it effectively. You must provide a means for users to get answers to these questions and to identify needs for further training.

As a member of the system development team that developed and implemented the new system, your job is winding down now that installation and conversion are complete. Chapter 18 completes the life cycle by focusing on maintenance. The end of implementation also marks the time for you to begin the process of project close-down. You read about project close-down in Chapter 3 when you learned about project management. At the end of this chapter, we will return to the topic of formally ending the systems development project.

After a brief overview of the coding, testing, and installation processes and the deliverables and outcomes from these processes, we will talk about software application testing. The following section presents the four types of installation: direct, parallel, single location, and phased. Afterwards, you will read about the process of documenting systems and training and supporting users as well as the deliverables from these processes. In the next section of the chapter, you will learn about the various types of documentation and numerous methods available for delivering training and support services. You will read about implementation as an organizational change process, with many organizational and people issues involved in the implementation effort. Finally, you will see how the implementation of an Internet-based system is similar to the implementation of more traditional systems.

SYSTEM IMPLEMENTATION

System implementation is made up of many activities. The six major activities we are concerned with in this chapter are coding, testing, installation, documentation, training, and support (see Figure 17-1). The purpose of these steps is to convert the physical system specifications into working and reliable software and hardware, document the work that has been done, and provide help for current and future users and caretakers of the system. These steps are often done by other project team members besides analysts, although analysts may do some programming. In any case, analysts are responsible for ensuring that all of these various activities are properly planned and executed. Next, we will briefly discuss these activities in two groups, first coding, testing, and installation, and then, documenting the system and training and supporting users.

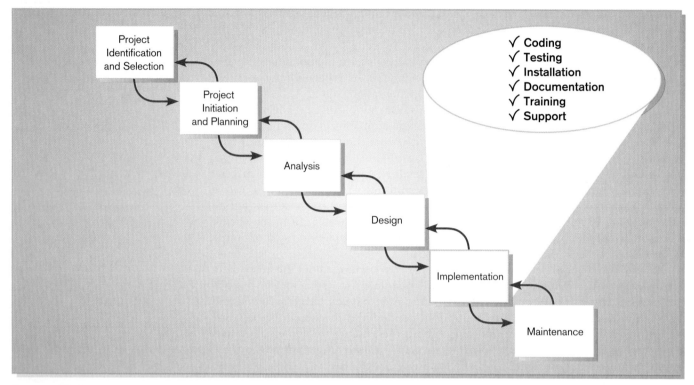

Figure 17-1
The systems development life cycle with the implementation phase highlighted

The Processes of Coding, Testing, and Installation

Coding, as we mentioned before, is the process whereby the physical design specifications created by the analysis team are turned into working computer code by the programming team. Depending on the size and complexity of the system, coding can be an involved, intensive activity. Once coding has begun, the testing process can begin and proceed in parallel. As each program module is produced, it can be tested individually, then as part of a larger program, and then as part of a larger system. You will learn about the different strategies for testing later in the chapter. We should emphasize that although testing is done during implementation, you must begin planning for testing earlier in the project. Planning involves determining what needs to be tested and collecting test data. This is often done during the analysis phase because testing requirements are related to system requirements.

Installation is the process during which the current system is replaced by the new system. This includes conversion of existing data, software, documentation, and work procedures to those consistent with the new system. Users must give up the old ways of doing their jobs, whether manual or automated, and adjust to accomplishing the same tasks with the new system. Users will sometimes resist these changes and you must help them adjust. However, you cannot control all the dynamics of user-system interaction involved in the installation process.

Deliverables and Outcomes from Coding, Testing, and Installation

Table 17-1 shows the deliverables from the coding, testing, and installation processes. The most obvious outcome is the code itself, but just as important as the code is documentation of the code. Some programming languages, such as COBOL, are said to be largely self-documenting because the language itself spells out much about the program's logic, the labels used for data variables, and the locations where data are accessed and output. But even COBOL code can be mysterious to maintenance pro-

TABLE 17-1   Deliverables for Coding, Testing, and Installation

1. Coding
   a. Code
   b. Program documentation
2. Testing
   a. Test scenarios (test plan) and test data
   b. Results of program and system testing
3. Installation
   a. User guides
   b. User training plan
   c. Installation and conversion plan
      i. Software and hardware installation schedule
      ii. Data conversion plan
      iii. Site and facility remodeling plan

grammers who must maintain the system for years after the original system was written and the original programmers have moved on to other jobs. Therefore, clear, complete documentation for all individual modules and programs is crucial to the system's continued smooth operation. Increasingly, CASE tools are used to maintain the documentation needed by systems professionals. The results of program and system testing are important deliverables from the testing process, as they document the tests as well as the test results. For example, what type of test was conducted? What test data were used? How did the system handle the test? The answers to these questions can provide important information for system maintenance as changes will require retesting and similar testing procedures will be used during the maintenance process.

The next two deliverables, user guides and the user training plan, result from the installation process. User guides provide information on how to use the new system, and the training plan is a strategy for training users so they can quickly learn the new system. The development of the training plan probably began earlier in the project and some training, on the concepts behind the new system, may have already taken place. During the early stages of implementation, the training plans are finalized and training on the use of the system begins. Similarly, the installation plan lays out a strategy for moving from the old system to the new, from the beginning to the end of the process. Installation includes installing the system (hardware and software) at central and user sites. The installation plan answers such questions as when the new system will be installed, which installation strategies will be used, who will be involved, what resources are required, which data will be converted and cleansed, and how long the installation process will take. It is not enough that the system is installed; users must actually use it.

As an analyst, your job is to ensure that all of these deliverables are produced and are done well. You may produce some of the deliverables, such as test data, user guides, and an installation plan; for other deliverables, such as code, you may only supervise or simply monitor their production or accomplishment. The extent of your implementation responsibilities will vary according to the size and standards of the organization you work for, but your ultimate role includes ensuring that all the implementation work leads to a system that meets the specifications developed in earlier project phases. See the box on the next page, The Future Programmer, for some insights into the changing role of programming and the nature of systems implementation in systems development.

The Processes of Documenting the System, Training Users, and Supporting Users

Although the process of documentation proceeds throughout the life cycle, it receives formal attention during the implementation phase because the end of implementation largely marks the end of the analysis team's involvement in system development. As the team is getting ready to move on to new projects, you and the other analysts need to prepare documents that reveal all of the important informa-

The Future Programmer

Where and by whom programming is done continues to change as the nature of programming languages evolves, resulting in improved programming productivity and the opening of programming to less highly skilled personnel. One prediction suggests that future programmers can be grouped into four categories.

- IS department programmers: the number of people who work in IS are in a clear decline, from nearly 2 million in 1994 to several hundred thousand by 2010; some people believe that these jobs are really being distributed out of the central IS function into business units, possibly under different titles.

- Software company programmers: These programmers work for consulting and packaged software companies, and the number will likely rise from roughly 600,000 in 1994 to several million by 2010.

- Embedded software programmers: These programmers produce code that is embedded in other products, like cars,

office equipment, and consumer electronics; this group will likely dramatically increase from several million in 1994 to over 10 million by 2010.

- Occasional programmers: Professionals and technicians (accountants, engineers, managers, and so forth) who program as part of their main duties; this group should rise from roughly 20 million in 1994 to over 100 million by 2010.

One theory is that standard business system components (objects in some terminologies) can be assembled into new systems by less skilled programmers. Thus, the job of what we call today a programmer will be to build these components and to ensure the quality of assembled systems. Although the number of occasional programmers (likely not trained in information systems) is exploding, the need for highly skilled programmers and programming work is far from diminishing.

Adapted from Bloor, 1994

tion you have learned about this system during its development and implementation. There are two audiences for this final documentation: (1) the information systems personnel who will maintain the system throughout its productive life and (2) the people who will use the system as part of their daily lives. The analysis team in a large organization can get help in preparing documentation from specialized staff in the information systems department.

Larger organizations also tend to provide training and support to computer users throughout the organization. Some of the training and support is very specific to particular application systems while the rest is general to particular operating systems or off-the-shelf software packages. For example, it is common to find courses on Microsoft Windows™ and WordPerfect™ in organization-wide training facilities. Analysts are mostly uninvolved with general training and support, but they do work with corporate trainers to provide training and support tailored to particular computer applications they have helped develop. Centralized information system training facilities tend to have specialized staff who can help with training and support issues. In smaller organizations that cannot afford to have well-staffed centralized training and support facilities, fellow users are the best source of training and support users have, whether the software is customized or off-the-shelf (Nelson and Cheney, 1987).

NET SEARCH
The nature of jobs in the IS field is ever changing. Visit http://www. prenhall.com/hoffer to complete an exercise related to this topic.

Deliverables and Outcomes from Documenting the System, Training Users, and Supporting Users

Table 17-2 shows the deliverables from documenting the system, training users, and supporting users. At the very least, the development team must prepare user documentation. The documentation can be paper-based, but it should also include computer-based modules. For modern information systems, this documentation includes any on-line help designed as part of the system interface. The development team should think through the user training process: Who should be trained? How much training is adequate for each training audience? What do different types of users need to learn during training? The training plan should be supple-

TABLE 17-2   Deliverables for Documenting the System, Training, and Supporting Users

1. Documentation	3. User training modules
a. System documentation	a. Training materials
b. User documentation	b. Computer-based training aids
2. User training plan	4. User support plan
a. Classes	a. Help desk
b. Tutorials	b. On-line help
	c. Bulletin boards and other support mechanisms

mented by actual training modules, or at least outlines of such modules, that at a minimum address the three questions stated previously. Finally, the development team should also deliver a user support plan that addresses such issues as how users will be able to find help once the information system has become integrated into the organization. The development team should consider a multitude of support mechanisms and modes of delivery. Each deliverable is addressed in more detail later in the chapter.

SOFTWARE APPLICATION TESTING

As we mentioned previously, analysts prepare system specifications that are passed on to programmers for coding. Although coding takes considerable effort and skill, the practices and processes of writing code do not belong in this text. However, as software application testing is an activity analysts plan (beginning in the analysis phase) and sometimes supervise, depending on organizational standards, you need to understand the essentials of the testing process.

Testing software begins earlier in the systems development life cycle, even though many of the actual testing activities are carried out during implementation. During analysis, you develop a master test plan. During design, you develop a unit test plan, an integration test plan, and a system test plan. During implementation, these various plans are put into effect and the actual testing is performed.

The purpose of these written test plans is to improve communication among all the people involved in testing the application software. The plan specifies what each person's role will be during testing. The test plans also serve as checklists you can use to determine whether all of the master test plan has been completed. The master test plan is not just a single document but a collection of documents. Each of the component documents represents a complete test plan for one part of the system or for a particular type of test. Presenting a complete master test plan is far beyond the scope of this book. Refer to Mosley's *Handbook of MIS Software Application Testing* for a complete test plan, which comprises a 101-page appendix. To give you an idea of what a master test plan involves, we present an abbreviated table of contents in Table 17-3.

A master test plan is a project within the overall system development project. Since at least some of the system testing will be done by people who have not been involved in the system development so far, the Introduction provides general information about the system and the needs for testing. The Overall Plan and Testing Requirements sections are like a baseline project plan for testing, with a schedule of events, resource requirements, and standards of practice outlined. Procedure Control explains how the testing is to be conducted, including how changes to fix errors will be documented. The fifth and final section explains each specific test necessary to validate that the system performs as expected.

TABLE 17-3   **Table of Contents of a Master Test Plan**

1. Introduction
   a. Description of system to be tested
   b. Objectives of the test plan
   c. Method of testing
   d. Supporting documents
2. Overall Plan
   a. Milestones, schedule, and locations
   b. Test materials
      1. Test plans
      2. Test cases
      3. Test scenarios
      4. Test log
   c. Criteria for passing tests
3. Testing Requirements
   a. Hardware
   b. Software
   c. Personnel
4. Procedure Control
   a. Test initiation
   b. Test execution
   c. Test failure
   d. Access/change control
   e. Document control
5. Test Specific or Component Specific Test Plans
   a. Objectives
   b. Software description
   c. Method
   d. Milestones, schedule, progression, and locations
   e. Requirements
   f. Criteria for passing tests
   g. Resulting test materials
   h. Execution control
   i. Attachments

Adapted from Mosley, 1993

Some organizations have specially trained personnel who supervise and support testing. Testing managers are responsible for developing test plans, establishing testing standards, integrating testing and development activities in the life cycle, and ensuring that test plans are completed. Testing specialists help develop test plans, create test cases and scenarios, execute the actual tests, and analyze and report test results.

Seven Different Types of Tests

Software application testing is an umbrella term that covers several types of tests. Mosley (1993) organizes the types of tests according to whether they employ static or dynamic techniques and whether the test is automated or manual. Static testing means that the code being tested is not executed. The results of running the code are not an issue for that particular test. Dynamic testing, on the other hand, involves execution of the code. Automated testing means the computer conducts the test while manual means that people do. Using this framework, we can categorize types of tests as shown in Table 17-4.

Let's examine each type of test in turn. **Inspections** are formal group activities where participants manually examine code for occurrences of well-known errors. Syntax, grammar, and some other routine errors can be checked by automated inspection software, so manual inspection checks are used for more subtle errors. Each programming language lends itself to certain types of errors that programmers make when coding, and these common errors are well-known and documented (for

Inspections: A testing technique in which participants examine program code for predictable language-specific errors.

TABLE 17-4   **A Categorization of Test Types**

	Manual	*Automated*
Static	Inspections	Syntax checking
Dynamic	Walkthroughs	Unit test
	Desk checking	Integration test
		System test

Adapted from Mosley, 1993

common coding errors in COBOL, see Litecky and Davis, 1976). Code inspection participants compare the code they are examining to a checklist of well-known errors for that particular language. Exactly what the code does is not investigated in an inspection. It has been estimated that code inspections have been used by organizations to detect from 60 to 90 percent of all software defects as well as to provide programmers with feedback that enables them to avoid making the same types of errors in future work (Fagan, 1986). The inspection process can also be used for such things as design specifications.

Unlike inspections, what the code does is an important question in a *walkthrough*. Using structured walkthroughs is a very effective method of detecting errors in code. As you saw in Chapter 6, structured walkthroughs can be used to review many systems development deliverables, including logical and physical design specifications as well as code. Whereas specification walkthroughs tend to be formal reviews, code walkthroughs tend to be informal. Informality tends to make programmers less apprehensive about walkthroughs and helps increase their frequency. According to Yourdon (1989), code walkthroughs should be done frequently when the pieces of work reviewed are relatively small and before the work is formally tested. If walkthroughs are not held until the entire program is tested, the programmer will have already spent too much time looking for errors that the programming team could have found much more quickly. The programmer's time will have been wasted, and the other members of the team may become frustrated because they will not find as many errors as they would have if the walkthrough had been conducted earlier. Further, the longer a program goes without being subjected to a walkthrough, the more defensive the programmer becomes when the code is reviewed. Although each organization that uses walkthroughs conducts them differently, there is a basic structure that you can follow that works well (see Figure 17-2).

It should be stressed that the purpose of a walkthrough is to detect errors, not to correct them. It is the programmer's job to correct the errors uncovered in a walkthrough. Sometimes it can be difficult for the reviewers to refrain from suggesting ways to fix the problems they find in the code, but increased experience with the process can help change a reviewer's behavior.

What the code does is also important in **desk checking,** an informal process where the programmer or someone else who understands the logic of the program works through the code with a paper and pencil. The programmer executes each instruction, using test cases that may or may not be written down. In one sense, the reviewer acts as the computer, mentally checking each step and its results for the entire set of computer instructions.

Desk checking: A testing technique in which the program code is sequentially executed manually by the reviewer.

GUIDELINES FOR CONDUCTING A CODE WALKTHROUGH

1. Have the review meeting chaired by the project manager or chief programmer, who is also responsible for scheduling the meeting, reserving a room, setting the agenda, inviting participants, and so on.
2. The programmer presents his or her work to the reviewers. Discussion should be general during the presentation.
3. Following the general discussion, the programmer walks through the code in detail, focusing on the logic of the code rather than on specific test cases.
4. Reviewers ask to walk through specific test cases.
5. The chair resolves disagreements if the review team cannot reach agreement among themselves and assigns duties, usually to the programmer, for making specific changes.
6. A second walkthrough is then scheduled if needed.

Figure 17-2
Steps in a typical walkthrough (Adapted from Yourdon, 1989)

Unit testing: Each module is tested alone in an attempt to discover any errors in its code.

Integration testing: The process of bringing together all of the modules that a program comprises for testing purposes. Modules are typically integrated in a top-down, incremental fashion.

System testing: The bringing together of all of the programs that a system comprises for testing purposes. Programs are typically integrated in a top-down, incremental fashion.

Stub testing: A technique used in testing modules, especially where modules are written and tested in a top-down fashion, where a few lines of code are used to substitute for subordinate modules.

Among the list of automated testing techniques in Table 17-4, there is only one technique that is also static, syntax checking. Syntax checking is typically done by a compiler. Errors in syntax are uncovered but the code is not executed. For the other three automated techniques, the code is executed.

The first such technique is **unit testing**, sometimes called module testing. In unit testing, each module is tested alone in an attempt to discover any errors that may exist in the module's code. But since modules coexist and work with other modules in programs and system, they must be tested together in larger groups. Combining modules and testing them is called **integration testing**. Integration testing is gradual. First you test the coordinating module (the root module in a structure chart tree) and only one of its subordinate modules. After the first test, you add one or two other subordinate modules from the same level. Once the program has been tested with the coordinating module and all of its immediately subordinate modules, you add modules from the next level and then test the program. You continue this procedure until the entire program has been tested as a unit. **System testing** is a similar process, but instead of integrating modules into programs for testing, you integrate programs into systems. System testing follows the same incremental logic that integration testing does. Under both integration and system testing, not only do individual modules and programs get tested many times. so do the interfaces between modules and programs.

Current practice calls for a top-down approach to writing and testing modules. Under a top-down approach, the coordinating module is written first. Then the modules at the next level in the structure chart are written, followed by the modules at the next level, and so on, until all of the modules in the system are done. Each module is tested as it is written. Since top-level modules contain many calls to subordinate modules, you may wonder how they can be tested if the lower level modules haven't been written yet. The answer is **stub testing**. Stubs are two or three lines of code written by a programmer to stand in for the missing modules. During testing, the coordinating module calls the stub instead of the subordinate module. The stub accepts control and then returns it to the coordinating module.

Figure 17-3 illustrates stub and integration system testing. Stub testing is depicted as the innermost oval. Here the Get module (where data are input and read) is being written and tested, but as none of its subordinate modules have been written yet, each one is represented by a stub. In the stub testing illustrated by Figure 17-3, Get is tested with only one stub in place, for its left-most subordinate module. You would of course write stubs for all of the Get module's subordinate modules, just as you would for the Make (where new information is calculated) and Put (where information is output) modules. Once all of the subordinate modules are written and tested, you would conduct an integration test of Get and its subordinate modules, as represented by the second oval. As stated previously, the focus of an integration test is on the interrelationships among modules. You would also conduct integration tests of Make and its subordinate modules, of Put and its subordinates, and of System and its subordinates, Get, Make, and Put. Eventually your integration testing would include all of the modules in the large oval that encompasses the entire program.

System testing is more than simply expanded integration testing where you are testing the interfaces between programs in a system rather than testing the interfaces between modules in a program. System testing is also intended to demonstrate whether a system meets its objectives. This is not the same as testing a system to determine whether it meets requirements—that is the focus of acceptance testing, which will be discussed later. To verify that a system meets its objectives, system testing involves using nonlive test data in a nonlive testing environment. *Nonlive* means the data and situation are artificial, developed specifically for testing purposes, although both data and environment are similar to what users would encounter in everyday system use. The system test is typically conducted by information systems personnel led by the project team leader, although it can also be conducted by users under MIS guidance. The scenarios that form the basis for system tests are prepared as part of the master test plan.

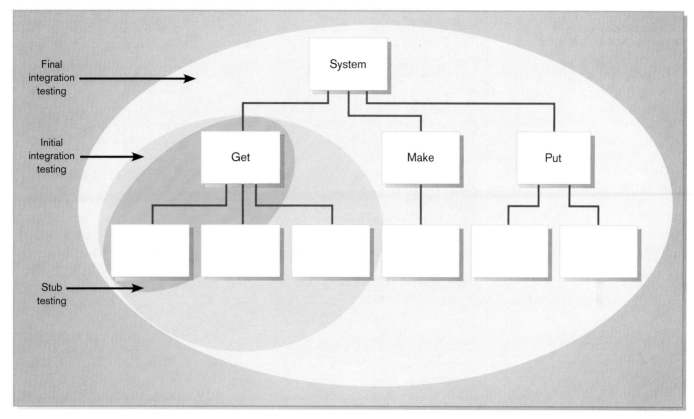

Figure 17-3
Comparing stub and integration testing

The Testing Process

Up to this point, we have talked about the master test plan and seven different types of tests for software applications. We haven't said very much about the process of testing itself. There are two important things to remember about testing information systems:

1. The purpose of testing is confirming that the system satisfies requirements.

2. Testing must be planned.

These two points have several implications for the testing process, regardless of the type of test being conducted. First, testing is not haphazard. You must pay attention to many different aspects of a system, such as response time, response to boundary data, response to no input, response to heavy volumes of input, and so on. You must test anything (within resource constraints) that could go wrong or be wrong about a system. At a minimum, you should test the most frequently used parts of the system as many other paths through the system as time permits. Planning gives analysts and programmers an opportunity to think through all the potential problem areas, list these areas, and develop ways to test for problems. As indicated previously, one part of the master test plan is creating a set of test cases, each of which must be carefully documented (see Figure 17-4 for an outline of a test case description).

A test case is a specific scenario of transactions, queries, or navigation paths that represent a typical, critical, or abnormal use of the system. A test case should be repeatable, so that it can be rerun as new versions of the software are tested. This is important for all code, whether written in-house, developed by a contractor, or purchased. Test cases need to determine that new software works with other existing software with which it must share data. Even though analysts often do not

Figure 17-4
Test case description form
(Adapted from Mosley, 1993)

Pine Valley Furniture Company
Test Case Description

Test Case Number:
Date:
Test Case Description:

Program Name:
Testing State:
Test Case Prepared By:

Test Administrator:

Description of Test Data:

Expected Results:

Actual Results:

do the testing, systems analysts, because of their intimate knowledge of applications, often make up or find test data. The people who create the test cases should not be the same people as those who coded and tested the system. In addition to a description of each test case, there must also be a description of the test results, with an emphasis on how the actual results differed from the expected results (see Figure 17-5). This description will indicate why the results were different and what, if anything, should be done to change the software. This description will then suggest the need for retesting, possibly introducing new tests necessary to discover the source of the differences.

One important reason to keep such a thorough description of test cases and results is so that testing can be repeated for each revision of an application. Although new versions of a system may necessitate new test data to validate new features of the application, previous test data usually can and should be reused. Results from the use of the test data with prior versions are compared to new versions to show that changes have not introduced new errors and that the behavior of the system, including response time, is no worse. A second implication for the testing process is that test cases must include illegal and out-of-range data. The system should be able to handle any possibility, no matter how unlikely; the only way to find out is to test.

If the results of a test case do not compare favorably to what was expected, the error causing the problem must be found and fixed. Programmers use a variety of

Pine Valley Furniture Company
Test Case Results

Test Case Number:
Date:

Program Name:
Module Under Test:

Explanation of difference between actual and expected output:

Suggestions for next steps:

Figure 17-5
Test case results form (Adapted from Mosley, 1993)

debugging tools to help locate and fix errors. A sophisticated debugging tool called a *symbolic debugger* allows the program to be run on-line, even one instruction at a time if the programmer desires, and allows the programmer to observe how different areas of data are affected as the instructions are executed. This cycle of finding problems, fixing errors, and rerunning test cases continues until no additional problems are found. There are specific testing methods that have been developed for generating test cases and guiding the test process (Mosley, 1993). See the box, Automating Testing, for an overview of tools to assist you in testing software.

N E T   S E A R C H
There are several software products used by system developers for software testing tasks. Visit http://www.prenhall.com/hoffer to complete an exercise related to this topic.

Automating Testing

Software testing tools provide the following functions, which improve the quality of testing:

- Record or build scripts of data entry, menu selections and mouse clicks, and input data, which can be replayed in exact sequence for each test run

- Compare the results of one test run with those from prior test cases to identify errors or to highlight the results of new features

- Supported unattended script playing to simulate high-volume or stress situations. Such tools can reduce the time for software testing by almost 80 percent.

Radically Combining Coding and Testing

Although coding and testing are in many ways part of the same process, it is not uncommon in large and complicated systems development environments to find the two practices separated from each other. Big companies and big projects often have dedicated testing staffs, who develop test plans and then use the plans to test software after it has been written. You have already seen how many different types of testing there are, and you can deduce from that how elaborate and extensive testing can be. A very different approach has been developed by Kent Beck as part of eXtreme programming (Beck, 2000). Under this approach, coding and testing are intimately related parts of the same process. The programmers who write the code also write the tests. The emphasis is on testing those things that can break or go wrong, not on testing everything. Code is tested very soon after it is written. The overall philosophy behind eXtreme programming is that code will be integrated into the system it is being developed for and tested within a few hours after it has been written. Code is written, integrated into the system, and then tested. If all the tests run successfully, then development proceeds. If not, the code is reworked until the tests are successful.

Another part of eXtreme programming that makes the code-and-test process work more smoothly is the practice of pair programming. All coding and testing is done by two people working together, writing code and writing tests. Beck says pair programming is not one person typing while the other one watches. Rather, the two programmers work together on the problem they are trying to solve, exchanging information and insight and sharing skills. Compared to traditional coding practices, the advantages of pair programming include: (1) more (and better) communication among developers, (2) higher levels of productivity, (3) higher quality code, and (4) reinforcement of the other practices in eXtreme programming, such as the code-and-test discipline (Beck, 2000). Although the eXtreme programming process has its advantages, just as with any other approach to systems development, it is not for everyone and not for every project.

Acceptance Testing by Users

Acceptance testing: The process whereby actual users test a completed information system, the end result of which is the users' acceptance of it.

Alpha testing: User testing of a completed information system using simulated data.

Beta testing: User testing of a completed information system using real data in the real user environment.

Once the system tests have been satisfactorily completed, the system is ready for **acceptance testing**, which is testing the system in the environment where it will eventually be used. Acceptance refers to the fact that users typically sign off on the system and "accept" it once they are satisfied with it. As we said previously, the purpose of acceptance testing is for users to determine whether the system meets their requirements. The extent of acceptance testing will vary with the organization and with the system in question. The most complete acceptance testing will include **alpha testing**, where simulated but typical data are used for system testing; **beta testing**, in which live data are used in the users' real working environment; and a system audit conducted by the organization's internal auditors or by members of the quality assurance group.

During alpha testing, the entire system is implemented in a test environment to discover whether or not the system is overtly destructive to itself or to the rest of the environment. The types of tests performed during alpha testing include the following:

- Recovery testing—forces the software (or environment) to fail in order to verify that recovery is properly performed.
- Security testing—verifies that protection mechanisms built into the system will protect if from improper penetration.
- Stress testing—tries to break the system (for example, what happens when a record is written to the database with incomplete information or what happens under extreme on-line transaction loads or with a large number of concurrent users).

Bugs in the Baggage

Testing a complex software system can be long and frustrating. A case in point was the software to control 4,000 baggage cars at the Denver International Airport. Errors in the software put the airport's opening on hold for months, costing taxpayers $500,000 a day and turning airport bonds into junk status. The airport was supposed to have opened in March 1994, but it did not open until February 1995 because of problems with the baggage-handling system. The system routinely damaged luggage and routed bags to the wrong flights. Various causes of the delay were identified, including last minute design change requests from airport officials and mechanical problems. The bottom-line lesson is that system designers must build in plenty of test and debugging time when scaling up proven technology into a much more complicated environment.

United Airlines, the major carrier at Denver International Airport, took over as systems integrator in October 1994. At the same time, the City of Denver commissioned a traditional conveyor-belt baggage-handling system in the airport for an additional $50 million. When the airport opened in 1995 (and as of this writing), only United used the automated baggage system, and then only to ferry bags to its flights. The system was not able to ferry bags from the planes back to the airport. All the other carriers used the traditional conveyor system. The City of Denver tried to recover $80 million of the system's $193 million cost from BAE Automated Systems Inc., the system's vendor. In 1996, BAE sued United for $17.5 million and the city for $4.1 million in withheld fees. United countersued. In September 1997, all sides settled, and no details were released.

Sources include Bozman, 1994; Leib, 1997; and Scheier, 1994.

- Performance testing—determines how the system performs on the range of possible environments in which it may be used (for example, different hardware configurations, networks, operating systems, and so on); often the goal is to have the system perform with similar response time and other performance measures in each environment.

In beta testing, a subset of the intended users run the system in their own environments using their own data. The intent of the beta test is to determine whether the software, documentation, technical support, and training activities work as intended. In essence, beta testing can be viewed as a rehearsal of the installation phase. Problems uncovered in alpha and beta testing in any of these areas must be corrected before users can accept the system. There are many stories systems analysts can tell about long delays in final user acceptance due to system bugs (see the box, Bugs in the Baggage, for one famous incident).

INSTALLATION

The process of moving from the current information system to the new one is called **installation**. All employees who use a system, whether they were consulted during the development process or not, must give up their reliance on the current system and begin to rely on the new system. Four different approaches to installation have emerged over the years: direct, parallel, single location, and phased (Figure 17-6). The approach an organization decides to use will depend on the scope and complexity of the change associated with the new system and the organization's risk aversion.

Installation: The organizational process of changing over from the current information system to a new one.

Direct Installation

The direct or abrupt approach to installation (also called "cold-turkey") is as sudden as the name indicates: the old system is turned off and the new system is turned on (Figure 17-6a). Under **direct installation**, users are at the mercy of the new system. Any errors resulting from the new system will have a direct impact on the users and how they do their jobs and, in some cases—depending on the centrality of the system to the organization—on how the organization performs its business. If the new system fails, considerable delay may occur until the old system can again be

Direct installation: Changing over from the old information system to a new one by turning off the old system when the new one is turned on.

Figure 17-6
Comparison of installation strategies
(a) Direct installation

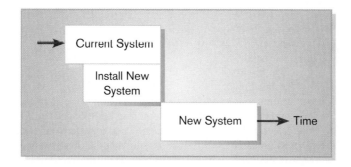

(b) Parallel installation

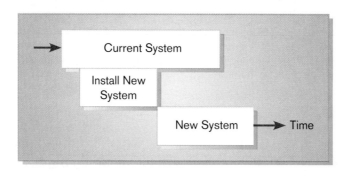

(c) Single location installation (with direct installation at each location)

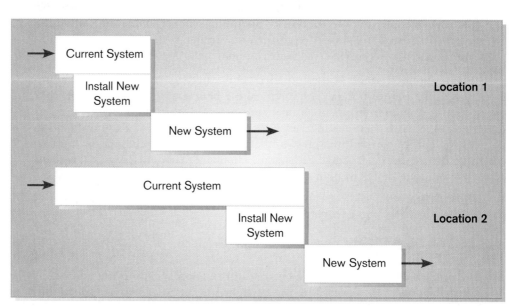

(d) Phased installation

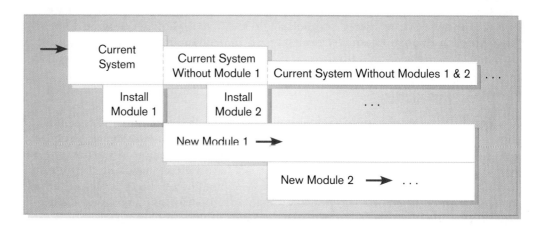

made operational and business transactions reentered to make the database up-to-date. For these reasons, direct installation can be very risky. Further, direct installation requires a complete installation of the whole system. For a large system, this may mean a long time until the new system can be installed, thus delaying system benefits or even missing the opportunities that motivated the system request. On the other hand, it is the least expensive installation method and it creates considerable interest in making the installation a success. Sometimes, a direct installation is the only possible strategy if there is no way for the current and new systems to coexist, which they must do in some way in each of the other installation approaches.

Parallel Installation

Parallel installation is as riskless as direct installation is risky. Under parallel installation, the old system continues to run alongside the new system until users and management are satisfied that the new system is effectively performing its duties and the old system can be turned off (Figure 17-6b). All of the work done by the old system is concurrently performed by the new system. Outputs are compared (to the greatest extent possible) to help determine whether the new system is performing as well as the old. Errors discovered in the new system do not cost the organization much, if anything, as errors can be isolated and the business can be supported with the old system. Since all work is essentially done twice, a parallel installation can be very expensive; running two systems implies employing (and paying) two staffs to not only operate both systems but also to maintain them. A parallel approach can also be confusing to users since they must deal with both systems. As with direct installation, there can be a considerable delay until the new system is completely ready for installation. A parallel approach may not be feasible, especially if the users of the system (such as customers) cannot tolerate redundant effort or the size of the system (number of users or extent of features) is large.

> **Parallel installation:** Running the old information system and the new one at the same time until management decides the old system can be turned off.

Single Location Installation

Single location installation, also known as location and pilot installation, is a middle-of-the-road approach compared to direct and parallel installation. Rather than convert all of the organization at once, single location installation involves changing from the current to the new system in only one place or in a series of separate sites over time (Figure 17-6c depicts this approach for a simple situation of two locations). The single location may be a branch office, a single factory, or one department, and the actual approach used for installation in that location may be any of the other approaches. The key advantage to single location installation is that it limits potential damage and potential cost by limiting the effects to a single site. Once management has determined that installation has been successful in one location, the new system may be deployed in the rest of the organization, possibly continuing with installation in one location at a time. Success at the pilot site can be used to convince reluctant personnel in other sites that the system can be worthwhile for them as well. Problems with the system (the actual software as well as documentation, training, and support) can be resolved before deployment to other sites. Even though the single location approach may be simpler for users, it still places a large burden on IS staff to support two versions of the system. On the other hand, because problems are isolated at one site at a time, IS staff can devote all its efforts to the success at the pilot site. Also, if different locations require sharing of data, extra programs will need to be written to synchronize the current and new systems; although this will happen transparently to users, it is extra work for IS staff. As with each of the other approaches (except phased installation), the whole system is installed; however, some parts of the organization will not get the benefits of the new system until the pilot installation has been completely tested.

> **Single location installation:** Trying out a new information system at one site and using the experience to decide if and how the new system should be deployed throughout the organization.

Phased Installation

Phased installation, also called staged installation, is an incremental approach. Under phased installation, the new system is brought on-line in functional components; different parts of the old and new systems are used in cooperation until the whole new system is installed (Figure 17-6d shows the phase-in of the first two modules of a new system). Phased installation, like single location installation, is an attempt to limit the organization's exposure to risk, whether in terms of cost or disruption of the business. By converting gradually, the organization's risk is spread out over time and place. Also, a phased installation allows for some benefits from the new system before the whole system is ready. For example, new data-capture methods can be used before all reporting modules are ready. For a phased installation, the new and replaced systems must be able to coexist and probably share data. Thus, bridge programs connecting old and new databases and programs often must be built. Sometimes, the new and old systems are so incompatible (built using totally different structures) that pieces of the old system cannot be incrementally replaced, so this strategy is not feasible. A phased installation is akin to bringing out a sequence of releases of the system. Thus, a phased approach requires careful version control, repeated conversions at each phase, and a long period of change, which may be frustrating and confusing to users. On the other hand, each phase of change is smaller and more manageable for all involved.

Planning Installation

Each installation strategy involves converting not only software but also data and (potentially) hardware, documentation, work methods, job descriptions, offices and other facilities, training materials, business forms, and other aspects of the system. For example, it is necessary to recall or replace all the current system documentation and business forms, which suggests that the IS department must keep track of who has these items so that they can be notified and receive replacement items. In practice you will rarely choose a single strategy to the exclusion of all others; most installations will rely on a combination of two or more approaches. For example, if you choose a single location strategy, you have to decide how installation will proceed there and at subsequent sites. Will it be direct, parallel, or phased?

Of special interest in the installation process is the conversion of data. Since existing systems usually contain data required by the new system, current data must be made error-free, unloaded from current files, combined with new data, and loaded into new files. Data may need to be reformatted to be consistent with more advanced data types supported by newer technology used to build the new system. New data fields may have to be entered in large quantities so that every record copied from the current system has all the new fields populated. Manual tasks, such as taking a physical inventory, may need to be done in order to validate data before they are transferred to the new files. The total data conversion process can be tedious. Furthermore, this process may require that current systems be shut off while the data are extracted so that updates to old data, which would contaminate the extract process, cannot occur.

Any decision that requires the current system to be shut down, in whole or in part, before the replacement system is in place must be done with care. Typically, off hours are used for installations that require a lapse in system support. Whether a lapse in service is required or not, the installation schedule should be announced to users well in advance to let them plan their work schedules around outages in service and periods when their system support might be erratic. Successful installation steps should also be announced, and special procedures put in place so that users can easily inform you of problems they encounter during installation periods. You should also plan for emergency staff to be available in case of system failure so that business operations can be recovered and made operational as quickly as possible. Another consideration is the business cycle of the organization. Most organizations face heavy workloads at particular times of year and relatively light loads at other times. A well-known example is the retail industry, where the busiest time of year is the fall, right before the year's major gift-giving holidays. You wouldn't want to schedule installation of a new point-of-sale system to begin December 1 for a department store. Make sure you understand the cyclical nature of the business you are working with before you schedule installation.

Planning for installation may begin as early as the analysis of the organization supported by the system. Some installation activities, such as buying new hardware, remodeling facilities, validating data to be transferred to the new system, and collecting new data to be loaded into the new system, must be done before the software installation can occur. Often the project team leader is responsible for anticipating all installation tasks and assigns responsibility for each to different analysts.

Each installation process involves getting workers to change the way they work. As such, installation should be looked at not as simply installing a new computer system, but as an organizational change process. More than just a computer system is involved—you are also changing how people do their jobs and how the organization operates.

DOCUMENTING THE SYSTEM

In one sense, every information systems development project is unique and will generate its own unique documentation. In another sense, though, system development projects are probably more alike than they are different. Each project shares a similar systems development life cycle, which dictates that certain activities be undertaken and each of those activities be documented. Bell and Evans (1989) illustrate how generic systems development life cycle maps onto a generic list of when specific systems development documentation elements are finalized (Table 17-5). As you compare the generic life cycle in Table 17-5 to the

TABLE 17-5   SDLC and Generic Documentation Corresponding to Each Phase

Generic Life-cycle Phase	Generic Document
Requirements specification	System requirements specification
	Resource requirements specification
Project control structuring	Management plan
	Engineering change proposal
System development	
Architectural design	Architecture design document
Prototype design	Prototype design document
Detailed design & implementation	Detailed design document
Test specification	Test specifications
Test implementation	Test reports
System delivery	User's guide
	Release description
	System administrator's guide
	Reference guide
	Acceptance sign-off

Adapted from Bell and Evans, 1989

life cycle presented in this book, you will see that there are differences, but the general structure of both life cycles is the same, as both include the basic phases of analysis, design, implementation, and project planning. Specific documentation will vary depending on the life cycle you are following, and the format and content of the documentation may be mandated by the organization you work for. However, a basic outline of documentation can be adapted for specific needs, as shown in Table 17-5. Note that this table indicates when documentation is typically finalized; you should start developing documentation elements early, as the information needed is captured.

We can simplify the situation even more by dividing documentation into two basic types, **system documentation** and **user documentation**. System documentation records detailed information about a system's design specifications, its internal workings, and its functionality. In Table 17-5, all of the documentation listed (except for System delivery) would qualify as system documentation. System documentation can be further divided into internal and external documentation (Martin and McClure, 1985). **Internal documentation** is part of the program source code or is generated at compile time. **External documentation** includes the outcome of all of the structured diagramming techniques you have studied in this book, such as data flow and entity-relationship diagrams. Although not part of the code itself, external documentation can provide useful information to the primary users of system documentation—maintenance programmers. In the past, external documentation was typically discarded after implementation, primarily because it was considered too costly to keep up-to-date, but today's CASE environment makes it possible to maintain and update external documentation as long as desired.

While system documentation is intended primarily for maintenance programmers (see Chapter 18), user documentation is intended primarily for users. An organization may have definitive standards on system documentation, often consistent with CASE tools and the system development process. These standards may include the outline for the project dictionary and specific pieces of documentation within it. Standards for user documentation are not as explicit.

User Documentation

User documentation consists of written or other visual information about an application system, how it works, and how to use it. An excerpt of on-line user documentation for Microsoft Access appears in Figure 17-7. Notice how the documentation has hot links to the meaning of important terms. The documentation lists the steps necessary to actually perform the task the user inquired about. The "notes" section that follows explains specific restrictions and constraints that will affect what the user is attempting. You should also notice how some words in the documentation are blue and that icons are used to represent Access buttons. This notation signifies that these particular words and buttons are hypertext links to related material elsewhere in the documentation. Hypertext techniques, rare in on-line PC documentation just a few years ago, are now the rule rather than the exception.

Figure 17-7 represents the content of a reference guide, just one type of user documentation (there is also a quick reference guide). Other types of user documentation include a user's guide, release description, system administrator's guide, and acceptance sign-off (Table 17-5). The reference guide consists of an exhaustive list of the system's functions and commands, usually in alphabetical order. Most on-line reference guides allow you to search by topic area or by typing in the first few letters of your keyword. Reference guides are very good for very specific information (as in Figure 17-7) but not as good for the broader picture of how you perform all the steps required for a given task. The quick reference guide provides essential

System documentation: Detailed information about a system's design specifications, its internal workings, and its functionality.

User documentation: Written or other visual information about an application system, how it works, and how to use it.

Internal documentation: System documentation that is part of the program source code or is generated at compile time.

External documentation: System documentation that includes the outcome of structured diagramming techniques such as data flow and entity-relationship diagrams.

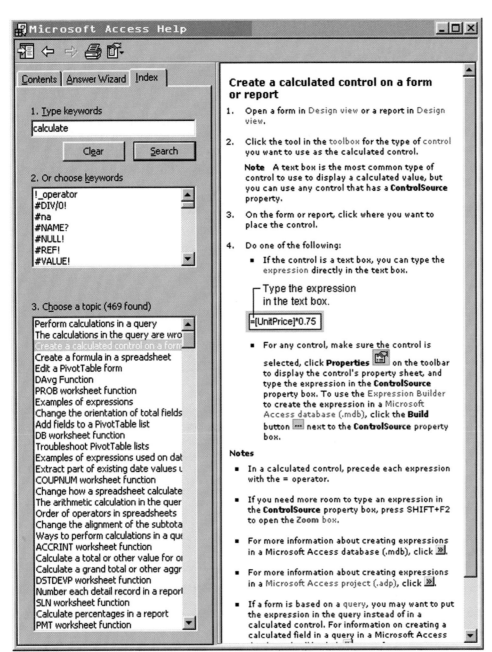

Figure 17-7
Example of on-line user documentation (from Microsoft Access™)

TABLE 17-6   Outline of a Generic User's Guide

Preface
1. Introduction
  1.1 Configurations
  1.2 Function flow
2. User interface
  2.1 Display screens
  2.2 Command types
3. Getting started
  3.1 Login
  3.2 Logout
  3.3 Save
  3.4 Error recovery
  3.n [Basic procedure name]
n. [Task name]
Appendix A—Error Messages
([Appendix])
Glossary
  Terms
  Acronyms
Index)

Adapted from Bell and Evans, 1989

information about operating a system in a short, concise format. Where computer resources are shared and many users perform similar tasks on the same machines (as with airline reservation or mail order catalog clerks), quick reference guides are often printed on index cards or as small books and mounted on or near the computer terminal. An outline for a generic user's guide (from Bell and Evans, 1989) is shown in Table 17-6. The purpose of such a guide is to provide information on how users can use computer systems to perform specific tasks. The information in a user's guide is typically ordered by how often tasks are performed and how complex they are.

In Table 17-6, sections with an "n" and a title in square brackets mean that there would likely be many such sections, each for a different topic. For example, for an accounting application, sections 4 and beyond might address topics such as entering a transaction in the ledger, closing the month, and printing reports. The items in parentheses are optional, included as necessary. An index becomes more important the larger the user's guide. Although a generic user's guide outline is helpful in providing an overview for you of what a user's guide might contain, we have included outlines of user guides from several popular PC software packages in Figure 17-8. Notice how different they are.

A release description contains information about a new system release, including a list of complete documentation for the new release, features and enhancements, known problems and how they have been dealt with in the new release, and information about installation. A system administrator's guide is intended primarily for those who will install and administer a new system and contains information about the network on which the system will run, software interfaces for peripherals such as printers, troubleshooting, and setting up user accounts. Finally, the acceptance sign-off allows users to test for proper system installation and then signify their acceptance of the new system with their signatures.

Figure 17-8
Outlines of user's guides from various popular PC software packages
(a) Microsoft Word™

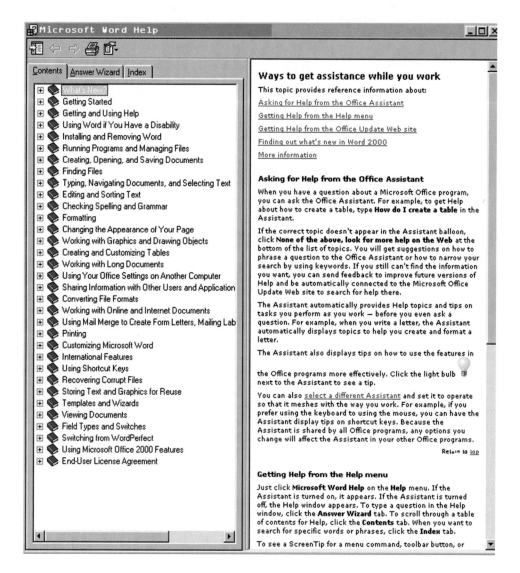

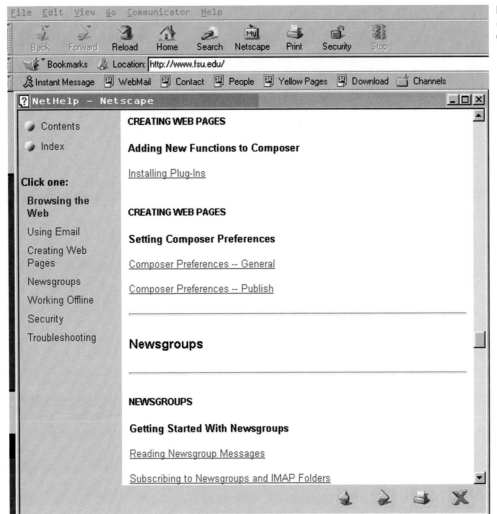

Preparing User Documentation

User documentation, regardless of its content or intended audience, was once provided almost exclusively in big, bulky paper manuals, and it was out of date almost as soon as it was printed. Most documentation is now delivered on-line in hypertext format. Regardless of format, user documentation is an upfront investment that should pay off in reduced training and consultation costs later (Torkzadeh and Doll, 1993). As a future analyst, you need to consider the source of documentation, its quality, and whether its focus is on the information system's functionality or on the tasks the system can be used to perform.

The traditional source of user documentation has been the organization's information systems department. Even though information systems departments have always provided some degree of user documentation, for much of the history of data processing the primary focus of documentation was the system. In a traditional information systems environment, the user interacted with an analyst and computer operations staff for all of his or her computing needs (Figure 17-9). The analyst acted as intermediary between the user and all computing resources. Any reports or other output that went to the user were generated by the operations staff, based on a regular reporting schedule. Since users were consumers and providers of data and infor-

Figure 17-9
Traditional information system environment and its focus on system documentation (Adapted from Torkzadeh and Doll, 1993)

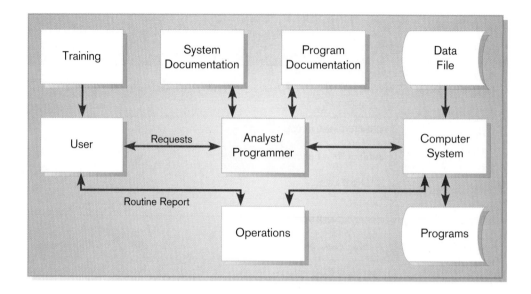

NET SEARCH
The ability to write information systems documentation is an important skill. Visit http://www.prenhall.com/hoffer to complete an exercise related to this topic.

mation, most documentation developed during a traditional information systems development project was system documentation for the analysts and programmers who had to know how the system worked. Although some user documentation was generated, most documentation was intended to assist maintenance programmers who tended to the system for years after the analysis team had finished its work.

In today's end-user information systems environment, users interact directly with many computing resources, users have many options or querying capabilities from which to choose when using a system, and users are able to develop many local applications themselves (Figure 17-10). Analysts often serve as consultants for these local end-user applications. For end-user applications, the nature and purpose of documentation has changed from documentation intended for the maintenance programmer to documentation for the end user. Application-oriented documentation, whose purpose is to increase user understanding and utilization of the organization's computing resources, has also come to be important. While some of this user-oriented documentation continues to be supplied by the information systems department, much of it now originates with vendors and with users themselves.

Figure 17-10
End user information system environment and its focus on user documentation (Adapted from Torkzadeh and Doll, 1993)

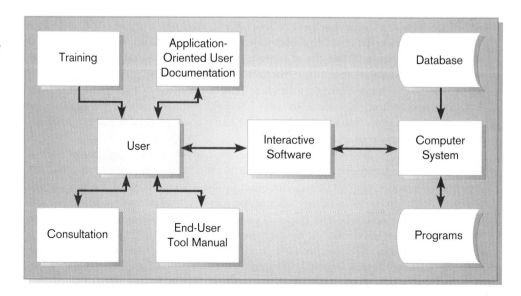

TRAINING AND SUPPORTING USERS

Training and **support** are two aspects of an organization's **computing infrastructure** (Kling and Iacono, 1989). Computing infrastructure is made up of all the resources and practices required to help people adequately use computer systems to do their primary work. It is analogous to the infrastructure of water mains, electric power lines, streets, and bridges that form the foundation for providing essential services in a city. Henderson and Treacy (1986) identify infrastructure as one of four fundamental issues IS managers must address. They suggest that training and support are most important in the early stages of end-user computing growth and less so later on. Rockart and Short (1989) cite the "development and implementation of a general, and eventually 'seamless,' information technology infrastructure" as a major demand on information technology. They list the creation of an effective information technology infrastructure as one of the five key issues for senior organizational managers in the 1990s. Thus, training and support are critical for the success of an information system. As the person whom the user holds responsible for the new system, you and other analysts on the project team must ensure that high-quality training and support are available.

Although training and support can be talked about as if they are two separate things, in organizational practice the distinction between the two is not all that clear, as the two sometimes overlap. After all, both deal with learning about computing. It is clear that support mechanisms are also a good way to provide training, especially for intermittent users of a system (Eason, 1988). Intermittent or occasional system users are not interested in, nor would they profit from, typical user training methods. Intermittent users must be provided with "point of need support," specific answers to specific questions at the time the answers are needed. A variety of mechanisms, such as the system interface itself and on-line help facilities, can be designed to provide both training and support at the same time.

The value of support is often underestimated. Few organizations invest heavily in support staff, which can lead to users solving problems for themselves or somehow working around them (Gasser, 1986). Adequate user support may be essential for successful information system development, however. One study found that user satisfaction with support provided by the information systems department was the factor most closely related to overall satisfaction with user development of computer-based applications (Rivard and Huff, 1988).

Training Information Systems Users

Computer use requires skills, and training people to use computer applications can be expensive for organizations (Kling and Iacono, 1989). Training of all types is a major activity in American corporations (they spent $54 billion on training in 2000), but information systems training is often neglected. Many organizations tend to underinvest in computing skills training. It is true that some organizations institutionalize high levels of information system training, but many others offer no systematic training at all. Still, of those U.S. companies that provide training, 99 percent teach employees how to use computer applications. Almost three-quarters of that computer application training is delivered in a classroom by a live, in-person instructor (Training, 2000).

The type of necessary training will vary by type of system and expertise of users. The list of potential topics from which you must determine if training will be useful include the following:

- Use of the system (e.g., how to enter a class registration request)
- General computer concepts (e.g., computer files and how to copy them)
- Information system concepts (e.g., batch processing)

Support: Providing ongoing educational and problem-solving assistance to information system users. For in-house developed systems, support materials and jobs will have to be prepared or designed as part of the implementation process.

Computing infrastructure: All the resources and practices required to help people adequately use computer systems to do their primary work.

TABLE 17-7   Seven Common Methods for Computer Training in 1987

1. Tutorial—one person taught at a time
2. Course—several people taught at a time
3. Computer-aided instruction
4. Interactive training manuals—combination of tutorials and computer-aided instruction
5. Resident expert
6. Software help components
7. External sources, such as vendors

Adapted from Nelson and Cheney, 1987

- Organizational concepts (e.g., FIFO inventory accounting)
- System management (e.g., how to request changes to a system)
- System installation (e.g., how to reconcile current and new systems during phased installation)

As you can see from this partial list, there are potentially many topics that go beyond simply how to use the new system. It may be necessary for you to develop training for users in other areas so that users will be ready, conceptually and psychologically, to use the new system. Some training, like concept training, should begin early in the project since this training can assist in the "unfreezing" element of the organizational change process.

Each element of training can be delivered in a variety of ways. Table 17-7 lists the most common training methods used by information system departments a few years ago. Despite the importance and value of training, most of the methods listed in Table 17-7 are underutilized in many organizations. Users primarily rely on just one of these delivery modes: more often than not, users turn to the resident expert and to fellow users for training, as shown in how often people use each mode listed in Table 17-7 (see Figure 17-11). One study reported that 89.4 percent of end users consulted their colleagues about how to use microcomputers, but only 48 percent consulted central information systems staff (Lee, 1986). Users are more likely to turn to local experts for help than to the organization's technical support staff because the local expert understands both the users' primary work and the computer systems they use (Eason, 1988). Given their dependence on fellow users for training, it should not be surprising that end users describe their most common mode of computer training as "self-training" (Nelson and Cheney, 1987). Self-training has been found to be associated with particular sets of user skills: using application development software, using packaged application software, data communication, using hardware, using operating systems, and graphics skills. The last four sets of skills are also highly associated with company-provided training. There appear to be areas of training best accomplished by centralized, company-provided training on the one hand and by self-training on the other.

One conclusion from the experience with user training methods is that an effective strategy for training on a new system is to first train a few key users and then organize training programs and support mechanisms which involve these users to provide further training, both formal and on demand. Often, training is most effective if you customize it to particular user groups, and the lead trainers from these groups are in the best position to do this.

Figure 17-11
Frequency of use of computer training methods (Nelson and Cheney, 1987)

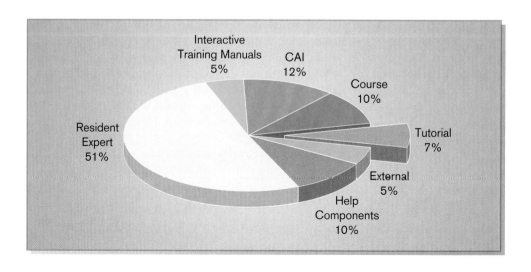

While individualized training is expensive and time-consuming, technological advances and decreasing costs have made this type of training more feasible. Similarly, the number of training modes used by information systems departments today has increased dramatically beyond what is listed in Table 17-7. Training modes now include videos, interactive television for remote training, multimedia training, on-line tutorials, and **electronic performance support systems (EPSS)**. These may be delivered via videotapes, CD-ROMS, company intranets, and the Internet.

Electronic performance support systems are on-line help systems that go beyond simply providing help—they embed training directly into a software package (Cole et al., 1997). An EPSS may take on one or more forms: they can be an on-line tutorial, provide hypertext-based access to context-sensitive reference material, or consist of an expert system shell that acts as a coach. The main idea behind the development of an EPSS is that the user never has to leave the application to get the benefits of training. Users learn a new system or unfamiliar features at their own pace and on their own machines, without having to lose work time to remote group training sessions. Furthermore, this learning is on-demand when the user is most motivated to learn, since the user has a task to do. EPSS is sometimes referred to as "just-in-time knowledge." One example of an EPSS with which you may be familiar is Microsoft's Office Assistant[TM]. Office Assistants (Figure 17-12) are animated characters that come up on top of such applications as Access[TM] and Word[TM]. You ask questions and the Office Assistants return answers that provide educational information, such as graphics, examples, and procedures, as well as hypertext nodes for jumping to related help topics. Microsoft's Office Assistants communicate with the application you are running to see where you are, so you can determine, by reading the context-sensitive information, if what you want to do is possible from your present location. Some EPSS environments actually walk the user step-by-step through the task, coaching the user on what to do or allowing the user to get additional on-line assistance at any point.

Training for information systems is increasingly being made available both over company intranets and over the Internet. International Data Corporation projects that the on-line training industry will grow to a $11.4 billion business by the year 2003 (Sambataro, 2000). Individual companies may prepare the training and make it available with the help of vendors who convert the content to work on the Internet. This is the case with PriceWaterhouseCoopers's Tax News Network, which is available to 2000 subscribers as well as 3000 company staffers. PriceWaterhouseCoopers features a two-hour training course on new tax legislation on its password-protected site. Alternatively, an organization may prepare its own training content using course authoring software. An example of this approach is Stanford University's Education Program for Gifted Youth, which Stanford provides over the Internet from its own servers. Still a third alternative is to access training provided by third party vendors. An example of this service is provided by Smart Force which offers hundreds of courses over the Internet. Smart Force specializes in training on information technology products from multiple vendors, including Microsoft, Oracle, Novell, Powersoft, and SAP. Accessing training over the Internet has the potential to save companies thousands of dollars each year in training costs, especially when it comes to information technology training. Instead of having to send personnel off-site for weeks and pay their travel expenses, companies can gain access to Internet training for lower cost and personnel can get the training at their desks.

As both training and support for computing are increasingly able to be delivered on-line in modules, with some embedded in software packages and applications (as is the case for EPSS), the already blurred distinction between them blurs even more. Some of the issues most particular to computer user support are examined in the next section.

Electronic performance support system (EPSS): Component of a software package or application in which training and educational information is embedded. An EPSS can take several forms, including a tutorial, an expert system shell, and hypertext jumps to reference material.

Figure 17-12
A Microsoft Office Assistant™ note

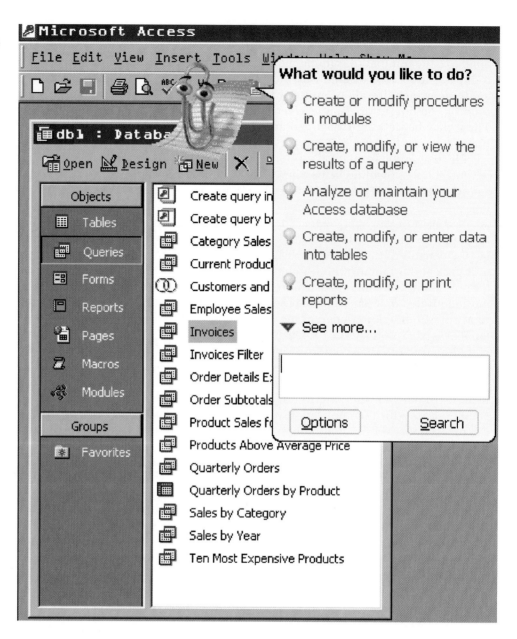

Supporting Information Systems Users

Historically, computing support for users has been provided in one of a few forms: paper, on-line versions of paper-based support, vendors, or by other people who work for the same organization. As we stated earlier, support, whatever its form, has often been inadequate for users' needs. Yet users consider support to be extremely important. A 1993 J. D. Power and Associates survey found user support to be the number one criterion contributing to user satisfaction with personal computing, cited by 26 percent of respondents as the most important factor (Schurr, 1993).

As organizations moved away from mainframe-based computing to increased reliance on personal computers in the 1980s, the need for support among users also increased. More support was made available on-line, but organizations also began to provide institutionalized user support in the form of information centers and help desks. An **information center** comprises a group of people who can answer questions

Information center: An organizational unit whose mission is to support users in exploiting information technology.

and assist users with a wide range of computing needs, including the use of particular information systems. Information center staff might do the following:

- Install new hardware or software
- Consult with users writing programs in fourth-generation languages
- Extract data from organizational databases onto personal computers
- Set up user accounts
- Answer basic on-demand questions
- Provide a demonstration site for viewing hardware and software
- Work with users to submit system change requests

When you expect an organizational information center to help support a new system, you will likely want to train the information center staff as soon as possible on the system. Information center staff will need additional training periodically as new system features are introduced, new phases of a system are installed, or system problems and workarounds are identified. Even if new training is not required, information center staff should be aware of system changes, since such changes may result in an increase in demand for information center services.

Personnel in an information center or help desk (we will discuss help desks in a later section in this chapter) were typically drawn from the ranks of information systems workers or knowledgeable users in functional area departments. There was rarely any type of formal training for people to learn how to work in the support area and, in general, this remains true today. As organizations move to client/server architectures, their needs for support have increased even more than during the introduction of PCs, and organizations find themselves in the position of having to rely more than ever on vendor support (From, 1993). This increased need for support comes in part from the lack of standards governing client/server products and the resulting need to make equipment and software from different vendors compatible. Vendors are able to provide the necessary support, but as they have shifted their offerings from primarily expensive mainframe packages to inexpensive off-the-shelf software, they find they can no longer bear the cost of providing the support for free. Most vendors now charge for support, and many have instituted 900 numbers or sell customers unlimited support for a given monthly or annual charge.

Automating Support In an attempt to cut the costs of providing support and to catch up with the demand for additional support services, vendors have automated much of their support offerings. Common methods for automating support include on-line support forums, bulletin board systems, on-demand fax, and voice-response systems (Schneider, 1993). On-line support forums provide users access to information on new releases, bugs, and tips for more effective usage. Forums are offered on on-line services such as America OnLine,™ over the Internet, and over company intranets. On-demand fax allows users to order support information through an 800 number and receive that information instantly over their fax machines. Finally, voice-response systems allow users to navigate option menus that lead to prerecorded messages about usage, problems, and workarounds. Organizations have established similar support mechanisms for systems developed or purchased by the organization. Internal e-mail, group support systems, and office automation can be used to support such capabilities within an organization.

Other organizations can follow the lead of vendors in automating some of their support offerings in similar ways. Some software vendors work with large accounts and corporations to set up or enhance their own support offerings. The cost to the corporation can go from $5,000 to as much as $20,000. Offerings include access to knowledge bases about a vendor's products, electronic support services, single point of contact, and priority access to vendor support personnel (Schneider, 1993).

Product knowledge bases include all of the technical and support information about vendor products and provide additional information for on-site personnel to use in solving problems. Some vendors allow users to access some of the same information that is available in on-line forums through bulletin boards which the vendors set up and maintain themselves and on Internet sites. Many vendors now supply complete user and technical documentation on CD-ROM, including periodic updates, so that a user organization can provide this library of documentation, bug reports, workaround notices, and notes on undocumented features on-line to all internal users. Electronic support services include all of the vendor support services discussed earlier, but tailored specifically for the corporation. The single point of contact is a system engineer who is often based on site and serves as a liaison between the corporation and the vendor. Finally, priority access means that corporate workers can always get help via telephone or e-mail from a person at the vendor company, usually within a prespecified response time of four hours or less.

Such vendor-enhanced support is especially appropriate in organizations where a wide variety of a particular vendor's products is in use, or where most in-house application development either utilizes vendor products as components of the larger system or where the vendor's products are themselves used as the basis for applications. An example of the former would be the case where an organization has set up a client/server architecture based on a particular vendor's SQL server and application programmer interfaces (APIs). Which applications are developed in-house to run under the client/server architecture depends heavily on the server and APIs, and direct vendor support dealing with problems in these components would be very helpful to the enterprise information systems staff. An example of the second would include order entry and inventory control application systems developed using Microsoft's Access™ or Excel™. In this case, the system developers and users, who are sometimes the same people for such package-based applications, can benefit considerably from directly questioning vendor representatives about their products.

Providing Support Through a Help Desk   Whether assisted by vendors or going it alone, the center of support activities for a specific information system in many organizations is the help desk. A **help desk** is an information systems department function, staffed by IS personnel, possibly part of the information center unit. The help desk is the first place users should call when they need assistance with an information system. The help desk staff either deals with the users' questions or refers the users to the most appropriate person.

Help desk: A single point of contact for all user inquiries and problems about a particular information system or for all users in a particular department.

For many years, a help desk was the dumping ground for people IS managers did not know what else to do with. Turnover rates were high because the position was sometimes little more than a complaints department, the pay was low, and burnout rates were high. In today's information systems-dependent enterprises, however, this situation has changed. Help desks are gaining new respect as management comes to appreciate the special combination of technical skills and people skills needed to make good help desk staffers. In fact, a recent survey reveals that the top two valued skills for help desk personnel are related to communication and customer service (Crowley, 1993).

Help desk personnel (as well as the personnel in the more general information center) need to be good at communicating with users, by listening to their problems and intelligently communicating potential solutions. These personnel also need to understand the technology they are helping users with. It is crucial, however, that help desk personnel know when new systems and releases are being implemented and when users are being trained for new systems. Help desk personnel themselves should be well trained on new systems. One sure recipe for disaster is to train users on new systems but not train the help desk personnel these same users will turn to for their support needs.

Support Issues for the Analyst to Consider

Support is more than just answering user questions about how to use a system to perform a particular task or about the system's functionality. Support also consists of such tasks as providing for recovery and backup, disaster recovery, and PC maintenance; writing newsletters and offering other types of proactive information sharing; and setting up user groups. It is the responsibility of analysts for a new system to be sure that all forms of support are in place before the system is installed.

For medium to large organizations with active information system functions, many of these issues are dealt with centrally. For example, users may be provided with backup software by the central information systems unit and a schedule for routine backup. Policies may also be in place for initiating recovery procedures in case of system failure. Similarly, disaster recovery plans are almost always established by the central IS unit. Information systems personnel in medium to large organizations are also routinely responsible for PC maintenance, as the PCs belong to the enterprise. There may also be IS unit specialists in charge of composing and transmitting newsletters, or overseeing automated bulletin boards and organizing user groups.

When all of these (and more) services are provided by central IS, you must follow the proper procedures to include any new system and its users in the lists of those to whom support is provided. You must design training for the support staff on the new system, and you must make sure that system documentation will be available to it. You must make the support staff aware of the installation schedule. And you must keep these people informed as the system evolves. Similarly, any new hardware and off-the-shelf software has to be registered with the central IS authorities.

When there is no official IS support function to provide support services, you must come up with a creative plan to provide as many services as possible. You may have to write backup and recovery procedures and schedules, and the users' departments may have to purchase and be responsible for the maintenance of their hardware. In some cases, software and hardware maintenance may have to be outsourced to vendors or other capable professionals. In such situations, user interaction and information dissemination may have to be more informal than formal: informal user groups may meet over lunch or over a coffee pot rather than in officially formed and sanctioned forums.

ORGANIZATIONAL ISSUES IN SYSTEMS IMPLEMENTATION

Despite the best efforts of the systems development team to design and build a quality system and to manage the change process in the organization, the implementation effort sometimes fails. Sometimes employees will not use the new system that has been developed for them or, if they do use the system, their level of satisfaction with it is very low. Why do systems implementation efforts fail? This question has been the subject of information systems research for the past 30 years. In this section, we will try to provide some answers, first by looking at what conventional wisdom says are important factors related to implementation success, then by investigating factor-based models and political models of systems implementation.

Why Implementation Sometimes Fails

The conventional wisdom that has emerged over the years is that there are at least two conditions necessary for a successful implementation effort: management support of the system under development and the involvement of users in the development process (Ginzberg, 1981b). Conventional wisdom holds that if both of these conditions are met, you should have a successful implementation. Yet, despite the support and active participation of management and users, information systems implementation sometimes fails (see the box, System Implementation Failures, for examples).

Management support and user involvement are important to implementation success, but they may be overrated compared to other factors that are also important. Research has shown that the link between user involvement and implementation success is sometimes weak (Ives and Olson, 1984). User involvement can help reduce the risk of failure when the system is complex, but user participation in the development process only makes failure more likely when there are financial and time constraints in the development process (Tait and Vessey, 1988). Information systems implementation failures are too common, and the implementation process is too complicated, for the conventional wisdom to be completely correct. The search for better explanations for implementation success and failure has led to two alternative approaches: factor models and political models.

System Implementation Failures

In 1985, the New Jersey Motor Vehicles Division implemented a new vehicle registration system. The new system had been developed by a major accounting firm in Applied Data Research's ideal, a 4GL. Although appropriate for management information systems or decision support systems, 4GLs are not well-suited for high-volume transaction processing systems. As might have been predicted, the vehicle registration system was a disaster. Using the system entailed as much as a five-minute response time, instead of the 2-second time the Division requested. System use resulted in a large number of incorrect vehicle registrations: over one million drivers in New Jersey had problems with their car registrations. The critical parts of the system had to be reprogrammed in COBOL at a cost of over two million dollars (which the accounting firm paid for) (Zwass, 1992).

Another state system has had an even more colorful history. In 1988, the Department of Health and Rehabilitative Services (HRS) began work on the development of the Florida On-Line Recipient Integrated Data Access system, or FLORIDA for short. The purpose of FLORIDA was to provide a single point of eligibility testing for the welfare services administered by the state, including food stamps, Aid to Families with Dependent Children (AFDC), and Medicaid. Access to FLORIDA would be possible from state welfare offices throughout the state, but FLORIDA and all of its files would reside in Tallahassee, the state capital. In 1989, EDS Federal Corporation won the bid for primary contractor for FLORIDA. One of the contract's stipulations was that the system be developed and implemented within 29 months.

FLORIDA was handed over to the state in June 1992. By the time the system came on-line, it had changed in at least two key aspects from the original design. First, the system was now running on a single mainframe system instead of three. Second, whereas the original design had called for eligibility rulings to be generated overnight, the final system allowed instant, on-line eligibility determinations.

From the time it was first handed over to the state, FLORIDA was a disaster. Response time often approached eight minutes. System crashes, lasting as long as eight hours, were common. Adding to the problems was the fact that, from July 1990 to July 1993, Florida's food stamps, AFDC, and Medicaid enrollments doubled, from 800,000 to 1.6 million people. FLORIDA was pushed to the limits of its capacity. IBM was commissioned to bring in a new, powerful mainframe system for FLORIDA to run on. As of September 1993, over $173 million had been spent on FLORIDA's development and implementation, almost double the original estimate, and the system was still not complete. Consultants estimated the system had only two more years left to operate, given FLORIDA's then current capacity and projections for the state's welfare enrollments.

The human and financial toll from FLORIDA has easily surpassed those that resulted from New Jersey's vehicle registration system. FLORIDA's use resulted in at least $263 million in welfare over- and underpayments. In 1992, EDS sued HRS for $46 million in payments EDS says it never received. Throughout 1992 and 1993, several key information systems personnel in HRS were fired or forced to resign. In the spring of 1993, the Governor demoted the secretary of HRS and sent the lieutenant governor to HRS to manage the FLORIDA disaster. In August, 1993, a grand jury indicted the former project leader and former deputy secretary of HRS for information systems, for filing false status reports and for falsifying payment reports. In the spring of 1994, all charges were dropped against the former deputy secretary of HRS. The FLORIDA system's former project manager went to trial in May, 1994, and was convicted of two misdemeanor counts of making false statements. The judge in the case overthrew the verdict in June of that year.

In January 1995, the state announced that FLORIDA had been considerably improved, with the error rate for issuing benefits cut to only 12.5 percent. Yet in October of the same year, the state sued EDS for $60 million for improper business practices in its development of FLORIDA. The state also wanted to ban EDS from ever doing business again in Florida. Two months later, a judge threw out the state's suit. In March 1996, EDS won its 1992 suit against the state. In May of that year, the State of Florida agreed to pay EDS $42.8 million in back payments.

(Compiled from 1993–1996 stories in the Tallahassee Democrat and Information Week.)

Factor Models of Implementation Success

Several research studies have found other factors that are important to a successful implementation process. Ginzberg (1981b) found three additional important factors: commitment to the project, commitment to change, and extent of project definition and planning. Commitment to the project involves managing the system development project so that the problem being solved is well understood and the system being developed to deal with the problem actually solves it. Commitment to change involves being willing to change behaviors, procedures, and other aspects of the organization. The extent of project definition and planning is a measure of how well planned the project is. The more extensive the planning effort, the less likely is implementation failure. In other research, Ginzberg (1981a) uncovered another important factor related to implementation success: user expectations. The more realistic a user's early expectations about a new system and its capabilities, the more likely it is that the user will be satisfied with the new system and actually use it.

Although there are many ways to determine if an implementation has been successful, the two most common and trusted are the extent to which the system is used and the user's satisfaction with the system (Lucas, 1997). Lucas, who has extensively studied information systems implementation, identified six factors that influence the extent to which a system is used (1997):

1. *User's personal stake.* How important the domain of the system is for the user; in other words, how relevant the system is to the work the user performs. The user's personal stake in the system is itself influenced by the level of support management provides for implementation, and by the urgency to the user of the problem addressed by the system. The higher the level of management support and the more urgent the problem, the higher the user's personal stake in the system.

2. *System characteristics.* Includes such aspects of the system's design as ease of use, reliability, and relevance to the task the system supports.

3. *User demographics.* Characteristics of the user, such as age and degree of computer experience.

4. *Organization support.* These are the same issues of support you read about earlier in the chapter. The better the system support infrastructure, the more likely an individual will be to use the system.

5. *Performance.* What individuals can do with a system to support their work will have an impact on extent of system use. The more users can do with a system and the more creative ways they can develop to benefit from the system the more they will use it. The relationship between performance and use goes both ways. The higher the levels of performance, the more use. The more use, the greater the performance.

6. *Satisfaction.* Use and satisfaction also represent a two-way relationship. The more satisfied the users are with the system, the more they will use it. The more they use it, the more satisfied they will be.

The factors identified by Lucas and the relationships they have to each other are shown in the model in Figure 17-13. In the model, it's easier to see the relationships among the various factors, such as how management support and problem urgency affect the user's personal stake in the system. Notice also that the arrows that show the relationships between use and performance and satisfaction have two heads, illustrating the two-way relationships use has with these factors.

It should be clear that, as an analyst and as someone responsible for the successful implementation of an information system, there are some factors you have more control over than others. For example, you will have considerable influence over the system's characteristics, and you may have some influence over the levels of support that will be provided for users of the system. You have no direct control over a user's

Figure 17-13
Lucas's model of implementation success, 1997 (Adapted with the permission of the McGraw-Hill Companies. All rights reserved.)

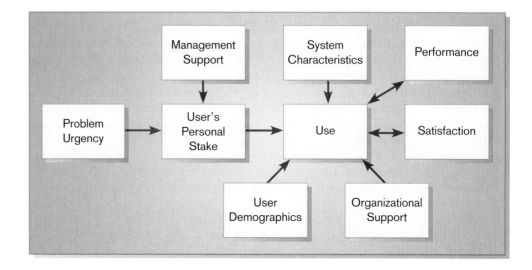

demographics, personal stake in the system, management support, or the urgency of the problem to the user. This doesn't mean you can ignore factors you can't change. On the contrary, you need to understand these factors very well, because you will have to balance them with the factors you *can* change in your system design and in your implementation strategy. You may not be able to change a user's demographics or personal stake in a system, but you can help design the system and your implementation strategy with these factors in mind.

Political Implementation Models

Factors model of implementation process have helped analysts better understand implementation and why it may or may not be successful. Certainly, taking the different factors into account and realizing how they work together to influence implementation is important. But just as the conventional wisdom about implementation could not explain the whole story, neither can factor models hope to completely explain the implementation process. Political models have been proposed as another perspective to help you understand how a systems development project can succeed. We will use two examples from the implementation literature in MIS to illustrate the usefulness of political models.

Political models assume that individuals who work in an organization have their own self-interested goals, which they pursue in addition to the goals of their departments and the goals of their organizations. Political models also recognize that power is not distributed evenly in organizations. Some workers have more power than others. People may act to increase their own power relative to that of their coworkers and, at other times, people will act to prevent coworkers with more power (such as bosses) from using that power or from gaining more.

Markus (1981) tells the story of a division of an organization where implementation appears to have succeeded. Workers in two manufacturing plants, called Athens and Capital City, were using the new work-in-progress (WIP) system, which made it possible for management to use a planning and forecasting system based in part on the output of the WIP system. Workers from both plants had been involved in the systems development process, especially workers from the Capital City plant. When workers at Athens resisted using the new WIP system, extensive management pressure seemed to force people at Athens to give in and begin using the new system. It appeared that the conventional wisdom about systems development had prevailed: Workers had participated in development, and management had been forcefully supportive. Implementation was a success.

Markus presents a political interpretation of the story that provides another explanation of events (see Figure 17-14). She begins by examining the history and power relationships within the division and the two plants. Although Athens was the plant that resisted using the new WIP system, they actually had superior information systems support at the beginning of the WIP system development process. Athens had once been a separate company and, until the division head decided new WIP and new planning systems were needed, Athens had been allowed a large amount of autonomy in how it operated. Further, the work performed at the two plants was tightly coupled. The Athens plant manufactured airplane parts, which were then refined and finished at Capital City. Athens' manufacturing process was unpredictable and unreliable and resulted in a high scrap rate. Capital City never knew where Athens was in the manufacturing process for any particular part, and this uncertainty complicated Capital City's efforts to complete its work and finish parts in time for promised delivery dates.

According to Markus, when the opportunity arose to develop a system that would lessen the dependency of Capital City on Athens, people at Capital City were anxious to participate in the system's development. With a new WIP system containing data on Athens' production process, data which Capital City could access, the Capital City plant could greatly reduce the uncertainty associated with its own work. Understandably, the people at Athens were not so enthusiastic, and resisted participating in the development process and using the new system. After the WIP system was completed and installed in both plants, the people at Athens continued to use their old WIP system, much to the chagrin of division management and Capital City. Division management applied a great deal of pressure, until finally Athens began to use the system. Markus states, however, that the people at Athens began using the new system not because of management pressure but because the economy began to improve, and managing the work at Athens became too difficult without the new system.

One lesson to be learned from Markus's case study is that factor-based analyses can only go so far in explaining successful implementation in organizations. History and power relationships are also important considerations. Sometimes political interpretations provide better explanations for the implementation process and why events took the course they did.

Once an information system has been successfully implemented, the importance of documentation grows. A successfully implemented system becomes part of the daily work lives of an organization's employees. Many of those employees will use the system, but others will maintain it and keep it running.

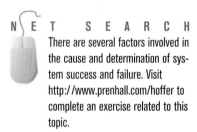

N E T   S E A R C H
There are several factors involved in the cause and determination of system success and failure. Visit http://www.prenhall.com/hoffer to complete an exercise related to this topic.

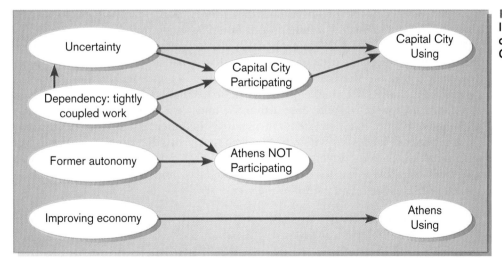

Figure 17-14
Illustrating the political explanation of system success at Athens and Capital City

ELECTRONIC COMMERCE APPLICATION: SYSTEM IMPLEMENTATION AND OPERATION FOR PINE VALLEY FURNITURE'S WEBSTORE

PINE VALLEY FURNITURE

Like many other analysis and design activities, system implementation and operation of an Internet-based electronic commerce application is no different than the processes followed for other types of applications. Previously, you read how Jim Woo and the Pine Valley Furniture development team transformed the conceptual data model for the WebStore into a set of normalized relations. Here we will examine how the WebStore system was tested before it was installed and brought on-line.

The programming of all WebStore software modules has been completed. The programmers extensively tested each unique module, and it was now time to perform a system-wide test of the WebStore. In this section, we will examine how test cases were developed, how bugs were recorded and fixed, and how alpha and beta testing were conducted.

Developing Test Cases for the WebStore

To begin the system-wide testing process, Jim and the PVF development team developed test cases to examine every aspect of the system. Jim knew that system testing, like all other aspects of the SDLC, needed to be a very structured and planned process. Before opening the WebStore to the general public, every module and component of the system needed to be tested within a controlled environment. Based upon his experience in implementing other systems, Jim felt that they would need to develop approximately 150–200 separate test cases to fully examine the WebStore. To help focus the development of test cases and to assign primary responsibility to members of his team to specific areas of the system, Jim developed the following list of testing categories:

- Simple functionality: Add to cart, list section, calculate tax, change personal data
- Multiple functionality: Add item to cart and change quantity, create user account and change address
- Function chains: Add item to cart, check out, create user account, purchase
- Elective functions: Returned items, lost shipments, item out of stock
- Emergency/Crisis: Missing orders, hardware failure, security attacks

The development group broke into five separate teams, each working to develop an extensive set of cases for each of the testing categories. Each team had one day to develop their test cases. Once developed, each team would lead a walkthrough, so that everyone would know the totality of the testing process, and to facilitate extensive feedback to each team so that the testing process would be as comprehensive as possible. To make this point, Jim stated, "What happens when a customer repeatedly enters the same product into the shopping cart? Can we handle that? What happens when the customer repeatedly enters and then removes a single product? Can we handle that? Although some of these things are unlikely to ever occur, we need to be confident that the system is robust to any type of customer interaction. We must develop every test case necessary to give us confidence that the system will operate as intended, 24-7-365!"

A big part of successful system testing is to make sure that no information is lost and that all tests are described in a consistent way. To achieve this, Jim provided all teams with standard forms for documenting each case and for recording the results of each test. This form had the following sections:

- Test Case ID
- Category/Objective of Test
- Description

- System Version
- Completion Date
- Participant(s)
- Machine Characteristics (processor, operating system, memory, browser, etc.)
- Test Result
- Comments

The teams also developed standard codes for each general type of test and this was used to create the Test Case ID. For example, all tests related to "Simply Functionality" were given an ID with SF as a prefix and a number as the suffix—for example, SF001. The teams also developed standards for categorizing, listing objectives, and writing other test form contents. Establishing these standards assured that the testing process would be consistently documented.

Bug Tracking and System Evolution An outcome of the testing process is the identification of system bugs. Consequently, in addition to setting a standard method for writing and documenting test cases, Jim and the teams established several other rules to assure a smooth testing process. Experienced developers have long known that an accurate bug tracking process is essential for rapid troubleshooting and repair during the testing process. You can think of bug tracking as creating a "paper trail" that makes it much easier for programmers to find and repair the bug. To make sure that all bugs were documented in a similar way, the team developed a bug tracking form that had the following categories:

- Bug Number (simple incremental number)
- Test Case ID that Generated the Bug
- Is the Bug Replicatible?
- Affects
- Description
- Resolution
- Resolution Date
- Comments

PVF development team agreed that bug fixes would be made in batches, because all test cases would have to be redone every time the software was changed. Redoing all the test cases each time the software is changed is done to assure that in the process of fixing the bug, no other bugs are introduced into the system. As the system moved along in the testing process—as batches of bugs are fixed—the version number of the software is incremented. During the development and testing phases, the version is typically below the "1.0" first release version.

Alpha and Beta Testing the WebStore

After completing all system test cases and resolving all known bugs, Jim moved the WebStore into the alpha testing phase, where the entire PVF development team as well as personnel around the company would put the WebStore through its paces. To motivate employees throughout the company to actively participate in testing the WebStore, several creative promotions and give-aways were held. All employees were given a tee shirt that said, "I shop at the WebStore, do you?" Additionally, all employees were given $100 dollars to shop at the WebStore and were offered a free lunch for their entire department if they found a system bug while shopping on the system. Also during alpha testing, the development team conducted extensive recovery, security, stress, and performance testing. Table 17-8 provides a sample of the types of tests performed.

TABLE 17-8   Sample of Tests Conducted on the WebStore During Alpha Testing

Test Type	Tests Performed
Recovery	• Unplug main server to test power backup system. • Switch off main server to test the automatic switching to backup server.
Security	• Try to purchase without being a customer. • Try to examine server directory files both within the PVF domain and when connecting from an outside Internet service provider.
Stress	• Have multiple uses simultaneously establish accounts, process purchases, add to shopping cart, remove from shopping cart, and so on.
Performance	• Examine response time using different connection speeds, processors, memory, browsers, and other system configurations • Examine response time when backing up server data.

After completing alpha testing, PVF recruited several of their established customers to help in beta testing the WebStore. As real-world customers used the system, Jim was able to monitor the system and fine tune the servers for optimal system performance. As the system moved through the testing process, fewer and fewer bugs were found. After several days of "clean" usage, Jim felt confident that it was now time to open the WebStore for business.

WebStore Installation

Throughout the testing process, Jim kept PVF management aware of each success and failure. Fortunately, because Jim and the development team followed a structured and disciplined development process, there were far more successes than failures. In fact, he was now confident that the WebStore was ready to go on-line and would recommend this to PVF's top management that it was now time to "flip the switch" and let the world enter the WebStore.

PROJECT CLOSE-DOWN

In Chapter 3 you learned about the various phases of project management, from project initiation to closing down the project. If you are the project manager and you have successfully guided your project through all of the phases of the systems development life cycle presented so far in this book, you are now ready to close down your project. Although the maintenance phase is just about to begin, the development project itself is over. As you will see in the next chapter, maintenance can be thought of as a series of smaller development projects, each with its own series of project management phases.

As you recall from Chapter 3, your first task in closing down the project involves many different activities, from dealing with project personnel to planning a celebration of the project's ending. You will likely have to evaluate your team members, reassign most to other projects, and perhaps terminate others. As project manager, you will also have to notify all of the affected parties that the development project is ending and that you are now switching to maintenance mode.

Your second task is to conduct post-project reviews with both your management and your customers. In some organizations, these post-project reviews will follow formal procedures and may involve internal or EDP (electronic data processing) auditors. The point of a project review is to critique the project, its methods, its deliverables, and its management. You can learn many lessons to improve future projects from a thorough post-project review.

The third major task in project close-down is closing out the customer contract. Any contract that has been in effect between you and your customers during the project (or as the basis for the project) must be completed, typically through the consent of all contractually involved parties. This may involve a formal "signing-off" by the clients that your work is complete and acceptable. The maintenance phase will typically be covered under new contractual agreements. If your customer is outside of your organization, you will also likely negotiate a separate support agreement.

As an analyst member of the development team, your job on this particular project ends during project close-down. You will likely be reassigned to another project dealing with some other organizational problem. Maintenance on your new system will begin and continue without you. To complete our consideration of the systems development life cycle, however, we will cover the maintenance phase and its component tasks in Chapter 18.

Summary

This chapter presented an overview of the various aspects of the systems implementation process. In the chapter, you studied seven different types of testing: (1) code inspections, where code is examined for well-known errors; (2) walkthroughs, where a group manually examines what the code is supposed to do; (3) desk checking, where an individual mentally executes the computer instructions; (4) syntax checking, typically done by a compiler; (5) unit or module testing; (6) integration testing, where modules are combined and tested together until the entire program has been tested as a whole; and (7) system testing, where programs are combined to be tested as a system and where the system's meeting of its objectives is examined. You also learned about acceptance testing, where users test the system for its ability to meet their requirements, using live data in a live environment.

You read about four types of installation: (1) direct, where the old system is shut off just as the new one is turned on; (2) parallel, where both old and new systems are run together until it is clear the new system is ready to be used exclusively; (3) single location, where one site is selected to test the new system; and (4) phased, where the system is installed bit by bit.

You learned about four types of documentation: system documentation, which describes in detail the design of a system and its specifications; internal documentation, that part of system documentation included in the code itself or emerges at compile time; external documentation, that part of system documentation which includes the output of diagramming techniques such as data flow and entity-relationship diagramming; and user documentation, which describes a system and how to use it for the system's users.

Training and support are both part of an organization's computing infrastructure. A computing infrastructure is analogous to a city's infrastructure of streets, water

mains, and electrical lines. The computing infrastructure allows an organization's computer users to continue to operate, just as a city's infrastructure allows the city to continue to operate. In many organizations, the computing infrastructure and the people charged with providing and maintaining it are underfunded.

Computer training has typically been provided in classes and tutorials. While there is some evidence lectures have their place in teaching people about computing and information systems, the current emphasis in training is on automated delivery methods, such as on-line reference facilities, multimedia training, and electronic performance support systems. The latter embed training in applications themselves in an attempt to make training a seamless part of using an application for daily operations. The emphasis in support is also on providing on-line delivery, including on-line support forums and bulletin board systems. As organizations move toward client/server architectures, they rely more on vendors for support. Vendors provide many on-line support services, and they work with customers to bring many aspects of on-line support in-house. An information center provides general support to all users, and a help desk provides aid to users in a particular department or for a particular system.

You saw how information systems researchers have been trying to explain what constitutes a successful implementation, using conventional wisdom, factor models, and political models. If there is a single main point in this chapter, it is that implementation is a complicated process, from managing programmer teams to the politics that influence what happens to a system after it has been successfully implemented to planning and implementing useful training and support mechanisms. Analysts have many factors to identify and manage for a successful systems implementation. Successful implementation rarely happens by accident or occurs in a totally predictable manner. The first step in a successful implementation effort may be realizing just that fact.

In many ways, the implementation of an Internet-based system is no different. Just as much careful atten-

tion, if not more, has to be paid to the details of an Internet implementation as is the case for a traditional system. Successful implementation for an Internet-based system is not an accident either.

Key Terms

1. Acceptance testing
2. Alpha testing
3. Beta testing
4. Computing infrastructure
5. Desk checking
6. Direct installation
7. Electronic performance support system (EPSS)
8. External documentation
9. Help desk
10. Information center
11. Inspections
12. Installation
13. Integration testing
14. Internal documentation
15. Parallel installation
16. Phased installation
17. Single location installation
18. Stub testing
19. System documentation
20. Sytem testing
21. Support
22. Unit testing
23. User documentation

Match each of the key terms above with the definition that best fits it:

_____ A testing technique in which participants examine program code for predictable language-specific errors.

_____ A testing technique in which the program code is sequentially executed manually by the reviewer.

_____ Component of a software package or application in which training and educational information is embedded. It may include a tutorial, expert system, and hypertext jumps to reference material.

_____ Written or other visual information about an application system, how it works, and how to use it.

_____ Changing over from the old information system to a new one by turning off the old system when the new one is turned on.

_____ Each module is tested alone in an attempt to discover any errors in its code; also called *module testing*.

_____ The organizational process of changing over from the current information system to a new one.

_____ System documentation that includes the outcome of structured diagramming techniques, such as data flow and entity-relationship diagrams.

_____ The process whereby actual users test a completed information system, the end result of which is the users' acceptance of it.

_____ Detailed information about a system's design specifications, its internal workings, and its functionality.

_____ Running the old information system and the new one at the same time until management decides the old system can be turned off.

_____ The process of bringing together all of the modules that a program comprises for testing purposes. Modules are typically integrated in a top-down, incremental fashion.

_____ An organizational unit whose mission is to support users in exploiting information technology.

_____ A technique used in testing modules, especially modules that are written and tested in a top-down fashion, where a few lines of code are used to substitute for subordinate modules.

_____ Changing from the old information system to the new one incrementally, starting with one or a few functional components and then gradually extending the installation to cover the whole new system.

_____ The bringing together of all the programs that a system comprises for testing purposes. Programs are typically integrated in a top-down, incremental fashion.

_____ System documentation that is part of the program source code or is generated at compile time.

_____ Providing ongoing educational and problem-solving assistance to information system users. Support material and jobs must be designed along with the associated information system.

_____ User testing of a completed information system using real data in the real user environment.

_____ Trying out a new information system at one site and using the experience to decide if and how the new system should be deployed throughout the organization.

_____ User testing of a completed information system using simulated data.

_____ A single point of contact for all user inquiries and problems about a particular information system or for all users in a particular department.

_____ All the resources and practices required to help people adequately use computer systems to do their primary work.

Review Questions

1. What are the deliverables from coding, testing, and installation?

2. Explain the testing process for code.

3. What are structured walkthroughs for code? What is their purpose? How are they conducted? How are they different from code inspections?

4. What are the four approaches to installation? Which is the most expensive? Which is the most risky? How does an organization decide which approach to use?

5. What is the conventional wisdom about implementation success?

6. List and define the factors that are important to successful implementation efforts.

7. Explain Lucas's model of implementation success.

8. How would you characterize political models of the implementation process? What can political models do that factor models don't?

9. What is the difference between system documentation and user documentation?

10. What were the common methods of computer training in 1987? What training methods have been added since then?

11. What is self-training?

12. What proof do you have that individual differences matter in computer training?

13. Why are corporations relying so heavily on vendor support as they move to client/server architectures?

14. Describe the delivery methods many vendors employ for providing support.

15. Describe the various roles typically found in a help desk function.

16. Describe the roles of each programmer in pair programming.

Problems and Exercises

1. Prepare a testing strategy or plan for Pine Valley Furniture's Purchasing Fulfillment System.

2. Which installation strategy would you recommend for PVF's Purchasing Fulfillment System? Which would you recommend for Hoosier Burger's Inventory Control System? If you recommended different approaches, please explain why. How is PVF's case different from Hoosier Burger's?

3. Develop a table that compares the four installation strategies, showing the pros and cons of each. Try to make a direct comparison when a pro of one is a con of another.

4. One of the difficult aspects of using the single location approach to installation is choosing an appropriate location. What factors should be considered in picking a pilot site?

5. You have been a user of many information systems including, possibly, a class registration system at your school, a bank account system, a word processing system, and an airline reservation system. Pick a system you have used and assume you were involved in the beta testing of that system. What would be the criteria you would apply to judge whether this system was ready for general distribution?

6. Why is it important to keep a history of test cases and the results of those test cases even after a system has been revised several times?

7. How much documentation is enough?

8. What is the purpose of electronic performance support systems? How would you design one to support a word processing package? A database package?

9. Discuss the role of a centralized training and support facility in a modern organization. Given advances in technology and the prevalence of self-training and consulting among computing end users, how can such a centralized facility continue to justify its existence?

10. Is it good or bad for corporations to rely on vendors for computing support? List arguments both for and against reliance on vendors as part of your answer.

11. Suppose you were responsible for establishing a training program for users of Hoosier Burger's inventory control system (described in previous chapters). Which forms of training would you use? Why?

12. Suppose you were responsible for establishing a help desk for users of Hoosier Burger's inventory control system (described in previous chapters). Which support system elements would you create to help users be effective? Why?

13. Your university or school probably has some form of microcomputer center or help desk for students. What functions does this center perform? How do these functions compare to those outlined in this chapter?

14. Suppose you were responsible for organizing the user documentation for Hoosier Burger's inventory control system (described in previous chapters). Write an outline that shows the documentation you would suggest be created, and generate the table of contents or outline for each element of this documentation.

Field Exercises

1. Interview someone you know or have access to who works for a medium to large organization. Ask for details on a specific instance of organizational change: What changed? How did it happen? Was it well planned or ad hoc? How were people in the organization affected? How easy was it for employees to move from the old situation to the new one?

2. Reexamine the data you collected in the interview in Field Exercise 1. This time, analyze the data from a political perspective. How well does the political model explain how the organization dealt with change? Explain why the political model does or does not fit.

3. Ask a systems analyst you know or have access to about implementation. Ask what the analyst believes is necessary for a successful implementation. Try to determine whether the analyst believes in factor models or political models of implementation.

4. Prepare a research report on successful and unsuccessful information system implementations. After you have found information on two or three examples of both successful and unsuccessful system implementations, try to find similarities and differences among the examples of each type of implementation. Do you detect any patterns? Can you add to either the factor or political models you read about in the chapter?

5. Talk with people you know who use computers in their work. Ask them to get copies of the user documentation they rely on for the systems they use at work. Analyze the documentation. Would you consider it good or bad? Support your answer. Whether good or bad, how might you improve it?

6. Volunteer to work for a shift at a help desk at your school's computer center. Keep a journal of your experiences. What kind of users did you have to deal with? What kinds of questions did you get? Do you think help desk work is easy or hard? What skills are needed by someone in this position?

7. Let's say your professor has asked you to help him or her train a new secretary on how to prepare class notes for electronic distribution to class members. Your professor uses word processing software and an e-mail package to prepare and distribute the notes. Assume the secretary knows nothing about either package. Prepare a user task guide that shows the secretary how to complete this task.

References

Beck, K. 2000. *eXtreme Programming eXplained.* Upper Saddle River, NJ: Addison-Wesley.

Bell, P., and C. Evans. 1989. *Mastering Documentation.* New York, NY: John Wiley & Sons.

Bloor, R. 1994. "The Disappearing Programmer." *DBMS* 7(9) (Aug.): 14–16.

Bozman, J. S. 1994. "United to Simplify Denver's Troubled Baggage Project." *ComputerWorld,* 10(10).

Cole, K., O. Fischer, and P. Saltzman, 1997. "Just-in-Time Knowledge Delivery." *Communications of the ACM* 40(7): 49–53.

Crowley, A. 1993. "The Help Desk Gains Respect." *PC Week* 10 (November 15): 138.

Eason, K. 1988. *Information Technology and Organisational Change.* London: Taylor & Francis.

Fagan, M. E. 1986. "Advances in Software Inspections." *IEEE Transactions on Software Engineering* SE-12(7) (July): 744–51.

From, E. 1993. "Shouldering the Burden of Support." *PC Week* 10 (November 15): 122–44.

Gasser, L. 1986. "The Integration of Computing and Routine Work." *ACM Transactions on Office Information Systems* 4 (July): 205–25.

Ginzberg, M. J. 1981a. "Early Diagnosis of MIS Implementation Failure: Promising Results and Unanswered Questions." *Management Science* 27(4): 459–78.

Ginzberg, M. J. 1981b. "Key Recurrent Issues in the MIS Implementation Process." *MIS Quarterly* 5(2) (June): 47–59.

Henderson, J. C., and M. E. Treacy. 1986. "Managing End-User Computing for Competitive Advantage." *Sloan Management Review* (Winter): 3–14.

Ives, B. and M. H. Olson. 1984. "User Involvement and MIS Success: A Review of Research." *Management Science* 30(5): 586–603.

Kling, R., and S. Iacono. 1989. "Desktop Computerization and the Organization of Work." In *Computers in the Human Context,* edited by T. Forester. Cambridge, MA: The MIT Press: 335–56.

Lee, D. M. S. 1986. "Usage Pattern and Sources of Assistance for Personal Computer Users." *MIS Quarterly* 10 (December): 313–25.

Leib, J. 1997. "Baggage Suite Quietly Settled." *The Denver Post Online,* Sept. 11.

Litecky, C. R., and G. B. Davis. 1976. "A Study of Errors, Error Proneness, and Error Diagnosis in COBOL." *Communications of the ACM* 19(1): 33–37.

Lucas, H. C. 1997. *Information Technology for Management.* New York, NY: McGraw-Hill.

Martin, J., and C. McClure. 1985. *Structured Techniques for Computing.* Upper Saddle River, NJ: Prentice Hall.

Markus, M. L. 1981, "Implementation Politics: Top Management Support and User Involvement." *Systems/Objectives/Solutions* 1(4): 203–15.

Mosley, D. J. 1993. *The Handbook of MIS Application Software Testing.* Englewood Cliffs, NJ: Yourdon Press.

Nelson, R. R., and P. H. Cheney. 1987. "Training End Users: An Exploratory Study." *MIS Quarterly* 11 (December): 547–59.

Rivard, S., and S. L. Huff. 1988. "Factors of Success for End-User Computing." *Communications of the ACM* 31 (May): 552–61.

Rockart, J. F., and J. E. Short. 1989. "IT in the 1990s: Managing Organizational Interdependence." *Sloan Management Review* (Winter): 7–17.

Sambataro, M. 2000. "Just-in-Time Learning." *ComputerWorld* (Apr. 3).

Schneider, J. 1993. "Shouldering the Burden of Support." *PC Week* 10 (November 15): 123, 129.

Schurr, A. 1993. "Support is No. 1." *PC Week* 10 (November 15): 126.

Tait, P., and I. Vessey. 1988. "The Effect of User Involvement on System Success: A Contingency Approach." *MIS Quarterly* 12(1) (March): 91–108.

Torkzadeh, G., and W. J. Doll. 1993. "The Place and Value of Documentation in End-User Computing." *Information & Management* 24(3): 147–58.

Training. 2000. "Industry Report 2000." October. www. trainingsupersite.com/publications/magazines/training/ trg_toc.htm.

Yourdon, E. 1989. *Managing the Structured Techniques*. 4th ed. Englewood Cliffs, NJ: Prentice-Hall.

BROADWAY ENTERTAINMENT COMPANY, INC.

Designing a Testing Plan for the Customer Relationship Management System

CASE INTRODUCTION

The students from Stillwater State University are eager to get reactions to the initial prototype of MyBroadway, the Web-based customer relationship management system for Carrie Douglass, manager of the Centerville, Ohio, Broadway Entertainment Company (BEC) store. Based on the user dialogue design (see BEC Figure 15–1 at the end of Chapter 15), the team divided up the work of building the prototype. Tracey Wesley accepted responsibility for defining the database, starting from the normalized relations they developed (see the BEC case at the end of Chapter 12), and then populating with sample data the portions of the database that in production would come from the BEC store and corporate systems. Because John Whitman and Aaron Sharp had the most Microsoft Access experience on the team, they were responsible for developing the menus, forms, and displays for specific subsets of the customer pages. Missi Davies accepted the role of developing and managing the process of testing the system. The team decided that it would be desirable to have someone who was not directly involved in developing the system take responsibility for all aspects of testing. Testing would include tests conducted by Missi herself as well as use of the prototypes by BEC store employees and customers. While Tracey, John, and Aaron were developing the prototype, Missi began organizing the testing plan.

PREPARING THE TESTING PLAN

Now that the database and human interface elements of the system had been reasonably well outlined, Missi has a general understanding of the likely functionality and operation of MyBroadway. Because MyBroadway has a natural modular design, Missi believes that a top-down, modular approach can be used as a general process for testing.

Missi decides that the testing plan must involve a sequence of related steps, in which separate modules and then combinations of modules are used. Missi will test the individual modules and will initially test combinations of modules. But, once she has tested all the customer pages in major categories of functionality, and is reasonably sure they work, then it will be time to test the prototype with store employees and customers.

After studying the dialogue design for MyBroadway (BEC Figure 15-1 at the end of Chapter 15), she determines that there are five major modules that could be independently tested by employees and customers. These testing modules correspond to the five pages on the third level of the dialogue diagram: pages 1.1, 1.2, 1.3, 2.1, and 2.2. Missi decides, however, that such a piecemeal testing will be too confusing for employees and customers, but these modules can drive the internal testing process. Before her independent tests of pages and modules can be done, she knows that she will have to test Tracey's work on building the database. Missi pulls out the E-R diagram the team developed for MyBroadway (see BEC Figure 10-1 at the end of Chapter 10). Data for the Comment, Pick, and Request entities will be entered by customers of MyBroadway. Product, Sale, and Rental data are fed from in-store BEC systems. So, Missi sees the steps of the testing plan emerging.

Missi determines that the first step is to have Tracey build the database and populate it with sample data simulating the feeds from in-store systems. For the prototype, the team won't actually build the feeds. Missi contacts Carrie to request printouts of data on products and sales and rental history. Missi asks Tracey to be prepared to test the loading of Product data first. It would appear that only Product data from the in-store data feeds are needed to test pages 1.1, 1.2, and 1.3. This approach will allow Missi to test Tracey's work on loading Product data before John needs a stable database for the Product Information module he is developing. Then Missi will test John's work on the Product Information module of MyBroadway. She can separately test Tracey's work of loading sales and rental history and Aaron's work to build the Rental Information module.

Missi will select some of the data Carrie provides to give to Tracey for use in Tracey's testing. Once she sees the data, Missi may create some other fictitious data to cover special circumstances (e.g., products that have missing field values or extreme field values). Missi will keep some of the data Carrie sends her for use in her own testing of the procedures Tracey builds for loading data. Missi decides that John and Aaron should do their own testing of pages until they believe that the pages are working properly. She wants them to use their own test data for this purpose, and she will develop separate test data when she looks at their pages.

Missi puts her ideas for a testing plan into a rough outline (BEC Figure 17-1). At this point, the outline does not show a time line or sequence of steps. Missi knows that she has to develop this time line before she can present the testing plan to her team members. After she reviews this outline and time line with the whole team, Aaron, who is maintaining the project schedule, will use Microsoft Project to enter these activities into the official project schedule. Professor Tann, the instructor of the

information systems project course at Stillwater, requires each team to maintain.

PREPARING A TEST CASE

Among all the work Missi must do to manage the testing process, she must develop a detailed test case for each of the testing steps assigned to her in the overall testing plan in BEC Figure 17-1. Having not tested a new system before, Missi believes that she should develop one case so that she can get feedback from Professor Tann before proceeding to develop the rest of the cases. Missi decides to develop one of the easiest test cases first, and a suitable candidate for this appears to be the test for page 1.1.1, the entry by a customer of a new comment on a product.

Missi reviews the normalized relations the team developed for the database (BEC Figure 12–2 at the end of Chapter 12). Page 1.1.1 deals with entry of data in the

BEC Figure 17-1
Outline of testing plan for MyBroadway

```
Collect Product, Sales, and Rental data from Carrie
    Select subset for Tracey
    Create extra sample data for Tracey's tests
    Design test documentation format for Tracey
    Give data to Tracey
    Create extra sample data for Missi's tests
Design test documentation format for John and Aaron
Design test documentation format for Missi's testing
Conduct module tests
    Conduct walkthrough with Tracey on Product data entry procedures
    Test Product data entry procedures from Tracey
        Provide Tracey with feedback on testing
    Test Product Information navigation pages from John and Aaron
    Test page 1.1.1
    Test page 1.1.2
    Do integration test of pages 1.1.1 and 1.1.2
    Test page 1.2.1
    Test page 1.3
    Conduct walkthrough with Tracey on Rental and Sales history data entry procedures
    Test Rental and Sales history data entry procedures from Tracey
        Provide Tracey with feedback on testing
    Test Rental Information navigation pages from John and Aaron
    Test pages 2.1.1 and 2.1.1.1
    Test page 2.2
        Provide John and Aaron with feedback on testing
Conduct tests of revisions made from initial module tests
    Cycle revising and retesting until system is ready for client testing
Conduct tests with employees
    Design employee feedback forms to collect test results
    Test pages 1 and below
    Test pages 2 and below
Review employee test results
    Cycle revising and retesting internally until system addresses employee concerns
Conduct tests with customers
    Design customer feedback forms to collect test results
    Test whole system with customers
Review customer test results
    Cycle revising and retesting internally until system addresses customer concerns
Summarize results of testing for inclusion in final report to professor and client
```

Comment table. The Comment table in the database will contain data for:

- *Membership_ID.* Indicates who is entering the comment. Missi assumes this datum will be collected in a prior page, so this datum will not be an integral part of testing page 1.1.1.

- *Comment_Time_Stamp.* Indicates when the comment is entered. John's procedures for the page will have to get this computer system value and store it in the table, but the user won't deal with the data. Whether the time is captured correctly can be tested during the integration testing of pages 1.1.1 and 1.1.2, because comments on a product are reviewed in chronological order.

- *Product_ID.* Indicates on which product the customer is entering a comment. This value will be entered or selected on page 1.1, so pages 1.1 and 1.1.1 should be tested together. Missi asks John how he is designing page 1.1. John has decided to have users select the product through a series of questions so that only a valid product is used in pages 1.1.1 and 1.1.2. Thus, tests on this field will check that only existing products appear among the values for a customer to select.

- *Parent/Child?* Indicates whether the comment is entered by a parent or a child. This field has two values, and John says he will use a pair of radio buttons for entry of this field, and the choice of "Parent" will be the default on the page. Thus, there is no meaningful data entry test of this, but when doing the integration testing with page 1.1.2, Missi will check that the proper value was recorded. Missi makes a note to tell Tracey that John assumes that "Parent" is the default value for this field in the database.

- *Member_Comment.* Free-form textual comment entered by the customer about the selected product. Special cases of this field are that the customer submits the comment before entering a value for this field or enters a comment longer than can be stored. Because the team chose to use the Memo data type for this field, a truncated field is unlikely.

Besides making sure MyBroadway can handle alternative values for each field, Missi considers other important tests she learned about in her classes at Stillwater. Because the prototype will be used on only one PC in the store, there are no issues of concurrent use of the Comment data. Also, stress testing is of no concern for the same reason. Because Carrie has never indicated that security was a concern, no security testing is to be done. A type of recovery test would be to turn off the power to the PC during the entry of data. Performance testing is also of no concern with the limited usage prototype the team is building.

CASE SUMMARY

Missi is fairly confident that she has a good start on detailing the testing plan for MyBroadway. Once she can put her ideas on the example test case into a form for Professor Tann to review along with the testing outline, Missi will be ready to set a time line for testing. She needs to check with her team members to see when they think each module of the system will be ready for testing, and when they will need the instructions from her on how they should do their individual alpha testing.

CASE QUESTIONS

1. Using Figure 17-4 as a guide, develop a test case description and summary form for the test Missi has designed for page 1.1.1.

2. Critically evaluate the outline of a testing plan Missi has developed (BEC Figure 17-1). Can you think of missing steps? Are there too many steps, and should some steps be combined?

3. The testing outline in BEC Figure 17-1 does not show sequencing of steps and what steps could be done in parallel. Develop a testing schedule from this figure, using Microsoft Project or other charting tool, that shows how you would suggest sequencing the testing steps. Make assumptions for the length of each testing step.

4. One element of the testing plan outline is cycling the alpha testing of pages as Missi tests and finds problems, and then the other team members have to rewrite the code for the problematic module. What guideline would you use to determine when to stop the alpha testing and release the module for beta testing with employees?

5. Design the test documentation format that Tracey, John, or Aaron is to use to explain how that student tested his or her code and the results of that testing.

6. Design the customer feedback form to be used to capture comments from customers during their use of the MyBroadway prototype. What measures of usability should be established for MyBroadway, and is a customer feedback form a sufficient means to capture all the usability measures you believe should be collected?

7. How would you suggest that the beta testing with customers be conducted? For example, should users use the system directly or through someone else at the keyboard and mouse? Should the customer be observed while he or she is using the system, either by a student team member watching or by videotape?

Chapter 18

Maintaining

Information Systems

LEARNING OBJECTIVES

After studying this chapter, you should be able to:

- Explain and contrast four types of maintenance.

- Describe several facts that influence the cost of maintaining an information system and apply these factors to the design of maintainable systems.

- Describe maintenance management issues including alternative organizational structures, quality measurement, processes for handling change requests, and configuration management.

- Explain the role of CASE when maintaining information systems.

INTRODUCTION

As the twentieth century draws to a close, software development and maintenance operations have become major corporate cost centers. In addition, the magnitude and replacement value of software portfolios have become significant tax liabilities. Finally, the on-rushing "Year 2000" problem brings on a set of new and exceptionally large software costs that will have a major financial impact on corporations and government agencies. (Capers Jones, 1997)

In this chapter, we discuss systems maintenance, the largest systems development expenditure for many organizations. In fact, more programmers today work on maintenance activities than work on new development. Your first job after graduation may very well be as a maintenance programmer/analyst. This dispropor-tionate distribution of maintenance programmers is interesting since software does not wear out in a physical manner as do buildings and machines.

There is no single reason why software is maintained; however, most reasons relate to a desire to evolve system functionality in order to overcome internal processing errors or to better support changing business needs. Thus, maintaining a system is a fact of life for most systems. This means that maintenance can begin soon after the system is installed. As with the initial design of a system, maintenance activities are not limited only to software changes but include changes to hardware and business procedures. A question many people have about maintenance relates to how long organizations should maintain a system. Five years? Ten years? Longer? There is no simple answer to this question, but it is most often an issue of economics. In other words, at what point does it make financial sense to discontinue evolving an older system and build or purchase a new one? The focus of a great deal of upper IS management attention is devoted to assessing the trade-offs between maintenance and new development. In this chapter, we will provide you with a better understanding of the maintenance process and describe the types of issues that must be considered when maintaining systems.

In this chapter, we also briefly describe the systems maintenance process and the deliverables and outcomes from this process. This is followed by a detailed discussion contrasting the types of maintenance, an overview of critical management issues, and a description of the role of CASE and automated development tools in the maintenance process. Finally, we describe the process of maintaining Web-based applications, including an example for the Pine Valley Furniture's WebStore application.

MAINTAINING INFORMATION SYSTEMS

Once an information system is stalled, the system is essentially in the maintenance phase of the SDLC. When a system is in the maintenance phase, some person within the systems development group is responsible for collecting maintenance requests from system users and other interested parties, like system auditors, data center and network management staff, and data analysts. Once collected, each request is analyzed to better understand how it will alter the system and what business benefits and necessities will result from such a change. If the change request is approved, a system change is designed and then implemented. As with the initial development of the system, implemented changes are formally reviewed and tested before installation into operational systems.

The Process of Maintaining Information Systems

Throughout this book, we have drawn the systems development life cycle as the waterfall model where one phase leads to the next with overlap and feedback loops. As shown in Figure 18-1, the maintenance phase is the last phase of the SDLC. Yet, a life cycle, by definition, is circular in that the last activity leads back to the first. This means that the process of maintaining an information system is the process of returning to the beginning of the SDLC (see Figure 18-2) and repeating development steps until the change is implemented.

As shown in Figure 18-1, four major activities occur within maintenance:

1. Obtaining Maintenance Requests
2. Transforming Requests into Changes
3. Designing Changes
4. Implementing Changes

Obtaining maintenance requests requires that a formal process be established whereby users can submit system change requests. Earlier in the book, we presented a user request document called a Systems Service Request (SSR), which is shown in Figure 18-3. Most companies have some sort of document like an SSR to request new development, to report problems, or to request new system features with an existing system. When developing the procedures for obtaining maintenance requests, organizations must also specify an individual within the organization to collect these requests and manage their dispersal to maintenance personnel. The process of collecting and dispersing maintenance requests is described in much greater detail later in the chapter.

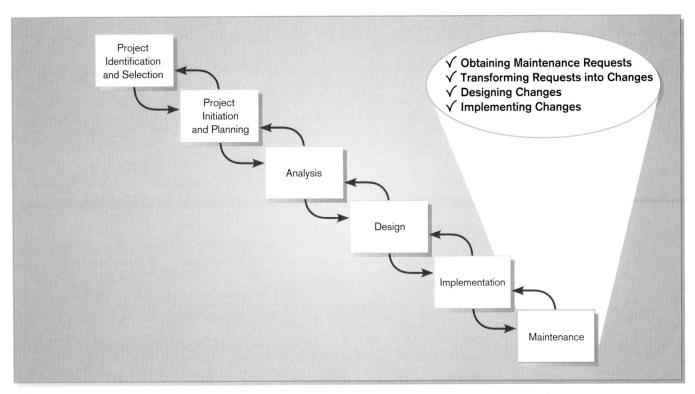

Figure 18-1
Systems development life cycle with maintenance phase highlighted

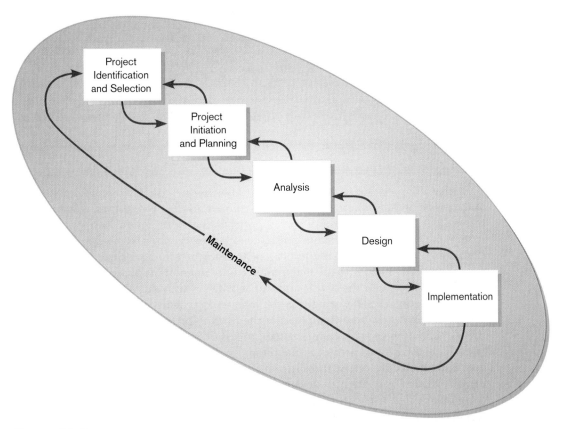

Figure 18-2
Maintenance phase makes the systems development process a life cycle

Figure 18-3
Systems Service Request for Purchasing Fulfillment System (Pine Valley Furniture)

Pine Valley Furniture
System Service Request

REQUESTED BY ___Juanita Lopez_____ DATE ___November 1, 2001___

DEPARTMENT _____Purchasing, Manufacturing Support_____

LOCATION _____Headquarters, 1-322_____

CONTACT _____Tel: 4-3267   FAX: 4-3270   e-mail: jlopez_____

TYPE OF REQUEST   URGENCY

[X]   New System   [   ]   Immediate – Operations are impaired or
                                                      opportunity lost
[   ]   System Enhancement   [   ]   Problems exist, but can be worked around
[   ]   System Error Correction   [X]   Business losses can be tolerated until new
                                                      system installed

PROBLEM STATEMENT

Sales growth at PVF has caused greater volume of work for the manufacturing support unit within Purchasing. Further, more concentration on customer service has reduced manufacturing lead times, which puts more pressure on purchasing activities. In addition, cost-cutting measures force Purchasing to be more agressive in negotiating terms with vendors, improving delivery times, and lowering our investments in inventory. The current modest systems support for manufacturing purchasing is not responsive to these new business conditions. Data are not available, information cannot be summarized, supplier orders cannot be adequately tracked, and commodity buying is not well supported. PVF is spending too much on raw materials and not being responsive to manufacturing needs.

SERVICE REQUEST

I request a thorough analysis of our current operations with the intent to design and build a completely new information system. This system should handle all purchasing transactions, support display and reporting of critical purchasing data, and assist purchasing agents in commodity buying.

IS LIAISON   Chris Martin (Tel: 4-6204   FAX: 4-6200   e-mail: cmartin)

SPONSOR   Sal Divario, Director, Purchasing

------------------------- TO BE COMPLETED BY SYSTEMS PRIORITY BOARD -------------------------
     [   ]   Request approved   Assigned to _____
                                     Start date _____
     [   ]   Recommend revision
     [   ]   Suggest user development
     [   ]   Reject for reason _____
                              _____
                              _____
                              _____

Once a request is received, analysis must be conducted to gain an understanding of the scope of the request. It must be determined how the request will affect the current system and the duration of such a project. As with the initial development of a system, the size of a maintenance request can be analyzed for risk and feasibility (see Chapter 6). Next, a change request can be transformed into a formal design change, which can then be fed into the maintenance implementation phase. Thus, many similarities exist between the SDLC and the activities within the maintenance process. Figure 18-4 equates SDLC phases to the maintenance activities described previously. The figure shows that the first phase of the SDLC—project identification and selection—is analogous to the maintenance process of obtaining a maintenance request (step 1). SDLC phases project initiation and planning and analysis are analogous to the maintenance process of transforming requests into a specific system change (step 2). The logical and physical design phases of the SDLC, of course, equate to the designing changes process (step 3). Finally, the SDLC phase implementation equates to step 4, implementing changes. This similarity between the maintenance process and the SDLC is no accident. The concepts and techniques used to initially develop a system are also used to maintain it.

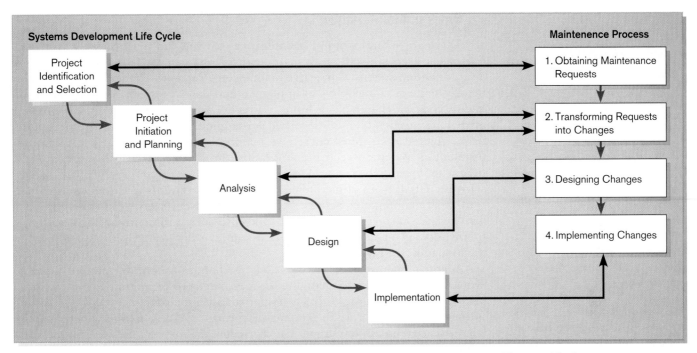

Figure 18-4
Maintenance activities in relation to the SDLC

Deliverables and Outcomes

Since the maintenance phase of the SDLC is basically a subset of the activities of the entire development process, the deliverables and outcomes from the process are the development of a new version of the software and new versions of all design documents developed or modified during the maintenance process. This means that all documents created or modified during the maintenance effort, including the system itself, represent the deliverables and outcomes of the process. Those programs and documents that did not change may also be part of the new system. Since most organizations archive prior versions of systems, all prior programs and documents must be kept to ensure the proper versioning of the system. This enables prior versions of the system to be recreated if needed. A more detailed discussion of configuration management and change control is presented later in the chapter.

Because of the similarities between the steps, deliverables, and outcomes of new development and maintenance, you may be wondering how to distinguish between these two processes. One difference is that maintenance reuses most existing system modules in producing the new system version. Other distinctions are that we develop a new system when there is a change in the hardware or software platform or when fundamental assumptions and properties of the data, logic, or process models change.

CONDUCTING SYSTEMS MAINTENANCE

A significant portion of the expenditures for information systems within organizations does not go to the development of new systems but to the maintenance of existing systems. We will describe various types of maintenance, factors influencing the complexity and cost of maintenance, alternatives for managing maintenance, and the role of CASE during maintenance. Given that maintenance activities consume the majority of information systems-related expenditures, gaining an understanding of these topics will yield numerous benefits to your career as an information systems professional.

TABLE 18-1   Types of Maintenance

Type	Description
Corrective	Repair design and programming errors
Adaptive	Modify system to environmental changes
Perfective	Evolve system to solve new problems or take advantage of new opportunities
Preventive	Safeguard system from future problems

Maintenance: Changes made to a system to fix or enhance its functionality.

Corrective maintenance: Changes made to a system to repair flaws in its design coding, or implementation.

Adaptive maintenance: Changes made to a system to evolve its functionality to changing business needs or technologies.

Perfective maintenance: Changes made to a system to add new features or to improve performance.

Preventive maintenance: Changes made to a system to avoid possible future problems.

Types of Maintenance

There are several types of **maintenance** that you can perform on an information system (see Table 18-1). By maintenance, we mean the fixing or enhancing of an information system. **Corrective maintenance** refers to changes made to repair defects in the design, coding, or implementation of the system. For example, if you had recently purchased a new home, corrective maintenance would involve repairs made to things that had never worked as designed, such as a faulty electrical outlet or misaligned door. Most corrective maintenance problems surface soon after installation. When corrective maintenance problems surface, they are typically urgent and need to be resolved to curtail possible interruptions in normal business activities. Of all types of maintenance, corrective accounts for as much as 75 percent of all maintenance activity (Andrews and Leventhal, 1993). This is unfortunate because corrective maintenance adds little or no value to the organization; it simply focuses on removing defects from an existing system without adding new functionality (see Figure 18-5).

Adaptive maintenance involves making changes to an information system to evolve its functionality to changing business needs or to migrate it to a different operating environment. Within a home, adaptive maintenance might be adding storm windows to improve the cooling performance of an air conditioner. Adaptive maintenance is usually less urgent than corrective maintenance because business and technical changes typically occur over some period of time. Contrary to corrective maintenance, adaptive maintenance is generally a small part of an organization's maintenance effort but does add value to the organization.

Perfective maintenance involves making enhancements to improve processing performance, interface usability, or to add desired, but not necessarily required, system features ("bells and whistles"). In our home example, perfective maintenance would be adding a new room. Many system professionals feel that perfective maintenance is not really maintenance but new development.

Preventive maintenance involves changes made to a system to reduce the chance of future system failure. An example of preventive maintenance might be to increase the number of records that a system can process far beyond what is currently needed or to generalize how a system sends report information to a printer so that the system can easily adapt to changes in printer technology. In our home example, preventive maintenance could be painting the exterior to better protect the home from severe weather conditions. As with adaptive maintenance, both perfective and preventive maintenance are typically a much lower priority than corrective maintenance. Over the life of a system, corrective maintenance is most likely to occur after initial system installation or after major system changes. This means that adaptive, perfective, and preventive maintenance activities can lead to corrective maintenance activities if not carefully designed and implemented.

Figure 18-5
Types of maintenance (Adapted from Andrews and Leventhal, 1993)

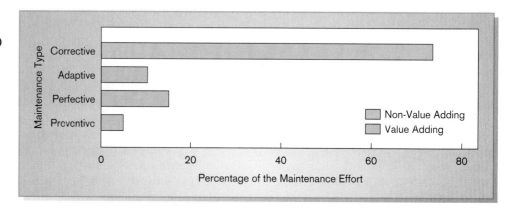

The Cost of Maintenance

Information systems maintenance costs are a significant expenditure. For some organizations, as much as 80 percent of their information systems budget is allocated to maintenance activities (Pressman, 1992). Additionally, the proportion of systems expenditures has been rising due to the fact that many organizations have accumulated more and more older systems that require more and more maintenance. For example, Figure 18-6 shows that in the 1970s, most of an organization's information systems' expenditures were allocated to new development rather than to maintenance. However, over the years, this mix has changed so that the majority of expenditures are now earmarked for maintenance. This means that you must understand the factors influencing the **maintainability** of systems. Maintainability is the ease with which software can be understood, corrected, adapted, and enhanced. Systems with low maintainability result in uncontrollable maintenance expenses.

Maintainability: The ease with which software can be understood, corrected, adapted, and enhanced.

Numerous factors influence the maintainability of a system. These factors, or cost elements, will determine the extent to which a system has high or low maintainability. Table 18-2 shows numerous elements that influence the cost of maintenance, many of which you can influence as a systems analyst. Of these factors, three are most significant: number of latent defects, number of customers, and documentation quality.

The number of latent defects refers to the number of unknown errors existing in the system after it is installed. Because corrective maintenance accounts for most maintenance activity, the number of latent defects in a system influences most of the costs associated with maintaining a system. If there are no errors in the system after it is installed, then maintenance costs will be relatively low. If there are a large number of defects in the system when it is installed, maintenance costs will likely be high.

A second factor influencing maintenance costs is the number of customers for a given system. In general, the greater the number of customers, the greater the maintenance costs. For example, if a system has only one customer, problem and change requests will come from only one source. A single customer also makes it much easier for the maintenance group to know how the system is actually being used and the extent to which users are adequately trained. If the system fails, the maintenance group is only notifying, supporting, and retraining a small group of people and updating their programs and documentation. On the other hand, for a system with thousands of users, change requests (possibly contradictory or incompatible) and error reports come from many places, and notification of problems, customer support, and system and documentation updating become a much more significant problem. For example, notifying all customers of a problem is fast and easy when you

Figure 18-6
New development versus maintenance as a percent of software budget (Adapted from Pressman, 1987)

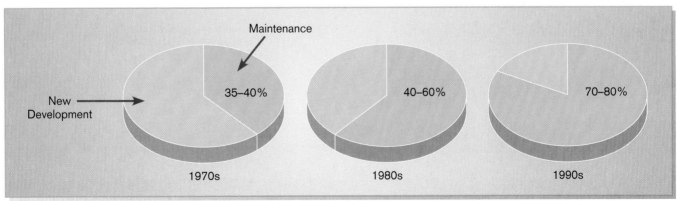

Maintenance

New Development   35–40%   40–60%   70–80%

1970s   1980s   1990s

TABLE 18-2   Cost Elements of Maintenance

Element	Description
Defects	Number of unknown defects in a system when it is installed
Customers	Number of different customers that a maintenance group must support
Documentation	Quality of technical system documentation including test cases
Personnel	Number and quality of personnel dedicated to the support and maintenance of a system
Tools	Software development tools, debuggers, hardware, and other resources
Software Structure	Structure and maintainability of the software

Adapted from Jones, 1986

have a single customer whom you can contact using a telephone, fax, or electronic mail message. Yet, it will be difficult and expensive to quickly and easily contact thousands of users about a catastrophic problem. In sum, the greater the number of customers, the more critical it is that a system have high maintainability.

A third major contributing factor to maintenance costs is the quality of system documentation. Figure 18-7 shows that without quality documentation, maintenance effort can increase exponentially. In conclusion, numerous factors will influence the maintainability and thus the overall costs of system maintenance. System professionals have found that the number of defects in the installed system drive all other cost factors. This means that it is important that you remove as many errors as possible before installation.

The quality of the maintenance personnel also contributes to the cost of maintenance. In some organizations, the best programmers are assigned to maintenance. Highly skilled programmers are needed because the maintenance programmer is typically not the original programmer and must quickly understand and carefully change the software. Tools, such as those that can automatically produce system documentation where none exists, can also lower maintenance costs. Finally, well-structured programs make it much easier to understand and fix programs.

N E T   S E A R C H

The Total Cost of Ownership has become very important to information systems managers. Visit http://www.prenhall.com/hoffer to complete an exercise related to this topic.

Figure 18-7
Quality documentation eases maintenance (Adapted from Hanna, 1992)

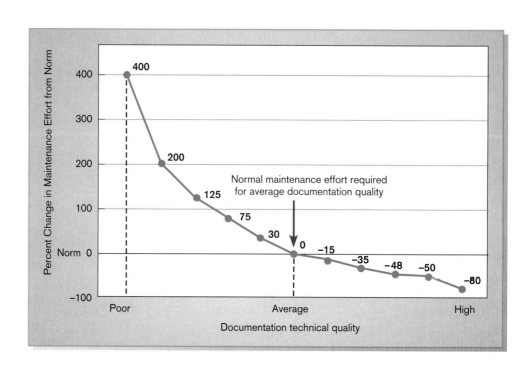

TABLE 18-3   Worldwide Totals of Programmers Working on New Development versus Maintenance

Year	Programmers on New Programs	Programmers on Maintenance
1950	90	10
1960	8,500	1,500
1970	65,000	35,000
1980	1,200,000	800,000
1990	3,000,000	4,000,000
2000	4,000,000	6,000,000

Adapted from Jones, 1986

Managing Maintenance

As maintenance activities consume more and more of the systems development budget, maintenance management has become increasingly important. For example, Table 18-3 shows the worldwide growth of programmers who are working on new development versus maintenance. These data show that the number of people working on maintenance has now surpassed the number working on new development. Maintenance has become the largest segment of programming personnel, and this implies the need for careful management. We will address this concern by discussing several topics related to the effective management of systems maintenance.

Managing Maintenance Personnel   One concern with managing maintenance relates to personnel management. Historically, many organizations had a "maintenance group" that was separate from the "development group." With the increased number of maintenance personnel, the development of formal methodologies and tools, changing organizational forms, end-user computing, and the widespread use of very high-level languages for the development of some systems, organizations have rethought the organization of maintenance and development personnel (Chapin, 1987). In other words, should the maintenance group be separated from the development group? Or, should the same people who build the system also maintain it? A third option is to let the primary end users of the system in the functional units of the business have their own maintenance personnel. The advantages and disadvantages to each of these organizational structures are summarized in Table 18-4.

In addition to the advantages and disadvantages listed in Table 18-4, there are numerous other reasons why organizations should be concerned with how they

TABLE 18-4   Advantages and Disadvantages of Different Maintenance Organizational Structures

Type	Advantages	Disadvantages
Separate	Formal transfer of systems between groups improves the system and documentation quality	All things cannot be documented, so the maintenance group may not know critical information about the system
Combined	Maintenance group knows or has access to all assumptions and decisions behind the system's original design	Documentation and testing thoroughness may suffer due to a lack of a formal transfer of responsibility
Functional	Personnel have a vested interest in effectively maintaining the system and have a better understanding of functional requirements	Personnel may have limited job mobility and lack access to adequate human and technical resources

manage and assign maintenance personnel. One key issue is that many systems professionals don't want to perform maintenance because they feel that it is more exciting to build something new rather than change an existing system (Martin, Brown, DeHayes, Hoffer, and Perkins, 1999). In other words, maintenance work is often viewed as "cleaning up someone else's mess." Also, organizations have historically provided greater rewards and job opportunities to those performing new development. thus making people shy away from maintenance-type careers. As a result, no matter how an organization chooses to manage its maintenance group—separate, combined, or functional—it is now common to rotate individuals in and out of maintenance activities. This rotation is believed to lessen the negative feelings bout maintenance work and to give personnel a greater understanding of the difficulties and relationships between new development and maintenance.

Measuring Maintenance Effectiveness A second management issue is the measurement of maintenance effectiveness. As with the effective management of personnel, the measurement of maintenance activities is fundamental to understanding the quality of the development and maintenance efforts. To measure effectiveness, you must measure these factors:

- Number of failures
- Time between each failure
- Type of failure

Measuring the number and time between failures will provide you with the basis to calculate a widely used measure of system quality. This metric is referred to as the **mean time between failures (MTBF)**. As its name implies, the MTBF measure shows the average length of time between the identification of one system failure until the next. Over time, you should expect the MTBF value to rapidly increase after a few months of use (and corrective maintenance) of the system (see Figure 18-8 for an example of the relationship between MTBF and age of a system). If the MTBF does not rapidly increase over time, it will be a signal to management that major problems

Mean time between failures (MTBF): A measurement of error occurrences that can be tracked over time to indicate the quality of a system.

Figure 18-8
How the mean time between failures should change over time

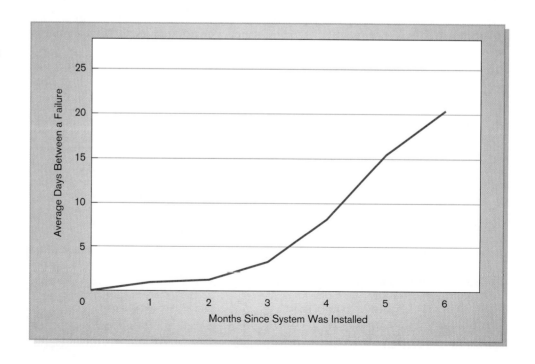

exist within the system that are not being adequately resolved through the maintenance process.

A more revealing method of measurement is to examine the failures that are occurring. Over time, logging the types of failures will provide a very clear picture of where, when, and how failures occur. For example, knowing that a system repeatedly fails logging new account information to the database when a particular customer is using the system can provide invaluable information to the maintenance personnel. Were the users adequately trained? Is there something unique about this user? Is there something unique about an installation that is causing the failure? What activities were being performed when the system failed?

Tracking the types of failures also provides important management information for future projects. For example, if a higher frequency of errors occurs when a particular development environment is used, such information can help guide personnel assignments, training courses, or the avoidance of a particular package, language, or environment during future development. The primary lesson here is that without measuring and tracking maintenance activities, you cannot gain the knowledge to improve or know how well you are doing relative to the past. To effectively manage and to continuously improve, you must measure and assess performance over time.

Controlling Maintenance Requests   Another maintenance activity is managing maintenance requests. There are various types of maintenance requests—some correct minor or severe defects in the system while others improve or extend system functionality. From a management perspective, a key issue is deciding which requests to perform and which to ignore. Since some requests will be more critical than others, some method of prioritizing requests must be determined.

Figure 18-9 shows a flow chart that suggests one possible method you could apply for dealing with maintenance change requests. First, you must determine the type of request. If, for example, the request is an error—that is, a corrective maintenance request—then the flow chart shows that a question related to the error's severity must be asked. If the error is "very severe," then the request has top priority and is placed at the top of a queue of tasks waiting to be performed on the system. In other words, for an error of high severity, repairs to remove it must be made as soon as possible. If, however, the error is considered "nonsevere," then the change request can be categorized and prioritized based upon its type and relative importance.

If the change request is not an error, then you must determine whether the request is to adapt the system to technology changes and/or business requirements or to enhance the system so that it will provide new business functionality. For adaptation requests, they too will need to be evaluated, categorized, prioritized, and placed in the queue. For enhancement-type requests, they must first be evaluated to see whether they are aligned with future business and information systems' plans. If not, the request will be rejected and the requester will be informed. If the enhancement appears to be aligned with business and information systems plans, it can then be prioritized and placed into the queue of future tasks. Part of the prioritization process includes estimating the scope and feasibility of the change. Techniques used for assessing the scope and feasibility of entire projects should be used when assessing maintenance requests (see Chapter 6).

The queue of maintenance tasks for a given system is dynamic—growing and shrinking based upon business changes and errors. In fact, some lower-priority change requests may never be accomplished, since only a limited number of changes can be accomplished at a given time. In other words, changes in business needs between the time the request was made and when the task finally rises to the top of the queue may result in the request being deemed unnecessary or no longer important given current business directions. Thus, managing the queue of pending tasks is an important activity.

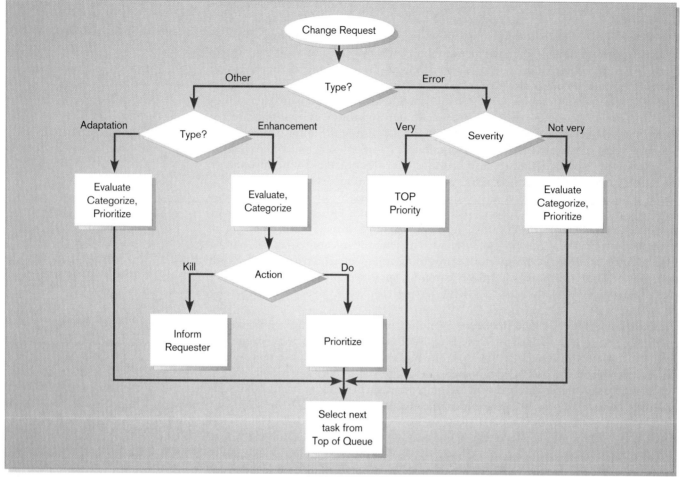

Figure 18-9
Flowchart of how to control
maintenance requests (Adapted
from Pressman, 1992)

To better understand the flow of a change request, see Figure 18-10. Initially, an organizational group that uses the system will make a request to change the system. This request flows to the project manager of the system (labeled 1). The project manager evaluates the request in relation to the existing system and pending changes, and forwards the results of this evaluation to the System Priority Board (labeled 2). This evaluation will also include a feasibility analysis that includes estimates of project scope, resource requirements, risks, and other relevant factors. The board evaluates, categorizes, and prioritizes the request in relation to both the strategic and information systems plans of the organization (labeled 3). If the board decides to kill the request, the project manager informs the requester and explains the rationale for the decision (labeled 4). If the request is accepted, it is placed in the queue of pending tasks. The project manager then assigns tasks to maintenance personnel based upon their availability and task priority (labeled 5). On a periodic basis, the project manager prepares a report of all pending tasks in the change request queue. This report is forwarded to the priority board where they reevaluate the requests in light of the current business conditions. This process may result in removing some requests or reprioritizing others.

Although each change request goes through an approval process as depicted in Figure 18-10, changes are usually implemented in batches, forming a new release of the software. It is too difficult to manage a lot of small changes. Further, batching changes can reduce maintenance work when several change requests affect the same or highly related modules. Frequent releases of new system versions may also confuse users if the appearance of displays, reports, or data entry screens changes.

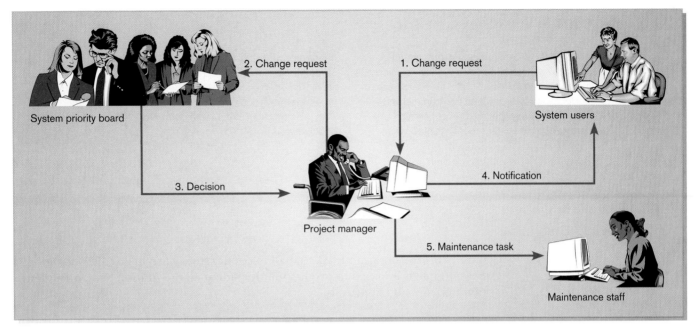

Figure 18-10
How a maintenance request moves through an organization (Adapted from Pressman, 1987)

Configuration Management   A final aspect of managing maintenance is **configuration management**, which is the process of assuring that only authorized changes are made to a system. Once a system has been implemented and installed, the programming code used to construct the system represents the **baseline modules** of the system. The baseline modules are the software modules for the most recent version of a system where each module has passed the organization's quality assurance process and documentation standards. A **system librarian** controls the baseline source code modules. If maintenance personnel are assigned to make changes to a system, they must first check out a copy of the baseline system modules because no one is allowed to directly modify the baseline modules. Only those that have been checked out and have gone through a formal check-in process can reside in the library. Before any code can be checked back in to the librarian, the code must pass the quality control procedures, testing, and documentation standards established by the organization.

When various maintenance personnel working on different maintenance tasks complete each task, the librarian notifies those still working that updates have been made to the baseline modules. This means that all tasks being worked on must now incorporate the latest baseline modules before being approved for check-in. Following a formal process of checking modules out and in, a system librarian helps to assure that only tested and approved modules become part of the baseline system. It is also the responsibility of the librarian to keep copies of all prior versions of all system modules, including the **build routines** needed to construct *any version* of the system that *ever* existed. It may be important to reconstruct old versions of the system if new ones fail, or to support users that cannot run newer versions on their computer system.

Special software systems have been created to manage system configuration and version control activities (see box "Configuration Management Tools"). This software is increasingly necessary as the change control process is complicated in organizations deploying several different networks, operating systems, languages, and database management systems in which there may be many concurrent versions of an application, each for a different platform. One function of this software is to control access to libraried modules. Each time a module is checked out or in, this activity is recorded, after being authorized by the librarian. The soft-

Configuration management: The process of assuring that only authorized changes are made to a system.

Baseline modules: Software modules that have been tested, documented, and approved to be included in the most recently created version of a system.

System librarian: A person responsible for controlling the checking out and checking in of baseline modules for a system when a system is being developed or maintained.

Build routines: Guidelines that list the instructions to construct an executable system from the baseline source code.

Configuration Management Tools

There are two general kinds of configuration management tools: revision control and source code control. With revision control tools, each system module file is "frozen" (unchangeable) to a specific version level—or "floating"—a programmer may check out, lock, and modify. Only the most recent version of a module is stored; previous versions are reconstructed when needed by applying changes in reverse order. Source code control tools extend the description to address inter-related files. These tools also help in rebuilding any historic version of a system by recompiling the proper source code modules. They also trace an executable code module back to its source code version.

ware helps the librarian track that all necessary steps have been followed before a new module is released to production, including all integration tests, documentation updates, and approvals.

Role of CASE and Automated Development Tools in Maintenance

N E T S E A R C H
There are many software tools available for configuration management. Visit http://www.prenhall.com/hoffer to complete an exercise related to this topic.

In traditional systems development, much of the time is spent on coding and testing. When software changes are approved, code is first changed and then tested. Once the functionality of the code is assured, the documentation and specification documents are updated to reflect system changes. Over time, the process of keeping all system documentation "current" can be a very boring and time-consuming activity that is often neglected. This neglect makes future maintenance by the same or *different* programmers difficult at best.

A primary objective of using CASE and other automated tools (see Chapter 4) for systems development and maintenance is to radically change the way in which code and documentation are modified and updated. When using an integrated development environment, analysts maintain design documents such as data flow diagrams and screen designs, not source code. In other words, design documents are modified and then code generators automatically create a new version of the system from these updated designs. Also, since the changes are made at the design specification level, most documentation changes such as an updated data flow diagram will have already been completed during the maintenance process itself. Thus, one of the biggest advantages to using CASE, for example, is its benefits during system maintenance.

In addition to using general automated tools for maintenance, two special-purpose tools, reverse engineering and reengineering tools (see Chapter 4), are primarily used to maintain older systems that have incomplete documentation or that were developed prior to CASE use. These tools are often referred to as *design recovery tools* since their primary benefit is to create high-level design documents of a program by reading and analyzing its source code.

Recall that reverse engineering tools are those that can create a representation of a system or program module at a design level of abstraction. For example, reverse engineering tools read program source code as input, perform an analysis, and extract information such as program control structures, data structures, and data flow. Once a program is represented at a design level using both graphical and textual representations, the design can be more effectively restructured to current business needs or programming practices by an analyst. Similarly, reengineering tools extend reverse engineering tools by automatically, or interactively with a systems analyst, altering an existing system in an effort to improve its quality or performance.

WEBSITE MAINTENANCE

All of the discussion on maintenance in this chapter applies to any type of information system, no matter on what platform it runs. There are, however, some special issues and procedures needed for Websites, due to their nature and operational status. These issues and procedures include:

- *24×7×365* Most Websites are never purposely down. In fact, an e-commerce Website has the advantage of continuous operation. Thus, maintenance of pages and the overall site usually must be done without taking the site off-line. However, it may be necessary to lock out use of pages in a portion of a Website while changes are made to those pages. This can be done by inserting a "Temporarily Out of Service" notice on the main page of the section being maintained and disabling all links within that segment. Alternatively, references to the main page of the section can be temporarily rerouted to an alternative location where the current pages are kept while maintenance is performed to create new versions of those pages. The really tricky nuance is keeping the site consistent for a user during a session; that is, it can be confusing to a user to see two different versions of a page within the same on-line session. Browser caching functions may bring up an old version of a page even when that page changes during the session. One precaution against confusing is locking, as explained above. Another approach is to not lock a page being changed, but to include a date and time stamp of the most recent change. This gives the page visitor an indication of the change, which may reduce confusion.

- *Check for broken links* Arguably the most common maintenance issue for any Website (besides changing the content of the site) is validating that links from site pages (especially for links that go outside the source site) are still accurate. Periodic checks need to be performed to make sure active pages are found from all links—this can be done via various software such as CyberSpider (www.CyberSpider.com), Doctor HTML (www.imagiware.com), or Linkbot (http://tetranetsoftware.com)—note the irony of any URL in a textbook! In addition, periodic human checks need to be performed to make sure that the content found at a still-existing referenced page is still the intended content.

- *HTML validation* Before modified or new pages are published, these pages should be processed by a code validation routine to insure that all the code, including applets, work. If you are using a HTML, XML, Perl, or other editor, such a feature is likely built into the software.

- *Re-registration* It may be necessary to re-register a Website with search engines when the content of your site significantly changes. Re-registration may be necessary for visitors to find your site based on the new or changed content.

- *Future editions* One of the most important issues to address to insure effective Website use is to avoid confusing visitors. Especially, frequent visitors can be confused if the site is constantly changing. To avoid confusion, you can post indications of future enhancements to the site and, as with all information systems, you can batch changes to reduce the frequency of site changes.

NET SEARCH
Besides the broken link chaser software listed in this section, there are many other tools and consulting services available to assist in Website maintenance. Visit http://www.prenhall.com/hoffer to complete an exercise related to this topic.

ELECTRONIC COMMERCE APPLICATION: MAINTAINING AN INFORMATION SYSTEM FOR PINE VALLEY FURNITURE'S WEBSTORE

In this chapter, we examined various aspects of conducting system maintenance. Maintenance is a natural part of the life of any system. In this section, we conclude our discussion of Pine Valley Furniture's WebStore by examining a maintenance activity for this system.

Maintaining Pine Valley Furniture's WebStore

Early on a Saturday evening, Jackie Judson, vice president of marketing at Pine Valley Furniture (PVF), was reviewing new product content that was recently posted on the company's electronic commerce Website, the WebStore. She was working on Saturday evening because she was leaving the next day for a long overdue two-week vacation to the Black Hills of South Dakota. Before she could leave, however, she wanted to review the appearance and layout of the pages.

Midway through the review process, pages from the WebStore began to load very slowly. Finally, after requesting detailed information on a particular product, the WebStore simply stopped working. The title bar on her Web browser reported the error:

Cannot Find Server

Given that her plane for Rapid City left in less than 12 hours, Jackie wanted to review the content and needed to figure out some way to overcome this catastrophic system error. Her first thought was to walk over to the offices of the information systems development group within the same building. When she did, she found no one there. Her next idea was to contact Jim Woo, senior systems analyst and the project manager for the WebStore system. She placed a call to Jim's home and found that he was at the grocery store but would be home soon. Jackie left a message for Jim to call her ASAP at the office.

Within 30 minutes, Jim returned the call and was on his way into the office to help Jackie. Although it is not a common occurrence, this is not the first time that Jim has gone into the office to assist users when systems have failed during off hours. Before leaving for the office, he connected to the Internet and also found the WebStore to be unavailable. Since PVF outsourced the hosting of the WebStore to an outside vendor, Jim immediately notified the vendor that the WebStore was down. The vendor was a local Internet service provider (ISP) that had a long-term relationship with PVF to provide Internet access, but had limited experience with hosting commercial Websites. Jim was informed that a system "glitch" was responsible for the outage and that the WebStore would be on-line within the next hour or so. Unfortunately, this was not the first time the WebStore failed and Jim felt powerless. More than ever before, he believed that PVF needed to a better way to learn about system failures and, more importantly, an improved confidence that the system would operate reliably.

On Monday morning, Jim requested a meeting with several senior PVF managers. At this meeting, he posed the following questions:

- How much is our Website worth?
- How much does it cost our company when our Website goes down?
- How reliable does our Website need to be?

These questions encouraged a spirited discussion. Everyone agreed that the WebStore was "critical" to PVF's future and unanimously agreed that it was "unacceptable" for the site to be down. One manager summarized the feeling of the group by stating, "I cannot think of a single valid excuse for the system to crash . . . our customers have incredibly high expectations of us . . . one major mishap could prove disastrous for PVF."

Jim outlined to the group what he felt the next steps needed to be. "I believe that the root of our problem is with our contract with our Web hosting company. Specifically, we need to renegotiate our contract, or find a different vendor, so that it includes wording to reflect our expectations of service. Our current agreement does not address how emergencies are responded to or what remedies we have for continued system failures. The question we must also address is the cost differences between having a site that operates 99 percent of the time versus one that operates 99.999 percent of the time. I believe, he continued, that it could increase our costs tens of thousands of dollars per month to guarantee extremely high levels of system reliability."

At the conclusion of the meeting, the senior managers unanimously agreed that Jim should immediately develop a plan for addressing the WebStore's service level problems. To begin this process, Jim prepared a detailed list of specific vendor services they desired. He felt a very specific list was needed so that the relative costs for different services and varying levels of service (e.g., response times for system failures and penalties for noncompliance) could be discussed.

When asked by a colleague what type of maintenance was being performed on the WebStore by delivering the system more reliability Jim had to pause and think. "Well, it is clearly *adaptive* in that we plan to migrate the system to a more reliable environment. It is also *perfective* and *preventive* . . . It is perfective in that we want to make some operational changes that will improve system performance and it is also preventive given that one of our objectives is to reduce the likelihood of system failures." Through this discussion it became clear to Jim that the system was much larger than simply the HTML used to construct the WebStore, but also the hardware, system software, procedures, and response team in place to deal with unforeseen events. Although he had heard it said many times before, Jim now understood that a successful system reflected all aspects of the system.

Summary

Maintenance is the final phase in the systems development life cycle where systems are changed to rectify internal processing errors or to extend the functionality of the system. Maintenance is where a majority of the financial investment in a system occurs and can span more than 20 years. More and more information systems professionals have devoted their careers to systems maintenance and, as more systems move from initial development into operational use, it is likely that even more professionals will in the future.

It is during maintenance that the SDLC becomes a life cycle since requests to change a system must first be approved, planned, analyzed, designed, and then implemented. In some special cases where business operations are impaired due to an internal system error, quick fixes can be made. This, of course, circumvents the normal maintenance process. After quick fixes are made, maintenance personnel must back up and perform a thorough analysis of the problem to make sure that the correction conforms to normal systems development standards for design, programming, testing, and documentation.

Maintenance requests can be one of four types: corrective, adaptive, perfective, and preventive. Corrective maintenance is used to repair design and programming errors. Adaptive maintenance is used to modify the system to changes in business informational or processing needs, or to migrate the system onto a different technology platform. Perfective maintenance is used to evolve the system so that it can solve new problems or take advantage of new opportunities; in other words, to extend the capabilities of the system beyond those initially intended for the system. Preventive maintenance is used to safeguard the system from future problems.

How a system is designed and implemented can greatly impact the cost of performing maintenance. The number of unknown errors in a system when it is installed is a primary factor in determining the cost of maintenance. Other factors, such as the number of separate customers and the quality of documentation, significantly influence maintenance costs.

Organizations can choose three general methods for managing maintenance personnel. One method is to have a separate maintenance group. A second approach is to have the same people who construct the system also maintain it. A third option is to have maintenance personnel housed within the functional areas of the business that use the system on a day-to-day basis. Each approach has its advantages and disadvantages and no approach is universally best. A second maintenance management issue relates to understanding how to measure the quality of the maintenance effort. Most organizations track the frequency, time, and type of each failure and compare performance over time. Since limited resources preclude organizations from performing all maintenance requests, some formal process for reviewing requests must be established to make sure that only those requests deemed consistent with organizational and information systems plans be performed. A central source, usually a project manager, is used to collect maintenance requests. When requests are submitted, this person forwards each request to a committee charged to assess its merit. Once assessed, the project manager assigns higher-priority activities to maintenance personnel.

Maintenance personnel must be controlled from making unapproved changes to a system. To do this, most organizations assign one member of the maintenance staff, typically a senior programmer or analyst, to serve as system librarian. The librarian controls the checking out and checking in of system modules to assure that appropriate procedures for performing maintenance such as adequate testing and documentation are adhered to.

CASE is actively employed during maintenance. The primary benefit to using CASE is its ability to enable maintenance to be performed on design documents rather than low-level source code. Reverse engineering and re-engineering CASE tools are used to recover design specifications of older systems that were not constructed using CASE or for systems with inadequate design specifications. Once recovered into a design specification, these older systems can then be changed at the design level rather than the source code level,

yielding a significant improvement in maintenance personnel productivity.

Website maintenance involves some special attention, including: 24×7×365 operation, checking for broken external links, validating code changes before publishing new or revised pages, re-registration of the Website for search engines, and avoiding visitor confusion by previewing future changes.

Business process reengineering is a major source for maintenance requests. BPR is often part of the formal strategic organizational and information systems planning effort where key business processes are redesigned using information technologies to radically improve business performance. Many of the tools and techniques used to design information systems are also used in the BPR process.

Key Terms

1. Adaptive maintenance
2. Baseline modules
3. Build routines
4. Configuration management
5. Corrective maintenance
6. Maintainability
7. Maintenance
8. Mean time between failures (MTBF)
9. Perfective maintenance
10. Preventive maintenance
11. System librarian

Match each of the key terms above with the definition that best fits it.

_____ Changes made to a system to fix or enhance its functionality.

_____ Changes made to a system to repair flaws in its design, coding, or implementation.

_____ Changes made to a system to evolve its functionality to changing business needs or technologies.

_____ Changes made to a system to add new features or to improve performance.

_____ Changes made to a system to avoid possible future problems.

_____ The ease with which software can be understood, corrected, adapted, and enhanced.

_____ A measurement of error occurrences that can be tracked over time to indicate the quality of a system.

_____ The process of assuring that only authorized changes are made to a system.

_____ Software modules that have been tested, documented, and approved to be included in the most recently created version of a system.

_____ A person responsible for controlling the checking out and checking in of baseline modules for a system when a system is being developed or maintained.

_____ Guidelines that list the instructions to construct an executable system from the baseline source code.

Review Questions

1. Contrast the following terms:
   a. adaptive maintenance, corrective maintenance, perfective maintenance, preventive maintenance
   b. baseline modules, build routines, system librarian
   c. maintenance, maintainability

2. List the steps in the maintenance process and contrast them with the phases of the systems development life cycle.

3. What are the different types of maintenance and how do they differ?

4. Describe the factors that influence the cost of maintenance. Are any factors more important? Why?

5. Describe three ways for organizing maintenance personnel and contrast the advantages and disadvantages of each approach.

6. What types of measurements must be taken to gain an understanding of the effectiveness of maintenance? Why is tracking mean time between failures an important measurement?

7. What managerial issues can be better understood by measuring maintenance effectiveness?

8. Describe the process for controlling maintenance requests. Should all requests be handled in the same way or are there situations when you should be able to circumvent the process? If so, when and why?

9. What is meant by configuration management? Why do you think organizations have adopted the approach of using a systems librarian?

10. How is CASE used in the maintenance of information systems?

11. What is the difference between reverse engineering and re-engineering CASE tools?

12. What are some special maintenance issues and procedures especially relevant for Websites?

Problems and Exercises

1. Maintenance has been presented as both the final stage of the SDLC (see Figure 18-1) and as a process similar to the SDLC (see Figure 18-4). Why does it make sense to talk about maintenance in both of these ways? Do you see a conflict in looking at maintenance in both ways?

2. In what ways is a request to change an information system handled differently from a request for a new information system?

3. According to Figure 18-5, corrective maintenance is by far the most frequent form of maintenance. What can you do as a systems analyst to reduce this form of maintenance?

4. What other or additional information should be collected on a System Service Request (see Figure 18-3) for a maintenance request?

5. Briefly discuss how a systems analyst can manage each of the six Cost Elements of Maintenance listed in Table 18-2.

6. Suppose you were a system librarian. Using entity-relationship diagramming notation, describe the database you would need to keep track of the information necessary in your job. Consider operational, managerial, and planning aspects of the job.

7. Suppose an information system were developed following a RAD approach (see Chapter 19). How might maintenance be different than if the system had been developed following the traditional life cycle? Why?

8. Software configuration management is similar to configuration management in any engineering environment. For example, the product design engineers for a refrigerator need to coordinate dynamic changes in compressors, power supplies, electronic controls, interior features, and exterior designs as innovations to each occur. How do such product design engineers manage the configuration of their products? What similar practices do systems analysts and librarians have to follow?

9. In the section on Pine Valley Furniture's WebStore, it is mentioned that Jim Woo will prepare a list of ISP Web hosting services. Prepare such a list for Website maintenance. Contrast the responsibilities of the ISP to those of PVF.

Field Exercises

1. Study an information systems department with which you are familiar or to which you have access. How does this department organize for maintenance? Has this department adopted one of the three approaches outlined in Table 18-4 or does it use some other approach? Talk with a senior manager in this department to discover how well this maintenance organization structure works.

2. Study an information systems department with which you are familiar or to which you have access. How does this department measure the effectiveness of systems maintenance? What specific metrics are used and how are these metrics used to effect changes in maintenance practices? If there is a history of measurements over several years, how can changes in the measurements be explained?

3. With the help of other students or your instructor, contact a system librarian in an organization. What is this person's job description? What tools does this person use to help him or her in the job? To whom does this person report? What previous jobs did this person hold, and what jobs does this person expect to be promoted into in the near future?

4. Interview the Webmaster at a company where you work or where you have a contact. Investigate the procedures followed to maintain this Website. Document these procedures. What differences or enhancement did you find to the special Website maintenance issues and procedures listed in this chapter.

References

Andrews. D. C., and N. S. Leventhal. 1993. *Fusion: Integrating IE, CASE, JAD: A Handbook for Reengineering the Systems Organization.* Englewood Cliffs, NJ: Prentice-Hall.

Chapin, N. 1987. "The Job of Software Maintenance." *Proceedings of the Conference on Software Maintenance.* Washington DC: IEEE Computer Society Press: 4–12.

Hanna, M. 1992. "Using documentation as a life-cycle tool." *Software Magazine* (December): 41–46.

Jones, C. 1986. *Programming Productivity.* New York, NY: McGraw-Hill.

Jones C. 1997. "How to Measure Software Costs." *Application Development Trends* (May): 32–36.

Martin, E. W., C. V. Brown, D. W. DeHayes, J. A. Hoffer, and W. C. Perkins, 1999. *Managing Information Technology: What Managers Need to Know.* 3rd Edition. New York, NY: MacMillan.

Pressman, R. S. 1987. *Software Engineering: A Practitioner's Approach.* New York, NY: McGraw-Hill.

Pressman, R. S. 1992. *Software Engineering: A Practitioner's Approach.* 2d ed. New York, NY: McGraw-Hill.

BROADWAY ENTERTAINMENT COMPANY, INC.

Designing a Maintenance Plan for the Customer Relationship Management System

CASE INTRODUCTION

Maintenance of MyBroadway, the Web-based customer relationship management system for Carrie Douglass, manager of the Centerville, Ohio, Broadway Entertainment Company (BEC) store, deals with a phase of systems development with which the students from Stillwater State University are unfamiliar. None of the systems they developed for programming and systems analysis and design classes ever reached this phase. And, MyBroadway, since it is a prototype, is not really designed to be maintained after the student team does its work. Professor Tann, instructor of their MIS project class, had encouraged all the student teams to consider the maintenance issue, even if the teams are not themselves involved in maintenance of the systems they build for the course.

PREPARING THE MAINTENANCE PLAN

The MyBroadway team members discussed the concept of maintenance for their information system. They concluded that each prototyping iteration is similar to a maintenance step. At each iteration, a flaw in the system may need to be repaired (corrective maintenance), a new business requirement may have been identified (adaptive maintenance), a desirable system improvement may be made (perfective maintenance), or the system may change in anticipation of future issues (preventive maintenance). Already, the team has gone through many iterations of MyBroadway, as additional modules are included or changes to the system interface are made. Prototyping is intended to be rather informal and user driven, so tight controls on making changes would not be appropriate. On the other hand, making many small changes takes considerable effort, can seem trivial at each iteration to the end user, and is open-ended and difficult to mesh with the team members busy class schedules. Thus, some systematic way to handle changes, even if not very formal, could make the team more productive.

The Stillwater students decided that preventive maintenance should be the lowest priority, to be done only when other changes are not requested or when the pre-

ventive change can be incorporated easily along with other changes. As with all maintenance, corrective maintenance is a top priority. Prototyping is driven by the identification of adaptive and to some degree perfective changes, so these must be important. The student concluded that maintenance for prototyping is the art of setting priorities and trying to cluster changes that can easily be done at the same time to get the most effect out of an iteration.

Given the above observations, the MyBroadway team members came up with a procedure for handling requests for changes (see BEC Figure 18-1). One source of requests would be the feedback forms submitted by customers and employees after trying an iteration of the prototype; another source would be suggestions from Carrie and store employees for adaptive and perfective features, and the final source would be ideas made by the team members. Any of these sources could generate any type of request, although suggestions from Carrie, store employees, and the team members would likely not be corrective maintenance requests. No preference is given to any particular source, but changes will be grouped by module: (1) customer comments and employee picks, (2) products requests and inventory review, (3) rental status and extension, (4) child purchase/rental history, (5) interface with Entertainment Tracker, and (6) the database. One team member is designated as the lead on each module. Corrective requests will be done as soon as possible without consideration of priority analysis or synergy with other requests. That is, whenever a team member has time to work on the system module for which he or she is responsible, he or she will try to make all corrective changes in the queue for that module and create a new iteration (version) of the prototype including those changes. If there are no corrective maintenance requests waiting and a team member has time, he or she will work on any adaptive requests for his or her module. Perfective requests are held until an adaptive request involving the same module is being handled. Because there is a desire to rapidly iterate as productively as possible in order to develop a working prototype that handles currently iden-

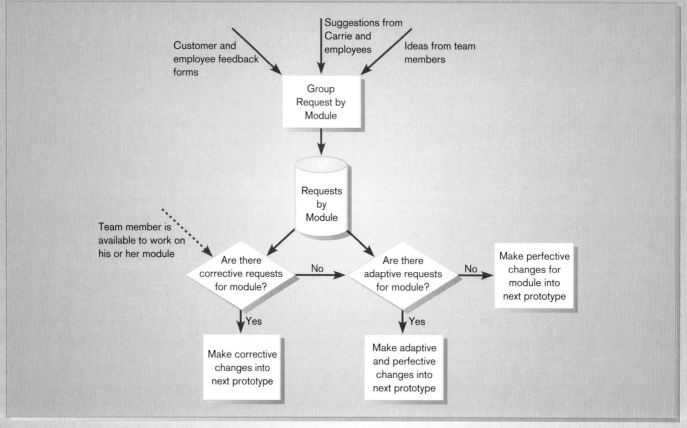

BEC Figure 18-1
MyBroadway prototyping and maintenance management procedure

tified needs, preventive requests receive the lowest priority and are not worked on unless no other request is being handled by a team member.

The BEC student team believes the above procedure will give them both discipline and an opportunity to put their efforts into making each prototype iteration have the greatest impact. By categorizing changes by module the student with the greatest familiarity with a module will be able to make the changes the most accurately and with the greatest motivation (it is "his" or "her" module).

said by seasoned systems developers, "the proof is in the details." The project must be well planned, the project must have proper resources that are well managed, user requirements must be realistic and met or exceeded, and implementation must go smoothly. You have had a chance to critique the work of the Stillwater students at each step in systems development. We trust you have found some flaws in their work and have suggested some significant improvements. Such critical review by peers is a cornerstone of professional systems analysis and design.

CASE SUMMARY

This is the last Broadway Entertainment Company case. Do you sense a conclusion? Not really. The results of the MyBroadway project are still to be determined by how well the Stillwater students handle all the changes for all the iterations of the system prototype. Their success is also to be determined by how well you have given them guidance in this and the preceding chapters by your answers to the case questions. Success on a systems development project comes from the project team members doing outstanding work in each phase of systems development. As is often

CASE QUESTIONS

1. Critically review the prototyping / maintenance procedure outlined in BEC Figure 18-1. What suggestions do you have for improvement? Why?

2. What do you consider to be the strengths of assigning one team member to each module? What are the weaknesses of this resource allocation model?

3. Does the procedure outlined in BEC Figure 18-1 adequately address the concerns of maintenance of a typical Website? What missing elements do you see?

4. What kinds of changes would be made that are assigned to the database module? What is the advantage of considering this a separate module?

5. What kinds of changes will be made to the Entertainment Tracker module? Would it be possible to combine the database and Entertainment Tracker modules under one team member? Justify your answer.

6. Maintenance includes not only the system itself but also all the documentation about the system. Since the MyBroadway team is using a prototyping methodology, what is the responsibility of each team member to update documentation produced from prior systems development phases when change requests relate to flaws or inadequacies in the requirements determination, systems analysis, or systems design of prior steps?

Part SIX

Advanced Analysis and Design Methods

● **Chapter 19**
Rapid Application Development

● **Chapter 20**
Object-Oriented Analysis and Design

An Overview of Part SIX

Advanced Analysis and Design Methods

Even after over half a century of business computer applications, systems professionals are still creating new and improved methods for systems development. Recent increased cost and competitive pressures in business have caused organizations to seek ways of doing business that take less human effort (the largest source of costs) and reduce the cycle time from idea conception to implementation. Although no foolproof, "silver-bullet" methodology exists, two methodologies show promise for improved systems development in at least some situations. These methodologies are Rapid Application Development (RAD) and object-oriented analysis and design (OOAD). Part VI explains these alternatives to the structured systems development approach, the approach emphasized in this text.

Rapid Application Development is an approach that promises better and cheaper systems and more rapid deployment by having systems developers and end users work together jointly in real-time to develop systems. As you will see in Chapter 19, RAD is not a single methodology but is more a general strategy of developing information systems. RAD relies on bringing together several systems development components that you have studied in this book. RAD depends on extensive user involvement. Much of end-user involvement takes place in the *prototyping* process, where users and analysts work together to design interfaces and reports for new systems. The prototyping is conducted in sessions that resemble traditional *Joint Application Development* (JAD) sessions. The primary difference is that in RAD the prototype becomes the basis for the new system. This is accomplished through reliance on *integrated CASE tools*, which include *code generators* for creating code from the designs end users and analysts create during prototyping. The code includes the interfaces and the application programs that use them.

Chapter 19 reports several RAD success stories, which highlight the diverse situations in which RAD may be applied. These success stories come from aerospace, government, and financial services, and point out the primary advantage of RAD: information systems developed in as little as a quarter the usual time. But shorter development cycles also mean cheaper systems, as fewer organizational resources need be devoted to developing any particular system. RAD also involves smaller development teams, which results in even more savings. Finally, because there is less time between the end of design and conversion, the new system is closer to current business needs and is therefore a higher quality than would be a similar system developed the traditional way. However, because of the shortcuts taken during RAD, it should be applied only in situations that require rapid deployment, where the system being developed is decoupled from other systems, the business model behind the system is relatively easy to discern, and system controls are not important.

Chapter 20 thoroughly covers OOAD. An object-oriented model, as you will learn in this chapter, is built around *objects*. An object *encapsulates* both data and *behavior*, implying that analysts can use the object-oriented approach for both data modeling and process modeling. By using a common representation and by offering benefits such as *inheritance* and code reuse, a systems analyst can use the object-oriented modeling approach to provide a powerful environment for developing complex systems. These principles of objects, encapsulation, behaviors, inheritance, and the like are the foundation for object-oriented systems development.

In Chapter 20 you will learn how to represent system requirements (use cases), the static and dynamic structure and behaviors of an information system (object and class models and state-transition diagrams, respectively), and the interaction between system components-objects (sequence and activity diagrams). In contrast to the SDLC, the object-oriented development life cycle is more like an onion than a waterfall. In the early stages (or core) of development, the model you build is abstract, focusing on external qualities of the application system. As the model evolves, it becomes increasingly detailed, shifting the focus to how the system will be built and how it should function—system architecture, data structures, and algorithms. Important in this process is the concept that the model of a system evolves. A major benefit of OOAD is the continuity of modeling techniques across life cycle steps.

You will also see guidelines on transforming the analysis models to design models. OOAD blends analysis with design by using the same type of diagram in different phases. As you move from analysis to design, you need to refine the diagrams based on the requirements of the implementation environment. During design, you need to consider technical details relating to the target hardware, operating system, programming language, and database management system, and refine the models that you developed earlier. OOAD truly blends analysis with design through the evolution of techniques, rather than changing from one set of techniques to another.

Chapter 19

Rapid Application Development

After studying this chapter, you should be able to:

- Explain the Rapid Application Development (RAD) approach and how it differs from traditional approaches to information systems development.

- Describe the system development components essential to RAD.

- Discuss the conceptual pillars that support the RAD approach.

- Explain the advantages and disadvantages of RAD as an exclusive systems development methodology.

I N T R O D U C T I O N

RAD [Rapid Application Development] has been demonstrated in many projects to be so superior to traditional development that it seems irresponsible to continue to develop systems the old way.

James Martin, 1991

James Martin is generally credited with inventing Rapid Application Development, or RAD, so it should be no surprise that he so enthusiastically supports the concept, as the above quote shows. Not everyone shares his enthusiasm, however, as you will see. In this chapter, you will learn about RAD: what it is, the components it relies on to succeed, software tools that can support RAD, and Martin's four-phase RAD life cycle. There are many different RAD approaches; in fact, most consulting firms and many corporations have

defined their own RAD methodologies. The ones used by information system consulting firms are often proprietary, so we cannot discuss them here. So that you can understand any RAD methodology you might encounter, we concentrate on the concepts of RAD and use several specific publicly known RAD approaches as examples. We end the chapter with examples of how the RAD approach can pay off handsomely and the advantages and disadvantages of RAD.

The topic of RAD is very extensive, and like object-oriented analysis and design (see Chapter 20), this chapter introduces you only to the major principles of RAD. To conduct a project with RAD requires applying skills and concepts discussed elsewhere in this text (such as interviewing, designing screens and reports, and using automated tools for systems development); thus, this is not a self-contained chapter on how to conduct an information systems development project via RAD. By carefully studying this chapter, you will, however, be able to see how to apply the skills learned in other parts of this text within the RAD approach.

RAPID APPLICATION DEVELOPMENT

Rapid Application Development (RAD) is an approach to developing information systems that promises better and cheaper systems and more rapid deployment by having systems developers and end users work together jointly in real-time to develop systems. RAD grew out of the convergence of two trends: the increased speed and turbulence of doing business in the late 1980s and early 1990s, and the ready availability of high-powered computer-based tools to support systems development and easy maintenance. As the conditions of doing business in a changing, competitive global environment became more turbulent, management in many organizations began to question if it made sense to wait two to three years to develop systems (in a methodical, controls-rich process) that would be obsolete upon completion. On the other hand, CASE tools and prototyping software were diffusing throughout organizations, making it relatively easy for end users to see what their systems would look like before they were completed. Why not use these tools to more productively address the problems of developing systems in a rapidly changing business environment? And so RAD was born.

The creation of RAD did not lead to its immediate adoption, however. To some, RAD was not an approach that could be taken seriously for enterprise-wide information systems. But increasing disenchantment with traditional systems development methods and the long development times associated with them led more and more firms to seriously consider RAD. The ready availability of increasingly powerful software tools created to support RAD also increased interest in this approach. RAD is becoming more and more a legitimate way to develop information systems. Today, the focus is increasingly on the rapid development of Web-based systems. RAD tools, and software created to support rapid development, almost all provide for the speedy creation of Web-based applications. For example, IBM has developed a suite of tools that enable the fast, easy development of e-business applications. These tools include VisualAge Generator, VisualAge for Java, WebSphere Studio, and WebSphere Application Server.

The Process of Developing an Application Rapidly

As you will see in this chapter, RAD is not a single methodology but is instead a general strategy of developing information systems. As such, there are many different approaches to developing applications rapidly. Some are special life cycles, such as the Martin RAD life cycle, which will be introduced soon. Others focus more on specific software tools and visual development environments that enable the process of rapidly developing and deploying applications. Whatever strategy is taken, however, the goal remains the same: to rapidly analyze a business problem, design a viable system solution through intense cooperation between users and developers, and to quickly get the finished application in the hands of users, saving time, money, and other resources in the process.

As Figure 19-1 shows, the same phases that are followed in the traditional SDLC are also followed in RAD, but the phases are shortened and combined with each other to produce a more streamlined development technique. Planning and design phases in RAD are shortened by focusing work on system functional and user interface requirements at the expense of detailed business analysis and concern for system performance issues. Also, usually RAD looks at the system being developed in

Rapid Application Development (RAD): Systems development methodology created to radically decrease the time needed to design and implement information systems. RAD relies on extensive user involvement, Joint Application Design sessions, prototyping, integrated CASE tools, and code generators.

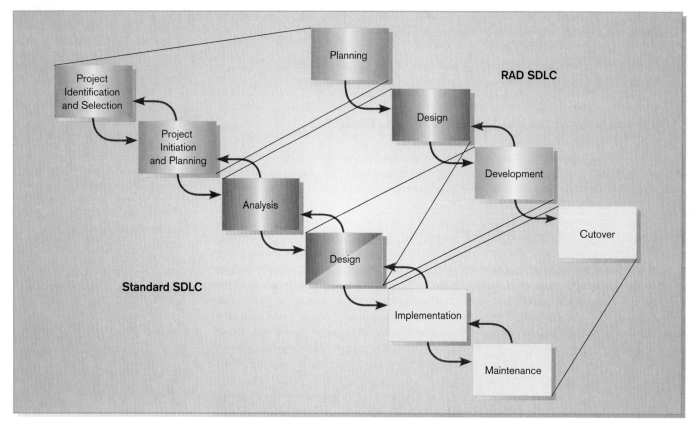

Figure 19-1
RAD systems development life
cycle compared to standard SDLC

isolation from other systems, thus eliminating the time-consuming activities of coordinating with existing standards and systems during design and development. The emphasis in RAD is generally less on the sequence and structure of processes in the life cycle and more on doing different tasks in parallel with each other and on using prototyping extensively. Notice also that the iteration in the RAD life cycle is limited to the design and development phases. This is where the bulk of the work in a RAD approach takes place. Although it is possible in RAD for there to be a return to planning once design has begun, it is rarely done. Similarly, although it is possible to return to development from the cutover phase, RAD is designed to minimize iteration at this point in the life cycle. The high level of user commitment and involvement throughout RAD implies that the system that emerges should be more readily accepted by the user community (and hence more easily implemented during cutover) than would be a system developed using traditional techniques.

Deliverables and Outcomes

The deliverables and outcomes of RAD are the same as for the traditional SDLC: a systems development plan, which includes the application being developed; a description of user and business process requirements for the application; logical and physical designs for the application; and the application's construction and implementation, with a plan for its continued maintenance and support. Whereas the traditional SDLC is indifferent as to the specific tools and techniques used to create and modify each of the deliverables listed above, RAD puts a heavy emphasis on the use of computer-based tools to support as much of the development process as possible. This emphasis on speed and the tools that make it possible is what sets RAD apart from traditional approaches.

In the next section, we describe the components most useful to a successful RAD effort. This is followed by sections on the foundations of the RAD approach, on the tools and techniques designed to support rapid application development and short-ening project schedules, on situations where RAD has been successfully employed, and on the advantages and disadvantages of RAD.

COMPONENTS OF RAD

To succeed, RAD relies on bringing together several systems development components found in this book. As you might have gathered from the definition, RAD depends on extensive user involvement. End users are involved from the beginning of the development process, where they participate in application planning; through requirements determination, where they work with analysts in system prototyping; and design and implementation, where they work with system developers to validate final elements of the system's design. Much of end-user involvement takes place in the *prototyping* process, where users and analysts work together to design interfaces and reports for new systems. The prototyping is conducted in sessions that resemble traditional *Joint Application Design (JAD)* sessions. The primary difference is that in RAD, the prototype becomes the basis for the new system—the screens designed during prototyping become screens in the production system. This is accomplished through reliance on *integrated CASE tools*, which include *code generators* for creating code from the designs end users and analysts create during prototyping. The code includes the interfaces as well as the application programs that use them. Alternatively, RAD may employ visual development environments instead of CASE tools with code generators, but the benefits from rapid prototyping are the same. In many cases, the basis for the production system is being built even as users are talking about the system during development workshops. In many cases, end users can get hands-on experience with the developing system before the design workshops are over. To further help speed the process, the reuse of templates, components, or previous systems described in the CASE tool repository is strongly encouraged.

APPROACHES TO RAD

Martin suggests there are four important supporting components or "pillars" of RAD. The first consists of software development tools. Tools alone, however, are not enough to make RAD work (see Figure 19-2). Tools are necessary but not sufficient. According to Martin, the other three pillars of RAD are people, who must be trained in the right skills; a coherent methodology, which spells out the proper tasks to be done in the proper order; and the support and facilitation of management (Martin, 1991).

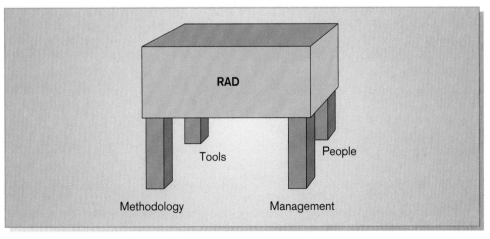

Figure 19-2
Martin's four necessary pillars for the RAD approach

Martin suggests that no information systems organization should be converted to the RAD approach overnight. Instead, a small group of well-trained and dedicated professionals, called a RAD cell, should be created to demonstrate the viability of RAD through pilot projects. Over time, the cell can grow, gradually adding more people, skills, and projects, until RAD is the predominant approach of the information systems unit.

Steve McConnell (1996) presents another approach to rapid development. Underlying his approach are the sound if somewhat conservative management practices of careful planning, efficient use of time, and the humane use of human resources. McConnell suggests his own set of four pillars fundamental to a successful RAD approach: avoiding classic mistakes, applying development fundamentals, managing risks to avoid catastrophic setbacks, and applying schedule-oriented practices (see Figure 19-3). McConnell lists 36 different classic development mistakes, 10 of which are included in Table 19-1. It should be pointed out that McConnell's classic mistakes apply to all systems development projects, not just RAD projects. Most of the mistakes listed in Table 19-1 are easily understood from their brief descriptions, such as unrealistic expectations, insufficient planning, and overestimated savings from new tools or methods. Some of the remaining items need additional explanation.

Under people-related mistakes, weak personnel refers to employees that are not as well trained in the skills necessary to a particular project. Weak has nothing to do with the employees' personalities or physical abilities. Since skills such as prototyping, fourth-generation language programming, and interaction with end users are needed for RAD, an excellent analyst without these skills will struggle on a RAD project. The second people-related mistake, adding people late to a project, may not seem like a mistake. Common sense would dictate that additional people should add to productivity, as the remaining work can now be divided among more people, leaving each person with less to do in the same amount of time. But as Brooks (1995) convincingly argued, adding more people also adds more need for coordination among all the people on the project, new and old. Adding more people actually ends up reducing productivity on a project. Brooks also identified the problem called the silver-bullet syndrome. The silver-bullet syndrome occurs when developers believe a new and usually untried technology is all that is needed to cure the ills of any development project. In application development, however, there is no silver bullet. No one technology can solve every problem, and those who believe it will be disappointed.

Two other mistakes from the list that need additional explanation are both product-related mistakes: feature creep and requirements gold-plating. Feature

Figure 19-3
McConnell's four necessary pillars for the RAD approach

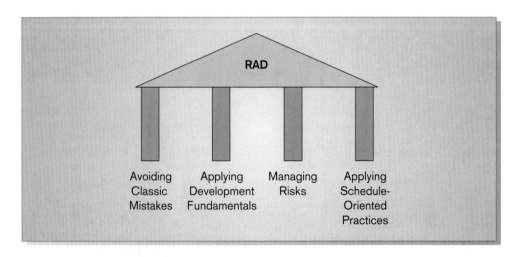

TABLE 19-1   Ten of McConnell's 36 Classic Development Mistakes

People-Related	Process-Related	Product-Related	Technology-Related
1. Weak personnel	1. Insufficient planning	1. Feature creep	1. Silver-bullet syndrome
2. Adding people to a project late	2. Overly optimistic schedules	2. Requirements gold-plating	2. Overestimated savings from new tools or methods
3. Unrealistic expectations	3. Planning to catch up later		

creep refers to the tendency of system requirements to change over the lifetime of the development project. It's called feature creep because more and more features that were not in the original specifications for the application "creep in" during the development process. The average project may see a 25 percent change in requirements, all of which can have a negative impact on keeping the project on schedule and containing project costs. Changes to an application typically cost 50 to 200 times less if they are made during requirements determination rather than during the physical design process. Requirements gold-plating means an application may have more requirements than it needs, even before the development project begins. Many of these requirements can be extreme and complex and not really necessary.

McConnell's second pillar is development fundamentals. Many of these fundamentals are stressed in other parts of this text and include such practices as proper project estimating, scheduling, planning, and tracking; measuring software quality and productivity; managing system requirements; engaging in good design techniques; and assuring quality. The third pillar is managing risk, a topic discussed in Chapters 2 and 3. Once the first three pillars have been successfully built, you have reached what McConnell calls "efficient development." Without efficient development, achieving the reduced schedules necessary for rapid development will be more difficult. Once achieved, however, you can go on to develop the fourth pillar: schedule-oriented practices, designed to improve development speed, reduce schedule risk, and make progress visible. McConnell lists over 25 best schedule-oriented practices in his book, everything from outsourcing to using a spiral life cycle to staged delivery. Many of his schedule-oriented practices are focused on choosing the appropriate software tools and life-cycle approaches, both of which are discussed in more detail in the next sections.

Software Tools for RAD

There are many different software tools that can be successfully used to support the RAD approach. Remember that there is no silver bullet for software development, and that includes software tools that support RAD. Some tools will work better for certain applications than others, and developers need to have a good understanding of the limits as well as the possibilities of any development environment they decide to use.

NET SEARCH
There are many different computer-based tools to support RAD. Visit http://www.prenhall.com/hoffer to complete an exercise on this topic.

We have previously mentioned how integrated CASE tools can be used to support RAD. Such CASE tools should have prototyping facilities as well as code generators in order to provide the best level of support. An example of a CASE tool with these capabilities is COOL:Gen from Computer Associates. COOL:Gen allows developers and users to work together closely to design windows and navigation paths between them (see Figure 19-4). It generates code in several languages, including C and C++.

Other software tools that can support RAD are types of visual development environments (see Chapter 4). Visual development environments are based on different underlying programming languages and allow developers to design windows and

Figure 19-4
Example CASE tool used during
RAD: COOL:Gen™ from Computer
Associates

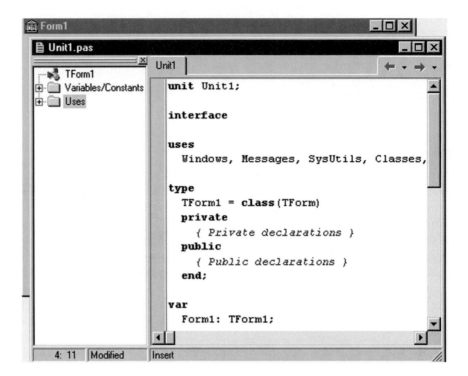

manipulate different aspects of graphical user interfaces (GUIs), such as push buttons and list boxes. Examples include Microsoft's Visual Basic, based on the Basic programming language; Inprise/Borland's Delphi, based on object-oriented Pascal (see Figure 19-5); PowerSoft's PowerBuilder; and IBM's VisualAge Generator. These are by no means the only development environments that can support RAD but are listed here only as examples of the many tool sets available.

Martin's RAD Life Cycle

Recall that Martin listed tools as just one of his four pillars necessary for RAD. Another pillar he listed was methodology. For methodology, Martin developed a specific RAD life cycle, which contains the basic phases of any RAD life cycle (see Figure 19-1)—planning, design, development, and cutover—although he calls the phases requirements planning, user design, construction, and cutover (see Figure 19-6).

Requirements planning incorporates elements of the traditional planning and analysis phases. During this phase, high-level managers, executives, and knowledgeable end users determine system requirements, but the determination is done in the context of a discussion of business problems and business areas. Once specific systems have been identified for development, users and analysts conduct a joint

Figure 19-5
Inprise/Borland's Delphi
development environment
(*Source:* Borland Corporation, 1998.
Reprinted by permission of
Inprise/Borland.)

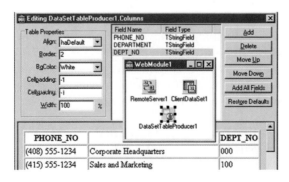

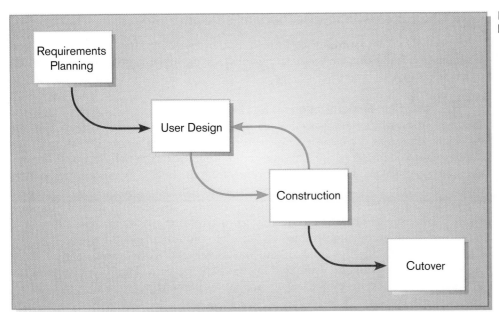

Figure 19-6
Martin's four-phase RAD life cycle

requirements planning workshop to reach agreement on system requirements. The overall planning process is not all that much different from planning in the traditional SDLC.

During user design, end users and information systems professionals participate in JAD workshops, where those involved use integrated CASE tools to support the rapid prototyping of system design. Users and analysts work closely and quickly to create prototypes that capture system requirements and that become the basis for the physical design of the system being developed. Users sign off on the CASE-based design—there are no paper-based specifications. Because user design ends with agreement on a computer-based design, the gap between the end of design and the handing over of the new system to users might only take three months instead of the usual 18.

In construction, the same information systems professionals who created the design now generate code using the CASE tools' code generator. End users also participate, validating screens and other aspects of the design as the application system is being built. Construction can be combined with user design into one phase for smaller systems.

Cutover means delivery of the new system to its end users. Because the RAD approach is so fast, planning for cutover must begin early in the RAD process. Cutover involves many of the traditional activities of implementation, including testing, training users, dealing with organizational changes, and running the new and old systems in parallel, but all of these activities occur on an accelerated basis.

According to Martin, RAD can produce a system in six months that would take 24 months to produce using the traditional systems development life cycle.

RAD SUCCESS STORIES

When used successfully, the results of RAD can be dramatic. Here are three stories about successful use of the RAD approach. The first story focuses on the use of a particular software tool suite, Inprise/Borland's Delphi. Both of the Delphi stories are based on postings to the Delphi Web pages (see References at the end of the chapter). The third story is based on the use of another RAD tool, IBM's VisualAge for Java.

RAD Success with Inprise/Borland's Delphi

Two RAD success stories with Inprise/Borland's Delphi, one from the U.S. government and one from the financial business sector, suggest the broad applicability of the RAD approach.

U.S. Navy Fleet Modernization   The U.S. Navy has as an ongoing challenge: the modernization of its fleet as well as its maintenance. Modernization has become especially crucial as the Navy has downsized from a fleet of almost 600 ships in the 1980s to just 260 in 1997. The Navy wanted to upgrade its fleet modernization management system at the same time it was renovating the fleet. The requirement was to move from three character-based systems to a unified, GUI-based system based on a single database. The Fleet Modernization program chose Delphi to build the new system. Two aspects of Delphi that were attractive to the Navy were its support of rapid prototyping and the promise of reuse of components in the object-oriented Pascal environment. The final system was developed with Delphi using a RAD approach in six months, an estimated savings of 50 percent. The Navy estimates the new system resulted in immediate savings of 20 percent per year due to reduced maintenance costs.

First National Bank of Chicago   First National Bank of Chicago is one of the 10 largest banks in the United States, with assets of over $122 billion. In October 1994, the Bank was awarded a bid to develop a new Electronic Federal Tax Payment System, intended to foster faster tax withholdings and eliminate all paper in the process. The contract, signed by a joint venture of the Bank with Mercantile Bank of St. Louis, specified exacting system requirements and very short deadlines. The venture chose Delphi for the job. The use of Delphi enabled rapid development and rapid prototyping of a GUI front-end to the system.

At one point, over 125 programmers organized into several teams were working on the project. Fifteen to twenty developers worked on the GUI front-end. Rapid prototyping for requirements elicitation allowed coding to begin in May 1995, only seven months after the contract was awarded. The system was completed and ready for testing in March 1996, only 10 months later. After three months of certification testing, the system went live in July 1996. The size of the database used in the Electronic Federal Tax Payment System was not trivial at 225 tables, with 250 million rows and 55 gigabytes in size. By the year 2000, there were 500 million rows maintained on-line. RAD techniques, with Delphi's support, allowed the project to come in on time and on budget, despite the extensive requirements and short deadlines specified in the contract.

RAD Success with VisualAge for Java

Comdata is the leading provider of financial and information services to the North American transportation industry. Over 3,000 trucking companies rely on Comdata's MOTRS (Modular Over The Road System) system to help manage their businesses and to monitor the spending of their truck drivers. These drivers use Compdata's Comchek card to purchase fuel, food, and lodging while on the road.

The original version of MOTRS was PC-based, written in VisualBasic. Comdata customers accessed the system through dial-up, uploaded account data, and downloaded reports. Long distance phone charges for access averaged $400 per month for customers. As Comdata customers ran different versions of MOTRS, written for different computer platforms, Comdata also incurred large costs for updating the system so it would run on different platforms. Comdata decided that it needed a new version of MOTRS by the end of 1999. The new version had to be Y2K compliant, eliminate the need for long distance dial-up access, and there had to be a single version of the system. These requirements pointed to an e-business solution. Comdata decided to deploy MOTRS on the Internet, and they chose IBM Global Services as the vendor to help them migrate their system.

To develop the Web-based MOTRS, IBM used VisualAge for Java and the IBM WebSphere Application Server software to create servlets. Servlets are programming modules, written in Java, that reside on the server and expand the functions of the Web server. VisualAge for Java was also used to create Java applets. Applets are embedded in HTML code on Web pages and are downloaded to the customer client machine and executed by the client Web browser. The WebSphere Application Server software was used to manage and control the servlets.

The MOTRS Internet project took only nine months to complete. This included three months for research and design, three months for coding, and three months for testing: one month each for quality assurance, acceptance testing by Comdata, and beta testing by customers. The entire project cost $500,000, including the server software and hardware, development, and everything else. The system became generally available in September 1998, long before the deadline for deployment set for the end of 1999. After the redevelopment effort, MOTRS was Y2K compliant, there was a single PC version of it, and customers now pay only $20 per month for an Internet Service Provider account through which to access the system.

N E T   S E A R C H

There are many success stories about RAD. Visit http://www. prenhall.com/hoffer to complete an exercise on this topic.

ADVANTAGES AND DISADVANTAGES OF RAD

As these success stories show, the primary advantage of RAD is obvious: information systems developed in as little as one-quarter the usual time. But shorter development cycles also mean cheaper systems, as fewer organizational resources need be devoted to develop any particular system. Martin (1991) points out that RAD also involves smaller development teams, which results in even more savings. Finally, because there is less time between the end of design and conversion, the new system is closer to current business needs and is therefore of higher quality than would be a similar system developed the traditional way.

Others point out, however, that although RAD works, it only works well for systems that have to be developed quickly (Gibson and Hughes, 1994). Such systems include those that are developed in response to quickly changing business conditions, perceived opportunities to gain a competitive edge by moving quickly in a market or industry, or new government regulations. If speed is a goal, however, then other aspects of an application must oftentimes be sacrificed (McConnell, 1996). In general, in a systems development effort, you can obtain two of three key characteristics of the development effort—speed, cost, and quality—but never all three. If hard-and-fast goals for speed and cost are set, quality of the final application will suffer. Similarly, if you must achieve a certain level of quality and you must do so in a fixed amount of time, the cost will be greater than it might otherwise have been. Finally, if you have set goals for cost and quality, you will take longer to develop the application than you might have wanted originally.

It should also be pointed out that taking too much time to develop a system can actually have a negative effect on the quality of the system. Too much time between the initial requirements determination and the delivery of the completed system adds many complications to the overall development process. It will be more difficult to accurately estimate requirements at the time the system will be delivered because business conditions will have changed so much during the intervening time. Although predicting the future is always difficult, predicting the near-term future is more certain than predicting a future that is three to five years away. Changing business conditions will also make it easier for system owners and developers to become side-tracked during the development process. Finally, too much time between the beginning and end of a project almost guarantees problems with feature creep, again due to changing business conditions. The key, then, is finding the right balance between time to completion and quality. Too much time can be just as bad as too little.

While RAD is the methodology of choice for all systems development, RAD's emphasis on what a system does and how it does it implies that essential information about the business models and processes is not included. In its highly accelerated analysis and design phases, RAD leaves little room for understanding the business area, what it does, what its functions are and who performs them, or what the people in the business area do in their jobs. The greater the reliance on RAD, the greater the risk that many systems will be out of alignment with the business.

Another drawback of RAD has to do with the very thing that makes it so attractive as a systems development methodology: its speed. Because the RAD process puts such an emphasis on speed, many important software engineering concepts, such as interface consistency, programming standards, module reuse, scalability, and systems administration, are overlooked (Bourne, 1994). For example:

- *Consistency* In their efforts to quickly paint screens, RAD analysts often ignore the need to be consistent both within an application and across a suite of related applications. Areas of concern include window size and color, consistent format masks, and using the same error message for the same offense.

- *Programming standards* Documentation standards and data naming standards should be established early in RAD or it may be difficult to implement them later.

- *Module reuse* Many times during prototyping, analysts forget that similar screen or report designs may have already been created. Often in RAD there is no mechanism in place that allows analysts to easily determine whether modules that can be reused are already in existence.

- *Scalability* If the system designed during RAD is useful, its use will gradually spread beyond those initial users who helped build it. Developers should build such growth into their initial system. Scalability also applies to hardware; system scope; the number, type, and users of reports; growth in the software team developing and maintaining the system as it grows; user training as system use expands; and security.

- *Systems administration* System administration is typically ignored altogether during RAD, as the emphasis is usually on the excitement of seeing a new application system develop before users' eyes. Important system administration issues include database maintenance and reorganization, backup and recovery, distribution or installation of application updates, and scheduling and implementing planned system downtime and restarts.

NET SEARCH
RAD may be changing the nature of software development. Visit http://www.prenhall.com/hoffer to complete an exercise on this topic.

Although RAD can reduce a project's cost, because fewer resources and less time go into the overall development effort, there are ways RAD can end up increasing costs. RAD is reliant on high levels of user commitment and participation, so it is crucial that key members of the user community be involved during the entire RAD process. The problem is that key users have their own work to do. Pulling key individuals away from their own work can come at a high cost to the organization. If the key user can't do his or her job because of a RAD effort, the organization must either wait for the user to get back to the job or find another way to accomplish the same work. Neither solution is optimal and the organization may ultimately suffer. Again, a suitable balance has to be found between a key user's commitment to his or her job and to the RAD effort.

Clearly, RAD is a powerful approach to information systems development, with the potential for many payoffs. Just as clearly, there are risks associated with using RAD, especially for every systems development project in an organization's portfolio of projects. The relative advantages and disadvantages of the RAD approach are summarized in Table 19-2.

TABLE 19-2   Advantages & Disadvantages of RAD

Advantages	Disadvantages
Dramatic time savings during the system development effort.	More speed and lower cost may lead to lower overall system quality (e.g., due to lack of attention to internal controls).
Can save time, money, and human effort.	Danger of misalignment of system developed via RAD with the business due to missing information on underlying business processes.
Tighter fit between user requirements and system specifications.	May have inconsistent internal designs within and across systems.
Works especially well where speed of development is important, as with rapidly changing business conditions or where systems can capitalize on strategic opportunities.	Possible violation of programming standards related to inconsistent naming conventions and insufficient documentation.
Ability to rapidly change system design as demanded by users.	Difficulties with module reuse for future systems.
System optimized for users involved in RAD process.	Lack of scalability designed into the system.
Concentrates on essential system elements from user viewpoint.	Lack of attention to later systems administration built into the system (e.g., not integrated into overall enterprise data model and missing system recovery features).
Strong user stake and ownership of system.	High costs of commitment on the part of key user personnel.

Summary

Rapid Application Development (RAD) is an alternative to the traditional systems development life cycle. All of the phases of the traditional life cycle are included in RAD, but the phases are executed at an accelerated rate. RAD relies on heavy user involvement, Joint Application Development sessions, prototyping, integrated CASE tools, and code generators to design and implement systems quickly. The abbreviated RAD life cycle, as defined by James Martin, begins with requirements planning, followed by user design, construction, and cutover.

The primary advantage of RAD is the quick development of systems, but quick development may also lead to cost savings and higher quality systems. RAD does have drawbacks: with its emphasis on developing systems quickly, the detailed business models that underlie information systems are often neglected, leading to the risk that systems may be out of alignment with the overall business. Similarly, the speed of development may lead to analysts overlooking systems engineering concepts such as consistency, programming standards, module reuse, scalability, and systems administration. If applied successfully, however, RAD may result in dramatic savings and improved performance. Systems may be designed and implemented in one-quarter to one-half the time needed for the traditional life cycle approach.

Key Terms

1. Code generators
2. JAD
3. Prototyping
4. Rapid Application Development (RAD)

Match each of the key terms above with the definition that best fits it.

_____ A structured process in which users, managers, and analysts work together for several days in a series of intensive meetings to specify or review system requirements.

_____ An iterative process of systems development in which requirements are converted to a working system that is continually revised through close work between an analyst and users.

_____ CASE tools that enable the automatic generation of program and database definition code directly from the

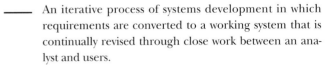

design documents, diagrams, forms, and reports stored in the repository.

—— Systems development methodology created to radically decrease the time needed to design and implement information systems. It relies on extensive user involvement, Joint Application Design sessions, prototyping, integrated CASE tools, and code generators.

Review Questions

1. List and briefly define the four phases of RAD, as defined by James Martin.

2. List and briefly explain two different views for the four pillars of RAD.

3. Explain the systems development components essential to RAD.

4. What trends in information systems encouraged the invention of the RAD approach to systems development?

5. Explain the concept of scalability and its influence on systems development.

6. Explain the advantages and disadvantages of RAD.

7. Given the relative advantages and disadvantages of RAD, why would a systems development group follow the RAD approach for a project? When is the best time to use RAD and when is the best time to use a more traditional method?

8. What is the role of JAD in Rapid Application Development?

9. Why is the RAD approach a good fit with Web-based development?

Problems and Exercises

1. Compare RAD and prototyping. How are these methodologies different from one another? In what ways are they similar?

2. How might RAD be used in conjunction with the structured systems development life cycle? Are RAD and the SDLC totally different approaches or could they complement each other?

3. What types of tools are necessary to do RAD? Is it possible to do RAD without fourth-generation languages and other tools?

4. One of the criticisms of RAD is that RAD may cause a system to be out of alignment with the direction of the business. Explain how this may occur and suggest what might be done to overcome this potential hazard of RAD.

5. Consider Figure 19-1. From what you read in this chapter, answer the following questions:

   a. How does the RAD planning phase differ from the corresponding SDLC phases?

   b. How does the RAD design phase differ from the corresponding SDLC phases?

   c. How does the RAD development phase differ from the corresponding SDLC phases?

   d. How does the RAD cutover phase differ from the corresponding SDLC phases?

6. Describe the characteristics of a situation requiring a new information system that would be ideal for the use of the RAD approach.

7. Consider Figures 19-2 and 19-3. Which of these two viewpoints on the essential pillars for RAD helps you the most to understand the keys to successful application of RAD? Why?

8. Consider Table 19-1. Explain what you would do during RAD to avoid the following classic development mistakes:

   a. Unrealistic expectations
   b. Insufficient planning
   c. Overly optimistic schedules

Field Exercises

1. Electronic Data Systems (EDS) uses a form of RAD which they call SLC–RISE (Systems Life Cycle–Rapid Iterative Systems Engineering). Contact a systems analyst at EDS or at a client organization and investigate the impact of this methodology on systems development projects. Have all systems professionals in the client organization embraced RISE? What does the RISE methodology do to address some of the potential RAD hazards outlined in this chapter?

2. Contact an organization that has done a RAD-based systems development project. Investigate why they chose that project using RAD as opposed to some other methodology also used in their organization. Explain what factors that organization considers when deciding to conduct a project using a RAD or an SDLC methodology. What factors do they consider contribute to the success of a RAD based project?

3. If you have access to a CASE tool as part of your course, study the features that might be helpful in support of RAD. Which components of the CASE tool would be helpful in which RAD life-cycle phases? What capabilities are missing?

If you do not have access to a CASE tool, contact an organization that uses a CASE tool and investigate how they use that tool in RAD or to accomplish some of the goals of RAD. Document your findings in a report.

4. Visit the Inprise/Borland Delphi Website (http://www. inprise.com/delphi/) and other Websites to find additional RAD success stories. From these stories or cases, what do you personally conclude are the keys to successful use of the RAD approach? Why?

References

Bourne, K. C. 1994. "Putting Rigor Back in RAD." *Database Programming & Design* 7 (8): 25–30.

Brooks, F. P., Jr. 1995. *The Mythical Man-Month, Anniversary Edition*. Reading, MA: Addison-Wesley.

Gibson, M. L. and C. T. Hughes, 1994. *Systems Analysis and Design: A Comprehensive Methodology with CASE*. Danvers, MA: Boyd & Fraser Publishing Company.

IBM VisualAge Generator Website, http://www-4.ibm.com/software/ad/visgen/

Inprise/Borland Delphi Website, http://www.inprise.com/delphi/

Martin, J. 1991. *Rapid Application Development*. New York: Macmillan Publishing Company.

McConnell, S. 1996. *Rapid Development*. Redmond, WA: Microsoft Press.

Chapter 20

Object-Oriented Analysis and Design*

LEARNING OBJECTIVES

After studying this chapter, you should be able to:

- Define the following key terms: *use case*, *object*, *class*, *state*, *behavior*, *operation*, *encapsulation*, *constructor operation*, *query operation*, *update operation*, *association*, *multiplicity*, *abstract class*, *concrete class*, *class-scope attribute*, *abstract operation*, *method*, *polymorphism*, *overriding*, *aggregation*, *composition*, *event*, *state transition*, and *sequence diagram*.

- Describe the concepts and principles underlying the object-oriented approach.

- Describe the activities in the different phases of the object-oriented development life cycle.

- State the advantages of object-oriented modeling versus traditional systems development approaches.

- Develop a requirements model using use-case diagrams.

- Develop an object model of the problem domain using class diagrams.

- Develop dynamic models using state, interaction, and activity diagrams.

- Model real-world applications using UML diagrams.

INTRODUCTION

In Part III of this book you learned how to analyze the requirements for an information system using structured techniques. We introduced you to a variety of diagrams for modeling purposes, such as context diagrams, data flow diagrams, decision tables, and entity-relationship diagrams. With the traditional systems development life cycle and structured techniques, analysis and design are separate but overlapping steps, and different techniques are used for analysis and for design. You learned about information system design

*The original version of this chapter was written by Professor Atish P. Sinha.

issues and techniques in Part IV. In this chapter, we will introduce you to the object-oriented model, which is becoming increasingly popular because of its ability to thoroughly represent complex relationships, as well as to represent data and data processing with a consistent notation. By a *consistent notation*, we mean that the same or compatible notation is used throughout the systems development process. Thus, you can more easily blend analysis and design in an evolutionary process.

A model, as you already know, is an abstraction of the real world. It allows you to deal with the complexity inherent in a real-world problem by focusing on the essential and interesting features of an application. An object-oriented model, as you will learn in this chapter, is built around *objects*. An object *encapsulates* both data and *behavior*, implying that analysts can use the object-oriented approach for both data modeling and process modeling. By allowing a systems analyst to capture them together within a common representation, and by offering benefits such as *inheritance* and code reuse, the object-oriented modeling approach provides a powerful environment for developing complex systems. The principles of objects, encapsulation, inheritance, and *polymorphism* (discussed later) are the foundation for object-oriented systems development.

In this chapter you will also learn a wide variety of techniques that systems analysts use for object-oriented analysis and design. These techniques and associated notations are incorporated into a standard object-oriented language called the Unified Modeling Language, or UML. This chapter is based on the UML standard from the Object Management Group (OMG) (see Fowler, 2000). The techniques and notations within UML include

- *Use cases*, which represent the functional requirements, or the "what" of the system
- *Class diagrams*, which show the static structure of data and the operations that act on the data
- *State diagrams*, which represent dynamic models of how objects change their states in response to events
- *Sequence diagrams*, which represent dynamic models of interactions between objects

You will also learn in this chapter how analysis and design activities are blended together in the object-oriented approach.

OBJECT-ORIENTED DEVELOPMENT LIFE CYCLE

The object-oriented development life cycle, depicted in Figure 20-1, consists of progressively developing an object representation through three phases—analysis, design, and implementation—similar to the heart of the systems development life cycle explained in Chapter 1. In contrast to the SDLC, the object-oriented development life cycle is more like an onion than a waterfall. In the early stages (or core) of development, the model you build is abstract, focusing on external qualities of the application system. As the model evolves, it becomes more and more detailed, shifting the focus to how the system will be built and how it should function—system architecture, data structures, and algorithms. Ultimately, like any information system, the systems developer must generate code and database access routines. The emphasis in modeling should be on analysis and design, focusing on front-end conceptual issues, rather than back-end implementation issues, which unnecessarily restrict design choices (Rumbaugh et al., 1991).

The Process of Object-Oriented Analysis and Design

In the analysis phase, a model of the real-world application is developed showing its important properties. It abstracts concepts from the application domain and describes *what* the intended system must do, rather than *how* it will be done. The

Figure 20-1
Phases of object-oriented systems development cycle

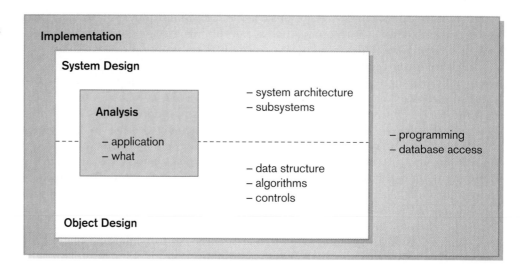

model specifies the functional behavior of the system, independent of concerns relating to the environment in which it is to be finally implemented. You need to devote sufficient time to clearly understand the requirements of the problem. The analysis model should capture those requirements completely and accurately. Remember that it is much easier and cheaper to make changes or fix flaws during analysis than during the later phases.

In the object-oriented design phase, you define *how* the application-oriented analysis model will be realized in the implementation environment. Jacobson et al. (1992) cite three reasons for using object-oriented design. These three reasons are:

- The analysis model is not formal enough to be implemented directly in a programming language. To move seamlessly into the source code requires refining the objects by making decisions on what operations an object will provide, what the inter-object communication should look like, what messages are to be passed, and so forth.

- The actual system must be adapted to the environment in which the system will actually be implemented. To accomplish that, the analysis model has to be transformed into a design model, considering different factors such as performance requirements, real-time requirements and concurrency, the target hardware and systems software, the DBMS and programming language to be adopted, and so forth.

- The analysis results can be validated using object-oriented design. At this stage, you can verify if the results from the analysis are appropriate for building the system, and make any necessary changes to the analysis model.

To develop the design model, you must identify and investigate the consequences that the implementation environment will have on the design (Jacobson et al., 1992). All strategic design decisions—such as how the DBMS is to be incorporated, how process communications and error handling are to be achieved, what component libraries are to be reused—are made. Next, you incorporate those decisions into a first-cut design model that adapts to the implementation environment. Finally, you formalize the design model to describe how the objects interact with one another.

Rumbaugh et al. (1991) separate the design activity into two stages: *system design* and *object design*. As the system designer, you propose an overall system architecture, which organizes the system into components called subsystems and provides the context to make decisions such as identifying concurrency; allocation of subsystems to processors and tasks; handling access to global resources; selecting the implementation of control in software; and more.

During object design, you build a design model by adding implementation details—such as restructuring classes for efficiency, internal data structures and algorithms to implement each class, implementation of control, implementation of associations, and packaging into physical modules—to the analysis model in accordance with the strategy established during system design. The application-domain object classes from the analysis model still remain, but they are augmented with computer-domain constructs, with a view toward optimizing important performance measures.

The design phase is followed by the implementation phase. In this phase, you implement the design using a programming language and/or a database management system. Translating the design into program code is a relatively straightforward process, given that the design model already incorporates the nuances of the programming language and/or the DBMS.

Deliverables and Outcomes

As with the structured modeling techniques you learned in Chapters 8 through 10, the deliverables from project activities using object-oriented modeling are diagrams and repository descriptions, which provide a full specification of the information system. A major characteristic of these diagrams is how tightly they are linked with each other. This is because all of the techniques and diagramming conventions for object-oriented modeling have been developed as a consistent set, rather than as relatively independent notations (as is true with the techniques of Chapters 8 through 10).

This consistency in notation yields certain outcomes for a systems development project using object-oriented modeling. Coad and Yourdon (1991b) identify several motivations and benefits of object-oriented modeling:

- The ability to tackle more challenging problem domains
- Improved communication among users, analysts, designers, and programmers
- Increased consistency among analysis, design, and programming activities
- Explicit representation of commonality among system components
- Robustness of systems
- Reusability of analysis, design, and programming results
- Increased consistency among all the models developed during object-oriented analysis, design, and programming

The last point needs further elaboration. In other modeling approaches, such as structured analysis and design, the models that are developed (e.g., data flow diagrams during analysis and structure charts during design) lack a common underlying representation and, therefore, are very weakly connected. In contrast to the abrupt and disjoint transitions that those approaches suffer from, the object-oriented approach provides a continuum of representation from analysis to design to implementation (Coad and Yourdon, 1991a), engendering a seamless transition from one model to another (Jacobson et al., 1992). For instance, moving from object-oriented analysis to object-oriented design entails expanding the analysis model with implementation-related details, not developing a whole new representation.

In this chapter, we show you how to analyze and design complex applications using the object-oriented approach. You will learn how to develop models by dealing with various aspects or views of a system. In general, each view will constitute multiple diagrams. Each view is developed using a different type of compatible diagram. Table 20-1 lists and briefly describes the types of object-oriented diagrams illustrated in this Chapter. We will show you how to draw those diagrams using a visual modeling language, described in the next section.

TABLE 20-1 Types of Object-Oriented Diagrams

Diagram	Explanation
Activity Diagram	Shows the sequence of data processing activities, including conditional and parallel activities, and how processing must be synchronized before further activities can happen.
Class Diagram	Shows the static structure of objects and their relationships.
Sequence Diagram	Shows the interactions among objects during a certain time period for a use case.
State Diagram	Shows the states of a single object and the events that cause the object to change from one state to another.
Use-Case Diagram	Shows the sequence of activities that a set of actors use to accomplish a major system function.

THE UNIFIED MODELING LANGUAGE

The Unified Modeling Language (UML) is "a language for specifying, visualizing, and constructing the artifacts of software systems, as well as for business modeling" (*UML Document Set*, 1997). It is a culmination of the efforts of three leading experts, Grady Booch, Ivar Jacobson, and James Rumbaugh, who have defined an object-oriented modeling language that is expected to become an industry standard in the near future. The UML builds upon and unifies the semantics and notations of the Booch (Booch, 1994), OOSE (Jacobson et al., 1992), and OMT (Rumbaugh et al., 1991) methods, as well as those of other leading methods.

The UML notation is useful for graphically depicting object-oriented analysis and design models. It not only allows you to specify the requirements of a system and capture the design decisions, but it also promotes communication among key persons involved in the development effort. A developer can use an analysis or design model expressed in the UML notation as a means to communicate with domain experts, users, and other stakeholders.

For representing a complex system effectively, it is necessary that the model you develop has a small set of independent views or perspectives. The UML allows you to represent multiple views of a system using a variety of graphical diagrams, such as the use-case diagram, class diagram, state diagram, sequence diagram, and collaboration diagram. The underlying model integrates those views so that the system can be analyzed, designed, and implemented in a complete and consistent fashion.

We will first show how to develop a use-case model during the requirements analysis phase. Next, we will show how to model the static structure of the system using class and object diagrams. You will then learn how to capture the dynamic aspects using state and interaction diagrams. Finally, we will provide a brief description of component and deployment diagrams, which are generated during the design and implementation phases.

USE-CASE MODELING

Jacobson et al. (1992) pioneered the application of use-case modeling for analyzing the functional requirements of a system. Because it focuses on *what* an existing system does or a new system should do, as opposed to *how* the system delivers or should deliver those functions, a use-case model is developed in the analysis phase of the object-oriented system development life cycle. Use-case modeling is done in the early stages of system development to help developers gain a clear understanding of the functional requirements of the system, without worrying about how those requirements will be implemented. The process is inherently iterative; developers need to involve the users in discussions throughout the model development process and finally come to an agreement on the requirements specification.

A use-case-driven design, in which use cases control the formation of all other models, promotes traceability among the different models. If user requirements change during the development life cycle, those changes are first made in the use-case model. Changes to the use cases then dictate what changes need to be made in the other models. The models are therefore traceable; for example, you can trace a set of requirements specified in a use case to elements in the other analysis, design, and test models. It is also possible to trace backward; for instance, you could see what effects, if any, a design change has on the use-case model.

A use-case model consists of actors and use cases. An **actor** is an external entity that interacts with the system (similar to an external entity in data flow diagramming). It is someone or something that exchanges information with the system. A **use case** represents a sequence of related actions initiated by an actor to accomplish a specific goal; it is a specific way of using the system (Jacobson et al., 1992). Note that there is a difference between an *actor* and a *user*. A user is anyone who uses the system. An actor, on the other hand, represents a role that a user can play. The actor's name should indicate that role. An actor is a type or class of users; a user is a specific instance of an actor class playing the actor's role. Note that the same user can play multiple roles. For example, if John Patton plays two roles, one as an instructor and the other as an advisor, we represent him as an instance of an actor called Instructor and as an instance of another actor called Advisor.

Because actors are outside the system, you do not need to describe them in detail. The advantage of identifying actors is that it helps you to identify the use cases they carry out. For identifying use cases, Jacobson et al. (1992) recommend that you ask the following questions:

- What are the main tasks performed by each actor?
- Will the actor read or update any information in the system?
- Will the actor have to inform the system about changes outside the system?
- Does the actor have to be informed about unexpected changes?

Actor: An external entity that interacts with the system (similar to an external entity in data flow diagramming).

Use case: A complete sequence of related actions initiated by an actor to accomplish a specific goal; it represents a specific way of using the system.

Developing Use-Case Diagrams

Use cases help you to capture the functional requirements of a system. During the requirements analysis stage, the analyst sits down with the intended users of the system and makes a thorough analysis of *what* functions they desire from the system. These functions are represented as use cases. For example, a university registration system has a use case for class registration and another for student billing. These use cases, then, represent the typical interactions the system has with its users.

In UML, a use-case model is depicted diagrammatically as in Figure 20-2. This **use-case diagram** is for a university registration system, which is shown as a box. Outside the box are four actors—Student, Registration Clerk, Instructor, and Bursar's Office—that interact with the system. An actor is shown using a stickman symbol with its name below. Inside the box are four use cases—Class registration, Registration for special class, Prereq courses not completed, and Student billing— which are shown as ellipses with their names underneath. These use cases are performed by the actors outside the system. Note that an actor does not necessarily have to be a human user. It could be anything, such as another system or a hardware device with which the system interacts or exchanges information.

A use case is always initiated by an actor. For example, Student billing is initiated by Bursar's Office. A use case can interact with actors other than the one that initiated it. The Student billing use case, although initiated by the Bursar's Office, interacts with the Student instances by mailing them tuition invoices. Another use case,

Use case diagram: A diagram that depicts the use cases and actors for a system.

Figure 20-2
Use-case diagram for a university registration system

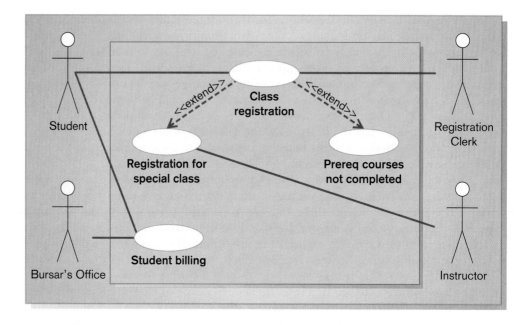

Class registration, is carried out by two actors, Student and Registration Clerk. This use case performs a series of related actions aimed at registering a student for a class.

A use case represents a complete functionality. You should not represent an individual action that is part of an overall function as a use case. For example, although submitting a registration form and paying tuition are two actions performed by users (students) in the university registration system, we do not show them as use cases because they do not specify a complete course of events; each of these actions is executed only as part of an overall function or use case. You can think of Submit registration form as one of the actions of the Class registration use case, and Pay tuition as one of the actions of the Student billing use case. A use case, therefore, is a complete sequence of related actions performed by an actor and the system during a dialog (Jacobson et al., 1992).

Relationships Between Use Cases

A use case may participate in relationships with other use cases. We will describe two types of relationships: extend and include. An *extend* relationship, shown as a dashed line arrow pointing toward the extended use case and labeled with the <<extend>> symbol, extends a use case by adding new behavior or actions (these new behaviors, or extension points, may be listed next to the <<extend>> symbol). In Figure 20-2, for example, the Registration for special class use case extends the Class registration use case by capturing the additional actions that need to be performed in registering a student for a special class. Registering for a special class requires prior permission of the instructor, in addition to the other steps carried out for a regular registration. You may think of Class registration as the *basic course*, which is always performed—independent of whether the extension is performed or not—and Registration for special class as an *alternative course*, which is performed only under special circumstances.

Another example of an extend relationship is that between the Prereq courses not completed and Class registration use cases. The former extends the latter in situations where a student registering for a class has not taken the prerequisite courses. In most cases, only Class registration would be performed, because students would have met the prerequisite requirements. However, in situations where a student has not taken one or more of the prerequisites, the Prereq courses not completed use case is performed, in addition to the Class registration use case. The extension use cases, Registration for special class and Prereq courses not completed, are inserted at appropriate points in the original use case. When the Class registration use case is initiated, it runs to each of those insertion points and checks if the conditions for extending the basic course are valid or not. If valid, the alternative course is performed; otherwise, the basic course continues till the next insertion point, or to its completion.

Figure 20-3 shows a use-case diagram for Hoosier Burger. Several actors and use cases can be identified. The actor that first comes to mind is Customer, which represents the class of all customers who order food at Hoosier Burger's; Order food is therefore represented as a use case. The other actor that is involved in this use case is Service Person. A specific scenario would represent a customer (an instance of Customer) placing an order with a service person (an instance of Service Person). At the end of each day, the manager of Hoosier Burger reorders supplies by calling suppliers. We represent this by a use case called Reorder supplies, involving the Manager and Supplier actors. A manager initiates the use case, which then sends requests to suppliers for various items.

Hoosier Burger also hires employees from time to time. We have therefore identified a use case called Hire employee, in which two actors, Manager and Applicant, are involved. When a person applies for a job at Hoosier Burger, the manager makes a hiring decision. Later, you will see how to capture all the details of interactions in this use case by drawing an interaction diagram.

So far you have seen one kind of relationship, extend, between use cases. Another kind of relationship is include, which arises when one use case references another use case. An include relationship is also shown diagrammatically as a dashed

N E T   S E A R C H
Use cases are becoming a widely used approach to process modeling even outside of the object-oriented world. Visit http://www.prenhall. com/hoffer to complete an exercise on this topic.

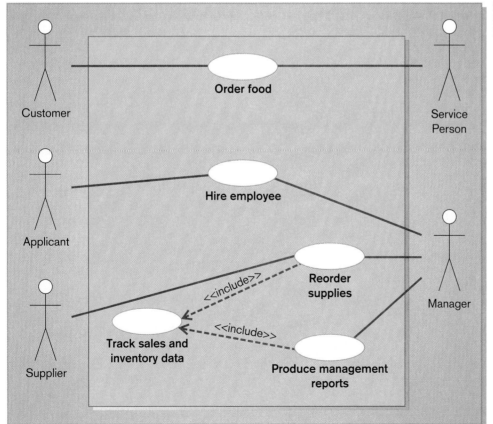

Figure 20-3
Use-case diagram for a Hoosier Burger's system

line pointing toward the use case that is being used; the line is labeled with the <<include>> symbol. In Figure 20-3, for example, the include relationship between the Reorder supplies and Track sales and inventory data use cases implies that the former uses the latter while executing. Simply put, when a manager reorders supplies, the sales and inventory data are tracked. The same data are also tracked when management reports are produced, so there is another include relationship between the Produce management reports and Track sales and inventory data use cases.

The Track sales and inventory data is a generalized use case, representing the common behavior among the specialized use cases, Reorder supplies and Produce management reports. When Reorder supplies or Produce management reports is performed, the entire Track sales and inventory data is used. Note, however, that it is only used when one of the specialized use cases is performed. Such a use case, which is never performed by itself, is called an abstract use case (Eriksson and Penker, 1998; Jacobson et al., 1992).

An Example of Use-Case Diagramming

Figure 20-4 shows a use-case model for different types of credit applications at a bank. The credit applications include those for home equity loans, home mortgage loans, auto loans, and credit cards. From the bank's perspective, therefore, the customers are home owners, home buyers, auto buyers, and credit card applicants. In the diagram, we have shown those different customer classes as actors—Home Owner, Home Buyer, Auto Buyer, and Credit Card Applicant—generalized into an

Figure 20-4
Use-case diagram for a bank lending system

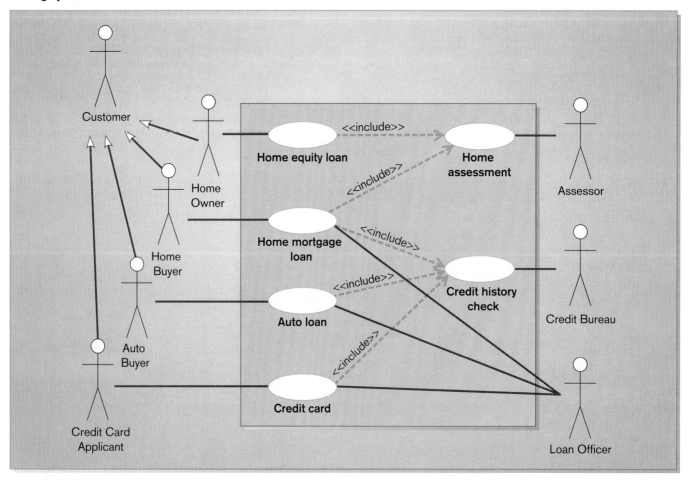

actor called Customer, which represents the class of all customers. Each of the four relationships between a specialized actor and Customer, shown by a straight line with a hollow triangle pointing toward the more general actor, is a generalization relationship (see discussion later in this chapter on class generalization).

To process most credit card applications, the bank needs to check the applicant's credit history, based on a report from the credit bureau. We, therefore, have a use case called **Credit history check**, which is used by three use cases relating to credit applications, that is, **Home mortgage loan**, **Auto loan**, and **Credit card**. The **Home assessment** use case models the common behavior among the **Home mortgage loan** and **Home equity loan** use cases; for both types of loans, the bank summons an assessor to assess the property value before making a decision.

If you are not quite sure about whether to employ *extend* or *include* to model a given relationship between two use cases, consider the following criteria suggested by Jacobson. If the intent is to model an extension to, or a variation of, a complete use case that exists in its own right, employ extend. In Figure 20-2, for example, **Class registration** is a complete use case. It is totally independent of the extension use cases—**Registration for special class** and **Prereq courses not completed**—that were inserted to model alternatives to its basic behavior. If the intent on the other hand is to factor the common behavior among two or more use cases into a single generalized use case, then you should employ the include relationship. The common behavior described by the generalized use case must be inserted into a specialized use case to complete it. For example, to carry out the **Home equity loan** or **Home mortgage loan** use case (see Figure 20-4), the **Home assessment** use case must be performed, too.

While a use-case diagram shows all the use cases in the system, it does not describe how those use cases are carried out by the actors. The contents of a use case are normally described in plain text. While describing a use case, you should focus on its external behavior, that is, how it interacts with the actors, rather than how the use case is performed inside the system (Eriksson and Penker, 1998). For example, we can describe the **Class registration** use case as follows:

1. A student completes a registration form by entering the following information:
   courseNumber
   sectionNumber
   term
   year

2. The student then takes the completed registration form to his or her advisor for signature. After checking the entries, the advisor signs the form.

3. The student then submits the form to a clerk at the registration office, who then enters the information into the computer.

4. The clerk provides the student with a computer printout of the classes in which he or she is successfully registered.

Note that the description includes the following three components:

1. The objective of the use case, which is to register the student for a class

2. The actor (Student) that initiates the use case

3. The exchange of information between the actors (Student and Registration Clerk) and the use case

In this section, we have shown how to develop a use-case model that captures the functional requirements of a system. A use-case diagram is an invaluable vehicle for communication between developers and end users. It also promotes modifiability, by allowing you to easily trace the effects of changes you make in the use-case model on other models (discussed in subsequent sections).

OBJECT MODELING: CLASS DIAGRAMS

In this section, we show how to develop object models by drawing class diagrams. We describe the main concepts and techniques involved in object modeling, including: objects and classes; encapsulation of attributes and operations; association, generalization, and aggregation relationships; cardinalities and other types of constraints; polymorphism; and inheritance. We show how you can develop class diagrams, using the UML notation, to provide a conceptual view of the system being modeled.

Representing Objects and Classes

In the object-oriented approach, we model the world in objects. Before applying the approach to a real-world problem, therefore, we need to understand what an object really is. An **object** is an entity that has a well-defined role in the application domain, and has state, behavior, and identity. An object is a concept, abstraction, or thing that makes sense in an application context (Rumbaugh et al., 1991). An object could be a tangible or visible entity (e.g., a person, place, or thing); it could be a concept or event (e.g., Department, Performance, Marriage, Registration, etc.); or it could be an artifact of the design process (e.g., User Interface, Controller, Scheduler, etc.).

Object: An entity that has a well-defined role in the application domain, and has state, behavior, and identity.

An object has a state and exhibits behavior, through operations that can examine or affect its state. The **state** of an object encompasses its properties (attributes and relationships) and the values those properties have, and its **behavior** represents how an object acts and reacts (Booch, 1994). An object's state is determined by its attribute values and links to other objects. An object's behavior depends on its state and the operation being performed. An operation is simply an action that one object performs upon another in order to get a response. You can think of an operation as a service provided by an object (supplier) to its clients. A client sends a message to a supplier, which delivers the desired service by executing the corresponding operation.

State: Encompasses an object's properties (attributes and relationships) and the values those properties have.

Behavior: Represents how an object acts and reacts.

Consider the example of a student, Mary Jones, represented as an object. The state of this object is characterized by its attributes, say, name, date of birth, year, address, and phone, and the values these attributes currently have. For example, name is "Mary Jones," year is "junior," and so on. Its behavior is expressed through operations such as calc-gpa, which is used to calculate a student's current grade point average. The Mary Jones object, therefore, packages both its state and its behavior together.

All objects have an identity; that is, no two objects are the same. For example, if there are two Student instances with the same name and date of birth, they are essentially two different objects. Even if those two instances have identical values for all the attributes, the objects maintain their separate identities. At the same time, an object maintains its own identity over its life. For example, if Mary Jones gets married and changes her name, address, and phone, she will still be represented by the same object.

Object class (class): A set of objects that share a common structure and a common behavior.

The term "object" is sometimes used to refer to a group of objects, rather than an individual object. The ambiguity can be usually resolved from the context. In the strict sense of the term, however, "object" refers to an individual object, not to a class of objects. That is the interpretation we follow in this text. If you want to eliminate any possible confusion altogether, you can use "object instance" to refer to an individual object, and **object class** (or simply **class**) to refer to a set of objects that share a common structure and a common behavior (just as we used entity type and entity instance in Chapter 10). In our example, therefore, *Mary Jones* is an object instance, while Student is an object class.

Class diagram: Shows the static structure of an object-oriented model: the object classes, their internal structure, and the relationships in which they participate.

You can depict the classes graphically in a class diagram as in Figure 20-5a. A **class diagram** shows the static structure of an object-oriented model: the object classes, their internal structure, and the relationships in which they participate. In UML, a class is represented by a rectangle with three compartments separated by horizontal lines. The class name appears in the top compartment, the list of attrib-

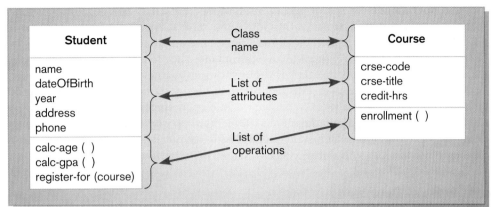

Figure 20-5
UML class and object diagrams
(a) Class diagram showing two classes

(b) Object diagram with two instances

utes in the middle compartment, and the list of operations in the bottom compartment of a box. The figure shows two classes, Student and Course, along with their attributes and operations.

The Student class is a group of Student objects that share a common structure and a common behavior. All students have in common the properties of name, date of birth, year, address, and phone. They also exhibit common behavior by sharing the calc-age, calc-gpa, and register-for(course) operations. A class, therefore, provides a template or schema for its instances. Each object knows to which class it belongs; for example, the Mary Jones object knows that it belongs to the Student class. Objects belonging to the same class may also participate in similar relationships with other objects; for example, all students register for courses and, therefore, the Student class can participate in a relationship called "registers-for" with another class called Course (see the later section on Association).

An **object diagram**, also known as instance diagram, is a graph of instances that are compatible with a given class diagram. In Figure 20-5b, we have shown an object diagram with two instances, one for each of the two classes that appear in Figure 20-5a. A static object diagram is an instance of a class diagram, providing a snapshot of the detailed state of a system at a point in time (*UML Notation Guide*, 1997).

In an object diagram, an object is represented as a rectangle with two compartments. The names of the object and its class are underlined and shown in the top compartment using the following syntax: objectname : classname. The object's attributes and their values are shown in the second compartment. For example, we have an object called Mary Jones that belongs to the Student class. The values of the name, dateOfBirth, and year attributes are also shown. Attributes whose values are not of interest to you may be suppressed; for example, we have not shown the address and telephone attributes for Mary Jones. If none of the attributes are of interest, the entire second compartment may be suppressed. The name of the object may also be omitted, in which case the colon should be kept with the class name as we have done with the instance of Course. If the name of the object is shown, the class name, together with the colon, may be suppressed.

Object diagram: A graph of instances that are compatible with a given class diagram.

Operation: A function or a service that is provided by all the instances of a class.

An **operation**, such as calc-gpa in Student (see Figure 20-5a), is a function or a service that is provided by all the instances of a class. It is only through such operations that other objects can access or manipulate the information stored in an object. The operations, therefore, provide an external interface to a class; the interface presents the outside view of the class without showing its internal structure or how its operations are implemented. This technique of hiding the internal implementation details of an object from its external view is known as **encapsulation** or information hiding (Booch, 1994; Rumbaugh et al., 1991). So while we provide the abstraction of the behavior common to all instances of a class in its interface, we encapsulate within the class its structure and the secrets of the desired behavior.

Encapsulation: The technique of hiding the internal implementation details of an object from its external view.

Types of Operations

Operations can be classified into three types, depending on the kind of service requested by clients: (1) constructor, (2) query, and (3) update (*UML Notation Guide*, 1997). A **constructor operation** creates a new instance of a class. For example, you can have an operation called create-student within Student that creates a new student and initializes its state. Such constructor operations are available to all classes and are therefore not explicitly shown in the class diagram.

Constructor operation: An operation that creates a new instance of a class.

Query operation: An operation that accesses the state of an object but does not alter the state.

A **query operation** is an operation without any side effects; it accesses the state of an object but does not alter the state (Fowler, 2000; Rumbaugh et al., 1991). For example, the Student class can have an operation called get-year (not shown), which simply retrieves the year (freshman, sophomore, junior, or senior) of the Student object specified in the query. Note that there is no need to explicitly show a query such as get-year in the class diagram since it retrieves the value of an independent, base attribute. Consider, however, the calc-age operation within Student. This is also a query operation because it does not have any side effects. Note that the only argument for this query is the target Student object. Such a query can be represented as a derived attribute (Rumbaugh et al., 1991); for example, we can represent "age" as a derived attribute of Student. Because the target object is always an implicit argument of an operation, there is no need to show it explicitly in the operation declaration.

Update operation: An operation that alters the state of an object.

An **update operation** has side effects; it alters the state of an object. For example, consider an operation of Student called promote-student (not shown). The operation promotes a student to a new year, say, from junior to senior, thereby changing the Student object's state (value of the year attribute). Another example of an update operation is register-for(course), which, when invoked, has the effect of establishing a connection from a Student object to a specific Course object. Note that, in addition to having the target Student object as an implicit argument, the operation has an explicit argument called "course," which specifies the course for which the student wants to register. Explicit arguments are shown within parentheses.

Scope operation: An operation that applies to a class rather than an object instance.

A **scope operation** is an operation that applies to a class rather than an object instance. For example, *avg-gpa* for the Student class (not shown with the other operations for this class in Figure 20-5) calculates the average gpa across all students (the operation name is underlined to indicate it is a scope operation).

Representing Associations

Association: A relationship among instances of object classes.

Parallel to the definition of a relationship for the E-R model (see Chapter 10), an **association** is a relationship among instances of object classes. As in the E-R model, the degree of an association relationship may be one (unary), two (binary), three (ternary), or higher (*n*-ary). In Figure 20-6, we illustrate how the object-oriented model can be used to represent association relationships of different degrees. An association is shown as a solid line between the participating classes. The end of an association where it connects to a class is called an **association role** (*UML Notation Guide*, 1997). Each association has two or more roles. A role may be explicitly named

Association role: The end of an association where it connects to a class.

Figure 20-6
**Examples of association
relationships of different degrees**

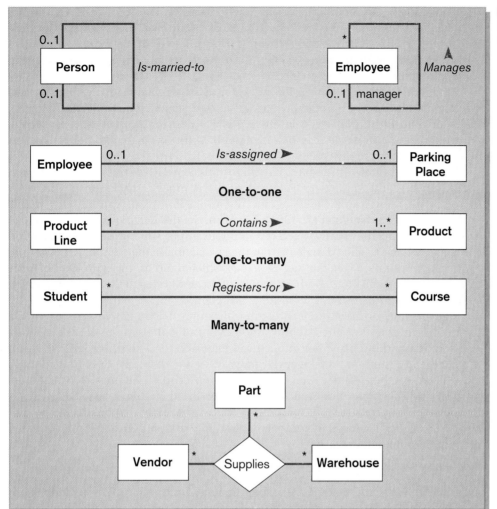

with a label near the end of an association (see the "manager" role in Figure 20-6a). The role name indicates the role played by the class attached to the end near which the name appears. Use of role names is optional. You can specify role names in place of or, in addition to, an association name.

Figure 20-6a shows two unary relationships, *Is-married-to* and *Manages*. At one end of the *Manages* relationship, we have named the role as "manager", implying that an employee can play the role of a manager. We have not named the other roles, but we have named the associations. When the role name does not appear, you may think of the role name as being that of the class attached to that end (Fowler, 2000). For example, you may call the role for the right end of the *Is-assigned* relationship in Figure 20-6b Parking Place.

Each role has a **multiplicity**, which indicates how many objects participate in a given association relationship. In a class diagram, a multiplicity specification is shown as a text string representing an interval (or intervals) of integers in the following format: lower-bound..upper-bound. The interval is considered to be closed, which means that the range includes both the lower and upper bounds. For example, a multiplicity of 2..5 denotes that a minimum of 2 and a maximum of 5 objects can participate in a given relationship. Multiplicities, therefore, are nothing but cardinality constraints, which you saw in Chapter 10. In addition to integer values, the upper bound of a multiplicity can be a star character (*), which denotes an infinite upper bound. If a single integer value is specified, it means that the range includes only that value.

Multiplicity: Indicates how many objects participate in a given relationship.

The most common multiplicities in practice are 0..1, *, and 1. The 0..1 multiplicity indicates a minimum of 0 and a maximum of 1 (optional one), while * (or equivalently, 0..*) represents the range from 0 to infinity (optional many). A single 1 stands for 1..1, implying that exactly one object participates in the relationship (mandatory one).

The multiplicities for both roles in the *Is-married-to* relationship are 0..1, indicating that a person may be single or married to one person. The multiplicity for the manager role in the *Manages* relationship is 0..1 and that for the other role is *, implying that an employee may be managed by only one manager, but a manager may manage many employees.

Figure 20-6b shows three binary relationships: *Is-assigned* (one-to-one), *Contains* (one-to-many), and *Registers-for* (many-to-many). A binary association is inherently bidirectional, though in a class diagram, the association name can be read in only one direction (Rumbaugh et al., 1991). For example, the *Contains* association is read from Product Line to Product. (Note: As in this example, you may show the direction explicitly by using a solid triangle next to the association name.) Implicit, however, is an inverse traversal of *Contains*, say, *Belongs-to*, which denotes that a product belongs to a particular product line. Both directions of traversal refer to the same underlying association; the name simply establishes a direction.

The diagram for the *Is-assigned* relationship shows that an employee is assigned a parking place or not assigned at all (optional one). Reading in the other direction, we say that a parking place has either been allocated for a single employee or not allocated at all (optional one again). Similarly, we say that a product line contains many products, but at least one, whereas a given product belongs to exactly one product line (mandatory one). The diagram for the third binary association states that a student registers for multiple courses, but it is possible that he or she does not register at all, and a course, in general, has multiple students enrolled in it (optional many in both directions).

In Figure 20-6c, we show a ternary relationship called *Supplies* among Vendor, Part, and Warehouse. As in an E-R diagram, we represent a ternary relationship using a diamond symbol and place the name of the relationship there. The relationship is many-to-many-to-many; it cannot be replaced by three binary relationships without loss of information.

The class diagram in Figure 20-7a shows binary associations between Student and Faculty, between Course and Course Offering, between Student and Course Offering, and between Faculty and Course Offering. The diagram shows that a student may have an advisor, while a faculty member may advise up to a maximum of 10 students. Also, while a course may have multiple offerings, a given course offering is scheduled for exactly one course. UML allows you to numerically specify any multiplicity. For example, the diagram shows that a course offering may be taught by one or two instructors (1,2). You can specify a single number (e.g., 2 for the members of a bridge team), a range (e.g., 11..14 for the players of a soccer team who participated in a particular game), or a discrete set of numbers and ranges (e.g., 3, 5, 7 for the number of committee members, and 20..32, 35..40 for the workload in hours per week of the employees of a company).

Figure 20-7a also shows that a faculty member plays the role of an instructor, as well as that of an advisor. While the advisor role identifies the Faculty object associated with a Student object, the advisees role identifies the set of Student objects associated with a Faculty object. We could have named the association, say, *Advises*, but, in this case, the role names are sufficiently meaningful to convey the semantics of the relationship.

Figure 20-7b shows another class diagram for a customer order. The corresponding object diagram is presented in Figure 20-8; it shows some of the instances of the classes and the links among them. (Note: Just as an instance corresponds to a class, a link corresponds to a relationship.) In this example, we see the orders placed by two

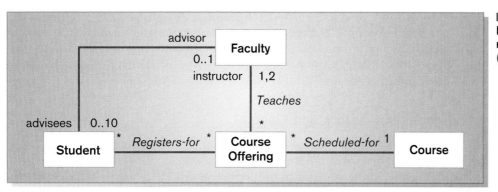

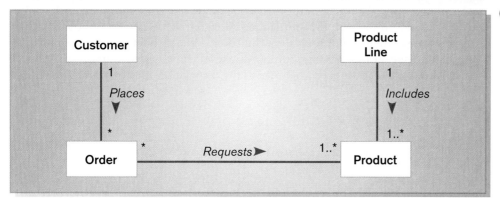

Figure 20-7
Examples of binary association relationships
(a) University example

(b) Customer order example

Figure 20-8
Object diagram for customer order example

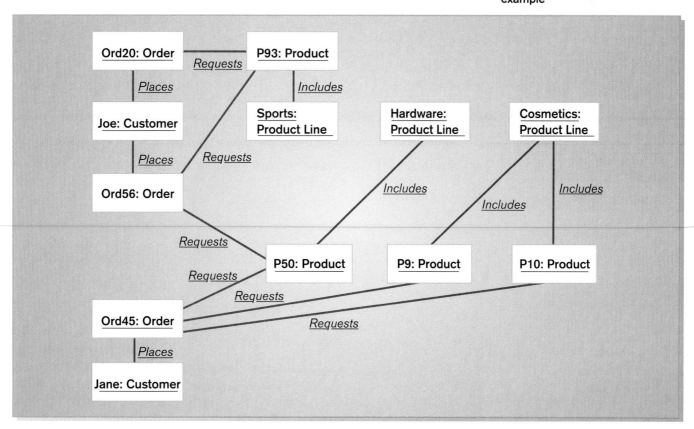

customers, Joe and Jane. Joe has placed two orders, Ord 20 and Ord 56. In Ord 20, Joe has ordered product P93 from the sports product line. In Ord 56, he has ordered the same sports product again, as well as product P50 from the hardware product line. Notice that Jane has ordered the same hardware product as Joe, in addition to two other products (P9 and P10) from the cosmetics product line.

Representing Association Classes

Association class: An association that has attributes or operations of its own, or that participates in relationships with other classes.

When an association itself has attributes or operations of its own, or when it participates in relationships with other classes, it is useful to model the association as an **association class** (just as we used an "associative entity" in Chapter 10). For example, in Figure 20-9a, the attributes term and grade really belong to the many-to-many association between Student and Course. The grade of a student for a course cannot be determined unless both the student and the course are known. Similarly, to find the term(s) in which the student took the course, both student and course must be known. The checkEligibility operation, which determines if a student is eligible to register for a given course, also belongs to the association, rather than to any of the

Figure 20-9
Association class and link object
(a) Class diagram showing association classes

(b) Object diagram showing link objects

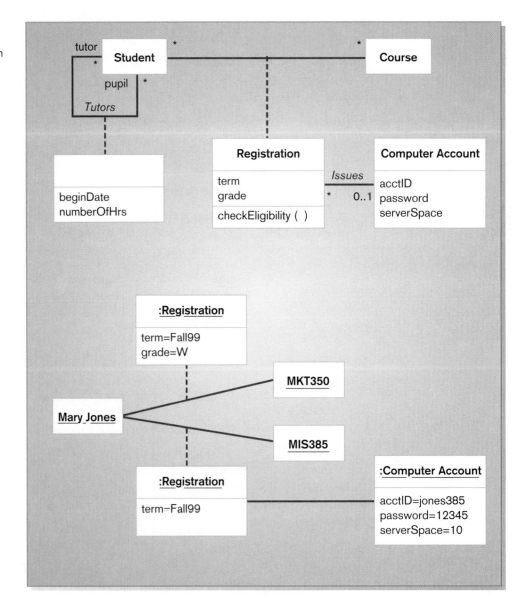

two participating classes. We have also captured the fact that, for some course registrations, a computer account is issued to a student. For these reasons, we model Registration as an association class, having its own set of features and an association with another class (Computer Account). Similarly, for the unary *Tutors* association, beginDate and numberOfHrs (number of hours tutored) really belong to the association, and, therefore, appear in a separate association class.

You have the option of showing the name of an association class on the association path, or the class symbol, or both. When an association has only attributes, but does not have any operations or does not participate in other associations, the recommended option is to show the name on the association path, but to omit it from the association class symbol, to emphasize its "association nature" (*UML Notation Guide*, 1997). That is how we have shown the *Tutors* association. On the other hand, we have displayed the name of the Registration association—which has two attributes and one operation of its own, as well as an association called *Issues* with Computer Account—within the class rectangle to emphasize its "class nature."

Figure 20-9b shows a part of the object diagram representing a student, Mary Jones, and the courses she has registered for in the Fall 1999 term: MKT350 and MIS385. Corresponding to an association class in a class diagram, link objects are present in an object diagram. In this example, there are two link objects (shown as :Registration) for the Registration association class, capturing the two course registrations. The diagram also shows that for the MIS385 course, Mary Jones has been issued a computer account with an ID, a password, and a designated amount of space on the server. She still has not received a grade for this course, but, for the MKT350 course, she received a grade of W because she withdrew from the course.

Figure 20-10 shows a ternary relationship among the Student, Software, and Course classes. It captures the fact that students use various software tools for different courses. For example, we could store the information that Mary Jones used Microsoft Access and Oracle for the Database Management course, Rational Rose and Visual C++ for the Object-Oriented Modeling course, and Level5 Object for the Expert Systems Course. Now suppose we want to estimate the number of hours per week Mary will spend using Oracle for the Database Management course. This process really belongs to the ternary association, and not to any of the individual classes. Hence we have created an association class called Log, within which we have

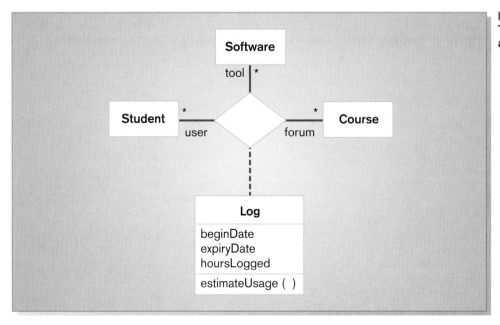

Figure 20-10
Ternary relationship with association class

declared an operation called estimateUsage. In addition to this operation, we have specified three attributes that belong to the association: beginDate, expiryDate, and hoursLogged.

Representing Derived Attributes, Derived Associations, and Derived Roles

A derived attribute, association, or role is one that can be computed or derived from other attributes, associations, and roles, respectively. A derived element (attribute, association, or role) is shown by placing a slash (/) before the name of the element. For instance, in Figure 20-11, age is a derived attribute of Student, because it can be calculated from the date of birth and the current date. Because the calculation is a constraint on the object class, the calculation is shown on this diagram within braces near the Student object class. Also, the *Takes* relationship between Student and Course is derived, because it can be inferred from the *Registers-for* and *Scheduled-for* relationships. By the same token, participants is a derived role because it can be derived from other roles.

Representing Generalization

In the object-oriented approach, you can abstract the common features (attributes and operations) among multiple classes, as well as the relationships they participate in, into a more general class. This is known as *generalization*. The classes that are generalized are called subclasses, and the class they are generalized into is called a superclass.

Consider the example shown in Figure 20-12a. There are three types of employees: hourly employees, salaried employees, and consultants. The features that are shared by all employees: empName, empNumber, address, dateHired, and printLabel, are stored in the Employee superclass, while the features that are peculiar to a particular employee type are stored in the corresponding subclass (e.g., hourlyRate and computeWages of Hourly Employee). A generalization path is shown as a solid line from the subclass to the superclass, with a hollow triangle at the end of, and pointing toward, the superclass. You can show a group of generalization paths for a given superclass as a tree with multiple branches connecting the individual subclasses, and a shared segment with a hollow triangle pointing toward the superclass. In Figure 20-12b, for instance, we have combined the generalization paths from Outpatient to Patient, and from Resident Patient to Patient, into a shared segment with a triangle pointing toward Patient. We also specify that this generalization is dynamic, meaning that an object may change subtypes.

Figure 20-11
Derived attribute, association, and role

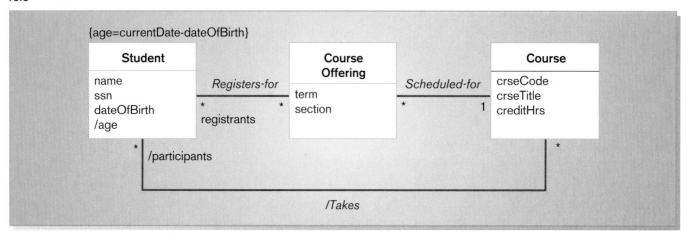

Figure 20-12
Examples of generalization, inheritance, and constraints
(a) Employee superclass with three subclasses

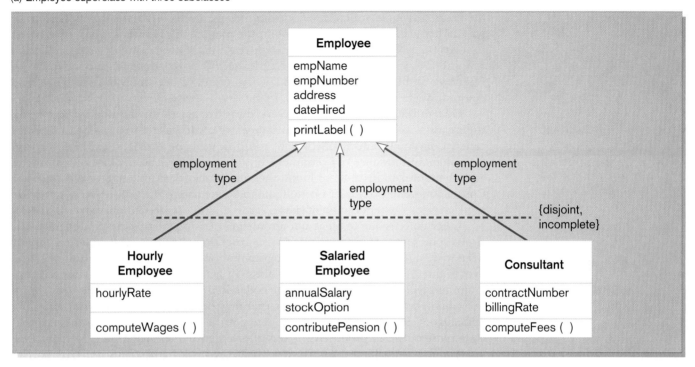

(b) Abstract Patient class with two concrete subclasses

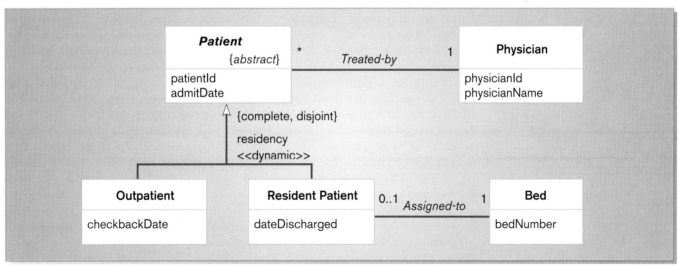

You can indicate the basis of a generalization by specifying a discriminator next to the path. A discriminator shows which property of an object class is being abstracted by a particular generalization relationship (Rumbaugh et al., 1991). You can discriminate on only one property at a time. For example, in Figure 20-12a, we discriminate the Employee class on the basis of employment type (hourly, salaried, consultant). To discriminate a group of generalization relationships, as in Figure 20-12b, we need to specify the discriminator only once. Although we discriminate the Patient class into two subclasses, Outpatient and Resident Patient, based on residency, we show the discriminator label only once next to the shared line.

An instance of a subclass is also an instance of its superclass. For example, in Figure 20-12b, an Outpatient instance is also a Patient instance. For that reason, a generalization is also referred to as an *Is-a* relationship. Also, a subclass inherits all the features from its superclass. For example, in Figure 20-12a, in addition to its own special features—hourlyRate and computeWages—the Hourly Employee subclass inherits empName, empNumber, address, dateHired, and printLabel from Employee. An instance of Hourly Employee will store values for the attributes of Employee and Hourly Employee, and, when requested, will apply the printLabel and computeWages operations.

Generalization and inheritance are transitive across any number of levels of a superclass/subclass hierarchy. For instance, we could have a subclass of Consultant called Computer Consultant, which would inherit the features of Employee and Consultant. An instance of Computer Consultant would be an instance of Consultant and, therefore, an instance of Employee, too. Employee is an ancestor of Computer Consultant, while Computer Consultant is a descendant of Employee; these terms are used to refer to generalization of classes across multiple levels (Rumbaugh et al., 1991).

Inheritance is one of the major advantages of using the object-oriented model. It allows code reuse: There is no need for a programmer to write code that has already been written for a superclass. The programmer only writes code that is unique to the new, refined subclass of an existing class. In actual practice, object-oriented programmers typically have access to large collections of class libraries in their respective domains. They identify those classes that may be reused and refined to meet the demands of new applications. Proponents of the object-oriented model claim that code reuse results in productivity gains of several orders of magnitude.

Abstract class: A class that has no direct instances, but whose descendants may have direct instances.

Concrete class: A class that can have direct instances.

Notice that in Figure 20-12b the *Patient* class is in italics, implying that it is an abstract class. An **abstract class** is a class that has no direct instances, but whose descendants may have direct instances (Booch, 1994; Rumbaugh et al., 1991). (Note: You can additionally write the word *abstract* within braces just below the class name. This is especially useful when you generate a class diagram by hand.) A class that can have direct instances (e.g., Outpatient or Resident Patient) is called a **concrete class**. In this example, therefore, Outpatient and Resident Patient can have direct instances, but *Patient* cannot have any direct instances of its own.

The *Patient* abstract class participates in a relationship called *Treated-by* with Physician, implying that all patients, outpatients and resident patients alike, are treated by physicians. In addition to this inherited relationship, the Resident Patient class has its own special relationship called *Assigned-to* with Bed, implying that only resident patients may be assigned to beds. So, in addition to refining the attributes and operations of a class, a subclass can also specialize the relationships in which it participates.

In Figure 20-12, the words "complete" and "disjoint" have been placed within braces, next to the generalization. They indicate semantic constraints among the subclasses (*complete* corresponds to total specialization in the EER notation, whereas *incomplete* corresponds to partial specialization). In UML, a comma-separated list of keywords is placed either near the shared triangle as in Figure 20-12b, or near a dashed line that crosses all of the generalization lines involved as in Figure 20-12a (*UML Notation Guide*, 1997). Any of the following UML keywords may be used: overlapping, disjoint, complete, and incomplete. According to the *UML Notation Guide* (1997), the definition of these terms are:

- *Overlapping*: A descendant may be descended from more than one of the subclasses (same as the overlapping rule in EER diagramming).

- *Disjoint*: A descendant may not be descended from more than one of the subclasses (same as the disjoint rule in EER diagramming).

- *Complete*: All subclasses have been specified (whether or not shown). No additional subclasses are expected (same as the total specialization rule in EER diagramming).

- *Incomplete*: Some subclasses have been specified, but the list is known to be incomplete. There are additional subclasses that are not yet in the model (same as the partial specialization rule in EER diagramming).

Both the generalizations in Figures 20-12a and 20-12b are disjoint. An employee can be an hourly employee, a salaried employee, or a consultant, but cannot, say, be both a salaried employee and a consultant. Similarly, a patient can be an outpatient or a resident patient, but not both at the same time. The generalization in Figure 20-12a is incomplete, specifying that an employee might not belong to any of the three types. In such a case, an employee will be stored as an instance of Employee, a concrete class. In contrast, the generalization in Figure 20-12b is complete, implying that a patient has to be either an outpatient or a resident patient, and nothing else. For that reason, *Patient* has been specified as an abstract class.

In Figure 20-13, we show an example of an overlapping constraint. The diagram shows that research assistants and teaching assistants are graduate students. The overlapping constraint indicates that it is possible for a graduate student to serve as both a research assistant and a teaching assistant. For example, Sean Bailey, a graduate student, has a research assistantship of 12 hours per week and a teaching assistantship of 8 hours per week. Also notice that Graduate Student has been specified as a concrete class so that graduate students without an assistantship can be represented. The ellipsis marks (. . .) under the generalization line based on the "level" discriminator does not represent an incomplete constraint. It simply indicates that there are other subclasses in the model that have not been shown in the diagram. For example, although Undergrad Student is in the model, we have opted not to show it in the diagram since the focus is on assistantships. You may also use ellipsis marks when there are space limitations.

In Figure 20-14, we represent both graduate and undergraduate students in a model developed for student billing. The calc-tuition operation computes the tuition a student has to pay; this sum depends on the tuition per credit hour (tuitionPerCred), the courses taken, and the number of credit hours (creditHrs) for each of those courses. The tuition per credit hour, in turn, depends on whether the student is a graduate or an undergraduate student. In this example, that amount is $300 for all graduate students and $250 for all undergraduate students. To denote that, we have underlined the tuitionPerCred attribute in each of the two subclasses, along with its value. Such an attribute is called a **class-scope attribute**, which specifies

Class-scope attribute: An attribute of a class which specifies a value common to an entire class, rather than a specific value for an instance.

Figure 20-13
Example of overlapping constraint

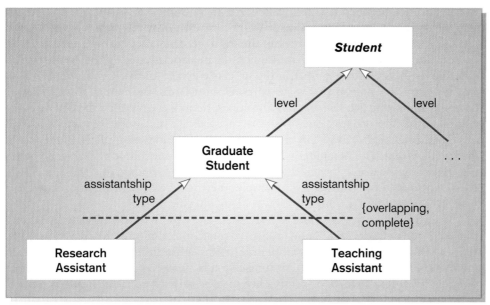

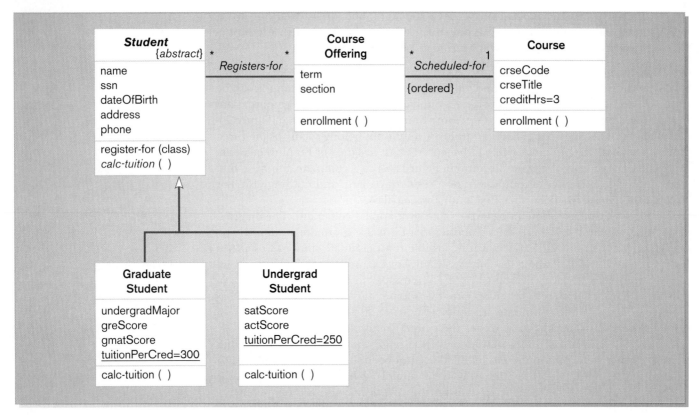

Figure 20-14
Polymorphism, abstract operation, class-scope attribute, and ordering

a value common to an entire class, rather than a specific value for an instance (Rumbaugh et al., 1991).

You can specify an initial default value of an attribute using an equals sign (=) after the attribute name. This is the initial attribute value of a newly created object instance. For example, in Figure 20-14, the creditHrs attribute has an initial value of 3, implying that when a new instance of Course is created, the value of creditHrs is set to 3 by default. You can write an explicit constructor operation to modify the initial default value. The value may also be modified later through other operations. The difference between an initial value specification and a class-scope attribute is that while the former allows the possibility of different attribute values for the instances of a class, the latter forces all the instances to share a common value.

In addition to specifying the multiplicity of an association role, you can also specify other properties, for example, if the objects playing the role are ordered or not. In the figure, we placed the keyword constraint "{ordered}" next to the Course Offering end of the *Scheduled-for* relationship to denote the fact that the offerings for a given course are ordered into a list—say, according to term and section. It is obvious that it makes sense to specify an ordering only when the multiplicity of the role is greater than one. The default constraint on a role is "{unordered};" that is, if you do not specify the keyword "{ordered}" next to the role, it is assumed that the related elements form an unordered set. For example, the course offerings are not related to a student who registers for those offerings in any specific order.

The Graduate Student subclass specializes the abstract Student class by adding four attributes—undergradMajor, greScore, gmatScore, and tuitionPerCred—and by refining the inherited *calc-tuition* operation. Notice that the operation is shown in italics within the Student class, indicating that it is an abstract operation. An **abstract operation** defines the form or protocol of the operation, but not its implementation (Rumbaugh et al., 1991). In this example, the Student class defines the protocol of

Abstract operation: Defines the form or protocol of the operation, but not its implementation.

the *calc-tuition* operation, without providing the corresponding **method** (the actual implementation of the operation). The protocol includes the number and types of the arguments, the result type, and the intended semantics of the operation. The two concrete subclasses, Graduate Student and Undergrad Student, supply their own implementations of the calc-tuition operation. Note that because these classes are concrete, they cannot store abstract operations.

Method: The implementation of an operation.

It is important to note that although the Graduate Student and Undergraduate Student classes share the same calc-tuition operation, they might implement the operation in quite different ways. For example, the method that implements the operation for a graduate student might add a special graduate fee for each course the student takes. The fact that the same operation may apply to two or more classes in different ways is known as **polymorphism**, a key concept in object-oriented systems (Booch, 1994; Rumbaugh et al., 1991). The enrollment operation in Figure 20-14 illustrates another example of polymorphism. While the enrollment operation within Course Offering computes the enrollment for a particular course offering or section, an operation with the same name within Course computes the combined enrollment for all sections of a given course.

Polymorphism: The same operation may apply to two or more classes in different ways.

Interpreting Inheritance and Overriding

We have seen how a subclass can augment the features inherited from its ancestors. In such cases, the subclass is said to use *inheritance for extension*. On the other hand, if a subclass constrains some of the ancestor attributes or operations, it is said to use *inheritance for restriction* (Booch, 1994; Rumbaugh et al., 1991). For example, a subclass called Tax-Exempt Company may suppress or block the inheritance of an operation called compute-tax from its superclass, Company.

The implementation of an operation can also be overridden. **Overriding** is the process of replacing a method inherited from a superclass by a more specific implementation of that method in a subclass. The reasons for overriding include extension, restriction, and optimization (Rumbaugh et al., 1991). The name of the new operation remains the same as the inherited one, but it has to be explicitly shown within the subclass to indicate that the operation is overridden.

Overriding: The process of replacing a method inherited from a superclass by a more specific implementation of that method in a subclass.

In *overriding for extension*, an operation inherited by a subclass from its superclass is extended by adding some behavior (code). For example, a subclass of Company called Foreign Company inherits an operation called compute-tax, but extends the inherited behavior by adding a foreign surcharge to compute the total tax amount.

In *overriding for restriction*, the protocol of the new operation in the subclass is restricted. For example, an operation called place-student(job) in Student may be restricted in the International Student subclass by tightening the job argument (see Figure 20-15). While students in general may be placed in all types of jobs during the summer, international students may be limited to only on-campus jobs because of visa restrictions. The new operation overrides the inherited operation by tightening the job argument, restricting its values to only a small subset of all possible jobs. This example also illustrates the use of multiple discriminators. While the basis for one set of generalizations is a student's "level" (graduate or undergraduate), that for the other set is his or her "residency" status (United States or international).

In *overriding for optimization*, the new operation is implemented with improved code by exploiting the restrictions imposed by a subclass. Consider, for example, a subclass of Student called Dean's List Student, which represents all those students who are on the Dean's list. To qualify for the Dean's list, a student must have a grade point average (gpa) greater than or equal to 3.50. Suppose Student has an operation called mailScholApps, which mails applications for merit- and means-tested scholarships to students who have a gpa greater than or equal to 3.00, and whose family's total gross income is less than $20,000. The method for the operation in Student will have to check both conditions, whereas the method for the same operation in the

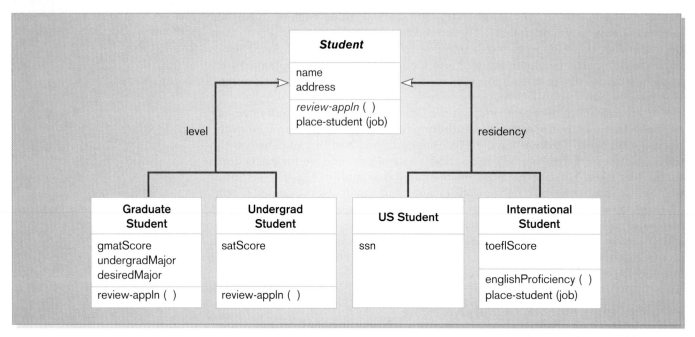

Figure 20-15
Overriding inheritance

Dean's List Student subclass can improve upon the speed of execution by removing the first condition from its code. Consider another operation called findMinGpa, which finds the minimum gpa among the students. Suppose the Dean's List Student class is sorted in ascending order of the gpa, but the Student class is not. The method for findMinGpa in Student must perform a sequential search through all the students. In contrast, the same operation in Dean's List Student can be implemented with a method that simply retrieves the gpa of the first student in the list, thereby obviating the need for a time-consuming search.

Representing Multiple Inheritance

So far you have been exposed to single inheritance, where a class inherits from only one superclass. But sometimes, as we saw in the example with research and teaching assistants, an object may be an instance of more than one class. This is known as **multiple classification** (Fowler, 2000; *UML Notation Guide*, 1997). For instance, Sean Bailey, who has both types of assistantships, has two classifications: one as an instance of Research Assistant and the other as an instance of Teaching Assistant. Multiple classification, however, is discouraged by experts. It is not supported by the ordinary UML semantics and many object-oriented languages.

Multiple classification: An object is an instance of more than one class.

To get around the problem, we can use multiple inheritance, which allows a class to inherit features from more than one superclass. For example, in Figure 20-16, we have created Research & Teaching Assistant, which is a subclass of both Research Assistant and Teaching Assistant. All students who have both research and teaching assistantships may be stored under the new class. We may now represent Sean Bailey as an object belonging to only the Research & Teaching Assistant class, which inherits features from both its parents, such as researchHrs and assignProject(proj) from Research Assistant and teachingHrs and assignCourse(crse) from Teaching Assistant.

Representing Aggregation

Aggregation: A *part-of* relationship between a component object and an aggregate object.

An **aggregation** expresses a *Part-of* relationship between a component object and an aggregate object. It is a stronger form of association relationship (with the added "part-of" semantics) and is represented with a hollow diamond at the aggregate end.

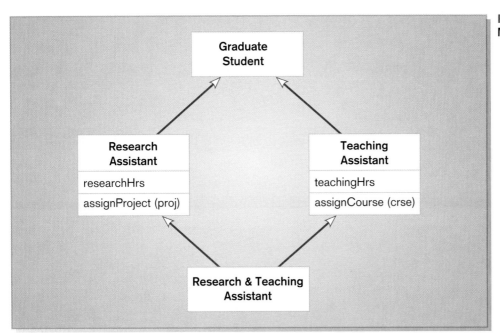

Figure 20-16
Multiple inheritance

For example, Figure 20-17 shows a personal computer as an aggregate of CPU (up to four for multiprocessors), hard disks, monitor, keyboard, and other objects. Note that aggregation involves a set of distinct object instances, one of which contains or is composed of the others. For example, a Personal Computer object is related to (consists of) CPU objects, one of its parts. In contrast, generalization relates object classes: An object (e.g., Mary Jones) is simultaneously an instance of its class (e.g., Graduate Student) and its superclass (e.g., Student). Only one object (e.g., Mary Jones) is involved in a generalization relationship. This is why multiplicities are indicated at the ends of aggregation lines, whereas there are no multiplicities for generalization relationships.

Figure 20-18a shows an aggregation structure of a university. The object diagram in Figure 20-18b shows how Riverside University, a University object instance, is related to its component objects, which represent administrative units (e.g., Admissions, Human Resources, etc.) and schools (e.g., Arts and Science, Business, etc.). A school object (e.g., Business), in turn, comprises several department objects (e.g., Accounting, Finance, etc.).

Notice that the diamond at one end of the relationship between Building and Room is not hollow, but solid. A solid diamond represents a stronger form of

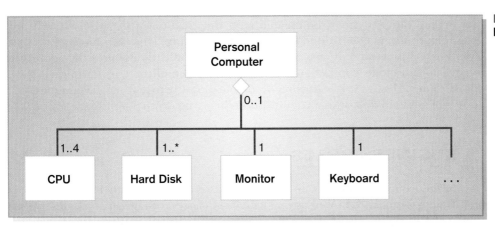

Figure 20-17
Example of aggregation

Figure 20-18
Aggregation and composition
(a) Class diagram

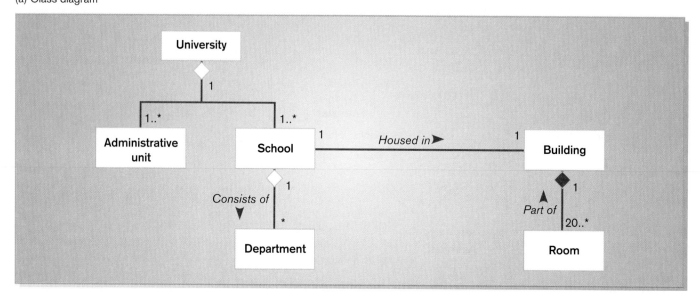

(b) Object diagram

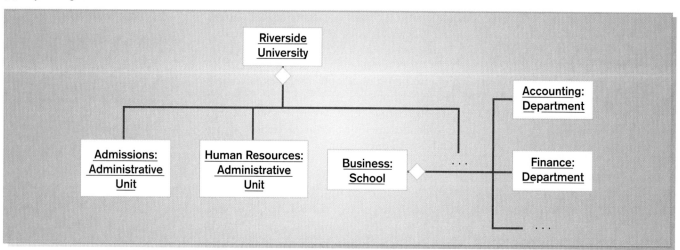

Composition: A part object that belongs to only one whole object, and that lives and dies with the whole.

aggregation, known as composition (Fowler, 2000; *UML Notation Guide*, 1997). In **composition**, a part object belongs to only one *whole* object; for example, a room is part of only one building. Therefore, the multiplicity on the aggregate end may not exceed one. Parts may be created after the creation of the whole object; for example, rooms may be added to an existing building. However, once a part of a composition is created, it lives and dies with the whole; deletion of the aggregate object cascades to its components. If a building is demolished, for example, so are all its rooms. However, it is possible to delete a part before its aggregate dies, just as it is possible to demolish a room without bringing down a building.

N E T   S E A R C H
There are a wide variety of software tools to assist in object-oriented analysis and design. Visit http://www. prenhall.com/hoffer to complete an exercise on this topic.

BUSINESS RULES

In the examples provided in this chapter, we captured many business rules as constraints—implicitly, as well as explicitly—on classes, instances, attributes, operations, relationships, and so forth. For example, you saw how to specify cardinality

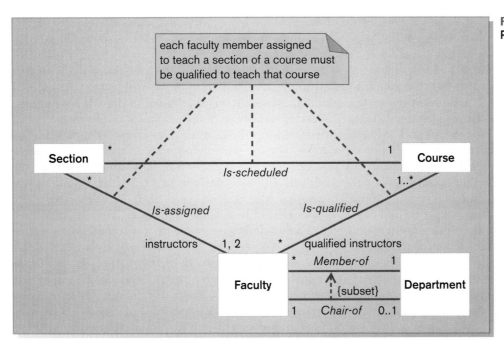

Figure 20-19
Representing business rules

constraints and ordering constraints on association roles. You also saw how to represent semantic constraints (e.g., overlapping, disjoint, etc.) among subclasses. Many of the constraints that were discussed so far in this chapter were imposed by including a set of UML keywords within braces—for example, {disjoint, complete} and {ordered}—and placing them close to the elements the constraints apply to. For example, in Figure 20-14, we expressed a business rule that offerings for a given course are ordered. But if you cannot represent a business rule using such a predefined UML constraint, you can define the rule in plain English, or in some other language such as formal logic.

When you have to specify a business rule involving two graphical symbols (such as those representing two classes or two associations), you can show the constraint as a dashed arrow from one element to the other, labeled by the constraint name in braces (*UML Notation Guide*, 1997). In Figure 20-19, for example, we have stated the business rule that the chair of a department must be a member of the department by specifying the *Chair-of* association as a subset of the *Member-of* association.

When you have to specify a business rule involving three or more graphical symbols, you can show the constraint as a note and attach the note to each of the symbols by a dashed line (*UML Notation Guide*, 1997). In Figure 20-19, we have captured the business rule that "each faculty member assigned to teach a section of a course must be qualified to teach that course" within a note symbol. Because this constraint involves all the three association relationships, we have attached the note to each of the three association paths.

DYNAMIC MODELING: STATE DIAGRAMS

You have seen how to model the static structure of a system using class diagrams. In this section, we show you how to model the dynamic aspects of a system from the perspective of state transitions. In UML, state transitions are shown using state diagrams. A state diagram depicts the various state transitions or changes an object can experience during its lifetime, along with the events that cause those transitions.

State transition: Changes in the attributes of an object or in the links an object has with other objects.

Event: Something that takes place at a certain point in time; it is a noteworthy occurrence that triggers a state transition.

From the perspective of dynamic modeling, a *state* is a condition during the life of an object during which it satisfies some condition(s), performs some action(s), or waits for some event(s) (*UML Notation Guide*, 1997). The state changes when the object receives some event; the object is said to undergo a **state transition**. The state of an object depends on its attribute values and links to other objects.

An **event** is something that takes place at a certain point in time. It is a noteworthy occurrence that triggers a state transition. Some examples of events are: a customer places an order; a student registers for a class; a person applies for a loan; and a company hires a new employee. For the purpose of modeling, an event is considered to be instantaneous, though, in reality, it might take some time. A state, on the other hand, spans a period of time. An object remains in a particular state for some time before transitioning to another state. For example, an Employee object might be in the Part-time state (as specified in its employment-status attribute) for a few months, before transitioning to a Full-time state, based on a recommendation from the manager (an event).

Not all attributes are important in determining the state of an object. For example, the address and phone attributes of a Student object may not be important. When a student changes apartments, the object does not undergo any state transition. (If, however, for the purposes of your model, you want to classify a student as being in one of two states, On-campus or Off-campus, then address could be an important attribute.) In determining the state of an object, only those attributes that affect its gross behavior are considered to be important.

The other thing to note is that the state of an object does not change whenever an attribute changes its value. For example, from a bank's perspective, the fluctuations in a customer's monthly debt payments will not normally trigger state transitions. However, if the debt-to-income ratio exceeds a certain threshold (say, 45 percent) after some event—say, the purchase of a new yacht—then the Customer object may transition from a Low-risk state to a High-risk state. Within a given state, an object's gross behavior will remain qualitatively the same in response to an event, while the behavior could be quite different in another state (Rumbaugh et al., 1991). For example, the bank's lending procedures for all low-risk customers are more or less the same, but the procedures could be very different for high-risk customers.

Modeling States and State Transitions

State diagram: A model of the states of an object and the events that cause the object to change from one state to another.

In UML, a state is shown as a rectangle with rounded corners. In Figure 20-20, for example, we have shown different states of a Student object, such as Inquiry, Applied, Approved, Rejected, and so forth. This **state diagram** shows how the object transitions from an initial state (shown as a small, solid, filled circle) to other states, when certain events occur or when certain conditions are satisfied. When a new Student object is created, it is in its initial state. The event that created the object, inquires, transitions it from the initial state to the Inquiry state. When a student in the Inquiry state submits an application for admission, the object transitions to the Applied state. The transition is shown as a solid arrow from Inquiry (the source state) to Applied (the target state), labeled with the name of the event, submits application.

A transition may be labeled with a string consisting of the event name, parameters of the event, guard condition, and action expression. A transition, however, does not have to show all the elements; it only shows those that are relevant to the transition. For example, we label the transition from Inquiry to Applied with simply the event name. But, for the transition from Applied to Approved, we show the event name (evaluate), the guard condition (acceptable), and the action taken by the transition (mail approval letter). It simply means that an applicant is approved for admission if the admissions office evaluates the application and finds it acceptable. If acceptable, a letter of approval is mailed to the student. The guard condition is shown within square brackets, and the action is specified after the "/" symbol.

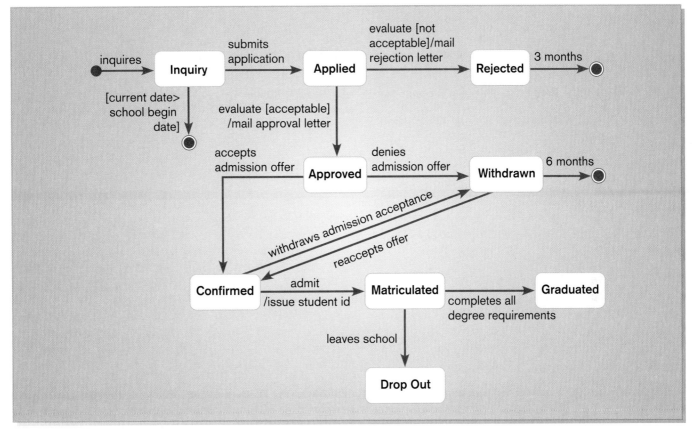

Figure 20-20
State diagram for the Student
object

If the **evaluate** event results in a not-acceptable decision (another guard condition), a rejection letter is mailed (an action) and the Student object undergoes a state transition from the Applied to the Rejection state. It remains in that state for three months, before reaching the final state. In the diagram, we have shown an elapsed-time event, 3 months, indicating the amount of time the object waits in the current state before transitioning. The final state is shown as a bull's eye: a small, solid, filled circle surrounded by another circle. After transitioning to the final state, the Student object ceases to exist.

Notice that the Student object may transition to the final state from two other states: Inquiry and Withdrawn. For the transition from Inquiry, we have not specified any event name or action, but we have shown a guard condition, current date > school begin date. This condition implies that the Student object ceases to exist beyond the first day of school unless, of course, the object has moved in the meantime from the Inquiry state to some other state.

The state diagram shown in Figure 20-20 captures all the possible states of a Student object, the state transitions, the events or conditions that trigger those transitions, and the associated actions. For a typical student, it captures the student's sojourn through college, right from the time when he or she expressed an interest in the college until graduation.

Diagramming Substates and Decomposing Events

We can expand a state or an event to show more details. In Figure 20-21, we have expanded the Matriculated state of a student into its substates. A matriculated student could be in any one of the substates: Freshman, Sophomore, Junior, or Senior. The ini-

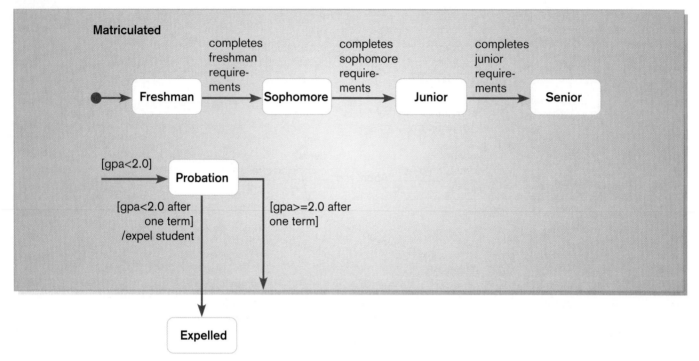

Figure 20-21
Nested state diagram for Matriculated state of Student object

tial state transitions automatically to the Freshman state, indicating the first substate of a matriculated student. When the student completes the freshman requirements, a transition fires and moves the object to the Sophomore state. Similar transitions move the Student object through the subsequent substates. Figure 20-21 is an example of a nested state diagram, where the substates are nested within the general state, Matriculated.

The diagram captures a further substate called Probation. The transition to this state is shown as an arrow from the enclosing rectangle, which represents the Matriculated state. What that means is that the transition to Probation could fire from any substate of Matriculated, whether it be Freshman, Sophomore, Junior, or Senior. However, there is a guard condition on the transition, which indicates that a matriculated student is placed on probation only when his or her grade point average falls below 2.0.

A Student object in the Probation state could go back to its previous state if the gpa equals or exceeds 2.0 after the next term (represented by an arrow from Probation to the outer rectangle). Otherwise, the object transitions to the Expelled state, indicating that the student is expelled from the college.

An event can also be expanded in a lower level diagram. In Figure 20-22, the accepts admission offer event shown in the high-level diagram of Figure 20-20 has been refined to show its fine structure. When a student verbally accepts the admission offer, the corresponding Student object transitions to the Preliminary Acceptance state. When the student deposits an amount of money, the object transitions to the Strong Acceptance state. If the deposit is not made by a given deadline, the object enters another state, which we have not named. Instead, we have shown the activity performed in that state, which is to follow up with the student.

An activity is an operation that is performed when an object is in a certain state. An ongoing activity is labeled by the keyword "do" followed by the "/" sign. It could be a continuous operation such as displaying the balance on a cash register, or it could be a sequential operation such as processing a customer order. This type of activity takes time to complete and can be interrupted by other events. On the other hand, an entry or an exit activity is an atomic operation performed on entry to or exit from a state.

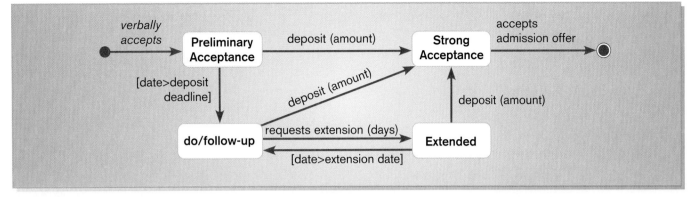

Figure 20-22
Nested state diagram for the accepts admission offer event

After the follow-up, if the student requests an extension for paying the deposit, the object enters the Extended state. Note that the requests extension event has an argument or parameter called "days," which indicates the number of days for which the extension is requested. You can attach multiple arguments to an event. If the student deposits the amount by the extension date, then the object transitions to the Strong Acceptance state. From there, it reaches the Confirmed state, which is implicitly represented by labeling the final state with the name of the top-level event, accepts admission offer.

A State Transition Diagram for Hoosier Burger

In Figure 20-23, we have shown how an object, representing a customer order at Hoosier Burger, dynamically changes states. When a customer places an order, a new Order object is created and the register is reset. The object enters the Checking state and it remains in that state until all the items have been checked. The state is shown with two compartments. The top compartment indicates the name of the state and the bottom one shows the activities associated with the state.

The transition from Checking back to itself is a self-transition. It fires every time the customer orders the next item. The entry activity, calculate line total, is performed on entry to the Checking state. This activity is performed even for the self-transition, capturing the fact that a new line total has to be calculated for each order line. The display line total is a continuous operation; it displays all the line totals for the order.

When all the items are checked, the Order object enters the Checked state. On entry, the total for the entire order is calculated, and the register displays that figure. When the customer pays, the object enters the Paid state, and the balance to be paid back to the customer is calculated and displayed. On exit from this state, the balance is paid back to the customer and the order enters the Processing state. The transition arrow is shown without an event name to indicate that the transition automatically fires when the activities in the source state (Paid) are completed. While in the Processing state, all the order items are prepared. Finally, when all the items have been served, the Order object transitions to the Delivered state.

Notice that there is a transition arrow from the outer rectangle surrounding the different order states to a state called Cancelled. That captures the transition from any of those states to the Cancelled state in the event that the customer cancels the order. Showing a common transition this way makes the diagram much simpler than showing a separate transition from each of the states.

In general, an action associated with an event can be modeled as an action or as entry activity; both are atomic operations. For example, we have shown adjust register as an entry activity within the Cancelled state, but we could have just as well represented it as an action associated with the **cancel order** event. Sometimes, however, you have to choose one over the other. The calculate line total operation is shown as an

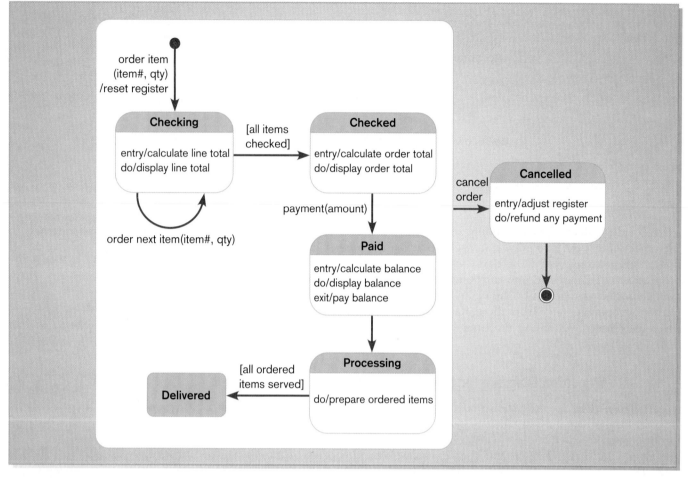

Figure 20-23
State diagram for the Order object in Hoosier Burger's system

entry activity within the Checking state. This operation will be performed for both transitions entering the state. On the other hand, we represented reset register as an action associated with the **order item** event, because this operation is performed only when the first item is ordered, not for subsequent items (shown by the self-transitions).

Summary of State Transition Diagramming

In this section, we showed examples of various state diagrams. Remember that a state diagram captures the states of a single class of objects. To develop a dynamic model for an application, you need to combine all the state diagrams you have drawn for different object classes. In such a model, the interaction among the component state diagrams are shown via shared events.

DYNAMIC MODELING: SEQUENCE DIAGRAMS

In this section we will show how to design some of the use cases we identified earlier in the analysis phase. A use-case design describes how each use case is performed by a set of communicating objects (Jacobson et al., 1992).

In UML, an interaction diagram is used to show the pattern of interactions among objects for a particular use case. There are two types of interaction diagrams: *sequence diagrams* and collaboration diagrams (*UML Notation Guide*, 1997). Both of them express similar information, but they do so in different ways. While sequence diagrams show the explicit sequencing of messages, collaboration diagrams show the relationship among objects. We will show you how to design use cases using sequence diagrams.

Components of a Sequence Diagram

A **sequence diagram** depicts the interactions among objects during a certain period of time. Because the pattern of interactions varies from one use case to another, each sequence diagram shows only the interactions pertinent to a specific use case. It shows the participating objects by their lifelines, and the interactions among those objects—arranged in time sequence—by the messages they exchange with one another.

A sequence diagram may be presented either in a generic form or in an instance form. The generic form shows all possible sequences of interactions, that is, the sequences corresponding to all the scenarios of a use case. For example, a generic sequence diagram for the **Class registration** use case (see Figure 20-2) would capture the sequence of interactions for every valid scenario of that use case. The instance form, on the other hand, shows the sequence for only one scenario. A scenario in UML refers to a single path, among possibly many different paths, through a use case (Fowler, 2000). A path represents a specific combination of conditions within the use case. In Figure 20-24, we have shown a sequence diagram, in

Sequence diagram: Depicts the interactions among objects during a certain period of time.

Figure 20-24
Sequence diagram for a class registration scenario with prerequisites

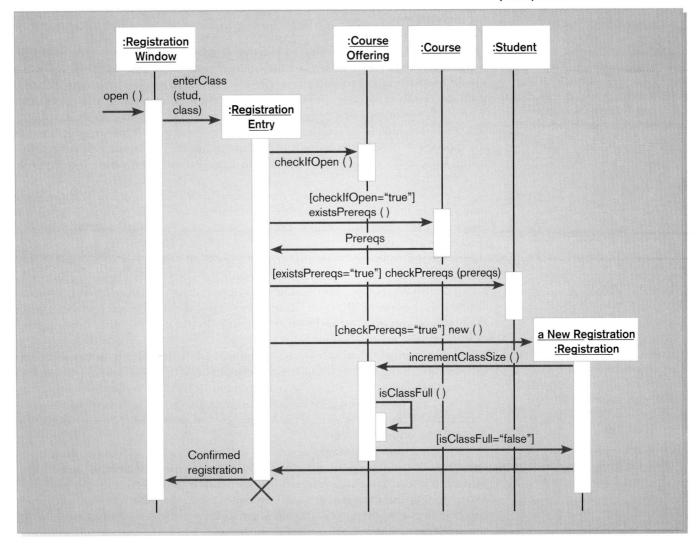

instance form, for a scenario where a student registers for a course that specifies one or more prerequisite courses as requirements.

The vertical axis of the diagram represents time and the horizontal axis represents the various participating objects. Time increases as we go down the vertical axis. The diagram has six objects, from an instance of Registration Window on the left, to an instance of Registration called "a New Registration" on the right. The ordering of the objects has no significance. However, you should try to arrange the objects in such a way that the diagram is easy to read and understand. Each object is shown as a vertical dashed line called the lifeline; the lifeline represents the object's existence over a certain period of time. An object symbol—a box with the object's name underlined—is placed at the head of each lifeline.

A thin rectangle, superimposed on the lifeline of an object, represents an activation of the object. An **activation** shows the time period during which the object performs an operation, either directly or through a call to some subordinate operation. The top of the rectangle, which is at the tip of an incoming message, indicates the initiation of the activation, and the bottom its completion.

Objects communicate with one another by sending messages. A message is shown as a solid arrow from the sending object to the receiving object. For example, the checkIfOpen message is represented by an arrow from the Registration Entry object to the Course Offering object. When the arrow points directly into an object box, a new instance of that object is created. Normally the arrow is drawn horizontally, but, in some situations (discussed later), you may have to draw a sloping message line.

Messages could be of different types (Eriksson and Penker, 1998; *UML Notation Guide*, 1997). Each type is indicated in a diagram by a particular type of arrowhead. A **synchronous message**, shown as a full, solid arrowhead, is one where the caller has to wait for the receiving object to complete executing the called operation before it itself can resume execution. An example of a synchronous message is checkIfOpen. When a Registration Entry object sends this message to a Course Offering object, the latter responds by executing an operation called checkIfOpen (same name as the message). After the execution of this operation is completed, control is transferred back to the calling operation within Registration Entry with a return value, "true" or "false."

A synchronous message always has an associated return message. The message may provide the caller with some return value(s), or simply acknowledge to the caller that the operation called has been successfully completed. We have not shown the return for the checkIfOpen message; it is implicit. We have explicitly shown the return for the existsPrereqs message from Registration Entry to Course. The tail of the return message is aligned with the base of the activation rectangle for the existsPrereqs operation. The message returns the list of prerequisites, if any, for the course in question. Return messages, if shown, unnecessarily clutter the diagram; you may just show the ones that help in understanding the sequence of interactions.

A **simple message** simply transfers control from the sender to the recipient without describing the details of the communication. In a diagram, the arrowhead for a simple message is drawn as a transverse tick mark. As we have seen, the return of a synchronous message is a simple message. The "open" message in Figure 20-24 is also a simple message; it simply transfers control to the Registration Window object.

An **asynchronous message**, shown as a half arrowhead in a sequence diagram, is one where the sender does not have to wait for the recipient to handle the message. The sender can continue executing immediately after sending the message. Asynchronous messages are common in concurrent, real-time systems, where several objects operate in parallel. We do not discuss asynchronous messages further in this chapter.

Activation: The time period during which an object performs an operation.

Synchronous message: A type of message in which the caller has to wait for the receiving object to finish executing the called operation before it can resume execution itself.

Simple message: A message that transfers control from the sender to the recipient without describing the details of the communication.

Asynchronous message: A message in which the sender does not have to wait for the recipient to handle the message.

Designing a Use Case with a Sequence Diagram

Let us now see how we can design use cases. We will draw a sequence diagram for an instance of the class registration use case, one in which the course has prerequisites. A description of this scenario is provided below.

1. Registration Clerk opens the registration window and enters the registration information (student and class).

2. Check if the class is open.

3. If the class is open, check if the course has any prerequisites.

4. If the course has prerequisites, then check if the student has taken all of those prerequisites.

5. If the student has taken those prerequisites, then register the student for the class, and increment the class size by one.

6. Check if the class is full; if not, do nothing.

7. Display the confirmed registration in the registration window.

The diagram of Figure 20-24 shows the sequence of interactions for this scenario. In response to the "open" message from Registration Clerk (external actor), the registration window pops up on the screen and the registration information is entered. This creates a new Registration Entry object, which then sends a checkIfOpen message to the Course Offering object (representing the class the student wants to register for). There are two possible return values: true or false. In this scenario, the assumption is that the class is open. We have therefore placed a guard condition, checkIfOpen = "true," on the message existsPrereqs. The guard condition ensures that the message will be sent only if the class is open. The return value is a list of prerequisites; the return is shown explicitly in the diagram.

For this scenario, the fact that the course has prerequisites is captured by the guard condition, existsPrereqs = "true." If this condition is satisfied, the Registration Entry object sends a checkPrereqs message, with "prereqs" as an argument, to the Student object to determine if the student has taken those prerequisites. If the student has taken all the prerequisites, the Registration Entry object creates an object called "a New Registration," which denotes a new registration.

Next, a New Registration sends a message called incrementClassSize to Course Offering in order to increase the class size by one. The incrementClassSize operation within Course Offering then calls upon isClassFull, another operation within the same object; this is known as self-delegation (Fowler, 2000). Assuming that the class is not full, the isClassFull operation returns control to the calling operation with a value of "false." Next, the incrementClassSize operation completes and relinquishes control to the calling operation within "a New Registration."

Finally, on receipt of the return message from a New Registration, the Registration Entry object destroys itself (the destruction is shown with a large X) and sends a confirmation of the registration to the registration window. Note that Registration Entry is not a persistent object; it is created on the fly to control the sequence of interactions and is deleted as soon as the registration is completed. In between, it calls several other operations within other objects by sequencing the following messages: checkIfOpen, existsPrereqs, checkPrereqs, and new. Hence Registration Entry may be viewed as a control object (Jacobson et al., 1992).

Apart from the Registration Entry object, "a New Registration" is also created during the time period captured in the diagram. The messages that created these objects are represented by arrows pointing directly toward the object symbols. For example, the arrow representing the message called "new" is connected to the object symbol for "a New Registration." The lifeline of such an object begins when the message that creates it is received (the dashed vertical line is hidden behind the activation rectangle).

As we discussed before, the Registration Entry object is destroyed at the point marked by X. The lifeline of this object, therefore, extends from the point of creation to the point of destruction. For objects that are neither created nor destroyed during the time period captured in the diagram—for example, Course Offering, Course, and Student—the lifelines extend from the top to the bottom of the diagram.

Figure 20-25 shows the sequence diagram for a slightly different scenario, where a student registers for a course without any prerequisites. Notice that the guard condition to be satisfied for creating a New Registration, existsPrereqs = "false," is different from that in the previous scenario. Also, because there is no need to check if the student has taken the prerequisites, there is no need to send the checkPrereqs message to Student. The Student object, therefore, does not participate in this scenario.

There is another difference between this scenario and the previous one. In this scenario, when the incrementClassSize operation within Course Offering calls isClassFull, the value returned is "true." Before returning control to "a New Registration," the incrementClassSize operation self-delegates again, this time calling setStatus to set the status of the class to "closed."

Both of the sequence diagrams we have seen so far are in instance form. In Figure 20-26, we present a sequence diagram in generic form. This diagram encompasses all possible combinations of conditions for the Prereq courses not completed

Figure 20-25
Sequence diagram for a class registration scenario without prerequisites

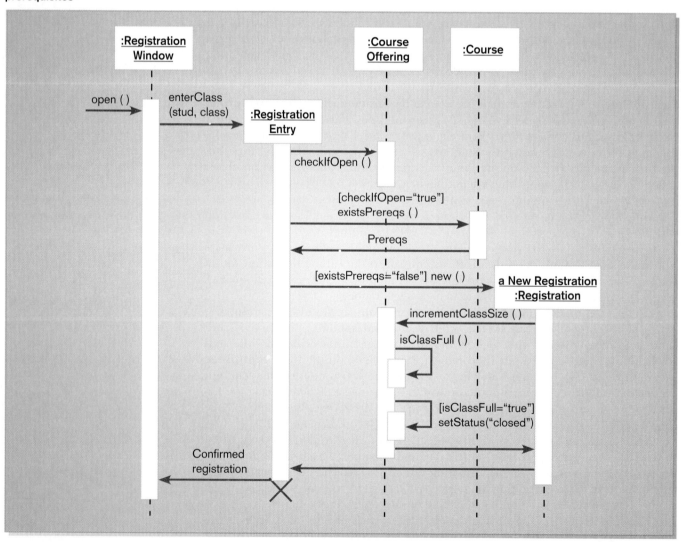

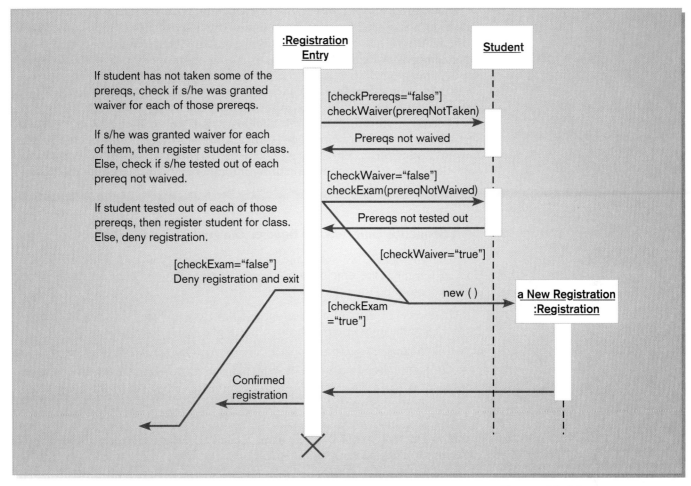

If student has not taken some of the prereqs, check if s/he was granted waiver for each of those prereqs.

If s/he was granted waiver for each of them, then register student for class. Else, check if s/he tested out of each prereq not waived.

If student tested out of each of those prereqs, then register student for class. Else, deny registration.

Figure 20-26
A generic sequence diagram for the "Prereq courses not completed" use case

use case (see Figure 20-2). Because this use case is an extension to the Class registration use case, we have not shown the Registration Window object. It is assumed that the Registration Entry object has already been created by the original use case. To improve understandability, we have provided textual description in the left margin. You may provide such description in either the left or the right margin, but try to align the text horizontally with the corresponding element in the diagram. The contents of the use case are described as follows:

1. If the student has not taken one or more of the prerequisites for the course he or she wants to register for, check if the student has been granted a waiver for each of those prerequisites.

2. If waiver was not granted for one or more of the prerequisites not taken, then check if the student tested out each of those prerequisites by taking an exam.

3. If the student did not test out for any of those prerequisites, then deny registration. Otherwise, register the student for the class and provide a confirmation.

Because this use-case extension pertains only to those registration situations where a student has not taken the prerequisite courses, we have placed a guard condition, checkPrereqs = "false," on the checkWaiver message from Registration Entry to Student. This message invokes the checkWaiver operation within Student to find if the student has been granted waivers on all the prerequisites he or she has not taken. Note that the operation has to be applied to each of the prerequisites not taken. The iteration is described in the text on the left margin.

The diagram also exhibits branching, with multiple arrows leaving a single point. Each branch is labeled by a guard condition. The first instance of branching is based on the value returned by the checkWaiver operation. If checkWaiver = "true," the system creates a New Registration object, bypassing other operations. If checkWaiver = "false"—meaning that some of the prerequisites in question were not waived—Registration Entry sends another message, checkExam, to Student to check if he or she tested out of each of the prerequisite courses not waived.

There is another instance of branching at this point. If checkExam = "false," Registration Entry sends a message (to Registration Window), denying the registration and exiting the system. We have deliberately bent the message line downward to show that none of the other remaining interactions takes place. If checkExam = "true," then "a New Registration" is created.

A Sequence Diagram for Hoosier Burger

In Figure 20-27, we show another sequence diagram, in generic form, for Hoosier Burger's Hire employee use case (see Figure 20-3). The description of the use case is given on the following page.

1. On receipt of an application for a job at Hoosier Burger, the data relating to the applicant is entered through the application entry window.

2. The manager opens the application review window and reviews the application.

3. If the initial review is negative, the manager discards the application and conveys the rejection decision to the applicant. No further processing of the application is involved.

4. If the initial review is positive, then the manager sets up a date and time to interview the applicant. The manager also requests the references specified in the application for recommendation letters.

5. The manager interviews the candidate and enters the additional information gathered during the interview into the application file.

6. When the recommendation letters come in, the manager is ready to make a decision. First, he or she prepares a summary of the application. Based on the summary, he or she then makes a decision. If the decision is to reject the candidate, the application is discarded and the decision is conveyed to the applicant. The processing of the application comes to an end.

7. If the decision is to hire the candidate, a potential employee file is created and all relevant information about the candidate (e.g., name, social security number, birthdate, address, phone, etc.) is entered into this file. The hiring decision is conveyed to the applicant.

In the sequence diagram for this use case, we have explicitly shown Manager as an external actor. The branching after the return value from the review message is received represents the two options Manager has. If review = "+ve," then an object called "an Interview" is created through the setup message shown in the upper branch. We have shown the arguments to the message—date and time—because their values are required to set up an interview. Notice that if review = "–ve" (lower branch), the discard message is sent to destroy the Application. The operations in between, for example, enterInfo, prepareSummary, and so forth, are completely bypassed.

Note that within the "an Interview" object created by the setup operation, there is another operation called collectInfo, which is invoked when the object receives the collectInfo message from Application Review Window. The operation collects all the relevant information during the interview, and enters this information into an Application. After "an Interview" receives a successful return message (not shown) from "an Application," it self destructs, as there is no need for it anymore.

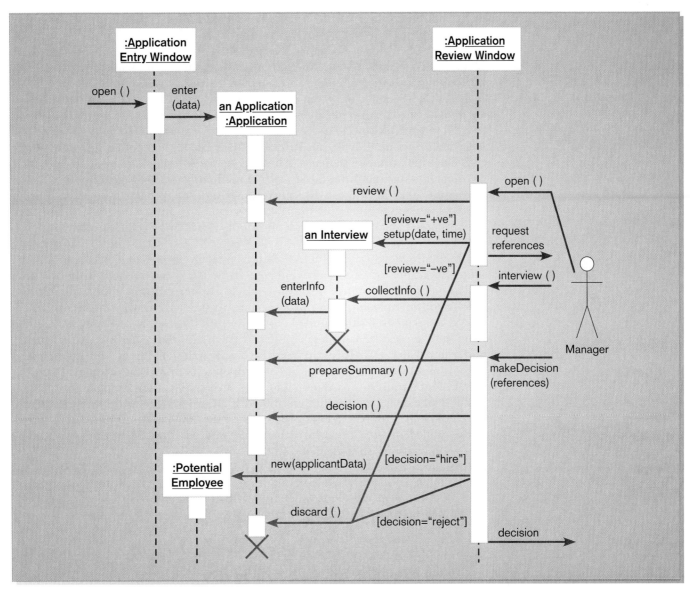

Figure 20-27
Sequence diagram for Hoosier Burger's Hire employee use case

Next, Manager sends a makeDecision message, which invokes a corresponding operation within Application Review Window. This operation first sends a prepareSummary message to "an Application," followed by another called "decision" to the same object. There is branching again at this point, depending on the return value. If decision = "hire," then a message called "new" is sent to create an instance of Potential Employee, which stores the relevant applicant data. If decision = "reject," the discard operation destroys "an Application." In either case, the decision is conveyed to the applicant.

Summary of Sequence Diagrams

In this section, we showed you how to design use cases by drawing sequence diagrams. A sequence diagram is an invaluable tool for specifying and understanding the flow of control. When coding the system, sequence diagrams help you to effectively and easily capture the dynamic aspects of the system by implementing the operations, messages, and the sequencing of those messages in the target programming language.

PROCESS MODELING: ACTIVITY DIAGRAMS

Activity diagram: Shows the conditional logic for the sequence of system activities needed to accomplish a business process.

As you have seen, a sequence diagram shows how objects interact over time to accomplish a specific system function or activity; and a state diagram shows how an object progresses through different states. An **activity diagram** shows the conditional logic for the sequence of system activities needed to accomplish a business process. An individual activity may be manual or automated, and often represents the actions needed to move an object between states. Further, each activity is the responsibility of a particular organizational unit. Thus, an activity diagram is another view or model of a system, which combines aspects of both sequence and state diagrams, and is similar to a data flow diagram from the structured methodology (see Chapter 8).

Figure 20-28 illustrates a typical customer ordering process for a stock-to-order business, such as a catalog or Internet sales company. Interactions with other business processes, such as replenishing inventory, forecasting sales, or analyzing profitability, are not shown.

In Figure 20-28 each column, called a swimlane, represents the organizational unit responsible for certain activities. The vertical axis is time, but without a time scale (that is, the distance between symbols implies nothing about the absolute amount of time passing). The process starts at the large dot. Activities are represented by rectangles with rounded corners. A fork means that several parallel, independent sequences of activities are initiated (such as after the Receive Order activity) and a join (such as before the Send Invoice activity) signifies that independent streams of activities now must all reach completion to move on to the next step.

A branch indicates conditional logic. For example, after available inventory is pulled from stock, it must be determined if all ordered items were found. If not, Purchasing must prepare a back order. After either the back order inventory arrives and is pulled or the original order could be filled completely, the process flow merges to continue on to the Send Order activity.

An activity diagram clearly shows parallel and alternative behaviors (Fowler, 2000), and is a good way to document work or document flows through an organization. However, objects are obscured and the links between objects are not shown. An activity diagram can be used to show the logic of a use case.

ANALYSIS VERSUS DESIGN

NET SEARCH
An advantage to the object-oriented approach is that UML is a standard for object-oriented systems modeling. Visit http://www.prenhall.com/hoffer to complete an exercise on this topic.

We have introduced you to a variety of diagramming techniques for capturing different aspects of a system. At the beginning of the chapter, we also described the activities involved in the various phases of the object-oriented development life cycle. You might wonder which diagrams belong to the analysis phase and which diagrams belong to the design phase. The answer is that, in general, the same type of diagram may apply to different phases. However, as you move from analysis to design, you need to refine the diagrams based on the requirements of the implementation environment. During design, you need to consider technical details relating to the target hardware, operating system, programming language, and database management system, and refine the models that you developed earlier accordingly.

All the diagrams you have seen so far are useful for analyzing a system. Use-case diagrams help you to generate a requirements analysis model, which captures the functionalities of the system as seen by external actors. Class diagrams help you to analyze the objects in the problem domain (business objects) and describe their static structure. During analysis, you model the dynamic aspects of the system using state and interaction diagrams.

When you move to design, you start with the existing set of analysis models and keep adding technical details. For example, you might add several interface classes to your class diagrams to model the windows that you will later implement using a

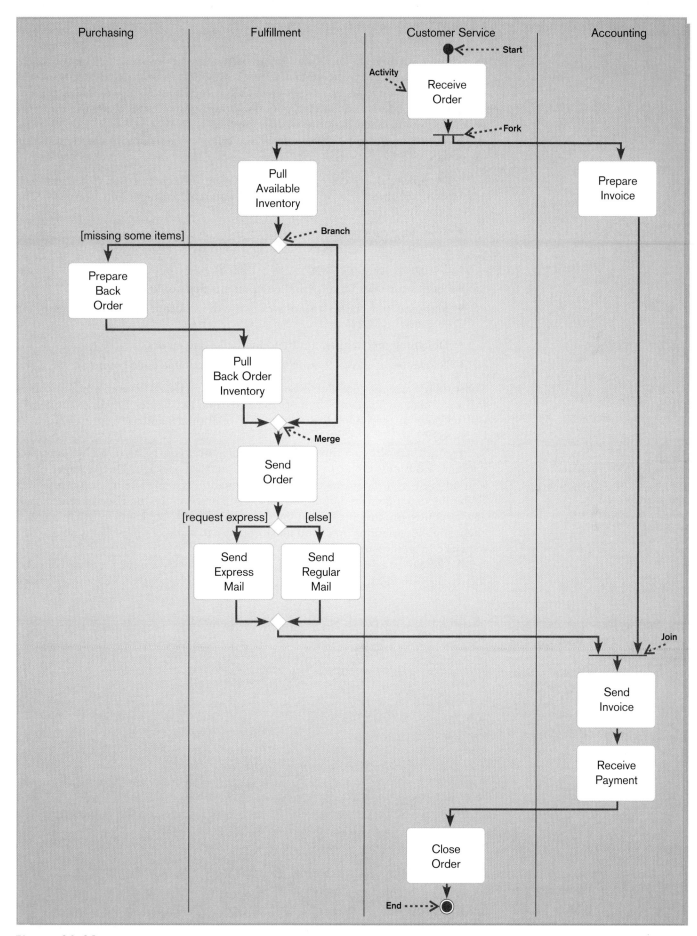

Figure 20-28
Activity diagram for customer order processing

GUI development tool. You would define all the operations in detail, specifying the procedures, signatures, and return values completely. If you decide to use a relational DBMS, you need to map the object classes and relationships to tables, primary keys and foreign keys. The models generated during the design phase will therefore be much more detailed than the analysis models.

Eriksson and Penker (1998) identify the following activities that are typical during design:

- Designing the architecture of the system in terms of functional *packages*, such as those for user interface, database storage and retrieval, and communication
- Modeling concurrent processes by using active classes, multiple threads of control, asynchronous messages, and so forth
- Designing user interfaces
- Identifying class libraries and components that could be reused
- Mapping the classes to constructs in the chosen DBMS for storage and retrieval
- Designing mechanisms for handling faults and errors
- Allocation of classes to source code and executable code components

These design activities parallel those performed using the structured methodologies, many of which are explained in more detail starting in Chapter 12. For example, you will see how to map an E-R model into a relational database representation in Chapter 12.

Figure 20-29 shows a three-layered architecture, consisting of a User Interface package, a Business Objects package, and a Database package. The **packages** represent different generic subsystems of an information system. The dashed arrows represent the dependencies among the packages. For example, the User Interface package depends on the Business Objects package; the packages participate in a client-supplier relationship. If you make changes to some of the business objects, the interface (e.g., screens) might change.

A package consists of a group of classes. Classes within a package are cohesive, that is, they are tightly coupled. The packages themselves should be loosely coupled

Package: A set of cohesive, tightly coupled classes representing a subsystem.

Figure 20-29
An example of UML Packages and dependencies

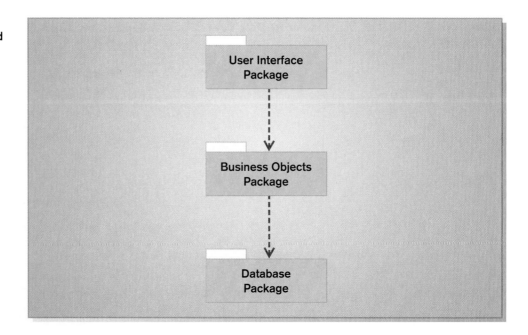

so that changes in one package do not affect the other packages a great deal. In the architecture of Figure 20-29, the User Interface package contains all the windows, the Business Objects package contains the problem domain objects that you identified during analysis, and the Database package contains a Persistence class for data storage and retrieval. In the university registration system that we considered earlier, the User Interface package could include Microsoft Windows class libraries for developing different types of Windows. The Business Objects package would include all the domain classes, such as Student, Course, Course Offering, Registration, and so forth. If you are using an SQL server, the classes in the Database package would contain operations for data storage, retrieval, and update (all using SQL commands).

During design, you would also refine the other analysis models. For example, you may need to show the interaction between a new window object you introduced during design and the other existing objects in a sequence diagram. Also, once you have selected a programming language, for each of the operations shown in the sequence diagram, you should provide the exact names that you would be using in the program, along with the names of all the arguments.

In addition to the types of diagrams you have seen so far, there are two other types of diagrams—component diagrams and deployment diagrams—that are pertinent during the design phase. A **component diagram** shows the software components or modules and their dependencies. For example, you can draw a component diagram showing the modules for source code, binary code, and executable code and their dependency relationships. Figure 20-30 shows a component diagram for the university registration system. In this figure, three software components have been identified: Class Scheduler, Class Registration, and GUI (graphical user interface). The small circles in the diagram represent interfaces. The "registration" interface, for example, is used to register a student for a class, and the schedule update interface is used for updating a class schedule. We have also shown two dependencies. The GUI component depends on the features of the registration interface. The Class Registration component depends on the existing class schedule, which is retrieved via the schedule retrieval interface.

Another type of diagram, a deployment diagram (not illustrated), shows how the software components, processes, and objects are deployed into the physical architecture of the system. It shows the configuration of the hardware units (e.g., computers,

Component diagram: Shows the software components or modules and their dependencies.

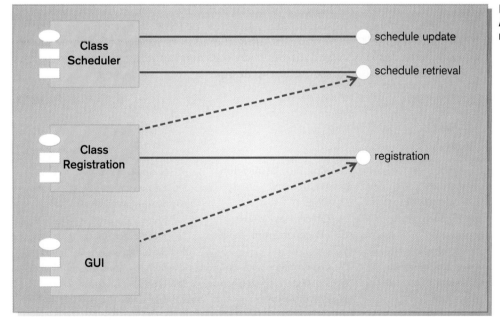

Figure 20-30
A component diagram for class registration

communication devices, etc.) and how the software (components, objects, etc.) is distributed across the units. For example, a deployment diagram for the university registration system might show the topology of nodes in a client-server architecture, and the deployment of the Class Registration component to a Windows NT Server and of the GUI component to client workstations.

When the design phase is complete, you move on to the implementation phase where you code the system. If you are using an object-oriented programming language, translating the design models to code should be relatively straightforward. Programming of the system is followed by testing. The system is developed after going through multiple iterations, with each new iteration providing a better version of the system. The models that you developed during analysis, design, and implementation are navigable in both directions. You should be able to easily trace how a class defined during analysis is coded in the implementation phase. Similarly, you should be able to reverse engineer from a piece of code into the corresponding object class.

Summary

In this chapter, we introduced you to the object-oriented modeling approach, which is becoming increasingly popular because it allows you to thoroughly model a real-world application—both in terms of its data and processes—using a common underlying representation. We described the activities involved in the different phases of the object-oriented development life cycle, and emphasized the seamless nature of the transitions that an object-oriented model undergoes as it evolves through the different phases, from analysis to design to implementation. This is in sharp contrast to other modeling approaches, such as structured analysis and design, which lack a common underlying representation and, therefore, suffer from abrupt and disjoint model transitions.

We showed how to develop different types of models using the Unified Modeling Language (UML) notation. You saw how to develop a requirements analysis model through use-case diagrams. A use-case diagram shows the interaction between external actors and actions performed within a system. You then saw how to model the static structure of objects in the problem domain using class diagrams. Objects have both states (conditions during the life of an object) and behavior (operations), which are encapsulated within. Associations exist between objects, just as relationships exist between entities in an entity-relationship diagram. Next, we introduced state diagrams for capturing the dynamic state transitions of an object. State transitions occur when events trigger changes in an object. We also showed how to develop a dynamic model of interactions among objects using sequence diagrams. A sequence diagram shows the passing of messages between objects. Messages activate operations within objects, thus causing the system to perform needed functions. An activity diagram shows the conditional logic for the sequence of system activities needed to accomplish a business process. We provided guidelines on transforming the analysis models to design models, and introduced other diagramming techniques that are useful during design.

Key Terms

1. Abstract class
2. Abstract operation
3. Activation
4. Activity diagram
5. Actor
6. Aggregation
7. Association
8. Association class
9. Association role
10. Asynchronous message
11. Behavior
12. Class diagram
13. Class-scope attribute
14. Component diagram
15. Composition
16. Concrete class
17. Constructor operation
18. Encapsulation
19. Event
20. Method
21. Multiple classification
22. Multiplicity
23. Object
24. Object class (class)
25. Object diagram
26. Operation
27. Overriding
28. Package
29. Polymorphism
30. Query operation
31. Scope operation
32. Sequence diagram
33. Simple message
34. State
35. State diagram
36. State transition
37. Synchronous message
38. Update operation
39. Use case
40. Use-case diagram

Match each of the key terms above with the definition that best fits it.

_____ An external entity that interacts with the system.

_____ A complete sequence of related actions initiated by an actor.

_____ A diagram that depicts the use cases and actors for a system.

_____ An entity that has a well-defined role in the application domain, and has state, behavior, and identity.

_____ Encompasses an object's properties (attributes and relationships) and the values those properties have.

_____ Represents how an object acts and reacts.

_____ A set of objects that share a common structure and a common behavior.

_____ Shows the static structure of an object-oriented model.

_____ A graph of instances that are compatible with a given class diagram.

_____ A function or a service that is provided by all the instances of a class.

_____ The technique of hiding the internal implementation details of an object from its external view.

_____ An operation that creates a new instance of a class.

_____ An operation that accesses the state of an object but does not alter the state.

_____ An operation that alters the state of an object.

_____ A relationship between object classes.

_____ An operation that applies to a class rather than an object instance.

_____ The end of an association where it connects to a class.

_____ Indicates how many objects participate in a given relationship.

_____ An association that has attributes or operations of its own, or that participates in relationships with other classes.

_____ A class that has no direct instances.

_____ A class that can have direct instances.

_____ An attribute of a class that specifies a value common to an entire class, rather than a specific value for an instance.

_____ Defines the form or protocol of the operation, but not its implementation.

_____ The implementation of an operation.

_____ The same operation may apply to two or more classes in different ways.

_____ The process of replacing a method inherited from a superclass by a more specific implementation of that method in a subclass.

_____ An object is an instance of more than one class.

_____ A *part-of* relationship between a component object and an aggregate object.

_____ A part object that belongs to only one whole object, and that lives and dies with the whole.

_____ Changes in the attributes of an object or in the links an object has with other objects.

_____ Something that takes place at a certain point in time.

_____ A model of the states of an object and the events that cause the object to change from one state to another.

_____ Depicts the interactions among objects during a certain period of time.

_____ The time period during which an object performs an operation.

_____ A type of message in which the caller has to wait for the receiving object to finish executing the called operation before it can resume execution itself.

_____ A message that transfers control from the sender to the recipient without describing the details of the communication.

_____ A message in which the sender does not have to wait for the recipient to handle the message.

_____ Shows the conditional logic for the sequence of system activities needed to accomplish a business process.

_____ A set of cohesive, tightly coupled classes representing a subsystem.

_____ Shows the software components or modules and their dependencies.

Review Questions

1. Contrast the following terms:
   a. actor; use case
   b. extends relationship; uses relationship
   c. object class; object
   d. attribute; operation
   e. state; behavior
   f. operation; method
   g. query operation; update operation
   h. abstract class; concrete class
   i. class diagram; object diagram
   j. association; aggregation
   k. generalization; aggregation

l. aggregation; composition

m. overriding for extension; overriding for restriction

n. state; event

o. event; action

p. entry action; exit action

q. state diagram; sequence diagram

r. generic sequence diagram; instance sequence diagram

s. synchronous message; asynchronous message

t. sequence diagram; activity diagram

2. State the activities involved in each of the following phases of the object-oriented development life cycle: object-oriented analysis, object-oriented design, and object-oriented implementation.

3. Compare and contrast the object-oriented analysis and design models with the structured analysis and design models.

4. State the conditions under which the "extend" relationship should be used between use cases.

5. State the conditions under which the "include" relationship should be used between use cases.

6. Give an example of an abstract use case. Your example should involve at least two other use cases and show how they are related to the abstract use case.

7. State the conditions under which a designer should model an association relationship as an association class.

8. Using a class diagram, give an example for each of the following types of relationships: unary, binary, and ternary. Specify the multiplicities for all the relationships.

9. Add role names to the association relationships you identified in Review Question 8.

10. Add operations to some of the classes you identified in Review Question 8.

11. Give an example of generalization. Your example should include at least one superclass and three subclasses, and a minimum of one attribute and one operation for each of the classes. Indicate the discriminator and specify the semantic constraints among the subclasses.

12. If the diagram you developed for Review Question 11 does not contain an abstract class, extend the diagram by adding an abstract class that contains at least one abstract operation. Also, indicate which features of a class are inherited by other classes.

13. Using (and, if necessary, extending) the diagram from your solution to Review Question 12, give an example of polymorphism.

14. Give an example of aggregation. Your example should include at least one aggregate object and three component objects. Specify the multiplicities at each end of all the aggregation relationships.

15. Give an example of state transition. Your example should show how the state of the object undergoes a transition based on some event.

Problems and Exercises

1. The use-case diagram shown in Figure 20-2 captures the student billing function, but does not contain any function for accepting tuition payment from students. Revise the diagram to capture this functionality. Also, express some common behavior among two use cases in the revised diagram by using "include" relationships.

2. Suppose that the employees of the university are not billed for tuition. Their spouses do not get a full tuition waiver, but pay for only 25 percent of the total tuition. Extend the use-case diagram of Figure 20-2 to capture these situations.

3. An auto rental company wants to develop an automated system that would handle car reservations, customer billing, and car auctions. Usually a customer reserves a car, picks it up, and then returns it after a certain period of time. At the time of pick up, the customer has the option to buy or waive collision insurance on the car. When the car is returned, the customer receives a bill and pays the specified amount. In addition to renting out cars, every six months or so, the auto rental company auctions the cars that have accumulated over 20,000 miles. Draw a use-case diagram for capturing the requirements of the system to be developed. Include an abstract use case for capturing the common behavior among any two use cases. Extend the diagram to capture corporate billing, where corporate customers are not billed directly; rather, the corporations they work for are billed and payments are made sometime later.

4. Draw a use-case diagram for the faculty tenure application and review problem as follows: A faculty member applies for tenure in his or her sixth year by submitting a portfolio summarizing his or her work. In rare circumstances he or she can come up for tenure earlier than the sixth year, but only if the faculty member has permission of the department chair and college dean. New professors, who have worked at other universities before taking their current jobs, rarely, if ever, come in with tenure. They are usually asked to undergo one "probationary" year during which they are evaluated and only can then be granted tenure. Top administrators coming in to a new university job, however, can often negotiate for retreat rights that enable them to become a tenured faculty member should their administrative post end. These retreat arrangements generally have to be approved by faculty. The tenure review process begins with an evaluation of the candidate's portfolio by a committee of faculty within the candidate's department. The committee then writes a recommendation on tenure and sends it to the department's chairperson who then makes a recommendation, and passes the portfolio and recommendation on to the next level, a college-wide faculty committee.

This committee does the same as the department committee and passes its recommendation, the department's recommendation, and the portfolio on to the next level, a university-wide faculty committee. This committee does the same as the other two committees and passes everything on to the provost. The provost then writes his or her own recommendation and passes everything to the president, the final decision maker. This process, from the time the candidate creates his or her portfolio until the time the president makes a decision, can take an entire academic year. The focus of the evaluation is on research, which could be grants, presentations, and publications, though preference is given for empirical research that has been published in top-ranked, refereed journals and where the publication makes a contribution to the field. The candidate must also do well in teaching and services (i.e., to the university, community, or to the discipline) but the primary emphasis is on research.

5. Draw a class diagram for the system to be developed for the auto rental company explained in Problem and Exercise 3. The diagram should identify all the relevant classes from the problem domain, such as cars, customers, and so forth, and show their attributes and operations, as well as the relationships among the classes. In drawing the diagram, you can make any assumptions, as long as they are realistic.

6. Draw a class diagram, showing the relevant classes, attributes, operations, and relationships, for each of the following situations (if you believe that you need to make additional assumptions, clearly state them for each situation):

   a. A company has a number of employees. The attributes of Employee include employeeID (primary key), name, address, and birthdate. The company also has several projects. Attributes of Project include projectName and startDate. Each employee may be assigned to one or more projects, or may not be assigned to a project. A project must have at least one employee assigned, and may have any number of employees assigned. An employee's billing rate may vary by project, and the company wishes to record the applicable billing rate for each employee when assigned to a particular project. At the end of each month, the company mails a check to each employee who has worked on a project during that month. The check amount is based on the billing rate and the hours logged for each project assigned to the employee.

   b. A university has a large number of courses in its catalog. Attributes of Course include courseNumber (primary key), courseName, and Units. Each course may have one or more different courses as prerequisites, or may have no prerequisites. Similarly, a particular course may be a prerequisite for any number of courses, or may not be a prerequisite for any other course. The university adds or drops a prerequisite for a course only when the director for the course makes a formal request to that effect.

   c. A laboratory has several chemists who work on one or more projects. Chemists also may use certain kinds of equipment on each project. Attributes of Chemist include name and phoneNo. Attributes of Project include projectName and startDate. Attributes of Equipment include serialNo and cost. The organization wishes to record assignDate—that is, the date when a given equipment item was assigned to a particular chemist working on a specified project—as well as totalHours, that is, the total number of hours the chemist has used the equipment for the project. The organization also wants to track the usage of each type of equipment by a chemist. It does so by computing the average number of hours the chemist has used that equipment on all assigned projects. A chemist must be assigned to at least one project and one equipment item. A given equipment item need not be assigned, and a given project need not be assigned either a chemist or an equipment item.

   d. A college course may have one or more scheduled sections, or may not have a scheduled section. Attributes of Course include courseID, courseName, and units. Attributes of Section include sectionNumber and semester. The value of sectionNumber is an integer (such as 1 or 2), which distinguishes one section from another for the same course, but does not uniquely identify a section. There is an operation called findNumSections, which finds the number of sections offered for a given course in a given semester.

   e. A hospital has a large number of registered physicians. Attributes of Physician include physicianID (primary key) and specialty. Patients are admitted to the hospital by physicians. Attributes of Patient include patientID (primary key) and patientName. Any patient who is admitted must have exactly one admitting physician. A physician may optionally admit any number of patients. Once admitted, a given patient must be treated by at least one physician. A particular physician may treat any number of patients, or may not treat any patients. Whenever a patient is treated by a physician, the hospital wishes to record the details of the treatment, by including the date, time, and results of the treatment.

7. A student, whose attributes include studentName, Address, phone, and age, may engage in multiple campus-based activities. The university keeps track of the number of years a given student has participated in a specific activity and, at the end of each academic year, mails an activity report to the student showing his participation in various activities. Draw a class diagram for this situation.

8. Draw a class diagram for some organization that you are familiar with—Boy Scouts/Girl Scouts, sports team, and so forth. In your diagram, indicate names for at least four association roles.

9. Draw a use-case diagram, as well as a class diagram, for the following situation (state any assumptions you believe you have to make in order to develop a complete diagram): Stillwater Antiques buys and sells one of a kind antiques of all kinds (for example, furniture, jewelry, china, and clothing). Each item is uniquely identified by an item number, and is also characterized by a description, asking price, condition, and open-ended comments. Stillwater works with many different individuals, called clients, who sell items to and buy items from the store. Some clients only sell items to Stillwater, some only buy items, and some others both sell and buy. A client is identified by a client number, and is also

described by a client name and client address. When Stillwater sells an item in stock to a client, the owners want to record the commission paid, the actual selling price, sales tax (tax of zero indicates a tax-exempt sale), and date sold. When Stillwater buys an item from a client, the owners want to record the purchase cost, date purchased, and condition at time of purchase.

10. Draw a use-case diagram, as well as a class diagram, for the following situation (state any assumptions you believe you have to make in order to develop a complete diagram): The H. I. Topi School of Business operates international business programs in ten locations throughout Europe. The School had its first of 9000 graduates in 1965. The School keeps track of each graduate's student number, name, country of birth, current country of citizenship, current name, current address, and the name of each major the student completed (each student has one or two majors). In order to maintain strong ties to its alumni, the School holds various events around the world. Events have a title, date, location, and type (for example, reception, dinner, or seminar). The School needs to keep track of which graduates have attended which events. For an attendance by a graduate at an event, a comment is recorded about information School officials learned from that graduate at that event. The School also keeps in contact with graduates by mail, e-mail, telephone, and fax interactions. As with events, the School records information learned from the graduate from each of these contacts. When a School official knows that he or she will be meeting or talking to a graduate, a report is produced showing the latest information about that graduate and the information learned during the past two years from that graduate from all contacts and events the graduate attended.

11. Draw a class diagram for the following problem. A nonprofit organization depends on a number of different types of persons for its successful operation. The organization is interested in the following attributes for all of these persons: social security number (ssn), name, address, and phone. There are three types of persons who are of greatest interest: employees, volunteers, and donors. In addition to the attributes for a person, an employee has an attribute called dateHired, and a volunteer has an attribute called skill. A donor is a person who has donated one or more items to the organization. An item, specified by a name, may have no donors, or one or more donors. When an item is donated, the organization records its price, so that at the end of the year, it can identify the top 10 donors.

There are persons other than employees, volunteers, and donors who are of interest to the organization, so that a person need not belong to any of these three groups. On the other hand, at a given time a person may belong to two or more of these groups (for example, employee and donor).

12. A bank has three types of accounts: checking, savings, and loan. Following are the attributes for each type of account:
CHECKING: Acct_No, Date_Opened, Balance, Service_Charge
SAVINGS: Acct_No, Date_Opened, Balance, Interest_Rate
LOAN: Acct_No, Date_Opened, Balance, Interest_Rate, Payment

Assume that each bank account must be a member of exactly one of these subtypes. At the end of each month the bank computes the balance in each account and mails a statement to the customer holding that account. The balance computation depends on the type of the account. For example, a checking account balance may reflect a service charge, whereas a savings account balance may include an interest amount. Draw a class diagram to represent the situation. Your diagram should include an abstract class, as well as an abstract operation for computing balance.

13. An organization has been entrusted with developing a Registration and Title system that maintains information about all vehicles registered in a particular state. For each vehicle that is registered with the office, the system has to store the name, address, telephone number of the owner, the start date and end date of the registration, plate information (issuer, year, type, and number), sticker (year, type, and number), and registration fee. In addition, the following information is maintained about the vehicles themselves: the number, year, make, model, body style, gross weight, number of passengers, diesel-powered (yes/no), color, cost, and mileage. If the vehicle is a trailer, diesel-powered and number of passengers are not relevant. For travel trailers, the body number and length must be known. The system needs to maintain information on the luggage capacity for a car, maximum cargo capacity and maximum towing capacity for a truck, and horsepower for a motorcycle. The system issues registration notices to owners of vehicles whose registrations are due to expire after two months. When the owner renews the registration, the system updates the registration information on the vehicle.

a. Develop a static object model by drawing a class diagram that shows all the object classes, attributes, operations, relationships, and multiplicities. For each operation, show its argument list.

b. Draw a state diagram that captures all the possible states of a Vehicle object, right from the time the vehicle was manufactured until it goes to the junkyard. In drawing the diagram, you may make any necessary assumptions, as long as they are realistic.

c. Select any state or event from the high-level state diagram that you have drawn and show its fine structure (substates and their transitions) in a lower-level diagram.

d. One of the use cases for this application is "Issue registration renewal notice," which is performed once every day. Draw a sequence diagram, in generic form, showing all possible object interactions for this use case.

14. Draw a state diagram showing all the states and transitions of a Car object for the auto rental system described in Problem and Exercise 3. The diagram should show the events that cause the transitions, along with any associated actions. If necessary, show the activities associated with a state.

15. One of the use cases for the auto rental system in Problem and Exercise 3 is "Car reservation." Draw a sequence diagram, in instance form, to describe the sequence of interactions for each of the following scenarios of this use case:

a. Car is available during the specified time period.

b. No car in the desired category (e.g., compact, midsize, etc.) is available during the specified time period.

16. Draw an activity diagram for the following employee hiring process. Compare your answer to this question to Problem and Exercise 11 in Chapter 8. In what ways are activity diagrams and DFDs different?

Projects, Inc., is an engineering firm with approximately 500 engineers of different types. New employees are hired by the personnel manager based on data in an application form and evaluations collected from other managers who interview the job candidates. Prospective employees may apply at any time. Engineering managers notify the personnel manager when a job opens and list the characteristics necessary to be eligible for the job. The personnel manager compares the qualifications of the available pool of applicants with the characteristics of an open job, then schedules interviews between the manager in charge of the open position and the three best candidates from the pool. After receiving evaluations on each interview from the manager, the personnel manager makes the hiring decision based upon the evaluations and applications of the candidates and the characteristics of the job, and then notifies the interviewees and the manager about the decision. Applications of rejected applicants are retained for one year, after which time the application is purged. When hired, a new engineer completes a nondisclosure agreement, which is filed with other information about the employee.

Field Exercises

1. Interview a friend or family member to elicit from them common examples of superclass/subclass relationships. You will have to explain the meaning of this term and provide a common example, such as: PROPERTY: RESIDENTIAL, COMMERCIAL; or BONDS: CORPORATE, MUNICIPAL. Use the information they provide to construct a class diagram segment and present it to this person. Revise if necessary until it seems appropriate to you and your friend or family member.

2. Visit two local small businesses, one in the service sector and one in manufacturing. Interview employees from these organizations to obtain examples of both superclass/subclass relationships and operational business rules (such as "a customer can return merchandise only if they have a valid sales slip"). In which of these environments is it easier to find examples of these constructs? Why?

3. Ask a database administrator or database or systems analyst in a local company to show you an EER (or E-R) diagram for one of the organization's primary databases. Translate this diagram into a class diagram.

4. Interview a systems analyst in a local company that uses object-oriented programming and system development tools. Ask to see any analysis and design diagrams they have drawn of the database and applications. Compare these diagrams to the ones in this chapter. What differences do you see? What additional features and notations are used and what are their purpose?

References

Booch, G. 1994. *Object-Oriented Analysis and Design with Applications.* 2nd. ed. Redwood City, CA: Benjamin/Cummings.

Coad, P., and E. Yourdon. 1991a. *Object-Oriented Analysis.* 2nd. ed. Englewood Cliffs, NJ: Prentice-Hall.

Coad, P., and E. Yourdon. 1991b. *Object-Oriented Design.* Englewood Cliffs, NJ: Prentice-Hall.

Eriksson, H., and M. Penker. 1998. *UML Toolkit.* New York: John Wiley.

Fowler, M. 2000. *UML Distilled: A Brief Guide to the Object Modeling Language.* 2nd. ed. Reading, MA: Addison-Wesley.

Jacobson, I., M., Christerson, P. Jonsson, and G. Overgaard. 1992. *Object-Oriented Software Engineering: A Use-Case Driven Approach.* Reading, MA: Addison-Wesley.

Rumbaugh, J., M. Blaha, W. Premerlani, F. Eddy, and W. Lorensen. 1991. *Object-Oriented Modeling and Design.* Englewood Cliffs, NJ: Prentice-Hall.

UML Document Set. 1997. Version 1.0 (January). Santa Clara: Rational Software Corp.

UML Notation Guide. 1997. Version 1.0 (January). Santa Clara: Rational Software Corp.

Glossary of Terms

Note: Chapter in which term is defined is listed parentheses after the definition.

Abstract class A class that has no direct instances, but whose descendants may have direct instances. (20) *See also* Concrete class.

Abstract operation Defines the form or protocol of the operation, but not its implementation. (20) *See also* Operation.

Acceptance testing The process whereby actual users test a completed information system, the end result of which is the users' acceptance of it. (17) *See also* System testing.

Action stubs That part of a decision table that lists the actions that result for a given set of conditions. (9)

Activation The time period during which an object performs an operation. (20)

Activity diagram Shows the conditional logic for the sequence of system activities needed to accomplish a business process. (20)

Actor An external entity that interacts with the system (similar to an external entity in data flow diagramming). (20) *See also* Use case.

Adaptive maintenance Changes made to a system to evolve its functionality to changing business needs or technologies. (18) *See also* Corrective maintenance, Perfective maintenance, Preventive maintenance.

Affinity clustering The process of arranging planning matrix information so the clusters of information with some predetermined level or type of affinity are placed next to each other on a matrix report. (5)

Aggregation A *part-of* relationship between a component object and an aggregate object. (20) *See also* Composition.

Alpha testing User testing of a completed information system using simulated data. (17) *See also* Beta testing, System testing.

Analysis The third phase of the SDLC in which the current system is studied and alternative replacement systems are proposed. (1)

Analysis tools CASE tools that enable automatic checking for incomplete, inconsistent, or incorrect specifications in diagrams, forms, and reports. (4)

Application independence The separation of data and the definition of data from the applications that use these data. (1)

Application program interface (API) Software building blocks that are used to assure that common system capabilities like user interfaces and printing are standardized as well as modules for facilitate the data exchange between clients and servers. (16)

Application server A "middle-tier" software and hardware combination that lies between the Web server and the corporate network and systems. (11) A computing server where data analysis functions primarily reside. (16)

Application service provider Organizations that host and run computer applications for other companies, typically on a per use or license basis. (11)

Application software Computer software designed to support organizational functions or processes. (1)

Association A relationship among instances of object classes. (20) *See also* Association role.

Association class An association that has attributes or operations of its own, or that participates in relationships with other classes. (20)

Association role The end of an association where it connects to a class. (20) *See also* Multiplicity.

Associative entity An entity type that associates the instances of one or more entity types and contains attributes that are peculiar to the relationship between those entity instances; also called a gerund. (10)

Asynchronous message A message in which the sender does not have to wait for the recipient to handle the message. (20) *See also* Sequence Diagram, Simple message, Synchronous message.

Attribute A named property or characteristic of an entity that is of interest to the organization. (10) *See also* Multivalued attribute.

Balancing   The conservation of inputs and outputs to a data flow diagram process when that process is decomposed to a lower level. (8)

Baseline modules   Software modules that have been tested, documented, and approved to be included in the most recently created version of a system. (18)

Baseline Project Plan (BPP)   A major outcome and deliverable from the project initiation and planning phase, which contains the best estimate of a project's scope, benefits, costs, risks, and resource requirements. (6)

Behavior   Represents how an object acts and reacts. (20)

Beta testing   User testing of a completed information system using real data in the real user environment. (17) *See also* Alpha testing.

Binary relationship   A relationship between instances of two entity types. This is the most common type of relationship encountered in data modeling. (10)

Bottom-up planning   A generic information systems planning methodology that identifies and defines IS development projects based upon solving operational business problems or taking advantage of some business opportunities. (5) *See also* Corporate strategic planning, Top-down planning.

Boundary   The line that marks the inside and outside of a system and that sets off the system from its environment. (2)

Build routines   Guidelines that list the instructions to construct an executable system from the baseline source code. (18)

Business case   The justification for an information system, presented in terms of the tangible and intangible economic benefits and costs, and the technical and organizational feasibility of the proposed system. (6)

Business Process Reengineering (BPR)   The search for, and implementation of, radical change in business processes to achieve breakthrough improvements in products and services. (7) *See also* Disruptive technologies, Key business processes.

Business rules   Specifications that preserve the integrity of a conceptual or logical data model. (10)

Candidate key   An attribute (or combination of attributes) that uniquely identifies each instance of an entity type. (10) *See also* Identifier, Primary key.

Cardinality   The number of instances of entity B that can (or must) be associated with each instance of entity A. (10)

Cascading style sheets (CSS)   A set of style rules that tells a Web browser how to present a document. (16)

Class diagram   Shows the static structure of an object-oriented model the object class, their internal structure, ad the relationships in which they participate. (20)

Class-scope attribute   An attribute of a class which specifies a value common to an entire class, rather than a specific value for an instance. (20)

Client   The (front-end) portion of the client/server database system that provides the user interface and data manipulation functions. (16)

Client/server architecture   A LAN-based computing environment in which a central database server or engine performs all database commands sent to it from client workstations, and application programs on each client concentrate on user interface functions. (16) *See also* File server.

Closed system   A system that is cut off from its environment and does not interact with it. (2) *See also* Open system.

Closed-ended questions   Questions in interviews and on questionnaires that ask those responding to choose from a set of specified responses. (7) *See also* Open-ended questions.

Code generators   CASE tools that enable the automatic generation of program and database definition code directly from the design documents, diagrams, forms, and reports stored in the repository. (4)

Cohesion   The extent to which a system or subsystem performs a single function. (2)

Command language interaction   A human-computer interaction method where users enter explicit statements into a system to invoke operations. (14)

Competitive strategy   The method by which an organization attempts to achieve its mission and objectives. (5)

Component   An irreducible part of aggregation of parts that make up a system, also called a subsystem. (2) *See also* Interrelated components.

Component diagram   Shows the software components or modules and their dependencies. (20) *See also* Package.

Composition   A part object that belongs to only one whole object, and that lives and dies with the whole. (20) *See also* Aggregation.

Computer-aided software engineering (CASE)   Software tools that provide automated support for some portion of the systems development process. (3) *See also* Cross lifecycle CASE, I-CASE, Lower CASE, Upper CASE.

Computing infrastructure   All the resources and practices required to help people adequately use computer systems to do their primary work. (17) *See also* Support.

Conceptual data model   A detailed model that captures the overall structure of organizational data while being independent of any database management system or other implementation consideration. (10) *See also* Entity-relationship data model, Logical data model.

Concrete class A class that can have direct instances. (20) *See also* Abstract class.

Condition stubs That part of a decision table that lists the conditions relevant to the decision. (9)

Configuration management The process of assuring that only authorized changes are made to a system. (18)

Constraint A limit to what a system can accomplish. (2)

Constructor operation An operation that creates a new instance of a class. (20)

Context development A method that helps you to better understand how a system fits within existing business activities and data. (16)

Context diagram An overview of an organizational system that shows the system boundaries, external entities that interact with the system, and the major information flows between the entities and the system. (8) *See also* Data flow diagram.

Cookie crumbs A technique for showing a user where they are in a Website by placing a series of "tabs" on a Web page that shows a user where they are and where they have been. (14)

Corporate strategic planning An ongoing process that defines the mission, objectives, and strategies of an organization. (5)

Corrective maintenance Changes made to a system to repair flaws in its design coding, or implementation. (18) *See also* Adaptive maintenance, Corrective maintenance, Perfective maintenance, Preventive maintenance.

Coupling The extent to which subsystems depend on each other. (2)

Critical path The shortest time in which a project can be completed. (3)

Critical path scheduling A scheduling technique whose order and duration of a sequence of task activities directly affect the completion date of a project. (3)

Cross life-cycle CASE CASE tools designed to support activities that occur *across* multiple phases of the systems development life cycle. (4) *See also* Lower CASE, Upper CASE.

Cross referencing A feature performed by a data dictionary that enables one description of a data item to be stored and accessed by all individuals so that a single definition for a data item is established and used. (4)

Customization Internet sites that allow a user to customize information to their personal preferences. (16) *See also* Personalization.

Data Raw facts about people, objects, and events in an organization. (1)

Data couple A diagrammatic representation of the data exchanged between two modules in a structure chart. (15) *See also* Flag.

Data dictionary The repository of all data definitions for all organizational applications. (4)

Data flow Data in motion, moving from one place in a system to another. (1)

Data flow diagram A picture of the movement of data between external entities and the processes and data stores within a system. (8)

Data mart A data warehouse that is limited in scope; whose data are obtained by selecting and (where appropriate) summarized data from the enterprise data warehouse. (16)

Data store Data at rest, which may take the form of many different physical representations. (8)

Data type A coding scheme recognized by system software for representing organizational data. (12)

Data warehouse A subject-oriented, integrated, time-variant, nonvolatile collection of data used in support of management decision making. (16) *See also* Data mart, Enterprise data warehouse (EDW).

Database A shared collection of logically related data designed to meet the information needs of multiple users in an organization. (1)

Database engine The (back-end) portion of the client/server database system running on the server and providing database processing and shared access function. (16)

Data-oriented approach An overall strategy of information systems development that focuses on the ideal organization of data rather than where and how data are used. (1) *See also* Process-oriented approach.

Decision table A matrix representation of the logic of a decision, which specifies the possible conditions for the decision and the resulting actions. (9) *See also* Action stubs, Condition stubs, Rules.

Decision tree A graphical representation of a decision situation in which decision situation points (nodes) are connected together by arcs (one for each alternative on a decision) and terminate in ovals (the action that is the result of all of the decisions made on the path leading to that oval). (9)

Default value A value a field will assume unless an explicit value is entered for that field. (12)

Degree The number of entity types that participate in a relationship. (10)

Deliverable An end product in a phase of the SDLC. (3)

Denormalization   The process of splitting or combining normalized relations into physical tables based on affinity of use of rows and fields. (12)

Design   The fourth phase of the SDLC in which the description of the recommended solution is converted into logical and then physical system specifications. (1)

Design strategy   A high-level statement about the approach to developing an information system. It includes statements on the system's functionality, hardware and system software platform, and method for acquisition. (11)

Desk checking   A testing technique in which the program code is sequentially executed manually by the reviewer. (17)

DFD completeness   The extent to which all necessary components of a data flow diagram have been included and fully described. (8)

DFD consistency   The extent to which information contained on one level of a set of nested data flow diagrams is also included on other levels. (8)

Diagramming tools   CASE tools that support the creation of graphical representations of various system elements such as process flow, data relationships, and program structures. (4)

Dialogue   The sequence of interaction between a user and a system. (14)

Dialogue diagramming   A formal method for designing and representing human-computer dialogues using box and line diagrams. (14)

Direct installation   Changing over from the old information system to a new one by turning off the old system when the new one is turned on. (17)

Discount rate   The rate of return used to compute the present value of future cash flows. (6)

Disjoint rule   Specifies that if an entity instance of the supertype is a member of one subtype, it cannot simultaneously be a member of any other subtype. (10) *See also* Overlap rule.

Disruptive technologies   Technologies that enable the breaking of long-held business rules that inhibit organizations from making radical business changes. (7) *See also* Business Process Reengineering (BPR).

Documentation generators   CASE tools that enable the easy production of both technical and user documentation in standard formats. (4)

Domain   The set of all data types and values that an attribute can assume. (10)

Domain naming system (BIND)   A method for translating Internet domain names into Internet Protocol (IP) addresses. (16)

Drop-down menu   A menu positioning method that places the access point of the menu near the top line of the display; when accessed, menus open by dropping down onto the display. (14) *See also* Pop-up menu.

Economic feasibility   A process of identifying the financial benefits and costs associated with a development project. (6)

Electronic commerce   Internet-based communication to support day-to-day business activities. (5)

Electronic data interchange (EDI)   The use of telecommunications technologies to directly transfer business documents between organizations. (5)

Electronic performance support system (EPSS)   Component of a software package or application in which training and educational information is embedded. An EPSS can take several forms, including a tutorial, and expert system shell, and hypertext jumps to reference material. (17)

Encapsulation   The technique of hiding the internal implementation details of an object from its external view. (20) *See also* Abstract operation, Constructor operation, Method, Polymorphism, Query operation, Scope operation, Update operation.

Enterprise data warehouse (EDW)   A centralized, integrated data warehouse that is the control point and single source of all data made available to end users for decision-support applications throughout the entire organization. (16) *See also* Data mart, Data warehouse.

Enterprise resource planning (ERP) systems   A system that integrates individual traditional business functions into a series of modules so that a single transaction occurs seamlessly within a single information system rather than several separate systems. (11)

Entity instance (instance)   A single occurrence of an entity type. (10)

Entity type   A collection of entities that share common properties or characteristics. (10)

Entity-relationship data model (E-R model)   A detailed, logical representation of the entities, associations, and data elements for an organization or business area. (10) *See also* Conceptual data model.

Entity-relationship diagram (E-R diagram)   A graphical representation of an E-R model. (10)

Environment   Everything external to a system that interacts with the system. (2)

Event   Something that takes place at a certain point in time; it is a noteworthy occurrence that triggers a state transition. (20)

Extensible style language (XSL)   A specification for separating style from content when generating HTML documents. (16)

External documentation System documentation that includes the outcome of structured diagramming techniques such as data flow and entity-relationship diagrams. (17) *See also* Internal documentation.

Extranet Internet-based communication to support business-to-business activities. (5) *See also* Internet, Intranet.

Feasibility study Determines if the information system makes sense for the organization from an economic and operational standpoint. (3)

Field The smallest unit of named application data recognized by system software. (12)

File organization A technique for physically arranging the records of a file. (12) *See also* Hashed file organization, Indexed file organization, Sequential file organization.

File server A device that manages file operations and is shared by each client PC attached to a LAN. (16) *See also* Client/server architecture.

Flag A diagrammatic representation of a message passed between two modules. (15) *See also* Data couple.

Foreign key An attribute that appears as a nonprimary key attribute in one relation and as a primary key attribute (or part of a primary key) in another relation. (12) *See also* Referential integrity.

Form A business document that contains some predefined data and may include some areas where additional data are to be filled in. An instance of a form is typically based on one database record. (13)

Form and report generators CASE tools that support the creation of system forms and reports in order to prototype how systems will "look and feel" to users. (4)

Form interaction A highly intuitive human-computer interaction method whereby data fields are formatted in a manner similar to paper-based forms. (14)

Formal system The official way a system works as described in organizational documentation. (7) *See also* Informal system.

Functional decomposition An iterative process of breaking the description of a system down into finer and finer detail, which creates a set of charts in which one process on a given chart is explained in greater detail on another chart. (8)

Functional dependency A particular relationship between two attributes. For a given relation, attribute B is functionally dependent on attribute A if, for every valid value of A, that value of A uniquely determines the value of B. The functional dependence of B on A is represented by A→B. (12) *See also* Partial dependency, Transitive dependency.

Functional hierarchy diagram A picture of the various tasks performed in a business and how they are related to each other. The tasks are broken down into their various parts, and all the parts are shown in the same representation. (8)

Gantt chart A graphical representation of a project that shows each task as a horizontal bar whose length is proportional to its time for completion. (3)

Gap analysis The process at discovering discrepancies between two or more sets of DFDs or discrepancies within a single DFD. (8)

Hashed file organization The address for each row is determined using an algorithm. (12)

Help desk A single point of contact for all user inquiries and problems about a particular information system or for all users in a particular department. (17) *See also* Computing infrastructure, Information center, Support.

Homonyms A single attribute name that is used for two or more different attributes. (12)

HTTP (hypertext transfer protocol) A communications protocol for exchanging information on the Internet. (16)

I-CASE An automated systems development environment that provides numerous tools to create diagrams, forms, and reports; provides analysis, reporting, and code generation facilities; and seamlessly shares and integrates data across and between tools. (4)

Icon Graphical pictures that represent specific functions within a system. (14) *See also* Object-based interaction.

Identifier A candidate key that has been selected as the unique, identifying characteristic for an entity type. (10) *See also* Candidate key, Primary key.

Implementation The fifth phase of the SDLC in which the information system is coded, tested, installed, and supported in an organization. (1)

Incremental Commitment A strategy in systems analysis and design in which the project is reviewed after each phase and continuation of the project is rejustified in each of these reviews. (5)

Index A table used to determine the location of rows in a file that satisfy some condition. (12)

Indexed file organization The rows are stored either sequentially or nonsequentially, and an index is created that allows software to locate individual rows. (12)

Indifferent condition In a decision table, a condition whose value does not affect which actions are taken for two or more rules. (9)

Informal system The way a system actually works. (7) *See also* Formal system.

Information center An organizational unit whose mission is to support users in exploiting information technology. (17) *See also* Computing infrastructure, Help desk, Support.

Information   Data that have been processed and presented in a form suitable for human interpretation, often with the purpose of revealing trends or patterns. (1)

Information repository   Automated tools used to manage and control access to organizational business information and application portfolio as components within a comprehensive repository. (4)

Information systems analysis and design   The complex organizational process whereby computer-based information systems are developed and maintained. (1)

Information systems planning (ISP)   An orderly means of accessing the information needs of an organization and defining the systems, databases, and technologies that will best satisfy those needs. (5) *See also* Corporate strategic planning, Top-down planning.

Informational systems   Systems designed to support decision making based on stable point-in-time or historical data. (16)

Inheritance   The property that occurs when entity types or object classes are arranged in a hierarchy and each entity type or object class assumes the attributes and methods of its ancestors; that is, those higher up in the hierarchy. Inheritance allows new but related classes to be derived from existing classes. (1)

Input   Whatever a system takes from its environment in order to fulfill its purpose. (2)

Inspections   A testing technique in which participants examine program code for predictable language-specific errors. (17)

Installation   The organizational process of changing over from the current information system to a new one. (17) *See also* Direct installation, Parallel installation, Phased installation, Single location installation.

Intangible benefit   A benefit derived from the creation of an information system that cannot be easily measured in dollars or with certainty. (6) *See also* Tangible benefit.

Intangible cost   A cost associated with an information system that cannot be easily measured in terms of dollars or with certainty. (6) *See also* Tangible cost.

Integrated CASE.   *See* I-CASE.

Integration depth   A measurement of how far into the existing technology infrastructure a system penetrates. (16)

Integration testing   The process of bringing together all of the modules that a program comprises for testing purposes. Modules are typically integrated in a top-down, incremental fashion. (17)

Interface   In systems theory, point of contact where a system meets its environment or where subsystems meet each other. (2) In human-computer interaction, a method by which users interact with information systems. (14)

Internal documentation   System documentation that is part of the program source code or is generated at compile time. (17)

Internet   A large worldwide network of networks that use a common protocol to communicate with each other; a global computing network to support business-to-consumer electronic commerce. (5) *See also* Extranet, Intranet.

Interrelated components   Dependence of one subsystem on one or more subsystems. (2)

Intranet   Internet-based communication to support business activities within a single organization. (5) *See also* Internet, Extranet.

JAD session leader   The trained individual who plans and leads Joint Application Design sessions. (7)

Joint-Application Design (JAD)   A structured process in which users, managers, and analysts work together for several days in a series of intensive meetings to specify or review system requirements. (1)

Key business processes   The structured, measured set of activities designed to produce a specific output for a particular customer or market. (7) *See also* Business Process Reengineering (BPR).

Legal and contractual feasibility   The process of assessing potential legal and contractual ramifications due to the construction of a system. (6)

Level-0 diagram   A data flow diagram that represents a system's major processes, data flows, and data stores at a high level of detail. (8)

Level-n diagram   A DFD that is the result of n nested decomposition of a series of subprocesses from a process on a Level-0 diagram. (8)

Lightweight graphics   The use of small simple images to allow a Web page to more quickly be displayed. (13)

Local area network (LAN)   The cabling, hardware, and software used to connect workstations, computers, and file servers located in a confined geographical area (typically within one building or campus). (16)

Logical design   The part of the design phase of the SDLC in which all functional features of the system chosen for development in analysis are described independently of any computer platform. (1) *See also* Physical design.

Logical system description   Description of a system that focuses on the system's function and purpose without regard to how the system will be physically implemented. (2)

Lower CASE   CASE tools designed to support the implementation and maintenance phases of the systems development life cycle. (4) *See also* Upper CASE.

Maintainability   The ease with which software can be understood, corrected, adapted, and enhanced. (18)

Maintenance The final phase of the SDLC in which an information system is systematically repaired and improved. (1) Changes made to a system to fix or enhance its functionality. (18) *See also* Adaptive maintenance, Corrective maintenance, Perfective maintenance, Preventive maintenance.

Mean time between failures (MTBF) A measurement of error occurrences that can be tracked over time to indicate the quality of a system. (18)

Menu interaction A human-computer interaction method where a list of system options is provided and a specific command is invoked by user selection of a menu option. (14) *See also* Drop-down menu, Pop-up menu.

Method The implementation of an operation. (20)

Middleware A combination of hardware, software, and communication technologies that bring together data management, presentation, and analysis into a three-tiered client/server environment. (16)

Mission statement A statement that makes it clear what business a company is in. (5)

Modularity Dividing a system up into chunks or modules of a relatively uniform size. (2) *See also* Cohesion, Coupling.

Module A self-contained component of a system, defined by function. (15)

Multiple classification An object is an instance of more than one class. (20)

Multiplicity Indicates how many objects participate in a given relationship. (20)

Multivalued attribute An attribute that may take on more than one value for each entity instance. (10)

Natural language interaction A human-computer interaction method whereby inputs to and outputs from a computer-based application are in a conventional speaking language such as English. (14)

Nominal Group Technique A facilitated process that supports idea generation by groups. At the beginning of the process, group members work alone to generate ideas, which are then pooled under the guidance of a trained facilitator. (7)

Normalization The process of converting complex data structures into simple, stable data structures. (12)

Null value A special field value, distinct from 0, blank, or any other value, that indicates that the value for the field is missing or otherwise unknown. (12)

Object A structure that encapsulates (or packages) attributes and methods that operate on those attributes. An object is an abstraction of a real-world thing in which data and processes are placed together to model the structure and behavior of the real-world object. (1) An entity that has a well-defined role in the application domain, and has state, behavior, and identity. (20) *See also* Object class, Object diagram.

Object class A logical grouping of objects that have the same (or similar) attributes and behaviors (methods). (1, 20) *See also* Abstract class, Concrete class.

Object diagram A graph of instances that are compatible with a given class diagram. (20)

Object-based interaction A human-computer interaction method where symbols are used to represent commands or functions. (14) *See also* Icon.

Objective statements A series of statements that express an organization's qualitative and quantitative goals for reaching a desired future position. (5)

Object-oriented analysis and design (OOAD) Systems development methodologies and techniques based on objects rather than data or processes. (1)

One-time cost A cost associated with project start-up and development, or system start-up. (6)

On-line analytical processing (OLAP) The use of graphical software tools that provide complex analysis of data stored in a database. (16) *See also* Data mart, Enterprise data warehouse.

On-line transaction processing (OLTP) The immediate automated responses to the requests of users. (16)

Open system A system that interacts freely with its environment, taking input and returning output. (2)

Open-ended questions Questions in interviews and on questionnaires that have no prespecified answers. (7) *See also* Closed-ended questions.

Operation A function or a service that is provided by all the instances of a class. (20)

Operational feasibility The process of assessing the degree to which a proposed system solves business problems or takes advantage of business opportunities. (6)

Operational systems Systems that are used to interact with customers and run a business in real time. (16)

Organizational breadth A measurement that tracks the core business functions affected by a system. (16)

Output Whatever a system returns to its environment in order to fulfill it purpose. (2)

Outsourcing The practice of turning over responsibility of some to all of an organization's information systems application and operations to an outside firm. (11)

Overlap rule Specifies that an entity instance can simultaneously be a member of two (or more) subtypes. (10) *See also* Disjoint rule.

Overriding The process of replacing a method inherited from a superclass by a more specific implementation of that method in a subclass. (20)

Package A set of cohesive, tightly coupled classes representing a subsystem. (20) *See also* Component diagram.

Parallel installation Running the old information system and the new one at the same time until management decides the old system can be turned off. (17)

Partial specialization rule Specifies that an entity instance of the supertype is allowed not to belong to any subtype. (10) *See also* Total specialization rule.

Perfective maintenance Changes made to a system to add new features or to improve performance. (18) *See also* Adaptive maintenance, Corrective maintenance, Preventive maintenance.

Personalization Providing Internet content to users based upon knowledge of that customer. (16) *See also* Customization.

PERT chart A diagram that depicts project tasks and their interrelationships. PERT stands for Program Evaluation Review Technique. (3)

Phased installation Changing from the old information system to the new one incrementally, starting with one or a few functional components and then gradually extending the installation to cover the whole new system. (17)

Physical design The part of the design phase of the SDLC in which the logical specifications of the system from logical design are transformed into technology-specific details from which all programming and system construction can be accomplished. (1) *See also* Logical design.

Physical file A named set of table rows stored in a contiguous section of secondary memory. (12)

Physical system description Description of a system that focuses on how the system will be materially constructed. (2)

Physical table A named set of rows and columns that specifies the fields in each row of the table. (12)

Pointer A field of data that can be used to locate a related field or row of data. (12)

Political feasibility The process of evaluating how key stakeholders within the organization view the proposed system. (6)

Polymorphism The same operation may apply to two or more classes in different ways. (20)

Pop-up menu A menu positioning method that places a menu near the current cursor position. (14) *See also* Drop-down menu.

Present value The current value of a future cash flow. (6)

Preventive maintenance Changes made to a system to avoid possible future problems. (18) *See also* Adaptive maintenance, Corrective maintenance, Perfective maintenance.

Primary key An attribute whose value is unique across all occurrences of a relation. (12) *See also* Candidate key, Identifier, Secondary key.

Primitive DFD The lowest level of decomposition for a data flow diagram. (8)

Process The work or actions performed on data so that they are transformed, stored, or distributed. (8)

Processing logic The steps by which data are transformed or moved and a description of the events that trigger these steps. (1)

Process-oriented approach An overall strategy to information systems development that focuses on how and when data are moved through and changed by an information system. (1) *See also* Data-oriented approach.

Project A planned undertaking of related activities to reach an objective that has a beginning and an end. (3)

Project closedown The final phase of the project management process that focuses on bringing a project to an end. (3)

Project execution The third phase of the project management process in which the plans created in the prior phases (project initiation and planning) are put into action. (3)

Project identification and selection The first place of the SDLC in which an organization's total information system needs are identified, analyzed, prioritized, and arranged. (1)

Project initiation The first phase of the project management process in which activities are performed to assess the size, scope, and complexity of the project and to establish procedures to support later project activities. (3)

Project initiation and planning The second phase of the SDLC in which a potential information systems project is explained and an argument for continuing or not continuing with the project is presented; a detailed plan is also developed for conducting the remaining phases of the SDLC for the proposed system. (1)

Project management A controlled process of initiating, planning, executing, and closing down a project. (3)

Project manager A systems analyst with a diverse set of skills—management, leadership, technical, conflict management, and customer relationship—who is responsible for initiating, planning, executing, and closing down a project. (62)

Project planning The second phase of the project management process, which focuses on defining clear, discrete activities and the work needed to complete each activity within a single project. (3)

Project workbook An on-line or hard-copy repository for all project correspondence, inputs, outputs, deliverables, procedures, and standards that is used for performing project audits, orienting new team members, communicating with management and customers, identifying future projects, and performing post-project reviews. (3) *See also* Repository.

Prototyping An iterative process of systems development in which requirements are converted to a working system that is continually revised through close work between an analyst and users. (1) *See also* Rapid Application Development (RAD).

Pseudocode A method for representing the instructions in a module with language very similar to computer programming code. (15)

Purpose The overall goal or function of a system. (2)

Query operation An operation that accesses the state of an object but does not alter the state. (20) See also Abstract operation, Query operation, Update operation.

Rapid Application Development (RAD) Systems development methodology created to radically decrease the time needed to design and implement information systems. RAD relies on extensive user involvement, Joint Application Design sessions, prototyping, integrated CASE tools, and code generators. (19) *See also* Prototyping.

Recurring cost A cost resulting from the ongoing evolution and use of a system. (6)

Recursive foreign key A foreign key in a relation that references the primary key values of that same relation. (12)

Reengineering Automated tools that read program source code as input, perform an analysis of the program's data and logic, and then automatically, or interactively with a systems analyst, alter an existing system in an effort to improve its quality or performance. (4) *See also* CASE.

Referential integrity An integrity constraint specifying that the value (or existence) of an attribute in one relation depends on the value (or existence) of the same attribute in another relation. (12) *See also* Foreign key.

Relational database model Data represented as a set of related tables or relations. (12)

Relations A named, two-dimensional table of data. Each relation consists of a set of named columns and an arbitrary number of unnamed rows. (12)

Relationship An association between the instances of one or more entity types that is of interest to the organization. (10)

Repeating group A set of two or more multivalued attributes that are logically related. (10)

Report A business document that contains only predefined data; it is a passive document used solely for reading or viewing. A report typically contains data from many unrelated records or transactions. (13)

Repository A centralized database that contains all diagrams, forms, and report definitions, data structure, data definitions, process flows and logic, and definitions of other organizational and system components; it provides a set of mechanisms and structures to achieve seamless data-to-tool and data-to-data integration. (4) *See also* Data dictionary, I-CASE, Information repository, Project workbook.

Request for proposal (RFP) An RFP is a document provided to vendors to ask them to propose hardware and system software that will meet the requirements of your new system. (11)

Resources Any person, group of people, piece of equipment, or material used in accomplishing an activity. (3)

Reusability The ability to design software modules in a manner so that they can be used again and again in different systems without significant modification. (4)

Reverse engineering Automated tools that read program source code as input and create graphical and textual representations of program design-level information such as program control structures, data structures, logical flows, and data flow. (4) *See also* CASE.

Rules That part of a decision table that specifies which actions are to be followed for a given set of conditions. (9)

Scalable The ability to seamlessly upgrade the capabilities of the system through either hardware upgrades, software upgrades, or both. (11)

Schedule feasibility The process of assessing the degree to which the potential time frame and completion dates for all major activities within a project meet organizational deadlines and constraints for affecting change. (6)

Scope operation An operation that applies to a class rather than an object instance. (20)

Scribe The person who makes detailed notes of the happenings at a Joint Application Design session. (7)

Second normal form (2NF) A relation is in second normal form if every nonprimary key attribute is functionally dependent on the whole primary key. (12) *See also* Functional dependency, Partial dependency.

Secondary key One or a combination of fields for which more than one row may have the same combination of values. (12) *See also* Primary key.

Sequence diagram Depicts the interactions among objects during a certain period of time. (20)

Sequential file organization The rows in the file are stored in sequence according to a primary key value. (12)

Simple message A message that transfers control from the sender to the recipient without describing the details of the communication. (20) *See also* Asynchronous message, Sequence diagram, Synchronous message.

Single location installation   Trying out a new information system at one site and using the experience to decide if and how the new system should be deployed throughout the organization. (17)

Slack time   The amount of time that an activity can be delayed without delaying the project. (3)

Source/sink   The origin and/or destination of data, sometimes referred to as external entities. (8)

Stakeholder   A person who has an interest in an existing or new information system. (1)

State   Encompasses an object's properties (attributes and relationships) and the values those properties have. (20) *See also* Behavior.

State diagram   A model of the states of an object and the events that cause the object to change from one state to another. (20)

State transition   Changes in the attributes of an object or in the links an object has with other objects. (20)

Statement of Work (SOW)   Document prepared for the customer during project initiation and planning that describes what the project will deliver and outlines generally at a high level all work required to complete the project. (6)

Structure chart   Hierarchical diagram that shows how an information system is organized. (15)

Structured English   Modified form of the English language used to specify the logic of information system processes. Although there is no single standard, Structured English typically relies on action verbs and noun phrases and contains no adjectives or adverbs. (9)

Stub testing   A technique used in testing modules, especially where modules are written and tested in a top-down fashion, where a few lines of code are used to substitute for subordinate modules. (17)

Subtype   A subgrouping of the entities in an entity type that is meaningful to the organization and that shares common attributes or relationships distinct from other subgroupings. (10) *See also* Supertype.

Supertype   A generic entity type that has a relationship with one or more subtypes. (10) *See also* Subtype.

Support   Providing ongoing educational and problem-solving assistance to information system users. For in-house developed systems, support materials and jobs will have to be prepared or designed as part of the implementation process. (17) *See also* Computing infrastructure, Help desk, Information center.

Synchronous message   A type of message in which the caller has to wait for the receiving object to finish executing the called operation before it can resume execution itself. (20) *See also* Asynchronous message, Sequence diagram, Simple message.

Synonyms   Two different names that are used for the same attribute. (12)

System   An interrelated set of components, with an identifiable boundary, working together for some purpose. (2) *See also* Closed system, Open system.

System documentation   Detailed information about a system's design specifications, its internal workings, and its functionality. (17)

System librarian   A person responsible for controlling the checking out and checking in of baseline modules for a system when a system is being developed or maintained. (18)

System testing   The bringing together of all of the programs that a system comprises for testing purposes. Programs are typically integrated in a top-down, incremental fashion. (17) *See also* Acceptance testing, Alpha testing, Beta testing, Integration testing, Stub testing.

Systems analysis   The organizational role most responsible for the analysis and design of information systems. (1)

Systems development life cycle (SDLC)   The traditional methodology used to develop, maintain, and replace information systems. (1)

Systems development methodology   A standard process followed in an organization to conduct all the steps necessary to analyze, design, implement, and maintain information systems. (1)

Tangible benefit   A benefit derived from the creation of an information system that can be measured in dollars and with certainty. (6) *See also* Intangible benefit.

Tangible cost   A cost associated with an information system that can be measured in dollars and with certainty. (6) *See also* Intangible cost.

Technical feasibility   A process of assessing the development organization's ability to construct a proposed system. (6)

Template-based HTML   Templates to display and process common attributes of higher-level, more abstract items. (13)

Ternary relationship   A simultaneous relationship among instances of three entity types. (10)

Thin client   A client device designed so that most processing and data storage occurs on the server. (16)

Third normal form (3NF)   A relation is in third normal form (3NF) if it is in second normal form and there are no functional (transitive) dependencies between two (or more) nonprimary key attributes. (12)

Three-tiered client/server   Advanced client/server architecture in which there are three logical and distinct applications—data management, presentation, and analysis—which are combined to create a single information system. (16)

Top-down planning　A generic information systems planning methodology that attempts to gain a broad understanding of the information system needs of the entire organization. (5) *See also* Bottom-up planning.

Total specialization rule　Specifies that each entity instance of the supertype *must* be a member of some subtype in the relationship. (10) *See also* Partial specialization rule.

Triggering operation (trigger)　An assertion or rule that governs the validity of data manipulation operations such as insert, update, and delete. (10)

Unary relationship (recursive relationship)　A relationship between the instances of one entity type. (10)

Unit testing　Each module is tested alone in an attempt to discover any errors in its code. (17)

Update operation　An operation that alters the state of an object. (20) *See also* Abstract operation, Constructor operation, Query operation.

Upper CASE　CASE tools designed to support the information planning and the project identification and selection, project initiation and planning, analysis, and design phases of the systems development life cycle. (4) *See also* Lower CASE.

Usability　An overall evaluation of how a system performs in supporting a particular user for a particular task. (13)

Use case　A complete sequence of related actions initiated by an actor to accomplish a specific goal; it represents a specific way of using the system. (20)

Use case diagram　A diagram that depicts the use cases and actors for a system. (20)

User documentation　Written or other visual information about an application system, how it works, and how to use it. (17)

Value chain analysis　Analyzing an organization's activities to determine where value is added to products and/or services and the costs incurred for doing so; usually also includes a comparison with the activities, added value, and costs of other organizations for the purpose of making improvements in the organization's operations and performance. (5)

Walkthrough　A peer group review of any product created during the systems development process. Also called structured walkthrough. (6)

Web server　A computer that is connected to the Internet and that stores files written in HTML—Hypertext Markup Language—that are publicly available through an Internet connection. (11)

Well-structured relation (or table)　A relation that contains a minimum amount of redundancy and allows users to insert, modify, and delete the rows without errors or inconsistencies. (12) *See also* Normalization.

Work breakdown structure　The process of dividing the project into manageable tasks and logically ordering them to ensure a smooth evolution between tasks. (3)

XML (extensible markup language)　An Internet authoring language that allows designers to create customized tags, enabling the definition, transmission, validation, and interpretation of data between applications. (16)

Glossary of Acronyms

Note: Some acronyms are abbreviations for entries in the Glossary of Terms. For these and some other acronym entities, we list in parenthesis the chapter or appendix in which the associated term is defined or introduced. Other acronyms are generally used in the information systems field and are included here for your convenience.

ACM	Association for Computing Machinery (2)
AITP	Association of Information Technology Professionals
API	Application Program Interface (16)
ASP	Application Service Provider
ATM	Automated Teller Machine
AT&T	American Telephone & Telegraph
BEA	Break Even Analysis (6)
BEC	Broadway Entertainment Company (4)
BPP	Baseline Project Plan (6)
BPR	Business Process Reengineering (7)
BSP	Business Systems Planning (5)
B-to-B	Business-to-Business (5)
B-to-C	Business-to-Consumer (5)
CASE	Computer-Aided Software Engineering (4)
CCP	Certified Computing Professional (2)
CD	Compact Disk
CD-ROM	Compact Disk-Read Only Memory
CIO	Chief Information Officer
COBOL	COmmon Business Oriented Language
COCOMO	COnstruction COst MOdel
CRT	Cathode Ray Tube
C/S	Client/Server (16)
CSS	Cascading Style Sheets (16)
CTS	Customer Tracking System (6)
CUA	Common User Access

DB2	Data Base 2
DBMS	Database Management System
DCR	Design Change Request
DFD	Data Flow Diagram (8)
DSS	Decision Support System
DVD	Digital Video Disk
EC	Electronic Commerce (5)
EDI	Electronic Data Interchange (5)
EDS	Electronic Data Systems
EDW	Enterprise Data Warehouse (16)
EER	Extended Entity Relationship
EFT	Electronic Funds Transfer
EIS	Executive Information System
EPSS	Electronic Performance Support System (17)
E-R	Entity-Relationship
ERD	Entity-Relationship Diagram (10)
ERP	Enterprise Resource Planning (11)
ES	Expert System
ESS	ExecutiveSupport System
ET	Estimated Time (3)
ET	Entertainment Tracker (4)
EUC	End-User Computing (17)
FHD	Functional Hierarchy Diagram (8)
FORTRAN	FORmula TRANSlater
GDSS	Group Decision Support System
GSS	Group Support System (7)
GUI	Graphical User Interface
HTML	HyperText Markup Language
HTTP	HyperText Transfer Protocol (16)
IBM	International Business Machines

I-CASE	Integrated Computer-Aided Software Engineering (4)		**PV**	Present Value (6)
ID	Identifier		**PVF**	Pine Valley Furniture (3)
IDSS	Integrated Development Support System		**RAD**	Rapid Application Development (19)
IE	Information Engineering		**RAM**	Random Access Memory
IEF	Information Engineering Facility		**R&D**	Research and Development
I/O	Input/Output		**RFP**	Request for Proposal (11)
IS	Information System		**RFQ**	Request for Quote (11)
ISA	Information Systems Architecture		**ROI**	Return on Investment (6)
ISP	Information Systems Planning (5)		**ROM**	Read Only Memory
IT	Information Technology		**SDLC**	Systems Development Life Cycle (1)
ITAA	Information Technology Association of America		**SDM**	Systems Development Methodology (1)
IU	Indiana University		**SNA**	System Network Architecture
JAD	Joint Application Design (1)		**SOS**	Simplicity, Organize, Show (14)
KLOC	Thousand Lines of Code		**SOW**	Statement of Work (6)
LAN	Local Area Network (16)		**SPTS**	Sales Promotion Tracking System
MIS	Management Information System		**SQL**	Structure Query Language
MOTRS	Modular Over the Road System (19)		**SSR**	System Service Request (3)
M:N	Many-to-Many		**TPS**	Transaction Processing System
MRP	Material Requirements Planning		**TQM**	Total Quality Management
MTBF	Mean Time Between Failures (18)		**TVM**	Time Value of Money (6)
MTTR	Mean Time to Repair Defect		**UML**	Unified Modeling Language
NGT	Nominal Group Technique (7)		**VB**	Visual Basic
NPV	Net Present Value (6)		**VBA**	Visual Basic for Applications
OA	Office Automation		**WAP**	Wireless Application Protocol (16)
ODBC	Open DataBase Connectivity (16)		**WIP**	Work in Process
OLAP	On-line Analytical Processing (16)		**WML**	Wireless Markup Language (16)
OLTP	On-line Transaction Processing (16)		**XML**	Extensible Markup Language (16)
OO	Object-Oriented		**XSL**	Extensible Style Language (16)
OOAD	Object-Oriented Analysis and Design (1)		**1:1**	One-to-One
PC	Personal Computer		**1:*M***	One-to-Many
PERT	Project Evaluation and Review Technique (3)		**1NF**	First Normal Form (12)
PIP	Project Initiation and Planning (1)		**2NF**	Second Normal Form (12)
POS	Point-of-Sale		**3NF**	Third Normal Form (12)
			4GL	Fourth-Generation Language

Index

Note: *Italicized* page number indicates that term is defined on that page.

A

Abdel-Hamid, T. K., 66
Abrupt installation. *See* Direct installation
Abstract class, *676*
Abstract operation, *678–679*
Acceptance testing, *582–583*
Action stubs, *288*
Activation, *690*
Activity diagram, 285, *696*, 697
Actor, *661*
Adaptive maintenance, 569, *620*
Affinity clustering, *149*
Aggregation, *680–682*
  composition, 682
Alexander, M. B., 561
Alpha testing, *582–583*, 605–606
Alternative design strategy, 348–375
  Baseline Project Plan update, 367–372
  for Broadway Entertainment Company, Inc., 379–381
  constraints and, 352–353
  deliverables and outcomes, 351
  features in, 352
  generation issues, 351–363
  hardware and systems software issues, 361–362
  implementation issues, 362
  for Internet-based e-commerce, 373–375
  organizational issues, 362–363
  outsourcing and, 353–354
  process of, 350–351
  selection process, 350–351, 366–367
  sources of software and, 354–361
  weighted approach and, 366–367
Alternative generation and selection, 200, 363–365. *See also* Alternative design strategy
Analysis, *21*. *See also* Object-oriented analysis and design (OOAD)
  alternative design strategies and, 348–375
  blending with design (OOAD), 696–700
  data flow diagrams in, 257–265
  of procedures/documents, 215–220
  requirements determination, 200–201
  requirements structuring and, 306
  structured, 24
  See also Systems analysis
Analysis tools, 104–106, *105*
Analytical skills, for systems analyst, 13
Andrews, D. C., 620
Applegate, L. M., 178n, 354
Application independence, *10–11*
Application program, in client/server architecture, 532–533
Application program interface (API), *533*

Application server, *374, 535*
Application Server/Object Framework architecture, 375
Application service provider (ASP), 7, *357–358*
Application software, *5*, 22
Approved development platforms, 49–50
Aranow, E. B., 309
Architecture. *See* Client/server architecture; Multi-tier system architecture; Three-level architecture; Two-level architecture
Artificial intelligence, 117
Askren, S., 535
Assessment. *See* Evaluation criteria; specific types of assessment
Association, *668–672*
  binary, 668
  derived, 674
  generalization, 674–679
  multiplicity, 669–670
  n-ary, 668
  part-of relationship (*See* Aggregation)
  role, 668–669
  ternary, 668
  unary, 668
Association class, *672–674*
Association for Computing Machinery (ACM), 49, 50–52
  code of ethics, 50–53
Association of Information Technology Professionals, 49
Association role, *668–669*
  multiplicity of, 669–670
Associative entity, 323–325, *324*, 415
Asynchronous message, *690*
Atkinson, R. A., 141
Attribute, *314*
  candidate key and, 315
  class-scope, 677–678
  defining, 314–315
  derived, 674
  identifier and, 315–316
  multivalued, 316–317
  naming, 314
  repeating group and, 316
Attributive entity, 316
Automated development tools, 628
Automated test types, 578

B

B-to-B applications, 5, 152
  extranet and, 152
B-to-C applications, 152
Balancing, 105–106, 239, *251–253*

Banker, R. D., 358
Baseline modules, *627*
Baseline Project Plan (BPP), 164, *166*, 181–189, 201
  changes in, 75
  development of, 73–74
  executing, 74
  monitoring project with, 74
  review of, 73–74, 185–189
  sections of, 181–185
  updating at end of analysis, 367–373
Beck, Kent, 518, 582
Behavior, 657, *666*
Bell, P., 587, 589
Benbasat, I., 442
Benson, R. J., 145
Beta testing, *582*, 583, 605–606
Bill-of-materials structure, 319, 320, 402
Binary relationship, *319*–320, 399–402
Bischoff, J., 16
Blattner, M., 472, 473
Bloor, R., 574
Bohm, C., 13
Booch, G., 660, 666, 668, 676, 679
Bottom-up approach, 310, 568
Bottom-up initiatives, 137
Bottom-up planning, *146–147*
Boundary, *33*
Bourne, K. C., 26, 652
Bozman, J. S., 583
BPR. *See* Business Process Reengineering (BPR)
Break-even analysis, 175–177
Broadway Entertainment Company, Inc. (case), 3
  company background, 123–129
  conceptual data modeling for, 345–347
  design strategy for, 379–381
  distributed and information systems design for, 562–565
  finalizing design specifications for, 525–526
  form and report design for, 457–459
  interface and dialogue design for, 496–498
  Internet systems design for, 562–565
  logic modeling for, 301–303
  maintenance plan for, 635–637
  physical database design for, 429–431
  process modeling for, 278–281
  project identification and selection, 158–162
  requirements determination for, 235–237
  testing plan for, 612–614
Brooks, F. P., Jr., 110
Brown, C. V., 624
Bruce, T. A., 315
Bug tracking, 605

721

Credits

Commander™ Execu-View is a trademark of Comshare®, Inc.

VisualBasic, Windows, and Windows NT are registered trademarks and Visual C++ is a trademark of Microsoft Corporation in the United States and other countries.

Netscape Communications Corporation has not authorized, sponsored, endorsed, or approved this publication and is not responsible for its content. Netscape and the Netscape Communications Corporate Logos are trademarks and the trade names of Netscape Communications Corporation. All other product names and/or logos are trademarks of their respective owners.

Oracle: Certain portions of copyrighted Oracle Corporation screen displays of software programs have been reproduced herein with the permission of Oracle Corporation.

Chapter 1   Page 20: Figure 1-8; The *Systems Development Life Cycle*, 1990–2000 Level 8 Technologies Inc. Reprinted by permission. Page 22: Figure 1-9; from www.tumyeto.com/tydu/skatebrd/organizations/plans/14pipe.jpg; www.tumyeto.com/tydu/skatebrd/organizations/iuscblue.html; accessed September 16, 1999. Reprinted by permission of the International Association of Skateboard Companies. Page 26: Figure 1-10; from "Prototyping: The New Paradigm for Systems Development" by J. D. Naumann and A.M. Jenkins, *MIS Quarterly* 6 (3): 29–44. Reprinted by permission.

Chapter 2   Page 47: Table 2-2; from *Rapid Development* by Steven C. McConnell. Reproduced by permission of Microsoft Press. All rights reserved. Page 49: Figure 2-9; adapted with permission from Option Technologies, Inc., Orlando, FL, 1992. Page 51: Figure 2-10; "ACM Code of Ethics and Professional Conduct," *Communications of the ACM*, Vol. 36:2, 1993. Reprinted by permission of Association for Computing Machinery. Page 52: Table 2-3; from "Ethics and Information Systems: The Corporate Domain" by J. H. Smith and J. Hasnas in *MIS Quarterly* 23 (1): 109–127. Reprinted by permission. Page 57: Figure 2-11; from *The Code of Ethics of The Association of Information Technology Professionals*.

Copyright © 1997. The Association of Information Technology Professions (www.aitp.org). Currently celebrating its 50th anniversary, it recently received the "Champions of Industry Award." Reprinted by permission. All rights reserved.

Chapter 4   Page 97: Figure 4-2; Imagix 4D. Copyright © 1994–2000. Reprinted by permission of Imagix Corp. Page 98: Table 4-2; from Chen M., and R. J. Norman. 1992. "Integrated Computer-aided Software Engineering (CASE): Adoption, Implementation, and Impacts." *Proceedings of the Hawaii International Conference on System Sciences*, edited by J. F. Nanamker, Jr. Los Alamitos, CA: IEEE Computer Society Press, Vol. 3: 362–73. Page 99: Table 4-5; from Chen M., and R.J. Norman. 1992. "Integrated Computer-aided Software Engineering (CASE): Adoption, Implementation, and Impacts." *Proceedings of the Hawaii International Conference on System Sciences*, edited by J. F. Nanamker, Jr. Los Alamitos, CA: IEEE Computer Society Press, Vol. 3: 362–73. Page 101: Table 4-7; from McClure, C.L. 1989. *CASE is Software Automation*. Upper Saddle River, NJ: Prentice Hall. Figure 4-12; from Hanna, M. 1992. "Using Documentation as a Life-Cycle Tool." *Software Magazine* 12(12) (Dec.): 41–51. Page 112: Figure 4-13; a form designed in Visual C++. Page 113: Figure 4-14; a form designed in Visual C++. Page 114: Figure 4-15a,b; building a menu with Visual Basic. Page 116: Figure 4-16; Visual Basic development environment within Visual Studio. Page 116: Figure 4-17; PowerBuilder development environment (Source Sybase, 2000; www.sybase.com). Used with permission of Sybase, Inc. Page 117: Figure 4-18; Borland's Delphi™ development environment (Source: Borland Software Corporation; 2000, 2001. www.borland.com). Page 118: Figure 4-19a,b: Courtesy of Allaire.

Chapter 5   Page 137: Table 5-1; from McKeen, J. D., T. Guimaraes, and J. C. Wetherbee. 1994. "A Comparative Analysis of MIS Project Selection Mechanisms." *Data Base* 25 (Feb.): 43–59. Page 139: Figure 5-2; adapted with the permission of The Free Press, a Division of Simon & Schuster, Inc., from *Competitive Advantage: Creating and Sustaining Superior Performance* by Michael E. Porter.

731